W9-BPP-049

Population Biology

POPULATION BIOLOGY

The Coevolution of Population Dynamics and Behavior

JOHN MERRITT EMLEN

Indiana University

MACMILLAN PUBLISHING COMPANY

New York

COLLIER MACMILLAN PUBLISHERS

London

This Book is Dedicated to Ellen

Copyright © 1984, Macmillan Publishing Company, a division of Macmillan, Inc.
Printed in the United States of America

All rights reserved. No part of this book may be reproduced or transmitted in any form or by any means, electronic or mechanical, including photocopying, recording, or any information storage and retrieval system, without permission in writing from the Publisher.

Macmillan Publishing Company
866 Third Avenue, New York, New York 10022

Collier Macmillan Canada, Inc.

Library of Congress Cataloging in Publication Data

Emlen, John Merritt.
Population biology.

Bibliography: p.
Includes index.
1. Population biology. I. Title. [DNLM:
1. Population dynamics. 2. Genetics, Population.
QH 455 E53p]
QH352.E46 1984 574.5'248 83-5401
ISBN 0-02-333660-9

Printing: 1 2 3 4 5 6 7 8 Year: 4 5 6 7 8 9 0 1

ISBN 0-02-333660-9

Preface

The last decade has seen dramatic splits and realignments among the fields of genetics, animal behavior, and what was once referred to as the study of ecology. The wedding of population genetics and evolutionary ecology, now matured to embrace the complexities of quantitative genetics, has given birth to rigorous theories of life history, dispersal, sociobiology, and the functional responses of one species to another. The dynamics of populations provide a template on which natural selection and genetic processes mold morphology, anatomy, physiology, and behavior; these adaptations, in turn, feed back on the population process. It is the cybernetic nature of these processes that provides the fascination of population biology for many of its practitioners. For all of its immense complexity, this interdisciplinary study of the interaction of population dynamics and the evolution of behavior may provide us with the best hope of understanding our environment and our own human nature. In the pages that follow I have tried not only to provide the reader with the basic tools, facts, and theories of population biology, but to describe the search for patterns of feedback among the often far-flung concepts and to convey some of the excitement accompanying discovery of these emerging relationships.

This book is designed for three purposes. First, it is a textbook suitable for advanced undergraduates, honor students, and graduate students. Indeed, any one of its three major sections comprises a full course in itself; the introductory section gives a treatment of the mathematics necessary to a study of population biology beyond the elementary level. Second, it is both a synthesis and a compendium of new ideas and suggestions for future research. Finally, the book provides an extensive review of the key literature through 1982 and so should serve as a useful reference source.

Acknowledgments

A large number of people contributed substantially to the writing of this book, including the students of two undergraduate honors courses in ecology and the members of a graduate course in population biology. Suggestions, critiques, and thoughtful reviews were graciously provided by Hal Caswell, Hseu-Wen Chang, T. J. Giesel, Peter Grant, Austin Hughes, Michael Lynch, Steve Martindale, Randolph Phillis, Ellen Pikitch, Eric Pianka, Robert Rockwell, Dolph Schluter, Robert Selander, and Alan Templeton. My particular thanks, for heroic efforts to set my thinking straight, for insightful discussions of the ideas, and for enthusiastic encouragement go to Robert Fagen, Craig Stewart, and Daniel Sulzbach. If this book is successful, that success will owe much to the efforts of these people.

J. M. Emlen

Contents

Part Four Adaptation: Fine-Tuning the Structural Framework *309*

Glossary of Symbols

$A = \{\alpha_{ij}\}$	"community matrix" of species interactions
$E(x)$	(1) properly, the "expected value" of x: the true, population mean of x; (2) also used to denote the mean of x over time
F_{IS}	coefficient of inbreeding arising from consanguineous or assortative mating
F_{IT}	$= 1 - (1 - F_{IS})(1 - F_{ST})$, the total inbreeding due to both of the processes F_{IS} and F_{ST}
F_{ST}	coefficient of inbreeding arising from genetic drift
k	$= 2W_2 - W_1 - W_3$, or $2m_2 - m_1 - m_3$
$k^{(x)}$	$= 2x_2 - x_1 - x_3$
K	carrying capacity; the equilibrium value of n for populations obeying the logistic equation
$\mathbf{L}$	Leslie matrix, used to project a population vector at time t onto a population vector at time $t + 1$
m_i	fitness (continuous measure); equal to $(1/n_i)(dn_i/dt)$ when there are no fecundity differences among genotypes, where n_i is the population density of the ith genotype
n	population density
n_e	effective population size
r	instantaneous rate of increase of a population
r_0	maximum value of r for a given environment; intrinsic rate of increase
R	ratio of population size at one time to that one time unit previous
V_A	additive genetic variance in fitness at a given locus
$V_A(x)$	additive genetic variance in x at a given locus
V_g, $V_g(x)$	additive genetic variances in fitness, x, over all loci
W	fitness (discrete measure); average number of offspring of a genotype per individual of that genotype born
x	(1) general term used to describe a variable; (2) displacement in n from population equilibrium
x^*	unless otherwise indicated, the geometric mean of x
$\hat{x}$	equilibrium value of x
$\bar{x}$	mean of x over individuals, space, and so on, at some point in time
α	$= p(W_1 - W_2) + q(W_2 - W_3)$, or $p(m_1 - m_2) + q(m_2 - m_3)$, where p and q are allele frequencies, and W and m represent fitnesses
$\alpha^{(x)}$	$= p(x_1 - x_2) + q(x_2 - x_3)$, where x represents trait (X) values for the three genotypes

α_{ij} species interaction (usually competition) coefficient; the effect of an individual of species j on the growth rate of species i, *relative to* the impact of an individual of species i on the growth rate of species I:

$$\frac{\alpha r_i/\alpha n_j}{\alpha r_i/\alpha n_i}$$

β $= q(W_1 - W_2) + p(W_2 - W_3)$, or $q(m_1 - m_2) + p(m_2 - m_3)$

$\beta^{(x)}$ $= q(x_1 - x_2) + p(x_2 - x_3)$

δ coefficient relating to death rate

Δx change, over a unit time, in the value of x

ϵ (1) ln $R_{max'}$ the discrete analog of r_0; (2) statistical error term, with mean zero

η (1) coefficient determining shape of the functional response curve; (2) the phenotypic correlation between mates

λ eigenvalue of a matrix

λ_1 Perron root; the real, positive eigenvalue of the Leslie matrix

ρ (1) deviation, in space, of n from the average; (2) contribution of an individual to its genotype's fitness; (3) size of a propagule

τ time lag

$\phi(x)$ probability density of the variable x; usually denotes the normal distribution

$\prod$ symbol denoting product:

$$\prod_{i=1}^{k} p_1 = p_1 p_2 \cdots p_k$$

$\sum$ symbol denoting summation:

$$\sum_{i=1}^{k} p_i = p_1 + p_2 + \cdots + p_k$$

Introduction

Largemouth bass are a favorite quarry for sports fishermen, who, every year, remove several hundred thousand of these fish from American lakes. Yet this species maintains a viable population; it has compensated its losses to man by somehow raising its fecundity or its natural survival until birthrate and death rate balance. We hear often of species which, under the onslaught of human disturbance, are declining toward extinction: several species of whales, for example, or certain heavily exploited oceanic fish. But what is most surprising is that, in spite of heavy exploitation, extensive habitat destruction, and in some cases, deliberate attempts at extermination, most species of animals and plants continue obstinately to persist. When death rate rises from one external cause, birthrate is raised by forces internal to the population and balances the losses. How and why does this compensation occur? How far can it be pushed? If we were to double the fishing intensity on bass, would their compensatory powers finally be overtaxed? Losses of individuals by exploitation of this sort, or from increased natural predation or disease, even in the presence of a consequent rise in birthrate or drop in mortality by other causes, might be expected to depress a population's average size. Do they? And if so, how much? What would be the effect of removing mostly young as opposed to old individuals? Suppose that fishermen take disproportionately more individuals of other species that share the bass's food supply. Might increased exploitation, then, by depressing these competitor populations, enhance the available food and actually increase the number of bass? What is the role of predators, or prey, in the dynamics of the bass population? Finally, how are the answers to these questions affected by the nature of the environment? What is the impact of climatic variation, fluctuations,in the abundance of food, or spatial heterogeneity in the distribution of resources?

These are the sorts of questions that population biologists seek to answer. How are we to address them? Clearly, we must first develop expressions capable of describing how population numbers change in time. In this book this is done in a series of steps. In Part Two we present simple, classical models, mathematically tractable and, because they ignore most biological complications, broadly applicable but imprecise. The logistic equation and the Ricker equation, pertaining to species that breed once in their lives and show no overlapping of generations, belong in this category, as do the Leslie matrices used for describing age-structured populations, and the Lotka–Volterra equations designed for describing competition among species or prey–predator relationships. These "models" of population dynamics form a basic, descriptive framework for the study of populations. They are useful as crude predictors, they are often powerful heuristically, they give us a foundation on which more complete and accurate models can be built, and they provide a departure point for further study.

The equations of the basic framework are seductive in their simplicity but can seldom be relied on for precise answers. Thus we must next begin to replace the approximations with specifics and to eliminate the simplifying assumptions. We must fine-tune the models. Unfortunately, even simple elaborations of the basic equations often prove prohibitively cumbersome mathematically. This increased complexity is coupled with loss of generality—usually the elaborations are species or situation specific—and an often minimal, or even unknown, increase in precision. As Oster and Takahashi (1974) put it: "We complicate our models step by step, hoping that before we pass into the abyss of mathematical intractability, a reasonable compromise will have been struck between precision and generality." This ubiquity of diminishing returns has led to two quite different approaches to the problem. Where accurate prediction is sought—for management decisions or measures of environmental impact, for example—it is often possible to construct equation systems that incorporate huge arrays of biological and other, environmental information. Human comprehension of interacting processes may become quite submerged in a towering babel of computer software, and the model, usually a theoretical–empirical hybrid, is generally too specialized to be of heuristic interest. Its purpose is strictly descriptive. Many population biologists, on the other hand, are concerned less with accurate description than with using models to generate ideas and as aids in understanding biological processes *in general*. These people are more inclined to fine-tune the simple models in a more qualitative fashion: Can a (reasonably) accurate expression for birthrate be written as a function of nutrient intake? How is nutrient intake related to availability of prey? Are there general rules which can be discovered relating degree of uncertainty in weather, say, to life history? How do differences in life history affect the dynamics of populations? Such questions have generated at least one whole field of study, sociobiology; they relate directly to nearly the entire field of ecology; and their pursuit often draws heavily on the disciplines of physiology and endocrinology. In particular, this branch of population biology is inextricably wedded to population genetics, for the questions asked are of an ultimate (evolutionary) rather than a proximate (mechanistic) nature. We are likely to inquire, given a set of environmental circumstances and biological constraints, what behavioral, morphological, and physiological patterns would be expected to evolve.

In this book, the second approach to fine-tuning is followed. Accordingly, Part Three deals with the genetic background necessary for addressing evolutionary questions. In this section we first examine the genetical structure of populations as affected by drift, consanguineous mating, and temporal and spatial heterogeneity. Then natural selection is discussed, first at the level of a single gene locus and then incorporating effects of linkage and epistasis. Quantitative genetics theory follows. In Chapter 9 we explore the effects of temporal and spatial heterogeneity, and the efficacy and importance of group selection. We also examine inclusive fitness as it pertains to the evolution of social organization.

Part Four applies the material of Part Two and Three, dealing with the interplay of population and genetic processes and the consequent evolutionary implications for fine-tuning the basic population models. Among the topics covered are the evolution of life histories—birth, death, and dispersal rates—social organization, foraging behavior, defense mechanisms, and the evolutionary consequences of competition.

Throughout the book we make extensive use of simple mathematics, including calculus and matrix algebra. The level of mathematical sophistication required is considerably less than may appear at first glance to one with little quantitative training, and is held at a fairly consistent level in most chapters. It represents a compromise between the knowledge of the average undergraduate student and that necessary to understand fully the recent literature. The rationale for this approach is straightforward: Texts dealing with other than strictly elementary material in population biology have generally dealt with the questions of this inherently quantitative field by either denying its nature or concentrating on subsets of the discipline which require less computational knowledge. This is reasonable for a low-level, survey course, but it is misleading and unfair to the student interested in pursuing advanced studies in the discipline. The recent rapid development of mathematically sophisticated procedures already threatens to split the field; it would be unfortunate to promulgate this schism by allowing potential researchers to remain ignorant of these contributions.

At the stage in a student's learning corresponding to this book there is no need to demand detailed knowledge of mathematical material, no need for proofs, and no need for high levels of computational skill. A basic *feeling* for the material is enough to permit exploration of many aspects of population biology not otherwise approachable. Students need not possess the insight necessary to invent a population model, but they should be able to understand its origin when it is derived for them. They need not necessarily be able to find the solution to an equation, but they should understand what the equation describes and why. An introduction to matrix algebra, and brief treatments of dispersion, partial derivatives, and the Taylor expansion, presented in this spirit, comprises Chapter 1. Students who feel they know the topic may choose to bypass the material. Or, on the other hand, by skipping the problem sets, readers may find this chapter a quick and easy way to review their knowledge.

Part One
Tools of the Trade

Just as one cannot write an essay without a knowledge of grammar, or learn about cell metabolism without an appreciation of chemistry, one cannot expect to understand population biology without a certain feeling for mathematics. Of course, the study of natural history, the exercise of field research or the amassing of interesting facts—all parts of population biology—do not require such a feeling. But the interweaving of these endeavors into a scientific framework from which we can draw an understanding of natural relationships is made vastly more simple with its use. In fact, population biology, as a discipline today, owes its existence, in large part, to the recent influx into ecology of mathematical ideas from geneticists, physicists, and mathematicians.

Historically, biologists have had rather uneven backgrounds in mathematics. Some of you will probably yawn at the trivialities of this first chapter. If so, skip them. Others will find some of the material quite unfamiliar. Do not be intimidated. This is not a math text and you will not be subjected to details. The emphasis is not on manipulating nasty equations or solving difficult problems, but rather on an understanding of where an equation comes from and what it means. Mathematics is a language that has the value of being both concise and precise, but it does not enjoy wide familiarity or affection. Therefore, it is properly used when words would be awkward or imprecise. This book is written in that spirit. However, circumstances in which words are inadequate arise more than occasionally in the study of population biology. It will also become apparent that the conciseness of a mathematical expression may lead us quite obviously to useful conclusions where verbal descriptions would leave us floundering in confusion. So do not be cavalier with this chapter—it is basic to what follows.

Chapter 1

Some Mathematics for Population Biologists

1.1 Matrices and Simple Matrix Operations

A *matrix* is defined as a rectangular array of numbers. If you were to display data gathered on four parameters over 20 replicates in a 4 (row) by 20 (column) table, that table would constitute a 4×20 matrix. Similarly, suppose that you made a list, across a page, of the coefficients

$$(a, b, c, d)$$

of the equation

$$y = a + bx + cx^2 + dx^3$$

This list is a 1 (row) by 4 (column) matrix. If the list had been written *down* the page,

$$\begin{bmatrix} a \\ b \\ c \\ d \end{bmatrix}$$

it would have been a 4 (row) by 1 (column) matrix. The number 5 is also a matrix—a 1 (row) by 1 (column) matrix. Matrices that have only one row or only one column are generally referred to as *vectors*; 1 by 1 matrices are called *scalars*. In general, matrices with n rows and m columns are referred to as "$n \times m$."

The value of the matrix concept will become apparent as you proceed through the book. For the moment, consider the following example. A population is comprised of individuals of several age classes. Each has its own, characteristic probability of survival to the next age class, and each contributes its own, characteristic number of offspring. If (say) old individuals do most of the reproducing, a population bulge of senior mothers will be followed by a glut of very young animals, followed, upon the death of the senior mothers, by a period of low reproduction during which mortality produces a shrinking population. Then, as the babies reach old age, the cycle repeats. Suppose that we wished to describe the population dynamics of all age groups. It would be possible to construct a series of simultaneous equations, one for each age group, for each period in time. But unless a computer is used, solution of the several equations becomes tedious, especially if the population is to be projected over many years. The use of matrices simplifies the problem vastly. One merely rewrites this *series* of equations as a *single* matrix equation—using matrices instead of simple numbers (scalars). Solution of the matrix equation gives a general description of the dynamics of all age groups over all time. The bulk of this chapter is devoted to the art of setting up matrix equations (which requires a knowledge of how to add, multiply, and divide matrices) and solving them.

Matrix Addition

It will be helpful to introduce some notation. Vectors and other matrices of size greater than 1×1 are often distinguished from scalars by boldface designation. Vectors usually are denoted by lowercase boldface letters, larger matrices by capital letters. Thus a 6×7 matrix might be called **A**, while a 1×10 vector could be referred to as **b**. This particular notation will be adhered to in this book.

Matrices can also be written

$$\mathbf{A} = \{a_{ij}\}$$

where a_{ij} denotes the element in the ith row, jth column, and the braces indicate that we are referring to the *set* of all a_{ij} values that comprise the matrix.

Using such notation it is now an easy matter to define matrix addition. The sum of two matrices, **A** and **B** (call it **Q**), is

$$\mathbf{Q} = \mathbf{A} + \mathbf{B} = \{a_{ij}\} + \{b_{ij}\} = \{a_{ij} + b_{ij}\} = \{q_{ij}\} \tag{1-1}$$

That is, to find the sum of two matrices, simply add the corresponding elements. Examples:

$$\begin{bmatrix} 2 & 3 & 1 \\ 0 & 0 & 2 \end{bmatrix} + \begin{bmatrix} 1 & 0 & 0 \\ 0 & 1 & 1 \end{bmatrix} = \begin{bmatrix} 2+1 & 3+0 & 1+0 \\ 0+0 & 0+1 & 2+1 \end{bmatrix} = \begin{bmatrix} 3 & 3 & 1 \\ 0 & 1 & 3 \end{bmatrix}$$

$$\begin{bmatrix} a & b \\ c & d \end{bmatrix} + \begin{bmatrix} e & f \\ g & h \end{bmatrix} = \begin{bmatrix} a+e & b+f \\ c+g & d+h \end{bmatrix}$$

If two matrices have different numbers of rows or different numbers of columns, corresponding elements do not exist and addition is not defined:

$$\begin{bmatrix} 1 & 3 \\ 2 & 0 \\ 1 & 1 \end{bmatrix} + \begin{bmatrix} 0 & 1 & 1 \\ 1 & 2 & 0 \\ 0 & 0 & 1 \end{bmatrix} = \text{undefined}$$

Add the following matrices.

(i) $[4 \quad 1 \quad 1] + [2 \quad 0 \quad 1]$ Answer: $[6 \quad 1 \quad 2]$

(ii) $\begin{bmatrix} 2 & 1 & 3 \\ 0 & 0 & 4 \\ 5 & 5 & 4 \end{bmatrix} + \begin{bmatrix} 2 & 0 & 0 \\ 0 & 3 & 0 \\ 0 & 5 & 0 \end{bmatrix}$ Answer: $\begin{bmatrix} 4 & 1 & 3 \\ 0 & 3 & 4 \\ 5 & 10 & 4 \end{bmatrix}$

(iii) $\begin{bmatrix} \lambda & 0 \\ 0 & \lambda \end{bmatrix} - \begin{bmatrix} 2 & 0 \\ 0 & 1 \end{bmatrix}$ Answer: $\begin{bmatrix} \lambda - 2 & 0 \\ 0 & \lambda - 1 \end{bmatrix}$

The matrix of all zeros, usually written **0**, behaves like the zero of scalar arithmetic. The sum of **0** and any other matrix, **A**, with the same number of rows and columns gives us simply **A**.

$$\mathbf{A} + \mathbf{0} = \mathbf{A}$$

Matrix Multiplication

The simplest form of matrix multiplication is scalar multiplication. To multiply a matrix by a scalar, we write

$$c \cdot \mathbf{A} = c\{a_{ij}\} = \{ca_{ij}\} \tag{1-2}$$

For example,

$$2\begin{bmatrix} 3 & 1 \\ 0 & 2 \end{bmatrix} = \begin{bmatrix} 2 \times 3 & 2 \times 1 \\ 2 \times 0 & 2 \times 2 \end{bmatrix} = \begin{bmatrix} 6 & 2 \\ 0 & 4 \end{bmatrix}$$

In contrast to the above, the multiplication of one matrix by another is *not* done, as one might guess, by multiplying corresponding elements. Rather, to find the (i, j)th element of the product, $\mathbf{Q} = \mathbf{A} \cdot \mathbf{B}$, one looks at the ith row of $\mathbf{A}$ and the jth column of $\mathbf{B}$, multiplies the first elements of each, multiplies the second elements of each, and so on, then adds these products. For example, suppose that

$$\mathbf{A} = \begin{bmatrix} 2 & 1 & 3 \\ 0 & 1 & 2 \end{bmatrix} \quad \text{and} \quad \mathbf{B} = \begin{bmatrix} a & d \\ b & e \\ c & f \end{bmatrix}$$

Then, diagrammatically,

$$q_{11} = \begin{bmatrix} \boxed{2 \quad 1 \quad 3} \\ 0 \quad 1 \quad 2 \end{bmatrix} \begin{bmatrix} \boxed{\begin{matrix} a \\ b \\ c \end{matrix}} & \begin{matrix} d \\ e \\ f \end{matrix} \end{bmatrix} = 2a + b + 3c$$

$$q_{12} = \begin{bmatrix} \boxed{2 \quad 1 \quad 3} \\ 0 \quad 1 \quad 2 \end{bmatrix} \begin{bmatrix} \begin{matrix} a \\ b \\ c \end{matrix} & \boxed{\begin{matrix} d \\ e \\ f \end{matrix}} \end{bmatrix} = 2d + e + 3f$$

$$q_{21} = \begin{bmatrix} 2 \quad 1 \quad 3 \\ \boxed{0 \quad 1 \quad 2} \end{bmatrix} \begin{bmatrix} \boxed{\begin{matrix} a \\ b \\ c \end{matrix}} & \begin{matrix} d \\ e \\ f \end{matrix} \end{bmatrix} = b + 2c$$

$$q_{22} = \begin{bmatrix} 2 \quad 1 \quad 3 \\ \boxed{0 \quad 1 \quad 2} \end{bmatrix} \begin{bmatrix} \begin{matrix} a \\ b \\ c \end{matrix} & \boxed{\begin{matrix} d \\ e \\ f \end{matrix}} \end{bmatrix} = e + 2f$$

so that

$$\mathbf{Q} = \{qij\} = \begin{bmatrix} 2a + b + 3c & 2d + e + 3f \\ b + 2c & e + 2f \end{bmatrix}$$

Suppose that we wished to find

$$\mathbf{X} = \begin{bmatrix} 2 & 0 \\ 0 & 4 \end{bmatrix} \begin{bmatrix} 1 & 2 \\ 3 & 4 \end{bmatrix}$$

Write

$$x_{11} = \begin{bmatrix} \boxed{2 \quad 0} \\ 0 \quad 4 \end{bmatrix} \begin{bmatrix} \boxed{\begin{matrix} 1 \\ 3 \end{matrix}} & \begin{matrix} 2 \\ 4 \end{matrix} \end{bmatrix} = 2 \times 1 + 0 \times 3 = 2$$

$$x_{12} = \begin{bmatrix} \boxed{2 \quad 0} \\ 0 \quad 4 \end{bmatrix} \begin{bmatrix} \begin{matrix} 1 \\ 3 \end{matrix} & \boxed{\begin{matrix} 2 \\ 4 \end{matrix}} \end{bmatrix} = 2 \times 2 + 0 \times 4 = 4$$

$$x_{21} = \begin{bmatrix} 2 \quad 0 \\ \boxed{0 \quad 4} \end{bmatrix} \begin{bmatrix} \boxed{\begin{matrix} 1 \\ 3 \end{matrix}} & \begin{matrix} 2 \\ 4 \end{matrix} \end{bmatrix} = 0 \times 1 + 4 \times 3 = 12$$

$$x_{22} = \begin{bmatrix} 2 \quad 0 \\ \boxed{0 \quad 4} \end{bmatrix} \begin{bmatrix} \begin{matrix} 1 \\ 3 \end{matrix} & \boxed{\begin{matrix} 2 \\ 4 \end{matrix}} \end{bmatrix} = 0 \times 2 + 4 \times 4 = 16$$

so that

$$\mathbf{X} = \begin{bmatrix} 2 & 4 \\ 12 & 16 \end{bmatrix}$$

Multiply the following matrices:

(iv)
$$\begin{bmatrix} 1 & 2 \\ 3 & 1 \end{bmatrix}\begin{bmatrix} 2 \\ 0 \end{bmatrix} \qquad \text{Answer: } \begin{bmatrix} 2 \\ 6 \end{bmatrix}$$

(v)
$$\begin{bmatrix} \lambda_1 & 0 \\ 0 & \lambda_2 \end{bmatrix}\begin{bmatrix} 3 & 1 \\ 1 & 2 \end{bmatrix} \qquad \text{Answer: } \begin{bmatrix} 3\lambda_1 & \lambda_1 \\ \lambda_2 & 2\lambda_2 \end{bmatrix}$$

Notice that the number of columns in **A** must be the same as the number of rows in **B**.

At this point two facts should be apparent. First, the product of a $(n \times k)$ and a $(k \times m)$ matrix is of size $(n \times m)$—for a verification of this, review the results of the problems above. Second, an $(n \times k)$ matrix can be multiplied by a $(k \times m)$ matrix because the number of columns in the first (k) equals the number of rows in the second (k). But it is *not*, in general, possible to reverse the order of these matrices and obtain a product. For example,

$$\begin{bmatrix} 1 & 2 \\ 3 & 1 \end{bmatrix}\begin{bmatrix} 2 \\ 0 \end{bmatrix} = \begin{bmatrix} 2 \\ 6 \end{bmatrix}$$

but

$$\begin{bmatrix} 2 \\ 0 \end{bmatrix}\begin{bmatrix} 1 & 2 \\ 3 & 1 \end{bmatrix} = \text{undefined}$$

In fact, even when the number of rows and columns is such that **AB** and **BA** are both defined, these products are generally *not* equal:

$$\begin{bmatrix} 1 & 2 \\ 0 & 3 \end{bmatrix}\begin{bmatrix} 2 & 1 \\ 4 & 0 \end{bmatrix} = \begin{bmatrix} 10 & 1 \\ 12 & 0 \end{bmatrix}$$

$$\begin{bmatrix} 2 & 1 \\ 4 & 0 \end{bmatrix}\begin{bmatrix} 1 & 2 \\ 0 & 3 \end{bmatrix} = \begin{bmatrix} 2 & 7 \\ 4 & 8 \end{bmatrix}$$

or

$$\begin{bmatrix} 1 & 2 & 3 \\ 2 & 1 & 0 \end{bmatrix}\begin{bmatrix} 0 & 0 \\ 0 & 1 \\ 3 & 0 \end{bmatrix} = \begin{bmatrix} 9 & 2 \\ 0 & 1 \end{bmatrix}$$

$$\begin{bmatrix} 0 & 0 \\ 0 & 1 \\ 3 & 0 \end{bmatrix}\begin{bmatrix} 1 & 2 & 3 \\ 2 & 1 & 0 \end{bmatrix} = \begin{bmatrix} 0 & 0 & 0 \\ 2 & 1 & 0 \\ 3 & 6 & 9 \end{bmatrix}$$

In more formal terms, matrix multiplication is not *commutative*.

With your growing experience in multiplying matrices, the following notation should now make sense. The (i,j)th element of the product of two matrices, $\mathbf{A} = \{a_{ij}\}$ and $\mathbf{B} = \{b_{ij}\}$, can be written

$$q_{ij} = \sum_k a_{ik} b_{kj} \tag{1-3}$$

That is, from the ith row of **A**, the jth column of **B**, take the product of the first $(k = 1)$ elements, $a_{i1}b_{1j}$, the product of the second $(k = 2)$ elements, $a_{i2}b_{2j}$, and so on, and sum them $(\sum_k)$.

We are now in a position to set up the matrix equation for that problem of a population with age classes mentioned earlier. Suppose that in our particular population individuals live to age 3. Let the number of individuals between ages 0 and 1 at time t be $n_1(t)$, the number between ages 1 and 2 be $n_2(t)$, and those between 2 and 3 be $n_3(t)$. Then the population at time t can be described by the vector $(3 \times 1$ matrix$)$

$$\begin{bmatrix} n_1(t) \\ n_2(t) \\ n_3(t) \end{bmatrix}$$

Now suppose that individuals in age class 1 (at time t) live to reach age class 2 (at time $t + 1$) with probability $\frac{1}{2}$, and that survival from age class 2 to age class 3 is $\frac{1}{4}$. Then

$$
\begin{aligned}
n_2(t + 1) &= \tfrac{1}{2}n_1(t) \\
n_3(t + 1) &= \tfrac{1}{4}n_2(t)
\end{aligned}
\tag{1-4}
$$

Finally, suppose that the numbers of offspring still alive one time unit later, born to individuals in age classes 1, 2, and 3, are, respectively, 0, 1, and 4. Since these offspring comprise the first age class, n_1, in that next time unit ($t + 1$), we can write

$$
n_1(t + 1) = n_1(t)\cdot 0 + n_2(t)\cdot 1 + n_3(t)\cdot 4 \tag{1-5}
$$

Equations (1-4) and (1-5) can now be combined in matrix form:

$$
\begin{bmatrix} n_1(t + 1) \\ n_2(t + 1) \\ n_3(t + 1) \end{bmatrix} = \begin{bmatrix} 0 & 1 & 4 \\ \frac{1}{2} & 0 & 0 \\ 0 & \frac{1}{4} & 0 \end{bmatrix} \begin{bmatrix} n_1(t) \\ n_2(t) \\ n_3(t) \end{bmatrix} \tag{1-6}
$$

Verify that this is so by multiplying the right-hand side of (1-6) and comparing the results with (1-4) and (1-5).

If the square (3×3) matrix is designated $\mathbf{L}$, and the vector of abundances $\mathbf{n}(t)$, then (1-6) can be written

$$
\mathbf{n}(t + 1) = \mathbf{L}\cdot n(t) \tag{1-7}
$$

Suppose that the population described above has the structure

$$
\mathbf{n}(0) = \begin{bmatrix} 200 \\ 0 \\ 0 \end{bmatrix}
$$

at time zero. Then, at time $t = 1$,

$$
\mathbf{n}(1) = \mathbf{L}\cdot\mathbf{n}(0) = \begin{bmatrix} 0 & 1 & 4 \\ \frac{1}{2} & 0 & 0 \\ 0 & \frac{1}{4} & 0 \end{bmatrix} \begin{bmatrix} 200 \\ 0 \\ 0 \end{bmatrix} = \begin{bmatrix} 0 \\ 100 \\ 0 \end{bmatrix}
$$

at time $t = 2$,

$$
\mathbf{n}(2) = \mathbf{L}\cdot\mathbf{n}(1) = \begin{bmatrix} 0 & 1 & 4 \\ \frac{1}{2} & 0 & 0 \\ 0 & \frac{1}{4} & 0 \end{bmatrix} \begin{bmatrix} 0 \\ 100 \\ 0 \end{bmatrix} = \begin{bmatrix} 100 \\ 0 \\ 25 \end{bmatrix}
$$

and so on. In this manner the population structure may be projected over any time period (provided that the elements of $\mathbf{L}$, the birth and survival rates, remain unchanged). The procedure, although tedious, is easy to carry out. Also, as we shall see shortly, it is possible to find explicit solutions for $n_1(t)$, $n_2(t)$, and $n_3(t)$ at any t, from (1-7). Such solutions remove even the tedium. But first, it is necessary to explore a few more basics.

Matrix Transposition

In (1-6) and (1-7), we dealt with a simple matrix problem involving column vectors

$$
\begin{bmatrix} n_1 \\ n_2 \\ n_3 \end{bmatrix}
$$

Suppose that we had wanted to write these n-values as a *row* vector, $[n_1 \quad n_2 \quad n_3]$. This new vector is really the same as the old one except that each row in the first has become a column in the second. We call the second, the row vector, the *transpose* of its corresponding column vector, $\mathbf{n}$. The transpose of $\mathbf{n}$ is usually written $\mathbf{n}^T$. We can also define the transpose of other matrices—simply switch rows and columns:

$$\begin{bmatrix} 1 & 2 & 3 \\ 0 & 1 & 2 \end{bmatrix}^T = \begin{bmatrix} 1 & 0 \\ 2 & 1 \\ 3 & 2 \end{bmatrix}$$

Transpose the following matrix.

(vi) $$\begin{bmatrix} 2 & 2 & 1 \\ 0 & 3 & 1 \end{bmatrix} \qquad \text{Answer: } \begin{bmatrix} 2 & 0 \\ 2 & 3 \\ 1 & 1 \end{bmatrix}$$

Making use of the notation in (1-3), the (i, j)th element of the product of $\mathbf{A}$ and $\mathbf{B}$, $\mathbf{Q} = \mathbf{A}\cdot\mathbf{B}$ is

$$q_{ij} = \sum_k a_{ik} b_{kj} \tag{1-8}$$

But note now that the (i, k)th element of $\mathbf{A}$ is, by definition, the (k, i)th element of $\mathbf{A}^T$ (rows and columns transposed). Similarly, the (k, j)th element of $\mathbf{B}$ is the (j, k)th element of $\mathbf{B}^T$. Equation (1-8) can therefore be altered to read

$$q_{ij} = \sum_k a^T_{ki}\, b^T_{jk} = \sum_k b^T_{jk}\, a^T_{ki}$$

The right side of this expression, by (1-3), is the (j, i)th element of $\mathbf{B}^T\cdot\mathbf{A}^T$. But it is also the (i, j)th element of $\mathbf{Q}$—or, equivalently, the (j, i)th element of $\mathbf{Q}^T$. Thus

$$\mathbf{Q}^T = \mathbf{B}^T\cdot\mathbf{A}^T \tag{1-9}$$

But since

$$\mathbf{Q}^T = (\mathbf{AB})^T$$

it follows that

$$(\mathbf{A}\cdot\mathbf{B})^T = \mathbf{B}^T\cdot\mathbf{A}^T \tag{1-10}$$

If we had really wished to attack the population problem above using row vectors, we could have converted

$$\mathbf{n}(t+1) = \mathbf{L}\cdot\mathbf{n}(t)$$

to

$$\mathbf{n}(t+1)^T = [\mathbf{L}\cdot\mathbf{n}(t)]^T = \mathbf{n}(t)^T\cdot\mathbf{L}^T$$

Verify that this is so by returning to the numerical calculations following (1-7) and redoing them using row vectors.

Calculate the following, first by multiplying and then transposing and, after that, by first transposing and then multiplying.

(vii) $$\left(4\begin{bmatrix} 0 & 1 \\ 1 & 0 \end{bmatrix}\right)^T \quad \text{(see equation (1-2))}$$

(viii) $$\left(\begin{bmatrix} 1 \\ 2 \\ 3 \end{bmatrix} [1 \quad 4]\right)^T$$

(ix) $$\left(3\begin{bmatrix} \lambda_1 & 0 \\ 0 & \lambda_2 \end{bmatrix}\right)^T$$

Before moving on, we assert the following without proof.

$$\mathbf{AB(C)} = \mathbf{A(BC)}$$

Verify this fact for the following problem.

(x)
$$[1 \quad 2 \quad 3] \begin{bmatrix} 1 & 0 \\ 2 & 1 \\ 2 & 1 \end{bmatrix} \begin{bmatrix} 3 & 0 \\ 1 & 0 \end{bmatrix} = ?$$

Determinants and the Matrix Inverse

Returning to the population problem in (1-7), suppose that we knew the elements of **L**, and the vector $\mathbf{n}(t)$, and wished to know the population vector in the *preceding* time unit, $\mathbf{n}(t-1)$. Equation (1-7) can be written

$$\mathbf{n}(t) = \mathbf{L} \cdot \mathbf{n}(t-1)$$

but so far we have not discussed a technique for working backward from $\mathbf{n}(t)$ to $\mathbf{n}(t-1)$. By analogy to the more familiar number system, we define an *identity matrix*, always denoted by **I**, which (like the number 1), when pre- or postmultiplied times any matrix, leaves it unchanged. That is,

$$\mathbf{I} \cdot \mathbf{A} = \mathbf{A} \cdot \mathbf{I} = \mathbf{A}$$

for any matrix, **A**. It is easy to verify that **I** must be uniquely of the form

$$\mathbf{I} = \begin{bmatrix} 1 & 0 & 0 & 0 & 0 & \cdots \\ 0 & 1 & 0 & 0 & 0 & \cdots \\ 0 & 0 & 1 & 0 & 0 & \cdots \\ 0 & 0 & 0 & 1 & 0 & \cdots \\ \cdots & \cdots & \cdots & \cdots & \cdots & \cdots \end{bmatrix}$$

—1's on the major diagonal (upper left to lower right), and 0's elsewhere.

Having defined **I**, we now define the *inverse*, $\mathbf{A}^{-1}$, of a matrix **A**, in such a way that

$$\mathbf{A} \cdot \mathbf{A}^{-1} = \mathbf{A}^{-1} \cdot \mathbf{A} = \mathbf{I} \tag{1-11}$$

Not all matrices have inverses, but suppose for the moment that an inverse can be found for **L**. Then, if

$$\mathbf{n}(t) = \mathbf{L} \cdot \mathbf{n}(t-1)$$

it follows that (premultiplying both sides by $\mathbf{L}^{-1}$),

$$\mathbf{L}^{-1}\mathbf{n}(t) = \mathbf{L}^{-1}(\mathbf{L} \cdot \mathbf{n}(t-1)) = (\mathbf{L}^{-1} \cdot \mathbf{L})\mathbf{n}(t-1) = \mathbf{I} \cdot \mathbf{n}(t-1) = \mathbf{n}(t-1)$$

Thus we have solved, at least in principle, the problem posed above. We now need to discover a method for finding inverses.

At several points in the following chapters we shall encounter simple matrix equations which can be solved by finding matrix inverses. To find them we need first to introduce a quantity known as a *determinant*. The determinant of a matrix is a scalar that is found by appropriate manipulation of the elements of a matrix. Determinants are defined *only for square* $(n \times n)$ *matrices*. In the case of a 1×1 matrix, the determinant

$$\det(\mathbf{A}) \qquad \text{or} \qquad |\mathbf{A}|$$

is simply the element itself. Thus if c is some number,

$$\det(c) = |c| = c$$

(Note that $|\cdot|$ does *not* indicate absolute value here as it does in scalar algebra.)

The determinant of a 2×2 matrix—say, $\begin{bmatrix} a & b \\ c & d \end{bmatrix}$—is

$$\det \begin{bmatrix} a & b \\ c & d \end{bmatrix} = \begin{vmatrix} a & b \\ c & d \end{vmatrix} = ad - bc \tag{1-12}$$

Find the determinants of the following matrices.

(xi) $\begin{bmatrix} 3 & 1 \\ 0 & 2 \end{bmatrix}$ Answer: $6 - 0 = 6$

(xii) $\begin{bmatrix} 3-\lambda & 0 \\ 0 & 1-\lambda \end{bmatrix}$ Answer: $(3-\lambda)(1-\lambda) - 0 = 3 - 4\lambda + \lambda^2$

(xiii) $\begin{bmatrix} 2-\lambda & 4 \\ 0 & -\lambda \end{bmatrix}$ Answer: $(2-\lambda)(-\lambda) - 4 \cdot 0 = \lambda(\lambda - 2)$

In most cases the determinant of a matrix has no easily interpretable meaning. It is, however, extremely useful for solving certain types of problems. Its value lies in the fact that matrices whose determinant equals zero have no inverse (see equation 1-14). Do the following matrices possess inverses?

(xiv) $\begin{bmatrix} 1 & 4 \\ 2 & 2 \end{bmatrix}$ Answer: Yes

(xv) $\begin{bmatrix} 4 & 2 \\ 2 & 1 \end{bmatrix}$ Answer: No

(xvi) $\begin{bmatrix} 1-\lambda & 0 \\ 0 & 2-\lambda \end{bmatrix}$ Answer: No if $\lambda = 1$ or 2, yes, otherwise

The determinant of any $m \times m$ matrix can be found by calculating:

$$\det(\mathbf{A}) = |\mathbf{A}| = \sum_k a_{ik}(-1)^{i+k}|\mathbf{A}_{ik}| \tag{1-13}$$

where i is the number of an arbitrary but fixed row of $\mathbf{A}$, and $\mathbf{A}_{ik}$ is the matrix $\mathbf{A}$ but with row i and column k deleted. $\mathbf{A}_{ik}$ matrices are referred to as *minors* of $\mathbf{A}$. Thus for $i = 1$

$$\begin{vmatrix} 2 & 1 & 3 \\ 4 & 0 & 1 \\ 2 & 2 & 1 \end{vmatrix} = 2\begin{vmatrix} 0 & 1 \\ 2 & 1 \end{vmatrix} - 1\begin{vmatrix} 4 & 1 \\ 2 & 1 \end{vmatrix} + 3\begin{vmatrix} 4 & 0 \\ 2 & 2 \end{vmatrix} = 2(0-2) - 1(4-2) + 3(8-0) = 18$$

or again for $i = 1$

$$\begin{vmatrix} 1 & 0 & 3 & 0 \\ 0 & 1 & 1 & 1 \\ 2 & 0 & 1 & 1 \\ 3 & 1 & 2 & 1 \end{vmatrix} = 1\begin{vmatrix} 1 & 1 & 1 \\ 0 & 1 & 1 \\ 1 & 2 & 1 \end{vmatrix} - 0\begin{vmatrix} 0 & 1 & 1 \\ 2 & 1 & 1 \\ 3 & 2 & 1 \end{vmatrix} + 3\begin{vmatrix} 0 & 1 & 1 \\ 2 & 0 & 1 \\ 3 & 1 & 1 \end{vmatrix} - 0\begin{vmatrix} 0 & 1 & 1 \\ 2 & 0 & 1 \\ 3 & 1 & 2 \end{vmatrix}$$

$$= 1\left(1\begin{vmatrix} 1 & 1 \\ 2 & 1 \end{vmatrix} - 1\begin{vmatrix} 0 & 1 \\ 1 & 1 \end{vmatrix} + 1\begin{vmatrix} 0 & 1 \\ 1 & 2 \end{vmatrix}\right) - 0$$

$$+ 3\left(0\begin{vmatrix} 0 & 1 \\ 1 & 1 \end{vmatrix} - 1\begin{vmatrix} 2 & 1 \\ 3 & 1 \end{vmatrix} + 1\begin{vmatrix} 2 & 0 \\ 3 & 1 \end{vmatrix}\right) - 0$$

$$= 1(-1 + 1 - 1) + 3(0 + 1 + 2) = -1 + 9 = 8$$

Which of the following matrices have inverses?

(xvii) $\begin{bmatrix} 2 & 0 & 3 \\ 1 & 1 & 2 \\ 0 & 4 & 0 \end{bmatrix}$ Answer: $\begin{vmatrix} 2 & 0 & 3 \\ 1 & 1 & 2 \\ 0 & 4 & 0 \end{vmatrix} = 2\begin{vmatrix} 1 & 2 \\ 4 & 0 \end{vmatrix} - 0\begin{vmatrix} 1 & 2 \\ 0 & 0 \end{vmatrix} + 3\begin{vmatrix} 1 & 1 \\ 0 & 4 \end{vmatrix}$

$$= 2(-8) - 0 + 3(4) = -4$$

This matrix has a nonzero determinant, so it *does* have an inverse.

(xviii) $\begin{bmatrix} -\lambda & 1 \\ 0 & 1-\lambda \end{bmatrix}$ Answer: $\begin{vmatrix} -\lambda & 1 \\ 0 & 1-\lambda \end{vmatrix} = -\lambda(1-\lambda) - 0 = \lambda^2 - \lambda$

Thus if $\lambda = 0$ or 1, the determinant equals zero and an inverse does not exist. For all other λ the matrix has an inverse.

(xix) $\begin{bmatrix} 1 & 2 & 3 \\ 0 & 1 & 1 \\ 1 & 4 & 5 \end{bmatrix}$ Answer: No

If all elements of a row or all elements of a column of a matrix are zero, the determinant of that matrix is zero. Other circumstances also guarantee a zero determinant. This occurs whenever one row (or column) can be written as a weighted sum of other rows (columns). For example, in problem (xix), the third row is the sum of the first row and twice the second row. The latter circumstances are not usually apparent, but learn to scan for zero rows or columns. A matrix that has a zero determinant is referred to as *singular*.

(xx) Consider the population problem (equations 1-6 and 1-7),

$$\mathbf{L} = \begin{bmatrix} 0 & 1 & 4 \\ \frac{1}{2} & 0 & 0 \\ 0 & \frac{1}{4} & 0 \end{bmatrix}$$

If $\mathbf{n}(t)$ is known, can $\mathbf{n}(t-1)$ be found? Answer: $\mathbf{n}(t-1)$ can be found if $\mathbf{L}$ has an inverse— that is, is nonsingular—has a nonzero determinant.

$$|\mathbf{L}| = \begin{vmatrix} 0 & 1 & 4 \\ \frac{1}{2} & 0 & 0 \\ 0 & \frac{1}{4} & 0 \end{vmatrix} = 0\begin{vmatrix} 0 & 0 \\ \frac{1}{4} & 0 \end{vmatrix} - 1\begin{vmatrix} \frac{1}{2} & 0 \\ 0 & 0 \end{vmatrix} + 4\begin{vmatrix} \frac{1}{2} & 0 \\ 0 & \frac{1}{4} \end{vmatrix}$$

$$= 0 - 0 + 4(\tfrac{1}{8}) = \tfrac{1}{2} \neq 0$$

so the answer is yes.

To calculate the inverse of any square matrix, write

$$\mathbf{A}^{-1} = \frac{\{(-1)^{i+j}|A_{ij}|\}}{|\mathbf{A}|^T} \tag{1-14}$$

where i and j denote rows and columns, T indicates transpose, $|\mathbf{A}|$ is the determinant, and $|A_{ij}|$ is the determinant of the (i,j)th minor of the matrix $\mathbf{A}$. Examples follow:

(a) $\mathbf{A} = \begin{bmatrix} 2 & 1 \\ 0 & 3 \end{bmatrix}$. Then

$$|A_{11}| = \det(3) = 3$$
$$|A_{12}| = \det(0) = 0$$
$$|A_{21}| = \det(1) = 1$$
$$|A_{22}| = \det(2) = 2$$
$$|A| = 6$$

So

$$\mathbf{A}^{-1} = \frac{1}{6}\begin{bmatrix} 3 & -0 \\ -1 & 2 \end{bmatrix}^T = \begin{bmatrix} \frac{3}{6} & -\frac{1}{6} \\ 0 & \frac{2}{6} \end{bmatrix}$$

(b) $\mathbf{A} = \begin{bmatrix} 1 & 0 & 0 \\ 0 & 2 & 0 \\ 0 & 1 & 1 \end{bmatrix}$. Then

$$|A_{11}| = \begin{vmatrix} 2 & 0 \\ 1 & 1 \end{vmatrix} = 2 \qquad |A_{23}| = \begin{vmatrix} 1 & 0 \\ 0 & 1 \end{vmatrix} = 1$$

$$|A_{12}| = \begin{vmatrix} 0 & 0 \\ 0 & 1 \end{vmatrix} = 0 \qquad |A_{31}| = \begin{vmatrix} 0 & 0 \\ 2 & 0 \end{vmatrix} = 0$$

$$|A_{13}| = \begin{vmatrix} 0 & 2 \\ 0 & 1 \end{vmatrix} = 0 \qquad |A_{32}| = \begin{vmatrix} 1 & 0 \\ 0 & 0 \end{vmatrix} = 0$$

$$|A_{21}| = \begin{vmatrix} 0 & 0 \\ 1 & 1 \end{vmatrix} = 0 \qquad |A_{33}| = \begin{vmatrix} 1 & 0 \\ 0 & 2 \end{vmatrix} = 2$$

$$|A_{22}| = \begin{vmatrix} 1 & 0 \\ 0 & 1 \end{vmatrix} = 1 \qquad \begin{aligned} |\mathbf{A}| &= 1|A_{11}| - 0|A_{12}| + 0|A_{13}| \\ &= 1(2) - 0 + 0 = 2 \end{aligned}$$

So

$$\mathbf{A}^{-1} = \frac{1}{2}\begin{bmatrix} 2 & 0 & 0 \\ 0 & 1 & -1 \\ 0 & 0 & 2 \end{bmatrix}^T = \frac{1}{2}\begin{bmatrix} 2 & 0 & 0 \\ 0 & 1 & 0 \\ 0 & -1 & 2 \end{bmatrix} = \begin{bmatrix} 1 & 0 & 0 \\ 0 & \frac{1}{2} & 0 \\ 0 & -\frac{1}{2} & 1 \end{bmatrix}$$

Check this result, and practice by inverting the following matrices.

(xxi)
$$\begin{bmatrix} 1 & 2 & 1 \\ 0 & 0 & 1 \\ 3 & 1 & 0 \end{bmatrix}$$

(xxii)
$$\begin{bmatrix} 1-\lambda & 0 & 0 \\ 0 & -\lambda & 0 \\ 0 & 1 & 1-\lambda \end{bmatrix}$$

(xxiii) Consider the matrix $\mathbf{L}$ in problem (xx). If

$$\mathbf{n}(t) = \begin{bmatrix} 20 \\ 20 \\ 0 \end{bmatrix},$$

what was $\mathbf{n}(t-1)$? What was $\mathbf{n}(t-2)$?

Answer: $\mathbf{L}^{-1} = \begin{bmatrix} 0 & 2 & 0 \\ 0 & 0 & 4 \\ \frac{1}{4} & 0 & -1 \end{bmatrix}$, so

$$\mathbf{n}(t-1) = \mathbf{L}^{-1}\cdot\mathbf{n}(t) = \begin{bmatrix} 0 & 2 & 0 \\ 0 & 0 & 4 \\ \frac{1}{4} & 0 & -1 \end{bmatrix}\begin{bmatrix} 20 \\ 20 \\ 0 \end{bmatrix} = \begin{bmatrix} 40 \\ 0 \\ 5 \end{bmatrix}$$

$$\mathbf{n}(t-2) = \mathbf{L}^{-1}\cdot n(t-1) = \begin{bmatrix} 0 & 2 & 0 \\ 0 & 0 & 4 \\ \frac{1}{4} & 0 & -1 \end{bmatrix} \begin{bmatrix} 40 \\ 0 \\ 5 \end{bmatrix} = \begin{bmatrix} 0 \\ 20 \\ 5 \end{bmatrix}$$

For an alternative way to obtain inverses (the Gauss–Jordan reduction method), the interested reader is referred to Anton (1977), Kolman (1980), or Williams (1978).

1.2 Eigenvalues and Eigenvectors

Eigenvalues

Consider once again the population process described earlier:

$$\mathbf{n}(t+1) = \mathbf{L}\cdot\mathbf{n}(t) \qquad (1\text{-}15)$$

and suppose that a stable age structure obtains—that is, each age class makes up an unchanging fraction of the total population. If the ratio of population sizes at time $t + 1$ and t is λ, we can then also write

$$\begin{bmatrix} n_1(t+1) \\ n_2(t+1) \\ n_3(t+1) \end{bmatrix} = \lambda \begin{bmatrix} n_1(t) \\ n_2(t) \\ n_3(t) \end{bmatrix},$$

or

$$\mathbf{n}(t+1) = \lambda\cdot\mathbf{n}(t) \qquad (1\text{-}16)$$

Subtracting (1-16) from (1-15), we have

$$0 = \mathbf{L}\cdot\mathbf{n}(t) - \lambda\cdot\mathbf{n}(t) = (\mathbf{L} - \lambda\mathbf{I})\mathbf{n}(t) \qquad (1\text{-}17)$$

This equation has an interesting property that is not immediately obvious but which is extremely useful: The determinant of $(\mathbf{L} - \lambda\mathbf{I})$ is *always* zero. To prove this assertion, suppose that $|\mathbf{L} - \lambda\mathbf{I}| \neq 0$. Then, since $\mathbf{L} - \lambda\mathbf{I}$ is a square matrix, it possesses an inverse. Thus

$$\begin{aligned} (\mathbf{L}-\lambda\mathbf{I})^{-1}(\mathbf{L}-\lambda\mathbf{I})\mathbf{n}(t) &= [(\mathbf{L}-\lambda\mathbf{I})^{-1}(\mathbf{L}-\lambda\mathbf{I})]\cdot\mathbf{n}(t) \quad \text{or} \quad (\mathbf{L}-\lambda\mathbf{I})^{-1}\cdot[(\mathbf{L}-\lambda\mathbf{I})\cdot\mathbf{n}(t)] \\ &= [\mathbf{I}]\cdot\mathbf{n}(t) \quad \text{or} \quad (\mathbf{L}-\lambda\mathbf{I})^{-1}[0] \qquad \text{by (1-17)} \\ &= \mathbf{n}(t) \quad \text{or} \quad 0 \end{aligned}$$

But if $\mathbf{n}(t)$ is not a zero vector—that is, if there exists a population at all—then what is shown above is a contradiction. Our supposition about $(\mathbf{L} - \lambda\mathbf{I})$ having a nonzero determinant must, therefore, be in error. We conclude that if

$$(\mathbf{L} - \lambda\mathbf{I})\cdot\mathbf{n}(t) = 0$$

then

$$|\mathbf{L} - \lambda\mathbf{I}| = 0$$

Why is this useful? Because $|\mathbf{L} - \lambda\mathbf{I}| = 0$ give us a polynomial equation in λ. Using the example in (1-6) gives us

$$\begin{aligned} 0 = |\mathbf{L} - \lambda\mathbf{I}| &= \left| \begin{bmatrix} 0 & 1 & 4 \\ \frac{1}{2} & 0 & 0 \\ 0 & \frac{1}{4} & 0 \end{bmatrix} - \lambda \begin{bmatrix} 1 & 0 & 0 \\ 0 & 1 & 0 \\ 0 & 0 & 1 \end{bmatrix} \right| = \begin{vmatrix} -\lambda & 1 & 4 \\ \frac{1}{2} & -\lambda & 0 \\ 0 & \frac{1}{4} & -\lambda \end{vmatrix} \\ &= -\lambda^3 + \tfrac{1}{2}\lambda + \tfrac{1}{2} \end{aligned}$$

whose solution is

$$\lambda = 1, \quad \frac{-1 + i}{2}, \quad \frac{-1 - i}{2} \qquad \text{where } i = \sqrt{-1}$$

The equation $|\mathbf{L} - \lambda\mathbf{I}| = 0$ is known as the *characteristic equation* of the matrix **L**, and the values of λ that satisfy the equation are called its *eigenvalues*. The second and third eigenvalues in the example above do not appear to make any sense—how can a population grow by a multiple of $(-1 + i)/2$ or $(-1 - i)/2$? Let us come back to this later. The first eigenvalue, $\lambda = 1$, tells us that *when the population is in stable age structure* it is constant in size (growth multiple of $\lambda = 1$ every generation).

If **L** happened to be,

$$\mathbf{L} = \begin{bmatrix} 0 & 2 & 3 \\ \frac{1}{2} & 0 & 0 \\ 0 & \frac{1}{4} & 0 \end{bmatrix}$$

how fast would the population be growing if it were in stable age structure?

Answer: When stable age structure holds, then

$$\mathbf{n}(t + 1) = \lambda \cdot \mathbf{n}(t) = \lambda \cdot \mathbf{I} \cdot \mathbf{n}(t)$$

Also,

$$\mathbf{n}(t + 1) = \mathbf{L} \cdot \mathbf{n}(t)$$

Therefore,

$$\mathbf{0} = \mathbf{L} \cdot \mathbf{n}(t) - \lambda \cdot \mathbf{I} \cdot \mathbf{n}(t) = (\mathbf{L} - \lambda\mathbf{I})\mathbf{n}(t),$$

so that

$$|\mathbf{L} - \lambda\mathbf{I}| = 0$$

Writing out this expression in full, we have

$$\mathbf{L} - \lambda\mathbf{I} = \left| \begin{bmatrix} 0 & 2 & 3 \\ \frac{1}{2} & 0 & 0 \\ 0 & \frac{1}{4} & 0 \end{bmatrix} - \lambda \begin{bmatrix} 1 & 0 & 0 \\ 0 & 1 & 0 \\ 0 & 0 & 1 \end{bmatrix} \right| = \begin{vmatrix} -\lambda & 2 & 3 \\ \frac{1}{2} & -\lambda & 0 \\ 0 & \frac{1}{4} & -\lambda \end{vmatrix} = -\lambda^3 + \lambda + \tfrac{3}{8}$$

whose solution is $\lambda = 1.151, -.497, -.654$. Hence (ignoring, for the moment, the negative roots) the population grows by a multiple of 1.151 each time unit.

Find the characteristic equations and eigenvalues for the following matrices.

(i) $\begin{bmatrix} 2 & 1 \\ 3 & 0 \end{bmatrix}$ Answer: $0 = \begin{vmatrix} 2 - \lambda & 1 \\ 3 & -\lambda \end{vmatrix} = \lambda^2 - 2\lambda - 3$, so $\lambda = 3, -1$

(ii) $\begin{bmatrix} 1 & 4 \\ 2 & 2 \end{bmatrix}$

(iii) $\begin{bmatrix} 1 & 2 & 0 \\ 0 & 1 & 0 \\ 0 & 0 & 1 \end{bmatrix}$

(iv) $\begin{bmatrix} 1 & 2 & 3 \\ 3 & 2 & 1 \\ 0 & -\frac{2}{3} & 0 \end{bmatrix}$

Note: Any cubic equation can be factored to give $(\lambda - \lambda_1)(\lambda - \lambda_2)(\lambda - \lambda_3) = 0$, where λ_1, λ_2, and λ_3 are the roots of the equation. So once any *one* root (say, λ_1) has been found, you can divide the characteristic equation by $(\lambda - \lambda_1)$—by long division—to find another, quadratic (and therefore solvable) equation for the remaining two roots.

Eigenvectors

We have shown that if

$$\mathbf{A}\cdot\mathbf{b} = \lambda\cdot\mathbf{b} \qquad \text{where } \mathbf{b} \neq 0 \tag{1-18}$$

where $\mathbf{A}$ is any square matrix and λ is a scalar, then $(\mathbf{A} - \lambda\mathbf{I})\,\mathbf{b} = 0$, whence

$$|\mathbf{A} - \lambda\mathbf{I}| = 0 \tag{1-19}$$

From (1-19) we can find the λ values for which (1-18) holds. But once the λ values characteristic for the matrix $\mathbf{A}$ have been found, can we simply plug them back in and expect (1-18) to hold for *any* vector $\mathbf{b}$? The answer is *no*. In fact, for each λ value calculated, there is a unique set of vectors that satisfy (1-18). For example, in the case of the population problem in (1-6),

$$\mathbf{L}\cdot\mathbf{n} = \lambda\mathbf{n} \qquad \text{where } \mathbf{L} = \begin{bmatrix} 0 & 1 & 4 \\ \frac{1}{2} & 0 & 0 \\ 0 & \frac{1}{4} & 0 \end{bmatrix} \tag{1-20}$$

when the population is at stable age distribution. We have already calculated the eigenvalues to be $\lambda = 1, (-1 + i)/2$, and $(-1 - i)/2$. The vector $\mathbf{n}$ which satisfies (1-20) *when* $\lambda = 1$ can be found by writing out $\mathbf{L}\cdot\mathbf{n} = \lambda\mathbf{n}$:

$$\begin{bmatrix} 0 & 1 & 4 \\ \frac{1}{2} & 0 & 0 \\ 0 & \frac{1}{4} & 0 \end{bmatrix} \begin{bmatrix} n_1 \\ n_2 \\ n_3 \end{bmatrix} = 1 \begin{bmatrix} n_1 \\ n_2 \\ n_3 \end{bmatrix}$$

so that

$$\begin{aligned} &(1) \qquad n_2 + 4n_3 = n_1 \\ &(2) \qquad \tfrac{1}{2}n_1 = n_2 \\ &(3) \qquad \tfrac{1}{4}n_2 = n_3 \end{aligned}$$

Solving these simultaneous equations, we get

$$\begin{aligned} \text{If } n_3 \text{ is set to } c, \text{ then} \quad n_2 &= 4c && \text{by (3)} \\ \text{and} \quad n_1 &= 2n_2 = 8c && \text{by (2)} \end{aligned}$$

The set of vectors corresponding to the eigenvalue $\lambda = 1$, then, is

$$c\begin{bmatrix} 8 \\ 4 \\ 1 \end{bmatrix}$$

where c is any nonzero number. The fact that we may multiply the eigenvector by any (nonzero) number should not be disturbing. It simply means that it is the ratios of the elements of the eigenvector, and not their absolute magnitudes, that are important. We can check the result above by writing

$$\mathbf{L}\cdot\mathbf{n} = \begin{bmatrix} 0 & 1 & 4 \\ \frac{1}{2} & 0 & 0 \\ 0 & \frac{1}{4} & 0 \end{bmatrix} \begin{bmatrix} 8c \\ 4c \\ c \end{bmatrix} = \begin{bmatrix} 8c \\ 4c \\ c \end{bmatrix} = 1\cdot\begin{bmatrix} 8c \\ 4c \\ c \end{bmatrix} = \lambda\begin{bmatrix} 8c \\ 4c \\ c \end{bmatrix} = \lambda\cdot\mathbf{n}$$

The interpretation of this vector is straightforward: **n** is the age vector that obtains when the population is at stable age distribution and growing at the rate λ. Again, the ratios $n_1 : n_2 : n_3$ in **n** are the ratios of the age groups when the population is at stable age distribution, no matter what the *absolute* numbers of individuals.

A vector **b** (or **n**) that satisfies an equation of type (1-18) [or (1-20)] for a given eigenvalue λ is referred to as an *eigenvector*.

Find the eigenvalues and corresponding eigenvectors associated with the following matrices.

(v) $\begin{bmatrix} 1 & 2 \\ 3 & 2 \end{bmatrix}$

Answer: $\begin{vmatrix} 1-\lambda & 2 \\ 3 & 2-\lambda \end{vmatrix} = \lambda^2 - 3\lambda - 4 = 0$, so $\lambda = 4, -1$.

For $\lambda = 4$, write

$$\begin{bmatrix} 1 & 2 \\ 3 & 2 \end{bmatrix}\begin{bmatrix} b_1 \\ b_2 \end{bmatrix} = 4\begin{bmatrix} b_1 \\ b_2 \end{bmatrix} = \begin{bmatrix} 4b_1 \\ 4b_2 \end{bmatrix}$$

so that

$$b_1 + 2b_2 = 4b_1$$
$$3b_1 + 2b_2 = 4b_2$$

Solving these, if b_1 is set to c, then b_2 must be $\frac{3}{2}c$, so the eigenvector is any multiple of $\begin{bmatrix} 1 \\ \frac{3}{2} \end{bmatrix}$. For $\lambda = -1$, write

$$\begin{bmatrix} 1 & 2 \\ 3 & 2 \end{bmatrix}\begin{bmatrix} b_1 \\ b_2 \end{bmatrix} = -1\begin{bmatrix} b_1 \\ b_2 \end{bmatrix} = \begin{bmatrix} -b_1 \\ -b_2 \end{bmatrix}$$

so that

$$b_1 + 2b_2 = -b_1$$
$$3b_1 + 2b_2 = -b_2$$

Solving these, if b_1 is set to c, b_2 must be $-c$, so the eigenvector is any nonzero multiple of $\begin{bmatrix} 1 \\ -1 \end{bmatrix}$.

(vi) $\begin{bmatrix} 1 & 2 \\ 0 & 3 \end{bmatrix}$

Answer: $\lambda = 1, 3$; eigenvectors are, respectively, any multiple of

$$\begin{bmatrix} 1 \\ 0 \end{bmatrix}, \begin{bmatrix} 1 \\ 1 \end{bmatrix}$$

Note: Singular matrices always have at least one zero root, and eigenvectors cannot be defined for zero roots.

Eigenvectors can be *row* as well as column vectors. For example, write

$$\mathbf{a}^T\begin{bmatrix} 1 & 2 \\ 3 & 2 \end{bmatrix} = \lambda\mathbf{a}^T$$

$\lambda = 4, -1$ [as in problem (v), using the same matrix]. For $\lambda = 4$,

$$[a_1 \quad a_2]\begin{bmatrix} 1 & 2 \\ 3 & 2 \end{bmatrix} = 4[a_1 \quad a_2]$$

so

$$a_1 + 3a_2 = 4a_1$$
$$2a_1 + 2a_2 = 4a_2$$

Solving, if we set a_1 to c, then a_2 must be c, so that the *row* eigenvector is any nonzero multiple of $[1 \quad 1]$. For $\lambda = -1$, the eigenvector is any nonzero multiple of $[1 \quad -\frac{2}{3}]$.
Notice that the row eigenvector is *not* simply the transpose of the column eigenvector! In fact, it is a totally different vector. Find *row* eigenvectors for problem (vi).

1.3 Diagonalizing Matrices

On those occasions when one absolutely must calculate eigenvectors, computers are generally used (it is simply not practical to calculate them for most matrices of size greater than 3×3). So there is no point in belaboring calculation methods. The foremost value of eigenvectors is in their *indirect* use to find general solutions to matrix equations. We address this use now.

Let $\mathbf{u}_i$ be the column eigenvector corresponding to the ith eigenvalue, λ_i, of a matrix $\mathbf{A}$. Then

$$\begin{aligned} \mathbf{A}\cdot\mathbf{u}_1 &= \lambda_1\mathbf{u}_1 \\ \mathbf{A}\cdot\mathbf{u}_2 &= \lambda_2\mathbf{u}_2 \\ \mathbf{A}\cdot\mathbf{u}_3 &= \lambda_3\mathbf{u}_3 \\ &\vdots \end{aligned} \tag{1-21}$$

If we stack $\mathbf{u}_1, \mathbf{u}_2, \mathbf{u}_3, \cdots$ next to each other to form a square matrix,

$$\mathbf{U} = [\mathbf{u}_1 \quad \mathbf{u}_2 \quad \mathbf{u}_3 \quad \cdots]$$

then

$$\begin{aligned} \mathbf{A}\cdot\mathbf{U} &= \mathbf{A}[\mathbf{u}_1 \quad \mathbf{u}_2 \quad \mathbf{u}_3 \quad \cdots] = [\mathbf{A}\cdot\mathbf{u}_1 \quad \mathbf{A}\cdot\mathbf{u}_2 \quad \mathbf{A}\cdot\mathbf{u}_3 \quad \cdots] = [\lambda_1\mathbf{u}_1 \quad \lambda_2\mathbf{u}_2 \quad \lambda_3\mathbf{u}_3 \quad \cdots] \\ &= [\mathbf{u}_1 \quad \mathbf{u}_2 \quad \mathbf{u}_3 \quad \cdots]\begin{bmatrix} \lambda_1 & 0 & 0 & \cdot \\ 0 & \lambda_2 & 0 & \cdot \\ 0 & 0 & \lambda_3 & \cdot \\ \cdot & \cdot & \cdot & \cdot \end{bmatrix} = \mathbf{U}\cdot\begin{bmatrix} \lambda_1 & 0 & 0 & \cdot \\ 0 & \lambda_2 & 0 & \cdot \\ 0 & 0 & \lambda_3 & \cdot \\ \cdot & \cdot & \cdot & \cdot \end{bmatrix} \end{aligned} \tag{1-22}$$

To make sure that you understand the machinations above, verify each step by writing out the $\mathbf{u}$ vectors in full.

The matrix

$$\begin{bmatrix} \lambda_1 & 0 & 0 & \cdot \\ 0 & \lambda_2 & 0 & \cdot \\ 0 & 0 & \lambda_3 & \cdot \\ \cdot & \cdot & \cdot & \cdot \end{bmatrix}$$

is usually written $\mathbf{\Lambda}$. Thus (1-22) can be rewritten in more tidy form as

$$\mathbf{A}\cdot\mathbf{U} = \mathbf{U}\cdot\mathbf{\Lambda}$$

from which

$$\mathbf{A} = (\mathbf{A})(\mathbf{U}\cdot\mathbf{U}^{-1}) = (\mathbf{A}\cdot\mathbf{U})\mathbf{U}^{-1} = (\mathbf{U}\cdot\mathbf{\Lambda})\mathbf{U}^{-1} = \mathbf{U}\cdot\mathbf{\Lambda}\cdot\mathbf{U}^{-1} \tag{1-23}$$

The fact that a square matrix $\mathbf{A}$ can be written as in (1-23) is of vital importance. Suppose that a biological process is described by a matrix equation of the form

$$\mathbf{x}(t+1) = \mathbf{A}\cdot\mathbf{x}(t)$$

(The population problem we have discussed repeatedly is a case in point.) Then we could write

$$\begin{aligned}\mathbf{x}(t) &= \mathbf{A}\cdot\mathbf{x}(t-1) = \mathbf{A}(\mathbf{A}\cdot\mathbf{x}(t-2)) = \mathbf{A}^2\cdot\mathbf{x}(t-2)\\ &= \mathbf{A}^2(\mathbf{A}\cdot\mathbf{x}(t-3)) = \mathbf{A}^3\cdot\mathbf{x}(t-3)\\ &= \cdots = \mathbf{A}^t\cdot\mathbf{x}(0)\end{aligned}$$

Having to multiply **A** times itself t times might be a rather horrible experience, especially if **A** were a large matrix and t a large number. But from (1-23) we can write

$$\begin{aligned}\mathbf{x}(t) = \mathbf{A}^t\cdot\mathbf{x}(0) &= (\mathbf{U}\Lambda\mathbf{U}^{-1})^t\cdot\mathbf{x}(0)\\ &= (\mathbf{U}\Lambda\mathbf{U}^{-1})(\mathbf{U}\Lambda\mathbf{U}^{-1})(\mathbf{U}\Lambda\mathbf{U}^{-1})\cdots\mathbf{x}(0)\\ &= \mathbf{U}(\Lambda\mathbf{U}^{-1}\mathbf{U})(\Lambda\mathbf{U}^{-1}\mathbf{U})(\Lambda\mathbf{U}^{-1}\mathbf{U})\cdots(\Lambda\mathbf{U}^{-1}\mathbf{U})\Lambda\mathbf{U}^{-1}\cdot\mathbf{x}(0)\\ &= \mathbf{U}(\Lambda\cdot\Lambda\cdot\Lambda\cdot\Lambda\cdot\Lambda\cdots)\Lambda\mathbf{U}^{-1}\cdot\mathbf{x}(0)\\ &= (\mathbf{U}\cdot\Lambda^t\cdot\mathbf{U}^{-1})\mathbf{x}(0)\end{aligned} \tag{1-24}$$

where

$$\Lambda^t = \begin{bmatrix}\lambda_1^t & 0 & 0 & \cdot\\ 0 & \lambda_2^t & 0 & \cdot\\ 0 & 0 & \lambda_3^t & \cdot\\ \cdot & \cdot & \cdot & \cdot\end{bmatrix} \tag{1-25}$$

Now consider any one element of $\mathbf{x}(t)$, noticing that it is a sum of the various λ_i^t, each multiplied by a value reflecting a combination of the elements of **U**, $\mathbf{U}^{-1}$, and $\mathbf{x}(0)$. That is,

$$x_i(t) = \sum_j C_{ij}\lambda_j^t \tag{1-26}$$

where $\{\lambda_j\}$ are the eigenvalues of **A** and the C_{ij} are scalar constants whose values can be calculated (if desired) by first finding the eigenvectors, then constructing **U**, and then multiplying out the expression, $\mathbf{U}\Lambda^t\mathbf{U}^{-1}\mathbf{x}(0)$. But all these calculations are really unnecessary, for if we know the values of $x_i(t)$ for the first few time periods, we can use them to find $\{C_{ij}\}$ in a much easier way. For example, suppose that

$$\mathbf{x}(t+1) = \mathbf{A}\cdot\mathbf{x}(t)$$

where

$$\mathbf{A} = \begin{bmatrix}1 & 2\\ 1 & 0\end{bmatrix} \quad \text{and} \quad \mathbf{x}(0) = \begin{bmatrix}3\\ 1\end{bmatrix} \quad \mathbf{x}(1) = \begin{bmatrix}5\\ 3\end{bmatrix} \tag{1-27}$$

Then

$$\mathbf{x}(t) = \mathbf{A}^t\cdot\mathbf{x}(0) = \mathbf{U}\cdot\begin{bmatrix}\lambda_1^t & 0\\ 0 & \lambda_2^t\end{bmatrix}\mathbf{U}^{-1}\cdot\mathbf{x}(0)$$

If we wished, we could find λ_1, λ_2, **U**, and $\mathbf{U}^{-1}$, and write out the expressions for $x_1(t)$ and $x_2(t)$. But these expressions are clearly just linear combinations (weighted sums) of the λ's. So, instead, write (equation 1-26)

$$\begin{aligned}x_1(t) &= C_{11}\lambda_1^t + C_{12}\lambda_2^t\\ x_2(t) &= C_{21}\lambda_1^t + C_{22}\lambda_2^t\end{aligned} \tag{1-28}$$

To find the numerical solution, find the eigenvalues $\lambda_1 = 2$ and $\lambda_2 = -1$ and substitute them into (1-28):

$$\begin{aligned}x_1(t) &= C_{11}(2)^t + C_{12}(-1)^t\\ x_2(t) &= C_{21}(2)^t + C_{22}(-1)^t\end{aligned} \tag{1-29}$$

But for age class 1 we know that the population is 3 at time 0, 5 at time 1. Therefore,

$$3 = x_1(0) = C_{11}(2)^0 + C_{12}(-1)^0 = C_{11} + C_{12}$$
$$5 = x_1(1) = C_{11}(2)^1 + C_{12}(-1)^1 = 2C_{11} - C_{12}$$

It is a straightforward matter to find

$$C_{11} = \tfrac{8}{3} \qquad C_{12} = \tfrac{1}{3} \tag{1-30}$$

Similarly, for the second age class

$$1 = x_2(0) = C_{21}(2)^0 + C_{22}(-1)^0 = C_{21} + C_{22}$$
$$3 = x_2(1) = C_{21}(2)^1 + C_{22}(-1)^1 = 2C_{21} - C_{22}$$

whence

$$C_{21} = \tfrac{4}{3} \qquad C_{22} = -\tfrac{1}{3} \tag{1-31}$$

The final expression is then [substituting (1-30) and (1-31) into (1-29)].

$$\begin{aligned} x_1(t) &= \tfrac{8}{3}\cdot 2^t + \tfrac{1}{3}(-1)^t \\ x_2(t) &= \tfrac{4}{3}\cdot 2^t - \tfrac{1}{3}(-1)^t \end{aligned} \tag{1-32}$$

Notice that the negative root (and negative roots in general) describe oscillations.

Use (1-26) to find solutions to the following matrix equations.

(i) $\mathbf{x}(t+1) = \begin{bmatrix} 1 & 1 \\ 0 & 2 \end{bmatrix} \mathbf{x}(t)$ where $\mathbf{x}(0) = \begin{bmatrix} 3 \\ 0 \end{bmatrix}$, $\mathbf{x}(1) = \begin{bmatrix} 3 \\ 0 \end{bmatrix}$

(ii) $\mathbf{y}(t) = \begin{bmatrix} -2 & 0 \\ 1 & 1 \end{bmatrix} \mathbf{y}(t-1)$ where $\mathbf{y}(0) = \begin{bmatrix} 2 \\ 2 \end{bmatrix}$. Find $\mathbf{y}(1)$ yourself.

(iii) $\mathbf{z}(t) = \begin{bmatrix} 1 & 1 \\ 9 & 1 \end{bmatrix}^t \cdot \mathbf{z}(0)$ where $\mathbf{z}(2) = \begin{bmatrix} 10 \\ 0 \end{bmatrix}$

(iv) $\mathbf{x}(t+1) = \begin{bmatrix} 1 & 0 & 0 \\ 2 & -1 & 0 \\ 3 & 0 & 2 \end{bmatrix} \mathbf{x}(t)$ where $\mathbf{x}(0) = \begin{bmatrix} 10 \\ 10 \\ 10 \end{bmatrix}$

The same approach, but with a twist, may be used to solve differential, matrix equations. Suppose that we have

$$\frac{d}{dt}\mathbf{x} = \mathbf{A}\cdot\mathbf{x} \tag{1-33}$$

In this case, define a new vector, $\mathbf{y}$:

$$\mathbf{y} = \mathbf{U}^{-1}\cdot\mathbf{x} \tag{1-34}$$

where, as before, $\mathbf{U}$ is the matrix of column eigenvectors. Then

$$\mathbf{x} = \mathbf{U}\cdot\mathbf{y}$$

and (1-33) may be written

$$\frac{d}{dt}\mathbf{x} = \frac{d}{dt}\mathbf{U}\mathbf{y} = \mathbf{U}\frac{d}{dt}\mathbf{y}$$

and

$$\mathbf{A}\cdot\mathbf{x} = \mathbf{A}(\mathbf{U}\cdot\mathbf{y})$$

so that

$$\mathbf{U}\frac{d}{dt}\mathbf{y} = \mathbf{A}\cdot\mathbf{U}\cdot\mathbf{y}$$

Premultiplying both sides by $\mathbf{U}^{-1}$, we obtain

$$\frac{d}{dt}\mathbf{y} = (\mathbf{U}^{-1}\cdot\mathbf{A}\cdot\mathbf{U})\mathbf{y} \tag{1-35}$$

which, as you recall, can be written

$$\begin{bmatrix} \frac{dy_1}{dt} \\ \frac{dy_2}{dt} \\ \frac{dy_3}{dt} \end{bmatrix} = \frac{d}{dt}\mathbf{y} = \mathbf{\Lambda}\cdot\mathbf{y} = \begin{bmatrix} \lambda_1 & 0 & 0 \\ \mathbf{0} & \lambda_2 & 0 \\ 0 & 0 & \lambda_3 \end{bmatrix} \mathbf{y} = \begin{bmatrix} \lambda_1 y_1 \\ \lambda_2 y_2 \\ \lambda_3 y_3 \end{bmatrix} \tag{1-36}$$

or, simply,

$$\begin{aligned} \frac{dy_1}{dt} &= \lambda_1 y_1 \\ \frac{dy_2}{dt} &= \lambda_2 y_2 \\ \frac{dy_3}{dt} &= \lambda_3 y_3 \end{aligned} \tag{1-37}$$

whose solutions are $y_1(t) = y_1(0)e^{\lambda_1 t}$, $y_2(t) = y_2(0)e^{\lambda_2 t}$, and $y_3(t) = y_3(0)e^{\lambda_3 t}$. What we have done is to separate the equations for y_i into independent parts. To find $x_i(t)$, note that

$$\mathbf{x} = \mathbf{U}\cdot\mathbf{y} \tag{1-38}$$

so that x_i is a linear combination of the y_i's:

$$x_i(t) = \sum_j k_{ij}y_j = \sum_j k_{ij}e^{\lambda_j t}y_j(0)$$

where k_{ij} are the elements of $\mathbf{U}^{-1}$. Letting $k_{ij}y_j(0) = C_{ij}$, the expression above finally becomes

$$x_i(t) = \sum_j C_{ij}e^{\lambda_j t} \tag{1-39}$$

This is a general solution identical in form to that of (1-26), except that, in the case of a differential equation, we deal with $e^{\lambda_j t}$ terms instead of λ_j^t terms.

Using (1-39), find solutions to the following equations.

(v) $$\frac{d}{dt}\mathbf{X} = \begin{bmatrix} 3 & 1 \\ 0 & 2 \end{bmatrix}\mathbf{X} \qquad \text{where } \mathbf{x}(0) = \begin{bmatrix} 2 \\ 0 \end{bmatrix}$$

(vi) $$\frac{d}{dt}\mathbf{X} = \begin{bmatrix} 1 & 2 \\ 2 & 2 \end{bmatrix}\mathbf{X} \qquad \text{where } \mathbf{x}(0) = \begin{bmatrix} 10 \\ 10 \end{bmatrix}$$

Answer to (v): For the first problem begin by finding $\lambda = 3, 2$, so that

$$x_i(t) = C_{i1}e^{3t} + C_{i2}e^{2t} \tag{1-40}$$

To find C_{i1} and C_{i2} we must find $x_i(t)$ for some $t \neq 0$. But this cannot be done directly; we must use the variable, y_i. Accordingly, we construct

$$\mathbf{U} = \begin{bmatrix} 1 & 1 \\ 0 & -1 \end{bmatrix} \qquad \mathbf{U}^{-1} = \begin{bmatrix} 1 & 1 \\ 0 & -1 \end{bmatrix}$$

and calculate

$$\mathbf{y}(0) = \mathbf{U}^{-1} \cdot \mathbf{x}(0) = \begin{bmatrix} 1 & 1 \\ 0 & -1 \end{bmatrix} \begin{bmatrix} 2 \\ 0 \end{bmatrix} = \begin{bmatrix} 2 \\ 0 \end{bmatrix}$$

Then

$$\frac{dy_1}{dt} = 3y_1 \qquad \frac{dy_2}{dt} = 2y_2 \qquad \text{(from 1-37)}$$

so that

$$y_1(t) = y_1(0)e^{3t} = 2e^{3t} \qquad y_2(t) = y_i(0)e^{2t} = 0$$

If $t = 1$, then

$$y_1(1) = 2e^3 \qquad y_2(1) = 0$$

and

$$\mathbf{x}(1) = \mathbf{U} \cdot \mathbf{y}(1) = \begin{bmatrix} 1 & 1 \\ 0 & -1 \end{bmatrix} \begin{bmatrix} 2e^3 \\ 0 \end{bmatrix} = \begin{bmatrix} 2e^3 \\ 0 \end{bmatrix}$$

This can now be used in (1-40)

$$x_1(0) = C_{11}e^0 + C_{12}e^0 = C_{11} + C_{12} = 2$$
$$x_1(1) = C_{11}e^3 + C_{12}e^2 = 2e^3 \qquad \text{so } C_{11}e + C_{12} = 2e$$

whence

$$C_{11} = 2, C_{12} = 0,$$

and

$$x_1(t) = 2e^{3t}$$

Similarly,

$$x_2(t) = 0$$

1.4 Interpreting Complex Roots

Return to the population problem.

$$\mathbf{n}(t + 1) = \begin{bmatrix} 0 & 1 & 4 \\ \frac{1}{2} & 0 & 0 \\ 0 & \frac{1}{4} & 0 \end{bmatrix} \mathbf{n}(t)$$

Applying the general solution technique of (1-26) to this case, we have

$$n_i(t) = C_{i1}\lambda_1^t + C_{i1}\lambda_2^t + C_{13}\lambda_3^t$$

where $\lambda_1 = 1$, $\lambda_2 = (-1 + i)/2$, and $\lambda_3 = (-1 - i)/2$. What are we to do with the *complex roots*—those containing a term involving the square root of a negative number?

Consider any complex number, written

$$u + iv$$

and designate this number as a point on a graph, the x-coordinate representing the magnitude of the *real* component (u), the y-coordinate giving the magnitude of the *imaginary* part (v) (see Fig. 1-1). Recall your basic rules of trigonometry. If the distance to the point (u, v) from the origin is called r, and the line connecting the point to the origin subtends an angle θ with the real axis, then

$$\frac{v}{r} = \sin\theta$$

$$\frac{u}{r} = \cos\theta \tag{1-41}$$

$$r = \sqrt{u^2 + v^2}$$

Thus $u + iv$ can be written $(r\cos\theta) + i(r\sin\theta) = r(\cos\theta + i\sin\theta)$, where r and θ are as defined above. The problem of raising $(u + iv)$ to some power (say t), then, is equivalent to the problem of finding $[r(\cos\theta + i\sin\theta)]^t$. To do this, recall another tenet of trigonometry:

$$\sin(\alpha + \beta) = \sin\alpha\cos\beta + \cos\alpha\sin\beta$$
$$\cos(\alpha + \beta) = \cos\alpha\cos\beta - \sin\alpha\sin\beta$$

Thus

$$\begin{aligned}(u + iv)^2 = [r(\cos\theta + i\sin\theta)]^2 &= r^2(\cos\theta\cos\theta + 2i\cos\theta\sin\theta + i^2\sin\theta\sin\theta)\\ &= r^2[(\cos\theta\cos\theta - \sin\theta\sin\theta)\\ &\quad + i(\sin\theta\cos\theta + \cos\theta\sin\theta)]\\ &= r^2[\cos(2\theta) + i\sin(2\theta)]\end{aligned}$$

Similarly,

$$\begin{aligned}(u + iv)^3 &= [r(\cos\theta + i\sin\theta)]^3 = r^3(\cos\theta + i\sin\theta)^2(\cos\theta + i\sin\theta)\\ &= r^3(\cos 2\theta + i\sin 2\theta)(\cos\theta + i\sin\theta)\\ &= r^3[(\cos 2\theta\cos\theta - \sin 2\theta\sin\theta) + i(\sin 2\theta\cos\theta + \sin\theta\cos 2\theta)]\\ &= r^3(\cos 3\theta + i\sin 3\theta)\end{aligned}$$

and so on. In the general case,

$$[r(\cos\theta + i\sin\theta)]^t = r^t(\cos\theta t + i\sin\theta t) \tag{1-42}$$

We now make use of a theorem (not proven here) that whenever an equation has a root of the form, $u + iv$, it also has a *complex conjugate* root of the form $u - iv$. Clearly, r is the same for both, and if ϕ is the angle for the second root,

$$\phi = \sin^{-1}\left(\frac{-v}{r}\right) = -\sin^{-1}\left(\frac{v}{r}\right) = -\theta$$

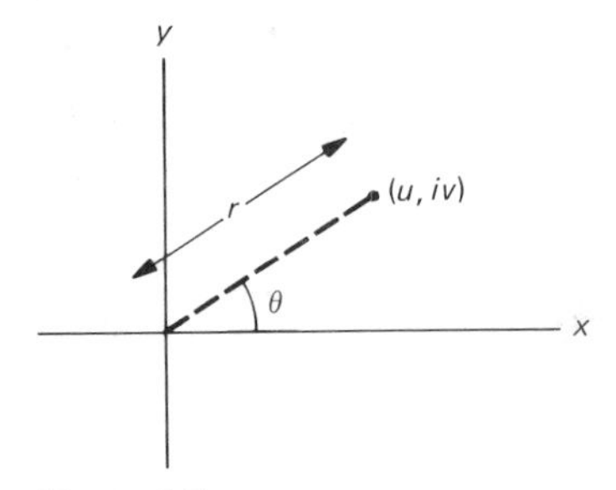

Figure 1-1

Designate the kth pair of conjugate roots with the subscripts $k1, k2$. The contribution to $x_i(t)$ of these roots (the corresponding terms in equation 1-26) is

$$\begin{aligned} C_{ik1}\lambda_{k1}^t t + C_{ik2}\lambda_{k2}^t t &= r_k^t C_{ik1}(\cos\theta_k t + i\sin\theta_k t) + r_k^t C_{ik2}(\cos\theta_k t - i\sin\theta_k t) \\ &= r_k^t[(C_{ik1} + C_{ik2})\cos\theta_k t + i(C_{ik1} - C_{ik2})\sin\theta_k t] \end{aligned} \tag{1-43}$$

If $x_i(t)$ is to be real—as it must be if it has biological meaning—then $(C_{ik1} + C_{ik2})$ and $i(C_{ik1} - C_{ik2})$ must both be real for all pairs of roots, k. This is always possible—it occurs when C_{ik1} and C_{ik2} are complex conjugates. Thus define

$$a_{ik} = (C_{ik1} + C_{ik2})$$
$$b_{ik} = i(C_{ik1} - C_{ik2})$$

whence the contribution to $x_i(t)$ of the kth pair of conjugate roots is

$$r_k^t(a_{ik}\cos\theta_k t + b_{ik}\sin\theta_k t) \tag{1-44a}$$

For differential equations,

$$e^{u_k t}(a_{ik}\cos\theta_k t + b_{ik}\sin\theta_k t) \tag{1-44b}$$

Example: Suppose that we have a discrete-time equation defined by a matrix with a characteristic equation with roots $\lambda = 1, -\frac{1}{2}, \frac{2}{3} + \frac{1}{3}i$, and $\frac{2}{3} - \frac{1}{3}i$. Then

$$x_i(t) = \sum_j C_{ij}\lambda_j^t = C_{i1}(1)^t + C_{i2}(-\tfrac{1}{2})^t + r^t(a_i\cos\theta t + b_i\sin\theta t)$$

where $r = \sqrt{u^2 + v^2} = \sqrt{\frac{4}{9} + \frac{1}{9}} = .746$, $\theta = \sin^{-1}(v/r) = \sin^{-1}(.333/.746) = \sin^{-1}(.446)$, and a_i, b_i, C_{i1} and C_{i2} depend on $x(0)$.

It is clear that complex roots, like negative roots, describe oscillations—$\cos\theta t$ and $\sin\theta t$ vary "sinusoidally" between -1 and $+1$ as θt increases with time.

In general we are less interested in specific solutions than in whether there are oscillations (are there negative or complex roots?), and whether the oscillations damp. With respect to the latter question, if the equation is a differential equation, then each cos, sin term has associated with it an e^{ut}, where u is the real part of the appropriate eigenvalue. If $u > 0$, the oscillations associated with that term clearly grow in amplitude. If $u < 0$, the oscillations damp out. In discrete-time equations [of the sort $\mathbf{x}(t + 1) = \mathbf{A}\cdot\mathbf{x}(t)$] each oscillatory term has associated with it an r^t value. Where $r > 1$, the oscillations grow. Where $r < 1$, they damp.

For the following problems, find all r (or u) values, and thus determine whether the oscillatory terms damp out.

(i) $\mathbf{x}(t + 1) = \begin{bmatrix} 1 & 3 \\ -2 & -3 \end{bmatrix}\mathbf{x}(t)$

Answer: $\lambda = -1 + 1.414i, -1 - 1.414i$. Thus

$$r = \sqrt{(-1)^2 + (1.414^2)} = \sqrt{3} > 1$$

and the oscillations grow in amplitude.

(ii) $\dfrac{d}{dt}\mathbf{x} = \begin{bmatrix} 3 & -1 \\ 2 & 2 \end{bmatrix}\mathbf{x}$

Answer: $\lambda = \dfrac{5 + i\sqrt{7}}{2}, \dfrac{5 - i\sqrt{7}}{2}$,

so the real part of the roots is $\frac{5}{2} > 0$—the oscillations grow in amplitude.

A caveat is in order. The general solutions given above (equations 1-26 and 1-39) are accurate only if all eigenvalues of the corresponding matrix are not identical. To mathematicians this is a legitimate concern; to biologists it is not. The chances of two identical eigenvalues occurring in nature is vanishingly small.

1.5 Dispersion

The average number of eggs laid by a given species of bird varies somewhat from female to female. Suppose that a given population of some species one year is characterized by 100 females laying three eggs, 1000 females laying four eggs, 2000 laying five, 3000 laying six, and 1000 laying seven. This pattern, known as a *frequency distribution*, may be illustrated as in Figure 1-2. Frequency distributions can be described with the use of a graph, as shown, or by word description. They can also be identified by *moments*. The first moment will be familiar to everyone. It is also known as the *mean*, usually designated μ. In the present example it may be easily found by noting that a fraction 100/7100 of the birds lay three eggs, 1000/7100 lay four, and so on. Thus

$$\mu = \frac{100}{7100}(3) + \frac{1000}{7100}(4) + \frac{2000}{7100}(5) + \frac{3000}{7100}(6) + \frac{1000}{7100}(7) = 5.535$$

In general, if we are dealing with some measure, x, and if $p(x)$ is the proportion of the total population that has value x, then

$$\mu = \sum_x xp(x) \tag{1-45}$$

μ is also, often written $E(x)$, the *expected value* of x. Note that the expected value of a constant, c, is that constant: $E(c) = c$.

The second moment is the mean of the x values squared, $\sum_x x^2p(x)$. It is generally convenient, though, to consider not the second moment per se, but rather the second moment "about the mean":

$$\sum_x (x - \mu)^2 p(x)$$

or

$$\begin{aligned} E(x - \mu)^2 &= E(x^2 - 2x\mu + \mu^2) \\ &= E(x^2) - 2E(x)\mu + \mu^2 \\ &= E(x^2) - \mu^2 \end{aligned} \tag{1-46}$$

This value is a measure of the width of the frequency distribution (Fig. 1-2), and is also known as the *variance*, usually written σ^2.

In many cases the mean (first moment) and variance (second moment about the mean) are sufficient to describe a frequency distribution reasonably well. However, if needed, we can also calculate a third moment about the mean,

$$\sum_x (x - \mu)^3 p(x)$$

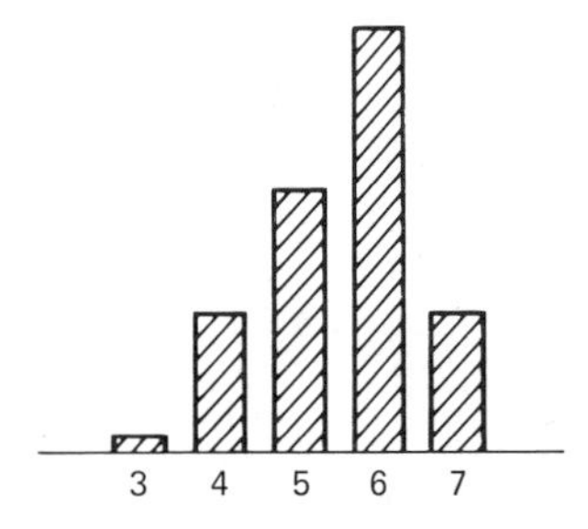

Figure 1-2 Bar graph of number of females laying 3, 4, . . . , 7 eggs.

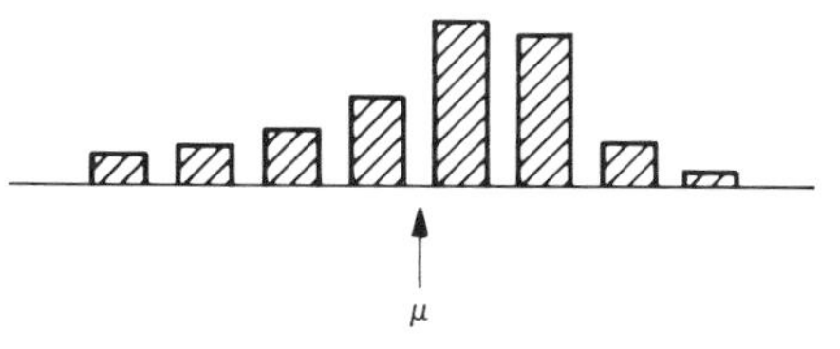

Figure 1-3

a fourth moment about the mean,

$$\sum_x (x - \mu)^4 p(x)$$

and so on. The quantity $\sum_x (x - \mu)^3 p(x)/\sigma^3$ is known as the *skew*. *Kurtosis* is given by $[\sum_x (x - \mu)^4 p(x) - 3]/\sigma^4$. A curve such as that in Figure 1-3 has more, larger, negative $(x - \mu)^3$ terms than positive, and so has negative skew. It is said to be skewed to the left. Curves shown in Figure 1-4a and b are, respectively *platykurtic* (having a large fourth moment about the mean) and *leptokurtic* (having a small fourth moment about the mean). For the moment let us worry only about the mean, μ, and the variance, σ^2.

Now suppose, as realistically must be the case, that we cannot examine all 7100 nests produced by the population above, but can only *sample* a few of them, say 100. If we take care to choose our sample in such a way that obvious bias toward large, small, or intermediate numbers of eggs is avoided (which may not always be easy), then we can argue that it is reasonably representative of the population as a whole. Suppose that we were to find two nests with three eggs, 18 with four, 22 with five, 51 with six, and 7 with seven eggs. We can calculate an average, or *sample mean*,

$$\bar{x} = \frac{2(3) + 18(4) + 22(5) + 51(6) + 7(7)}{100} = 5.43$$

We simply add up all the values and divide by the sample size,

$$\bar{x} = \frac{\sum_{i=1}^{n} x_i}{n} \qquad \text{where } n = \text{sample size}$$

Because we have only *sampled* the population, not recorded it in full, we cannot necessarily expect our sample mean $\bar{x}$ to equal the true mean μ. However, if we were to sample the population over and over, and find the mean of all our average x values, we would find

$$\begin{aligned} E(\bar{x}) &= E\left(\frac{\sum_{i=1}^{n} x_i}{n}\right) = \frac{1}{n}\sum_{i=1}^{n} E(x_i) \\ &= \frac{1}{n}\sum_{i=1}^{n} \mu = \frac{1}{n} n\mu = \mu \end{aligned} \tag{1-47}$$

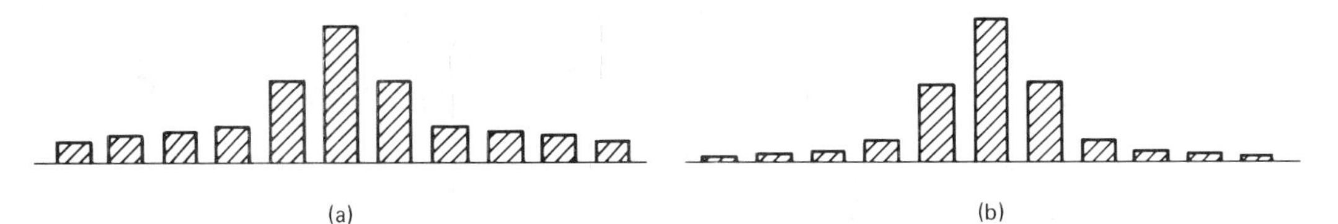

Figure 1-4

Because the expected value of the average, $E(\bar{x})$, equals the true mean, we say that $\bar{x}$ is an *unbiased estimator* of μ.

Variance, too, can be estimated from a sample. Because, from a sample, we cannot even known μ for sure, we must estimate μ with $\bar{x}$, and write an expression analogous to (1-46):

$$\text{sample variance—usually called } s^2 = \frac{\sum_{i=1}^{n} (x - \bar{x})^2}{n - 1}$$

Why should the denominator used in the expression to calculate sample variance be $n - 1$, not n? The reason is that dividing by $n - 1$ gives us an *unbiased estimate* of σ^2.

It is often convenient to use the notation $\text{var}(x)$ to represent either σ^2 or s^2. Whether $\text{var}(x)$ refers to the true or the sample variance should be clear from context.

We now derive two rules that lie at the heart of a powerful, statistical technique known as *analysis of variance*. These rules will also be useful to us from time to time in this book.

1. If we wished to calculate the variance of a series of values measured in meters, but our x values were in centimeters, we would need to use a conversion factor and calculate

$$\text{var}(cx) \qquad \text{where } c = \tfrac{1}{100} \text{ in this case}$$

But

$$\text{var}(cx) = E(cx - c\mu)^2 = c^2 E(x - \mu)^2 = c^2 \,\text{var}(x) \tag{1-48}$$

2. If we sample twice (or more times), adding the results, what is the variance of these sums?

$$\begin{aligned}\text{var}(x_1 + x_2) &= E[(x_1 + x_2) - (\mu + \mu)]^2 \\ &= E[(x_1 - \mu)^2 + 2(x_1 - \mu)(x_2 - \mu) + (x_2 - \mu)^2]\end{aligned}$$

But because x_1 and x_2 are sampled randomly from the same population the value of x_2 does not depend on the value of the previously measured x_1—the values are independent. But this means that the value of $(x_1 - \mu)$ varies independently of the value of $(x_2 - \mu)$. Therefore, for any value of x_1 (say), the expected value of $(x_1 - \mu)(x_2 - \mu)$ is

$$(x_1 - \mu)E(x_2 - \mu) = (x_1 - \mu)\cdot 0 = 0$$

Clearly, the expected value of these zeros over all x_1 must also be zero. Thus the middle term in the equation above drops out and we have

$$\text{var}(x_1 + x_2) = E(x_1 - \mu)^2 + E(x_2 - \mu)^2 = \text{var}(x_1) + \text{var}(x_2) \tag{1-49}$$

In the general case,

$$\text{var}\left(\sum_i x_i\right) = \sum_i \text{var}(x_i) \tag{1-50}$$

Equation (1-50) is the one utilized in deriving (1-51):

3. Suppose that we sampled (sample size n) many times from a population, each time calculating an average, $\bar{x}$. What is the variance of these averages? Write

$$\text{var}(\bar{x}) = \text{var}\left(\frac{1}{n}\sum_{i=1}^{n} x_i\right) = \frac{1}{n^2} n \,\text{var}(x) = \frac{1}{n} \text{var}(x) \tag{1-51}$$

We now summarize. Whenever observations of interest can be described in terms of so many observations with one value, so many with another, and so on, the corresponding description of these observations is a frequency distribution. A frequency distribution may pertain to a whole population of values. In this case it can be characterized by a mean,

$$\mu = E(x)$$

and a variance,

$$\sigma^2 = E(x - \mu)^2 = E(x^2) - \mu^2$$

and "higher-order" moments about the mean.

If a frequency distribution pertains to a sample (of size n) of the whole population, then the average, or sample mean value, $\bar{x}$, can be used as an unbiased estimate of the true mean, and the sample variance,

$$s^2 = \frac{\sum_{i=1}^{n} (x_i - \bar{x})^2}{n - 1}$$

can be used as an unbiased estimator of the true variance.

We saw that the variance of $x_1 + x_2$, where the value of x_2 is independent of the value of x_1, is simply $\text{var}(x_1) + \text{var}(x_2)$. But what if two values, x and y, say are *not* independent? For example, suppose that we had sample data on female weights and the weights of their offspring and wished to calculate a variance in biomass (weight) among family units. If x is female weight and y is combined offspring weight, family unit weight is $x + y$, and we wish to find $\text{var}(x + y)$. It is clear in this case that y may very much depend on x. Larger females, for example, might a priori be expected to produce larger or more numerous young. So x and y are not independent, and we cannot simply write

$$\text{var}(x + y) = \text{var}(x) + \text{var}(y)$$

But we can write the problem out the hard way:

$$\begin{aligned}
\text{var}(x + y) &= \frac{\sum_i [(x_i + y_i) - (\bar{x} + \bar{y})]^2}{n - 1} \\
&= \frac{\sum_i [(x_i - \bar{x})^2 + 2(x_i - \bar{x})(y_i - \bar{y}) + (y_i - \bar{y})^2]}{n - 1} \\
&= \frac{\sum_i (x_i - \bar{x})^2}{n - 1} + \frac{2\sum_i (x_i - \bar{x})(y_i - \bar{y})}{n - 1} + \frac{\sum_i (y_i - \bar{y})^2}{n - 1} \\
&= \text{var}(x) + 2\,\text{cov}(x, y) + \text{var}(y)
\end{aligned} \tag{1-52}$$

The term $\text{cov}(x, y)$ is known as the *covariance* of x and y.

To interpret covariance, look at its definition:

$$\text{cov}(x, y) = \frac{\sum_i (x_i - \bar{x})(y_i - \bar{y})}{n - 1} \tag{1-53}$$

Suppose that x and y covary in the *same direction*. That means that when a given x exceeds the average x, the corresponding y is also likely to exceed the average y; $(x - \bar{x})(y - \bar{y})$ is positive; if x is less than $\bar{x}$, then y is usually less than $\bar{y}$—again, $(x - \bar{x})(y - \bar{y})$ is positive. If large mothers, by and large, have larger litter weights and smaller mothers produce smaller litter weights, x and y are said to positively covary; the covariance is positive. If, on the other hand, large mothers generally produce smaller litter weight, and vice versa, $(x - \bar{x})$ and $(y - \bar{y})$ will usually be of opposite sign, so $\text{cov}(x, y)$ will be negative. If there is no relationship whatsoever between maternal and litter weight, there is no reason to expect $(x - \bar{x})(y - \bar{y})$ to be generally positive or negative, and the average of these products has an expected value of zero; $\text{cov}(x, y) = 0$.

Returning for a moment to (1-52), note that if x and y vary *identically*, so that $\text{cov}(x, y) = \text{cov}(x, x) = \text{var}(x)$, $\text{var}(y) = \text{var}(x)$, then $\text{var}(x + y) = \text{var}(x) + 2\,\text{cov}(x, y) + \text{var}(y) = \text{var}(x) + 2\,\text{var}(x) + \text{var}(x) = 4\,\text{var}(x)$. This makes good sense: In this case $\text{var}(x + y) = \text{var}(2x) = 4\,\text{var}(x)$. On the other hand, if x and y varied identically, but in opposite directions, an increase in x would be balanced by a corresponding drop in y, and $\text{var}(x + y) = \text{var}(0) = 0$. Equivalently, $\text{cov}(x, y) = \text{cov}(x, -x) = -\text{var}(x)$, so that $\text{var}(x + y) = \text{var}(x) - 2\,\text{var}(x) + \text{var}(x) = 0$. If x and y are independent, $\text{cov}(x, y) = 0$ and $\text{var}(x + y) = \text{var}(x) + \text{var}(y)$.

In the popular vernacular we are likely to associate positive, negative, or zero covariance with the expressions "x and y are correlated positively, or negatively, or x and y are not correlated." This terminology is also quite proper in more formal contexts, but we must provide a clear, rigorous definition of exactly what we mean by it. The formal definition of *correlation* is simply

$$\text{correlation}(x, y) = \frac{\text{cov}(x, y)}{\sqrt{\text{var}(x)\,\text{var}(y)}} \tag{1-54}$$

Obviously, the sign of a correlation is the same as that of the corresponding covariance.

As with mean and variance, covariance and correlation are written using different symbols depending on whether we are referring to a whole population or to a sample of that population. We have

$$\begin{aligned} \text{true covariance}(x, y) &= \sigma_{x,y} = E(x - \mu_x)(y - \mu_y) \\ \text{sample covariance}(x, y) &= s_{x,y} = \frac{\sum(x - \bar{x})(y - \bar{y})}{n - 1} \end{aligned} \tag{1-55}$$

$$\begin{aligned} \text{true correlation}(x, y) &= \rho_{x,y} = \frac{\sigma_{x,y}}{\sqrt{\sigma_x^2 \sigma_y^2}} = \frac{\sigma_{x,y}}{\sigma_x \sigma_y} \\ \text{sample correlation}(x, y) &= r_{x,y} = \frac{s_{x,y}}{\sqrt{s_x^2 s_y^2}} = \frac{s_{x,y}}{s_x s_y} \end{aligned} \tag{1-56}$$

In summary, whenever one parameter is associated with another parameter (e.g., mother weight and litter weight, wing length and bill length, number of species A in a quadrat sample and number of species B in the same quadrat sample, body fat content and hair color), it is possible to calculate a covariance or correlation. If the two parameters vary in the same direction—when the value of one is large that of the other is also generally large, and vice-versa—the covariance and correlation are positive. If the two parameters vary in opposite ways, the covariance and correlation are negative. A zero covariance and correlation are found when the two parameters vary in value independently of one another.

The example immediately above will serve as an illustration of the value of using correlation coefficients instead of covariance values in some instances. When the two parameters (e.g., body fat content and hair color) are measured in different units, interpretation of a covariance is a bit uncertain. The correlation, though, is a unitless quantity:

$$r_{(\text{weight, color})} = \frac{\text{cov}(\text{weight, color})}{\sqrt{\text{var}(\text{weight})\,\text{var}(\text{color})}}$$

which has units

$$\frac{(\text{weight})(\text{color})}{(\text{weight})(\text{color})} = \text{unitless}$$

Even when the two parameters in question (say, wing length and bill length) are measurable in the same units, correlation coefficients are often more useful than covariance values. In this case we might measure length in millimetres, centimetres, inches, and so on. The covariance is unit specific; the correlation is not:

$$\text{cov (mm, mm) has dimension mm}^2$$

$$r_{\text{mm, mm}} \text{ is unitless}$$

The measures presented here, population and sample means, variances, covariances, and correlations, are used extensively in statistical analyses, which we shall not explore at all in this book. But they are also extremely useful in describing data, and in this sense will be used quite often in this book.

In later chapters we shall adopt a convention with respect to notation for the mean that is slightly different from that used above. Often we shall have reason to refer to the mean or average of some trait X *over individuals* in a population. We shall call this mean $\bar{x}$, even though at times, we are speaking of a true mean. The expectation notation $E(\bar{x})$ will be used to denote means *over time*.

1.6 Partial Derivatives

Consider a hypothetical animal whose reproductive output, C in a given breeding season depends on its size w in a linear manner:

$$C \propto w$$

Suppose that C also depends on the quality of parental care, which comes with experience, and hence age, x:

$$C \propto (1 + ax)$$

We have, then,

$$C = bw(1 + ax) \tag{1-57}$$

where b is the appropriate proportionality constant. If there is a one-to-one relationship between w and x, (1-57) could be rewritten either as a function of w alone, or of x alone, and we could ask: What is the effect of age on $C(dC/dx = ?)$, or what is the effect of size on $C(dC/dw = ?)$? But suppose that size is a function not only of age, but of environmental conditions—for example, fish grow more slowly in crowded conditions or in situations of low food availability. In this case, we may have individuals all one age, but varied in size, or vice versa. It now makes sense to ask: For individuals of a particular size, how does reproduction vary with age? What we need to do to answer this question is to find the derivative of C with respect to x, as before, but *as if w were constant*—that is, we ignore the fact that, in reality, w varies as x varies. Such a derivative is called a *partial derivative* and is written

$$\frac{\partial C}{\partial x}$$

In the case above (equation 1-57),

$$\frac{\partial C}{\partial x} = \frac{\partial}{\partial x}[bw(1 + ax)] = abw$$

Similarly, we might ask, for individuals of a particular age, how does reproduction vary with size? Differentiate C with respect to w, but as if x were constant:

$$\frac{\partial C}{\partial w} = \frac{\partial}{\partial w}[bw(1 + ax)] = b(1 + ax)$$

Partial derivatives are related to standard derivatives through the following rule. If y is a function of (say) three variables, x_1, x_2, and x_3, then infinitesimal changes in the x values, dx_1, dx_2, and dx_3, lead to an infinitesimal change in y of

$$dy = \frac{\partial y}{\partial x_1} dx_1 + \frac{\partial y}{\partial x_2} dx_2 + \frac{\partial y}{\partial x_3} dx_3 \tag{1-58}$$

In more general terms,

$$df(x_1, x_2, x_3, \ldots) = \sum_i \frac{\partial f}{\partial x_i} dx_i \tag{1-59}$$

This rule may be usefully applied even when the x_i values are not independent. For example, suppose that a breeding pair of mayflies produce m males and f females in a population consisting of a total of M males and F females. It can be shown (Chapter 14) that the relative genetic contribution to future generations of that pair is

$$C = \frac{m}{M} + \frac{f}{F} \tag{1-60}$$

What sex ratio, $f:m$, will maximize C? From the rule above,

$$dC = \left(\frac{\partial C}{\partial m}\right) dm + \left(\frac{\partial C}{\partial f}\right) df$$

C is maximum when dC/dm, dc/df, and therefore, dC is zero. Write

$$0 = dC = \left(\frac{\partial C}{\partial m}\right) dm + \left(\frac{\partial C}{\partial f}\right) df \tag{1-61}$$

It follows that,

$$\frac{dm}{df} = -\left(\frac{\partial C/\partial f}{\partial C/\partial m}\right) \tag{1-62}$$

or

$$\frac{dm}{df} = -\frac{1/F}{1/M} = -\frac{M}{F}$$

and so

$$\int dm = \int -\frac{M}{F} df$$

or

$$m + c' = -\left(\frac{M}{F}\right) f + c''$$

where c' and c'' are constants. Letting $c'' - c' = c$, we have

$$m = c - \left(\frac{M}{F}\right) f$$

If the breeding pair can produce a total of n offspring, regardless of sex, then when $f = 0$, m must equal n. Thus

$$m = c - \left(\frac{M}{F}\right) \cdot 0 = n$$

and c must equal n. Finally, then,

$$m = n - \left(\frac{M}{F}\right)f \tag{1-63}$$

An interesting observation is now possible. If mayflies all produce young in the same sex ratio, then $M/F = m/f$, and (1-63) becomes

$$m = n - \left(\frac{m}{f}\right)f = n - m = f$$

That is, the sex ratio that maximizes genetic contribution is 1:1.

Partial derivatives will arise occasionally in this book. For the moment it is sufficient simply that you know what they are and remember that they can be strung together in the manner shown in (1-58) and (1-59).

1.7 The Taylor Expansion

Consider the following simple model for population growth:

$$\frac{dm}{dt} = Bm - Dm \tag{1-64}$$

where m is the biomass, B the birthrate, and D the death rate. The solution for this equation can be found by writing

$$\frac{dm}{dt} = (B - D)m$$

so that

$$\int_{m(0)}^{m(t)} \frac{dm}{m} = \int_0^t (B - D)dt$$

$$\ln \frac{m(t)}{m(0)} = (B - D)t$$

$$m(t) = m(0)e(B - D)t \tag{1-65}$$

But suppose that individuals require a time period τ to mature. Then the number of individuals producing offspring *now* (at time t) is not $m(t)$, because some of these will not yet be mature, but rather the surviving members of the population that existed τ units in the past (all of these will, by now, be mature) that is, $am(t - \tau)$, where a is the proportion that survived the interval τ. Equation (1-64) then becomes

$$\frac{dm(t)}{dt} = aBm(t - \tau) - Dm(t) \tag{1-66}$$

To solve this equation we need an expression for $m(t - \tau)$ in terms of $m(t)$ and τ.

A similar problem arises in the following case. Suppose that the number of offspring, Y, a male bird can produce in a breeding season is some function of his territory size, S,

$$Y = f(S) \tag{1-67}$$

If this function is known, then we can assess Y for any male for which S is known. In particular, we can find Y for that hypothetical, average male with territory size $\bar{S}$. (A bar denotes the *mean* value for all males.) But what is the average value of Y? It is *not* necessarily true that $\bar{Y} = \overline{f(S)}$ is the same as $f(\bar{S})$. For example, if $f(S)$ happened to be S^2, and S took on values of

1, 2, 2, 4, then $\overline{f(S)} = \overline{S^2} = (1^2 + 2^2 + 2^2 + 4^2)/4 = 25/4 = 6.25$, while $f(\bar{S})$ is $f(\frac{9}{4}) = (\frac{9}{4})^2 = \frac{81}{16} = 5.0625$. Suppose that we write

$$Y = f(S) = f(\bar{S} + \delta) \tag{1-68}$$

where δ is the deviation of S for a particular bird from the mean value. If we could evaluate $f(\bar{S} + \delta)$ as a function of $f(\bar{S})$ and δ, we would be able to average all such values to obtain the desired $\bar{Y}$.

In both of the examples above we encounter expressions of the form $f(x + \delta)$, which we wish to write as functions of $f(x)$ and δ. This goal can be attained through the use of the Taylor series. Because its complexity is well beyond the point of diminishing returns for this book's purposes, we omit the derivation. The expression is

$$f(x + \delta) = f(x) + \delta \frac{df}{dx} + \frac{1}{2}\delta^2 \frac{d^2f}{dx^2} + \frac{1}{6}\delta^3 \frac{d^3f}{dx^3} + \cdots = \sum_{i=0}^{\infty} \frac{\delta^i}{i!} \frac{d^if}{dx^i} \tag{1-69}$$

Equation (1-69) is also known as the *Taylor expansion.*

Now apply this expression to the problems (equations 1-66 and 1-68) above. Equation (1-66) becomes

$$\frac{dm(t)}{dt} = aB \sum_{i=0}^{\infty} \left[\frac{(-\tau)^i}{i!} \frac{d^i m(t)}{dt^i} \right] - Dm(t)$$

If τ is small so that the higher-order terms (terms with third, fourth, and so on, derivatives) can be ignored without severely affecting the solution to the equation, (1-66) becomes

$$\frac{dm(t)}{dt} \simeq (aB - D)m - (a\tau B)\frac{dm}{dt} + \left(\tau^2 \frac{aB}{2}\right)\frac{d^2m}{dt^2} = 0 \tag{1-70}$$

For a method of solving this rather unpleasant looking differential equation, try supposing that $m = Ke^{pt}$, where K is some constant. Differentiating this quantity and substituting back into (1-70), you will find that you can solve for p. To interpret what you find, see Section 1.4.

Equation (1-68) can be written

$$Y = f(S) = f(\bar{S} + \delta) = f(\bar{S}) + (\delta)\frac{df}{dS} + \frac{1}{2}(\delta)^2 \frac{d^2f}{dS^2} + \frac{1}{6}(\delta)^3 \frac{d^3f}{dS^3} + \cdots$$

Ignoring higher-order terms again, and noting that δ denotes individual variations from the mean,

$$\bar{Y} = f(\bar{S}) + (\bar{\delta})\frac{df}{dS} + \tfrac{1}{2}\overline{(\delta)^2}\frac{d^2f}{dS^2}$$

But $\bar{\delta}$ is the average of the individual deviations from the mean, $\bar{S}$, so it equals zero; and $\overline{\delta^2}$ is the average, *squared* deviation, which is the variance. Hence

$$\bar{Y} = f(\bar{S}) + \tfrac{1}{2}\,\mathrm{var}\,(S)\frac{d^2f}{dS^2} \tag{1-71}$$

If breeding success, $Y = f(S)$, increases with territory size at a decreasing rate (diminishing returns), then $dY/dS > 0$,

$$\frac{d}{dS}\left(\frac{dY}{dS}\right) = \frac{d^2Y}{dS^2} < 0$$

So $\bar{Y}$ *decreases* as the variance in territory size among males increases.

The Taylor expansion will be used frequently in this book. Remember it.

Part Two

The Structural Framework: Population Dynamics

An ultimate goal of the population biologist is to describe accurately the dynamics of one or more populations. Because the dynamics of any one population depend not only on an immense and complex array of features intrinsic to that population, but also on the dynamics of interacting species with their own collection of intrinsic features, this goal is not easily reached. One approach used commonly by applied ecologists is to build large, compound models that can be used to simulate dynamics on a computer. But when individual models, each describing some component part of a whole, interactive system are linked, even minute deviations from reality may snowball into huge errors in the final predictions. Thus accuracy depends, in part, on knowing and understanding in detail the limitations of each constituent model, the biological relationship of each of a population's intrinsic features to its birth, death, and migration processes. The next five chapters present a structural framework, a set of general, constituent models. These models are designed, inasmuch as possible, to be heuristic and intellectually manageable in their scope. As such the biological meaning and implications of any modifications in them is fairly easily understood. As they stand they are not to be considered precise descriptors of nature. They are skeletons to which the flesh of reality must be appended, general constructs that must be fine-tuned on a case-by-case basis. Some of the criteria by which fine-tuning can be accomplished are presented in Chapters 11 through 17.

Chapter 2

Populations Without Age Structure

By July the redwinged blackbirds of Wisconsin have finished breeding, there is no further recruitment of young, and death is the only remaining influence on abundance. The population enters a period of decline which will last until the following spring, when, again, males establish territories in the marshes and females begin to lay eggs. If one were to plot bird numbers against time, over several years, a sawtooth pattern would emerge, the peaks recurring every year in late spring. Most species show similar, seasonal fluctuations. In the case of the redwing, where fledging of young is spread over several weeks and most adults survive the winter, the teeth are rounded and the intervening troughs not very marked. In the mayfly, where egg laying is followed almost immediately by death and the adults emerge nearly synchronously a year later, the pattern is pronounced and the peaks sharp.

Although seasonal patterns in abundance are interesting in their own right, it is the year-to-year trends that are generally of concern to population biologists. Accordingly, it is often convenient to pick a certain time of year and look at the population size each year at that specific time. Since the population at any given time is some multiple of the population at the previous time, we can write

$$n(t + 1) = R(t)n(t) \tag{2-1}$$

where $n(t)$ is the population size or, more properly, *density* (number of individuals per unit area) at time t, and $R(t)$ is defined by

$$R(t) = \frac{n(t + 1)}{n(t)} \tag{2-2}$$

But this really does not tell us much. What we need to know is how $R(t)$ changes over time. This change is a function of the environmental conditions experienced by the population at time t, perhaps of past conditions as well, and as we shall see shortly, it is surely also a function of n, the population density itself.

The function $R(t)$ can be dissected as follows. If $b(t)'$ is the average fecundity (number of offspring produced per individual in the year t to $t + 1$), and $d(t)'$ is the death rate (proportion dying in the year = average probability of an individual dying over the year), then the number of young produced by the population between times t and $t + 1$ must be $b(t)'n(t)$, and the number of the original population remaining at $t + 1$ must be $[1 - d(t)']n(t)$. The new population density, then (in the absence of immigration or emigration), is

$$n(t + 1) = b(t)'n(t) + [1 - d(t)']n(t) \tag{2-3}$$

$$= [1 + b(t)' - d(t)']n(t) \tag{2-4}$$

To understand the function $R(t)$, we must understand what determines fecundity $b(t)'$ and mortality $d(t)'$. (The prime indicates that we are defining a discrete variable; the lack of a prime designates the continuous analogues—equations 2-6 and 2-7.)

Any detailed understanding of births and deaths depends on an in-depth knowledge of the organism under consideration. For the moment, however, we shall satisfy ourselves with some rules of thumb. Note that (2-1) can be extended to read

$$n(t) = R(t-1)n(t-1) = R(t-1)[R(t-2)n(t-2)]$$
$$= \prod_{i=0}^{t-1} R(i)n(0) = R^{*t}n(0) \tag{2-5}$$

where R^* is the geometric mean, between time 0 and t, of all the $R(i)$ values. Clearly, if $R^* < 1$, $n(t)$ will be smaller than $n(0)$, and that over a long period of time (t very large), unless R^* is very close indeed to 1.0, $n(t)$ will practically vanish. On the other hand, if $R^* > 1$, then as t becomes very large, the population will reach arbitrarily high densities. Since populations generally persist for fairly long times and do not approach infinite sizes, we must conclude that for every persisting species, R^* averages, in fact, very nearly 1.0. But the likelihood that this should be so by pure chance is astronomically small unless R exceeds 1.0 when n is small and drops below 1.0 as n becomes large. So on the basis of considerations of the simplest kind we know that, for n sufficiently large, R must be a declining function of n, and must eventually fall below 1.0. How R varies with n is not known precisely for any species, but reasonable approximations have been made and can be used to investigate some aspects of population behavior.

Equations 2-1 through 2-5 are population descriptions in discrete time—that is, they indicate values of n at specified points in time. They provide no information on n during the intervals between these points in time and thus do not describe seasonal changes. Of course, one could apply these same expressions, using a time interval of (say) 1 week instead of a year. This would provide a much more detailed population description, although still an incomplete one. A thorough description can be obtained only by choosing time intervals of infinitesimal length, dt. When this is done, we obtain

$$n(t+dt) = n(t) + \text{number of individuals entering the population in the interval } t + dt$$
$$- \text{ number dying in the interval } t \text{ to } t + dt \tag{2-6}$$

If $b(t)$ is the number births per individual per unit time, the number of newborn between t and $t + dt$ is $n(t)b(t)\,dt$; if $d(t)$ is the death rate, the number succumbing in the interval is $n(t)d(t)dt$. Equation (2-6) becomes

$$n(t+dt) = n(t) + n(t)b(t)dt - n(t)d(t)dt$$

from which

$$\frac{dn(t)}{dt} = [b(t) - d(t)]n(t) \tag{2-7}$$

The right-hand term in brackets is usually referred to as $r(t)$, the population growth rate. Hence

$$\frac{dn(t)}{dt} = r(t)n(t) \tag{2-8}$$

Just as (2-1) describes changes in n over discrete intervals, (2-8) describes changes in n over continuous time. Of course, changes in n are not really continuous—individuals come in integer numbers—and even if n is put in units of biomass, newborns come in discrete packages. But continuous approximations are often useful for describing population changes when breeding is more or less continuous. Discrete expressions are often more appropriate when populations breed only for short periods of time at specific intervals.

Equation (2-8) can be partially solved to give

$$\ln\frac{n(t)}{n(0)} = t\left[\int_0^t \frac{r(t')dt'}{t}\right] = tE(r)$$

where $E(r)$ is the arithmetic mean of r, over time. Finally,

$$n(t) = n(0)e^{tE(r)} \tag{2-9}$$

The relationship between the analogous, discrete, and continuous parameters can now be seen by writing (equations 2-5 and 2-9)

$$n(t) = n(0)R^{*t} = n(0)e^{tE(r)}$$

or

$$R^* = e^{E(r)} \tag{2-10}$$

If r is constant over time [so that $E(r) = r$, $R^* = R$], then (2-10) can be written

$$R = e^r$$

or, from (2-4) and (2-7),

$$1 + b' - d' = e^{b-d} \tag{2-11}$$

2.1 Basic Models

The first and most basic task of the population biologist is to describe how populations grow, or decline, to determine their equilibrium values, if such exist, and how they fluctuate in response to environmental variation. Complete, detailed descriptions of population behavior would be difficult or even impossible. For that reason we construct simplistic approximations, known as *models*. Models, at least if skillfully constructed, allow us to analyze systems so complex that they would be impossible to treat otherwise. In Maynard Smith's (1978) words: "In population biology we need simple models that make predictions that hold qualitatively in a number of cases, even if they are contradicted in detail in all of them." Some examples of commonly used models follow.

The Logistic Equation

Using the continuous approach of (2-8), we write,

$$\frac{dn}{dt} = rn$$

From simple considerations already discussed, we know that r, like R, must be a declining function of n as n becomes sufficiently large, and that for n large enough, the growth rate must become negative. Thus, while r may be a function of a variety of environmental parameters such as temperature, humidity, or abundance of food, it must also vary with n. Since we do not, a priori, know how r varies with n, we begin with the simplest possible relationship and see if the resulting population equation predicts real, observed population trends reasonably well. We let

$$r = r_0\left(\frac{1 - n}{K}\right) \tag{2-12a}$$

This is equivalent to supposing either that birthrate b drops linearly with n,

$$b = b_0(1 - \xi n)$$

or that death rate d rises with n,

$$d = d_0(1 + \nu n)$$

or both, so that

$$r - b - d = b_0(1 - \xi n) - d_0(1 + \nu n)$$

$$= r_0\left(1 - \frac{n}{K}\right) \tag{2-12b}$$

where $r_0 = b_0 - d_0$ and $K = (b_0 - d_0)/(b_0\xi + d_0\nu)$.

In this situation, growth rate is maximum when n is very small, drops off linearly as n rises, and finally, when n reaches K, becomes zero. We may think of K as the number of individuals the environment can support. Suppose, for example, that the environment factor which determines this number is food. Then, if $n < K$, there is an excess of food, individuals have high survival, produce more than the replacement number of young, and the population grows. If $n > K$, there is less than basic sustenance, mortality rises, births decline, and the population drops. When $n = K$, the food supply is exactly sufficient just to maintain the population at a steady level. In this example, food is said to be the *limiting factor*. K is known as the *carrying capacity*. The quantity r_0, the maximum attainable population growth rate, is the *intrinsic rate of increase* and is a function both of environmental conditions and attributes of the species in question.

Substitution of (2-12) into (2-8) gives us the *logistic equation*:

$$\frac{dn}{dt} = r_0 n\left(1 - \frac{n}{K}\right) \tag{2-13}$$

It is also worth noting that (2-13) represents the first three terms of the Taylor expansion of a much more general expression:

$$\frac{dn}{dt} = f(n)$$

where $f(n)$ is the true (unknown) descriptor of population growth. Then

$$\frac{dn}{dt} = f\left[0 + \left(\frac{n-0}{K}\right)\right] = f(0) + \frac{n}{K}\left(\frac{df(x)}{dx}\right)_{x=0} + \frac{1}{2}\left(\frac{n}{K}\right)^2\left(\frac{d^2f(x)}{dx^2}\right)_{x=0} + \cdots$$

$$\simeq f(0) + a\frac{n}{K} + b\left(\frac{n}{K}\right)^2$$

But a nonexistent population cannot grow. Hence $f(0) = 0$, and

$$\frac{dn}{dt} \simeq a\left(\frac{n}{K}\right) + b\left(\frac{n}{K}\right)^2 = r_0 n\left(1 - \frac{n}{K}\right)$$

where $r_0 = a/K$ and $b = -1$.

This differential equation is easily solved through the use of a simple substitution. (Note this substitution carefully, for we shall use it again.) Define the quantity

$$u = \frac{1}{n}$$

Then, by rules of differentiation,

$$\frac{du}{dt} = \frac{d(1/n)}{dn}\frac{dn}{dt} = -\frac{1}{n^2}\frac{dn}{dt} = -u^2\frac{dn}{dt}$$

Substituting into (2-13), we obtain, after some algebraic rearrangements,

$$-\frac{du}{dt} = \frac{r_0}{K}(Ku - 1) \tag{2-14}$$

Thus

$$\frac{1}{K}\ln\frac{Ku(t) - 1}{Ku(0) - 1} = -\frac{r_0 t}{K}$$

$$\ln\frac{[K/n(t)] - 1}{[K/n(0)] - 1} = -r_0 t$$

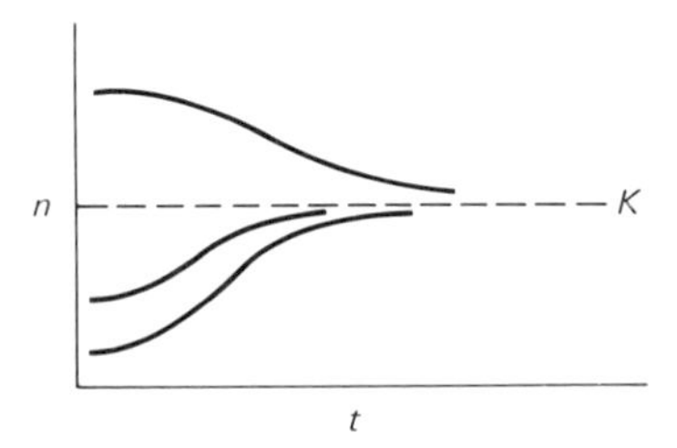

Figure 2-1 Hypothetical graph of population size as a function of time. K, carrying capacity.

or

$$n(t) = n(0)\frac{K}{[K - n(0)]e^{-r_0 t} + n(0)} \tag{2-15}$$

Plots of $n(t)$ versus time t for a variety of initial population sizes are pictured in Figure 2-1.

Before proceeding to test whether (2-15) reasonably describes population behavior in the real world, stop and ask what assumptions have been made in its derivation and for what, if anything, it will be useful. In addition to supposing that b, d, and therefore r change linearly with n, we have assumed:

1. *b and d are constant, for a given n, regardless of the age structure of the population.* This is vitiated for many populations in which several age groups occur. For example, suppose that there are two age groups, one of which produces many young, the other only very few. Surely, b depends not only on the total value of n, but also on its age composition. Thus we would expect the logistic equation to be valid only where all age groups have the same mortality and fecundity rates (so age composition does not matter), or the proportion of individuals in the different age groups remains constant in time (see Chapter 3). If individuals breed only once in their lives and then die, there is no overlapping of generations, no two age groups breeding at the same time, and the growth parameter does not depend on age structure (since there *is* no age structure). But populations of this sort vary in a discrete manner and are best described using the discrete model, unless the time between generations is very small (not a bad approximation for viruses, bacteria, and some protists).

2. *We have assumed the linear decline in dn/dt as n rises to be instantaneous.* Consider a population whose carrying capacity K is determined by available food. It will grow until there is only enough food to sustain its present members at which point, according to the logistic equation, its growth will stop. But suppose that at this point, most of its members have not yet reached their full, adult size. Even if numbers cease to increase, this population will suddenly find, as its members mature, that food is in short supply. Because of the time lag associated with growing up, the population will have overshot its carrying capacity and must now begin to decline. Or suppose that population growth is limited by waste products in an aquatic species' medium: As more individuals are generated, more waste is produced and waste concentration rises, feeding back negatively on survival or reproduction until growth stops. Gause (1932) found population growth of certain yeasts was limited in this manner in his laboratory. If the negative impact occurs only after a certain period of exposure, the population will overshoot K. Intuitively, such time lags should lead to oscillations. A straightforward, mathematical approach to this problem is as follows. If the lag in feedback is τ, then

$$\frac{dn(t)}{dt} = r_0 n(t)\left[1 - \frac{n(t - \tau)}{K}\right] \tag{2-16}$$

Using the Taylor expansion (Chapter 1), we next obtain

$$n(t - \tau) = \sum_{i=0}^{\infty} \frac{(-\tau)^i}{i!} \frac{d^i n(t)}{dt^i} \tag{2-17}$$

If (2-17) is substituted back into (2-16), the resulting expression is quite intractable. But if we are willing to settle for an approximate solution (or if the time lag is very short), we can ignore higher-order terms and write

$$\frac{dn}{dt} = r_0 n\left(1 - \frac{n - \tau\, dn/dt + \frac{1}{2}\tau^2\, d^2n/dt^2}{K}\right)$$

This expression is still rather unpleasant. But we can simplify it again if we consider only cases where fluctuations in n are small. Write

$$n = K + x$$

so that x is the deviation in n, at any time, from equilibrium, K. Then $dn/dt = d(K + x)/dt = dx/dt$, and the equation above becomes

$$\frac{dx}{dt} = r_0(K + x)\left(1 - \frac{K + x - \tau\, dx/dt + \frac{1}{2}\tau^2\, d^2x/dt^2}{K}\right)$$

Because x is very small (so that the x^2, $x\, dx/dt$, and $x\, d^2x/dt^2$ terms disappear), the expression becomes

$$\frac{dx}{dt} = -r_0\left(x - \tau\frac{dx}{dt} + \frac{1}{2}\tau^2\frac{d^2x}{dt^2}\right) \tag{2-18}$$

Write this in the form

$$b\frac{dx}{dt} + a\frac{d^2x}{dt^2} + cx = 0$$

where $a = r_0\tau^2/2$, $b = 1 - r_0\tau$, and $c = r_0$. This form of differential equation has a standard solution of the form

$$x = e^{-bt/2a}\left[C_1 \exp\left(\frac{\sqrt{b^2 - 4ac}}{2a}t\right) + C_2 \exp\left(\frac{-\sqrt{b^2 - 4ac}}{2a}t\right)\right] \tag{2-19}$$

which is oscillatory if $4ac > b^2$ (see Chapter 1). Substituting back for a, b, and c, we see that oscillations will occur if

$$2r_0^2\tau^2 > (1 - r_0\tau)^2$$

or, equivalently, if

$$0 > 1 - 2(r_0\tau) - (r_0\tau)^2 \tag{2-20}$$

or, finally, if $r_0\tau > 2.414$. Note that oscillations are more likely if either the intrinsic rate of increase r_0 is large, or if the time lag is long.

With all the problems above, have we disposed of the logistic equation by caveat? Not at all. It turns out to be a reasonably good predictor of population behavior for bacteria, yeasts, and protozoans (Pearl, 1927; Gause, 1931, 1934) (Fig. 2-2a). For laboratory populations of water fleas, *Daphnia* (Slobodkin, 1954), flour beetles (Chapman, 1928), *Drosophila* (Pearl, 1927), and even for sheep (Davidson, 1938) it gives a fair fit to data (Fig. 2-2b). In both the latter cases the real populations fluctuate—this may be due to time lags, or since these species possess overlapping generations, perhaps to other complications imposed by age structure (see Chapter 3). The logistic equation is far from perfect, but, often in modified form, it will serve our interests well later in this chapter and in subsequent chapters.

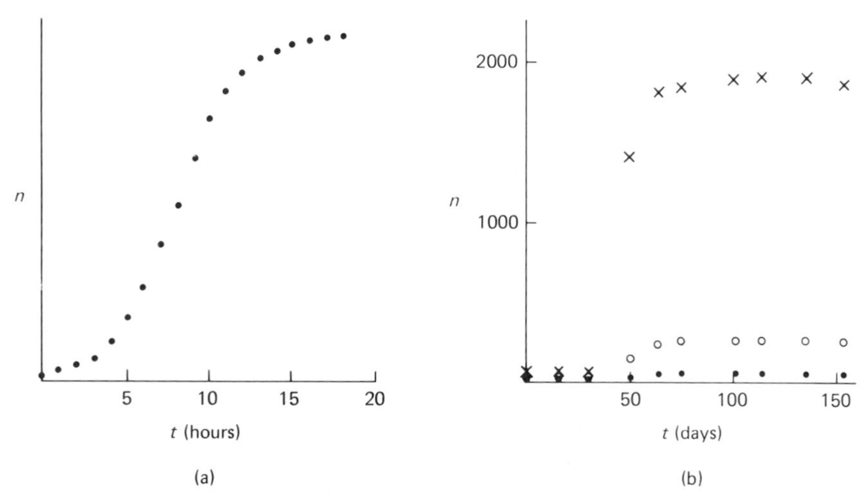

Figure 2-2 (a) Population growth in yeast as a function of time. (Data from Pearl, 1927.) (b) Growth of a flour beetle population in (solid circle) 4 g of flour, (open circle) 16 g of flour, (cross) 128 g of flour. (Data from Chapman, 1928.)

It is useful to point out that discrete analogs to the logistic equation can be written. For example, replace dn/dt with Δn, a change in n over one unit of time (a year, perhaps). Then (2-13) becomes

$$\Delta n = \epsilon n\left(1 - \frac{n}{K}\right) \tag{2-21}$$

where ϵ is the discrete analog of r_0 or

$$n(t + 1) = n(t) + \epsilon n(t)\left[1 - \frac{n(t)}{K}\right] = n(t)\left[1 + \epsilon\left(1 - \frac{n(t)}{K}\right)\right] \tag{2-22}$$

The Ricker Model

Invented by W. E. Ricker (1954), this model has been used extensively by fisheries biologists. In its original form it applies to species without overlapping generations, such as pacific salmon. The more common of the two derivations is given here.

Let P be the parental stock, the number of spawning adults, and let $R(t)$ be the number of *recruits*, the number of young produced in that spawning, at time t after they were spawned. (Do not confuse this R with the R defined earlier in the chapter!) Assume now that the young recruits, as they grow larger, pass through a vulnerable stage during which they experience a *density-dependent* mortality (related to the population density) which is directly proportional to the number of stock (via cannibalism, perhaps). After some critical age T_C is reached, these fish are presumed to have passed beyond such a vulnerable state, or, alternatively, the spawning stock has died off. So, considering only this source of mortality, the recruits R die off at a rate proportional to P.

$$\frac{dR(t)}{dt} = -\delta_D R(t)P \tag{2-23}$$

where δ_D is the density-dependent mortality rate and P, the spawning stock, is constant. The solution to (2-23) at the critical age T_C, is

$$R(T_C) = R(0)e^{-\delta_D T_C P} \tag{2-24}$$

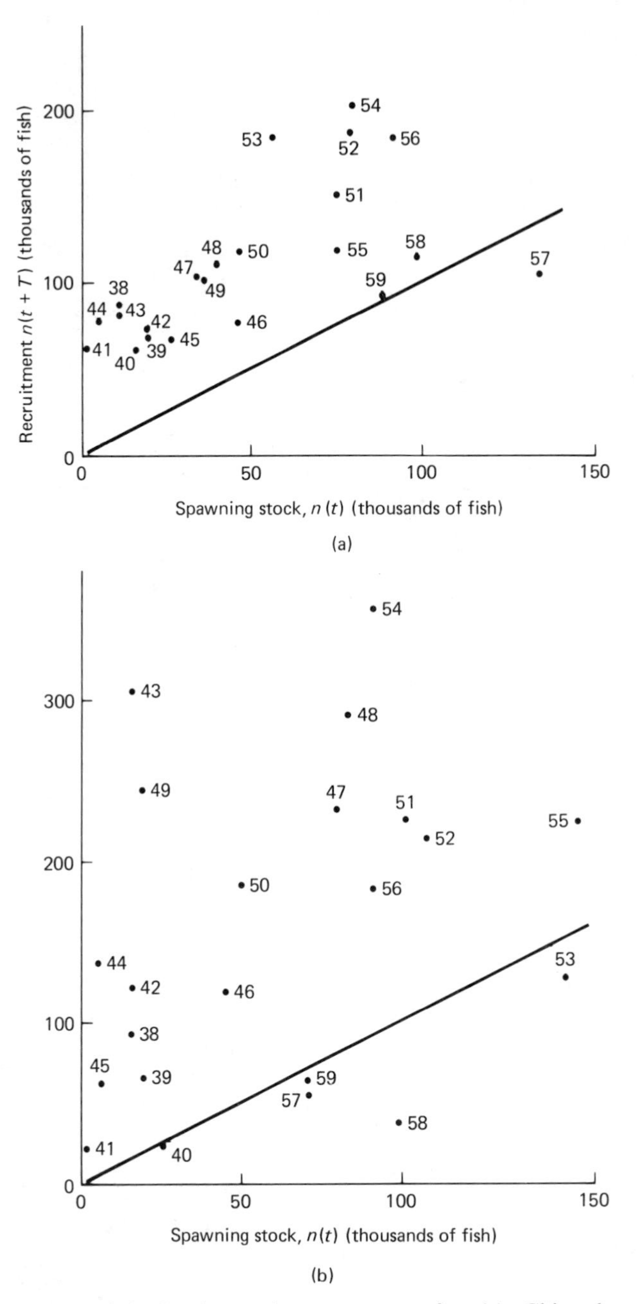

Figure 2-3 Stock-recruitment curves for (a) Chinook salmon and (b) sockeye salmon. The number of individuals at one time unit is plotted against the number at the previous time unit. The numbers refer to year class, for example, "58" indicates fish hatched in 1958. (From Van Hyning, 1973.)

Of course, there are other sources of mortality not related to P (accidental heat shock, biting a worm with a hook inside it), the *density-independent* mortality, which acts on R throughout. If the proportion of the above survivors living through the latter stresses to T_C is l_{TC}, then

$$R(T_C) = R(0)l_{TC}e^{-\delta_D T_C P} \tag{2-25}$$

and if the proportion of these subsequently surviving to spawning age T is l_T/l_{TC}, then

$$R(T) = R(T_C)\frac{l_T}{l_{TC}} = R(0)\frac{l_T l_{TC} e^{-\delta_D T_C P}}{l_{TC}} = R(0)l_T e^{-\delta_D T_C P} \tag{2-26}$$

Of course, $R(0)$, the number originally spawned, will be very nearly proportional to P. If that proportionality factor is called c, we have, finally,

$$R(T) = cPl_T e^{-\delta_D T_C P} = \alpha P e^{-\beta P} \tag{2-27}$$

where $\alpha = cl_T$ and $\beta = \delta_D T_C$. This is the *Ricker equation.*

Fisheries biologists prefer to use the symbols P and R as in (2-27). But to avoid confusion over the meaning of R and to maintain consistency, this equation will henceforth be written

$$n(t + T) = \alpha n(t)e^{-\beta n(t)} \tag{2-28}$$

The values of α and β can, of course, vary with time.

The Ricker model is a discrete population model. How accurately does it describe the dynamics of fish populations? One way to find out is to plot $n(t + T)$ against $n(t)$ as predicted by the model (called the *Ricker curve*, see Figure 2-4) and compare that curve against one generated by data. The results of such tests are mixed. Fairly typical examples are those of the Chinook salmon (*Onchorhynchus tshawytscha*) and sockeye salmon (*O. nerka*) pictured in Figure 2-3. The scatter in data points, arising (presumably) largely from year-to-year variations in density-independent environmental effects on mortality or fecundity obscure the nature of the underlying curve and make comparison with the theoretical Ricker curve difficult. Nevertheless, the general trends agree. The Ricker curve is *not* a very good predictor of $n(t + T)$ from $n(t)$, but if we are interested in the underlying relationship between population sizes in subsequent generations, with the noise stripped away, it may be very useful (Pikitch, personal communication). The scatter in the data provide an important lesson: We must remember that $R = n(t + 1)/n(t)$ is a function of environmental variables as well as a function of n, and unless we take these variables into account, we will never achieve a really good fit of theory and reality.

As long as environmental fluctuations produce scatter about the curve, there can be, technically speaking, no population equilibrium (the same is obviously also true of the logistic

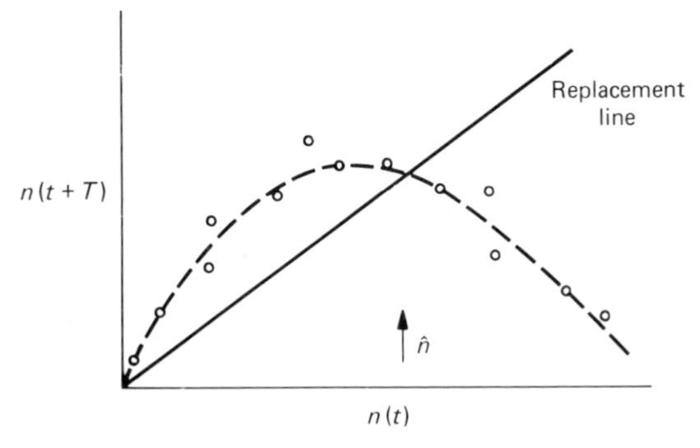

Figure 2-4 Hypothetical data fit to a Ricker stock-recruitment curve showing the replacement line and point of population equilibrium.

model). But ignore such fluctuations for the moment. The 45° line in Figure 2-4 is known as the *replacement line*—along this line $n(t + T) = n(t)$, so the population is at equilibrium. To the left of the point where this line crosses the Ricker curve, $n(t + T) < n(t)$, so the population is growing; to the right the population is declining. Thus the point of crossing represents a "stable" equilibrium; the population, if displaced from this density, returns to it. Equilibrium population size is given by $\hat{n}$, where

$$\hat{n} = \alpha\hat{n}e^{-\beta\hat{n}}$$

Thus

$$\hat{n} = \frac{\ln \alpha}{\beta} \tag{2-29}$$

This is equivalent to the K of the logistic model.

Other population models abound, but will not be covered here. The interested reader is referred to Larkin (1973).

2.2 Density Feedback

The logistic model, the Ricker model, and indeed all biologically reasonable models of population dynamics include a *density-feedback* term. In the case of the logistic equation,

$$r = \frac{1}{n}\frac{dn}{dt} = r_0\left(1 - \frac{n}{K}\right) = r_0 - \frac{r_0}{K}n$$

it is r_0/K: The addition of one individual to the population lowers r by an amount r_0/K. Formally, $dr/dn = -r_0/K$. With respect to the Ricker equation,

$$R = \frac{n(t + T)}{n(t)} = \alpha e^{-\beta n(t)}$$

the effect of adding an individual can be assessed by writing

$$\frac{1}{R}\left(\frac{dR}{dn}\right) = \frac{d \ln R}{dn} = \frac{d}{dn}(\ln \alpha - \beta n) = -\beta$$

Clearly, feedback must exist. It remains to discuss what biological characteristics of the population assure its existence.

It is clear that if a population increases in size long enough, a time will come when there will no longer be enough food to go round. Energy and nutrients are not in infinite supply, and as the available goods are spread among more and more individuals, the proportion of those individuals obtaining sufficient amounts both to sustain life and produce young must begin to drop. Eventually, either birthrate will fall or death rate will rise (or both) to the point where $r = b - d$ is zero, and population growth ceases. The population is said to be *food limited*. Of course, some other factor(s) may intervene to stop population growth below the level where food becomes critical. For example, barnacles growing on intertidal rocks may run out of space, while the waves continue to lavish planktonic food on them (the barnacles are said to be *space limited*); flour beetles may develop high egg cannibalism rates, overcrowded rats may develop pathological aggressiveness as a (not fully understood) consequence of high incidence of individual contact, so that population growth stops short of running low on resources (*stress limited*); some aquatic organisms may inadvertently limit their population growth by self-poisoning with waste products. Finally, high population levels mean abundant food for predators, whose numbers may consequently increase at a sufficient rate to stop the prey from further growth (*predator limited*).

Populations are rarely limited by any one factor—seldom is $\hat{n}$ ($= K$ in the logistic model) determined solely by a single resource, social stress, or predation. In general, many factors impinge on birth and death rates, and in addition, these factors interact. In some cases, however, we may be justified in at least strongly implicating one or another of these factors. This, of course, is not surprising in a laboratory situation. Slobodkin (1954), for example, grew *Daphnia obtusa* on algal food in 50 cc of filtered pond water and found an almost perfectly linear relation between equilibrium population levels, $\hat{n}$, and food supply. Clearly, neither space nor stress arising from individual interactions, nor waste products, were significant in determining $\hat{n}$, and there were no predators in the system. In the field, Southwood and Cross (1969) found that 70% of the variance in breeding success—ratio of young to adults, and thus a measure of birthrate—of the European partridge (*Perdix perdix*) could be accounted for by insect (food) abundance over the breeding season. Data of a more direct nature showed that as the winter minimum density (g dry weight/m^2) of invertebrate stock rose from 20 to 70 to 120 to 140, the winter survival of coal tits (*Parus ater*) rose, correspondingly, from 30 to 50 to 60 to 75% (Gibb, 1960). Jansson et al. (1981) added extra food to an experimental plot in Sweden and found that winter survival of two species of tit (*Parus montanus* and *P. cristatus*) nearly doubled. By the first week in April both species exhibited population densities 2.2 times that on a nearby control plot. Similar manipulations produced similar results in deermice (*Peromyscus maniculatus*) in western North America (Tait, 1981). In the latter case, the added food also promoted late winter breeding, absent in most natural populations. Southern (1970), in a study of the tawny owl (*Strix aluco*), found that when prey animals (the wood mouse, *Apodemus sylvaticus*, and the bank vole, *Clethrionomys glareolus*) were very scarce, no breeding took place. With increasing prey abundance the numbers of young successfully fledged rose asymptotically.

Perhaps the most clear-cut case of food limitation to be found in the literature is that of the pond snail *Lymnaea elodes* (Eisenberg, 1966). Eisenberg constructed twenty 3- by 15-ft enclosures along the edge of a pond in southern Michigan. In eight he altered the early June snail population to fivefold its initial level (which was roughly 1000 snails). In another eight he thinned the population to one-fifth. Four pens served as controls. On the basis of control pen mortality rates, he then predicted the autumn population levels, excluding new young, from initial levels in the 16 experimental enclosures and plotted the log of the observed values against the log of the predicted values. If density feedback were operating via summer mortality (in winter the snails are inactive, so density effects presumably are not present), the slope of the curve should be less than 1.0 (Fig. 2-5). Actual results were in accord with this hypothesis, but statistically, a null hypothesis of no density effects on mortality could not be rejected. Thus predation may contribute to population limitation, but is not sufficiently important to be statistically detectable. On the other hand, counts of young snails over the summer revealed that, within a week, new recruits in the dense pens had dropped well below five times the control level, and the number of young animals in the sparse pens was disproportionately high. By September the numbers of new generation snails bore no relation to the densities of their parental populations. The populations appeared to have been regulated back toward control levels by virtue of density feedback operating on the birthrate. But density effects on the birthrate suggest either stress or shortage of some resource—food or space. To test this, Eisenberg added artificial food in the form of spinach and observed a 25-fold increase in production of young.

Any resource may be a potential, limiting factor; food—encompassing energy supply, water, protein, sodium, essential vitamins, and so on—is only one. Others include cover (protection from predators or the elements), appropriate nesting sites, or space for settling larvae and growing adults. Is there any reason to believe that the feedback arising from a limited supply of such resources is linear as assumed in the logistic equation? That is, as we add more consumers, one at a time, is the consequent drop in r always the same? Clearly, the answer depends on the answer to several subsidiary questions:

1. What relation does r bear to the amount of resource used?

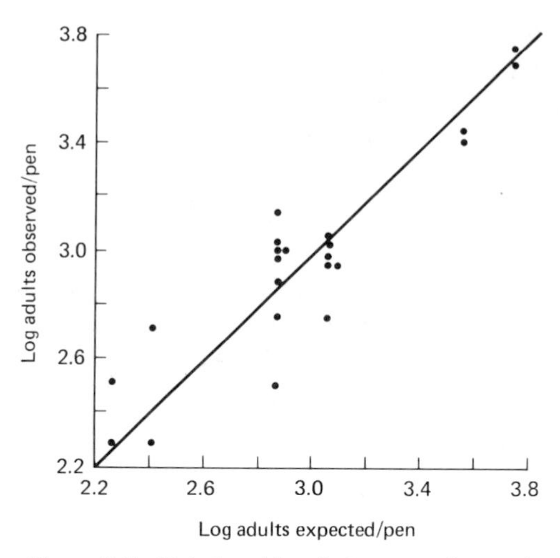

Figure 2-5 Relationship between observed and expected density in a population of pond snails. (From Eisenberg, 1966.)

2. What relation does amount of resource used bear to the amount of resource available?
3. What effect does resource use have on the amount of resource available?

Where the resource is food, these questions appropriately belong in a treatment of predator–prey relationships—the reader is referred to Chapter 4. Where the resource is space, the question of linearity, at first, seems quite simple: The probability that a new individual, randomly settling onto a substrate, finds space to live is clearly proportional to the amount of available space remaining, which declines linearly with n. But this is true only if new individuals settle randomly in space (which they do not—most seeds are dispersed short enough distances that clumps of related individuals are the rule; the planktonic larvae of barnacles and mussels show marked, substrate choice behavior when they are ready to settle and metamorphose). Furthermore, space occupied by one individual is *not* necessarily unavailable to another. Barnacles growing in dense clusters alter their growth form, becoming relatively taller and narrower; one individual may settle and grow on top of another without (at least for a time) suffocating it, and crowded barnacles may even share portions of each others' exoskeletons, thus saving space and reducing the cost of shell production (Brian Hazlett, personal communication).

The lesson in the paragraphs immediately above is straightforward. Resource limitation is not nearly as simple as implied in the logistic or the Ricker model. Where resource levels remain constant, as is often the case when the limiting factor is space, or where they are held constant, as in the laboratory, we can take comfort in the fact that the logistic equation proves reasonably accurate. Where resource levels vary as a function of n, as in most cases where the resource is a food (another species) supply, we must look to two-species or m-species interactive models for a reasonable understanding of population processes.

A population may be limited by competition among its members for a resource. It may also be affected by the individuals' behavior or by chemical interference which increases with n. Examples of the latter implicate, among various factors, waste products (alcohol in the case of yeasts), growth- or germination-inhibiting (called *allelopathic*) substances in plants and some animals (Jackson and Buss, 1975), and oxygen depletion for small pond organisms.

Population limitation by behavioral interference has never been proven to exist in the field, but evidence that behavior interactions *contribute* to slowed population growth abounds, especially in the small mammal literature. Calhoun (1952) found that as Norway rat (*Rattus norvegicus*) populations grew, the number of fights among their members increased, leading to infected wounds and higher mortality. Similar results have been reported for such diverse species as house mice (*Mus musculus*) (Southwick, 1955; Lloyd and Christian, 1967) and hermit crabs (*Pagurus bernhardus*) (Hazlett, 1968). Increased numbers of aggressive encounters should be expected. In the simplest case, where encounters are random in time, if one population is k times as big as another, each individual would be expected to encounter approximately k times as many other individuals, or to encounter other animals k times as often. If the same proportion of encounters are aggressive in each case, there will be k times the number of fights per individual. In fact, if each fight carries the same probability of leading to a death, the death rate (arising from aggressive encounters) would be k-fold. Using this rather simplistic assumption of random encounters and constant chance of aggression, we now construct the following model. Suppose that there were no population feedback other than individual aggression. Let b be the birthrate and d the rate of death *not* associated with aggression. The aggression-caused death rate is proportional to population size—we shall call it δn. Then

$$\begin{aligned}\frac{dn}{dt} &= n(b - d - \delta n) \\ &= (b - d)n\left(1 - \frac{\delta}{b - d} n\right) \\ &= (b - d)n\left[1 - \frac{n}{(b - d)/\delta}\right]\end{aligned}$$

This is the logistic equation with $b - d = r_0$, and $(b - d)/\delta = K$, and the population would come to equilibrium at

$$\hat{n} = \frac{b - d}{\delta}$$

Of course life is not really this simple. Calhoun found, in high-density rat populations, that dominant males not only injured other individuals, but also fiercely guarded local food supplies, lowering availability to others. Also, high population densities typically led to physiologically detectable stress reactions including hypertrophied adrenal glands and spleens, perhaps by way of increased frequency in aggressive contacts (Christian, 1956, 1961; Christian and Davis, 1956, 1964; Warnock, 1965). These physiological changes often led to increased aggression and abnormal behavior. For example, Southwick (1955), who kept house mice in 25- by 6-ft pens with unlimited food and water, found the number of fights per individual to rise faster than population size, and in some instances his males ran amock, destroying nests and young. Population equilibrium depended greatly on the "personalities" of the dominant individuals present. Generally, death rates rise and birthrates fall with increased stress.

There is some evidence for the role of stress in the regulation of wild populations. Herrenkohl (1979) has shown that daughters of stressed female rats experience lower fertility and fecundity than those of unstressed mothers, and that their sons might be demasculinized. In fact, young of stressed mothers display even the physiological symptoms of their female parent. This results in a stress feedback which has a time delay. The alert reader will immediately recognize this as a possible perpetrator of population cycling (Chitty, 1957); see also Krebs et al., 1969; Krebs, 1970; Krebs and Myers, 1974; Boonstra and Krebs, 1979).

Effects of resource shortage or stress on population growth need not be direct. In many fish species, for example, these factors act to slow body growth. For two quite different formulations of growth rate as a function of population density, see Walter (1934, cited in Gerking, 1978, Chap. 11), and Shuter and Koonce (1977). But smaller fish generally produce fewer eggs. Schopka and Hempel (1973) review relationships between herring (*Clupea harengus*) and cod (*Gadua morhua*) body size and fecundity in a variety of populations. The following relationship for Norwegian herring is a fairly typical example:

$$\text{number of eggs produced} = (.3323)(\text{length}^{3.405})$$

But clearly, if population density slows growth and smaller fish lay fewer eggs, there is a feedback relationship between n and birthrate (see also Weatherley, 1972). Also, in aquatic and marine systems there is often a predation-safe size. As organisms grow they pass through size ranges which are ideal for capture by various predators, but may eventually reach a size at which potential predators can no longer handle them. If growth rate is lowered, the fish spend more time in the critical, highly predated size ranges, and the death rate rises (Beverton and Holt, 1957).

The first argument above suggests one way in which resource shortage or stress may interact with predation to result in population limitation. Interactions of this sort, which make it impossible to define clearly what limits a population, are not uncommon. Some other interrelationships are as follows. If food is in short supply, animals may find it necessary to increase the proportion of their time spent foraging. But foraging generally requires leaving shelter and thus increases exposure to predators. Hunger may lead to weakness, making individuals more prone to disease (the same is true of plants growing in nutrient-deficient soils), or more susceptible to predation. On the other side of the coin, increased predation may force prey to spend more time in hiding or in other avoidance behavior, limiting their time to feed. Social stress also may weaken individuals or consume time beneficially spent in feeding. A particularly interesting interaction between social stress and food (actually sodium availability) is reported by Aumann and Emlen (1965). Records of 55 *Microtus pennsylvanicus* populations showed a striking, positive relationship between population peak densities and soil sodium content. To find out why, the authors ran a series of experiments with this species in the laboratory. In one experiment, eight 3- by 6-ft pens and eight 3- by 3-ft pens were seeded with mice and supplied, ad libidum, with food in the form of alfalfa and oats. Half of the pens in each group were given distilled water, the other half a 0.5% sodium chloride solution as well as distilled water. Over 15 to 19 weeks, during which the populations were undisturbed, the pens with access to the salt solution produced between half again and twice as many young ($p < .05$). In a second experiment, high-density groups (32 animals per pen) and low-density groups (eight animals per pen), half male and half female, were offered a choice between distilled water and the sodium chloride solution. The high-density populations used 70% (males) and 90% (females) more of the salt solution than did the low-density populations. A third experiment followed sodium choice in animals subjected to two weeks of low-, then intermediate-, then high-, and then again low-density pressures. Sodium choice increased markedly as density rose, and then fell with the return to low-density conditions. Because adrenalectomized animals are known to show increased appetite for sodium in the laboratory, Aumann and Emlen hypothesized that adrenal stress arising from crowded conditions increased the animals' need for sodium. The better that need could be satisfied, the greater stress the population would be able to withstand. Thus easy access to sodium permitted high population levels. If Aumann and Emlen's interpretation is correct, the 55 populations which inspired their study provide good evidence for the importance of social stress in microtine population dynamics.

2.3 Changing Environments

Environments are not constant. Food supply varies, predation levels vary, climate varies. Furthermore, different individuals come and go in time, and with them the distinctive characteristics they contribute to a population. These changes can be categorized in two ways. First there is a rough dichotomy between density-dependent and density-independent changes. By the first is meant changes in environmental parameters whose effects on r are density related. That is, if $S(t)$ is a measure of such a parameter at time t, a density-dependent change is one where

$$\frac{\partial r(t)}{\partial S(t)}$$

is a function of n. A density-independent change involves a parameter whose effect on r is independent of n—that is, one for which the derivative is not a function of n. Changes in population-limiting factors such as food supply are always density dependent. To see this, consider the fact that the limiting factor determines $\hat{n}$. In the case of the logistic model, $\hat{n} = K$, so that

$$r = r_0\left(1 - \frac{n}{\hat{n}}\right)$$

Thus

$$\frac{\partial r}{\partial S} = \frac{\partial r}{\partial \hat{n}}\frac{\partial \hat{n}}{\partial S} - \frac{r_0 n}{\hat{n}^2}\frac{\partial \hat{n}}{\partial S}$$

which is a function of n. The effects of density-dependent changes on population dynamics are best simulated by allowing $\hat{n}$ to vary.

Density-independent fluctuations are best incorporated into population models simply by adding a term $S(t)$ to the expression for r. Thus, if we were to use the logistic model, we would write

$$r = r_0\left(1 - \frac{n}{K}\right) - \gamma S(t)$$

Here r_0 is the rate of increase under ideal conditions and γS is the lowering of r_0 due to less than ideal conditions (see below).

Environmental fluctuations can also be categorized as deterministic, meaning that the values of $S(t)$ follow a predictable path; or capricious, indicating that the values of $S(t)$ are at least partly unpredictable, based on chance events. In Lewontin's (1966) words: "A system is capricious if there is less than perfect information . . . and if repeated sampling . . . fails to increase the information about the system." The response of a *population* to environmental fluctuations can also be categorized as deterministic or capricious. If the fluctuations are capricious, so are the population dynamics. However, capriciousness of population behavior does not require similar behavior in the environment; sometimes the environmental influences are deterministic but the population responds in *apparently* random fashion anyway.

In the first two parts of this section we treat two kinds of genuine randomness—chance variations in n arising from the fact that individuals are discrete units, and the effects on n of random, environmental changes (density dependent and density independent). The third part examines the stability of populations. In the fourth part we take up apparently capricious behavior in n which is *not* truly random.

The Importance of Individuals as Discrete Units

In deriving (2-7),

$$\frac{dn}{dt} = (b - d)n$$

we made the implicit assumption that for a given value of n, b, and d were fixed. But, of course, population processes are not really continuous. Individuals do not fade in and out of populations; they arrive and depart in discrete packages. In any instantaneous time interval the vastly most probable event is no change in n. An increase of one due to a birth or a decrease of one due to a death are less likely. Finally, unless births and deaths are nearly synchronized the probability of two or more births or deaths over such a minute time interval is vanishingly small.

Therefore, even in the absence of capricious changes in the external environment, the dynamics of a population cannot be determined with complete certainty (even if we knew our models to be complete and precise). We *can*, however, decide the *probability* that the population will take on some particular size. A model that uses laws of probability to deal with capricious influences is said to be *stochastic*. A model that ignores capricious influences, or describes a process that can be precisely determined, is called *deterministic*. The logistic model, for example, ignores chance events and so is deterministic. Let us examine a simple stochastic model of population dynamics.

Let b_i and d_i be the birth and death rates associated with a population of size i. Choose the time interval over which we describe population changes, Δt, to be infinitesimally small, so that (1) b_i and d_i are continuous rates as in the logistic model, and (2) in any one time interval there is virtually no chance of observing more than one birth or death. Let $p_i(t)$ be the probability distribution of n at time t—$\mathbf{p}(t)$ is the vector giving the probabilities that the population is of size $0, 1, 2, \ldots$. Finally, let $\mathbf{Q} = \{q_{ij}\}$ be the matrix which gives the probability that a population changes from size i to size j in a given (infinitesimal) time interval. Then

$$\mathbf{p}(t + dt)^T = \mathbf{p}(t)^T \cdot \mathbf{Q}$$

describes the relationship between the probability vector at time t and at time $t + dt$. This may be rewritten

$$\frac{d\mathbf{p}(t)}{dt} = \frac{\mathbf{p}(t + dt)^T - \mathbf{p}(t)^T}{dt} = \frac{\mathbf{p}(t)^T\mathbf{Q} - \mathbf{p}(t)^T}{dt} = \mathbf{p}(t)^T \frac{\mathbf{Q} - \mathbf{I}}{dt} \tag{2-30}$$

The elements of $\mathbf{Q}$ are found as follows. If the population is of size $i > 2$, then the probability that a birth occurs in the interval dt is $b_i dt$ (reproduction is assumed to require two individuals). Similarly, the probability that a death occurs is $d_i dt$. Thus the population increases to $i + 1$ with probability $b_i dt$, and decreases to $i - 1$ with probability $d_i dt$. It remains at size i with probability $1 - b_i dt - d_i dt$. If $i = 1$, the population can only remain at size 1 or decline to zero (with probability $d_1 dt$). If $i = 0$, it must remain there. The matrix $\mathbf{Q}$ is therefore,

$$\mathbf{Q} = \begin{bmatrix} 1 & 0 & 0 & 0 & \cdot & \cdot \\ d_1\,dt & 1 - d_1\,dt & 0 & 0 & \cdot & \cdot \\ 0 & d_2\,dt & 1 - d_2\,dt - b_2\,dt & b_2\,dt & \cdot & \cdot \\ 0 & 0 & d_3\,dt & 1 - d_3\,dt - b_3\,dt & b_3\,dt & \cdot \\ 0 & 0 & 0 & d_4\,dt & 1 - d_4\,dt - b_4\,dt & \cdot \\ \cdot & \cdot & \cdot & \cdot & \cdot & \cdot \end{bmatrix}$$

Before proceeding, recall that the general solution to an equation of the sort above (equation 2-30) is (Chapter 1)

$$p_i(t) = \sum_j C_{ij} e^{\lambda_j t} \tag{2-31}$$

where the C_{ij} are constants determined by the initial conditions and the λ_j are the eigenvalues of $(\mathbf{Q} - \mathbf{I})/dt$. Since the elements of $p_i(t)$ must always add to 1.0, for any t, we know immediately that there can be no λ_j greater than zero, (else $e^{\lambda_j t}$ approaches infinity as t becomes very large) and also that there must be at least one λ_j *not less* than zero (else all terms go eventually to zero). Now once the population falls to zero there is no recovery, so that size zero (extinction) is an *absorbing state*. Eventually, no matter how big the initial population, it will reach that state, and $p_i(t)$ must drop to zero for all $i \neq 0$. But because $p_i(t) = C_{i1}e^0 + C_{i2}e^{\lambda_2 t} + \cdots$, $p_i(t)$ can fall to zero only if $C_{i1} = 0$. Thus $C_{i1} = 0$ for all $i \neq 0$. Finally, we are in a position to answer an interesting question: at what rate do chance events lead to extinction? We write

$$\text{probability population is } \textit{not} \text{ extinct} = 1 - p_0(t) = \sum_{i \neq 0} p_i(t) = \sum_i \left(\sum_j C_{ij} e^{\lambda_j t} \right) = \sum_j C_j e^{\lambda_j t} \tag{2-32}$$

where $C_j = \sum_i C_{ij}$. Clearly, the rate at which this quantity decays is governed by the size of the largest, nonzero λ_j (which, we have seen, must be less than zero). An example follows.

Suppose that death rates are fixed, and birthrates fall linearly with n, such that $d = .5$, $b = .6(1 - .25n)$, except that $b_1 = 0$ because we need two individuals for reproduction. Then

$$\mathbf{Q} = \begin{bmatrix} 1 & 0 & 0 & 0 & 0 \\ .5\,dt & 1 - .5\,dt & 0 & 0 & 0 \\ 0 & .5\,dt & 1 - .8\,dt & .3\,dt & 0 \\ 0 & 0 & .5\,dt & 1 - .65\,dt & .15\,dt \\ 0 & 0 & 0 & .5\,dt & 1 - .5\,dt \end{bmatrix}$$

The eigenvalues of $(\mathbf{Q} - \mathbf{I})/dt$ are $\lambda = 0, -.5, -.6, -.675 + .425i$, and $-.675 - .425i$. From (2-32) (and since we know that $C_1 = 0$),

$$\begin{aligned} \text{probability population } \textit{not} \text{ extinct at time } t &= C_2 e^{-.5t} + C_3 e^{-.6t} + C_4 e^{-.675+.425i} \\ &\quad + C_5 e^{-.675-.425i} \\ &= C_2 e^{-.5t} + C_3 e^{-.6t} + Ce^{-.23t}(a \operatorname{Cos} \theta t \\ &\quad + b \operatorname{Sin} \theta t) \qquad \text{(see Chapter 1)} \end{aligned}$$

and the rate at which probability of extinction grows is limited by the term $e^{-.23t}$. This random drift toward extinction is *not* due to environmental fluctuations or to death rate exceeding birthrate! It is an inherent property of the system, a by-product of the fact that individuals are discrete units.

One important attribute of this random drift is that larger populations take much longer to decline than do small populations. To envision why, consider the following. The extinction of a population of size n will require an excess of n deaths over births. If births and deaths are equally likely and n is initially (say) 4, the extinction process is akin to tossing a fair coin until number of heads exceeds number of tails by 4. This should not take very long. On the other hand, if n is initially 40, we must toss the coin until heads exceed tails by 40. This will require many more tosses (i.e., time units). For an alternative approach to this question of chance extinction, see MacArthur (1972) or Leigh (1975).

The biological importance of the foregoing arguments is straightforward. Ultimately, *all* populations must die off. For small, local populations, this extinction may be rapid, and will therefore be a common phenomena. Species that form small isolated groups can be expected to be irregularly distributed as a result of local extinctions and recolonization (see Section 2.4). This will be particularly true when the species exhibit high birth and death rates. Large populations, either homogeneous groups of individuals or collections of smaller subgroups, any one of which may die off quite readily, may have such a tremendous life expectancy that the rule of ultimate extinction is of no immediate significance.

Changes in the External Environment

Over time there are likely to be many changes in the values of environmental parameters—the temperature regime of one year is not the same as that of another; drought in one year may give rise to poor forage the next; storms may kill more young in one year, less in another. Also, individuals differ from generation to generation in their response to the same environmental conditions. Some of these changes are predictable, some not; some or all will influence population dynamics. Three general sorts of questions can be asked. First, given the fact of environmental fluctuations, is it realistic or useful to assume populations are at or near their equilibria? Second, what kind of relationship can we expect between the average carrying capacity and the average value of n? Third, if population sizes deviate commonly from their equilibrium values, do they nevertheless bear a high correlation with the carrying capacity of the moment—do they "track" the environment?

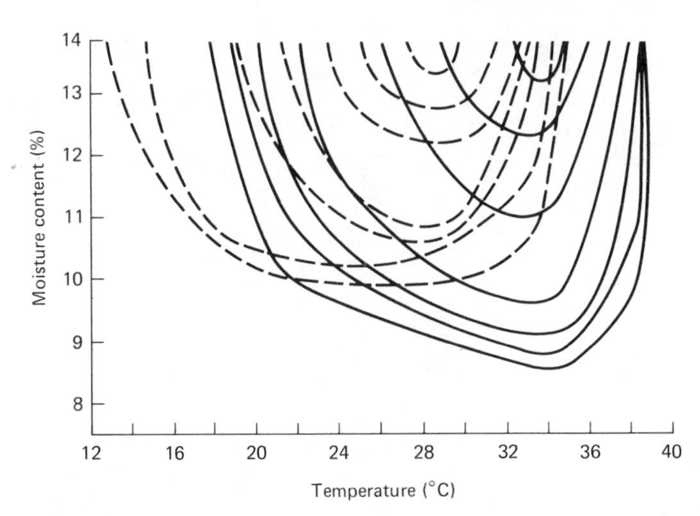

Figure 2-6 Topographic map of the finite rate of increase (λ) of *Calandra oryzae* (small strain) (solid line) and *Rhizopertha dominica* (dashed line) living in wheat of different moisture contents and at various temperatures. (From Birch, 1953a.)

In 1953, L. C. Birch (1953a,b) performed a series of elegant laboratory experiments which showed that population growth rate could be altered by climatic factors. Working with three species of grain beetle, he measured intrinsic rates of increase under various combinations of temperature and grain moisture content, plotting the results as a series of isoclines—as a kind of topographic map (Fig. 2-6). Temperature and moisture would seem to have no direct effect on resource levels in the environment and so should not affect K. In fact, the proportional effect of a change in these variables on r might be expected to be independent of whether the population is large or small. Temperature and moisture are, therefore, density-independent variables.

Actually, temperature and moisture, as well as hailstorms and late frosts are *not* entirely density independent, because (for example) mortality arising from them may be alleviated if low population levels enable individuals more easily to find a limited number of shelters. Nevertheless, it is convenient to think in terms of ideally, density-independent events if for no other reason than to distinguish their effects from clearly density-dependent events such as changes in food supply.

Before answering the more specific questions posed above, note that like random variation in birth and death events, density-independent fluctuations can lead to extinction. Extinction of populations obeying simple, exponential growth models was investigated by Sawyer and Slatkin (1981). Their arguments and calculations will not be presented here; but among their conclusions was the assertion that expected time to extinction should vary with the logarithm of the initial population size.

Let us examine persisting populations. What is the effect on population dynamics of fluctuations in density-independent variables? Suppose that under ideal conditions a population obeys the relationship

$$\Delta n(t) = \epsilon n(t)\left[1 - \frac{n(t)}{K}\right]$$

and that certain density-independent environmental events lead to fluctuations in population growth about this value. Then

$$\Delta n(t) = \epsilon n(t)\left[1 - \frac{n(t)}{K}\right] + \gamma n(t)\, S(t)$$

or

$$n(t + 1) = n(t) + \epsilon n(t)\left[1 - \frac{n(t)}{K}\right] + \gamma n(t)\, S(t) \tag{2-33}$$

where $S(t)$ represents the (changing) state of the environment, and γ measures the species' sensitivity to changes in S. Many density-independent influences, such as those contributed by continuous changes in temperature or humidity, may be viewed as either positive or negative. Were only these influences acting, S would be a continuous variable. We could let the expected value of S over time, $E(S)$, be zero, and define ϵ and K such that

$$\Delta n = \epsilon n\left(1 - \frac{n}{K}\right)$$

under hypothetical, average conditions. But there are other density-independent influences that are more appropriately described as discrete events (hailstorms, for example), and which are strictly deleterious to population growth. When ϵ and K are defined as above, the mean influence of both kinds of density-independent factors together must be negative. Hence $E(S) < 0$. For further discussion of the form of the fluctuation term, see Ricker (1958).

Note that equations (2-33) are either deterministic or stochastic, depending on whether $S(t)$ follows a known, predictable pattern in time or an unpredictable one. Realistically, though, $S(t)$ will have a predictable component (seasonality is largely predictable), but also a capricious element. Therefore, we shall treat the population process described in equations (2-33) as stochastic.

The following derivation is presented for the benefit of those students who are unwilling to accept textbook conclusions without knowing their rationale. For others, considerable time and effort may be saved by turning directly to (2-37) and (2-38). To avoid messy-looking equations, I write $E(S)$ in the form $\bar{S}$. Let n take on the value $\hat{n}$ when $S = \bar{S}$. Then

$$0 = \Delta n = \epsilon\hat{n}\left(1 - \frac{\hat{n}}{K}\right) + \hat{n}\gamma\bar{S}$$

so that

$$\hat{n} = K\left(1 + \frac{\gamma\bar{S}}{\epsilon}\right) \tag{2-34}$$

Now suppose that fluctuations are small (or consider only cases where this is so), so that $(n - \hat{n})^2$, $(n - \hat{n})(S - \bar{S})$ terms can be ignored. Setting $n - \hat{n} = x$, $S - \bar{S} = u$, we have, (equation 2-33)

$$\begin{aligned} n(t + 1) &= \hat{n} + x(t + 1) \\ &= [\hat{n} + x(t)] + [\epsilon(\hat{n} + x(t))]\left[1 - \frac{\hat{n} + x(t)}{K}\right] + \gamma(\hat{n} + x(t))[\bar{S} + u(t)] \end{aligned}$$

or

$$x(t + 1) = bx(t) + cu(t) \tag{2-35}$$

where $b = 1 + \epsilon - 2\epsilon\hat{n}/K + \gamma\bar{S} = 1 - \epsilon - \gamma\bar{S} = 1 - \dfrac{\epsilon\hat{n}}{K}$ and $c = \gamma\hat{n}$.

This can also be written,

$$\begin{aligned} x(t) &= b\big(x(t - 1)\big) + cu(t - 1) \\ &= b[bx(t - 2) + cu(t - 2)] + cu(t - 1) \\ &= \cdots b^t x(0) + b^{t-1}cu(0) + b^{t-2}cu(1) + \cdots cu(t - 1) \\ &= b^t x(0) + c\sum_{i=0}^{t-1} b^{t-1-i}u(i) \end{aligned}$$

Since b must be less than 1.0, the b^t terms eventually vanish, and we obtain

$$x(t) = c \sum_{i=0}^{t-1} b^{t-1-i} u(i) \tag{2-36}$$

whose expected value [because $E(u) = 0$] is zero. But if $E(x) = 0$, then

$$E(n) = E(\hat{n} + x) = \hat{n} + E(x) = \hat{n} = K\left[\frac{1 + \gamma\bar{S}}{\epsilon}\right] \tag{2-37}$$

Thus the mean influence of the fluctuations, $\bar{S} < 0$, depresses the average value of n below K. The amplitude of fluctuations—the variance in S—does not enter (2-37), so would seem to have no effect on average n. This conclusion, however, is in part an artifact of making fluctuations very small [i.e., of making var$(S) \to 0$]. We shall deal with this matter shortly.

The variance in n, over a long time period, is messy to calculate but is a useful quantity. It is given by

$$\lim_{t\to\infty} \text{var}(n) = \frac{K^2(1 + \gamma\bar{S})^2\sigma^2}{\epsilon^2[1 - (1 - \epsilon - \gamma\bar{S})^2]}\left[1 + 2\sum_k (1 - \epsilon + \gamma\bar{S})^k \rho_k\right] \tag{2-38}$$

where $\sigma^2 = \text{var}(u) = \text{var}(S)$ and ρ_k is the correlation between S values k time intervals apart (autocorrelation). (Remember, this is accurate only when fluctuations are very small.)

By looking at (2-38), we can answer the first of the questions posed at the beginning of this section: Is it realistic or useful to assume that populations are at or near their equilibria? If the environment fluctuates only very slightly, so that $\gamma^2\sigma^2$ is very small, or if $\epsilon - \gamma\bar{S}$ is small, and if the ρ_k are near zero or, even better, negative, then var(n) will be small, indicating a fairly constant population. If these conditions are not met, var(n) is apt to be considerable. What kind of species, then, can be thought of as being near population equilibrium? Clearly, it is those for which $\sigma^2 = \text{var}(\bar{S})$ is small—and those for which γ is small—those whose mobility enables them to escape deleterious circumstances, whose behavioral and physiological plasticity allow them to adapt to changing environmental conditions. Species that possess low reproductive potential and low susceptibility to density-independent fluctuations will also show more consistent population densities. As we shall see in Chapter 11, animal species with low b, d, and γ values are likely to be one and the same with large, mobile and adaptable species—terrestrial vertebrates, for example, as opposed to insects, large as opposed to small species. Humans, elephants, and caribou display much steadier numbers than do mosquitoes, houseflies, or field mice. When conclusions drawn from population models require assumptions of equilibrium or near-equilibrium conditions, they are most likely to be valid for the first class of organisms.

We can also answer the second question posed at the start of this section: What is the relationship between $E(n)$ and K? The answer is that $E(n)$ is depressed below K by the average level of punishment by density-independent disturbance. According to (2-37), the variance in S has no effect on the average population size. However, the formulation above was derived under the assumption that fluctuations in n were very small. In nature we are much more likely to see noninfinitesimal, even extremely large amplitude variations. The effect of large oscillations is to alter the conclusion of (2-37)—$E(n)$ is, in fact, depressed below K, even when $E(S) = 0$. This is easily demonstrated for a simple system.

Suppose that the population shifts regularly, back and forth, between two values, in response to regular changes in $S(t)$. We let the two values of $S(t)$ be $+u$ and $-u$, and write $n(t + 1) = \hat{n} + x$, $n(t) = n(t + 2) = \hat{n} - x$. Then (equation 2-33)

$$\begin{aligned} n(t+1) &= (\hat{n} + x) = (\hat{n} - x)(1 + \epsilon + \gamma u) - \frac{\epsilon}{K}(\hat{n} - x)^2 \\ n(t+2) &= (\hat{n} - x) = (\hat{n} + x)(1 + \epsilon - \gamma u) - \frac{\epsilon}{K}(\hat{n} + x)^2 \end{aligned} \tag{2-39}$$

Adding both sides of these equations and rearranging terms, we have

$$\frac{\epsilon}{K}\hat{n}^2 - \epsilon\hat{n} + \left(\frac{\epsilon}{K}x^2 + \gamma ux\right) = 0$$

so that

$$\frac{\hat{n}}{K} = \frac{1}{2} \pm \sqrt{\frac{1}{4} - \frac{x^2}{K^2} - \frac{\gamma ux}{K\epsilon}} < 1 \tag{2-40}$$

Thus even when $E(S) = 0$, density-independent fluctuations *may* depress $E(n)$.

Consider now the case of density-*dependent* fluctuations—where the degree to which environmental change affects birth and death rates depends on population density at the time. As noted previously, the simplest way to envision the impact of such change is to view K as a random variable. Either environmental conditions alter K directly—as, for example, by providing periods of food plenty or famine—or indirectly, by forcing organisms to seek shelter, effectively lowering K by curtailing foraging time. Roughgarden (1975) has developed such a model for the case where fluctuations in K are minute, and arrives at the expressions

$$E(n) = E(K) \tag{2-41}$$

$$\operatorname{var}(n) = \frac{\epsilon^2\sigma^2}{1 - (1 - \epsilon)^2}\left[1 + 2\sum_k (1 - \epsilon)^k \rho_k\right] \tag{2-42}$$

where ϵ, σ^2, and ρ_k are as in (2-38). Thus, as in the previous model, very small fluctuations do not depress $E(n)$ as long as the average influence of the environmental fluctuation, $E(S)$, is neutral. Also, as before, the variance in S does not affect $E(n)$. Boyce and Daley (1980) present a more general set of calculations. Instead of considering K to fluctuate randomly and minutely about a mean of $E(K)$, they consider random fluctuations of unspecified magnitude, in the quantity $Q = 1/K$. Their model yields

$$\lim_{\epsilon \to 0} E(n) = \frac{1}{E(Q)} \leqslant E(n) \leqslant E(K) = \lim_{\epsilon \to \infty} E(n) \tag{2-43}$$

Thus the expected population density is less than the expectation of K (as in equation 2-40), and becomes higher when the intrinsic rate of increase, ϵ, increases.

Are there data available to test these predictions? Because it is difficult to assess the cause of population fluctuations, in many cases it is not possible to give clear-cut illustrations. But some general observations can be relayed. Insects and other groups of organisms whose members are small are more prone to the vagaries of density-independent events—small organisms are less capable of defense against predators, more prone to such indignities (unheard of for elephants) as immobilization by surface tension, more susceptible to rain or hail, less buffered, by virtue of their large surface area/volume ratio, from drought or temperature changes, and more likely to be displaced by winds. Thus their numbers may be not only highly erratic, but also, at most times, well below the carrying capacity. Note, however, that under some circumstances, small size may enhance an organism's chances of finding shelter from environmental stresses. The sensitivity of such organisms makes it highly unlikely that they would survive periods with prolonged, particularly inclement weather; most are quite seasonal. But the very fact that these species operate mostly below their carrying capacities and "escape" the off-seasons (via aestivation, overwintering egg or pupal stages, etc.) enables them also to escape the effects of density-dependent fluctuations (changes in K). Thus the numerical gyrations of these species may be great, and may be almost entirely the result of density-independent events; in fact, insect population densities often display little or no density response, a fact that generated quite a heated controversy in the 1950s over the existence of density feedback (for a battlefront view, see Davidson and Andrewartha, 1948; Smith, 1961).

Species comprised of large individuals, more buffered against weather and accident, experience smaller, density-independent depression and view a given fluctuation in weather as relatively less important than their smaller, more sensitive relations. The value, γ^2 var(S), is perceived as being smaller. Also, such species generally have smaller reproductive potential ϵ. Thus we are more likely to see evidence of density feedback in these species, and will more likely find relatively stable population densities with values near K. Most large mammal populations vary from year to year by less than a factor of 2, birds by factors of roughly 2, meadow voles by as much as a factor of 10, but many insects may vary 100 or even 1 000-fold.

The formulations and predictions above relate to populations growing in discrete time. To treat the continuous-time case we must turn to *diffusion equations.* Because diffusion equations are coming into more and more frequent use in population biology, the following treatment is included for the benefit of the more mathematically advanced student. Others among you may, without loss of face or information prerequisite for later sections, skip the next few pages.

Consider the following, continuous-growth model:

$$\frac{dn(t)}{dt} = r_0 n(t)\left[1 - \frac{n(t)}{K(t)}\right] \tag{2-44}$$

and define $Q(t) = \hat{Q}(1 + S(t)) = 1/K(t)$, where $E(S) = 0$, var$(S) = \sigma^2$. Defining $u = 1/n$ (see equation 2-14), we obtain

$$\frac{du}{dt} = -r_0 u + r_0\hat{Q}(1 + S(t)) \tag{2-45}$$

Because $S(t)$ varies in a not entirely known manner it will be impossible to predict $u(t)$ from $u(0)$ with certainty. However, given $u(0)$ we can predict the probability that $u(t)$ will take a particular value. We define $\phi(u, t)$ as the probability density, so that the probability that $u(t)$ lies between u and $u + du$ at t is given by $\phi(u, t)dt$. Then we make use of the Fokker–Planck diffusion equation (see Kempthorne, 1973, for a simple derivation), which states that if $S(t)$ is *white noise*—$S(t)$ is uncorrelated with $S(t - \tau)$, for any τ,

$$\frac{\partial\phi(u, t)}{\partial t} = -\frac{\partial}{\partial u}[M(u)\phi(u, t)] + \frac{1}{2}\frac{\partial^2}{\partial u^2}[V(u)\phi(u, t)] \tag{2-46}$$

where

$$M(u) = E\left(\frac{du}{dt}\bigg| u\right) \qquad V(u) = \text{var}\left(\frac{du}{dt}\bigg| u\right)$$

the vertical slash indicates that we are looking at $E(du/dt)$ or var(du/dt) for a given value of w. Next, we suppose that the probability distribution will eventually stabilize so that $\partial\phi(u, t)/\partial t = 0$. Then (2-46) becomes

$$0 = -\frac{\partial}{\partial u}[M(u)\phi(u)] + \frac{1}{2}\frac{\partial^2}{\partial u^2}[V(u)\phi(u)]$$

whose solution is

$$\phi(u) = \frac{c}{V(u)}\exp\left[\int \frac{2M(u)}{V(u)}\partial u\right] \tag{2-47}$$

To apply (2-47) we find $M(u)$, by noting that, for a given value of u the expected value of du/dt is (from equation 2-46)

$$\begin{aligned} M(u) = E\left(\frac{du}{dt}\bigg| u\right) &= E[-r_0 u + r_0\hat{Q}(1 + S(t))] \\ &= -r_0 u + r_0\hat{Q} \end{aligned} \tag{2-48}$$

Next

$$V(u) = \text{var}[-r_0 u + r_0 \hat{Q}(1 + S)] = \text{var}(-r_0 u) + \text{var}(r_0 \hat{Q}(1 + S))$$

But, for given u, u is, by definition, constant. Thus

$$V(u) = 0 + r_0^2 \hat{Q}^2 \text{var}(1 + S) = r_0^2 \hat{Q}^2 \sigma^2 \tag{2-49}$$

Substituting these values and integrating, we have

$$\phi(u) = k \exp\left[-\frac{(u - \hat{Q})^2}{2(r_0 Q^2 \sigma^2/2)}\right] \tag{2-50}$$

where $k \propto c/r_0^2 Q^2 \sigma^2$. The expression in (2.50) is the normal distribution function with mean $= \hat{Q}$ and variance $= r_0 \hat{Q}^2 \sigma^2/2$. That is, when $1/K$ fluctuates in a white noise fashion, u will, in time, become normally distributed with a mean of $\hat{Q}$ and a variance of $r_0 \hat{Q}^2 \sigma^2/2$.

To find the expected value of n (rather than $u = 1/n$) in relation to the average $K(= 1/Q)$, we use the Taylor expansion (Chapter 1). Suppose that fluctuations in u, and thus presumably Q as well, are very small, so that terms of order higher than 3 can be conveniently ignored. We write

$$\begin{aligned} E(n) &= E[f(u)] \qquad \text{where } f(u) = 1/u \\ &= E[f(E(u) + \Delta u)] \end{aligned}$$

where Δ indicates the deviation, at any time, in u from its expected value. Using the Taylor expansion, this is

$$E(n) \simeq E(f) + E(\Delta u)\left(\frac{\partial f}{\partial u}\right) + \frac{1}{2}E(\Delta u^2)\left(\frac{\partial^2 f}{\partial u^2}\right) \qquad \text{where } f \text{ is evaluated at } E(u)$$

$$\frac{1}{E(u)} + \frac{\text{Var}(u)}{E(u)^2} = \frac{2 + r_0 \sigma^2}{2\hat{Q}}$$

Similarly,

$$\begin{aligned} E(K) &= E(g(Q)) \qquad \text{where } g(Q) = 1/Q \\ &= E[g(\hat{Q} + \Delta Q)] = \frac{1 + \sigma^2}{\hat{Q}} \end{aligned}$$

so that $\hat{Q} = E(K)/(1 + \sigma^2)$. Therefore,

$$E(n) = \frac{2 + r_0 \sigma^2}{2(1 + \sigma^2)} E(K) \tag{2-51}$$

Also,

$$\begin{aligned} \text{var}(n) = E(n^2) - E(n)^2 &= E[h(u)] - E(n)^2 \qquad \text{where } h(u) = 1/u^2 \\ &\simeq E\left[h + \Delta u \frac{\partial h}{\partial u} + \frac{1}{2}\Delta u^2 \frac{\partial^2 h}{\partial u^2}\right] - E(n)^2 \quad \text{where } h \text{ is evaluated at } E(u) \\ &= \frac{1}{\hat{Q}^2} + E\left[\Delta u\left(\frac{-2}{\hat{Q}^3}\right)\right] + E\left[\left(\frac{\Delta u^2}{2}\right)\left(\frac{6}{\hat{Q}^4}\right)\right] - E(n)^2 \\ &= \cdots \\ &= \frac{r_0 \sigma^2 (2 - r_0 \sigma^2)}{4\hat{Q}^2} \\ &= \frac{r_0 \sigma^2 (2 - r_0 \sigma^2) E(K)^2}{4(1 + \sigma^2)^2} \end{aligned} \tag{2-52}$$

As with the earlier models for both density-dependent and density-independent fluctuations, var(n) is least when σ^2, r_0 are smallest. Since we have assumed it out of existence in this formulation we cannot address the question of autocorrelation. But, unlike Roughgarden's model above, unless $r_0 \geqslant 2$, $E(n)$ will be *less* than $E(K)$. Furthermore, the greater the degree of fluctuation in $K(\sigma^2)$, the more depressed the average population size will be. There is a drawback to its use. There exist two slightly different rationales for defining the $M(x)$ and $V(x)$ for differential equations in x. Each provides us a solution, the solutions are generally not the same, and yet it is often not at all clear which rationale is appropriate (for further discussion, see Feldman and Roughgarden, 1975; Turelli, 1977). In the case derived above both rationales give the same conclusions, but this is not generally so.

Before leaving continuous models we will look briefly at one proposed by Nisbet and Gurney (1976). These authors were concerned with deterministic, seasonal changes in r_0 or K, and modified the logistic to their purposes:

$$\begin{aligned} r_0 &= \hat{r}_0[1 + a\cos(2\pi f t + \phi)] \\ K &= \hat{K}[1 + b\cos(2\pi f t)] \\ \frac{dn(t)}{dt} &= r_0 n(t)\left[1 - \frac{n(t-\delta)}{K}\right] \end{aligned} \tag{2-53}$$

Here a and b are coefficients denoting the importance of the fluctuations to r_0 and K, respectively, f is the reciprocal of the seasonal cycle period (i.e., $f = 1/T = 1$ if the cycle is one year in length), and ϕ and δ are time lags. We shall present here only some of the general conclusions, obtained via computer simulation. Small oscillations in K alone result in population fluctuations, small oscillations in r_0 only do not. Large fluctuations in K result in a drop in $E(n)$ below $\hat{K}$ which becomes increasingly pronounced for *small* r_0. Variance in n, as might be expected from our previous discussions, varies in a positive manner with r_0.

A value of $\phi = 0$ implies that r_0 and K vary in phase; that is, if summer is the season of maximal population growth potential (r_0), it is also the time of greatest equilibrium population size. As ϕ is increased to π, however, this correlation is reversed. Such a reversal, according to the model, results in a dramatic drop in $E(n)/K$ and var(n). Intuitively this makes sense, since large growth potential cannot be realized if there is an insufficient resource base to support its potential consequences; and a large K will, nevertheless, not assure a large n if the growth potential is nonexistent.

Returning to Roughgarden's model we can finally answer the third question posed at the beginning of this section: How well do populations track the environment—that is, how well are n and K correlated? Roughgarden (1975) calculated

$$\text{cov}(n, K) = \eta\frac{\sigma^2\epsilon}{1-\epsilon} \qquad \text{where } \eta = \sum_i (1-\epsilon)^k \rho_k \tag{2-54}$$

where ϵ and ρ_i are as before (equation 2-38). This value rises as η increases. In simple words, as the autocorrelation in environmental events rises, we will see a less obvious show of n chasing a changing K but never quite catching up. Note that tracking also improves as ϵ rises until $\epsilon = 1$, after which the covariance suddenly turns negative—the reproductive potential is so high that the population overshoots, leading to erratic behavior. Such overshooting cannot occur with continuous growth unless time lags occur.

More generally, the influence of K, k time units past, on the present value of K (expressed via autocorrelation ρ_k), is expressed as $(1-\epsilon)^k\rho_k$. If the environment is fluctuating erratically, ρ_k may be near zero, or negative for most or all k, and η will be zero or negative. Regular but rapid fluctuations are indicated by positive values of ρ_k for small time lags (k), but smaller, perhaps even negative values over longer lags. The value of η will be positive but not strongly positive. If the fluctuations are both regular and of long wavelength, then ρ_k is positive over a larger range in k and η is large. The autocorrelation function thus becomes increasingly positive with both regularity of fluctuations and with the time span of a single cycle. Combining this fact with earlier comments on population tracking (equation 2-54), the following conclusions emerge:

1. Populations will better track regularly changing (i.e., cycling) environments.
2. Populations will better track slowly changing environments.
3. Populations whose density-feedback operates at discrete time intervals (so that the discrete model is a better descriptor than the continuous model) and whose reproductive potential is high may overcompensate for environmental fluctuations and display erratic behavior. This will be true particularly when fluctuations are rapid or unpredictable.

Evidence in support of the conclusions above is largely anecdotal; it is difficult to assess changes in K, and often difficult to distinguish density-dependent from density-independent influences. Some laboratory experiments, however, are strongly suggestive. Luckinbill and Fenton (1978) raised *Paramecium primaurelia, P. tetraurelia,* and *Colpidium campylum* on daily rations of the bacterium *Enterobacter aerogenes.* Preliminary tests showed that, as expected, populations adjusted to an equilibrium density reflecting food level. This adjustment was essentially complete within about 14 days. Experiments were then run in which the food supply was varied every 21 days (slow cycle), 14 days (intermediate cycle), or 7 days (fast cycle—adjustment of n to new K cannot be accomplished in this period). As predicted, all three species tracked the changing food supply best in the slow-cycle experiment. In the 14-day cycle test, populations showed significant time lags and oscillations. These oscillations were most pronounced in the 7-day cycle experiment. Furthermore, *Colpidium,* a small organism with high population growth potential, showed extreme population instability in the latter experiments in which it fluctuated to extinction. *P. primaurelia,* the species with the lowest intrinsic growth rate, showed the least amplitude in its gyrations.

The Stability of Populations

The word "stability" has several possible meanings. In this book, unless otherwise stated, we shall be concerned with three.

1. A population is *neighborhood stable* if there exists an equilibrium population value, and if, when displaced by a *minute* amount, the population returns to that equilibrium value. It is conceivable—and such situations will be encountered later—that the population may come to an equilibrium at either of two or more values. Any or all such equilibria may or may not be neighborhood stable.
2. If there exists one equilibrium value and the population, after *any* perturbation, large or small, returns to that value, the population is said to be *globally stable.*
3. If, whenever, a population becomes very small, its growth rate is positive, then—barring chance events—it will never become extinct. The population is said to be *protected.*

A population that is globally stable is, of necessity, also protected. This is not necessarily true of neighborhood stable populations, which may or may not be globally stable. A protected population that is not neighborhood stable is one that stays within finite bounds (infinite populations are biologically unrealistic) but does *not* return to an equilibrium. It therefore cycles indefinitely. Generally, these cycles converge to a specific, repeated path. They are then known as *limit cycles.* These definitions of stability are illustrated using the logistic model.

Neighborhood Stability
We are given the relation

$$\frac{dn}{dt} = r_0 n\left(\frac{1 - n}{K}\right) \tag{2-55}$$

Equilibrium is defined when $dn/dt = 0$, and thus when $1 - n/K = 0$ (i.e., $n = K$). There is only one nonzero equilibrium in this case. Now suppose that we begin with $n = K$, and displace the population by an infinitesimal amount, x. At this new value, the rate of population change is given by

$$\frac{d(K + x)}{dt} = \frac{dx}{dt} = r_0(K + x)\left(1 - \frac{K + x}{K}\right) = -r_0 x - r_0 x^2$$

But since $x \to 0$, the x^2 term is vanishingly small and

$$\frac{dx}{dt} = -r_0 x \tag{2-56}$$

This gives us an expression for the rate of change in the displacement, x, of the population from equilibrium. Equation (2-56) is solved to give

$$x(t) = x(0)e^{-r_0 t} \tag{2-57}$$

Thus the displacement decays exponentially, meaning that the population returns to K—the population is neighborhood stable.

Global Stability
There are two methods for proving global stability. The usual approach is that described by the Russian mathematician Liapunov in his 1892 doctoral thesis. The object is to find a function, rather like a potential energy function, $V(n)$, which has a single minimum at $n = n_{\text{equil}}$, and for which $dV/dt < 0$ for all values of $n \neq n_{\text{equil}}$. If such a function exists, the population is globally stable. To see why, look at Figure 2-7. This graph has a single minimum at $n = n_{\text{equil}}$. [Formally, this is described by the conditions $(\partial V/\partial n)n_{\text{equil}} = 0$ and $(\partial^2 V/\partial n^2)n_{\text{equil}} > 0$.] If $dV/dt < 0$ for all $n \neq n_{\text{equil}}$, V will decrease back to its minimum (and thus n approaches n_{equil}) after any displacement.

In the case of the logistic equation it is easy to find a *Liapunov function.* For example,

$$V(n) = n - K \ln\frac{n}{K} \tag{2-58}$$

will suffice. To check this, write

$$0 = \frac{\partial V(n)}{\partial n} = 1 - K\frac{K}{K}K = 1 - \frac{K}{n}$$

so that $V(n)$ takes a single, extreme value when $n = K$ (which is n_{equil}). Also, $\partial^2 V(n)/\partial n^2 = \partial(1 - K/n)/\partial n = K/n^2 = 1/K > 0$, when $n = K$. Thus $n = K$ describes a minimum. Finally,

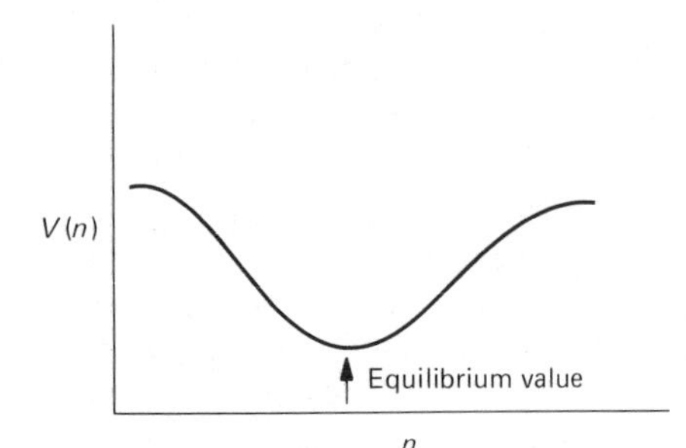

Figure 2-7 Graph of a hypothetical Liapunov function.

$$\frac{dV(n)}{dt} = d\left(n - K \ln \frac{n}{K}\right) = \frac{n - K}{n} \frac{dn}{dt}$$

$$= \frac{r_0 n(n - K)(K - n)}{nK} = \frac{-r_0(K - n)^2}{K} < 0$$

Therefore, $V(n) = n - K \ln(n/K)$ is a Liapunov function and we conclude that a population whose dynamics can be described with the logistic equation is globally stable. The difficulty with the Liapunov function approach to global stability is that there exist no rules, guides, or shortcuts for finding such functions. Just because we cannot discover one does not mean one does not exist. Thus we may be able to prove a system stable but we can never prove a system unstable.

The other approach, also easily applicable to the logistic model, is simply to rewrite the equation in such a way that $dx/x\,dt$, the proportional change in the displacement of n from K, can be shown always to be negative. In (2-55) we did this for x very small, but not for the general case. To accomplish this more general goal, write the logistic equation in the form of (2-14):

$$\frac{du}{dt} = \frac{-r_0}{K}(Ku - 1) \qquad \text{where } u = 1/n \tag{2-59}$$

and let the displacement from equilibrium ($u_{\text{equil}} = 1/K$) be y. Then

$$\frac{d(u_{\text{equil}} + y)}{dt} = \frac{dy}{dt} = \frac{-r_0}{K}\left[K\left(\frac{1}{K} + y\right) - 1\right] = -r_0 y$$

so

$$\frac{1}{y}\frac{dy}{dt} = -r_0 \tag{2-60}$$

Clearly, a displacement y of *any* size is followed by a return to equilibrium.

Protected Populations

An example of a protected population which displays limit cycles is one that obeys the logistic model with a very short time lag (see equations 2-16 through 2-18). We write

$$\lim_{n\to 0} \frac{dn}{dt} \simeq \lim_{n\to 0} r_0 n\left[1 - \frac{n - \tau\, dn/dt + (\tau^2/2) d^2n/dt^2}{K}\right]$$

$$= \lim_{n\to 0} \frac{r_0 n}{1 - \tau/K}$$

which is positive unless the time lag τ is extraordinarily long (and we have assumed it very short). So the population is protected. But (equation 2-20) if

$$r_0 \tau > 2.414$$

then an infinitesimal displacement x oscillates with increasing amplitude. So there is no neighborhood stable equilibrium, and the population cycles without damping.

Overcompensation and Chaos

Suppose that density feedback on a population occurs over a very short period of time each generation. Then the population size at time $t + 1$ depends on n at time t but not on events between t and $t + 1$, and a discrete model accurately describes the dynamics. Populations of the various Pacific salmon, where the spawning stock in one year is believed to be density influenced

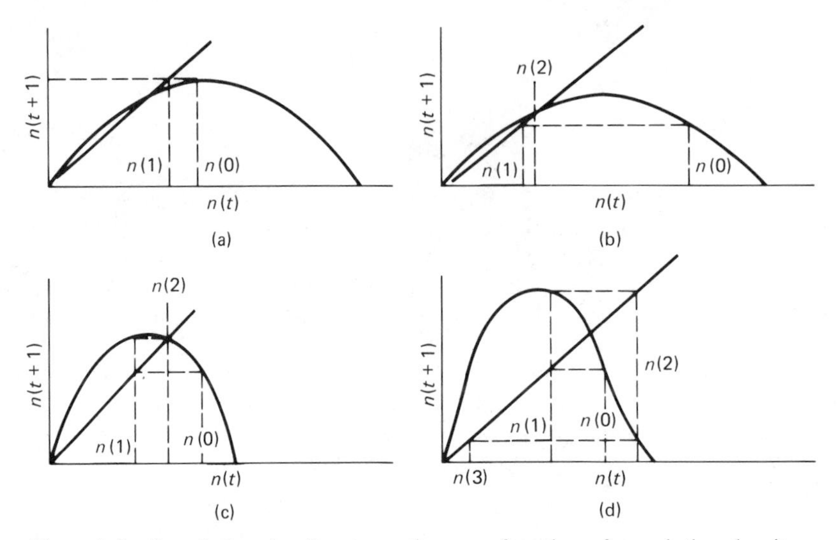

Figure 2-8 Population density at one time as a function of population density one time unit past when (a, b) the intrinsic rate ϵ is less than 1.0; when (c, d) ϵ exceeds 1.0. The sequence $n(0)$, $n(1)$, ... describes density changes over time.

over only a short period of time early in life (see the derivation of the Ricker equations in Section 2.1) would be of this sort, as might those of some insect species in which density feedback occurs only, or at least primarily, during one brief stage in the life cycle. Such populations may, at least theoretically, exhibit rather interesting behavior which is neither neighborhood nor globally stable, and which is much more complex than simple cycling. For other discussion on this issue, see May (1973a,b, 1974) and May and Oster (1976). Consider the discrete, logistic analog

$$\Delta n = \epsilon n\left(\frac{1 - n}{K}\right)$$

or, equivalently,

$$n(t + 1) = \left[1 + \epsilon - \frac{\epsilon}{K}n(t)\right]n(t) \tag{2-61}$$

When $\epsilon < 0$ the population inevitably moves toward extinction, so we shall only consider situations where $\epsilon > 0$. A plot of $n(t + 1)$ versus $n(t)$ is given in Figure 2-8; equilibrium population size is given by $n = K$, the value of n where the graph crosses the 45° "replacement" line—$n(t + 1) = n(t)$. Consider Figure 2-8a. To predict population trends, begin with time $= 0$ and a hypothetical population of size $n(0)$ on the $n(t)$ axis and, from the $n(t + 1)$ versus $n(t)$ curve, read off the value of $n(1)$ on the $n(t + 1)$ axis. Then use the replacement line to find this new value of n back on the $n(t)$ axis and repeat the process to find $n(2)$—and then $n(3)$, and so on. In Figure 2-8a, the slope of the curve is positive where it crosses the replacement line, so if $n(0)$ is not too far above K, $n(1)$ can also be seen to lie above K—and also $n(2)$ and $n(3)$. In fact, n gradually converges on K, without overshooting. The same occurs if $n(0)$ is taken anywhere to the left of K. In Figure 2-8b, the slope is again positive when $n = K$, but this time we have chosen $n(0)$ far enough to the right that $n(1)$ lies *below* K. Thus if the initial displacement of n from K is both positive and sufficiently large, the population will drop, overshooting K, and then converge back on K from below. The population in Figure 2-8a and b, distinguished by a positive slope where the curve crosses the replacement line, is clearly globally stable. But the slope is positive when

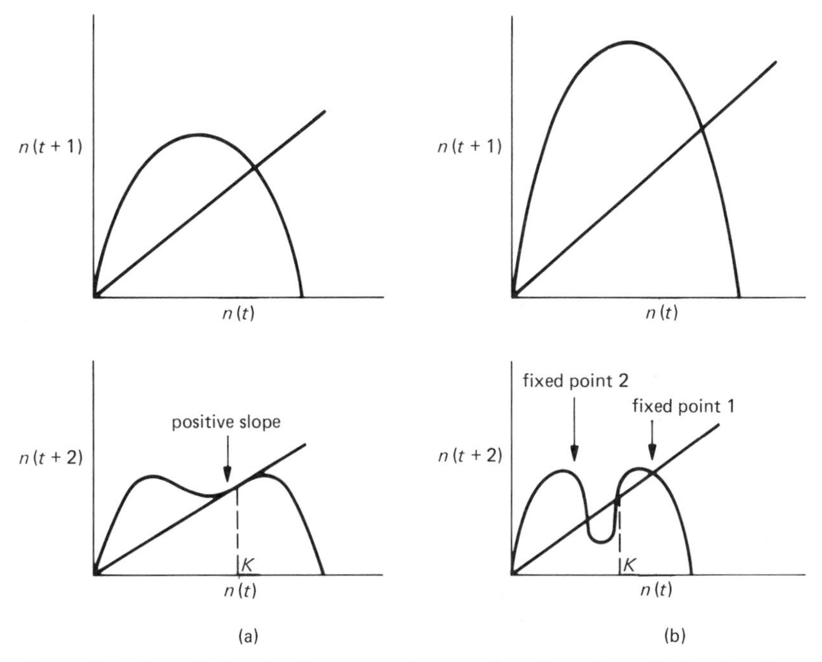

Figure 2-9 Population density at one time as a function of population density one or two time units past, for (a) $2.0 > \epsilon > 1.0$, and (b) $\epsilon > 2.0$.

$$\frac{\partial n(t+1)}{\partial n(t)} = \frac{\partial}{\partial n(t)}\left[1 + \epsilon - \frac{\epsilon a(t)}{K} n(t)\right]_{n(t)=K} = 1 - \epsilon > 0 \tag{2-62}$$

or, equivalently, when $\epsilon < 1.0$. A general conclusion, then: If $\epsilon < 1.0$, the population described in 2-61 is globally stable.

But what if $\epsilon > 1.0$, so that intersection with the replacement line occurs on the *downward* slope? In Figure 2-8c, n appears to converge on K as before, but in an oscillatory manner. In Figure 2-8d, the oscillations seem to increase in amplitude. These figures do not, as exhibited, give us the insight to predict the behavior of n. But consider a plot of $n(t + 2)$, the population *two* time units later, versus $n(t)$. In such a plot it is possible to look at n every *other* time unit. So if the population oscillates between $n(t)$ and $n(t + 1)$, we can look at the top (or the bottom) n values in the oscillation and ask whether the sequence $n(0), n(2), n(4), \ldots$ converges, as appears to be the case in Figure 2-8c, or diverges as in Figure 2-8d. To find where this new curve (Fig. 2-9) intersects the replacement line, set $n(t + 2) = n(t)$:

$$n(t) = n(t+2) = \left[1 + \epsilon - \epsilon\frac{n(t+1)}{K}\right] n(t+1)$$

$$= \left[1 + \epsilon - \epsilon\left(\frac{1 + \epsilon - \epsilon[n(t)/K]}{K}\right) n(t)\right]\left[1 + \epsilon - \epsilon\frac{n(t)}{K}\right]$$

The solutions (other than the trivial case where $n = 0$) are

$$n = K, \quad \frac{K}{2}\left[\left(\frac{1+2}{\epsilon}\right) \pm \sqrt{\frac{1+4}{\epsilon^2}}\right] \tag{2-63}$$

That is, if we start a population at size $\hat{n}$, where $n = \hat{n}$ satisfies (2-63), the population returns to $\hat{n}$ every second time unit. The values $\hat{n}$ are, in a sense, equilibrium values; they are known as *fixed points*.

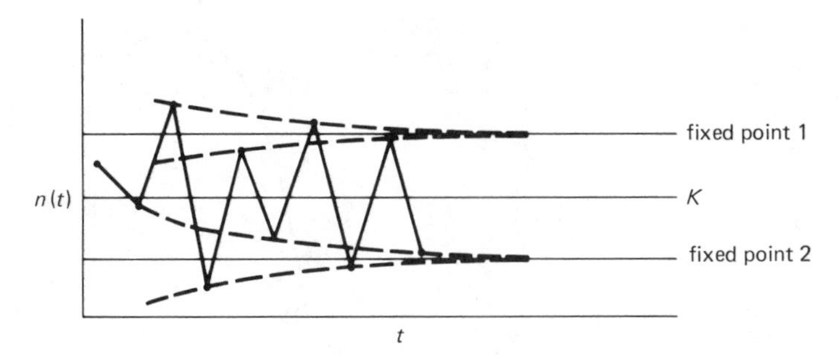

Figure 2-10 Trajectory of population density over time where $\epsilon >$ 1.0, showing oscillations about the two "fixed points."

If $1 < \epsilon < 2$, then the last two roots in (2-63) are imaginary, so the $n(t + 2)$ versus $n(t)$ graph has only one real fixed point—at $n = K$. Moreover, the slope of the curve at this fixed point is

$$\frac{dn(t + 2)}{dn(t)} = (1 - \epsilon)^2 > 0$$

The appropriate graph is shown in Figure 2-9a. Since the curve is symmetrical—the bulge on the left is no higher than that on the right—it is easy to verify, using the same argument as in Figure 2-8a and b, that the sequence $n(0)$, $n(2)$, $n(4)$, . . . converges on K.

If $\epsilon > 2$, then (2-63) shows there to be two real fixed points in addition to K (see Fig. 2-9b). Furthermore, substituting the lowest and highest fixed-point values into the derivation of (2-62) yields the same, *negative* slope. From an initial population value just to the right of K, the population diverges from K. The same occurs if we take $n(0) < K$. The point $n = K$ is, in fact, unstable. On the other hand, slight displacements from either of the extreme fixed points shows them to be "attracting." Thus we know that the population bounces, every alternate time unit, between the extreme fixed points; it does *not* converge on K anymore (Fig. 2.10). The logical

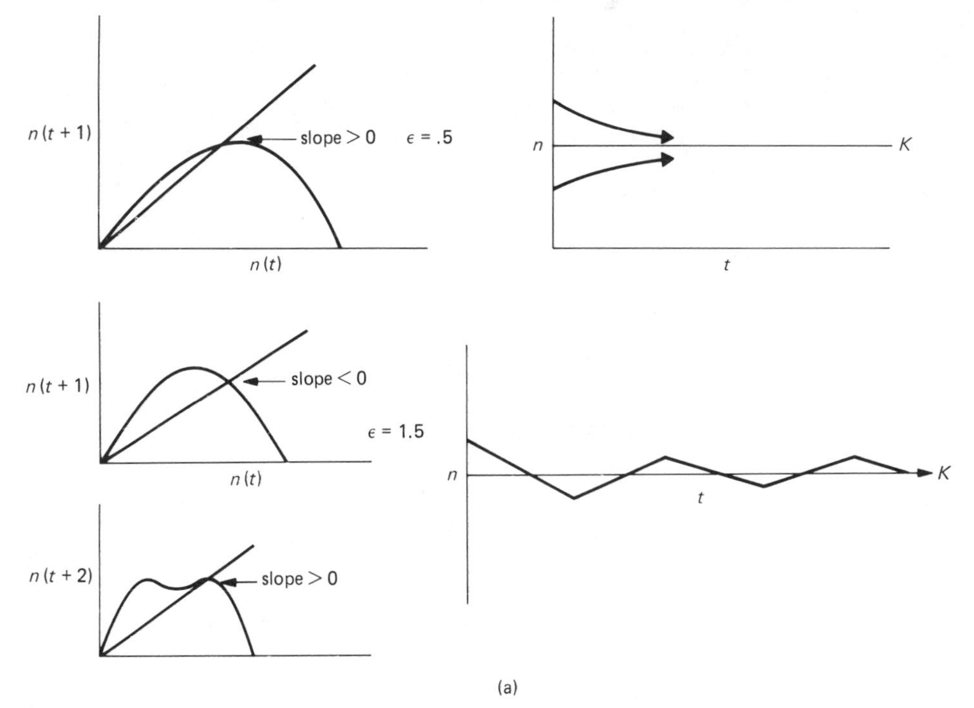

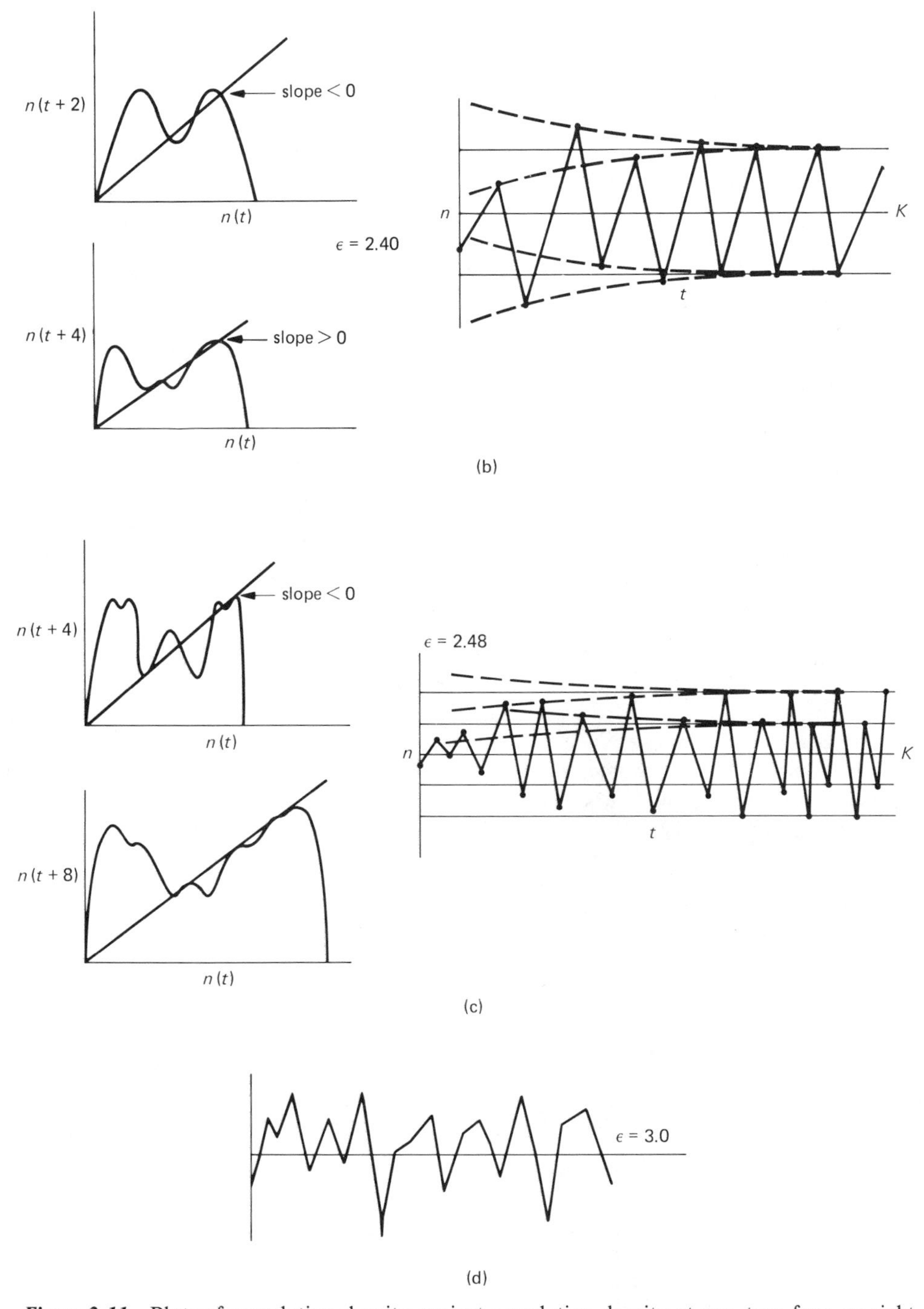

Figure 2-11 Plots of population density against population density at one, two, four, or eight time units past, and corresponding population trajectories for various values of the intrinsic rate ϵ. (a) Convergence on one fixed point, K; (b) convergence on two fixed points; (c) convergence on four fixed points; (d) no apparent pattern.

question at this point is whether, as ϵ increases still more, the sequences $n(0)$, $n(2)$, ... and $n(1)$, $n(3)$, ... continue to converge on the two fixed points.

Because, in Figure 2-9b, the graph intersects the replacement line (at its extreme fixed points) on the *downward* slope, and we are in the same situation as described in Figure 2-8c and d. In that instance we looked at every other time unit in the sequence $n(0)$, $n(1)$, $n(2)$, ... In this case, the equivalent will be to look at every other time unit in the sequence $n(0)$, $n(2)$, $n(4)$, ... ; that is, we will examine a graph of $n(t + 4)$, *four* time units delay, versus $n(t)$. The argument at this point becomes quite messy and will not be pursued further in any detail. It is not too difficult, though, by extension of the foregoing lines of reasoning, to intuit the results. As ϵ continues to rise, $n(0)$, $n(2)$, $n(4)$, ... ceases to converge, but a graph of $n(t + 4)$ versus $n(t)$ shows a positive slope, indicating convergence every *fourth* time unit (Fig. 2-11c) around four fixed points. As ϵ rises even further, we move on to $n(t + 8)$ versus $n(t)$ in order to find convergence every *eighth* time unit around *eight* fixed points, and so on. Before ϵ reaches 3.0, the number of fixed points has doubled, redoubled, and so on, and approaches infinity (Fig. 2-11d). This is a situation which May (1973b, 1974) has termed *chaos*.

The potential existence of chaos in the real world is of significance for several reasons. First, it means that even populations obeying very simple, descriptive models can exhibit extremely complex behavior. In fact, with an infinite number of fixed points, with cycles of infinite length, the oscillations clearly cannot form a repeated pattern. Thus the observed population trends cannot be completely described without prior knowledge of the exact form of the population equation—the dynamics are statistically indistinguishable from random motion. Second, the apparent chaos makes it impossible to investigate the relative roles of density feedback (which is responsible for this state of affairs) and density-independent influences, and thus extremely difficult to make any sorts of predictions about population dynamics at all.

A final few words on chaos are in order. For such behavior to occur in single species systems it is necessary, as initially stated, that $n(t + 1)$ be a *discrete* function of $n(t)$—that is, $n(t + 1)$ is not affected by n between times t and $t + 1$. It seems more reasonable, biologically, that density feedback be continuous. Thus the number of species for which this particular condition is met may be quite limited. Second, after density-independent influences are taken into account, ϵ, for any given species, is likely to rise to 2 or 3 only on occasion, not remain there consistently. Thus insect outbreaks clearly involve very large ϵ values, but in most intervening years ϵ may be quite small. Finally, structuring of populations in space is likely to vitiate the chaos argument because of the tendency for migration to average out the erratic behavior of subpopulations. It is a matter of conjecture whether chaos, at least as a result of the mechanism above, ever really exists in nature.

2.4 Spatially Structured Populations

To any reasonably thoughtful observer of nature it must be apparent that individuals of most species do not distribute themselves randomly about their habitat. In an open field, any portion of which we must, a priori, suppose is suitable for growth, we find milkweeds in clusters. Where fields occur as patches distributed through a matrix of forest, each may have its own, mostly isolated, population of meadow voles. Periwinkles form dense colonies on intertidal rocks, the individuals on one rock rarely or never making contact with those on an adjacent rock only 20 ft away. Populations are not, generally, continuous units with freely interacting members; they are spatially structured, consisting of several, partly connected, subpopulations. What are the implications of such spatial structure for the basic population models described above? What is its influence on rates of extinction, on the tracking of environmental fluctuations, on population stability? Consider the work of Smith (1980) on the pika *Ochotona princeps*.

Spatial Structure as a Buffer to Extinction

Pikas are small mammals which live at fairly high altitude in western North America, making their homes in the rocks of talus slopes. They are territorial and prefer areas along the talus edges where there is ready access to vegetation. Thus the carrying capacity of a given talus slope is roughly proportional to its perimeter. Smith chose to examine a series of 77 talus slopes, the results of mining activities near Bodie, California. He studied the pika populations on these areas in 1972 and returned to examine them again in 1977, slightly more than a generation later. In both years, about 40% of the slopes were vacant, but which areas were unoccupied differed among years. Eight of the areas empty in 1972 were populated in 1977, and 11 of the areas occupied in 1972 had been vacated in 1977. These facts indicate a high rate of local extinction and also the recolonization of empty areas from adjacent colonies. It suggests that we should view the pika population as a collection of interdependent units whose internal dynamics are governed, in part, by chance events and whose individual survival depends on occasional replenishment by migration. This is a somewhat different picture of population dynamics than has been painted so far in this book.

Of course, it would be a mistake to break too enthusiastically with such standbys as the logistic model. Local population dynamics certainly follow rules; they are not entirely prisoners of serendipity. In fact, the equations describing mean population densities over collections of subpopulations may be the same, or at least very similar, in form to those describing single, isolated populations. For example, suppose that

$$\frac{dn_h}{dt} = f(n_h) = r_0 n_h \left(1 - \frac{n_h}{K}\right) + [i - e(n_h)] \tag{2-64}$$

describes the dynamics of a subpopulation h, where i is immigration, and e describes net emigration. Then, using the Taylor expansion, and letting δ denote local deviations from the mean population density,

$$\frac{d\bar{n}}{dt} = \overline{f(n)} = f(\bar{n} + \delta) = \sum_{i=0}^{\infty} \overline{\frac{\delta^i}{i!}\left(\frac{d^i f}{dn^i}\right)}_{n=\bar{n}} \simeq \left[r_0 \bar{n}\left(1 - \frac{\bar{n}}{K}\right) + i - e(\bar{n})\right] + [0] + \frac{1}{2}\sigma_n^2\left(-\frac{2r_0}{K}\right) \tag{2-65}$$

If individuals are, say, Poisson distributed so that $\sigma_n^2 = \bar{n}$, then

$$\frac{d\bar{n}}{dt} = r_0\bar{n}\left(1 - \frac{\bar{n}}{K}\right)\left[1 - \frac{1}{K(1 - \bar{n}/K)}\right] + [\overline{i - e(n)}] = r_0^*\bar{n}\left(1 - \frac{\bar{n}}{K^*}\right) + [\overline{i - e(n)}]$$

where $r_0^* = r_0(1 - 1/K)$ and $K^* = K(1 - 1/K)$. With the exception of the final, bracketed term (of which more will be said later), the form of (2-65) is identical to that of (2-64), although the effective, intrinsic rate of increase and the effective carrying capacity are both altered downward. The appropriate word of caution, then, is not to berate simplistic population models (at least for reasons of existing spatial structure), but to note that the dynamics of any *one*, local subpopulation may bear little resemblance to that of another subpopulation or to the population as a whole.

Vance (1980) has used a sophisticated version of the foregoing approach, incorporating random density-independent terms. His mathematical treatment is a bit too high level for this book, but his results are worth noting. Vance was able to show that, in a continuous model, dispersal invariably decreased the variance in n among subpopulations and increased $\bar{n}$, the average density within subpopulations. Interestingly, when he analyzed a discrete analog to the equation above, this result was sometimes violated; this is, dispersal could be *destabilizing*.

An alternative approach to describing the dynamics of spatially structured populations is to use the method of (2-30) to (2-32). This provides for chance, local extinctions, and is satisfyingly descriptive of the kind of dynamics found by Smith for pikas.

The important point here is that the zero population class is no longer absorbing. A subpopulation that has gone extinct can, in a sense, reappear. Ashby (1960) gives an analogy. Suppose that we were to construct a board in which were displayed 100 light bulbs. We start the experiment with all lights turned on and suppose that each second there is a probability of $\frac{1}{2}$ that any given bulb will turn itself off. Suppose, however, that it were somehow possible to connect the light bulbs in such a way that if a bulb, which has just gone out, is connected to one that is still on, it will relight with a probability of $\frac{1}{2}$. The analogy is to a collection of local subpopulations, each of which has a chance, during any time unit, of becoming extinct, but a probability of reemerging as a result of colonization from adjacent extant populations. With no connections the expected time to extinction of all 100 bulbs is just 2 seconds. As the number of connections increases, so does the life expectancy of the board. And when all possible connections exist, the expected time to global extinction is about 10^{22} years (considerably greater than the estimated age of the universe). Dispersal among subpopulations of a divided population may often result in the *appearance* of great population stability (Ehrlich and Birch, 1967), and Vance (1980) has shown that "dispersal may reduce the impact of stochastic disturbances on population variation." When the subpopulations fluctuate out of phase, local buffering will occur by virtue of greater emigration from dense subpopulations. In particular, it may be that populations exhibiting no density feedback whatsoever on birth or death rates may nevertheless be density limited by virtue of density-dependent emigration among subpopulations. Suppose that

$$\frac{dn_h}{dt} = r_0 n_h + [i - e(n_h)] \tag{2-66}$$

as in (2-64), but without the density-feedback term. Then

$$\frac{d\bar{n}}{dt} = r_0\bar{n} + \overline{i - e(n)} \tag{2-67}$$

Now the value of i, for any subpopulation, is directly proportional to the number of migrants in the population as a whole, which, in turn, is proportional to $e(n)$. Also, since there will surely be some mortality resulting from migration, $\bar{i} < \overline{e(n)}$. Suppose that the survival fraction is β. Then $\bar{i} = \beta e(n)$. Thus

$$\overline{i - e(n)} = (\beta - 1)\overline{e(n)}$$

where β is some proportionality constant less than 1.0. Equation (2-67) is now

$$\begin{aligned}\frac{d\bar{n}}{dt} &= r_0\bar{n} - (1 - \beta)\overline{e(n)} \\ &\simeq r_0\bar{n} - (1 - \beta)\left[e(\bar{n}) + \frac{1}{2}\sigma^2\frac{\partial^2 e(n)}{\partial n^2}\right]\end{aligned} \tag{2-68}$$

But clearly the population in (2-68) is density limited if $e(n)$ is of order greater than 1; that is, if the second derivative of $e(n)$ with respect to n is positive. Suppose, for example, that the emigration rate is proportional to the population density ($= \lambda n$). That is, the probability that any given individual leaves a subpopulation rises with the density of that subpopulation. Then the number emigrating is

$$\overline{e(n)} = \overline{(\lambda n)n} = \overline{\lambda n^2} \quad \text{and} \quad \frac{d\bar{n}}{dt} = r_0\bar{n} - (1 - \beta)\lambda\bar{n}^2 - (1 - \beta)\lambda\sigma_n^2 \tag{2-69}$$

Population equilibrium occurs when $d\bar{n}/dt = 0$, or (for the case where individuals are Poisson distributed—$\sigma_n^2 = \bar{n}$),

$$0 = \frac{d\bar{n}}{dt} = [r_0 - (1 - \beta)\lambda]\bar{n} - (1 - \beta)\lambda\bar{n}^2 \tag{2-70}$$

so that

$$\bar{n}_{\text{equil}} = \frac{r_0 - (1 - \beta)\lambda}{(1 - \beta)\lambda} \tag{2-71}$$

The higher the migration-related mortality (the lower β is), the lower is the equilibrium population density. As intrinsic rate of increase rises, $\bar{n}$, rises. And as emigration rate rises (λ goes up), $\bar{n}$ decreases. On the other hand, suppose that emigration is a purely random process (rate $= \lambda =$ constant). Then

$$e(n) = (\lambda)n$$

and

$$\frac{d\bar{n}}{dt} = r_0\bar{n} - (1 - \beta)\lambda\bar{n} - 0 = [r_0 - (1 - \beta)\lambda]\bar{n} \tag{2-72}$$

Clearly, n is not density limited. An early discussion of this subject is to be found in the seminal paper of Lidicker (1962). For a much more sophisticated treatment of this problem, see Gurney and Nisbet (1975). See also Gadgil (1971), Lidicker (1975), and Nakano (1981).

Biologically, the buffering effects of spatial structure are probably most important in those species, such as insects, with high birth and death rates (high r_0) and high susceptibility to density-independent influences. Here local populations will often pass through bottlenecks of low numbers which make chance extinctions likely. Here also, we may find incidents of local extinction by catastrophe or predation (Chapter 4). Indeed, the continued existence of some such species may depend on an ability to use their high fecundity, coupled with great dispersal powers, to continually recolonize new areas as old ones are lost to the population. In these species, the mechanism of density feedback may be difficult to describe, for there will be little evidence even of its existence. Population dynamics are perhaps best described by probabilistic models. We shall discuss these "fugitive" species at more length in Chapters 4, 5, and 13.

The Evidence for Population Limitation by Emigration

Populations that approach their carrying capacities before emigration becomes significant cannot be said to be limited by emigration. Dispersal must occur before such saturation levels are reached. One can certainly argue that animals *ought* to disperse before saturation is reached. Golley et al. (1975), for example, point out that "saturation dispersal" would, by definition, involve destitute individuals, those that are unlikely to survive the migration to new areas. If an animal is to leave, it would do best to leave while it is still healthy, to "anticipate" the crunch and leave before it occurs (Lidicker, 1962). But by what mechanism might individuals recognize and respond to population pressure before limiting factors make themselves felt? One possible answer, clearly, is social stress. We previously argued that microtine populations experienced considerable stress during the peak phase, and that neither food nor predation was limiting. Indeed, fenced populations rise to considerably higher levels than unfenced populations, suggesting strongly that presaturation dispersal occurs (Krebs et al., 1969). Krebs et al. (1973) report on experiments with *Microtus pennsylvanicus* in which immigration to a trapped-out area was monitored. Immigration (arising from emigration elsewhere) was highest during the increase phase of the surrounding population. Indeed, the number of individuals dispersing was linearly related to the population growth parameter r, and apparently independent of population density; growing populations lost more individuals to emigration than did declining populations of the same size. This suggests that the emigrants were responding not to density pressures, but to anticipated future pressures. Similar behavior has been found in the related species, *M. townsendii* (Krebs et al., 1975). Grant (1978) studied *M. pennsylvanicus* in southern Quebec, removing preexisting residents of all small mammal species from .4 ha of grassland and deciduous woods in April, and introducing 16 study animals. These animals, which showed preference for the grassland portion of the plot, multiplied and, when their numbers reached 170 to 200 per hectare, began to spread (emigrate) into the woods. Was this dispersal prompted by food shortage? Or did social interactions promulgate presaturation dispersal? Grant, noting that his animals ate virtually nothing but grass, sampled their food to measure abundance and chemical composition. Calculations showed that to meet minimal energy requirements, the voles would have had to eat .1 to 1.0% of the available grass. This level of consumption, moreover, would serve to satisfy water, magnesium, calcium, and potassium needs. But sodium and phosphorus needs would require eating up to 10 times this quantity. That is, at worst, the animals would have had to consume between about 1 and 10% of the available grass. In actuality, they bothered to eat about 1.6%. In winter, for about 17 weeks, the animals would need to rely on old growth, with lower nutrient content. The voles would be harder put to meet nutritional needs, but would still easily survive in large numbers, eating only a very small fraction of the available grass. Grant concluded that food played a very minor role in what seemed a clear case of presaturation dispersal.

To those who would construct simple models of population dynamics for presaturation dispersers, a few cautions are in order. For example, the linear relation between number of dispersers and r found by Krebs et al. (1973) suggests a first-approximation model for population limited by emigration. Without other density feedback, the growth parameter r is simply r_0, minus the rate of dispersal. But if number of dispersers rises linearly with r, we obtain

$$\text{rate of dispersal} = \frac{a + br}{n}$$

where a and b are constants, so that

$$r = r_0 - \frac{a + br}{n} = \frac{r_0 n - a}{b + n}$$

which *increases* with n—$dr/dn > 0$. In fact, if dispersal is to be the limiting factor, we must have dr/dn negative. So our model is clearly inadequate.

Tamarin (1977) regressed numbers of dispersers against both r and n for populations of *M. pennsylvanicus* and *M. breweri*. In both cases, *both* independent variables contributed about equally to the dependent variable. We might, therefore, modify our model to read

$$r = r_0 - \frac{a + br}{n} - \frac{cn}{n} = \frac{(r_0 - c)n - a}{b + n}$$

But still, $dr/dn > 0$. Then is Krebs's system, and perhaps the others mentioned above really not limited by dispersal after all? Krebs noted clear evidence of depression in longevity, later maturation, and higher juvenile mortality in peak and declining populations (see also the discussion in Section 2.2), which indicated the existence of social stress. Dispersal may (or may not) be a response to stress, but it appears inadequate to alleviate stress fully. Perhaps dispersal works *in concert* with stress to limit these populations. Or perhaps the impact of dispersal is somehow magnified by the nature of the dispersers. For example, if the dispersers consisted of the healthier, more fecund members of the population, the impact of dispersal on r of the parental population would be greater than indicated simply by numbers of emigrants. Indeed, dispersing individuals *are* qualitatively different from the stay-at-homes. Dispersing males (*M. pennsylvanicus*) are more aggressive (Krebs et al., 1973), and dispersing females of *Mus musculus* (Crowcroft and Rowe, 1958; Lidicker, 1965), *Peromyscus leucopus* (Sheppe, 1965), *M. pennsylvanicus* (Krebs et al., 1973), and *M. townsendii* (Beacham, 1981) are more often in breeding condition. In addition, the emigrant portion of vole populations, and perhaps other species, consist disproportionately of juveniles and subadults (Gaines et al., 1979), the age groups that display the greatest reproductive value (see Chapter 12). Finally, there seem to be genetic differences between resident animals and emigrants (for a discussion, see Chapter 10). The case on population limitation by emigration is not closed.

Review Questions

1. How do we arrive at a population model? Do we just pick a formulation out of the air and see if it "predicts" real, observed population trends reasonably well? Or do we make a set of assumptions about things that might affect the rate of change of a population, develop an equation, and solve it to find out what behavior is predicted by the assumptions made? What are the values and pitfalls of each of these approaches?
2. In what way do we use the predictions of our models as data sets? Does this pose any serious problems for the way we think about populations or view the purpose and utility of model building?
3. Wynne-Edwards (1962) proposed that animals detect the imminence of density feedback before it occurs by means of social cues and react by withholding or curtailing reproductive activity. If some interference impeded the transmission of this information, how might we model population growth rate?
4. In the logistic equation K defines the population density supportable by the environment. Thus K must be a function of the various population-limiting factors. What information would we need in order to express K as an explicit function of, say, food supply, and available space for some particular species of animal or plant?
5. Is K also a function of social stress?
6. In analyzing the effect of density-independent fluctuations on population dynamics (equations 2-33) we used the terms S, a measure of environmental state, and γ, the species' sensitivity to S. Exactly what do S and γ represent in the real world? Are they more than just "fudge factors?" Are they, in themselves, amenable to analysis?
7. In light of questions 4 and 5, justify (or condemn) the use of variations in K to simulate fluctuations in density-dependent factors.
8. If a population is density regulated, it is expected that a density increase during one year is more likely than not to be followed the next year by a decrease, and vice versa. Suppose that you examined two populations for 10 years and found the average density to be $n = 1000$ for both. In the first (population A) you find that, indeed, increases are likely to be followed by decrease, and vice versa. But in species B no apparent pattern emerges; increases and decreases appear to follow in random order. Which species is likely to have the higher variance in density? Which probably has the higher carrying capacity?
9. Two species breed with the same frequency, produce identical-size litters, have identical survivorship schedules, and have identical responses to density feedback. In short, they are ecologically identical in all respects but one—species A occurs in rigidly structured social groups that vary between 18 and 20 individuals; species B is more loosely organized, forming groups of 2 to 100. If the two species coexisted, without interacting, in the same area, which would you expect to be more numerous? (*Hint:* Using the logistic equation as an approximate descriptor of the pertinent dynamics, the mean population growth rate per group is $\overline{dn/dt}$. Expand this expression.)
10. Naylor (1959) presents evidence that as *Tribolium confusum* populations approach saturation, dispersers from the overcrowded population include substantially more males than females. Any of the following might be responsible for this phenomenon: (a) males are more likely to disperse, while population sex ratio remains constant at 1 : 1; (b) more male offspring are produced at high population densities; (c) more males survive to adulthood at high densities. In the light of the discussion on dispersion in this chapter, what effect might these various possibilities have on r?

Chapter 3

Age-Structured Populations

Many species of plant and animal breed once in their lifetime and then die. Some, such as the various species of Pacific salmon mentioned in Chapter 2, live several years before mating; many go through a generation in one year (annuals); some show several, complete population turnovers in a year. Consider the case of the pink salmon (*Onchorhynchus gorbuscha*), which breeds every second year, and therefore consists of odd-year and even-year classes. If, as is generally believed, density feedback occurs almost entirely on the very young, who are still in fresh water and so isolated from their alternate-year fellows, it is possible to treat odd- and even-year classes as isolated populations. Effectively, there is no overlapping of generations, no important interactions among different age groups. With sockeye salmon (*O. nerka*), the situation is more complicated. These fish also breed but once in their lives, so in a sense, there is no generation overlap. But some individuals return to fresh water to spawn at four years of age; others wait until they are five. As a result, fish spawned in different years are *not* entirely isolated. If the density feedback exerted by the spawning stock is different for fish aged 4 and fish aged 5, survival of the young is a function of the *relative* abundances of these age classes as well as of the population density as a whole. This clearly complicates the notion of density feedback presented in Chapter 2, where r is a declining function of n without regard to the age distribution of the individuals constituting that n. Similar complications arise in the case of annual insect species whose eggs hatch or whose larvae emerge over a wide range of days, who breed only once, but where different individuals breed at different times in the year. Again, in a sense, there is no overlapping of generations. But the potential for interactions among different age groups certainly exists.

The material of Chapter 2, although by and large robust in that it obtains *qualitatively* for populations in general, refers, strictly speaking, to populations without interage class interactions. That is, it applies to the simplest kinds of annuals in which all members breed synchronously, or to populations in which there are only insignificant differences in birthrate, death rate, density impact, or susceptibility to density impact among the ages. In this chapter we deal with the opposite extreme, populations that breed more than once, and whose individuals vary with age in their potential contribution to the population dynamics.

It is fairly easy to see how changes in age structure (the fraction of individuals at various ages) might drive fluctuations in total abundance, and vice versa. Suppose, for example, that maximum longevity is 4 years, that individuals under 2 years old have a very low reproductive rate (insufficient to balance mortality), and that individuals over 2 have a high birthrate. If we begin with a young population there will be little recruitment; growth will be slow. But when the initial group has reached ages three and four there will be a surge of reproduction, and n will rise rapidly. This surge will then be followed by death of the oldsters, and the new population will consist primarily of young, low-fecundity individuals. Death rate will then exceed birthrate and the population will fall. But then the young grow up and there will be another reproductive bonanza—and so on. This simple example suggests that stable (unchanging) age distributions are not likely to be common in nature. Nevertheless, it is convenient, as a heuristic exercise, to consider hypothetical populations which *are* at stable age distribution. Basic relationships, obscured by complexities of the real universe, often present themselves lucidly in idealized worlds.

3.1 Basic Models

Life Tables

Consider first organisms whose individuals breed synchronously, in well-defined breeding seasons. Let

$n_x(t)$ = number of individuals of exactly age x at time t
b'_x = fecundity of individuals of age x
p_x = probability of surviving the interval $x, x + 1$
l_x = "survivorship," the proportion of individuals reaching age x
d_x = proportion of a cohort (all individuals born in the same breeding season) that die in the interval $x, x + 1$
$q_x = 1 - p_x$ be the proportion of individuals of age x that will die before reaching age $x + 1$
e_x = life expectancy of individuals now age x (actually, the expected number of future breeding seasons they will see)

The relationships among some of these parameters are fairly obvious. For example, l_x is the probability that an individual lives to age x, which is simply

$$l_x = p_0 p_1 p_2 \cdots p_{x-1} = \prod_{i=0}^{x-1} p_i \tag{3-1}$$

The survivorship can also be calculated in the following way:

$$\begin{aligned} l_0 &= 1.0 \qquad \text{(by definition)} \\ l_1 &= 1.0 - d_0 \\ l_2 &= 1.0 - d_0 - d_1 = l_1 - d_1 \\ l_3 &= 1.0 - d_0 - d_1 - d_2 = l_2 - d_2 \\ l_x &= l_{x-1} - d_{x-1} \end{aligned} \tag{3-2}$$

The age-specific mortality q_x is the probability that an individual will die before age $x + 1$, given that it is now age x—that is, given that it lives to age x. Therefore, the unconditional probability that an individual dies in the interval, d_x, is q_x times the probability that it survived to age x, l_x:

$$d_x = q_x l_x \qquad \text{or} \qquad q_x = \frac{d_x}{l_x} \tag{3-3}$$

The life expectancy can be written

$$\begin{aligned} e_x &= \left(\begin{matrix}\text{probability that the individual lives to reach} \\ \text{exactly one more age class}\end{matrix}\right)(1.0) \\ &\quad + \left(\begin{matrix}\text{probability that the individual lives to reach} \\ \text{exactly two more age classes}\end{matrix}\right)(2.0) \\ &\quad + \text{etc.} \\ &= \frac{l_x - l_{x+1}}{l_x}(1) + \frac{l_{x+1} - l_{x+2}}{l_x}(2) + \cdots \\ &= \frac{1}{l_x}(l_{x+1} + l_{x+2} + l_{x+3} + \cdots) = \frac{1}{l_x}\left(\sum_{i=x+1}^{\infty} l_i\right) \end{aligned} \tag{3-4}$$

Some examples will be useful.

Adolph Murie (1944) examined the skulls of Dall mountain sheep (*Ovis dalli*) in Alaska, using the horns to estimate age at death. Because bones decay only very slowly on the tundra, his data were representative of a long-time average population. Because of this fact and the fact that mountain sheep populations do not fluctuate greatly, the data can be thought of as characteristic of a sheep population with R very nearly 1.0. But this, in turn, means that there is no consistent change in the size of cohorts over time. Thus if (say) 19.9% of all the skulls found were from individuals in their first year (age 0 to 1), we would know that (approximately) 19.9% of any cohort, on average, die in their first year—$d_0 = .199$. Using this reasoning, Murie calculated d_x for all age classes. He then found survivorship values (equation 3-2), and then q_x (equation 3-3), and finally, e_x (equation 3-4). The results, collectively forming a *life table*, are shown in Table 3-1. Note the pattern followed by age-specific mortality, q_x, with age. This downward trend in q_x during early life, followed by a rise later, is almost universal in animals and plants—more will be said of it later (Chapter 12).

Among the more ingenious attempts to build life tables is the work of Edmondson (1945), who studied sessile rotifers. The species of interest, *Floscularia conifera*, lives in tubes which it builds from debris it gathers in the water and fastens to aquatic plants. Edmondson sprinkled carmine powder on the water surface on one day, allowing his animals to gather the particles and incorporate them into their tubes. Twenty-four hours later he sprinkled carbon powder over the same area. Later investigation of rotifer tubes showed red and black bands, the distance between them indicating growth over a 24-hour period. In this way the size–age relationship could be constructed, allowing Edmondson to ascertain the age of any individual he might sample. Also, the frequency of animals with red, but no black band directly gave him the mortality rates of different-aged individuals (see Table 3-2). Life tables have been constructed for a wide variety of organisms (see Deevey, 1947 and Caughley, 1966, for reviews).

Merely constructing life tables may be fun, offering a use for moldering field data, but it is of more than idle interest only for devotees of the life insurance business. However, more relationships among the various parameters of an age-structured population are useful. For example, let us again delve into the idyllic realm of stable age distributions. Where $n_x(t)$ is the population of individuals age x, at time t, we can write

$$n_x(t) = n_0(t - x)l_x \tag{3-5}$$

the individuals now (time t) aged x were age 0, x time units past, and a fraction l_x of them survived the intervening period. Furthermore, if the population is in stable age distribution, and growing by a multiple R each time period, every age group must also be growing by a multiple R.

Table 3-1

Age (yr)	d_x	l_x	q_x	e_x
0	.199	1.000	.199	6.759
1	.012	.801	.015	7.438
2	.013	.789	.0165	6.551
3	.012	.776	.0155	5.661
4	.030	.764	.0393	4.750
5	.046	.734	.0626	3.944
6	.048	.688	.0699	3.208
7	.069	.640	.108	2.136
8	.132	.571	.231	1.394
9	.187	.439	.426	.813
10	.156	.252	.619	.417
11	.090	.096	.937	.094
12	.003	.006	.500	.500
13	.003	.003	1.000	.000
$\geqslant 14$	—	.000	—	—

Table 3-2

Age (days)	d_x	l_x	q_x
0	.020	1.000	.020
1	.200	.980	.204
2	.060	.780	.077
3	.000	.720	.000
4	.300	.720	.416
5	.140	.420	.333
6	.060	.280	.214
7	.140	.220	.636
8	.040	.080	.500
9	.020	.040	.500
10	.020	.020	1.000

Hence

$$n_0(t) = n_0(t - x)R^x \qquad \text{or} \qquad n_0(t - x) = n_0(t)R^{-x} \tag{3-6}$$

Substituting (3-6) into (3-5) yields

$$n_x(t) = n_0(t)R^{-x}l_x \tag{3-7}$$

If R and l_x are constant in time (and, as we shall soon see, we cannot have stable age distribution if this is not so), then (3-7) holds for all time, and the variable t can be dropped. From (3-7) we can find the stable age distribution:

$$\left(\begin{array}{c}\text{Proportion of the population}\\ \text{at age } x\end{array}\right) = \frac{n_x}{\sum_x n_x} = \frac{n_0 R^{-x}l_x}{\sum_x n_0 R^{-x}l_x} = \frac{R^{-x}l_x}{\sum_x R^{-x}l_x} \tag{3-8}$$

or, more simply,

$$c_x \propto R^{-x}l_x \tag{3-9}$$

Equation (3-9) is more informative than it may appear at first glance. If $R > 1$ (the population is growing), then R^{-x} declines as x increases. Thus, for a given set of l_x values, $R^{-x}l_x$ in a growing population will be proportionately larger for small x, smaller for large x, than in a stable or declining population; growing populations have proportionately more young individuals. For example, to the extent that the general, age-specific mortality values q_x are similar (at least proportionately), most developing nations, with rapidly growing populations, should display disproportionate numbers of young and fewer old individuals than the industrialized countries. Figure 3-1 shows the age distribution for Mexico and the United States in 1974. In human beings, massive shifts in age structure accompanying population stable state have significant implications for a country's politics, economics, and tax structure.

The Leslie Matrix

In Chapter 2 we opened with some basic definitions and a simple expression for unlimited population growth:

$$\frac{dn}{dt} = r_0 n$$
$$n(t) = n(0)e^{r_0 t} \tag{3-10}$$

The Leslie matrix approach (Leslie, 1945) in its basic form is the age-structured population

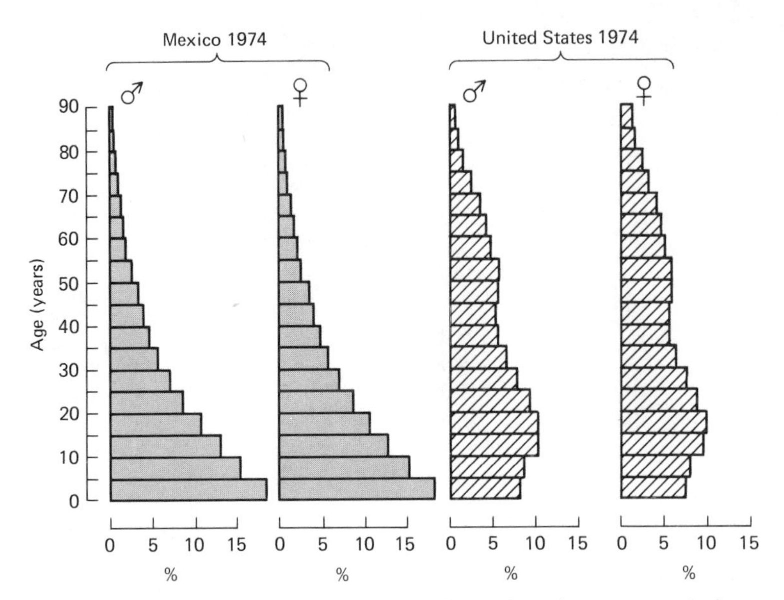

Figure 3-1 Age distribution in U.S. and Mexican human populations, 1974. (From the United Nations' *Demographic Year Book*, 1974.)

equivalent to (3-10). Nowhere does it incorporate density feedback, a caveat that has often been omitted in texts and in applications to real populations. Later in Section 3.1 we shall discuss ways in which the model can be modified to include density-feedback terms.

We begin with two simple relationships: In a population whose last reproductive age is k, and which breeds at discrete intervals (well-defined short breeding season),

$$n_x(t) = n_{x-1}(t-1)p_{x-1} \qquad \text{where } x > 1 \tag{3-11}$$

$$n_0(t) = \sum_{x=1}^{k} n_x(t)b'_x = \sum_{x=1}^{k} n_{x-1}(t-1)p_{x-1}b'_x \tag{3-12}$$

Equation (3-11) should be obvious; (3-12) gives the number of individuals in the first age class as the sum of newborn by parents of all ages. These relationships can now be put into matrix form:

$$\begin{bmatrix} n_0(t) \\ n_1(t) \\ n_2(t) \\ \vdots \\ n_{k-1}(t) \end{bmatrix} = \begin{bmatrix} p_0b'_1 & p_1b'_2 & p_2b'_3 & \cdot & p_{k-1}b'_k \\ p_0 & 0 & 0 & \cdot & \cdot \\ 0 & p_1 & 0 & \cdot & \cdot \\ 0 & 0 & p_2 & \cdot & \cdot \\ \cdot & \cdot & \cdot & \cdot & \cdot \\ 0 & 0 & 0 & p_{k-2} & 0 \end{bmatrix} \begin{bmatrix} n_0(t-1) \\ n_1(t-1) \\ n_2(t-1) \\ \vdots \\ n_{k-1}(t-1) \end{bmatrix} \tag{3-13}$$

or, more conveniently,

$$\mathbf{n(t)} = \mathbf{L} \cdot \mathbf{n(t} - 1) \tag{3-14}$$

The matrix **L** is the Leslie matrix. It belongs to a class of *projection matrices*, named for the fact that it projects the population vector at one time onto a new population vector in the next time unit. We shall deal with other population projection matrices later.

Recall that the general solution to an equation such as (3-13) is

$$n_i(t) = \sum_j C_{ij}\lambda_j^t \tag{3-15}$$

where $\{\lambda_j\}$ are the eigenvalues of **L** and $\{C_{ij}\}$ are constants whose values depend on the initial population vector. To illustrate, consider the hypothetical case of a population with the following life table:

x	p_x	b'_x
0	.4	0
1	.5	1.25
2	0.0	4.0

The corresponding Leslie matrix is

$$\mathbf{L} = \begin{bmatrix} .5 & 2.0 \\ .4 & 0 \end{bmatrix} \tag{3-16}$$

Suppose that we begin with a population of 100 individuals, all of age 0. What is the population vector nine time units later?

First, find the eigenvalues of **L**:

$$0 = \det\begin{bmatrix} .5 - \lambda & 2.0 \\ .4 & -\lambda \end{bmatrix} = \lambda^2 - .5\lambda - .8 \tag{3-17}$$

so

$$\lambda = 1.1787, \quad -.6787$$

Then, using (3-15),

$$\begin{aligned} 100 = n_0(0) &= C_{11}(1.1787)^0 + C_{12}(-.6787)^0 = C_{11} + C_{12} \\ n_0(1) &= C_{11}(1.1787)^1 + C_{12}(-.6787)^1 = 1.1787C_{11} - .6787C_{12} \end{aligned} \tag{3-18}$$

$$\begin{aligned} 0 = n_1(0) &= C_{21}(1.1787)^0 + C_{22}(-.6787)^0 = C_{21} + C_{22} \\ n_1(1) &= C_{21}(1.1787)^1 + C_{22}(-.6787)^1 = 1.1787C_{21} - .6787C_{22} \end{aligned} \tag{3-19}$$

To solve for the C_{ij}'s we need to know $n_0(1)$ and $n_1(1)$. These are easily found by direct calculation (equation 3-13):

$$\begin{bmatrix} n_0(1) \\ n_1(1) \end{bmatrix} = \begin{bmatrix} .5 & 2.0 \\ .4 & 0 \end{bmatrix}\begin{bmatrix} n_0(0) \\ n_1(0) \end{bmatrix} = \begin{bmatrix} .5 & 2.0 \\ .4 & 0 \end{bmatrix}\begin{bmatrix} 100 \\ 0 \end{bmatrix} = \begin{bmatrix} 50 \\ 40 \end{bmatrix}$$

Substituting $n_0(1) = 50$ and $n_1(1) = 40$ back into (3-18) and (3-19), we find

$$\begin{aligned} C_{11} &= 63.46 \qquad & C_{12} &= 36.54 \\ C_{21} &= 21.54 \qquad & C_{22} &= -21.54 \end{aligned}$$

so that, finally,

$$\begin{aligned} n_0(t) &= 63.46(1.1787)^t + 36.54(-.6787)^t \\ n_1(t) &= 21.54(1.1787)^t - 21.54(-.6787)^t \end{aligned}$$

At $t = 9$,

$$n_0(9) = 277.6 \qquad n_1(9) = 93.9$$

The total population is $277.6 + 93.9 = 371.5$.

A number of interesting generalizations can be made about the Leslie matrix.

1. The eigenvalues are given by the solutions to an equation of the form

$$1 = \sum_x \lambda^{-x} l_x b'_x \tag{3-20}$$

This is easily proven. Write

$$0 = \begin{vmatrix} p_0 b'_1 - \lambda & p_1 b'_2 & p_2 b'_3 & \cdots \\ p_0 & -\lambda & 0 & \cdots \\ 0 & p_1 & -\lambda & \cdots \\ \cdot & \cdot & \cdot & \cdots \end{vmatrix}$$

But, by (1-13), this is

$$0 = (-1)^{1+1}(p_0 b'_1 - \lambda) \begin{vmatrix} -\lambda & 0 & 0 & \cdot \\ p_1 & -\lambda & 0 & \cdot \\ 0 & p_2 & -\lambda & \cdot \\ \cdot & \cdot & \cdot & \cdot \end{vmatrix} + (-1)^{1+2}(p_1 b'_2) \begin{vmatrix} p_0 & 0 & 0 & \cdot \\ 0 & -\lambda & 0 & \cdot \\ 0 & p_2 & -\lambda & \cdot \\ \cdot & \cdot & \cdot & \cdot \end{vmatrix} + \text{etc.}$$

$$\begin{aligned} &= (+1)(p_0 b'_1 - \lambda)(-\lambda)^{k-1} + (-1)(p_1 b'_2)[p_0(-\lambda)^{k-2}] + \cdots \\ &= (p_0 b'_1 - \lambda)(-1)^{k-1}\lambda^{k-1} - (p_1 b'_2 p_0(-1)^{k-2}\lambda^{k-2} + (p_2 b'_3)p_0 p_1(-1)^{k-3}\lambda^{k-3} - \cdots \\ &= (-1)^{k-1}[\lambda^k - p_0 b'_1 \lambda^{k-1} - p_0 p_1 b'_2 \lambda^{k-2} - p_0 p_1 p_2 b'_3 \lambda^{k-3} - \cdots] \end{aligned}$$

Thus

$$\begin{aligned} 0 &= \lambda^k - p_0 b'_1 \lambda^{k-1} - p_0 p_1 b'_2 \lambda^{k-2} - p_0 p_1 p_2 b'_3 \lambda^{k-3} - \cdots \\ &= \lambda^k - l_1 b'_1 \lambda^{k-1} - l_2 b'_2 \lambda^{k-2} - l_3 b'_3 \lambda^{k-3} - \cdots \end{aligned}$$

Finally,

$$\lambda^k = l_1 b'_1 \lambda^{k-1} + l_2 b'_2 \lambda^{k-2} + l_3 b'_3 \lambda^{k-3} + \cdots$$

and dividing through by λ^k gives us

$$1 = \lambda^{-1} l_1 b'_1 + \lambda^{-2} l_2 b'_2 + \lambda^{-3} l_3 b'_3 + \cdots = \sum_x \lambda^{-x} l_x b'_x$$

2. Equations of the form (3-20) can be shown to have one and only one real, positive root (for a simple proof, see J. M. Emlen, 1973). All other roots are either negative or complex and, therefore (Chapter 1), describe population oscillations. The real, positive root, λ_1 (sometimes written λ_0), is also known as the *Perron root.*

3. If the oscillations described by the negative and complex roots die out, population growth is thereafter described by the Perron root alone. Thus

$$\mathbf{n}(t) = \mathbf{C}\lambda_1^t$$

But this says that once oscillations die out, each age group grows at the same rate as every other—that is, the population is in stable age distribution. Furthermore, the vector **n**, for which the above is true,

$$\mathbf{n}(\mathbf{t}+1) = \lambda_1 \cdot \mathbf{n}(\mathbf{t})$$

is the eigenvector associated with λ_1. Therefore, the stable age distribution is given by the eigenvector associated with the Perron root. Finally, when the population is in stable age distribution, so that

$$n_1(t+1) = n_i(t)\lambda_1$$

for all *i*, it follows that

$$n_{\text{total}}(t) = \sum_i n_i(t+1) = \lambda_1 \sum_i n_i(t) = \lambda_1 n_{\text{total}}$$

Thus $R = \lambda_1$. But be careful. R equals λ_1 *only* at stable age distribution.

4. The value of the Perron root is equal to or greater than the absolute values of all the other roots. Thus it would be hypothetically possible to obtain (for example)

$$\lambda_1 = 1.2$$
$$\lambda_2 = -.4 \qquad \text{absolute value} = .4$$
$$\lambda_3 = .5 + .6i \qquad \text{absolute value} = .5^2 + .6^2 = .78$$
$$\lambda_4 = .5 - .6i$$

but not

$$\lambda_1 = .8 \qquad\qquad \lambda_1 = 1.1$$
$$\lambda_2 = -.9 \quad \text{or} \quad \lambda_2 = -.7 + .9i$$
$$\lambda_3 = -.7 - .9i$$

The implications of this fact are enormous, for if a population is not steadily growing—so that $\lambda_1 \leqslant 1.0$, then $\lambda_j \leqslant 1.0$ for all other roots ($j \neq 1$), and population cycles cannot grow in amplitude; they must remain steady or damp out.

5. Can we predict, without calculating the eigenvalues, whether or not cycles will damp out? The answer is yes. Examine the top row of the Leslie matrix, containing the fecundity values $p_{i-1}b_i$. Note all the $p_{i-1}b_i$ values with *nonzero* values and find the highest common denominator of the corresponding indices, *i*. This number, the highest common denominator, gives the number of eigenvalues with absolute value equal to λ_1. For example, suppose that

$$\mathbf{L} = \begin{bmatrix} 0 & 2 & 4 & 10 \\ .4 & 0 & 0 & 0 \\ 0 & .1 & 0 & 0 \\ 0 & 0 & .1 & 0 \end{bmatrix}$$

The nonzero $p_{i-1}b'_i$ values are $p_1b'_2$, $p_2b'_3$, and $p_3b'_4$, with indices 2, 3, and 4. The highest common denominator of 2, 3, and 4 is *one*. Hence there is only one eigenvalue with absolute value equal to λ_1, the Perron root itself. All other roots have $|\lambda_j| < \lambda_1$, and since, in this case, $\lambda_1 = 1.0$, $|\lambda_j| < 1.0$ for all *j*, and any cycling that may occur must damp out. Suppose that

$$\mathbf{L} = \begin{bmatrix} 0 & 2 & 0 & 5 \\ .4 & 0 & 0 & 0 \\ 0 & .2 & 0 & 0 \\ 0 & 0 & .5 & 0 \end{bmatrix}$$

where, again, λ_1 happens to be 1.0. The nonzero $p_{i-1}b'_i$ elements are $p_1b'_2$ and $p_3b'_4$, with indices $i = 2, 4$. The highest common denominator of 2 and 4 is 2, so that there are *two* eigenvalues with absolute value $= \lambda_1$, one in addition to the Perron root. This population will experience sustained cycles.

The biological interpretation of this point is fairly straightforward. It is highly unlikely that we should find multiple-age breeders skipping perfectly good breeding seasons on a regular basis. So if we look at the highest common denominator of $p_{i-1}b'_i$ indices on a year-to-year basis, we are unlikely to find anything other than 1, implying that (if $\lambda_1 = 1$) most populations invariably approach stable age distribution on a year-to-year basis. Of course, any fluctuation in the environment will affect b_i and p_i values and alter λ_1, at least momentarily. Hence, in spite of the rather clean mathematics, it is unlikely that any other than the highly buffered species, operating consistently in the neighborhood of population equilibrium, $\hat{n}$, will really approach this ideal condition.

Now suppose that our age classes are based on 6-month intervals (say), for a species that breeds once a year. In this case we will obtain $p_{i-1}b'_i = 0$ every alternate time interval, and a highest common denominator of 2, implying (if $\lambda_1 = 1$) a sustained cycle. This is hardly surprising—populations do rise in numbers following reproduction, and fall back, via mortality, during the nonbreeding season.

For more information the reader with mathematical interests is referred to Demetrius (1971), Sykes (1969) and Pollard (1973).

Not all populations breed synchronously at some instantaneous point in time. Not even crudely. Indeed, some spread their reproductive season over much or most of a year. When this happens it makes no sense to use b'_x values. Instead, we must define a new measure of fecundity:

f_x = the number of young who live at least one time interval, produced per individual in the age range $x - 1, x$

If we consider the number of individuals contributing f_x young to be the number at age $x - 1$, we will overestimate reproductive output because some will die before reaching age x and so will not contribute their full potential in offspring. If we consider only those that live to age x, we will underestimate output. Thus we compromise, and approximate the number of individuals contributing f_x young during the interval by the number that reach some mean point in the interval. In practice this is taken to be the midpoint of the interval. It is at this effective breeding point that $n_x(t)$ is defined.

Let P_x be the probability of surviving from the effective breeding point in age interval $x - 1$ to x to the same time in the interval, x to $x + 1$. We can now write, analogous to (3-11) and (3-12),

$$n_x(t) = n_{x-1}(t-1)P_{x-1} \qquad \text{where } x \geqslant 2$$
$$n_1(t) = \sum_{x=1} n_x(t-1)f_x \tag{3-21}$$

so that

$$\mathbf{n(t)} = \begin{bmatrix} f_1 & f_2 & \cdot & \cdot & \cdot \\ P_1 & 0 & \cdot & \cdot & \cdot \\ 0 & P_2 & \cdot & \cdot & \cdot \\ \cdot & \cdot & \cdot & \cdot & \cdot \end{bmatrix} \mathbf{n(t-1)} \tag{3-22}$$

The characteristic equation is given by

$$1 = \lambda^{-1}f_1 + \lambda^{-2}P_1f_2 + \lambda^{-3}P_1P_2f_3 + \cdots$$

Suppose that survivorship is now taken to mean the probability of surviving from the effective breeding point of the age interval 0 to 1, to the corresponding time in the age interval $x - 1$ to x. This is estimated, in practice, with

$$L_x = \frac{l_{x-1} + l_x}{2}$$

Then $L_1 = 1.0$, $L_2 = P_1$, $L_3 = P_1P_2$, and so on, and the expression above becomes

$$1 = \lambda^{-1}L_1f_1 + \lambda^{-2}L_2f_2 + \cdots = \sum_x \lambda^{-x}L_xf_x \tag{3-23}$$

The equations and generalizations are exactly analogous in all ways to those describing populations with instantaneous breeding seasons.

Stage Projection Matrices

It is often not simple to ascertain organisms' ages in the field. By contrast it is usually quite easy to categorize them as infants, juveniles, and adults; as eggs, larvae, pupae, and adults; or as instars number 1, 2, 3, and so on. In such cases Leslie matrices, describing *age* structure, would be difficult or impossible to apply. But a matrix describing the *stage* structure might be constructed. Sometimes, as in the case of many insects, fecundity and survival are directly related to stage and only indirectly to age; in many fish, fecundity is a function of size, related to age only in that age and size are (often quite loosely) correlated. In these cases a Leslie matrix would actually be quite inappropriate, whereas a stage projection matrix might be useful. We will not cover this topic in any depth; the interested reader is referred to Lefkovitch (1965) and Usher (1972).

Consider a hypothetical situation in which a holometabolous insect spends 2 days as an egg, 5 as a larva, 4 as a pupa, and 10 as an adult. We can construct a matrix equation of the form

$$\begin{bmatrix} n_1(t) \\ n_2(t) \\ n_3(t) \\ \cdot \end{bmatrix} = \begin{bmatrix} g_{11} & g_{12} & g_{13} & \cdots \\ g_{21} & g_{22} & g_{23} & \cdots \\ g_{31} & g_{32} & g_{33} & \cdots \\ \cdot & \cdot & \cdot & \cdots \end{bmatrix} \begin{bmatrix} n_1(t-1) \\ n_2(t-1) \\ n_3(t-1) \\ \cdot \end{bmatrix} \qquad (3\text{-}24)$$

where n_i refers to the population of the ith *stage*, and g_{ij} gives the "contribution" of the ith to the jth stage over some defined time interval (say 1 day). Because each stage probably has a characteristic survival per time, and requires a different amount of time to develop before passing on to the next stage, assigning g_{ij} values is not at all straightforward. Note that (3-24) can be written

$$n_1(t) = \sum_j g_{1j} n_j(t-1)$$

$$n_2(t) = \sum_j g_{2j} n_j(t-1) \qquad (3\text{-}25)$$

etc.

Because each $n_i(t)$ is a linear combination of the $n_j(t-1)$ terms, it is possible to estimate g_{ij} values by standard, multiple regression.

A very simple, illustrative example, using least-squares analysis follows. Consider a species of animal that can be classified into two life history stages (say, a larval, dispersal stage and a sessile, adult stage), each of which has fairly consistent survival and fecundity characteristics. Let us suppose that we are extremely ignorant of this creature, lacking knowledge even of the duration of the two stages. However, suppose that we are able accurately to assess the populations of each stage over several months, with the results shown in Table 3-3.

Table 3-3

Month	n_1	n_2
1	100	10
2	200	10
3	200	30
4	200	50
5	200	40
6	300	40
7	300	50
8	200	30
9	100	10
10	100	10

To find g_{11} and g_{12}, let

$\mathbf{N}_i$ = a column vector giving the various, observed values of $n_i(t) - \bar{n}_i$ over months 2 to 10

$$\mathbf{N}_1 = \begin{bmatrix} 0 \\ 0 \\ 0 \\ 0 \\ 100 \\ 100 \\ 0 \\ -100 \\ -100 \end{bmatrix} \qquad \mathbf{N}_2 = \begin{bmatrix} -20 \\ 0 \\ 20 \\ 10 \\ 10 \\ 20 \\ 0 \\ -20 \\ -20 \end{bmatrix} \tag{3-26}$$

$\mathbf{X}$ = the matrix whose columns contain values of $n_1(t-1) - \bar{n}_1$ and $n_2(t-1) - n_2$, whose rows are the months from 1 to 9

$$\mathbf{X} = \begin{bmatrix} -100 & -20 \\ 0 & -20 \\ 0 & 0 \\ 0 & 20 \\ 0 & 10 \\ 100 & 10 \\ 100 & 20 \\ 0 & 0 \\ -100 & -20 \end{bmatrix} \tag{3-27}$$

$$\mathbf{g}_1 = \text{the vector} \begin{bmatrix} g_{11} \\ g_{12} \end{bmatrix} \tag{3-28}$$

Then, according to (3-24),

$$\mathbf{N}_1\,(\text{predicted}) = \mathbf{X}\mathbf{g}_1$$

and

$$\mathbf{N}_1(\text{predicted}) - \mathbf{N}_1 = \mathbf{X}\mathbf{g}_1 - \mathbf{N}_1 = \text{vector of deviations from expected values, or "errors"}$$

The sums of the errors squared is then

$$\text{SS} = (\mathbf{X}\mathbf{g}_1 - \mathbf{N}_1)^T(\mathbf{X}\mathbf{g}_1 - \mathbf{N}_1) = \mathbf{g}_1^T\mathbf{X}^T\mathbf{X}\mathbf{g}_i - 2\mathbf{g}_1^T\mathbf{X}^T\mathbf{N}_1 + \mathbf{N}_1^T\mathbf{N}_1 \tag{3-29}$$

The least-squares technique of statistics is to find the values g_{11} and g_{12} that minimize SS. This is done by differentiating SS with respect to g_{11} and g_{12} and setting the derivatives equal to zero. A little algebraic experimentation will show that this can be written

$$0 = \begin{bmatrix} \dfrac{d\text{SS}}{dg_{11}} \\ \dfrac{d\text{SS}}{dg_{12}} \end{bmatrix} = 2\mathbf{X}^T\mathbf{X}\mathbf{g}_1 - 2\mathbf{X}^T\mathbf{N}_1$$

so that

$$\mathbf{g}_1 = (\mathbf{X}^T\mathbf{X})^{-1}\mathbf{X}^T\mathbf{N}_1 \tag{3-30}$$

Substituting from (3-26) and (3-27) and performing the calculations, we obtain

$$[g_{11} \quad g_{12}] = [.41 \quad .51]$$

Following the same procedure gives us

$$[g_{21} \quad g_{22}] = [.11 \quad .26]$$

Thus we have

$$\mathbf{n(t)} = \mathbf{G}\cdot\mathbf{n(t} - 1) \qquad \text{where} \qquad \mathbf{G} = \begin{bmatrix} .41 & .51 \\ .11 & .26 \end{bmatrix} \tag{3-31}$$

whose general solution is

$$n_i(t) = \sum_j C_{ij}\lambda_j^t$$

where $\lambda_1 = 0.58$ and $\lambda_2 = 0.09$. This population is in a decline phase, and is not cycling (neither eigenvalue is negative).

Unlike the Leslie matrix, other stage projection matrices do *not* necessarily have only one real, positive root. In fact, rules applying to Leslie matrices do not in general apply to stage projection matrices.

Stage projection matrices have been applied in a number of biological studies. The interested reader is referred to Werner and Caswell (1977) for an application of the technique to teasel (*Dipsacus sylvestris*), a weedy plant.

Before leaving stage projection matrices, one word of caution. Suppose that using least-squares or some other regression technique, we were to find negative g_{ij} values. Clearly, these would make no sense biologically. So what might we conclude? Possibly these results might reflect statistical error arising from inadequate or inaccurate data. Possibly the $n_i(t)$ values are really density dependent and therefore nonlinear functions of the $\{n_j(t-1)\}$; possibly the system displays time lags, or depends on environmental parameters which are changing in value. On the other hand, g_{ij} values obtained by the least-squares method are purely abstract quantities providing a best, empirical fit; they are not really "contributions" of stage i to stage j. Thus **G** matrices obtained through regression are merely predictive, not heuristic tools. Further discussion of stage projection matrices may be found in Lefkovitch (1965) and Usher (1972). A method for combining both age and stage in a single projection matrix is provided by Law (1983).

The Hubbell–Werner Model

It is always possible to describe populations by means of stage projection matrices. Because there is an average fecundity and survival probability for individuals at any age, it is also always possible to write a Leslie matrix. On the other hand, construction of the appropriate matrices may be far from straightforward.

The application of network theory to age-structured population problems was first suggested by Lewis (1976), and then again by Hubbell and Werner (1979). Basically, the method serves as an illustrative tool for finding the eigenvalues of the appropriate Leslie or stage projection matrix without the necessity of actually setting up the matrix. Consider, as a first example, the following population.

A hypothetical creature survives to age 1 with probability p_0, at the end of which time a fraction, z, will reproduce, generating b_1' young apiece. These then die. The remaining $1-z$ do not reproduce immediately, but live to age 2 with probability p_1 and then produce b'_2 young apiece. This life history can be diagrammed as shown in Figure 3-2. The arrows pointing back to age 0 indicate the full cycle of young (age 0) to young. Each node indicates the passage of a year. In Figure 3-2 there are two young-to-young pathways, labeled A and B. Consider pathway A, and let the total number of young be n_0, the number that will follow pathway A be $n_0(\text{A})$. Then the number of young produced via pathway A (which runs *one* year) is

$$n_0^{(\text{A})}(t + 1) = n_0(t)\, p_0 z b'_1 \tag{3-32}$$

(Simply multiply the string of all the probabilities followed by the fecundity, which appear along the pathway.) Now suppose that n_0 is growing by a multiple λ each year. Thus

$$n_0^{(\text{A})}(t + 1) = \lambda n_0^{(\text{A})}(t)$$

which we have just decided was equal to $n_0(t)p_0 z b'_1$. So

$$\lambda n_0^{(\text{A})}(t) = n_0(t)p_0 z b'_1$$

or

$$n_0^{(\text{A})}(t) = \lambda^{-1} n_0(t)p_0 z b'_1 \tag{3-33}$$

The number of young produced via pathway B, by the same argument, only with a *two*-year time span, is (again, multiplying the string of probabilities and the fecundity value for this pathway)

$$n_0^{(\text{B})}(t + 2) = n_0(t)p_0(1 - z)p_1 b'_2$$

But over the 2 years,

$$n_0^{(\text{B})}(t + 2) = \lambda^2 n_0^{(\text{B})}(t)$$

Thus

$$n_0^{(\text{B})}(t) = \lambda^{-2} n_0(t)p_0(1 - z)p_1 b'_2 \tag{3-34}$$

Because both pathways lead to and depart from a common pool $n_0^{(\text{A})}$ and $n_0^{(\text{B})}$ must sum to n_0. Thus adding (3-33) and (3-34) together gives us

$$n_0(t) = n_0^{(\text{A})}(t) + n_0^{(\text{B})}(t) = n_0(t)\lambda^{-1} p_0 z b'_1 + n_0(t)\lambda^{-2} p_0(1 - z)p_1 b'_2$$

or

$$1 = \lambda^{-1} p_0 z b'_1 + \lambda^{-2} p_0(1 - z)p_1 b'_2 \tag{3-35}$$

Do not be surprised if (3-35) looks like the characteristic equation of a Leslie matrix—that is exactly what it is, for what we have done is to describe a population's bahavior on an age-specific basis. The only difference between this approach and the Leslie matrix method is that a *network diagram* (Fig. 3-2) has been substituted for a matrix as the visual aid underlying the

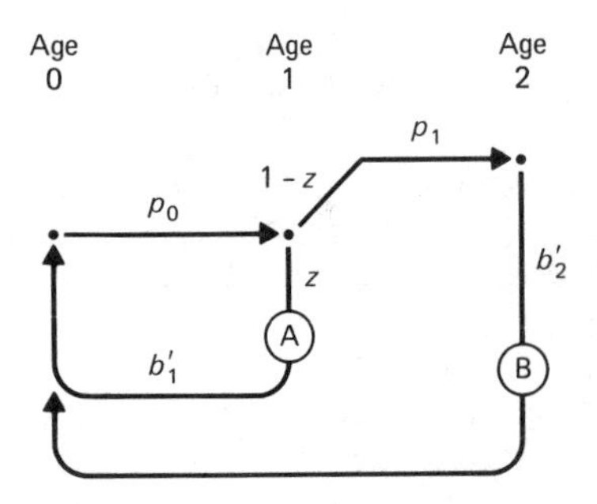

Figure 3-2 Hypothetical life history network.

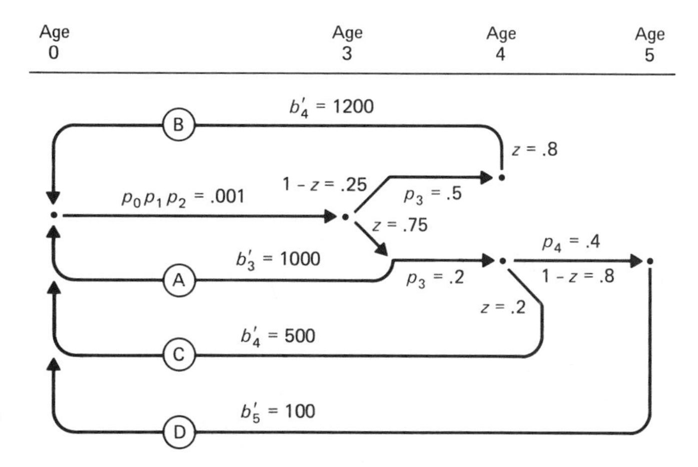

Figure 3-3 Hypothetical life history network.

calculation of λ values. The general solution to the problem above is

$$n_i(t) = C_{ij}\lambda_1 t + C_{i2}\lambda_2 t$$

where λ_1 and λ_2 are the roots of (3-35).

Consider the network diagram of Figure 3-3. Again, each node denotes the passage of a year. Following the notation of the example above, and multiplying all the probability values and the fecundity along the various pathways, we obtain

$$\begin{aligned} n_0^{(A)}(t) \quad & (.001)(.75)(1000)\lambda^{-3}n_0(t) \\ + n_0^{(B)}(t) \quad & (.001)(.25)(.5)(.8)(1200)\lambda^{-4}n_0(t) \\ + n_0^{(C)}(t) = \; & (.001)(.75)(.2)(.2)(500)\lambda^{-4}n_0(t) \\ + n_0^{(D)}(t) \quad & (.001)(.75)(.2)(.8)(.4)(100)\lambda^{-5}n_0(t) \\ n_0(t) = \; & (.75\lambda^{-3} + .135\lambda^{-4} + .0048\lambda^{-5})n_0(t) \end{aligned}$$

so that

$$1 = .75\lambda^{-3} + .135\lambda^{-4} + .0048\lambda^{-5}$$

Solving, we obtain $\lambda_1 = .96$, so that the population in question is declining very slowly. Because all other roots must be equal to or less than λ_1 in absolute value, any oscillations the population may experience will damp out.

Sometimes it may be useful to view the network not as a graph of connections among ages, but rather between critical points in the life history. This can be made clear by beginning with a trivial example and then elaborating. Consider a population of organisms that live to 1 year with probability $p_0 = .3$, at which point 60% become mature, but reproductively suppressed, and the other 40% breed, producing three offspring each. Both groups then survive to age 2 with probability .5 and have two young. We can graph this particular life history (Fig. 3-4). But this life history can also be graphed as in Figure 3-5.

By following the procedure of multiplying the strings of probability values and the fecundity for the different pathways, it is easy to see that the two graphs give equivalent characteristic equations. What is the advantage of the second approach? By using states rather than ages it becomes possible to treat individuals who regress to earlier states. Suppose that the example above, referred to some hypothetical creature who, if unsuccessful in finding a mate during the sexually active state in cycle C, reverted the next year, with survival probability p_2, back to a sexually repressed state. The network would then take the form shown in Figure 3-6.

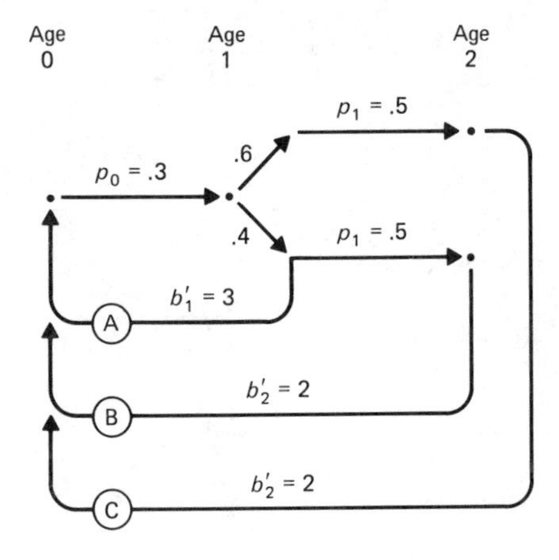

Figure 3-4 Hypothetical life history network.

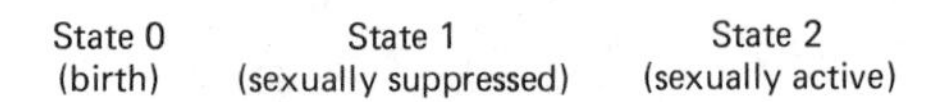

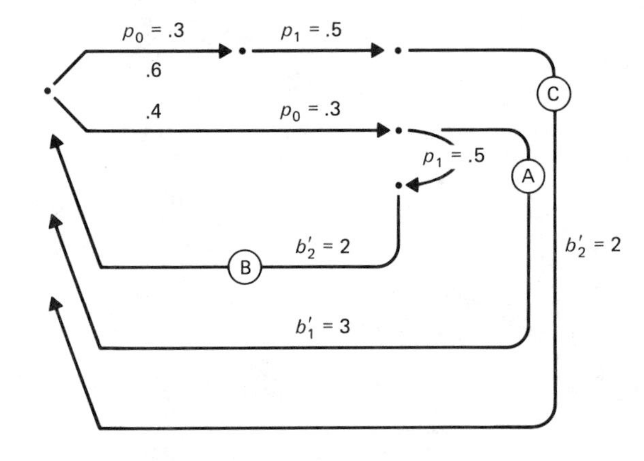

Figure 3-5 Hypothetical life history network.

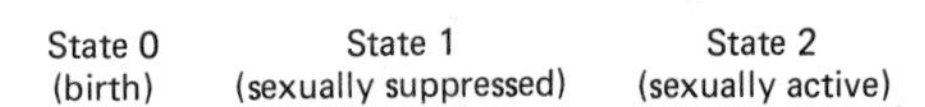

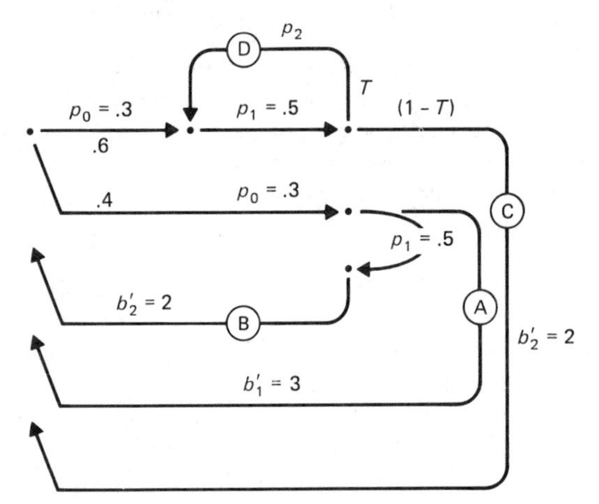

Figure 3-6 Hypothetical life history network. T is the probability of reproductive failure by individuals following pathway C or D.

The contributions of pathways A and B are

$$n_0^{(A)}(t+1) = n_0(t)(.3)(.4)(3) = .36n_0(t)$$
$$\text{so } n_0^{(A)}(t) = .36\lambda^{-1}n_0(t)$$
$$n_0^{(B)}(t+2) = n_0(t)(.3)(.4)(.5)(2) = .12n_0(t)$$
$$\text{so } n_0^{(B)}(t) = .12\lambda^{-2}n_0(t)$$

But pathway C is really many pathways, for an organism can bypass loop D, making its reproductive contribution in 2 years, or pass through loop D once before continuing along C, making its contribution after 4 years, or pass through loop D twice before continuing on to contribute after 6 years, and so on. Write

$$n_0^{(C)}(t+2) = n_0(t)(.3)(.6)(.5)(1-T)(2) = .18(1-T)n_0(t)$$
$$\text{so } n_0^{(C)}(t) = .18(1-T)\lambda^{-2}n_0(t)$$
$$n_0^{(C \text{ and } D)}(t+4) = n_0(t)(.3)(.6)(.5)(1-T)[T(p_2)(.5)](2) = .18(1-T)[.5p_2T]n_0(t)$$
$$\text{so } n_0^{(C+D)}(t) = .18(1-T)[.5p_2T]\lambda^{-4}n_0(t)$$
$$n_0^{(C \text{ and } D \text{ twice})}(t+6) = .18(1-T)[.5p_2T]^2n_o(t)$$
$$\text{so } n_0(C+D+D)(t) = .18(1-T)[.5p_2T]^{2\lambda-6}n_0(t)$$

etc.

Adding the various expressions, we obtain

$$n_0(t) = .36\lambda^{-1}n_0(t) + .12\lambda^{-2}n_0(t) + .18(1-T)\lambda^{-2}n_0(t)[1 + (.5p_2T\lambda^{-2}) + (.5p_2T\lambda^{-2})^2 + \cdots]$$

So, dividing by $n_0(t)$ and recalling that the series $1 + x + x^2 + x^3 + \cdots$, where $|x| < 1$, can be written $1/(1-x)$, the expression above finally becomes

$$1 = .36\lambda^{-1} + .12\lambda^{-2} + \frac{.18(1-T)\lambda^{-2}}{1 - .5p_2T\lambda^{-2}} \qquad (3\text{-}36)$$

Whenever a pathway contains a subloop, its contribution to the characteristic equation is enhanced, as shown in (3-36).

One more simple example should drive home this last point—which will later become very useful. Examine the network of Figure 3-7. This time there is only one full pathway (A) with one subloop (B). We can write

$$n_1^{(A)}(t+2) = n_1(t)p_0Tp_1b'$$
$$\text{so } n_1^{(A)}(t) = (p_0Tp_1b')\lambda^{-2}n_1(t)$$
$$n_1^{(A+B)}(t+3) = n_1(t)p_0[(1-T)p_1]Tp_1b'$$
$$\text{so } n_1^{(A+B)}(t) = (p_0Tp_1)\lambda^{-2}[(1-T)p_1\lambda^{-1}]n_1(t)$$
$$n_1^{(A+B+B)}(t+4) = n_1(t)p_0[(1-T)p_1']^2Tp_1b'$$
$$\text{so } n_1^{(A+B+B)}(t) = (p_0Tp_1)\lambda^{-2}[(1-T)p_1\lambda^{-1}]^2n_1(t)$$

etc.

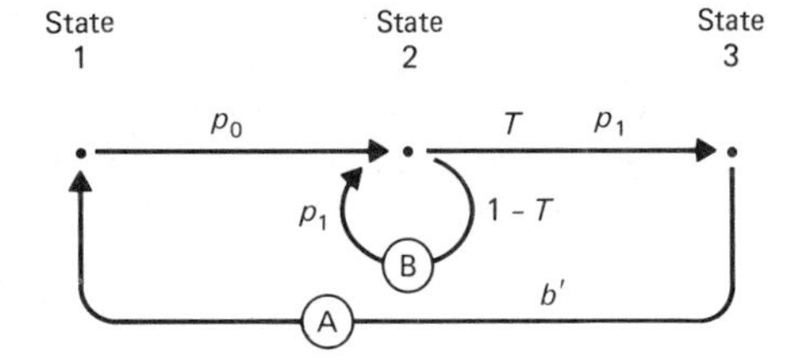

Figure 3-7 Hypothetical life history network.

Adding the expressions, and dividing both sides by $n_1(t)$ gives us

$$1 = \frac{p_0 T p_1 b' \lambda^{-1}}{1 - (1 - T)p'_1 \lambda^{-1}} \tag{3-37}$$

A Multiple-Age Ricker Model

Although the Ricker equation (Chapter 2) is sometimes used with age-structured fish populations, it applies, strictly speaking, only to cases with no generation overlap (some Pacific salmon, for example). The basic assumptions underlying its derivation, though, can be used to construct a legitimate analogous model of age-structured populations. The following example is taken loosely from J. Lawler (Lawler, Matusky and Skelley, Engineers, personal communication).

Let

$R_x(t)$ = number of recruits, age x, at time t

$\text{PEP}(t)$ = total egg production by parental stock of all ages, at time t(parental egg production)

$P(t)$ = total parental stock at time t

$\text{REP}(t)$ = total egg production of recruits of the above spawning at a specific time.

Suppose that the mortality of young arising from density feedback by parent stock is γP. If, as in Chapter 2, T_C is the critical age at which density feedback ceases, then the proportion of young, spawned at time $t - x$, surviving density feedback is

$$e^{-\gamma P(t-x) T_C}$$

Finally, let the proportion surviving all other mortality sources, to age x, be α''_x. Then the number of x-year-olds at time t is

$$R_x(t) = \text{PEP}(t - x)\alpha''_x e^{-\gamma P(t-x)T_C} \tag{3-38}$$

But $\text{PEP}(t - x)$ must be proportional to $P(t - x)$, so let β be a number such that

$$\text{PEP}(t - x) = \frac{\gamma T_C P(t - x)}{\beta} \tag{3-39}$$

Also, the total egg production at time t is

$$\text{REP}(t) = \sum_x \kappa_x R_x(t) \tag{3-40}$$

where κ_x is fecundity of x-year-olds. Define $\alpha'_x = \alpha''_x \kappa_x$, and substitute (3-39) and (3-40) into (3-38) to obtain

$$\text{REP}(t) = \sum_x \alpha'_x \text{PEP}(t - x) e^{-\beta' \text{PEP}(t-x)} \tag{3-41}$$

It is not likely that we will know the total egg production at various times in the past—if our knowledge of a population were this good, we would probably have little need for (3-41). Thus the formulation above, as written, is not terribly useful. We therefore begin with what may, at first, appear to be an outlandish assumption—and then try to justify it. Suppose that the total reproductive population varies only slightly from year to year, so that $\text{PEP}(t - x)$ is approximately equal for all x. We choose some measure of generation time T and approximate (3-41) by

$$\text{REP}(t) = \alpha'(T)\text{PEP}(t - T)e^{-\beta' \text{PEP}(t-T)} \tag{3-42}$$

Expression (3-42) is simply the original Ricker equation, but stated in terms of total egg production rather than number of individuals.

Can we justify the assumption that we have just made? As we might expect, there is good evidence that populations of fish in any given age group fluctuate considerably from year to year. Every now and then there is a year in which either spawning is particularly high or survival of young is high, resulting in a *strong year class*. An example is shown in Figure 3-8; one strong year class is marked with arrows. The appearance of strong or weak year classes is common in fish, and is a general phenomenon among the percids, where a dominant influence on year

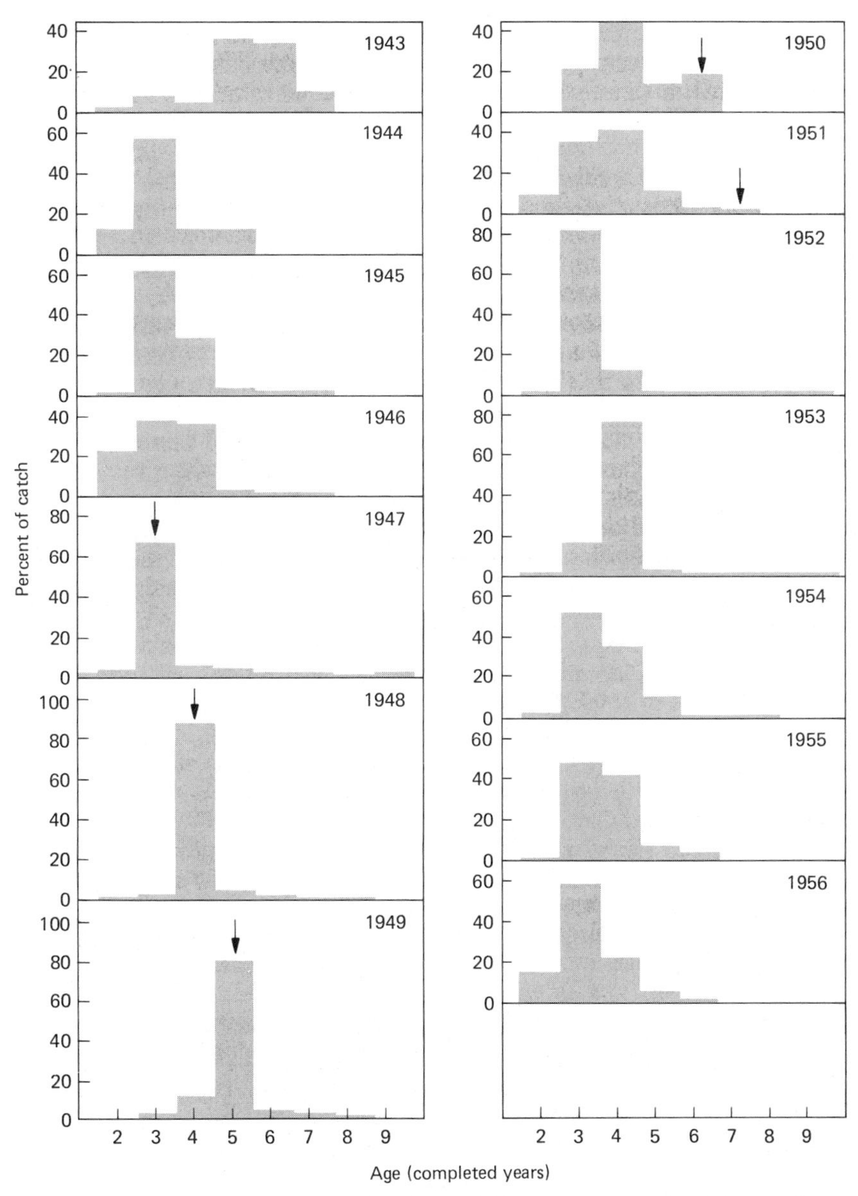

Figure 3-8 Age composition of samples of Lake Erie commercial whitefish catch for 1943–1956. (From Lawler, 1965, Fig. 2.)

Table 3-4

Year Class	Estimated Recruitment at Age 3 (Relative Numbers in Thousands per Unit of Fishing Effort)	Total Catch (tons/yr)
1951	5.9	
1952	21.9	
1953	8.3	
1954	8.2	
1955	2.5	5.23
1956	20.6	5.20
1957	5.6	3.87
1958	3.9	3.43
1959	5.5	3.77
1960	24.3	2.83
1961	24.8	3.98
1962	10.3	2.46
1963	8.3	4.44
1964	1.5	2.21
1965		2.88
1966		3.03
1967		2.57
1968		2.34
	var/mean^2 .575	.079

class strength seems to be water temperature (Koonce et al., 1977). In the European walleye *Stizostedion lucioperca*, a relative of the American walleye, year class strength varies up to 20-fold, and 100-fold differences have been reported for *Perca fluviatalis* (Willemsen, 1977). But (3-42) depends not on a constancy of year classes, but rather a constancy—or at least proportionately small variations—in the *sum* of year classes over all reproductive ages. Because means show less fluctuation than the individual values that collectively define them, and because density feedback is expected to ease population values toward some central tendency, this assumption may be less unlikely that at first appears. The data in Table 3-4 for North Sea herring, taken from Burd and Parnell (1970), show that whereas recruitment varies considerably, the total catch (an approximation to P) is quite steady.

Density Feedback

The Leslie matrix, stage projection matrix, and Hubbell–Werner models discussed above do not include terms for density feedback. Attempts to incorporate density feedback do a certain violence to the simple elegance of the Leslie matrix approach. But we can at least take comfort in the fact that density feedback will affect the f_i and P_i elements in such manner as to push λ_1 toward 1. Thus species which are somewhat buffered from density-independent fluctuations, whose populations hover near $\hat{n}$, can be counted on to have fairly fixed Leslie matrices from year to year. Species that are in rapid growth phases, without density influences, may to some extent be thought of similarly. Leslie (1959) was the first to investigate the influence of feedback in a matrix model. Supposing that the number of individuals of age x is depressed by feedback from the entire current population and from the entire population in the year of these individuals' births, he wrote

$$\hat{n}_x(t) = \frac{n_x(t) \text{ calculated from } \mathbf{Ln}(t-1)}{1 + \alpha n_{\text{total}}(t-x-1) + \beta n_{\text{total}}(t)} \tag{3-43}$$

At equilibrium this becomes

$$\hat{n} = \frac{\lambda_1 \hat{n}}{1 + \alpha\hat{n} + \beta\hat{n}}$$

so that

$$\hat{n} = \frac{\lambda_1 - 1}{\alpha + \beta} \tag{3-44}$$

Pennycuick et al. (1968) performed computer simulations using the Leslie matrix, and defining

$$b'_x = \frac{B_x a}{c + n_{\text{tot}}}$$

$$p_x = \frac{P_x}{1 + \exp(n_{\text{tot}}/d - f)}$$

where B_x is fecundity, P_x survival in the absence of density feedback, and a, c, d, and f are constants. Depending on the values of B_x, P_x, a, c, d, and f chosen, they obtained sigmoid growth with no oscillations, damped oscillations, or stable oscillations. In itself, this information is not terribly useful, although if we knew the true form of feedback on b'_x and p_x, this simulation approach might be of considerable value.

Smouse and Weiss (1975) have also done simulations using Leslie matrices whose elements are subject to density feedback. They find, when λ_1 (in the absence of feedback) is large ($\lambda_1 = 1.946$), that the equilibrium age structure depends on the nature of the feedback. If its impact is primarily to suppress reproduction, there is a larger equilibrium proportion of older individuals, whereas if density effects act mostly to lower survival, the equilibrium population is, on average, younger. This makes good, intuitive sense; if survival is relatively lower, fewer individuals live to old age. If fecundity drops, input to the youngest level is relatively less. Smouse and Weiss report that these results nearly disappeared when $\lambda_1 = 1.002$. Because the density-induced drops in b or p values necessary to bring population growth to a halt are minute in this case, their results, again, are intuitively sensible.

Continuous Growth Models

All four of the models described above are discrete. Although we have no intention of pursuing continuous age-structured population models in this text, we refer the reader, in the interest of completeness, to the "Von Foerster" equations (see Streifer, 1974).

Autoregression Models

With the exception of the stage projection matrix, all of the population models discussed in this chapter and in Chapter 2 are theoretical models; each is built from basic considerations of birthrate, death rate, and reasonable (or easy-to-handle) assumptions about density feedback, rather than constructed directly from observed, population dynamic data. But all of these models are admittedly simplistic, and because of stochastic fluctuations in the environment are difficult to assess for accuracy. We thus return to the statistical regression approach of the stage projection matrix and broaden it; can empirical models improve our accuracy in population prediction?

To find the elements of the stage projection matrix we regressed each $n_x(t)$ on $n_y(t-1)$ for all ages, y. To handle density feedback we must include higher-order [e.g., $n_y(t-1)^2$] terms in the regression. But note that such formulation cannot take into account environmental fluctuations. To incorporate predictable environmental changes (unpredictable changes must be treated with stochastic models) into a population model we must write $n(t)$ not only as a function of $n(t-1)$ but also of events in the past. Of course, one indicator of conditions in the past (T time units, say) is the population itself at that time. Thus it may be possible to improve on the stage projection model by including in the regression $n(t-T)$ terms for various T. In fact, the inclusion of such terms makes sense for quite another reason as well. The number of individuals present at time t consists of 1-year-olds born at time $t-1$, 2-year-olds born at time $t-2$, and so on, and thus is related to the breeding population at times $t-1, t-2$, and so on. Finally even linear functions may, implicitly, describe density feedback. Thus the expression

$$x(t) = \beta_1 x(t-1) + \beta_2 x(t-2) + \beta_3 x(t-3) + \cdots + \beta_k x(t-k) \qquad (3\text{-}45)$$

where x is some function of n [say $\log(n)$], may at once include age-structured effects, the impact of predictable environmental changes, and density feedback. The statistical theory and efficacy of using (3-45) to describe real populations has been championed and explored extensively by Poole (1972, 1976, 1978).

3.2 Changing Environments

In Chapter 2 we discussed three kinds of environmental change. First, individuals were noted to be discrete entities whose survival probabilities and fecundities might display different values by chance. Thus from generation to generation, birth and death might vary, resulting in a random drift in population size. A population size of zero (or 1 in a sexual, diploid species), which could be reached by chance, via this drift, is absorbing—a population reaching this size cannot recover. In populations with overlapping generations the elements of the matrix $(\mathbf{Q} - \mathbf{I})/dt$ will vary depending on age structure, but the qualitative conclusions remain the same. The second kind of change described in Chapter 2 was fluctuation in the external environment, and the third was the capricious behavior of a population with a high intrinsic rate of increase breeding at discrete intervals.

Fluctuations in the External Environment

The effect of environmental fluctuations, both density dependent and density independent, is to alter the values of $\{p_x, b'_x\}$, the elements of the Leslie matrix. Thus, in changing environments we can no longer treat $\mathbf{L}$ as a constant matrix, even if density feedback does not occur. Boyce (1977) was the first to simulate population behavior using a Leslie matrix with randomly varying elements. Recall that the underlying growth of a population, emerging through the various cycles described by the secondary eigenvalues, is given by λ_1, the Perron root. Boyce found that a random fluctuation in any Leslie matrix element around some mean value depressed this underlying growth multiple. The larger the variance, the lower the population growth (Fig. 3-9). Now if the population approached stable age distribution, then $\ln(R) = \ln(\lambda_1)$, and we could write

$$\lim_{t\to\infty} n(t) = \lim_{t\to\infty} \lambda_1(t-1)n(t-1) = \lim_{t\to\infty} \lambda_1(t-1)\lambda(t-2)n(t-2) = \cdots$$

$$= \lim_{t\to\infty} \prod_{T=0}^{t-1} \lambda_1(T)n(0) = (\lambda_1^*)n(0)$$

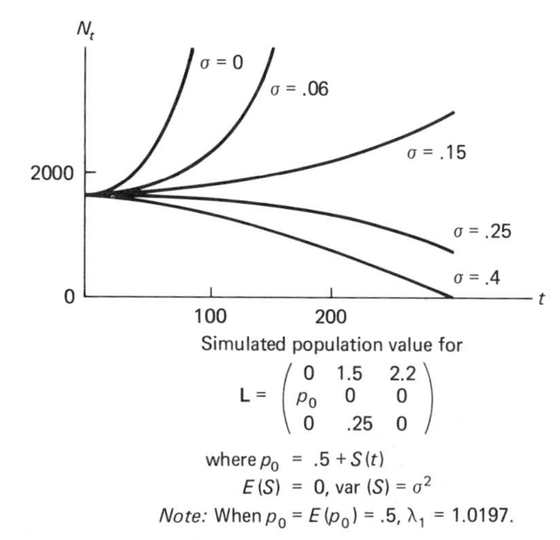

Figure 3-9 Population trajectory over time as a function of standard deviation in p_0. (From Boyce, 1977, Fig. 1.)

where λ_1^* is the geometric mean of the λ_1's over time. If $\lambda_1^* > 1.0$ or, equivalently, if $E(\ln \lambda_1) > 0$, the population would show a net growth. To find the effect of variation in Leslie matrix elements on population growth, we could then write (using the Taylor expansion)

$$E(\ln \lambda_1) \simeq E\left[(\ln \lambda_1) + \Delta x\left(\frac{d \ln \lambda_1}{dx}\right) + \frac{1}{2}\Delta x^2\left(\frac{d^2 \ln \lambda_1}{dx^2}\right)\right]$$

$$= \ln \lambda_1 + \frac{1}{2}\,\mathrm{var}(x)\frac{d^2 \ln \lambda_1}{dx^2}$$

where all expressions are evaluated at the mean value of x, the Leslie matrix element of interest. If the second derivative were negative ($\ln \lambda_1$ is a "concave" function of x), then $E(\ln \lambda_1) < \ln \lambda_1$ evaluated at mean x, and we would know that fluctuations in x decreased the population growth rate (see Daley, 1979). Unfortunately, the population does *not* reach a stable age distribution when x fluctuates, so the approach above may give misleading predictions. For example, Boyce (1979) shows that even when $\ln \lambda_1$ is a convex function of x, fluctuations will decrease population growth rate. This trend appears quite general. It is also interesting, and not unexpected, that simulations with Leslie matrices exhibiting density feedback indicate depressed values of $E(n)$ as a result of variation in the matrix elements (Boyce, 1979). Thus fluctuations have the same effects in both age-structured and non-age-structured populations.

Cohen (1977a,b), in apparent contradiction to Boyce's conclusions, proved in quite elegant fashion that the geometric mean of R, given sufficient time, should equal the Perron root, $\lambda_1(\bar{\mathbf{L}})$, of the Leslie matrix of mean fecundities and survivals, $\bar{\mathbf{L}}$. That is, variation in Leslie matrix elements should *not* depress the long-term, effective growth multiple. The confusion has been nicely resolved by Tuljapurkar and Orzack (1980), who proved that (in the case analyzed by Cohen, where fluctuations in the matrix elements is Markovian—there is a fixed probability an element will take on a particular value given its immediately past value)

$$\lim_{t \to \infty} \frac{n(t)}{n(0)}$$

is lognormally distributed with mean $\lambda_1(\bar{\mathbf{L}})t$ and variance $\sigma^2 t$. Thus the geometric mean growth multiple is indeed $\lambda_1(\bar{\mathbf{L}})$, but the *modal* value is $\lambda_1(\bar{\mathbf{L}}) - \sigma^2$. That is, the large values of R that maintain a geometric mean of $\lambda_1(\bar{\mathbf{L}}) \geqslant 1$ are rare events; the population may show an erratic decline for many time units (as found by Boyce) before one of these rare events occurs. In fact, the decline could well be of sufficient duration to cause extinction, even when $\lambda_1(\bar{\mathbf{L}})$ is considerably larger than 1.0. Interestingly, $\lambda_1(\bar{\mathbf{L}})$ is completely insensitive, whereas σ^2 is extremely sensitive to the nature of environmental autocorrelation.

The degree to which population densities are buffered against environmental fluctuations varies considerably between age-structured and simpler populations. It is not difficult to see that a catastrophe which destroys all young of the year would wipe out a species that breeds only once and then dies. But an age-structured population will respond to such an event with the loss of one age class, suffering only a momentary blow. Tracking of the environment by the population is complicated by the fact that any change in p_x, b'_x values alters the age structure and restarts oscillations. Cycles generated internally by a population's age structure may be expected to interact with cycles generated by environmental fluctuations, resulting in highly complex patterns.

Stability and Chaos

Without specific expressions for density feedback effects by and on the various age classes, it is impossible to assess the neighborhood stability of an age-structured population. In any case, examining the population for conditions leading to chaotic behavior would be very difficult. Suffice it to say that for populations with sufficiently high intrinsic growth rates (large enough λ_1 in the absence of density feedback), we can imagine unstable overshooting of equilibrium values at some or all age classes in response to perturbation.

An interesting variation on the question of population stability is the question of age-structure stability. The standard Leslie matrix, without density feedback, generally predicts (in a constant environment) convergence on a predictable, stable age distribution. But suppose that density feedback exists, that its intensity is a function of age structure, and that the age classes are affected differentially. For example, consider a population with two age classes in which only the older group reproduces, only the younger group is affected, via survival, by feedback, and where the feedback is generated only by older individuals. We can envision the following scenario. Begin with many older individuals (n_2 large) and a small number of juveniles. The many adults produce young and inhibit the juveniles from living to maturity. Hence, in the next time unit, n_1 will be large and n_2 small. This, in turn, leads to low recruitment of young and high survival of juveniles, so that again, n_2 becomes large, n_1 small. The density cycles in both age groups—and, probably, the total population—are sustained, generated by competition among the age classes. Population cycles arising from interage class competition have been mentioned by Slobodkin (1961) and J. M. Emlen (1973), but their potential significance and possibly common occurrence have been largely ignored.

Without specifically referring to interage class competition, Beddington (1974) (see also Cooke and Leon, 1976) has provided a framework for determining whether density feedback will result in population cycling. His interest in age-structure equilibrium led to the following neighborhood stability analysis (see Section 2.3). Write

$$\mathbf{n}(\mathbf{t}+1) = \mathbf{M}\cdot\mathbf{n}(\mathbf{t}) \tag{3-46}$$

where $\mathbf{M}$ represents the Leslie matrix analog with density-feedback terms. $\mathbf{M}$ itself is a function of $\mathbf{n}$, and at equilibrium (stable or unstable),

$$\hat{\mathbf{n}} = \mathbf{M}(\hat{\mathbf{n}})\cdot\hat{\mathbf{n}} \tag{3-47}$$

so that

$$0 = (\mathbf{M}(\hat{\mathbf{n}}) - I)\cdot\hat{\mathbf{n}}$$

and, therefore,

$$|\bar{M}(\hat{\mathbf{n}}) - I| = 0 \tag{3-48}$$

We now perturb the vector $\mathbf{n}$ by a minute amount, $\mathbf{x}$. Then, from (3-46),

$$\mathbf{n}(\mathbf{t}+1) = \hat{\mathbf{n}} + \mathbf{x}(\mathbf{t}+1) = \mathbf{M}(\hat{\mathbf{n}} + \mathbf{x}(\mathrm{t}))\cdot(\hat{\mathbf{n}} + \mathbf{x}(\mathrm{t})) \tag{3-47}$$

Applying the Taylor expansion (Section 1.7) to $\mathbf{M}(\mathbf{n} + \mathbf{x})$, and deleting higher-order terms, we obtain

$$\mathbf{M}(\hat{\mathbf{n}} + \mathbf{x}) \simeq \mathbf{M}(\hat{\mathbf{n}}) + \sum_i x_i \left(\frac{d\mathbf{M}}{dn_i}\right)_{\hat{n}_i} \qquad (3\text{-}48)$$

Therefore,

$$\hat{\mathbf{n}} + \mathbf{x}(t + 1) \simeq \mathbf{M}(\hat{\mathbf{n}}) + \left[\sum_i x_i(t)\left(\frac{d\mathbf{M}}{dn_i}\right)_{\hat{n}_i}\right](\hat{\mathbf{n}} + \mathbf{x}(t))$$

or, dropping x^2 and higher-order terms because they are vanishingly small, we obtain

$$\mathbf{x}(t + 1) \simeq \mathbf{M}(\hat{\mathbf{n}})\cdot\mathbf{x}(t) + \sum_i x_i(t)\left(\frac{d\mathbf{M}}{dn_i}\right)_{\hat{n}_i}\cdot\hat{\mathbf{n}} = \left[\mathbf{M}(\hat{\mathbf{n}}) + \sum_i \left(\frac{d\mathbf{M}}{dn_i}\right)_{\hat{n}_i}\cdot\mathbf{H}_i\right]\cdot\mathbf{x}(t) \qquad (3\text{-}49)$$

where $\mathbf{H}_i$ is the square matrix with n in the ith column and zeros elsewhere. If we call the bracketed matrix $\mathbf{M}_x$, then (3-49) is

$$\mathbf{x}(t + 1) = \mathbf{M}_x\cdot\mathbf{x}(t) \qquad (3\text{-}50)$$

and neighborhood stability can be assessed by looking at the eigenvalues of $\mathbf{M}_x$. If they are all less than 1.0 in absolute value, the system is stable. If any $\lambda < 0.0$, the system cycles.

We now apply Beddington's approach to the foregoing verbal example of two competing age classes. Quantifying that situation, suppose that $f_1 = 0, f_2 = 2, p_1 = .8(1 - n/190)$. Then

$$\mathbf{n}(t + 1) = \mathbf{M}\cdot\mathbf{n}(t) = \begin{bmatrix} 0 & 2 \\ .8\left(1 - \dfrac{n_2}{100}\right) & 0 \end{bmatrix} \mathbf{n(t)} \qquad (3\text{-}51)$$

The eigenvalues are given by

$$0 = |\mathbf{M} - \lambda I| = \lambda^2 - 1.6\left(1 - \frac{n_2}{100}\right) \qquad (3\text{-}52)$$

But equilibrium occurs when $\lambda_1 = 1.0$. Thus, from (3-52),

$$0 = 1 - 1.6\left(1 - \frac{\hat{n}_2}{100}\right) = -.6 + .016\hat{n}_2$$

so that

$$\hat{n}_2 = 37.5 \qquad (3\text{-}53)$$

and

$$\mathbf{M}(\hat{\mathbf{n}}) = \begin{bmatrix} 0 & 2 \\ .8\left(1 - \dfrac{n_2}{100}\right) & 0 \end{bmatrix} = \begin{bmatrix} 0 & 2 \\ .5 & 0 \end{bmatrix} \qquad (3\text{-}54)$$

Then

$$\begin{bmatrix} \hat{n}_1 \\ \hat{n}_2 \end{bmatrix} = \hat{n} = \mathbf{M}(\hat{\mathbf{n}})\cdot\hat{\mathbf{n}} = \begin{bmatrix} 0 & 2 \\ .5 & 0 \end{bmatrix}\begin{bmatrix} \hat{n}_1 \\ \hat{n}_2 \end{bmatrix} = \begin{bmatrix} 2\hat{n}_2 \\ .5\hat{n}_1 \end{bmatrix}$$

which gives us

$$\hat{n}_1 = 2\hat{n}_2 = 75.0 \qquad \text{(from equation 3-53)} \qquad (3\text{-}55)$$

We still need to find $\mathbf{M}_x$. To do so, recall that $\mathbf{H}_i$ is the matrix with $\hat{\mathbf{n}}$ in the ith column and zeros elsewhere:

$$\mathbf{H}_1 = \begin{bmatrix} \hat{n}_1 & 0 \\ \hat{n}_2 & 0 \end{bmatrix} = \begin{bmatrix} 75.0 & 0 \\ 37.5 & 0 \end{bmatrix}$$
$$\mathbf{H}_2 = \begin{bmatrix} 0 & \hat{n}_1 \\ 0 & \hat{n}_2 \end{bmatrix} = \begin{bmatrix} 0 & 75.0 \\ 0 & 37.5 \end{bmatrix} \tag{3-56}$$

Next, note that

$$\left(\frac{d\mathbf{M}}{dn_1}\right)_{\hat{\mathbf{n}}} = \frac{d}{dn_1}\begin{bmatrix} 0 & 2 \\ .8\left(1 - \frac{n_2}{100}\right) & 0 \end{bmatrix} = \mathbf{O} \tag{3-57}$$

$$\left(\frac{d\mathbf{M}}{dn_2}\right)_{\hat{\mathbf{n}}} = \frac{d}{dn_2}\begin{bmatrix} 0 & 2 \\ .8\left(1 - \frac{n_2}{100}\right) & 0 \end{bmatrix} = \begin{bmatrix} 0 & 0 \\ -.008 & 0 \end{bmatrix}$$

Finally,

$$\mathbf{M}_x = \mathbf{M}(\hat{\mathbf{n}}) + \sum_i \left(\frac{d\mathbf{M}}{dn_i}\right)_{\hat{\mathbf{n}}} \cdot \mathbf{H}_i = \begin{bmatrix} 0 & 2 \\ .5 & 0 \end{bmatrix} + \left(\frac{d\mathbf{M}}{dn_1}\right)_{\hat{\mathbf{n}}} \cdot \mathbf{H}_1 + \left(\frac{d\mathbf{M}}{dn_2}\right)_{\hat{\mathbf{n}}} \cdot \mathbf{H}_2$$
$$= \begin{bmatrix} 0 & 2 \\ .5 & -.6 \end{bmatrix} \tag{3-58}$$

The eigenvalues are

$$\lambda = -1.344, .744 \tag{3-59}$$

Because one of these eigenvalues is negative, we know that the system cycles: in fact, that eigenvalue exceeds 1.0 in absolute value, indicating that the cycles grow in amplitude; the system is unstable.

3.3 Spatially Structured Populations

Quoting Reddingius and den Boer (1970), spatial structure "spreads the risk of extinction over several subpopulations and stabilizes the population as whole." As we showed in Chapter 2, for simple populations without overlapping generations, age-structured populations can go locally extinct without dying out; division of populations into several, semi-independent parts with cross-migration renders the zero population state, locally nonabsorbing. There is little to be added to the discussion in Chapter 2 except to point out that both the Leslie matrix and the network model can be extended to apply to open populations. For example, consider the case of a population with two age classes and a constant (i.e., not density dependent) migration rate of m, for both age groups, existing in two, identical environments. Define the vector

$$\mathbf{n} = \begin{bmatrix} n_1(1) \\ n_1(2) \\ n_2(1) \\ n_2(2) \end{bmatrix} \tag{3-60}$$

where $n_x(y)$ is the population density of age class x in environment y. Because individuals born in one environment will end up in the other in the next time unit with probability m, remain at home with probability $1 - m$, the appropriate Leslie matrix is

$$\mathbf{L}^* = \begin{bmatrix} f_1(1-m) & f_1 m & f_2(1-m) & f_2 \\ f_1 m & f_1(1-m) & f_2 m & f_2(1-m) \\ p_1 & 0 & 0 & 0 \\ 0 & p_1 & 0 & 0 \end{bmatrix} \tag{3-61}$$

The corresponding Leslie matrix for either environment, alone, is simply

$$\mathbf{L} = \begin{bmatrix} f_1 & f_2 \\ p_1 & 0 \end{bmatrix} \tag{3-62}$$

Suppose that $f_1 = 0$, $f_2 = 2$, $p_1 = .5$, and $m = .2$. Then the eigenvalues of matrix $\mathbf{L}$ are $\lambda = \pm 1.0$; and the eigenvalues of the open system, $\mathbf{L}^*$, are ± 1.0, .775, $-.775$. Spatial structure may buffer the population from chance extinction (an issue not addressed by the Leslie matrix), but it apparently does not necessarily act to dampen population cycles (both systems have an additional eigenvalue with absolute value = 1.0). That this should be so is not surprising, though. Damping of cycles certainly will occur if migration is density dependent (this can be verified by application of Beddington's method; see Section 2.2).

Consider the following network model example. A plant population is divided into two parts, each occupying a different area. In habitat I, seeds germinate and live to age 0 with probability $p_0(\text{I})$, maturing and producing $b'_1(\text{I})$ seeds, a fraction m of which are dispersed to habitat II. In habitat II the corresponding values are $p_0(\text{II})$, $b'_1(\text{II})$, and m. The network diagram is shown in Figure 3-10. We follow the pathways leaving and returning to age 0 in either habitat (say habitat I), pathways are:

A: over one time unit
BC: over two time units
BDC: over three time units
BDDC: over four time units
etc.

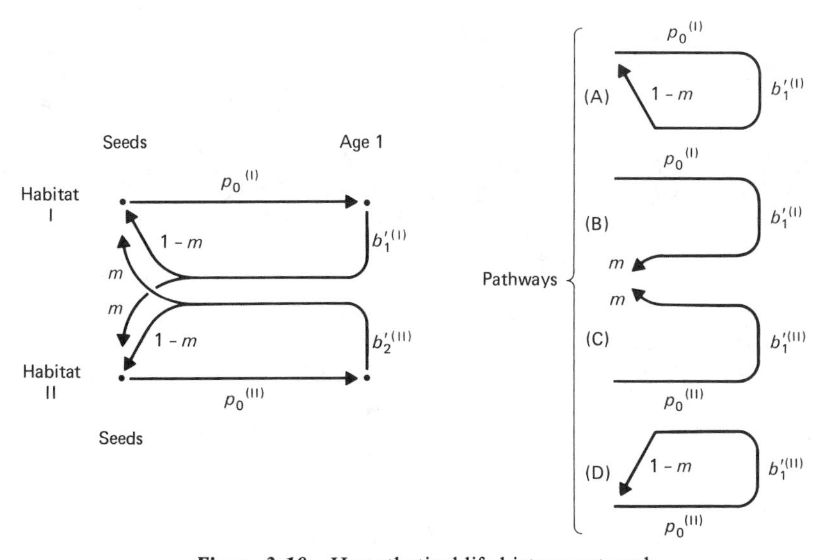

Figure 3-10 Hypothetical life history network.

Thus, using the notation of (3-32),

$$\begin{aligned} n_0^{(A)}(t+1) &= n_0(t)p_0^{(I)}(1-m)b_1'^{(I)} \\ n_0^{(BC)}(t+2) &= n_0(t)p_0^{(I)}mb_1'^{(I)}p_0^{(II)}mb_1'^{(II)} \\ n_0^{(BDC)}(t+3) &= n_0(t)p_0^{(I)}mb_1'^{(I)}[p_0^{(II)}(1-m)b_1'^{(II)}]p_0^{(II)}mb_1'^{(II)} \\ &\text{etc.} \end{aligned} \tag{3-63}$$

If n_0 grows by a multiple, λ, each time unit, equations (3-63) can be written,

$$\begin{aligned} n_0^{(A)}(t) &= n_0(t)p_0^{(I)}b_1'^{(I)}(1-m)\lambda^{-1} \\ n_0^{(BC)}(t) &= n_0(t)p_0^{(I)}p_0^{(II)}b_1'^{(I)}b_1'^{(II)}m^2\lambda^{-2} \\ n_0^{(BDC)}(t) &= n_0(t)p_0^{(I)}p_0^{(II)}b_1'^{(I)}b_1'^{(II)}m^2[p_0^{(II)}b_1'^{(II)}(1-m)]\lambda^{-3} \\ &\text{etc.} \end{aligned}$$

Adding these expressions and dividing through by $n_0(t)$ and rearranging terms yields

$$1 = (1-m)[p_0^{(I)}b_1'^{(I)} + p_0^{(II)}b_1'^{(II)}]\lambda^{-1} + (2m-1)[p_0^{(I)}b_1'^{(I)}p^{(II)}b_1'^{(II)}]\lambda^{-2} \tag{3-64}$$

The corresponding expression for habitat (I) alone is

$$1 = p_0^{(I)}b_1'^{(II)}\lambda^{-1} \tag{3-65}$$

The population above consists, in effect, of only one age class (age 0 does not contribute to population growth), but has been included as an illustrative case. With patience and algebraic care, the network model can be applied to multiple ages and multiple habitats. Because the network model is merely an alternative presentation of the Leslie matrix model, the conclusions are the same. Damping of cycles by migration in populations occurs when migration is density dependent.

Review Questions

1. If the Alaskan Dall sheep population referred to in Section 3-1 had been growing steadily for some time, how might Murie have changed his estimates of d_x? How might d_x have been calculated if the population had been oscillating due to a shifting age structure? What kinds of data would be needed to find d_x values or construct a projection matrix?
2. Why does the Leslie matrix described in (3-13) fail to include the last age class, n_k?
3. Make up a simple Leslie matrix and calculate the solutions for $n_x(t)$ for all x. Calculate R directly for each time period. How do these values relate to the eigenvalues, and particularly, λ_1?
4. Suppose that you travel to a remote area of Borneo where there is no seasonality and no disturbance (i.e., a constant environment) and release a number of hypothetical Bornean jungle mice. The life table after one year of data gathering (with x in units of, say, 6 months) looks like

x	d_x	b'_x
0	.5	0
1	.2	0
2	.1	4
3	.1	0
4	.1	2

Over the next 30 years will the population cycles grow or damp out? Can you answer this question without data on the nature of density feedback?

5. Suppose that:

x	d_x	b'_x
0	.5	0
1	.3	$4(1 - n/100)$
2	.2	$2(1 - n/100)$

What is the equilibrium population density?

6. A biennial occurs in three stages: seed, rosette, and mature plant. It takes a year for a seed to become a rosette. Suppose that density feedback operates via seed mortality so that the probability of becoming a rosette is

$$\frac{.01(1 - n_0)}{100{,}000}$$

where n_0 is the number of seeds per hectare. Half of *all* rosettes in any one year mature in the next (survival = .4), set 10,000 seeds, and die. The others survive the year with probability .8, but remain rosettes. At equilibrium how many *rosettes* are there per hectare?

7. In certain animals (e.g., some aquatic invertebrates, and poeciliid fishes) the population follows an annual cycle, with a summer peak and winter trough. In these species, females breed several times a year and their numbers of offspring depend both on the availability of food (which varies seasonally in a fairly predictable fashion) and on the female's size. Such populations cannot be easily modeled by equations such as those in Chapter 2. How might the stage projection matrix approach be adapted to model such populations?

8. Can the analysis presented in this chapter for age class stability be applied to cyclic changes in year class strength in, for example, sockeye salmon, in which run strengths exhibit a 4-year periodicity (Larkin, 1973)? Or is such behavior perhaps more appropriately modeled by treating each year class as a separate population and assuming the existence of multiple stable equilibrium points?

Chapter 4

Prey–Predator Interactions

In Chapters 2 and 3, populations were treated as isolates. Where fluctuations were examined there was little discussion of the biological *raison d'être* of these disturbances. There was no consideration at all of the possibility that environmental fluctuations might not only affect population dynamics but also be the direct result of population dynamics. But, in fact, the most important density-dependent influences on a population are generally its food supply (prey) and its predators, both of which are closely and dynamically related to it. A drop in carrying capacity is, more than occasionally, a direct reflection of a drop in food supply, which, in turn, may have come about through overexploitation due to high population density. Increased predation is apt to arise from the increased growth of a predator population or the attraction of predators from adjacent areas, due to high population densities. What this means, of course, is that the dynamics of one population cannot be understood except in the context of its prey's and its predator's dynamics.

As we have seen, the population models of Chapters 2 and 3 are not terribly realistic, and attempts to incorporate realism in the form of stochasticity, age structure, and spatial heterogeneity often lead to fearsomely intractable mathematics. So we may be excused for viewing the congestion caused by linking together several, interrelated systems of such equations with some apprehension. But population biologists are not (as a class) masochists. They either further sacrifice accuracy by simplifying their models, or they address more general questions whose *modus quaerendi* is less opaque. They have hoped for a perhaps nonexistent simplicity in nature. In 1960, Hairston et al. presented an argument whose basic conclusion was that terrestrial plants, as a group, are herbivore limited, that terrestrial herbivores, as a group, are predator limited, and that all terrestrial predators, taken together, are limited by their supply of prey. Since many terrestrial systems have only three trophic levels—no secondary predator—this conclusion, if true, means that, to a good approximation, herbivore–predator systems can be treated in isolation from plants. If the various herbivores are lumped into a single entity, and the predators also, the dynamics of the animal part of a terrestrial ecosystem can be largely treated as the interaction of only two compartments. However, Hairston et al.'s argument was roundly attacked by Murdoch (1966) and Ehrlich and Birch (1967), and only rather ineffectually defended by Slobodkin et al. (1967). It seems unlikely that the universe is either so consistent, green, or simple as Hairston et al. would like to believe. Of course, there will undoubtedly be systems in which secondary predation is of minor importance; in some instances a two-species model may depict nature reasonably accurately. In this chapter we commence in this optimistic spirit and investigate two-species prey–predator interactions. Confounding effects, including complications of including more species, are very briefly discussed. A more comprehensive investigation of multispecies interactions is postponed to Chapter 6.

Before embarking on a discussion of prey–predator interactions it is reasonable to ask just how much impact predation has on prey populations. In many cases it may be very little. Errington (1946, 1963), for example, found that predation on muskrats is limited almost entirely to vagrants, individuals which have left crowded population pockets and are destined to die in any case. He stressed that, in general, predators attack mostly the very young, the very old, the sick—those whose deaths have minimum impact on the preys' population dynamics. (This may be true as a general but not absolute rule. See, for example, Hornocker, 1970.) Errington (1956) suggests that density feedback from other sources should serve to limit severely the increase in prey populations whose predators have been removed. Howard (1974) in an excellent review paper argues strongly that predation may merely skim off the "excess" prey—those that would not contribute to prey population growth—and thus have little impact on equilibrium density.

Glasgow (1963) reports that spider (*Hersilia* sp.) webs alone each week capture 17% of a local tsetse fly population. If effects of other predators are added, the mortality rate is surely much higher. But are these flies that, if not predated, would die anyway at the hand of (other) density-dependent influences?

The common sea star *Pisaster ochraceus* eats about 25% of the available turban snails (*Tegula funebralis*) each year at Neah Bay, Washington. Total yearly mortality is 28%, very little more than 25% (R. T. Paine, personal communication). But if sea stars were removed, would mortality be only 3%? Or would density-dependent factors raise it back to somewhere near 28%? Predation data of the sort given above are easy to find (e.g., Connell, 1970, or Odum et al., 1964), but tell us little about the impact of predation. To find convincing evidence, we need a more direct approach.

Brooks and Dodson (1965) present strong evidence that predation by alewives (*Alosa pseudoharengus* or *A. aestivalis*), planktivorous fish which show dietary preferences for large food items, significantly alters the size composition of aquatic crustacea and rotifers. Wells (1970), working on lake Michigan, supports their assertion. In the 1950s there was an invasion and massive population explosion of alewives in Lake Michigan accompanied by a marked decline in the density of the larger zooplankton species. These prey reappeared in 1968, following a die-off of their fish predators in 1967.

Paine (1971) removed the carnivorous starfish, *Stichaster australis*, from a rocky, intertidal area in New Zealand and observed a resulting increase in tide level range of the mussel *Perna canaliculus* of about 40%. In addition, another (gastropod) predator, *Neothais* sp., nearly doubled in density and the number of intertidal species dropped from 20 to 14.

In northeastern Minnesota, wolves lay out a mosaic of pack territories with buffer zones up to 2 km wide between them. Pack members seldom venture into these zones except in times of food shortage. An immediate impact of this behavior is that deer can find refuge. One subsequent consequence is that high survival in these buffer zones leads to an age distribution skewed toward older individuals (see Chapter 3) (Mech, 1977).

To investigate the impact of insect "predation" on plants, Morrow and LaMarch (1978) worked with subalpine *Eucalyptus* trees in Australia. For each of two species they chose three individuals, each with two or more stems and sprayed one stem weekly, for a year, with insecticide. During that year they noticed significantly heavier foliage on the treated stems. In the next year, even without continued treatment, *all* stems were more luxuriant. Ring width, indicating growth rate, increased tremendously for all stems, especially the treated ones. Normalcy returned in subsequent years without further application of the insect poison.

Predation takes a variety of forms. A crude classification would first list the killing (and usually devouring) of prey. Herbivory is included here; while the victim may not die, it is usually damaged. Parasitism (including disease) is another form. The parasite may or may not ultimately kill the host, but may render it dead in evolutionary terms by destroying the gonads. Social parasitism, in which individuals of one species force the task of caring for their young on individuals of another species, is yet a third form of predation. The latter includes the habit of laying eggs in other animal's nests (well studied in birds; see Hamilton and Orians, 1965; Lack, 1968; Meyerriecks, 1972), and slave-making in social insects (Talbot and Kennedy, 1940; E. O. Wilson, 1971). The emphasis in this chapter is on the first variety of predation, although parasitism is occasionally addressed. Social parasitism will not be further discussed.

4.1 Basic Models

The Lotka–Volterra Model

We begin by supposing that the rate at which prey are eaten by predators is proportional to the encounter rate between individuals of these two species. In the simplest of worlds the frequency with which a given predator comes across prey will be proportional to the prey population density, n_1. Similarly, the number of predators confronting a prey in a given time interval will rise proportionately with the predator density, n_2. Hence the number of prey killed per unit time is proportional to $n_1 n_2$. Then, if the only source of prey mortality is predation, we can write

$$\frac{dn_1}{dt} = b_1 n_1 - d_1 n_1 = b_1 n_1 - \delta n_1 n_2 = n_1(b_1 - \delta n_2) \tag{4-1a}$$

where b_1 is birthrate and δ is a proportionality constant. Furthermore, if the birthrate of the predator is directly (and linearly) tied to its food intake per individual,

$$\frac{dn_2}{dt} = b_2 n_2 - d_2 n_2 = \beta \frac{n_1 n_2}{n_2} n_2 - d_2 n_2 = n_2(\beta n_1 - d_2) \tag{4-1b}$$

where β measures the efficiency with which food is converted to young predators. Equations (4-1a) and (4-1b) are collectively referred to as the *Lotka–Volterra predation equations*, after their originators (Lotka, 1925; Volterra, 1926).

These equations suggest an interesting possibility for prey–predator systems. Suppose that we begin by depicting the system as a point in prey–predator phase space (Fig. 4-1), and use equations (4-1) to increment n_1 and n_2 over time, plotting the results as we go. What appears is a repeated loop; the system cycles. Graphed another way (Fig. 4-2), we observe regular fluctuations in both species, those of the predator slightly lagging those of the prey. This pattern is easily explained. A high prey population results in a high predator birthrate and, correspondingly, a growth in the predator population. As the predator increases in density it exerts ever increasing pressure on the prey, eventually forcing prey numbers down. Then when the prey becomes sufficiently scarce, the predator birthrate drops and the predator population declines. Predation pressure falls, allowing the prey population once again to begin rising. The cycle is repeated over and over.

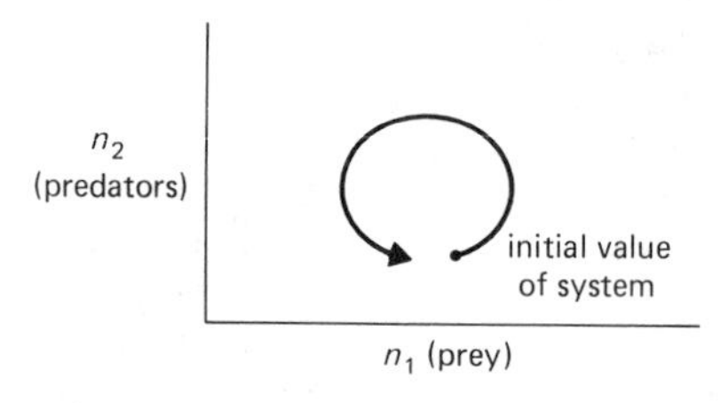

Figure 4-1 Plot of predator against prey population density. The arrow shows the path followed over time by the two populations.

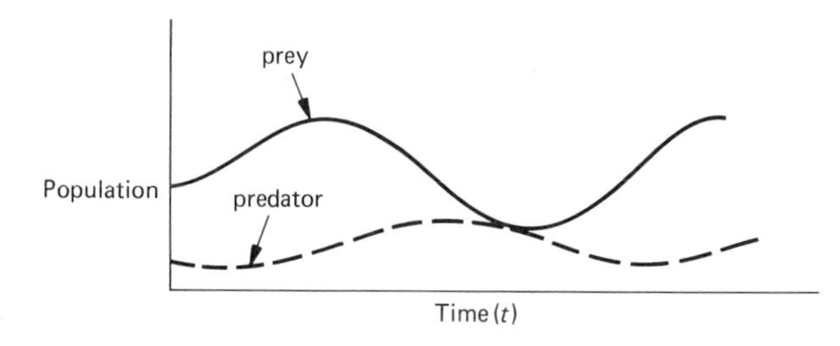

Figure 4-2 Hypothetical prey and predator population trajectories over time.

But while the Lotka–Volterra equations suggest the possibility of inherent prey–predator cycles, they are not terribly satisfactory. They inherit all the weaknesses of the logistic equation; they rely on simplistic assumptions about rate of prey capture and, in addition, fail even to contain a self-damping density-feedback term. Since it seems likely that prey numbers would not rise to fill the universe were the predators removed, nor predators increase without bound even with ad libitum food, the latter fault really needs correcting. This is easily done, as shown by the equations

$$\frac{dn_1}{dt} = n_1(b_1 - \delta' n_2 - \gamma_1 n_1) \equiv f_1(n_1, n_2) \tag{4-2a}$$

$$\frac{dn_2}{dt} = n_2(\beta' n_1 - d_2 - \gamma_2 n_2) \equiv f_2(n_1, n_2) \tag{4-2b}$$

But the system described in (4-2) does not cycle. This can be seen easily by performing a neighborhood stability analysis (recall that instability in the neighborhood of equilibrium implies either limit cycles or global instability (Chapter 2). First, set the expressions above equal to zero to find equilibrium values:

$$\hat{n}_1 = \frac{b_1\gamma_2 + \delta' d_2}{\gamma_1\gamma_2 + \delta'\beta} \qquad \hat{n}_2 = \frac{\beta' b_1 - \gamma_1 d_2}{\gamma_1\gamma_2 + \delta'\beta} \tag{4-3}$$

Then, perturbing n_1 and n_2 by minute amounts x_1 and x_2, we can use the Taylor expansion (Chapter 1) to find

$$\begin{aligned}\frac{dx_1}{dt} &= \frac{d(\hat{n}_1 + x_1)}{dt} = f_1(\hat{n}_1 + x_1, \hat{n}_2 + x_2)\\ &\simeq f_1(\hat{n}_1, \hat{n}_2) + x_1\frac{\partial f_1(\hat{n}, \hat{n}_2)}{\partial \hat{n}_1} + x_2\frac{\partial f_1(\hat{n}_1, \hat{n}_2)}{\partial \hat{n}_2}\\ &= x_1(b_1 - \delta'\hat{n}_2 - 2\hat{n}_1\gamma_1) + x_2(-\delta'\hat{n}_1) = -\hat{n}_1\gamma_1 x_1 - \delta'\hat{n}_1 x_2 \qquad \text{(see equations 4-2)}\end{aligned}$$

Similarly,

$$\frac{dx_2}{dt} \simeq x_2(\beta'\hat{n}_2) + x_2(\beta'\hat{n}_1 - d_2 - 2\gamma_2\hat{n}_2) = \beta'\hat{n}_2 x_1 - \gamma_2\hat{n}_2 x_2$$

In matrix form

$$\frac{d}{dt}\begin{bmatrix} x_1 \\ x_2 \end{bmatrix} = \begin{bmatrix} -\gamma_1\hat{n}_1 & -\delta'\hat{n}_1 \\ \beta'\hat{n}_2 & -\gamma_2\hat{n}_2 \end{bmatrix}\begin{bmatrix} x_1 \\ x_2 \end{bmatrix} \tag{4-4}$$

The system is stable if the real parts of all eigenvalues are negative (Chapters 1 and 2). In this case the eigenvalues are given by

$$\lambda^2 + (\hat{n}_1\gamma_1 + \hat{n}_2\gamma_2)\lambda + (\hat{n}_1\hat{n}_2\gamma_1\gamma_2 + \hat{n}_1\hat{n}_2\delta'\beta') = 0$$

Because all coefficients are positive, the real parts of λ must be negative.

Does this, then, mean that prey–predator systems do *not* cycle, after all? Let's look at some experimental evidence. Gause et al. (1936) (see also Gause, 1934) performed a series of laboratory experiments involving a single prey and a single predator species. In one set of studies the predatory mite *Cheyletus eruditus* was raised in glass tubes with its prey, *Aleuroglyphus agilis*, another mite. Several experiments were run, with a variety of food media for the prey species. In all cases cycles appeared, leading to decimation of the prey, followed by a crash of the predator population. Similar results were obtained with cultures of *Didinium nasutum* preying on *Paramecium caudatum* (both protozoans). When *P. bursaria* was used as a predator on the yeast *Saccharomyces exiguus*, all experimental runs converged to a limit cycle whose nature was independent of initial conditions. Other workers have also found cycles, in some cases very unstable (e.g., Huffaker, 1958), in others (fairly) well-behaved limit cycles (e.g., Utida, 1955). And there is evidence for prey–predator cycling in nature (see below). Clearly, equations (4-2) are not valid descriptors of the real world.

Functional Response Curves

Another correction to equations (4-1), taken together with inclusion of the self-damping terms of (4-2), may yield more satisfactory results. Specifically, we wish more accurately to describe the rate of prey capture per predator as a function of prey population density—what Holling (1965) has called the *functional response*. Holling described three forms of functional response curves (Fig. 4-3), which have since become known as type I, II, or III. All three curves describe an ultimate drop in prey mortality as prey become more abundant, a relationship usually referred to as *depensatory mortality*. Type I curves are typical of filter feeders and organisms which obtain nutrients from their medium by diffusion—wherever the food items are small enough, or their capture method is such, that there is essentially no "handling time" involved in feeding. The curve levels off abruptly when the food-gathering apparatus is saturated. Type II curves are fairly typical of invertebrates; the form of the relationship is easily obtained. Let

n_1 = prey density

m_1 = number of prey eaten over a time period T, per predator

T_H = time required to catch and devour a prey (handling time)

T = total time available for feeding

Then, since the eating of m_1 prey requires $m_1 T_H$ amount of time, the time available for search is $T - m_1 T_H$. We expect, to a good approximation, that the number of prey items found will be proportional to the prey density and the available search time. Thus

$$m_1 = a(T - m_1 T_H)n_1$$

where a is a proportionality constant. Then

$$m_1 = \frac{aTn_1}{1 + aT_H n_1} \quad \text{(type II curve)} \tag{4-5}$$

For alternative derivations of basically similar curves, see Watt (1959) and Ivlev (1961). See also

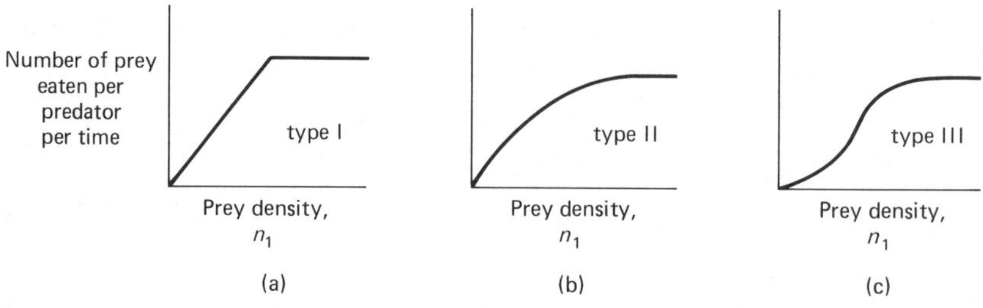

Figure 4-3 The three functional response curves (number of prey eaten per predator versus prey density) described by Holling (1965).

the modifications of this curve below, (equations 4-7, 4-8) dictated by spatial heterogeneity in the environment.

Type III curves are characteristic of many vertebrates and may arise in several ways as modifications of a basic, underlying type II relationship. Holling (1965) suggested that predators with a learning capability would discover the value of a prey species and so begin treating it as a staple food only if it were above some critical abundance. Thus intake would rise with increasing abundance only slowly at first, then disproportionately fast after the critical level was reached. Holling used the term *search image* to refer to the predator's response. Another concept of search image (Tinbergen, 1960) has the predator forming a mental image of the food it encounters more often, tending to overlook scarce food types and concentrate on common ones. Clearly, this behavior also would lead to a type III functional response curve. Alternatively, optimal diet choice (see Chapter 16) in the presence of alternative foods dictates a sigmoid functional response curve (J. M. Emlen, 1973, pp. 159–160). A sigmoid shape also arises when animals switch food preferences toward an increasingly common food (Murdoch, 1969—see also, below, Chapters 6 and 16) or when they exhibit *training effects* (become more adept at finding, subduing or handling commonly used items with practice—McNair, 1979). Finally, a population might display a type III curve, although its members individually respond in type II fashion. Referring to Figure 4-3b, suppose that some predator individuals fail to notice or respond to a food when it is very scarce, and that the level at which a response begins varies among individuals. The cumulative (or average) response of all individuals is sigmoid (Fig. 4-4). For purposes of description, purely empirical, the type III curve can be written

$$m_1 = \frac{aTn_1^\eta}{1 + acn_1^\eta} \qquad \text{(type III curve)} \qquad (4\text{-}6)$$

(Note the relationship to equation 4-5.) Generally η is given a value of 2.0, although clearly other values are possible (Real, 1979). Finally, there is no a priori reason to expect the coefficients, a, η, and c to be independent of n_1. For example, as food per predator becomes more common, predators may become more particular in which items they choose (see Chapter 16). If more easily handled items are preferred, c should decline with n_1/n_2. The coefficient a is not defined in the case of (4-6), but in (4-5) it reflects the search and capture efficiency of the predator. As n_1 rises, so may the available number of easily found or easily caught prey individuals. Thus perhaps a will rise as n_1 rises. In addition, as predators become more abundant they may affect each other's foraging efficiency—via aggressive interactions, perhaps, thereby *lowering a*. A discussion of this matter may be found in Hassell et al. (1976), who present data showing drops in food intake per predator (m_1) with an increase in predator density.

Considerations of the spatial distribution of prey complicate the foregoing formulations a bit. Suppose that n_1 prey individuals are randomly distributed in space and that we figuratively divide the environment into n_1 equal-size plots. Then each plot must have exactly 0, 1, 2, ... prey individuals, and, on average, there is one prey individual per plot. We now picture a predator which searches and, if it encounters a plot with prey, is capable of attacking one prey individual.

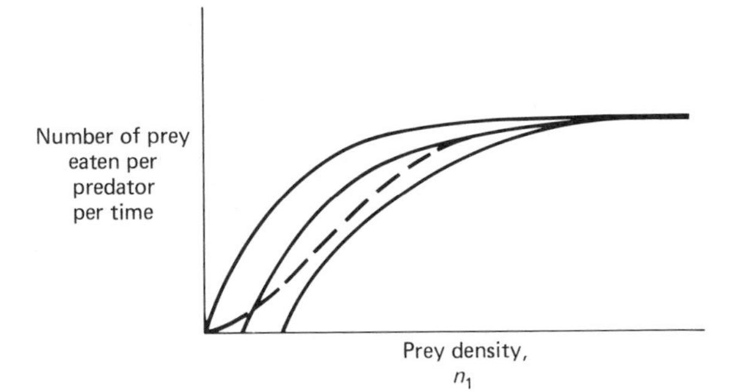

Figure 4-4 Average functional response over all individuals in a population (dashed line) and several individual responses (solid lines).

Suppose the distribution is such that n_1^* plots contain prey. Then the probability that a randomly encountered (by the predator) plot contains prey is n_1^*/n_1, and the probability that such a plot contains exactly x prey is

$$\Pr(x) = \frac{(n_1^*/n_1)^x e^{-n_1^*/n_1}}{x!}$$

Specifically, the proportion of plots in which the predator finds no prey at all is

$$\Pr(0) = e^{-n_1^*/n_1}$$

It follows that a fraction

$$1 - \Pr(0) = 1 - e^{-n_1^*/n_1}$$

of the encountered plots contain prey, and that if, over some time interval, a fraction α of the plots are searched, the number of prey consumed will be

$$m_1 = \alpha(\text{number of plots})(1 - e^{-n_1^*/n_1}) = \alpha n_1(1 - e^{-n_1^*/n_1}) \qquad (4\text{-}7)$$

(See also Nicholson, 1933.) Since each prey-containing plot encountered requires the same handling time, it is reasonable, if the predator takes only one prey per plot, that n_1^* versus n_1 should be a type II or type III functional response curve. Thus

$$m_1 = \alpha n_1\left[1 - \exp\left(\frac{-aTn_1^\eta}{1 + acn_1^\eta}\Big/n_1\right)\right] = \alpha n_1\left[1 - \exp\left(\frac{-aTn_1^{\eta-1}}{1 + acn_1^\eta}\right)\right] \qquad (4\text{-}8)$$

The shape of the curve generated by (4-8) is very close to that generated by (4-5) or (4-6), and, for the purposes of the very general discussion in this book, its increased accuracy does not justify its greater complexity. We shall continue to rely on the simpler forms of the standard type II and III curves. For an excellent discussion of the derivation and usefulness of (4-8) and related forms, the reader is referred to Hassell (1978).

So far we have merely indicated that type II and III response curves are observed in nature. We can be more specific. Holling (1965), in addition to reviewing data in the literature, experimented by burying sawfly (*Neodiprion sertifer*) cocoons in sand and allowing mice (*Peromyscus leucopus*) to search for them. Alternate food was supplied, ad libitum, in the form of dog biscuits. His results are shown in Table 4-1 and Figure 4-5. Figure 4-5 is clearly of the type III form. A cursory (and far from complete) survey of recent work reveals the following. Salt (1974) experimented with protozoan systems, using *Didinium nasutum* as a predator on *Paramecium aurelia*, and varying prey density. The result is a beautiful type II curve. Swenson (1977) looked at food consumption in walleye (*Stizostedion vitreum vitreum*) and Sauger (*S. canadense*) in Minnesota lakes, and found a clear-cut type II response curve for the former. Statistical scatter makes the picture less clear for Sauger.

Table 4-1

Number of Cocoons/m^2	Number Eaten per Time[a]	Mean Number Eaten per Time
24.02	8, 10, 13, 12, 3	9.2
48.04	12, 12, 15, 20, 20, 23	17.0
67.25	23, 26, 28, 29, 29, 33, 38	29.4
84.47	36, 40, 44, 50, 51, 59	46.7
105.68	51, 58, 58, 62, 74	60.6
124.90	63, 70, 70, 75, 78, 82	73.0
156.12	78, 84, 88, 92, 93	87.0
192.15	88, 90, 96, 99, 102	95.0
249.80	92, 92, 97, 100, 104	97.0

[a] Values are from several replicates.

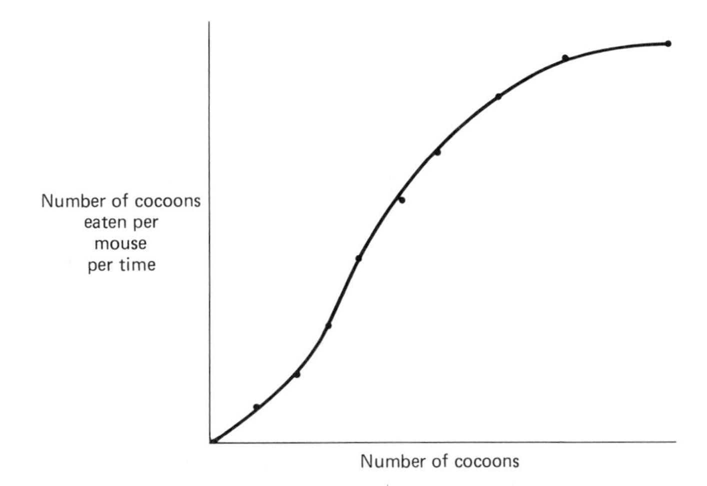

Figure 4-5 Example of a functional response curve (type III): number of sawfly cocoons eaten per mouse as a function of cocoon density. (After Holling, 1965.)

The most elegant work has probably been done on arthropods. Lawton et al. (1975), Beddington et al. (1976), and Hassell et al. (1976) provide extensive data for a wide variety of arthropod taxa, including both standard prey–predator systems and host–parasitoid systems. Both type II and III curves abound (see also Burnett, 1951; D. Rogers, 1972). Good reviews of functional response data available prior to 1975 can also be found in Murdoch and Oaten (1975).

Of course, while the nature of functional response curves gives us a notion of prey death rate, it does not necessarily tell us very much about predator population dynamics. It would be convenient if predator birthrate rose proportionately with food intake—and so also followed a type II or III curve. How good, or disastrous, would such an assumption be? Amazingly enough, it seems fairly accurate! Burnett (1951) looked at parasitism of sawfly (*Neodiprion sertifer*) cocoons by the chalcid wasp *Dahlbominus fuscipennis*, and obtained the results shown in Table 4-2 and Figure 4-6. Both prey death rate and predator birthrate show a clear, type II curve. Lawton et al. (1975) and Beddington et al. (1976) also find type II (or III) curves when predator (parasite) birthrates are plotted against prey (host) density—a good, linear relationship between birthrates and food consumption—for a variety of arthropods. Maker (1970), examining predation by pomarine jaegers (*Stercorarius pomarinus*) on their primary prey, the brown lemming (*Lemmus trimucronatus*) in Alaska, found similar results. The jaegers lay an invariant number of eggs, so reproductive response to food supply is proportional to the number of jaegers that nest. A plot of jaeger nests per area versus number of lemmings per area yields a type III curve.

Table 4-2[a]

Number of Host Cocoons/in.^{2}	Number of Wasp Eggs Laid (Predator Birth Rate)	Number of Cocoons Parasitized (Prey Death Rate)
.04	84.07	1.80
.08	143.87	3.13
.18	197.00	5.73
.32	268.20	9.80
.50	296.67	12.93
.72	344.93	15.63
1.00	349.27	17.53
1.28	361.73	18.40

[a] Numbers represent means over several replicates.

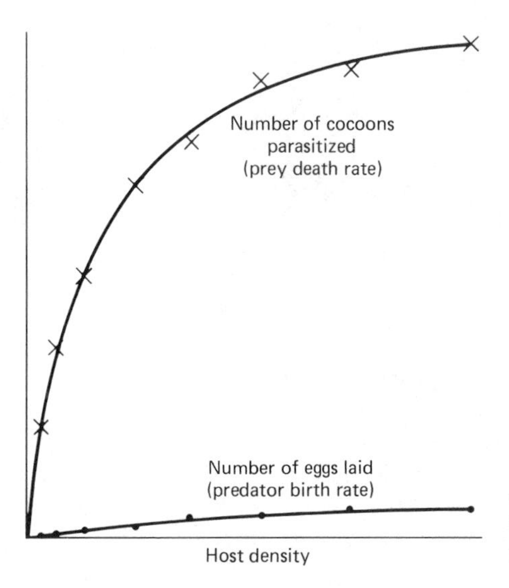

Figure 4-6 Example of a type II functional response curve (dots) and its effect on prey death rate (crosses): response of wasp parasites to sawfly cocoons.

The upshot of the previous few pages of discussion is that we may, to a reasonable approximation in many cases, describe both prey mortality due to predation and predator birthrate with a functional response equation (4-6). We shall make use of this convenient fact in a moment. But first, candor requires a caveat: Such simple relationships do not necessarily hold true, even crudely, for some prey–predator interactions. For example, herbivory does not necessarily kill the eaten plant. Thus plant death rate does not follow (4-6). Furthermore, light grazing not only reduces density feedback on the plant population, but actually stimulates individual growth, leading to greater seed production (e.g., see Mattson and Addy, 1976; McNaughton, 1979; Hilbert et al., 1981). The same phenomenon may occur in fish (predator)–bottom fauna (prey) associations (Hayne and Ball, 1956). For now we ignore such complex cases and return to simpler systems. The loss of prey to predation is then

$$\frac{\delta n_1^\eta n_2}{1 + \xi \delta n_1^\eta}$$

where $\eta = 1$ for type II curve, δ incorporates a, T from (4-5), $\xi \propto T_H/T$ (or c/T). Thus (4-2a), for prey dynamics, becomes

$$\frac{dn_1}{dt} = b_1 n_1 - \frac{\delta n_1^\eta n_2}{1 + \xi \delta n_1^\eta} - \gamma_1 n_1^2 \tag{4-9a}$$

The birthrate of the predator, rather than $\beta' n_1$, can now be more accurately written as

$$\frac{\beta \delta n_1^\eta}{1 + \xi \delta n_1^\eta}$$

where β reflects the efficiency with which prey is converted to predator biomass, so that

$$\frac{dn_2}{dt} = \frac{\beta \delta n_1^\eta}{1 + \xi \delta n_1^\eta} n_2 - d_2 n_2 - \gamma_2 n_2^2 \tag{4-9b}$$

Graphical Models

Equations (4-9) are not jointly solvable with the greatest of ease. Hence we shall apply them to a qualitative, graphical approach to prey–predator dynamics. The prey population possesses an equilibrium when $dn_1/dt = 0$. This occurs, for a type II response ($\eta = 1$), when (equation 4-9a)

$$n_2 = \frac{(b_1 - \gamma_1 n_1)(1 + \xi \delta n_1)}{\delta} \tag{4-10a}$$

This relationship (the prey zero-growth isocline) is graphed in Figure 4-7. In the shaded area both predation and density feedback are below the levels at which $dn_1/dt = 0$. Hence this area is one of growth of the prey (see arrows). Outside the curve, the prey population must decline from overexploitation or self-limiting factors.

A similar figure, describing dynamics for the predator, can be constructed. Set (4-9b) equal to zero. Then the predator zero-growth isocline is given by

$$n_2 = \left(\frac{\beta \delta n_1}{1 + \xi \delta n_1} - d_2\right) \Big/ \gamma_2 \tag{4-10b}$$

The corresponding graph is shown in Figure 4-8. To the right of (and below) the isocline there is sufficient food and insufficient density feedback to halt predator growth (see arrows); Above the line, n_2 declines.

Before combining Figures 4-7 and 4-8 to the purpose of describing prey–predator dynamics, note that these curves can be directly generated from laboratory (or field) data. Gause et al. (1936), in their previously mentioned experiments with *Paramecia* and yeast, initiated a number of trials with different values of n_1, n_2, and observed population changes in the following days. From these trials, the directions and magnitudes of change in n_1, and n_2 for a wide variety of initial points (n_1, n_2) could be determined. Rosenzweig (1969) has done the same for a rotifer (*Asplanchna*) (predator)–*Paramecium* (prey) system, using data from Maly (1969), and for a prey–predator system of mites, using data from Huffaker (1958).

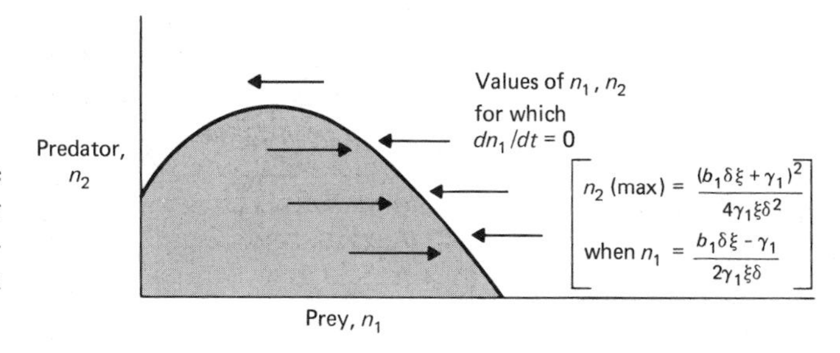

Figure 4-7 Prey isocline. The curve describes values of predator (n_2) and prey (n_1) population densities at which the prey population is neither growing nor declining.

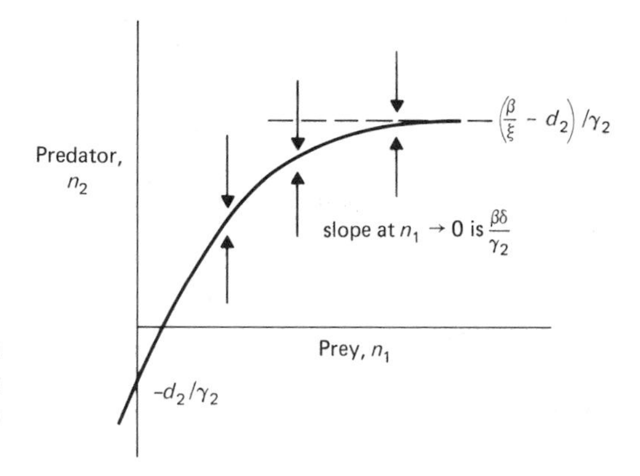

Figure 4-8 Predator isocline. The curve describes values of predator (n_2) and prey (n_1) population densities at which the predator population is neither growing nor declining.

The use of prey and predator zero-growth isoclines to make predictions about prey–predator dynamics was first suggested by Rosenzweig and MacArthur (1963). The following is both a review and extension of their method (see also J. M. Emlen, 1973, pp. 290–293).

First let us ask what the time courses of n_1, and n_2 are, and investigate the effects of high versus low efficiency of search and capture by the predator. An increase in such efficiency (or, equivalently, a drop in escape abilities by the prey) is described by an increase in the parameter δ. A system with high δ displays a prey isocline (Fig. 4-7) with low n_2 intercept $(= b_1/\delta)$, and lower $n_2(\max)$ $[= (b_1\delta\xi + \gamma_1)^2/4\delta^2\xi\gamma_1]$, occurring at a larger value of n_1 $[= (b_1\delta\xi - \gamma_1)/2\gamma_1\xi\delta]$. Also, the predator isocline (Fig. 4-8) will rise more rapidly at first (slope $= \beta\delta/\gamma_2$). These differences are depicted in Figures 4-9 and 4-10. Combining the predator and prey isoclines and drawing in the arrows indicating population growth or decline, we obtain Figure 4-11. It is now possible to estimate the dynamics of the system as a whole by starting at any point on the n_1–n_2 graphs and drawing in the vector sums of the arrows. For example, if n_1 is increasing (arrows pointing to the right), and n_2 is declining (arrows pointing down), the net direction of the system is down and to the right. Remember that as the path of the system crosses the predator isocline movement can only be to the right or left (exactly on the isocline n_2 is, by definition, not changing—no net movement up or down). Similarly, when the path crosses the prey isocline, movement can be only up or down (since, by definition, there is no change in n_1—no right or left movement). The dynamics of the systems in Figure 4-11 are shown in Figure 4-12. Obviously, the system with inefficient predation damps to a stable equilibrium while that with efficient predation cycles to extinction. Presumably, a system might exist with intermediate predator efficiency that would display stable limit cycles. We can conclude:

1. *Prey–predator systems are more apt to be stable, both neighborhood and globally, if the predator is not highly efficient at finding and capturing prey* (see Rosenzweig and MacArthur, 1963).

Good experimental evidence for this prediction is supplied by Luckinbill (1973), who experimented with prey–predator systems of *Paramecium aurelia* and *Didinium nasutum*. Luckinbill added methyl cellulose to the medium in some of his replicates, a procedure that has no effect on nonpredatory death rates or the birthrate of the prey, but by making the medium more viscous, slows down the movements of both species, thus reducing the predation rate (less

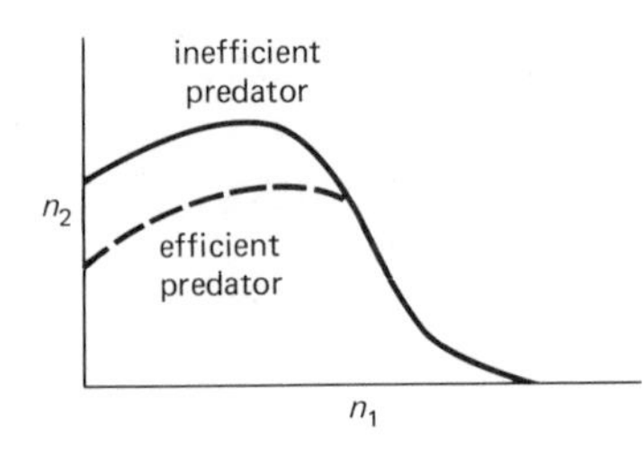

Figure 4-9 Prey isoclines for systems with an efficient and an inefficient predator.

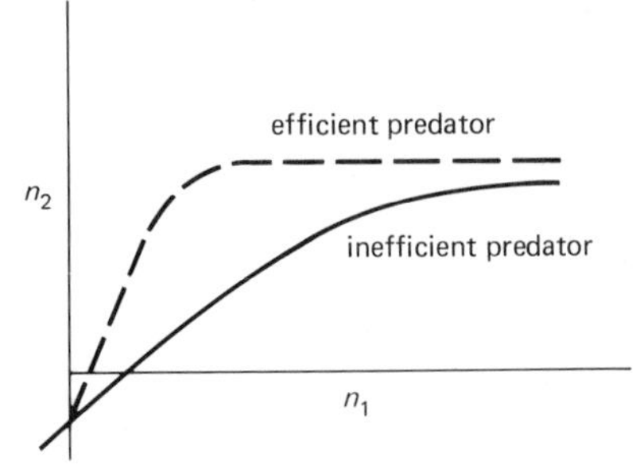

Figure 4-10 Predator isoclines for systems with an efficient and an inefficient predator.

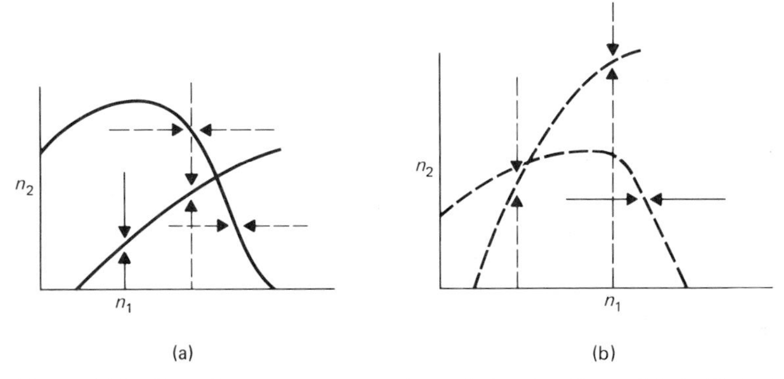

Figure 4-11 Superimposed prey and predator isoclines: (a) inefficient predator; (b) efficient predator. Arrows show the direction of population change.

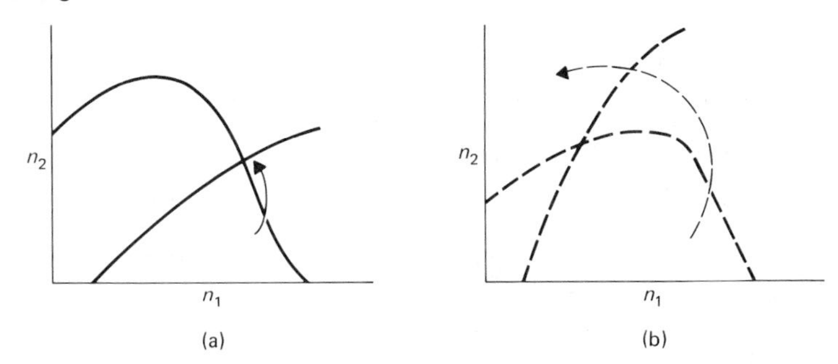

Figure 4-12 Superimposed prey and predator isoclines: (a) inefficient predator; (b) efficient predator. Arrows show the direction of population change.

efficient predator). The result is a stabilizing of cycles which otherwise lead to a crash in the system.

Suppose that the predator were more efficient not at finding and capturing prey, but at handling them once they are caught. Then ξ will be low, resulting in a higher rise in the predator isocline and a lower prey isocline with its peak shifted to the left (Fig. 4-13). The dynamics are shown in Figure 4-14, where, again, the system with the efficient predator is less stable.

2. *Prey–predator systems are more apt to be stable if the predator is not highly efficient at handling its prey.*

If the predator is efficient at converting prey biomass to predator biomass, then β is high and the predator isocline rises more steeply and reaches a higher point before leveling off. A graphical representation of the dynamics, again, will show that

3. *Prey–predator systems are more apt to be stable if the predator is not highly efficient at converting food to growth and reproduction.*

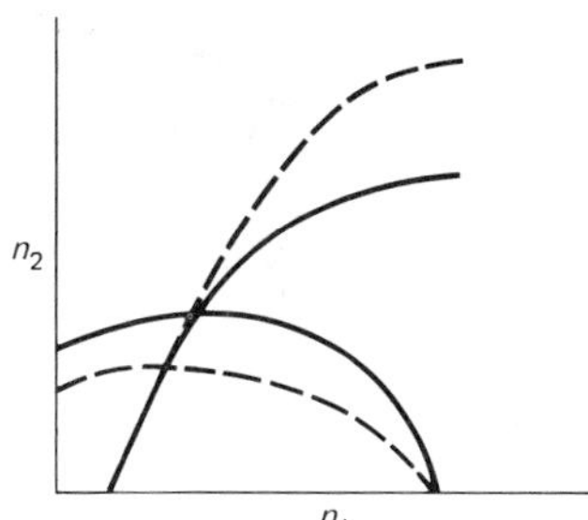

Figure 4-13 Superimposed prey and predator isoclines. Dashed lines, efficient predator; solid lines, inefficient predator.

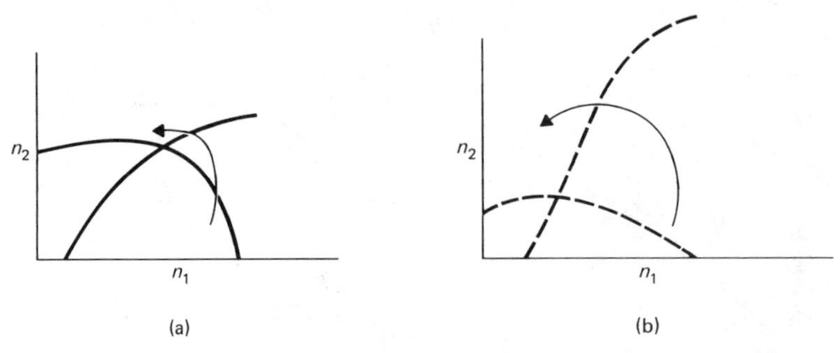

Figure 4-14 Superimposed prey and predator isoclines: (a) inefficient predator; (b) efficient predator. Arrows show the direction of population change.

Suppose that two prey–predator systems were identical in all respects except that one was relatively enriched—more food for the prey and hence less self-limitation for the prey (lower γ_1 value). A drop in γ_1 has no effect on the predator isocline, but raises the prey isocline, moves its peak to the right, and increases the value of the n_1 intercept (Fig. 4-15). It is not hard to see that the enriched system allows the predator population to grow to much higher levels before the prey ceases to grow, thus destabilizing the system (Fig. 4-16). If the system is enriched in the alternative sense of supplying a steady, alternative food source for the predator, γ_2 is lowered and the predator isocline rises more steeply. The effects of a steeper curve can be seen in Figure 4-12. Again, the system is destabilized. This fact was first noted and discussed by Rosenzweig (1971).

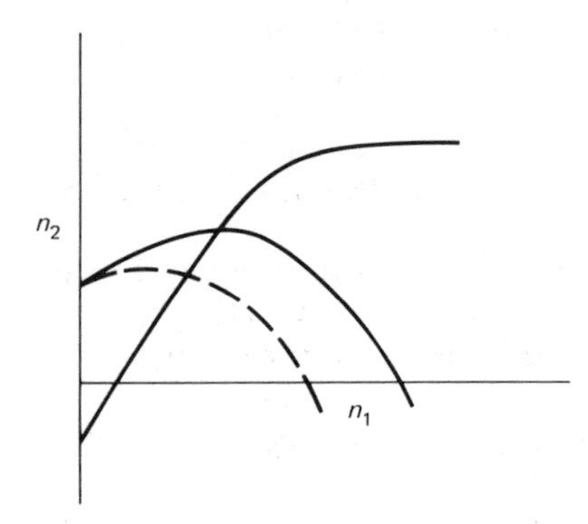

Figure 4-15 Superimposed prey and predator isoclines: enriched prey system versus unenriched predator system.

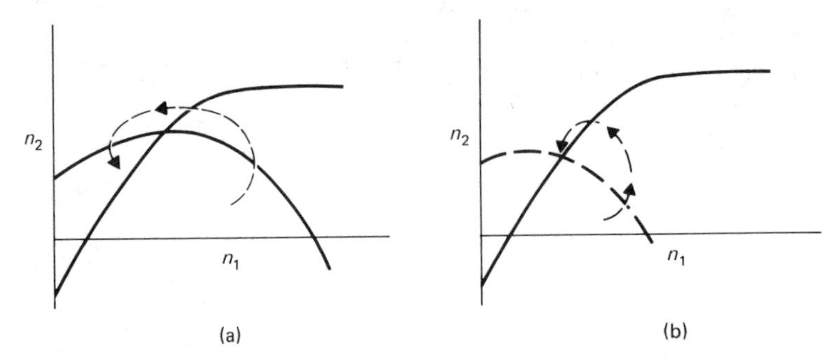

Figure 4-16 Superimposed prey and predator isoclines: (a) enriched system; (b) unenriched system. Arrows show the direction of population change.

4. *Enrichment of a prey–predator system by addition of food for the prey or alternative sustenance for the predator destabilizes that system.*

Evidence for the destabilization by enrichment comes, again, from experiments of Luckinbill (1973), who added enough bacteria to the medium that the *Paramecium* population density, in the absence of predation, would have risen by about two-thirds. The result was a crash in the system.

Many prey–predator systems have prey refugia built into them. In the simplest case, a limited number of safe hiding places (may exist) to which threatened prey may escape. The result is a pool of uncatchable individuals which may perpetuate the prey population even at the worst of times, and prevent, or at least greatly depress, the chance of extinction. But refugia will more likely come in forms other than hideaways of absolute safety. As food items become scarce it may be unfeasible, energetically, for predators to search and catch them. Thus the predator population may run out of reproductive energy long before it decimates the prey. This consideration can be incorporated into the predator dynamics equation (4-9b), simply by noting that not all the energy obtainable from food [$\propto \beta\delta n_1/(1 + \delta\xi n_1)$, for $\eta = 1.0$] is available for reproduction. If the cost of homeostasis for the predator is E_2, then (4-9b) (for $\eta = 1$) should be written

$$\frac{dn_2}{dt} = \left(\frac{\beta\delta n_1}{1 + \xi\delta n_1} - E_2\right)n_2 - d_2 n_2 - \gamma_2 n_2^2 \tag{4-11}$$

The effect on the predator isocline curve is shown in Figure 4-17 and is, clearly, stabilizing.

Still other forms of refugia exist. Barnacles (*Balanus cariosus*) live in the marine, rocky intertidal, where they are subject to predation by starfish and various predatory gastropods. Both sets of predators require time to penetrate a barnacle's defenses, to open it, and actually begin destruction of the soft tissues; the time increases with the size of the barnacle. Both appear loathe to feed when exposed to the air. Thus feeding generally takes place only when the tide is high enough to submerge the predator where it is attacking its prey. But at any given tide level, there is a limit to the time of submersal, and thus a limit to how long a predator can extend its attack on a barnacle. At low tide, the predators generally move to shelter. So a barnacle that escapes predation long enough may become sufficiently large that the number of minutes required to open it exceeds the contiguous number of minutes it is submerged. At this point it is safe from attack and joins the legions of the invulnerable. Paine (1976) describes a similar case involving the sea star *Pisaster ochraceus*, and its intertidal prey, the mussel *Mytilus californianus*, we conclude:

5. *Many (most?) prey species have available to them some form of "refuge."* The existence of such refugia prevents, or at least makes less likely, extinction by predation. Thus systems that are globally unstable in the absence of refugia may be prevented from crashing. They may, however, be expected to show cycling of greater amplitude than other systems.

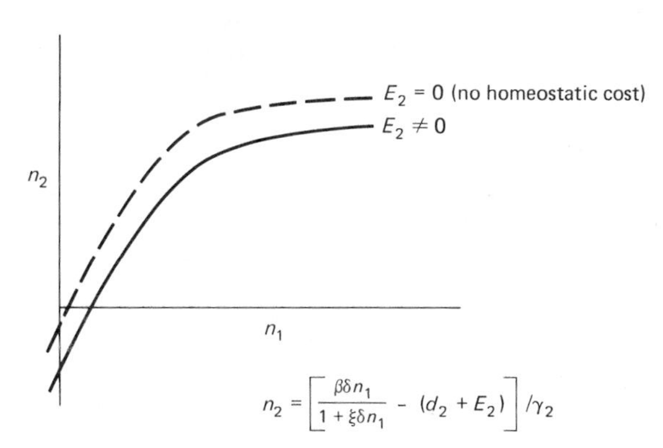

Figure 4-17 Predator isocline, showing the effect of homeostatic costs.

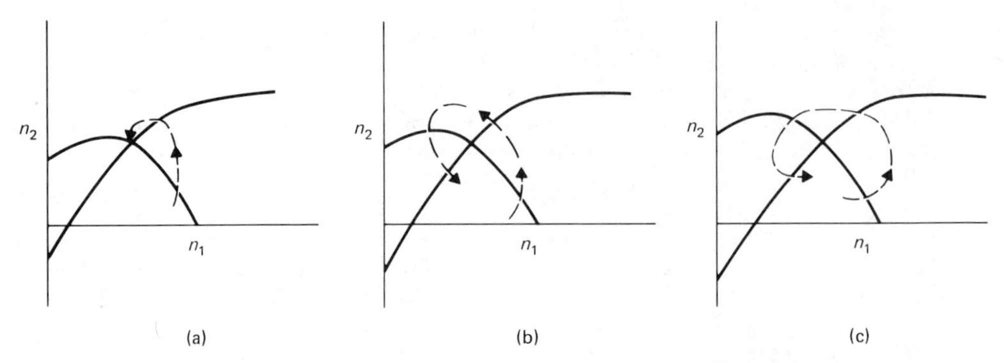

Figure 4-18 Superimposed prey and predator isoclines showing the effect of time lags on systems dynamics: (a) no time lags; (b) time lag in predator response; (c) time lag in prey response.

In Chapters 2 and 3 we discussed the impact of time lags on the dynamics of single populations. How will time lags affect a prey–predator system? The question is easily answered using the isocline graphs. Simply note that a lag in the population response of the prey to either predator pressure or self-limiting density feedback will result in the path of the system being *not yet quite vertical* when it crosses the prey isocline. With respect to lags in the predator population, the system's path will be *not yet quite horizontal* when it crosses the predator isocline. This is demonstrated in Figure 4-18. Clearly,

6. *Time lags in the population responses of either species to self-limiting density feedback, of prey to predation pressure, or of predators to prey availability, are destabilizing.* For further discussion of the effects of time lags on prey–predator systems, see Arditi et al. (1977).

We have yet to explore the implications for prey–predator dynamics of type III versus type II functional response behavior. But this is easily done. Referring back to (4-9) and (4-11), and setting $\eta = 2$, we see that the appropriate isoclines are:

Prey isocline:

$$n_2 = \frac{(b_1 - \gamma_1 n_1)(1 + \delta\xi n_1^2)}{\delta n_1} \tag{4-12a}$$

Predator isocline:

$$n_2 = \frac{\beta\delta n_1^2}{1 + \xi\delta n_1^2} - (d_2 + E_2) \tag{4-12b}$$

The corresponding graphs are shown in Figure 4-19, together with the graphs for a type II response systems with identical parameters. The change in shape of the prey isocline as we move from a type II to a type III functional response curve can be explained as follows. Refer to Figure 4-3c. At very low prey densities, predatory response is very slight. Because each predator individual does miniscule damage to its food supply, many are required to hold the prey population in check. At larger values of n_1 each predator is much more actively pursuing the prey. Thus fewer are needed to block prey population growth. Finally, at high prey densities each predator encounters more prey than it can use, the probability that a given prey gets eaten by a given predator drops, and so there must again be many predators to control the prey population. The prey isocline drops finally as prey approaches its carrying capacity. It is not clear whether systems with type II response curves are more or less stable than those with type III response curves. The predator isocline is much steeper in the latter case, suggesting greater instability. But there is a kind of built-in refugium in the prey isocline. However, the size of the invulnerable prey class may be small enough that random extinction (Chapters 2 and 3) is likely, largely negating the refugium effect. No generalities are in order here (but see Oaten and Murdoch, 1975, and Armstrong, 1976, for a different approach).

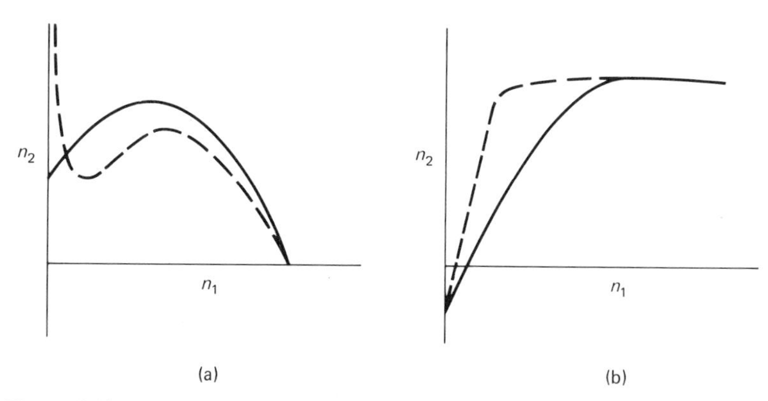

Figure 4-19 (a) Prey and (b) predator isoclines for systems with type II (solid lines) and type III (dashed lines) functional response curves.

One very interesting consequence of type III curves needs mentioning. If the sigmoid nature of the functional response is sufficiently pronounced [the parameter η in (4-6) is large enough], the prey isocline may intersect the predator isocline twice (Fig. 4-20). If the movement of the system in Figure 4-20b is plotted, it can be seen that two stable equilibria exist, one at P_1, the other at P_2. As long as n_1 remains to the right of P_3 it is in the *domain of attraction* of the higher equilibrium. But if, due to poor environmental conditions, disease, or other disturbance, n_1 is knocked below P_3, it enters the domain of attraction of P_1, the lower equilibrium. Peterman (1977) was one of the first to notice possible examples of this phenomenon and to explain it. He cites the case of the pink salmon (*Onchorynchus gorbuscha*) in the Atnarko river, British Columbia. This species is spawned in freshwater tributaries and, like other Pacific salmon, the young fish migrate to the ocean. Then, in precisely two years they return to breed and die. The result is odd- and even-year populations which very rarely exchange genes and which have minimal influence on each other. For many years these two, isolated (in time) populations fluctuated about a value of roughly 2 to $6\frac{1}{2}$ million individuals. But in 1961 a combination of heavy fishing pressure and inclement weather greatly reduced the stock of the odd-year group. Subsequently, while the even-year breeders remain in number between 2 and $6\frac{1}{2}$ million, the odd-year breeders have been consistently lower, at about 800,000. Apparently, the bad conditions in 1961 reduced the stock below the critical density level (P_3) and into the lower domain of attraction.

Peterman (1977) provides equations to demonstrate why populations may be perturbed into lower equilibria. However, the derivation used by Pikitch (in preparation) is more

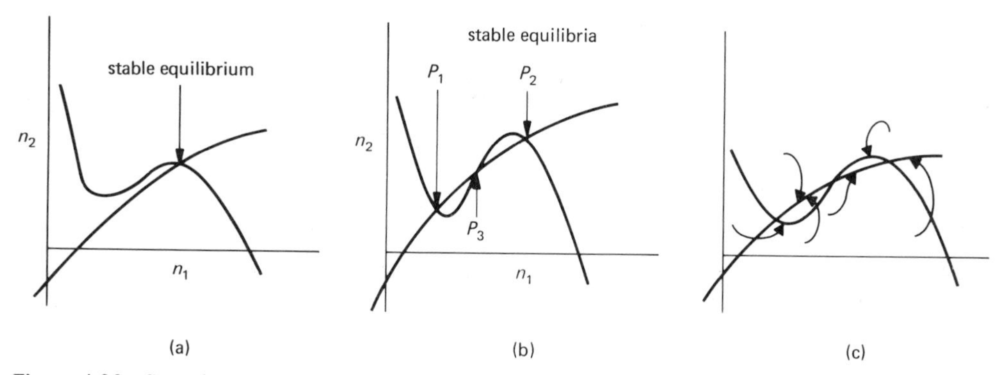

Figure 4-20 Superimposed prey and predator isoclines showing equilibria in two possible systems exhibiting type III functional response curves: (a) η is small; (b) η is large; (c) path of the system.

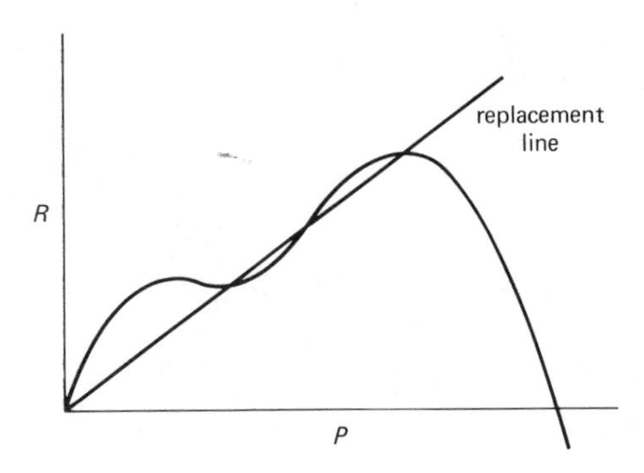

Figure 4-21 Stock (P)–recruitment (R) curve for a prey species experiencing extreme type III predation.

straightforward. Refer to Chapter 2 and the derivation of the Ricker curve (p. 43), and note that in addition to the linear, density feedback by parental stock, there should be a density-related mortality arising from predation—interspecific, cannibalistic, or both. Thus returning to fisheries notation (P is parental stock, R_x is number of recruits at age x), and referring to (4-6), we can write

$$\text{probability a recruit is predated in one unit time} = \frac{\text{number predated/time}}{\text{number of recruits}} \propto \frac{aTR_x^\eta}{1 + acR_x^\eta} \Big/ R_x$$

and

$$\text{density-dependent mortality rate} = \beta P + \frac{a'TR_x^{\eta-1}}{1 + acR_x^\eta} \tag{4-13}$$

If, as assumed for the Ricker model, this mortality occurs over a short period of early life, and the number of very young recruits is proportional to the parental stock, (4-13) becomes (for $\eta = 2$),

$$\text{density-dependent mortality} = \beta P + \frac{a''P}{1 + acP^2}$$

whence probability of surviving density-dependent pressures is

$$\exp\left[-\left(\beta P + \frac{a''P}{1 + acP^2}\right)\right] \tag{4-14}$$

This term should replace $e^{-\beta P}$ in the Ricker equation. Finally, then, the modified Ricker model is

$$R = \alpha P \exp\left[-\left(\beta P - \frac{a''P}{1 + acP^2}\right)\right] \tag{4-15}$$

A modified Ricker curve taken from (4-15) is shown in Figure 4-21. Clearly, there are two stable equilibria.

Although it is difficult to demonstrate the observation, statistically, it appears that bimodal Ricker curves may indeed occur. Pikitch (in preparation), for example, has demonstrated such a form for yellow perch (*Perca flavescens*) in Lake Erie, where this species is preyed on by walleye (*Stizostedion vitreum vitreum*). For further discussion of prey–predator isocline graphs and their possible applications, see Noy Meir (1975).

The Evidence for Prey–Predator Cycles

The preceding discussion indicates that prey–predator systems may display either neighborhood-stable equilibria, or limit cycles (or unstable cycles leading to extinction), and that the magnitude of cycles, when they appear, depends on a variety of factors, ranging from predator efficiency and food abundance for the prey to the presence (and nature) of refugia and time lags. Prey–predator cycles can certainly be generated in the laboratory. Do they occur in nature?

One difficulty with demonstrating prey–predator cycles is that, whereas cycling is common in natural populations, causes are almost impossible to pin down; cycling can also arise from time lags inherent in the species in question, from changing age structure, from environmental fluctuations, or as a result of genetical phenomena (Chapter 13). Evidence for the importance of predation in generating cycles comes in the form of coordinated population changes in prey and predator, the path of the predator lagging behind that of the prey. The following examples are classic, but it should not blindly be assumed that the dynamics described are entirely the result of interactions between consumer and consumee.

Pearson (1966) followed population fluctuations of the California meadow vole (*Microtus californicus*) through what appeared to be one and a half full cycles. Predators (raccoons, *Procyon lotor*; gray foxes, *Urocyon cinereargenteus*; skunks, *Spilogale putorius*; plus various owls, hawks, and feral cats) were monitored for abundance by counting the number of droppings they left behind. By dissecting these droppings and identifying the bits of bones to be found, Pearson also was able to estimate the rate at which voles were being predated. What he found was a large-amplitude fluctuation in the rodents, and a less marked, slightly lagging fluctuation in the number of rodents eaten and total number of predators. All signs point to a classic, simple prey–predator cycle. In support of the concept of predator-induced cycling, in general, Pearson also points out that where *Microtus* has invaded Brooks Island, in San Francisco Bay, it shows seasonal changes but no regular cycling. There are no predators on the island.

Data garnered from the number of pelts collected by the fur industry (mostly from the Hudson Bay Co., 1821–1952; and Dominion Bureau of Statistics, 1919–1951) show quite regular cycles of about 10 years for several species in the Canadian boreal forest. Predators include the wolf (*Canis lupus*), red fox (*Alopex lagopus*), coyote (*Canis latrans*), lynx (*Lynx canadensis*), mink (*Mustela vixen*), and various hawks and owls. Prey, among others, include the varying hare (*Lepus americanus*), muskrat (*Ondatra zibethica*), and ruffed grouse (*Bonasa umbellus*). The most thoroughly examined data are those pertaining to the lynx and the hare. The former shows a period of 9.7 years, the latter 9.6 years (the difference seems likely to be nothing but statistical noise in the data). Population densities are correlated ($r = .55 \pm 0.5$), and the lynx population lags, appropriately, behind the number of hares. (Various discussions and treatments of the data can be found in MacLulich, 1937; Elton and Nicholson, 1942; Butler, 1953; and Keith, 1963.)

Although it is generally accepted that these fluctuations represent true prey–predator cycles, a number of objections and confounding facts have surfaced. Butler (1953) and Watt (1968) note that fox and lynx populations lag changes in the number of hares by one year—since young of the year enter the adult population the next year, this is reasonable. Fisher (*Martes pennanti*) show a 2-year lag which, in light of this species' habit of delayed implantation, is also reasonable. But coyotes and wolves show even greater lags. Why? Then, on Anticosti Island, Canada, where lynx are absent, an introduced population of hares cycles in phase with its mainland counterparts. Elton and Nicholson (1942) feel that this strongly implicates some endogenous factor. Perhaps the predator population responds to prey numbers, but the prey species cycles as a result of some intrinsic quality—age structure, lags in response to population stress, and so on. But still more seems to be going on. The synchrony of the cycles of mink and muskrat (who do not eat hares) with those of the larger predators (who do) suggests that exogenous events are also important. Moran (1953a and b) feels that the observed cycles are, in fact, driven by prey–predator interactions, but perhaps triggered by exogenous events. Watt (1968, 1969) agrees and presents evidence in support of the idea. For example, statistical analyses of the lynx cycles turns up three clusters of years where the residual (statistically unexplained deviations) errors are huge. One occurs between 2 and 7 years after the eruption of Cosequina, one 3 to 5 years after the eruption of Krakatao, and one 5 to 6 years after the eruption of Katmai. Watt also finds a highly (statistically) significant correlation between this residual error in lynx numbers and temperature (departure from normal) 3 years previous (the latter is strikingly related to major volcanic events). Finally, there is at least rough synchrony of cycling over most of Canada (but see Section 4.3), suggesting, again, the role of major, external influences. Leslie (1959) had already shown, some time before, that independent (noninteracting) populations, initially cycling out of phase, could be brought into synchrony by random environmental perturbations.

More recently, believers of the boreal prey–predator model were jolted by a paper entitled "Do hares eat lynx?" (Gilpin, 1973). If prey population (n_1) is plotted against predator population (n_2) as in Figure 4-1, the direction of motion must be counterclockwise, reflecting the lag in predator numbers after those of the prey. But Gilpin found that over some periods of years the direction appeared to be clockwise for the lynx and hares. Perhaps the hares had become, in a sense, the predator by virtue of passing on epidemics to the lynx? He also suggested that the population response by lynx, even where it appropriately lagged its prey, was too quick and too strong. One possible explanation is that the prey–predator cycles are really an interaction between the fur bearers (prey) and the fur trappers (predators). Correlations in lynx and rabbit pelt numbers might arise in the following way. When hares were abundant, trappers might feel free to spend more time trapping the more profitable lynx, but would sit out the rabbit-poor years (Gilpin, 1973). If the cycles were in fact generated in this manner, the rabbit oscillations on lynx-free Anticosti Island would make more sense.

Weinstein (1977) takes a somewhat different tack. He finds no evidence of epidemics, and suggests that trappers could ill afford to bypass the poor hare years. But in those poor years, need for food might drive the trappers into areas of low lynx densities. Thus, although a fur bearer–fur trapper cycle is possible, it is also possible that the cycles are not even real, but merely artifacts of the fur records. But Finerty (1979) defends the original premise of MacLulich, and Elton and Nicholson, and feels that the awkward fact of occasional, clockwise plots of prey versus predator are statistical artifacts. Such misleading noise may arise from the fact that cycles are *not* perfectly synchronized over all of Canada and that the data used by Gilpin were not area-consistent. He cautions that the classical model should not yet be abandoned. The plot thickens with the observation that some Canadian fish populations cycle very nicely in unison with the lynx population (Rowen, 1950).

To the north and along parts of Hudson Bay to the east, the boreal forest gives way to tundra, and the 10-year cycle is replaced by a 3- to 4-year cycle (Keith, 1963). Again, the correlation of predator and prey populations, the former lagging the latter, suggest straightforward prey–predator cycles. It is tempting to think that the shift in period is a function of the different species involved, but that would be simplistic; in some cases the species are the same. It is difficult to see how climatic differences, or differences in vegetative dominants, might change the internal dynamics of prey–predator interactions. Why we find a within-biome consistency and an among-biome difference is, at this point, obscure. Let us examine one, closely studied system.

In northern Alaska, the brown lemming (*Lemmus trimucronatus*) undergoes regular 4-year cycles. Lagging by a year are the numbers of owls and jaegers, both lemming predators. Close examination of the dynamics reveals the following details (Pitelka, 1957). Consider a summer of low lemming abundance. High populations of nesting owls and jaegers impose heavy predation, and a sparse grass crop provides little food and little cover. By winter, effects of the low prey density on predator nesting success is apparent, predation declines, and the lemmings escape further decimation. By the following summer lemming numbers are slightly up, predators have not yet had a chance to reproduce, and the grass supply, spared ravages by the still sparse rodents, is recovering. Because lemmings are not yet common, raptor reproduction is mediocre and the next winter finds a continuing increase in both rodents and grass. By the second summer after our starting point, lemmings are abundant and devouring large quantities of grass, and owls and jaegers are having a very successful nesting season. The result is a winter of less grass food, less cover, and heavy predation. The lemmings begin to decline, and continue to do so through the next year, another year of raptor nesting success and decimation of lemming food and cover. The fourth summer finds us back at the initial situation of low lemming numbers, high predator numbers, and minimal grass cover.

The alert reader will have noticed that a third trophic level, grass, has been inserted into the prey–predator system. What is the effect of this complication? Pitelka feels that the important prey–predator interaction, that which drives the cycle, is the grass–lemming relationship. In fact, it appears that the top predator, the owls and jaegers, actually serve to damp the oscillations. Perhaps in the boreal forest, too, it is the plant–herbivore interactions that drive the cycles (see Pease et al., 1979).

Complications in the Prey–Predator Interaction

Although the lemming fluctuations serve as an example of prey–predator cycling, they are perhaps better as an illustration of the fact that models of two-species interactions do not necessarily provide an adequate picture of most population systems. How can we explore the dynamics of three-trophic-level systems? Also, what of systems with more than one predator species or alternate prey species? Hassell (1978) argues that prey species can be lumped together if only total prey numbers are of interest. But predicting the dynamics of all involved species, individually, is far more difficult. Of course, the general approach of using functional response curves to modify Lotka–Volterra equations can be used to construct models of such systems. The only complication in building the appropriate equations is in deriving the necessary functional response curves for situations where several prey species occur. Using the same symbols from (4-5), write

$$m_i = a_i n_i (T - \sum_j T_{Hj} m_j) \tag{4-16}$$

In matrix form,

$$\mathbf{m} = T \cdot \mathbf{A} \cdot \mathbf{N} \cdot \mathbf{1} - \mathbf{A} \cdot \mathbf{N} \cdot \mathbf{H} \cdot \mathbf{m}$$

where

$$\mathbf{A} = \begin{bmatrix} a_1 & 0 \\ 0 & a_2 \end{bmatrix} \qquad \mathbf{N} = \begin{bmatrix} n_1 & 0 \\ 0 & n_2 \end{bmatrix}$$

$$\mathbf{H} = \begin{bmatrix} T_{H1} & T_{H2} \\ T_{H1} & T_{H2} \end{bmatrix} \qquad \mathbf{m} = \begin{bmatrix} m_1 \\ m_2 \end{bmatrix} \qquad \mathbf{1} = \begin{bmatrix} 1 \\ 1 \end{bmatrix}$$

whence

$$\mathbf{m} = T \cdot (I + \mathbf{ANH})^{-1} \mathbf{AN1} \tag{4-17}$$

For (example) a two-prey system,

$$m_1 = \frac{a_1 n_1 (1 + a_2 n_2 T_{H2}) - a_2^2 n_2^2 T_{H2}}{(1 + a_1 n_1 T_{H1})(1 + a_2 n_2 T_{H2}) - a_1 a_2 n_1 n_2 T_{H1} T_{H2}}$$

$$m_2 = \frac{a_2 n_2 (1 + a_1 n_1 T_{H1}) - a_1^2 n_1^2 T_{H1}}{(1 + a_1 n_1 T_{H1})(1 + a_2 n_2 T_{H2}) - a_1 a_2 n_1 n_2 T_{H1} T_{H2}} \tag{4-18}$$

[Murdoch (1973) uses a simpler form which is an approximation of (4-18).]

It is probably unnecessary to point out that the resulting equations are difficult to deal with. Indeed, when various time lags and other considerations are included, the difficulties in dealing with the equations and the inherent inaccuracies make some analyses less than worthwhile. Multidimensional isocline graphs are impossibly cumbersome. For the moment, most population dynamics questions of multispecies systems can be answered only by gathering data specific to the system of interest. For example, Pitelka (1957) can state with some conviction that top predators stabilize a cycle driven by a grass–lemming interaction, and Pearson (1966) can assert that the presence of prey other than to *Microtus* acts to destabilize the cycles observed. The sorts of generalities arising from the simpler prey–predator models can be applied to multispecies systems only with extreme caution and skepticism. Questions of multispecies systems whose answers are generally reliable must be those whose answers do not depend on the precise form of the dynamics equations. Such questions are dealt with in Chapter 6.

Age Structure

In most discussions of prey–predator interactions, age structure is conveniently ignored. Considering the complications involved, this is hardly astounding. Pennycuick et al. (1968) published results of computer simulations based on Leslie matrices whose elements were functions of the densities of the species and age groups involved—the general result being oscillations. When various time lags were introduced, they still found oscillations. Smith and Mead (1974) constructed models of the interaction of a prey (two age groups) and a parasite (one age group) and found, via computer simulation, that age structure need not, but might, delay time to extinction. Beddington and Free (1976) studied a model of a predator with two age groups and a prey with one, finding that age structure could either stabilize or destabilize the interaction. If equilibrium prey density were held not too far below the value that would obtain in the absence of predation ($n_1 > .2K_1$), the system was almost always stable. That is, stability occurs if the predator is not too efficient at depressing the prey population (see earlier discussion). A useful addition to our knowledge arising from this study is that the occurrence of a very inefficient *younger* age class of predator introduces a kind of time lag which destabilizes the system.

A few generalities, probably true if intuition is any guide, are first, that age structure will introduce irregularities into the regular oscillations of single-aged prey–predator systems, and second, that as long as prey–predator cycles occur, neither prey nor predator can reach stable age distribution. Furthermore, changes in age structure seem destined to change the characteristics of both populations over time and thus alter the nature of the species-interactive cycles. Thus one process will drive the other, which, in turn, affects the first. The two kinds of cycles are also apt to be of different period, giving rise to "beats" in the population pattern (Oster and Takahashi, 1974).

4.2 Changing Environments

In Chapters 2 and 3 we noted that all populations have a finite probability of becoming extinct through the operation of chance events. Small populations disappear, on average, more quickly than larger ones, for whom the impact of such chance events may be all but negligible. If a predator drives its prey to very low levels, it increases the likelihood that it will be left with nothing to eat and so die off itself. Thus we need not project a movement across an axis of the isocline graphs to predict extinction; the systems are not as stable as those diagrams suggest. In fact, even prey–predator systems which should be highly static under constant conditions may cycle endlessly because of the perturbations produced by changing prey and predator personnel or random alterations in the number of offspring left or the number dying in any given time interval. Small local populations are more influenced by such events and so should exhibit greater tendencies to cycle.

Changes in the External Environment

Before submerging ourselves in models, some general observations are in order. In Chapter 2 it was demonstrated that density-independent changes could be expected to drop $E(n)$ below the carrying capacity. Since $E(n_1)$, the expected value of the prey, defines the expected carrying capacity of the predator, and since $E(n_2)/E(K_2)$ also drops with variability in K_2, it follows that $E(n_2)$ is proportionately diminished even more strongly than $E(n_1)$. The effect is exaggerated if the predator also is affected by density-independent variation. So the ratio $E(n_2)/E(n_1)$ will be highest in steady environments, lowest in erratic environments. That is, predation pressure (efficiency of predator) will be lowest where environments are the most changeable. It follows from our earlier discussion, then, that prey–predator cycles are less likely and, if they occur, of lower amplitude in fluctuating environments. Also, where environmental fluctuations are fairly regular, interactions between them and the inherent cycling of the prey–predator system will interact to promote oscillations of longer or shorter periodicity than expected on the basis of either driving force alone. Oster and Takahashi (1974), using mathematics beyond the scope of this book, have, in fact, demonstrated this latter prediction.

Maynard Smith and Slatkin (1973) attempted to model a prey–predator system in which the predator exhibits two age classes (young and reproductive adults), to determine stability conditions in a periodic environment. In accord with the study of Beddington and Free (1976) discussed in "Age Structure" (p. 122), they determined that persistence was possible when prey density was not forced substantially below what it would have been without presence of the predator (i.e., when predation is inefficient). They also found that differences in hunting abilities between young and old predators might promote coexistence (in partial contradiction to the conclusions of Beddington and Free), and that prey refugia buffered prey from extinction but did *not* make coexistence more likely. For methods used, the reader is referred to the original article.

Stability and Chaos

The neighborhood stability of a prey–predator system whose dynamic equations are known can be assessed in the usual way—by figuratively perturbing n_1 and n_2 from their equilibria by minute amounts and asking whether the perturbations damp or grow in magnitude over time (see Chapter 2 and equations 4-3 and 4-4). This technique can be useful for making qualitative comparisons of the stability of different prey–predator models (May et al., 1981). However, since we do *not* really know the exact form of the equations in question, the results will be somewhat ambiguous. The algebraic busy work simply does not justify the end. The conclusions reachable from an examination of isocline graphs is similarly nonquantitative, but probably no less useful.

Global stability, as stated in Chapter 2, is often difficult to prove, and impossible to disprove. To my knowledge no one has found a Liapunov function for the prey–predator system described in (4-9). However, Goh (1977) and Case and Casten (1979) have provided such a function for the Lotka–Volterra equations with self-limiting terms (equations 4-2). It is

$$V = \sum_{i=1}^{2} C_i \left[n_i - \hat{n}_i - \hat{n}_i \ln\left(\frac{n_i}{\hat{n}_i}\right) \right] \tag{4-19}$$

where $C_1 > 0$ and $C_2 > 0$. To prove that this is a Liapunov function, we must show that V takes its minimum value when $n_i = \hat{n}_i$ for both species, and that $dV/dt < 0$. First, differentiate V with respect to n_i:

$$\frac{\partial V}{\partial n_i} = C_i - C_i \hat{n}_i \frac{\hat{n}_i}{n_i}\left(\frac{1}{\hat{n}_i}\right) = C_i\left(1 - \frac{\hat{n}_i}{n_i}\right) \qquad \text{for } i = 1, 2$$

Setting this equal to zero, we see that V has an extreme point at $n_i = \hat{n}_i$. The second derivative, evaluated at $n_i = \hat{n}_i$, is

$$\frac{C_i\hat{n}_i}{\hat{n}_i^2} = \frac{C_i}{\hat{n}_i}$$

which is positive since C_i is positive. Thus $n_i = \hat{n}_i$ defines $V_{\min}$. Next,

$$\frac{dV}{dt} = \sum_{i=1}^{2} C_i\left(\frac{dn_i}{dt} - \frac{\hat{n}_i}{n_i}\frac{dn_i}{dt}\right) = \sum_{i=1}^{2} C_i\frac{dn_i}{dt}\left(1 - \frac{\hat{n}_i}{n_i}\right)$$

But, by definition, dn_i/dt is positive when $n_i < \hat{n}_i$, and vice versa. Thus $C_i(dn_i/dt)(1 - \hat{n}_i/n_i)$ is negative for $i = 1, 2$, and, unless $n_i = \hat{n}_i$, dV/dt is negative. Unfortunately, it is precisely when populations are far from their equilibrium, and thus prone to blow up or die off, that simplified models are most apt to be in error. Hence it is not at all clear that the existence of this Liapunov function tells us very much (but see Case and Casten, 1979, and the discussion of stability in Chapter 6).

A more general and perhaps more useful approach to global stability has been taken by Hastings (1978). Write, for any two interacting species,

$$\frac{dn_1}{dt} = n_1 f(n_1, n_2) \qquad \frac{dn_2}{dt} = n_2 g(n_1, n_2) \tag{4-20}$$

Drawing from mathematical theorems beyond the scope of this book, Hastings proves that the species pair above is globally stable if:

1. There exists a unique, biologically feasible equilibrium.
2. This equilibrium is neighborhood stable.
3. Both species sustain density-dependent mortality at all densities—that is,

$$\frac{\partial f}{\partial n_1} < 0 \quad \text{for all } n_1 > 0 \qquad \frac{\partial g}{\partial n_2} < 0 \quad \text{for all } n_2 > 0$$

4. There exist positive numbers A and B such that

$$\text{for any } n_2 > B, \quad f(C, n_2) < 0 \qquad \text{for some } C > 0$$
$$\text{for any } n_1 > A, \quad g(n_1, D) < 0 \qquad \text{for some } D > 0$$

This last condition may be restated as follows. Where the symbol "⇒" designates implication,

$$(n_2 > B) \Rightarrow \frac{dn_1}{dt} < 0 \qquad (n_1 > A) \Rightarrow \frac{dn_2}{dt} < 0$$

Equivalently,

$$\frac{dn_1}{dt} \geqslant 0 \Rightarrow (n_2 < B) \qquad \frac{dn_2}{dt} \geqslant 0 \Rightarrow (n_1 < A)$$

The first part of condition 4 can now be understood as follows. Suppose that the system is perturbed from equilibrium such that the prey population (n_1) begins to increase. Then $dn_1/dt > 0$, so that n_2, the predator population, must be less than B. But as long as $n_2 < B$, growth in n_1 is limited only by condition 3. Suppose that n_1 rises above A. But then $n_1 > A$, so that $dn_2/dt < 0$. This violates the basic nature of the prey–predator interaction. Thus the first part of condition 4, to be consistent with the nature of the interaction, must be translated: "The equilibrium value of the prey population is $\hat{n}_1 \leqslant A$. In particular, in the limit as $n_2 \to 0$, $\hat{n}_1 = A$." With respect to the latter part of condition 4, suppose that a perturbation promotes an increase in the predator population, n_2. Then $dn_2/dt > 0$, so that n_1 must lie below A. If n_2 rises above B, then n_1 begins to fall. In this case there is nothing inconsistent between this implication and the nature of the prey–predator system. Indeed, n_1 will continue to fall to extinction, followed by the crash of the predator. But clearly this violates condition 2, which requires neighborhood stability. Indeed, the point (n_1, B) is unstable because a slight disturbance might send n_2 *above* B. Therefore, the latter part of condition 4 must be translated, "$n_2 < B$. This is equivalent to requiring that density feedback be sufficiently strong—that is, $\partial g/\partial n_2 \leqslant \alpha < 0$, where α is some critical value that depends on the nature of the functions f and g."

The discussion above can be linked to the earlier discussion of prey–predator isocline graphs. Condition 1 states that the isoclines must cross in the first quadrant $(n_1, n_2 > 0)$. Condition 2 requires that the predator isocline cross the prey isocline at a sufficiently small angle. Condition 3 says that the prey curve must curve downward to the right and cross the n_1 axis (at some value A), and that the predator curve must also rise at a decreasing rate. Finally, condition 4 assures that the decrease in the slope of the predator curve occurs at a sufficiently rapid rate.

Populations of individuals that experience density feedback once, over a short period in their lives, and which breed synchronously at discrete intervals, are properly described with discrete-time equations (the occurrence of such creatures was discussed briefly in Chapter 2). We have already noted that species like this, if they possess high reproductive potential, may fluctuate chaotically. Will this also occur if the species interact as prey and predator? Systems appropriately modeled in this way may be found among seasonally breeding insects and their parasitoids. Beddington et al. (1975) have, in fact, examined these systems, using the difference equations of Nicholson and Bailey (1935), and adding a linear density-feedback term for the host. As intrinsic rate of increase for the host rises, the system first begins behaving erratically—the spiral of Figure 4-1 grows very rough, chaotic edges, although it remains closed. With still further increase in this parameter, violent density overcompensation leads to extinction.

4.3 Spatially Structured Populations

It is not a profound intellectual leap to conclude, from the discussions of Sections 2.3 and 3.3, that division of a prey–predator system into subsystems, loosely connected in space, promotes stability. We should expect lower rates of extinction and lower-amplitude fluctuations for connected, local systems than for isolated systems, and an even more pronounced effect on the collective assemblage of subsystems. This conjecture has been investigated directly, using several approaches.

Hillborn (1979) constructed hypothetical, multiple subsystems, using the model presented in (4-9) (with $n = 1$—type II function response curves):

$$\frac{dn_1}{dt} = g_1(n_1, n_2)$$

$$\frac{dn_2}{dt} = g_2(n_1, n_2)$$

and parameter values that he considered biologically reasonable. What he found, using computer simulations, was large oscillations within the subsystems. But when connections between the subsystems, in the form of immigration and emigration, were permitted, interesting results emerged. Hillborn wrote, for the ith member of a linear array of subsystems, $i = 1, 2, \ldots, 100$,

$$\frac{dn_{1i}}{dt} = g_1(n_{1i}, n_{2i}) + (I_1 n_{1,i+1} + I_1 n_{1,i-1}) - 2E_1 n_{1i}$$

$$\frac{dn_{2i}}{dt} = g_2(n_{1i}, n_{2i}) + (I_2 n_{2,i+1} + I_2 n_{2,i-1}) - 2E_2 n_{2i}$$

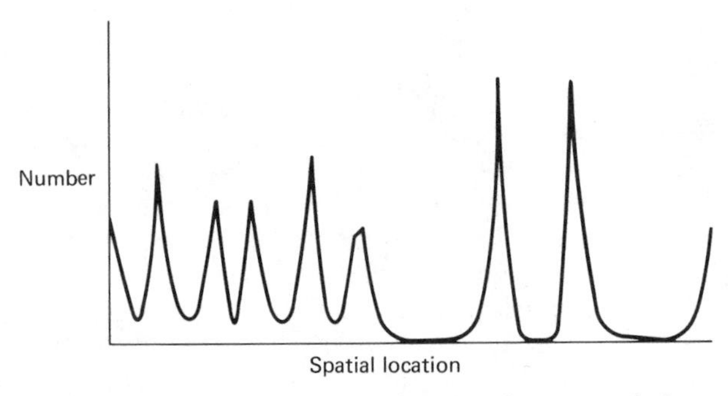

Figure 4-22 Dispersion pattern in a predator population occupying a linear habitat. (After Hillborn, 1979, Fig. 4.)

He initialized the system with $n_{1i} = 1.0$ for all subsystems i; and $n_2 = 1.0$ for one subsystem, zero for all others; that is, prey numbers were initially equal in all subsystems, predators were introduced into only one. Over time, the predator population propagated over the array of subsystems, in both directions, like a wave until, by generation 1000, it was spread, with random irregularities, over all 100 (see Fig. 4-22). Over the entire system the predator population still oscillated, but with markedly reduced amplitude. Local subsystems still occasionally crashed (extinction was assumed to occur if n_{1i} or n_{2i} fell below .001), and population distribution of both species was patchy over space. Hillborn did not introduce environmental variation into his model, but there is no reason to believe that its inclusion would alter the general conclusions.

One important point to emerge from Hillborn's paper is the following. It is possible to find values for b_1, γ_1, δ, and so on, which describe the observed behavior of the whole system according to (4-9). But these parameters have explicit ecological meanings and their true values, which determine the *local* dynamics, bear little or no numerical relation to the apparent values, determined as above. Thus, in nature, unless the size of the patch over which data are gathered corresponds to the extent of a single subsystem, analyses of the nature of the species interaction could be very misleading.

Stability of spatially structured prey–predator systems, in the sense of persistence, has also been explored by Hillborn (1975). Subsystems were pictured as occurring in patches spread over space, in two dimensions. Dispersal was supposed proportional to donor population density, as before, and random in direction, to any other patch. The underlying equations were of the general Lotka–Volterra type with a self-damping term for the prey. The results of a computer

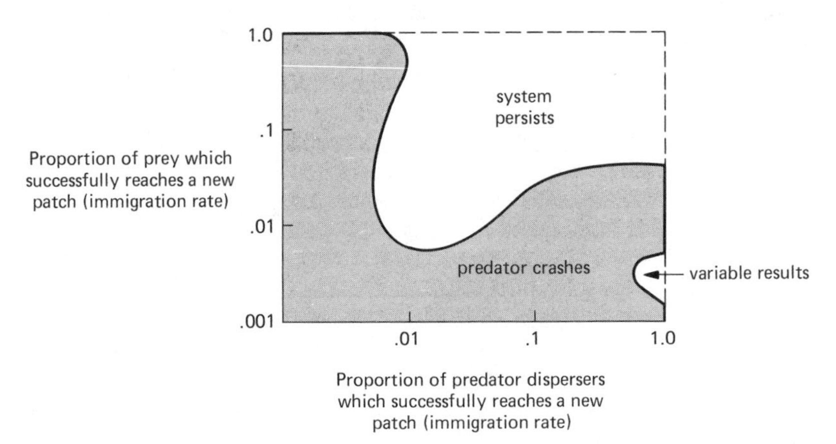

Figure 4-23 Relationship between immigration rate of prey, immigration rate of predators, and prey–predator system stability. (After Hillborn, 1975.)

simulation are shown in Figure 4-23. A qualitative interpretation is fairly straightforward. For a given rate of prey immigration, if the predator is too much slower, it will not be able to find sufficient food and will die out. If it disperses too successfully, the prey will not be able to escape to new areas fast enough and will be destroyed.

Crowley (1981) applied the technique of isocline graphs to prey–predator dynamics with dispersal. He confirmed the prediction that increasing migration rates of both prey and predator at first stabilize the system (in the sense of persistence), then destabilize it. He was also able to show that (a) an increase in the number of semi-isolated subpopulations ("cells") stabilizes the system so long as migration is not so high as to permit synchronization among them; (b) as the system gets very large, the various cells cluster into hypercells with a sufficiently low exchange rate that synchrony is avoided and stability enhanced even with quite high dispersal rates; (c) environmental fluctuations, if on a time scale less than that of the population system, creates asynchrony and so enhances stability; and (d) density-dependent migration is not necessary to produce stability.

Similar results have been obtained by workers using diffusion models (e.g., Steele, 1974; Levin and Segal, 1976; Hastings, 1977; Mimura and Murray, 1978). The mathematics involved are beyond the scope of this book. For further discussion, see Hassell and May (1973), Comins and Blatt (1974), and Levin (1977).

One general prediction has been made: Unless predator dispersal is too slow or extremely rapid, spatial structuring stabilizes prey–predator systems. In addition, the wavelike spread found in the simulation model of Hillborn (1979) may be of significance. Perhaps the rise in numbers of a prey and then its predator will tend to spread outward from a central focal point, like ripples on a water surface. In Hillborn's model the waves quickly damped. But if local systems were sufficiently unstable or if dispersal of both prey and predator were sufficiently low—perhaps due to great distances—might such waves be persistent? Might the boundaries of the involved species' ranges, at least if well defined, act as reflective barriers, and cause standing population waves? Let us examine the evidence.

The first prediction has been supported by observations both in the laboratory and in the field. Huffaker (1958) studied the interactions of a predaceous mite, *Typhlodromus occidentalis*, and its mite prey, *Eotetranychus sexmaculatus*. The latter infests oranges, so Huffaker set up an artificial ecosystem consisting of oranges. A series of experiments were run, each one involving more environmental heterogeneity than the last. Heterogeneity was introduced by increasing the number of oranges (but concurrently, partially covering them with wax to avoid undue enrichment of the system), and by interspersing them with rubber balls. However, the system immediately crashed. The predators rose in number, easily dispersed among the oranges and decimated the prey and then, themselves, starved. When the environment was further complicated by using more than one tray of oranges, each connected to another by a small bridge, and each consisting of many oranges, many rubber balls, and a system of Vaseline barriers, the prey rose, fell, and began to rise again before crashing. Finally, with the introduction of scattered wooden pegs and a fan blowing gently over the entire array, Huffaker succeeded in maintaining three complete oscillations before prey extinction. The impact of this final bit of complexity was that the prey species could mount the pegs, spin threads, and be blown over the Vaseline barriers, thus keeping a step ahead of the prey. Until this final step was added, the predators were able to disperse with sufficient speed relative to the prey that no escape was possible.

A classic field example is that of the prickly pear cactus, *Opuntia* spp., and its "predator," the *Cactoblastis cactorum* moth. The cactus, which is native to the Americas, was widely introduced and had become a pest in many parts of the world in the nineteenth century. By 1880 it had become a serious problem in Australia. By 1900 it covered an estimated 10 million acres, by 1920 about 60 million acres. In many areas, immense patches of cacti, 3 to 6 ft high, grew so densely that it was impossible to walk through them. Control came in the form of the *Cactoblastis* moth, imported from Argentina. This moth lays its eggs on cactus pads, which are tunneled through and eaten by the larvae. In addition to these direct effects, the insect's activities expose the plant to bacterial and fungal infection. So critical is the damage and so high is the moth's reproductive and dispersal abilities that by 1930 the cacti had almost disappeared from Australia. This drop in the prey population was, of course, followed by a decline in the moth population, and a partial recovery of the cactus. The present pattern is one of extreme local instability—a rise in cactus density followed by discovery and decimation by the moths. But over the continent as a whole, cactus and insect numbers are fairly constant. The plant manages to disperse, via scattering of seeds or small pads stuck to animals, to a new area and to grow before being rediscovered. By dispersing slightly more rapidly than their predator, the prey avoid global extinction (Dodd, 1940, 1959). This simple game of hide and seek, however, is less than the whole story. If the moth spread its eggs over all cactus plants in a patch, the speed with which it is capable of devouring cactus might well spell doom for the patch so rapidly that dispersal would be prevented. If so, the moth would quickly eliminate itself and become useless as a means of cactus control. In fact, *Cactoblastis* lays clusters of eggs on only a few plants in a patch, lending stability to the system by being an inefficient predator (Monro, 1967).

It has been suggested that spatial structure of populations might lead to prey–predator population waves. Do such waves occur? The answer is yes, although the waves are not necessarily sustained. In 1943 there was a large spruce budworm outbreak over an area of approximately 1000 square miles of northern Ontario. The spruce budworm, a parasite on trees, causes heavy defoliation, often leading to death of the host. Belyea (1952) describes the outward, wavelike spread of tree damage and recovery.

Population waves have been thoroughly discussed by Watt (1968), who calls them *epizootic waves*. Watt concentrated his attention on the boreal forest 10-year cycle, which he points out is *not* perfectly synchronized over all of Canada. In the nineteenth century the lynx population appeared to peak first in the Lake Athabasca region and then spread outward. Since the 1920s the epicenter has shifted to northern Manitoba or Saskatchewan (Keith, 1963). Hares, foxes, muskrats, and mink seem to increase first (except in the Maritime provinces) in the northern sections of the prairie provinces (Keith, 1963). The lynx and hare epicenters do *not* always occur in the same area, but they are generally close, so this fact does not necessarily contradict the theory of epizootic waves. Butler (1953) notes that there often seem to be two or more epicenters, and that the apparent waves grow, coalesce, and continue to spread outward. This observation complicates matters but, again, is not necessarily contradictory to theory. We do know that the dispersal necessary to the existence of epizootic waves exists. Prey species disperse when population densities are high (Chapter 2), and predators tend to follow their prey and to leave areas of low prey density. For example, Cox (1936) reports seeing vast numbers of snowshoe hares in mass migration from population dense areas, and infested with ticks, followed by raptors. Shelford (1945) documents the appearance of snowy owls (*Nyctea scandiaca*) great distances to the south in years of lemming (their major prey) population crashes to the north.

But whether the boreal cycles exhibit epizootic wave patterns or not, it is clear that prey–predator cycles, or at least outbreaks of prey or predators, can occur over wide areas with almost perfect synchrony. For example, in 1957 and 1958 there occurred a tremendous rise in the California vole (*Microtus californicus*) population over much of California, south-central Oregon, and parts of Nevada (Murray, 1965). There was no evidence for an epicenter and a spread. Also, we could argue that predator dispersal is apt to buffer the prey–predator cycles unless it is so rapid and widespread that subsystems merge into a homogeneous whole. Thus where dispersal conditions are such as to promote epizootic waves, there is apt to be little or no cycling to be propagated, and where dispersal is great, cycling but no waves. This view is argued by Galushin (1974), who checked predatory bird-banding records in Europe and the USSR, reporting that those species which disperse great distances (e.g., 1955 $\pm$ 1079 km per year for *Buteo lagopus*) exhibit strong, synchronized population fluctuations, whereas those which disperse only small distances (e.g., 39 $\pm$ 14 km per year for *Falco subbuteo*) show weak fluctuations. Sustained epizootic waves are an intriguing but so far unproven phenomenon.

Review Questions

1. What should the prey and predator isoclines look like if the predator is unaffected by its prey—that is, if it has several alternative prey so that the influence of the prey species in question on its population dynamics is miniscule? Can cycling occur in such cases? If the predator density changes upward or downward (due to some factor other than the prey in question) and restabilizes, what happens to the prey equilibrium? What difference will it make if the predator's functional response is type II? Type III?
2. Draw a series of hypothetical functional response curves and, without using numbers or equations, deduce the shape of the prey and predator isoclines. How might these curves appear for a plant–herbivore system in which light grazing stimulates plant productivity?
3. Consider two species whose adults do not interact interspecifically, but which eat the young of the other. Draw appropriate isoclines (for both type II and III responses) and analyze them. Under what conditions will stable coexistence occur?
4. Under what conditions will dispersal of prey and predators among environmental patches be stabilizing to the prey–predator system? Destabilizing to the system? Why?

Chapter 5

Competitive Interactions

If populations were consistently at or near equilibrium, there would be little point to investigating prey–predator interactions. The fixed depression, or altered age structure of prey density by predators, could never be ascertained except in the artificial atmosphere of a laboratory and there would be no dynamics to understand or predict. If virtually all fluctuations in the populations of prey or predator species could be accounted for by environmental factors other than the prey–predator interactions, the same would be true. What makes the interaction significant and interesting is that a change in one population has a discernible and, to a large extent, predictable effect on the birth and death rates of the other. Other kinds of population interactions are also detectable and lead to qualitatively, predictable effects.

1. Consider two species both of which experience density feedback via their use of a particular resource or combination of resources. For example, species whose equilibrium densities are influenced by food might exhibit strongly overlapping diets. Then whatever amount of the resources is used by one species i is, ipso facto, unavailable to the other, j. It follows that an increase in n_i feeds back negatively not only on r_i, but also on r_j:

$$\frac{\partial r_i}{\partial n_i} < 0 \quad . \quad \frac{\partial r_j}{\partial n_i} < 0$$

The species are said to be in interspecific competition—specifically *resource competition* or *exploitation competition.*

The intensity of competition, the magnitude of $\partial r_j/\partial n_i$ (and $\partial r_i/\partial n_j$), will vary with the amount of resource use overlap between the two species. With no overlap there can be no exploitation competition. Thus two barnacles whose population densities are limited by available space and which show extreme dietary overlap will be competing, exploitatively, for space (unless their microhabitat preferences dictate use of different space), but not for food. If m_k denotes the availability of the kth resource, interspecific competition (exploitative) intensity of species j on species i can be written

$$\sum_k \frac{\partial r_i}{\partial m_k}\frac{\partial m_k}{\partial n_j}$$

If either r_i is unaffected by m_k (the kth resource is unimportant to population growth of species i), or species j has no impact on that resource, the contribution of the resource to interspecific exploitation competition is zero.

2. If two species interfere with each other's *access* to a resource, surely $\partial r_i/\partial n_j < 0$. The species are clearly in competition, but not exploitation competition. Suppose that two species are, at least partially, predator limited. If one increases in density, predators may respond by population growth or local migration, thereby increasing their impact on the death rate of the other. Again, this time via an intermediary organism, the two species are in competition, but not exploitation competition. Any negative effect of one species on the rate of growth of another, arising from any cause *other than* the using up of shared resources, is called *interference competition.* The intensity of interspecific competition, both exploitation and interference, can still be written $\partial r_i/\partial n_j$.

In this chapter, interference resulting via interactions with third species will be ignored. Such complications are better left to Chapter 6, which deals with multispecies interactions. For our present purposes, we consider exploitation competition and direct interference arising through aggression or habitat disturbance.

3. Species may interact in a mutually beneficial or "facilitative" manner. This has long been recognized to occur among certain, distantly related taxa. But among closely related organisms, its possible importance has been acknowledged only recently. In light of the importance of intermediary species in many facilitative interactions, and in deference to the importance of competition as (historically) a major topic in itself, further discussion of facilitation will be postponed to Chapter 6.

The early experiments of Gause (1932, 1934) indicate that two ecologically similar species are unlikely to coexist indefinitely, at least in the laboratory. Among his competition studies Gause (1932) pitted two species of yeast, *Saccharomyces cervisiae* and *Schizosaccharomyces kephir*, against one another in an anaerobic medium of brewer's yeast extract, sugar, and water. Both produced ethyl alcohol as a waste product, and both were inhibited in growth by the alcohol. *Saccharomyces* was least susceptible to the toxin and so "won" the (interference) competition, driving *Schizosaccharomyces* to extinction. Park (1954) and his co-workers (e.g., Neyman et al., 1956) experimented with the flour beetles *Tribolium castaneum* and *T. confusum.* The animals were kept in flour-filled vials. Contents were sifted each month, beetle eggs, larvae, pupae, and adults were counted, and new flour was provided. At 34, 29 and 24°C, and at both 70% and 30% relative humidity, either species alone could persist indefinitely (except for *T. castaneum* at 24°C, 30% humidity). But when raised together, one or the other species always won, excluding the other. *T. confusum*, a more tropical species, won consistently at the highest temperature and humidity, and most of the time at 29°C, 70% humidity. *T. castaneum* did better at lower temperatures and lower humidities. The mechanism of competition among flour beetles is complex, involving a negative impact of population density on egg-laying rate and rate of larval development, "conditioning" of the flour, and cannibalism, both intra- and interspecific. The nature of the competition seems to be interference (see also the experiments of Birch 1953b). More recently, the *Tribolium* system has been elegantly elaborated in both experiments and theory; the interested reader is referred to Ghent (1966) and Mertz (1972). When one species excludes another by virtue of competitive superiority, the result is known as *competitive exclusion.*

But is there evidence for competition in the field? In particular, does competitive exclusion occur? In the Sierra Nevada, four chipmunks, *Eutamias alpinus*, *E. speciosus*, *E. amoenus*, and *E. minimus*, occupy mostly nonoverlapping, altitudinal zones. Here the aggressive *E. alpinus* and *E. amoenus*, in their preferred haunts, exclude the other two species. The losers, whose altitudinal ranges are considerably greater in areas of allopatry (where they occur *without* the other species) settle for what's left (Heller, 1971). Work by Sheppard (1971) and Meredith (1977) shows similar patterns for *E. amoenus* and *E. minimum* in Alberta. In all of the cases above there is clear evidence of interference competition.

DeLong (1966) compared the population ecology of *Mus musculus* (house mouse) populations in open field grids with or without *Microtus californicus.* Extra food was supplied so as to head off exploitation competition (at least for that particular resource). There were no differences in fecundity, body growth, percent of females lactating, or adult mortality in the two types of grid, but recruitment of young *Mus* into the population was greatly reduced in the presence of *Microtus.* Furthermore, over two years, the number of juveniles per lactating female dropped off with a rise in *Microtus* population density. The implication, that *Microtus* destroy young *Mus*, was borne out by laboratory studies. This is clearly a case, albeit somewhat artificial, of interference competition.

Of course, it is not only rodents that compete. In Big Bend National Park, the lizard species *Sceloporus merriami* and *Urosaurus ornatus* are sympatric (live together in the same area). Dunham (1980) marked individuals in six plots of similar size and vegetation profile, isolated by areas of unsuitable habitat, and censused the populations at intervals for 4 years. In two of those years dry weather and sparse insect (prey) populations led to lower rates of food capture, loss of weight, and lower prehibernation lipid levels in the lizard population as a whole. However, in plots where *S. merriami* had been removed, *U. ornatus* actually showed an *increase* in these measures over the same two years. In plots where *U. ornatus* had been removed there were no such dramatic effects, although in one of the years, *S. merriami* showed higher survival probabilities. In the wet years, removal of one species had no apparent effect on the other. It seems, then, that the interspecific interaction is more severe when food is scarce (dry years), implicating exploitation competition as either the agent of the conflict or as a motivating force behind (an unobserved) aggressive interference.

Haven (1973), working in the United States West Coast intertidal, erected small exclosures containing individuals of either one, both, or neither species of the limpets *Acmaea scabra* and *A. digitalis*. He found that if one species were absent, individuals of the remaining species grew more rapidly, and that an algal film began to build up. It appears as though these algal grazers compete for food, and that one species alone, until its population responds to the release of competition and increases, cannot keep ahead of the algal growth. It is a reasonable guess that faster individual growth rates mean larger individuals, on average, and a larger birthrate. Thus population growth of one species is expected to increase in the absence of the other.

Interference competition was implicated by studies of Connell (1961a,b) for the two barnacle species, *Balanus balanoides* and *Chthamalus stellatus*. After only one month in predator-excluding experimental enclosures mixed populations were dramatically shifting toward pure *Balanus* composition as individuals of this species overgrew and suffocated *Chthamalus* individuals or undercut them, flipping them off the substrate.

Among plants it is an easy matter to demonstrate density feedback due to shortage of certain soil nutrients. That is what the fertilizer industry is all about. All else being equal, plants that utilize available minerals more efficiently than others are competitively superior. Plants that can grow faster and so spread leaves and shade over their neighbors will win in the struggle to obtain sunlight for photosynthesis.

Plants competing for sunlight, for nutrients, and especially for water (in arid regions) are really in exploitation competition for air and soil space. But interference competition also occurs. If a plant can obtain exclusive rights to space by excluding or discouraging growth in its vicinity, it will protect its resource supply. Thus the existence of *allelopathy*, the inhibition of germination or growth by chemical agents, is adaptive. In forests, the area to be "defended" may be quite large, making it likely that the toxin will be found in and ultimately leached from falling leaves. This is true in, for example, walnuts (*Juglans nigra*) and various species of *Eucalyptus*. The barrenness of *Eucalyptus* forest understory is familiar to many people, and there will be few midwestern farmers who have not noticed that planting certain crops near walnut trees can be a very poor idea. In deserts, the advantage of exclusive space for gathering water is obvious and, indeed, allelopathy is particularly common in such areas. Clumps of sage (*Salvia leucophylla*), for example, are generally surrounded by vegetation bare zones about 1 m wide. The proximate causes of allelopathy are beyond the scope of this book. But a series of fascinating stories awaits the interested reader. See, for example, Davis (1928), Muller (1966), Wilson (1968), DelMoral and Cates (1971), and Einhellig and Rasmussen (1973). For examples of allelopathy among animals see Jackson and Buss (1975).

All of the examples above involve phylogenetically or ecologically similar species. But competitive interactions also occur among vastly different species. Dayton (1971) demonstrated the important interference that wave-whipped algal fronds had on the settlement and subsequent survival of intertidal barnacles. Brown and Davidson (1977) and Brown et al. (1979) showed that desert rodents (*Dipodomys merriami*, *Perognathus penicillatus*, and *P. amplus*) and ants (*Pheidole* spp.) fed on seeds of similar size and species and gathered them from similar microhabitats. Clearly, if seeds are a limiting factor, the potential for competition is present. Four sets of replicated circular plots 36 m in diameter were marked in *Larrea tridentata–Ambrosia deltoides* desert scrub in Arizona. In one set, rodents were trapped, removed, and fenced out. The authors estimate 90% success in ridding the area of these animals. In a second set, about 95% of the ants were killed with repeated doses of insecticide. In the third set both groups of animals were removed. The fourth acted as a control. Four years of data showed a

71% increase in density of ant colonies where rodents were missing. In the insect-free areas, rodents increased 20% in number, 29% in biomass (and 17 plots had more individuals, 5 the same or fewer, $p < .01$). In the set of plots with both ants and rodents gone, soil analyses showed a 5.5% increase in number of seeds over the control plots.

In spite of the plethora of evidence for competition, the reader should not forge onward in the belief that competition necessarily occurs between any two species sharing a trophic level or even between any two very ecologically and phylogenetically similar species. The view that competition is a pervasive and important interaction in biological systems has been strongly challenged (see Sections 5.2 and 5.3 and Chapters 6 and 7); it is quite easy to find examples of situations where competition "ought" to occur, but appears to be lacking. Shroder and Rosenzweig (1975) studied the kangaroo rats *Dipodomys merriami* and *D. ordii* on unenclosed trapping grids in New Mexico. Removal of two-thirds of all individuals of one or the other species from experimental grids resulted in virtually no migratory or population response by the other when compared with undisturbed, control grids. Cameron (1977) worked with natural populations of *Sigmodon hispidus* and *Reithrodontomys flavescens*, but did *not* find evidence for competition. Indeed, he did not find a tendency for increased immigration by either species when the other was removed. In fact, *Sigmodon* individuals were larger in the control plots than in plots where *Reithrodontomys* had been removed, and had higher survivorship and larger litters. Perhaps these experimental-control plot differences were due to the fact that the control plots were, by chance, richer in preferred *Sigmodon* habitat—but they are suspiciously close to implicating facilitation.

5.1 Basic Models

The Lotka–Volterra Model

Few people would deny the existence of interspecific competition in nature or its pervading importance in *structuring* communities (but see Chapter 17). This chapter reviews the impact of interspecific competition on population *dynamics*. To assess this impact, we begin, as in Chapters 2, 3, and 4, by constructing simple, mathematical models.

As with prey–predator interactions, the first models of competition were developed by Lotka and Volterra. Consider the logistic equation for some species i:

$$r_i = \frac{1}{n_i}\frac{dn_i}{dt} = r_{i0}\left(1 - \frac{n_i}{K_i}\right) = r_{i0}\left(1 - n_i\frac{1}{K_i}\right) \tag{5-1}$$

A quick glance should suffice to show that the addition of one individual of species i depresses the growth rate by a fraction $1/K_i$. What if we were to add an individual of *another* species, j? It is unlikely that the magnitude of depression in r_i evoked by an individual of species j should be identical to that produced by an individual of species i. Let us suppose, then, that the relative impact on r_i of individuals of species j and species i, is α_{ij}. It follows that the addition of one member of species j will depress r_i by a fraction $(1/K_i)\alpha_{ij}$. The impact of n_j individuals of the competing species is then $(r_{i0}/K_i)\,\alpha_{ij}n_j$, and the expression for r_i must become

$$r_i = \left(r_{i0} - \frac{r_{i0}}{K_i}n_i - \frac{r_{i0}}{K_i}\alpha_{ij}n_j\right) = r_{i0}\left(1 - \frac{n_i + \alpha_{ij}n_j}{K_i}\right) \tag{5-2a}$$

Similarly,

$$r_j = r_{j0}\left(1 - \frac{n_j + \alpha_{ji}n_i}{K_j}\right) \tag{5-2b}$$

Notice that the density feedback on the growth rate of species i by its competitor,

$$\frac{\partial r_i}{\partial n_j} = -\frac{r_{i0}\alpha_{ij}}{K_i}$$

divided by the density feedback arising from conspecifics,

$$\frac{\partial r_i}{\partial n_i} = -\frac{r_{i0}}{K_i}$$

gives us α_{ij}:

$$\frac{\partial r_i/\partial n_j}{\partial r_i/\partial n_i} = \frac{-r_{i0}\alpha_{ij}/K_i}{-r_{i0}/K_i} = \alpha_{ij} \tag{5-3}$$

(see also Gill, 1974). The quantities, α_{ij} and α_{ji}, are referred to as the *competition coefficients.*

The Lotka–Volterra competition equations suffer from all the same ailments as the underlying logistic model. In addition, there is an unspoken assumption that α_{ij} and α_{ji} are invariant with n_i and n_j, a simplification that is almost surely inaccurate. But these equations have a long history and are heuristically useful. We begin with the work of Gause and Witt (1935), who first suggested the isocline-graph approach that we discussed for prey–predator systems in Chapter 4.

At equilibrium, $r_i = 0$ and $r_j = 0$. Solving (5-2a) for n_j gives us

$$n_j = \frac{K_i}{\alpha_{ij}} - \frac{n_i}{\alpha_{ij}} \tag{5-4a}$$

A plot of n_j versus n_i, from (5-4*a*), is shown in Figure 5-1. Because equation (5-4*a*) describes the dynamics of species i, the line in Figure 5-1 gives us the values of n_i and n_j for which $dn_i/dt = 0$, the species i isocline. Note the values of the intercepts. When species j is absent ($n_j = 0$), n_i comes to equilibrium at its carrying capacity, K_i. As species i becomes very rare ($n_i \to 0$), a population of over K_i/α_{ji} individuals of species j is sufficient to halt its growth. Below the isocline there are sufficiently few individuals of the two species combined that species i can grow. Above the line, the combined species' density feedback forces a decline in n_i (note the arrows in Fig. 5-1).

Species j can be similarly treated. When $dn_j/dt = 0$, (5-2b) gives us

$$n_j = K_j - \alpha_{ji}n_i \tag{5-4b}$$

(see Fig. 5-2).

Figures 5-1 and 5-2 can be combined in four biologically reasonable ways (Fig. 5-3), depending on the relative values of the intercepts. The arrows designate, qualitatively, the dynamics of each species. In Figure 5-3a, no matter where the system begins (the initial values of n_i, n_j), the arrows point inexorably to the lower right. Species i "wins" the competition and species j is excluded. In Figure 5-3b, the arrows lead to the upper left. In this case species j

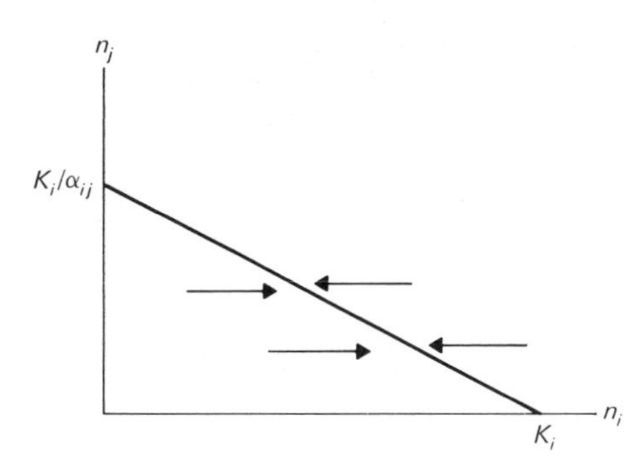

Figure 5-1 Competitor i's isocline graph. The line describes values of the two competitor populations n_1 and n_2 at which n_i neither increases nor decreases.

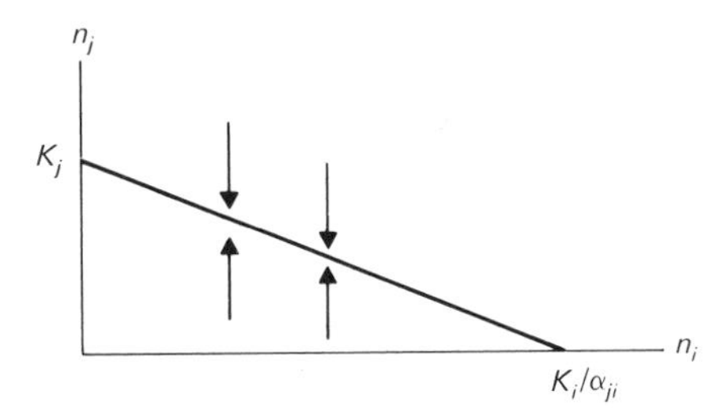

Figure 5-2 Competitor j's isocline graph. The line describes values of the two competitor populations n_1 and n_2 at which n_j neither increases nor decreases.

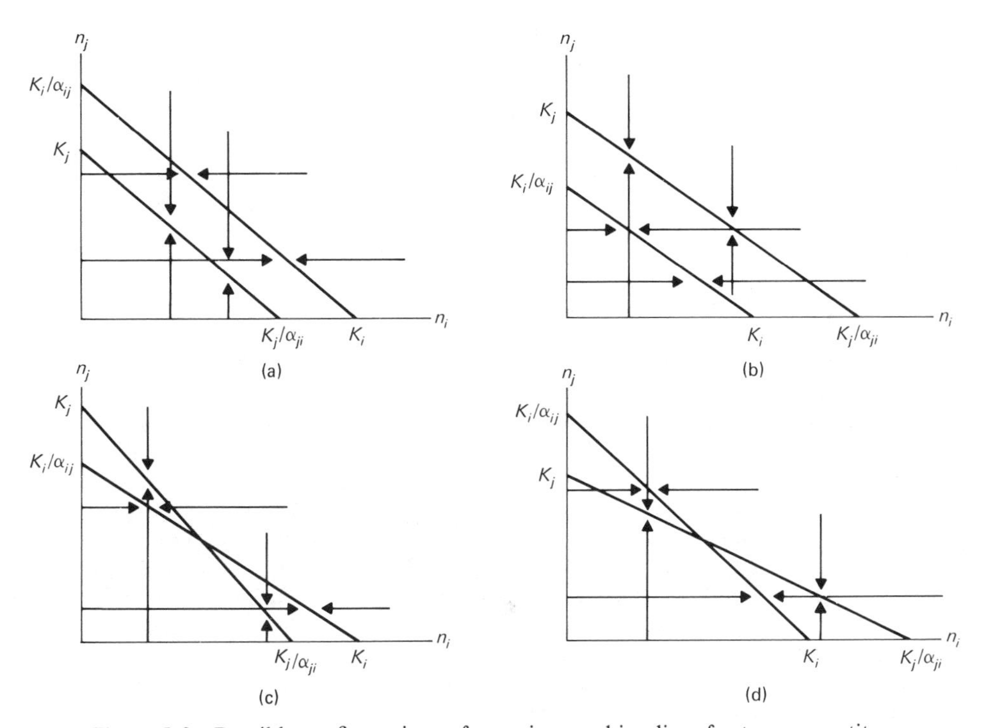

Figure 5-3 Possible configurations of superimposed isoclines for two competitors.

excludes species i. In Figure 5-3c, the system follows the arrows to either the upper left or to the lower right, depending on where it begins. If species i is, initially, at a strong enough numerical advantage, or if its intrinsic rate of increase is large relative to that of j so that the system initially moves strongly down and to the right, it will oust species j, and vice versa. Only if the configuration of isoclines is as in Figure 5-3d can coexistence occur, equilibrium occurring at the node, where the two isoclines cross. That this point represents a stable equilibrium can be seen easily by imagining a displacement. Suppose that the system is perturbed downward from the node. The arrows indicate an increase in both populations. If n_j rises enough faster than n_i that the system rises to the upper left of the node, the arrows eventually shift down and to the right. If the system moves to the right of and below the node, it changes to an upward and leftward direction. If either case it is carried back to the node. A displacement above the node has the same result.

What is unique about the isocline arrangement in Figure 5-3d that permits coexistence of the competitors? Because the four graphs differ only in respect to their intercept relationships, we can write: Coexistence is possible only if $K_i/\alpha_{ij} > K_j$, and $K_j/\alpha_{ji} > K_i$. Inversion of these inequalities yields the equivalent relationships

$$\frac{1}{K_j} > \frac{\alpha_{ij}}{K_i} \qquad \frac{1}{K_i} > \frac{\alpha_{ji}}{K_j} \tag{5-5}$$

The equilibrium solution can be found by setting $r_i = r_j = 0$ in (5-2):

$$\hat{n}_i = \frac{K_i - \alpha_{ij}K_j}{1 - \alpha_{ij}\alpha_{ji}} \tag{5-6}$$

Taken together with (5-5) it is easy to show that a necessary (but not sufficient) condition for coexistence is that $\alpha_{ij}\alpha_{ji} < 1.0$. In the spirit of testing this prediction, Gause (1932) experimented with two yeast species grown in anaerobic culture. He rearranged equations (5-2):

$$\alpha_{ij} = \frac{K_i[1 - (dn_i/dt)/r_{i0}n_i] - n_i}{n_j}$$

measured $n_i, n_j, dn_j/dt$, and dn_i/dt, and calculated α values at various points in time. The average values were 3.15, and 0.439. Clearly, $\alpha_{ij}\alpha_{ji}$ is not less than 1.0 and, indeed, coexistence did not occur.

Returning to (5-2) and (5-5), notice that $1/K_j$ is the proportional drop in r_j (species j) due to the addition of an individual of species j. α_{ij}/K_i is the proportional drop in r_i (species i) due to that extra individual of species j. So the inequality $1/K_j > \alpha_{ij}/K_i$ (equations 5-5) can be interpreted to read: The addition of one individual of species j has a greater (proportional) impact on r_j than on r_i. Equivalently, species j depresses its own population growth rate more than it depresses that of its competitor. The second inequality says the same with respect to the impact of adding an individual of species i. Finally, then, competitive coexistence is possible only if both species exhibit greater density feedback on their own growth rate than on that of their competitor. Under all other circumstances one species becomes extinct. Superficially, if the two species exploit the environment in identical ways (their "niches" are identical), their impact on each other should be the same as their impact on their own population. Thus condition (5-5) cannot be met. Only if their niches are *not* perfectly confluent (they differ somewhat in how they make a living) can they persist in sympatry (together, in the same area). The rule that species with the same niche cannot coexist has become known as *Gause's principle* or the *law of competitive exclusion* (Hardin, 1960).

"Laws" have a way of becoming lost in the labyrinth of ecological complexity, and Gause's principle has proven no exception. Cole (1960), in a reply to Hardin, pointed out myriad extenuating circumstances under which the principle would not hold. Indeed, because it is virtually impossible to determine whether two species have identical niches (presumably with respect to the parameters important to their interaction), or even to say with certainty what to measure when addressing the matter, the principle would seem to be of little use anyway. Nevertheless, a massive array of articles, over many intervening years, has raised and reraised the query of how different two species must be in order to coexist. We shall now proceed to discuss this question in some detail.

First, niche similarities or differences are not enough to describe competition. Two species may exploit the environment in identical ways, but if one converts the available resources into biomass more efficiently or at a different rate than the other, there is no reason to expect α_{ij} to equal α_{ji}. Even more important niche overlap does not take interference into consideration. If exploitation competition is such that interspecific overbalances intraspecific impacts [so that inequalities (5-5) are not met], coexistence may nevertheless occur if *interference* is stronger within than between species. Keep this critical fact in mind while reading the following material in this section.

The competition concept and Gause's principle became solidly engrained in ecology for many years largely on the basis of benign acceptance. Hundreds of papers have been published, their contents dutifully reporting on what differences had been found between species that presumably enabled them to coexist. But there were a few rumblings of discontent. Miller (1964), for example, experimentally estimated $K_i = K_j$, $\alpha_{ij} = \alpha_{ji} = 1.0$, for two laboratory *Drosophila* populations which, nevertheless, stubbornly refused to exhibit competitive exclusion.

The modern phase of doubt probably can be dated to the experiments of Ayala (1969). Note that the conditions (equations 5-5) can be multiplied by $\hat{n}_j, \hat{n}_i$, respectively, and then added to give

$$\frac{\hat{n}_i}{K_i} + \frac{\hat{n}_j}{K_j} > \frac{\alpha_{ji}\hat{n}_i}{K_j} + \frac{\alpha_{ij}\hat{n}_j}{K_i} \tag{5-7}$$

But (equations 5-2), at equilibrium,

$$\frac{\hat{n}_i}{K_i} = 1 - \frac{\alpha_{ij}\hat{n}_j}{K_i} \qquad \frac{\hat{n}_j}{K_j} = 1 - \frac{\alpha_{ji}\hat{n}_j}{K_j}$$

Hence (5-7) can be written

$$\frac{\hat{n}_i}{K_i} + \frac{\hat{n}_j}{K_j} > \left(1 - \frac{\hat{n}_j}{K_j}\right) + \left(1 - \frac{\hat{n}_i}{K_i}\right)$$

or

$$\frac{\hat{n}_i}{K_i} + \frac{\hat{n}_j}{K_j} > 1 \tag{5-8}$$

Ayala, working with *Drosophila serrata* and *D. pseudoobscura* in population cages, measured K_i and K_j in single-species populations, and $\hat{n}_i$ and $\hat{n}_j$ in his persistently coexisting mixed populations. Condition (5-8) was clearly violated. Further experiments showed that *D. serrata* and *D. nebulosa*, too, coexist. In this case (Ayala, 1972),

$$\frac{\hat{n}_i}{K_i} + \frac{\hat{n}_j}{K_j} < 1$$

Ayala (1971) reports on still further experiments with *D. pseudoobscura* and *D. willistoni*. A series of mixed populations were started with different numbers of individuals of the two species (input). After 5 weeks the numbers emerging were counted (output). Some of the results are shown in Table 5-1 and Figure 5-4. As the input ratio rises, the output ratio also rises, but not as fast. Where the graph crosses the 45° replacement line, a stable equilibrium is indicated; and indeed, a stable equilibrium occurs. But coupled with the violation of inequality (5-8), stable

Table 5-1

Species	Input	Output
D. willistoni, strain M11		
Alone	200	491 ± 25 (S.E.)
	500	725 ± 38
	800	728 ± 55
	1000	768 ± 28
In competition with *D. pseudoobscura*	200	287 ± 3
	500	315 ± 19
	800	501 ± 23
D. pseudoobscura		
In competition with *S. willistoni*	800	164 ± 15
	500	183 ± 14
	200	168 ± 1

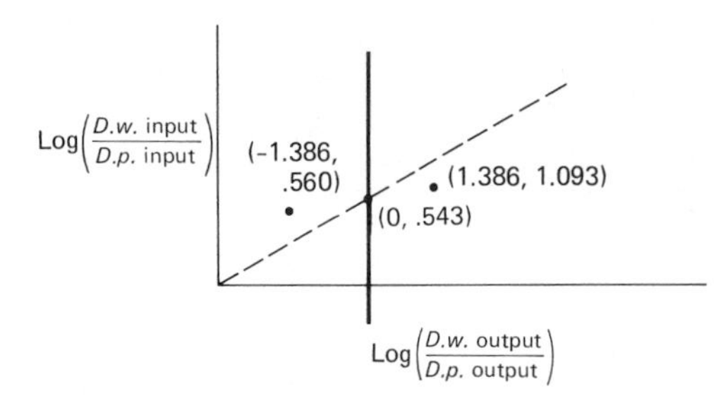

Figure 5-4 Relationship between relative population abundance of two fly species in one generation and that in the previous generation, indicating frequency-dependent competition. (From Ayala, 1971.)

coexistence, supposedly, cannot occur—unless the competition zero isoclines somehow differ from those in Figure 5-3. Ayala suggests that the isoclines are curved, reflecting a (negative) frequency-dependent interaction. Frequency-dependent competition clearly would be stabilizing. A losing species, as it declines, becomes a better and better competitor until (possibly) its disadvantage is reversed. That frequency-dependent competition, or at least density-dependent competition, occurs in nature has been demonstrated by Smith-Gill and Gill (1978), who showed that competition intensity between tadpoles of the frogs *Rana sylvatica* and *R. pipiens* was closely related to population densities of both species.

Other Models

Gilpin and Ayala (1976) have attempted to construct a general yet accurate competition model. After conjuring up a fair array of contestants, they settled on the following:

$$\frac{dn_i}{dt} = r_{i0}n_i\left[1 - \left(\frac{n_i}{K_i}\right)^{\theta_i} - \alpha'_{ij}\left(\frac{n_j}{K_i}\right)\right] \tag{5-9a}$$

where K_i, α'_{ij} are constants. Notice that

$$\alpha_{ij} = \frac{\partial r_i/\partial n_j}{\partial r_i/\partial n_i} = \frac{(-r_{i0}/K_i)\alpha'_{ij}}{(-r_{i0}/K_i)\theta_i(n_i/K_i)^{\theta_i - 1}} = \frac{\alpha'_{ij}}{\theta_i(n_i/K_i)^{\theta_i - 1}} \tag{5-9b}$$

which is clearly density dependent (a function of n_i). Because an increase in n_i generally accompanies a decrease in n_j, α_{ij} is also generally frequency dependent, and if $\theta_i < 1$, an increase in n_i results in a rise in α_{ij} (lowered competitive ability for the increasing species). Gilpin and Ayala used (5-9) to construct isocline graphs from Ayala's input/output data on *Drosophila* and found them to agree well with those predicted by (5-9). Theta values were

$$\theta_{D.\ willistoni} = .35 < 1.00 \qquad \theta_{D.\ pseudoobscura} = .12 < 1.00$$

The inequalities necessary for coexistence are given in form identical to those for the Lotka–Volterra model (equations 5-5):

$$\frac{1}{K_j} > \frac{\alpha'_{ij}}{K_i} \qquad \frac{1}{K_j} > \frac{\alpha'_{ji}}{K_j} \tag{5-10}$$

but because the isoclines are no longer straight lines, the interpretation of these inequalities is a bit different. To make expressions (5-10) equivalent to expressions (5-5), for similar interpretation, we must substitute for α'_{ij} and α'_{ji} (equation 5-9b). Thus

$$\frac{1}{K_j} > \frac{\alpha_{ij}\theta_i(n_i/K_i)^{\theta_i - 1}}{K_i} \qquad \frac{1}{K_i} > \frac{\alpha_{ji}\theta_j(n_j/K_j)^{\theta_j - 1}}{K_j}$$

The assessment of relative density-feedback intensities has suddenly become a bit complex. Other models include those of Hassell and coworkers (Hassell and Comins, 1976; Hassell, 1978).

The Michaelis–Menten Model

One weakness of the models described above is that they all treat K_i and K_j as constants. This may, to some degree, be justified in space-limited species such as many plants, or sessile animals, but for most other species K will reflect food supply at least slightly, and food supply is an interactive, not a passive quantity. In effect, we should be coupling the competition equations with prey–predator equations. One simple way of accomplishing this is to suppose birthrate of the competitors to be proportional to rate of resource utilization. If utilization follows a type II curve (Chapter 4), we can then write

$$\begin{aligned} \frac{dn_1}{dt} &= n_1\left(\frac{\beta_1 S}{a_1 + S} - d_1\right) \\ \frac{dn_2}{dt} &= n_2\left(\frac{\beta_2 S}{a_2 + S} - d_2\right) \\ \frac{dS}{dt} &= S\left[r_{SO}\left(1 - \frac{S}{K}\right) - \frac{\delta_1 n_1}{a_1 + S} - \frac{\delta_2 n_2}{a_2 + S}\right] \end{aligned} \qquad (5\text{-}11)$$

Here S is the resource (food), which is presumed to have a legitimate constant carrying capacity. Equations (5-11) are also the expressions for describing growth in a controlled chemostat, and the type II birth response terms are identical to the Michaelis–Menten terms used in the description of enzyme-controlled reactions: hence the term *Michaelis–Menten model* (or, sometimes, Monod model). Notice that its identity to the Michaelis–Menten form suggests that the type II curve might also apply to competition for inorganic nutrients. This possibility has been explored with very interesting results by Tilman (see Tilman, 1976).

More Competitors Than Resources?

As written, equations (5-11) assume the existence of only a single resource. Thus the question of niche overlap is moot in this case; both competitors must use the identical resource. Appropriately expanded equations for more than one resource are easily constructed using the form given in Chapter 4 (equations 4-17 and 4-18). For the moment, however, we shall keep things simple. This allows us to ask directly about the competitive exclusion principle. Is it possible, according to the Michaelis–Menten model, for two competitors to coexist on one resource (if their resource niches are identical)?

Equations (5-11) do not lend themselves easily to calculating competition coefficients, so the conditions for competitive coexistence are not immediately apparent. The problem has been tackled by Hansen and Hubbell (1980), who showed that only one species can persist, and that which one it is depends (when $d_1 = d_2 = d$) heavily on the quantities

$$a_i \frac{d}{\beta_i - d}, \qquad a_j \frac{d}{\beta - d}$$

However, an interesting caveat arises. The theoretical part of Hansen and Hubbell's analysis, like our earlier analysis of the Lotka–Volterra system, defined an equilibrium and asked whether there was coexistence *at that equilibrium*. Equations (5-11), though, allow for prey–predator interactions and can, therefore, cycle. What effect does cycling have on the question of coexistence?

DeJong (1976) offers the following argument. Suppose that one species, A, is the more efficient forager and so has the higher population growth rate *at low food densities*, but that its demographic performance improves less rapidly than its competitor's (B) as food conditions improve. Then when food is abundant, B "wins". But in the process of winning, B depresses the food supply to a level where A becomes the superior competitor. If this level is low enough that the combined population of A and B is declining before it is reached, food eventually becomes density dependent and negatively frequency dependent, permitting coexistence of A and B.

Hsu et al. (1978), in an elegant treatment of the subject, showed that, indeed, coexistence on a single resource can occur provided that the system, as in DeJong's argument, exhibits sustained limit cycles. Coexistence requires also that one species be an efficient user of the resource (high carrying capacity—that is, a dense population at the hypothetical equilibrium, in the absence of the competitor), and that the other have a high intrinsic rate of increase (maximum r). Armstrong and McGehee (1976, 1980) present a less general discussion with a slightly modified form of the equations. But their approach is mathematically simpler:

Consider the expressions

$$\frac{dn_1}{dt} = n_1\left(\frac{\beta_1 S}{a_1 + S} - d_1\right)$$

$$\frac{dn_2}{dt} = n_2(\beta_2 S - d_2) \tag{5-12}$$

$$\frac{dS}{dt} = S\left[r_{SO}\left(1 - \frac{S}{K}\right) - \frac{\delta_1 n_1}{a_1 + S} - \delta_2 n_2\right]$$

Now suppose that species 2 is not present. Then, at equilibrium ($n_1 = \hat{n}_1, S = \hat{S}$), so that when $dn_1/dt = 0$,

$$\hat{S} = \frac{a_1 d_1}{\beta_1 - d_1} \tag{5-13a}$$

If species 1 is absent, then at equilibrium ($n_2 = \hat{n}_2, S = S^*$), and $dn_2/dt = 0$. Thus

$$S^* = \frac{d_2}{\beta_2} \tag{5-13b}$$

Now species 2 can invade the first system if its growth rate, when rare, is positive:

$$\lim_{n_2 \to 0} \frac{dn_2/dt}{n_2} = \left(\frac{dn_2/dt}{n_2}\right)_{\hat{n}_1, \hat{S}} > 0 \tag{5-14a}$$

That is, if

$$\beta_2 \hat{S} - d_2 > 0,$$

or equivalently,

$$\hat{S} > S^* \tag{5-14b}$$

species 1 can invade the second system if

$$\lim_{n_1 \to 0} \frac{dn_1/dt}{n_1} = \left(\frac{dn_1/dt}{n_1}\right)_{\hat{n}_2, S^*} > 0 \tag{5-15a}$$

that is, if

$$S^* > \hat{S} \tag{5-15b}$$

Clearly, both conditions (5-14b) and (5-15b) cannot be met. Suppose, though, that the first system cycles. Then inequality (5-14a) becomes

$$\lim_{n_2 \to 0} \overline{\frac{dn_2/dt}{n_2}} = \beta_2 \bar{S} - d_2 > 0,$$

or

$$\bar{S} > S^* \qquad \text{[which replaces (5-14b)]} \tag{5-16}$$

So if the first system cycles such that $\bar{S} > S^*$, species 2 can invade. If cycling ceases after invasion of species 2, one of the competitors must disappear. If species 2 is the weaker competitor, it will decline until cycling begins again. Hypothetically, then, if species 2 is the weaker competitor, and the system cycles, both species will persist as long as

$$\bar{S} > S^* > \hat{S}$$

That this inequality can hold is shown as follows. Cycling or not, over a long period of time species 1, if persistent, must show an average r_1 of 0.0. Therefore,

$$0 = \overline{\frac{1}{n_1}\frac{dn_1}{dt}} = \overline{\frac{\beta_1 S}{a_1 + S}} - d_1 \simeq \frac{\beta_1 \bar{S}}{a_1 + \bar{S}} - d_1 - \frac{a_1 \beta_1}{(a_1 + \bar{S})^3} \operatorname{var}(S)$$

from the Taylor expansion. This expression requires that $\bar{S}$ increase as var(S) increases. Thus $\bar{S}$ takes its minimum value, $\hat{S}$, when var$(S) = 0$—no fluctuations— and increases as fluctuations increase. That is, always, $\bar{S} > \hat{S}$. Hence we require only that S^* lie between $\bar{S}$ and $\hat{S}$, a matter of the values of a_1, β_1, β_2, d_1, and d_2.

Equations (5-11) were used by Koch (1974b) to simulate competition between two species on a single prey, and not surprisingly, the predictions above were born out. Coexistence required that d of the weaker competitor be low or its b value be high, and usually occurred only when cycles were of high amplitude.

It is a pity that natural experiments are not available to test the foregoing theory; it is utterly impossible to control environmental variables sufficiently to know whether all other causative factors for coexistence have been eliminated. This is particularly true since the predictions of coexistence or exclusion in competitive systems with one resource seem depressingly dependent on the form of the mathematical model used. The Lotka–Volterra model, as well as some others, do not permit coexistence; the Michaelis–Menten equations do, under certain, cyclic conditions.

If the question of whether two species can coexist on a single, shared resource is important, it is important to ask what constitutes a single resource. If two animals both feed on only sunflower seeds, do they share one food? Or do they subdivide the seeds by (say) size, thereby making the apparently single food type into two? Haigh and Maynard Smith (1972) have produced a model of two competitors feeding on one prey which has larval and adult stages. It turns out, at least on the basis of the equations they choose to use, that the two stages act effectively as two resources and that the system is stable. They show the same to be true for competitors eating the same plant species, but with one concentrating on the roots, the other on shoots. Examples of the coexistence of highly similar species that subdivide a food in this manner are very common. One of the best involves the golden-mantled ground squirrel (*Citellus lateralis*), and the least chipmunk (*Eutamias minimus*) in Colorado. Although both species feed on a variety of plants, their diets are strongly overlapping and the dominant food (more than 80% for both species) is the dandelion. The former shows a pronounced preference for the stems; the latter eats the heads and seeds (Carleton, 1965).

Although the question of one resource supporting more than one competitor has received much attention and has contributed to our understanding of competitive processes, it may be largely academic. As the preceding paragraph implies, what constitutes a single resource depends on the competitors subdividing it. It is not possible even to define what we mean by a single resource without reference to its consumers. Second, the entire discussion above cleverly ignores interference. If two species depress their own growth rates more than those of their competitors by virtue of interference *plus* exploitation, the matter of number of discrete resources becomes moot. This point was realized by Schoener (1976), who constructed a hybrid model of the form

$$\frac{dn_i}{dt} = \xi_i n_i \left[\left(\frac{I_{0i}}{n_i + \beta n_j} + \frac{I_{Ei}}{n_i} \right) - (\gamma_{ii} n_i + \gamma_{ij} n_j) - C_i \right] \tag{5-17}$$

Here, I_{0i} is the amount of resource used exclusively by species i, I_{Ei} the amount shared with species j, β the likelihood that species j grabs an item of the resource before an individual of species i can use it, ξ_i a measure of the rate of conversion of food into biomass, γ_{ii} and γ_{ij} are interference coefficients, and C_i is the metabolic cost of living. This formulation, like those in the preceding section, does not consider the resource dynamics. Nevertheless, it does examine both exploitative and interference competition simultaneously. Analyzing this model, Schoener showed what we noted above—that a high enough ratio of intra- to interspecific interference can promote coexistence.

Age Structure

In addition to the example given above, where age could be used to make one prey species effectively two with respect to its predators, the existence of age structure has some important implications for competition theory. Competition pressure between individuals of different species may well vary depending on the age (or stage) of the individuals involved. Glass and Slade (1980), for example, report that the cotton rat (*Sigmodon hispidus*) has a negative feedback on the meadow vole (*Microtus ochrogaster*) but only during the former's reproductive period. This fact is interesting because it leads to the conclusion that competing populations can cycle (see also Pennycuick et al., 1968). This is illustrated with a simple example below. Using the Leslie matrix notation, where f_i is fecundity, P_i survival, suppose that two species with two age classes have identical population parameters and that

$$\begin{aligned} f_1 &= \hat{f}_1 \\ f_2 &= \hat{f}_2(1 - a_1 n_1 - a_2 m_1) \\ P_1 &= \hat{P}_1 \end{aligned}$$

where n_i, and m_i are the populations of age class i for species 1 and 2, and f_i, and p_i refer to the fecundity and survivial of both species. Then

$$\begin{aligned} n_1(t+1) &= n_1(t)\hat{f}_1 + \hat{n}_2(t)\hat{f}_2(1 - a_1 n_1(t) - a_2 m_1(t)) \\ n_2(t+1) &= n_1(t)\hat{P}_1 \\ m_1(t+1) &= m_1(t)\hat{f}_1 + m_2(t)\hat{f}_2(1 - a_1 m_1(t) - a_2 n_1(t)) \\ m_2(t+1) &= m_1(t)\hat{P}_1 \end{aligned} \tag{5-18}$$

Equilibrium values can be calculated from (5-18):

$$\begin{aligned} \hat{n}_1 = \hat{m}_1 &= \frac{\hat{f}_1 + \hat{f}_2 \hat{P}_1 - 1}{\hat{f}_2(a_1 + a_2)\hat{P}_1} \\ \hat{n}_2 = \hat{m}_2 &= \frac{\hat{f}_1 + \hat{f}_2 \hat{P}_1 - 1}{\hat{f}_2(a_1 + a_2)} \end{aligned} \tag{5-19}$$

Checking for neighborhood stability, we now perturb $\hat{n}_1, \hat{n}_2, \hat{m}_1$, and $\hat{m}_2$ by x_1, x_2, y_1, and y_2, respectively, and employ the Taylor expansion to obtain

$$\begin{bmatrix} x_1(t+1) \\ x_2(t+1) \\ x_3(t+1) \\ x_4(t+1) \end{bmatrix} = \begin{bmatrix} \hat{f}_1 - \hat{f}_2\hat{n}_2 a_1 & \hat{f}_2(1 - a\hat{n}_1) & -a_2\hat{f}_2\hat{n}_2 & 0 \\ \hat{P}_1 & 0 & 0 & 0 \\ -a_1\hat{f}_2\hat{n}_2 & 0 & \hat{f}_1 - \hat{f}_2\hat{n}_1 a_1 & \hat{f}_2(1 - a\hat{n}_1) \\ 0 & 0 & \hat{P}_1 & 0 \end{bmatrix} \begin{bmatrix} x_1(t) \\ x_2(t) \\ x_3(t) \\ x_4(t) \end{bmatrix} \tag{5-20}$$

where a is $(a_1 + a_2)$. Suppose now that we choose the following parameters,

$$\hat{P}_1 = .5$$
$$\hat{f}_1 = 0$$
$$\hat{f}_2 = 4$$
$$a_1 = a_2 = 1.0$$

Substituting these values into the matrix, and evaluating $\hat{n}_1$, and $\hat{n}_2$ with (5-19), the characteristic equation becomes

$$4\lambda^4 + 6\lambda^3 - 7\lambda^2 - 6\lambda + 4 = 0 \tag{5-21}$$

whose solution is $\lambda = .904, .548, -1.85, -1.10$. Thus the system cycles when, in the absence of age structure no cycling would take place.

We can easily imagine much more complex age-specific interactions. For example, in many fish, the adults of one species prey on the young of others, and both the young and the adults may compete, but perhaps for different resources. Or perhaps young compete only with young, adults with adults, but competition favors one species in early life, the other later on. General all-inclusive models can be built (e.g., see Travis et al., 1980), but are applicable only if quantitative information is available on the effects of species and age-specific abundances on age-specific birth and death rates.

5.2 Changing Environments

Individuals as Discrete Units

In our isocline-graph analyses we said nothing of the role of intrinsic rates of increase in determining the outcome of competition. In the case pictured in Figure 5-5, either species 1 or species 2 will "win," depending on initial conditions, but exactly what initial conditions will lead to victory by (say) species 1 is not clear. Species 1 must be sufficiently numerous relative to species 2, but how much is "sufficiently"? If $r_{10} \simeq r_{20}$ and the system begins at the asterisk (Fig. 5-5), it seems likely that the species 2 isocline will be crossed first, leading to the elimination of species 2. But if r_{20} is much larger than r_{10}, then, perhaps, the vector sum of the arrows is sufficiently upward that the system moves into the "domain of attraction" of species 2, leading to a reversed outcome. Because individuals come in discrete units, with genetic and experiential variations, r_0 values may vary in time. The result of this largely random variation is that it may be difficult to predict the outcome of competition when initial populations are either quite small (thus allowing maximum expression of individual variation) or quite near that critical

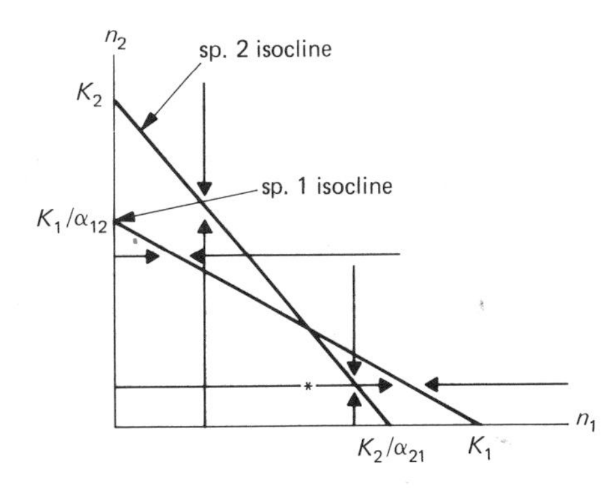

Figure 5-5 Superimposed competitor isoclines.

breakpoint which, in an ideal, deterministic world, would separate winner from loser (see also Chapter 17). Similarly, in communities where coexistence is indicated, if one species is relatively uncommon, chance variations may kill it off (see the discussion of the work by Hubbell, 1979, and Caswell, 1976, in Section 6.3). Thus in small local populations, the predictions of the models above apply only very loosely.

Fluctuations in the External Environment

Recall, from Chapters 2 and 3, that environmental fluctuations depress average population sizes below their carrying capacities. Species that are particularly prone to climatic vagaries, such as many insects, may spend most of their time in populations well below levels where resources become limiting. But if this is so, both intra- and interspecific feedback (i.e., competition) must be very slight. MacArthur in his 1972 book makes a good deal of this fact, pointing out that many species may experience only negligible competition pressures except on uncommon occasions. There are two implications for population dynamics.

1. Any spatial separation of species arising from exploitation competition (each species shifts more to unshared resources because there is a reduced supply of the shared ones) or interference (the species avoid interspecific harrassment) is apt to be slight. Distribution of either species in space will be nearly independent of the distribution of the other.
2. Relative population growth rates during periods of lull in competitive pressures (r_0) become extremely important in determining the outcome of competition—either by adjusting initial population densities before the crunch comes, or by altering the angle of the arrow-vector sums in the isocline graphs. We shall consider both.

Picture a pair of insect species, i and j, which occupy very similar habitat. Suppose, for example, that they occupy leaves of two types of plant in a prairie. If M_A, and M_B are the numbers of A and B leaves, then

$$\frac{n_{iA}}{M_A} = \frac{\text{number of insects of species } i \text{ on leaves of plant species } A}{\text{number of leaves of plant species } A} \qquad (5\text{-}22)$$

is the expected number of i-insects per A leaf. If there is no competition within or among species i and j, and both species are randomly (Poisson) and independently distributed among the leaves, expression (5-22) gives the Poisson parameter, λ_{Ai}. It follows that the probability that an A leaf carries x individuals of species i is

$$\frac{e^{-\lambda_{Ai}}\lambda_{Ai}^x}{x!}$$

In particular, when $n_{Ai}/M_A = \lambda_{Ai} \ll 1$, the probability that an A leaf carries two or more individuals of i is

$$\begin{aligned}1 - \frac{e^{-\lambda_{Ai}}\lambda_{Ai}^0}{0!} - \frac{e^{-\lambda_{Ai}}\lambda_{Ai}^1}{1!} &= 1 - e^{-\lambda_{Ai}}(1 + \lambda_{Ai}) \simeq 1 - (1 - \lambda_{Ai})(1 + \lambda_{Ai}) \\ &= 1 - (1 - \lambda_{Ai}^2) \\ &= \lambda_{Ai}^2\end{aligned}$$

and the expected number of A leaves with two or more is

$$M_A\lambda_{Ai}^2 = M_A\left(\frac{n_{Ai}}{M_A}\right)^2 = \frac{n_{Ai}^2}{M_A} \qquad (5\text{-}23)$$

The probability that an A leaf carries at least one insect of *both* species is

$$\left(1 - \frac{e^{-\lambda_{Ai}}\lambda_{Ai}^{0}}{0!}\right)\left(1 - \frac{e^{-\lambda_{Aj}}\lambda_{Aj}^{0}}{0!}\right) = (1 - e^{-\lambda_{Ai}})(1 - e^{-\lambda_{Aj}})$$

$$\simeq \lambda_{Ai}\lambda_{Aj}$$

$$= \frac{n_{Ai}n_{Aj}}{M_A^2}$$

and the expected number of A leaves with mixed species inhabitants is

$$M_A\left(\frac{n_{Ai}n_{Aj}}{M_A^2}\right) = \frac{n_{Ai}n_{Aj}}{M_A} \qquad (5\text{-}24)$$

The expected total number of leaves (A plus B) occupied by two or more individuals of species i (or j), and by individuals of both species i and j, is

$$\frac{n_{Ai}^2}{M_A} + \frac{n_{Bi}^2}{M_B} \quad \text{or} \quad \frac{n_{Aj}^2}{M_A} + \frac{n_{Bj}^2}{M_B}$$

$$\frac{n_{Ai}n_{Aj}}{M_A} + \frac{n_{Bi}n_{Bj}}{M_B} \qquad (5\text{-}25)$$

In the general case of many plant species, h, expressions (5-25) become

$$\sum_h \frac{n_{hi}^2}{M_h} \qquad \sum_h \frac{n_{hj}^2}{M_h}$$

$$\sum_h \frac{n_{hi}n_{hj}}{M_h} \qquad (5\text{-}26)$$

If there is intraspecific competition, the expected number of leaves of *either* plant species with two individuals of species i (or j) will be less than indicated in (5-26). If there is interspecific competition, the expectation of joint occupancy falls below that in (5-26).

A *guild* (Root, 1967) is a collection of species which exploit the environment in a similar manner. Thus all species in an area that eat insects in a forest would belong to a forest insect-feeder guild. Rathcke (1976) has used this approach to investigate competition in a guild of stem-boring insects, 48 species in all. She classified microhabitats, h, on the basis of plant species, stem size, and location (top, middle, and bottom of the stem). Since fewer than 10% of all stems were occupied at any time, (5-22) through (5-26) should be quite accurate. She then calculated the expectations (5-26) for all insects, i, and insect species pairs, i, j. These two quantities she called, respectively, f_{ii} and f_{ij}. The next step was to compare observed values to these expectations. The results were as follows: Of 53 pairs, over 2 years, for which she felt data were adequate for statistical analysis, 42 showed f_{ij}(expected) larger than f_{ij}(observed) for $i \neq j$; seven showed the reverse. The chance of finding a ratio of 42 : 7 or a more skewed ratio under the null hypothesis of half and half is only about 3×10^{-8}, so we must conclude that interspecific competition occurs. Also, the difference between f_{ij}(expected) and f_{ij}(observed) was generally larger when $i = j$ than when $i \neq j$. So *intraspecific* competition is stronger than *interspecific* competition, the condition necessary for coexistence (if competition is of importance in the system). On the other hand, of the 53 pairs, f_{ij}(expected) and f_{ij}(observed) for $i \neq j$ differed significantly ($p < .05$) in only 17 cases, not all of them differing in the same direction. Thus interspecific competition, if it is occurring, is usually very hard to detect—it is very weak for most species pairs. In the case of insects of this sort, Rathcke's results are, of course, exactly what we would predict. (*A cautionary word:* These same results might be promulgated by intraspecific clumping of individuals in space, even in the complete absence of competition.)

If competition is sporadic and the "winner" (were competition a constant influence) is at a disadvantage during competition-free periods (has a lower r_0), is an altered outcome possible? In particular, can such alteration of advantage lead to the coexistence of two species on a single resource (in the absence of interference)? Nineteen seventy-three was a popular year for addressing these questions (Grenney et al., 1973; Stewart and Levin, 1973), but the clearest exposition, although not the most quantitative, is that of Koch (1974a). Koch considered two hypothetical species with equal carrying capacities, obeying the Lotka–Volterra model, which started with populations well below K and so, for a time, grew in an almost unrestricted environment. As densities rose, the system entered a period of intense competition. At the end of the year, Koch invoked a 40% mortality across the board (winter die-off), bringing both populations back to low levels and allowing the cycle to begin again. One species, i, had a competitive advantage ($\alpha_{ij} < \alpha_{ji}$); the other possessed a higher growth potential ($r_{j0} > r_{i0}$).

Digress for a moment. If competition is purely exploitative, for a single resource, and species i uses the resource at x times the rate of species j, then species i impacts r_j x times as strongly as species j. Thus

$$\alpha_{ij} = \frac{1}{x} \qquad \alpha_{ji} = x \qquad \text{and} \qquad \alpha_{ij} = \frac{1}{\alpha_{ji}}$$

Using reciprocal α values, Koch found that environmental fluctuations could result in a reversal of competitive outcome. But he could never get coexistence. However, when $\alpha_{ij} \neq 1/\alpha_{ji}$ (meaning that there is either more than one resource, or interference), coexistence could be promoted when, in the absence of fluctuations, competitive exclusion was inevitable.

That fluctuations can promote coexistence may be more easily understood if the problem is looked at as follows. Suppose, were coexistence possible, that the two species began each year with populations of size n_i^*, n_j^*. Consider a year in which the better competitor is more numerous, the other less numerous than usual. Because the rate of approach to competitive population density levels is largely controlled by the faster grower, j, and because species j begins the year at lower density than normal, the competitive period is reached later, giving a greater amount of time for species j to exert its superior powers of growth. If the start of a year finds species i depressed and species j more common, the competitive period is reached sooner (since n_j begins already fairly high) and species j shows less increase than usual. Thus the fluctuations have introduced a negative frequency dependence into the system which buffers it, at least partly, from crashing.

A quite different approach has been taken by Abrams (1976). Using the Lotka–Volterra model with an added, randomly generated environmental variability term, $S(t)$—with expected value zero and variance σ^2—

$$\frac{dn_i}{dt} = r_{i0}n_i\left[1 - \frac{n_i + \alpha_{ij}n_j}{K_i} + S_i(t)\right] \tag{5-27}$$

he simulated two-species competition with extremely interesting results. To keep the problem simple he supposed that $K_i = K_j = 300$, $r_{i0} = r_{j0} = .03$, $\alpha_{ij} = \alpha_{ji} = \alpha$, and let $S_1(t)$ vary independently of $S_2(t)$. Extinction was deemed to occur when a species' population fell below some critical density. Some of his results are shown in Table 5-2. For any given extinction threshhold and α value, coexistence was always *discouraged* by environmental fluctuation.

At first glance Abrams's results seem to contradict the work of Koch and others cited above. But this is not the case. As Abrams set up his equations there was no net density-independent suppression of population growth (see Chapter 2). Koch began with two species which could not coexist in a constant environment and showed that an average density-independent depression $[E(S) < 0]$ might promote coexistence. Abrams started with two species which *could* coexist and showed, for $E(S) = 0$, that a greater variance in environmental state might bring one of them low enough that chance extinction could occur. In effect, what Abrams is saying is that avoidance of accidental random destruction requires smaller α values in fluctuating environments.

Table 5-2

Critical Density	α Value					
	0	*.25*	*.50*	*.75*	*.9*	*1.0*
10	144	120	48	24	24	0
5	168	144	72	48	24	0
1	240	216	144	96	48	0
.5	264	240	168	120	48	0
.01	320	280	240	160	80	0

Note: Critical density is the population level below which extinction occurs. Entries give the permissible variances [$1{,}000\,\sigma^2 = 1{,}000\,\mathrm{var}(S)$] below which values coexistence is possible.

Abrams also raised the question of what happens when the two competitors experience random environmental fluctuations in different ways. Suppose, for example, that both species are affected similarly by temperature. Then changes of warm to cool affect both in the same direction. But if one species survives better in warm weather, the other in cool weather, an increase in $S(t)$ for one species will be felt as a decrease by the other. Recall the paper of Birch (1953a), cited in Chapter 2, where R values of three grain beetles were differentially affected by temperature and humidity. Birch (1953b) also found temperature and humidity to affect competitive outcome between grain beetles. Similar results obtained by Park and co-workers have already been discussed in this chapter. To investigate this possibility, Abrams described a different $S(t)$ for each species. He wrote

$$\begin{aligned} S_i(t) &= \pm\gamma A(t) + (1-\gamma)B(t) \\ S_j(t) &= \pm\gamma A(t) + (1-\gamma)C(t) \end{aligned} \tag{5-28}$$

where $A(t)$, $B(t)$, and $C(t)$ are independently and randomly varying numbers. The similarity of $S_i(t)$ and $S_j(t)$ can be described by their covariance and their variances,

$$\begin{aligned} \mathrm{cov}(S_i, S_j) &= \pm\gamma^2\,\mathrm{var}(A) \\ \mathrm{var}(S_i) &= \gamma^2\,\mathrm{var}(A) + (1-\gamma)^2\,\mathrm{var}(B) \\ \mathrm{var}(S_j) &= \gamma^2\,\mathrm{var}(A) + (1-\gamma)^2\,\mathrm{var}(C) \end{aligned} \tag{5-29}$$

Choosing γ, $\mathrm{var}(B)$, and $\mathrm{var}(C)$ such that $\mathrm{var}(C) = \mathrm{var}(B) = [(1+\gamma)/(1-\gamma)]\,\mathrm{var}(A)$, the variance values above simplify to

$$\mathrm{var}(S_i) = \mathrm{var}(S_j) = \mathrm{var}(A) = \sigma^2$$

and the correlation coefficient of S_i and S_j becomes

$$\mathrm{cor}(S_i, S_j) = \frac{\mathrm{cov}(S_i, S_j)}{\mathrm{var}(S_i)\,\mathrm{var}(S_j)} = \gamma^2 \tag{5-30}$$

Table 5-3 shows results of further simulations. As the correlation coefficient drops (the species respond less and less similarly to the fluctuations), the value of α becomes increasingly critical, and populations become more and more susceptible to chance extinction by virtue of greater population variance.

Other authors have addressed Abrams's problem. May (1973a), using a diffusion equation approach, came to the conclusion that coexistence required lower α values in fluctuating environments. However, his definitions and his mathematics have been severely criticized by

Table 5-3

Correlation, γ		α Value				
		0	*.25*	*.5*	*.75*	*.9*
Positive	.25	216	168	144	96	24
	.5	192	192	192	144	48
	.75	216	192	192	192	168
	.9	216	216	216	216	192
Negative	.9	216	144	72	24	0

Note: Entries give the permissible variances [$1{,}000\,\sigma^2 = 1{,}000\,\text{var}(S)$] below which values coexistence is possible.

Turelli (1978). Roughgarden (1975) set up discrete-time equations for two or more competitors in a fluctuating environment, assuming minute amplitudes of population response to the fluctuations. His equations are complex, in spite of their crudeness, and will not be covered here. His conclusions are similar to those above.

There is still a wide array of questions to be answered respecting the effects of environmental fluctuations on competition. In the cases above, fluctuations are of a density-independent nature. Are the effects of changes in K similar? What if the fluctuations involve not the total resource availability (K) but the *relative* availability of different resources or different *classes* of resources (e.g., food versus shelter)? What if competitive dominance itself fluctuates, as appears to be the case in some aquatic arthropod communities (Lynch, 1978)?

Stability

We have already discussed some aspects of stability in competition pairs. Two things remain to be said. First, as might be expected, if the competitors are properly described with discrete equations and their intrinsic rates of increase are sufficiently high, chaotic fluctuations ensue (Hassell and Comins, 1976). Also, when chaos occurs, competitors, contrary to intuition, may oscillate *in phase* (Gilpin et al., 1982). A more interesting observation is that the outcome of competition between two species may be altered by time lags (Caswell, 1972; Pielou, 1974).

A general picture of the effects of time lags may be seen by writing the Lotka–Volterra equations and supposing feedback on species i and j to have delay times of τ_i, and τ_j, respectively.

$$\frac{dn_i}{dt} = \frac{r_{io}n_i(t)}{K_i}[K_i - n_i(t - \tau_i) - \alpha_{ij}n_j(t - \tau_j)] \qquad (5\text{-}31)$$

If the lags are very small, so that, using the Taylor expansion,

$$n(t - \tau) \simeq n(t) - \frac{\tau\, dn(t)}{dt}$$

then (5-31) becomes

$$\frac{dn_i}{dt} = r_{io}n_i(t)\frac{[K_i - n_i(t) - \alpha_{ij}n_j(t)] + \alpha_{ij}\tau_j\, dn_j(t)/dt}{K_i - r_{io}\tau_i n_i(t)} \qquad (5\text{-}32)$$

Writing the same for $dn_j(t)/dt$ and substituting into the above, we obtain, after some tedious

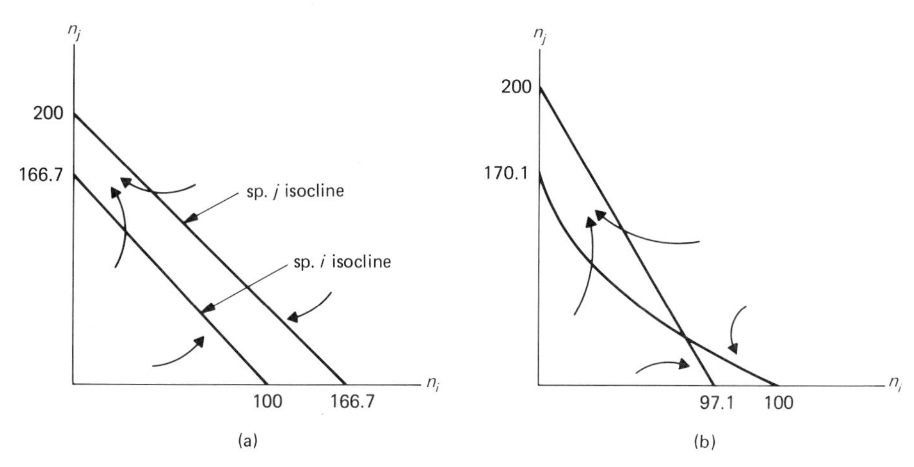

Figure 5-6 Superimposed competitor isoclines showing the effect of time lags: (a) no time lags (species j wins); (b) time lags.

rearrangement,

$$\frac{dn_i}{dt} = \frac{r_{io}n_i[(K_i - n_i - \alpha_{ij}n_j)(K_j - r_{jo}\tau_j n_j) + (K_j - n_j - \alpha_{ji}n_j)(r_{jo}\alpha_{ij}\tau_j n_j)]}{(K_i - r_{io}\tau_i n_i)(K_j - r_{jo}\tau_j n_j) - r_{io}r_{jo}\tau_i\tau_j n_i n_j \alpha_{ij}\alpha_{ji}} \tag{5-33}$$

The corresponding isocline for species i is

$$n_i = \frac{K_iK_j - K_j\alpha_{ji}n_j - r_{jo}\tau_j(K_i - \alpha_{ij}K_j)n_j}{K_j - r_{jo}\tau_j n_j(1 - \alpha_{ij}\alpha_{ji})} \tag{5-34a}$$

or

$$n_j = \frac{K_j(K_i - n_i)}{\alpha_{ij}K_j + r_{jo}\tau_j[(K_i - \alpha_{ij}K_j) - n_i(1 - \alpha_{ij}\alpha_{ji})]}$$

and for species j,

$$n_j = \frac{K_iK_j - K_i\alpha_{ij}n_i - r_{io}\tau_i(K_j - \alpha_{ji}K_i)n_i}{K_i - r_{io}\tau_i n_i(1 - \alpha_{ij}\alpha_{ji})} \tag{5-34b}$$

Consider the following simple example. Let $K_i = 100$, $K_j = 200$, $\alpha_{ij} = .6$, $\alpha_{ji} = 1.2$, $r_{io}\tau_i = 1.2$, and $r_{jo}\tau_j = 1.2$. The corresponding isocline graphs are shown in Figure 5-6. The insertion of a time lag has changed the system from one which species j always wins to one which species i could win if initially abundant enough.

5.3 Spatially Structured Populations

Most texts and a large fraction of even the most recent literature persist in treating competition in the context of Lotka–Volterra equations at or near equilibrium. As should be clear from the preceding pages, such treatment falls a bit short of adequately describing competition dynamics. In addition, temporal change is apt to obscure evidence of competition. Competition is likely to be an important influence on population dynamics only when populations track the environment fairly well, so that density feedback is consistent. The latter complication has led Wiens (1977) to suggest rather strongly that ecologists' preoccupation with interspecific competition may be, at best, misplaced. In fact, he points out that direct evidence for its existence is sparse—almost all the supporting observations are of an indirect nature—niche separations, for example (see Chapter 17). Spatial heterogeneity muddies the waters still more.

There is considerable evidence that in heterogeneous environments, slight preferences by competitors for different microhabitats act to separate the species in space. For example, Koplin and Hoffman (1968), examining *Microtus pennsylvanicus* and *M. montanus* in a zone of sympatry in the National Bison Range of western Montana, find the latter species in dry areas. If *M. pennsylvanicus* populations are trapped out, *M. montanus* extends its range into the more mesic, vacated areas. Cameron (1964) notes that *M. pennsylvanicus* is generally a field species, while its relative, *Clethrionomys gapperi*, is a forest denizen. However, on Newfoundland, where *Clethrionomys* has never successfully invaded, *Microtus* has extended its range to include forests, the preferred habitat of its mainland competitor. Working with these same two species, Grant (1969) observed habitat selection in enclosures which contained both field and forest. When both species occur together, they segregate as expected. When *Clethrionomys* is absent, *Microtus* occupies forest as well as grassland areas. When the latter species is absent, *Clethrionomys* will occasionally make use of the grassland.

Of course, differential use of microhabitats will likely (although not necessarily) result in differential use of limiting resources and so alleviate interspecific exploitation competition. Another, more certain consequence is that some separation in space will ameliorate interference. It is likely, although again not necessary (see, e.g., Roughgarden, 1978; or Chapter 17), that microhabitat preference differences have evolved *because* of the population growth-depressing effects of competition. If so, competition has played an important role in the spatial structuring of the environment. But once separation occurs, one can argue that the effects of competition on the population *dynamics* of the species involved may be very small. Thus spatial heterogeneity, by allowing microhabitat preferences to express themselves, buffers the effects of interspecific competition on population dynamics.

Effective spatial separation, with the same consequences, can be achieved by species utilizing the same resources in slightly different ways—in effect making two resources out of one by differences in exploitation technique. One example is offered by the tropical hummingbirds studied by Feinsinger (1976). In this case six species, feeding primarily on nectar, secondarily on small arthropods, coexist in Costa Rica. The large species, aggressively dominant by virtue of size, defend territories from which the smaller birds are excluded. Interspecific agonistic behavior serves to separate these species from each other in space. Territories are, as one might expect, situated around rich clumps of flowers. The smaller species are excluded from the rich flower clumps, but their slightly smaller metabolic needs permit them to make a living by "traplining"—following regular paths between the nonclumped flowers which are an energetically unfeasible food source for the larger species. The small species segregate in space by specializing on inner, or outer, upper or lower levels in the trees they visit. A very similar situation exists for flower-foraging bees of the genus *Trigonia* (Johnson and Hubbell, 1975).

Sometimes spatial separation may occur via chance events. Like separation by any other cause, this too can ameliorate competition. Levin (1974) provides the following theoretical scenario. Figuratively divide an area inhabited by two competitors into halves, and suppose that one species happens, by chance, to have a numerical advantage in one half, the other in the other half. Suppose that a sufficient numerical advantage leads to competitive exclusion of the other species. Then if the numerical asymmetry can be maintained, coexistence may occur. If migration between the halves is too great, the whole populations behave as if the area were homogeneous and one species "wins." If migration is too slow, both species may persist, but alone and in separate halves; they do not really coexist except in the sense that an occasional migrant induces a temporary mixing. But if migration is within certain definable limits, a persistent species mixture is maintained over the entire area.

A possible example of Levin's scheme can be found in the coexistence of two species of mites in Costa Rica (Colwell, 1973). These creatures live on the flowers of four plant species, and disperse among flowers by jumping on and off of the bills of three species of hummingbirds which forage on those flowers. By making observations on the foraging patterns of the birds, measuring the probabilities that a visit to a flower of one species would be followed by a visit to one of another, Colwell was able to calculate the "transition probability" that if a mite hopped on the bill of a bird visiting plant species i, it would end up being carried to species j. The transition matrix, rows denoting the donor plant, columns the recipient, is as follows:

$\mathbf{P} = \{p_{ij}\} =$

Donor Species \ Recipient Species	*1*	*2*	*3*	*4*
1	.916	.058	.006	.020
2	.062	.931	.002	.005
3	.020	.010	.880	.090
4	.005	.002	.012	.980

Notice that the mites are most apt to move among flowers of the same species, but that where cross-species movements occur, plant species 1 and 2, and species 3 and 4, form groups within which some movement occurs but between which almost no movement occurs. Thus if a mite starts out on species 1, it is apt to spend its life on species 1 or, perhaps, species 1 and 2, but is not very likely to spend much time on either 3 or 4. Interference and a very slight degree of flower species preference (as well as chance) shuttles the mites onto different species, where their (partial) isolation is perpetuated by hummingbird foraging behavior. All plant species host both mite species, but one mite is concentrated on plant species 1 and 2, the other on 3 and 4.

While spatial separation may occur by differential habitat selection, by interference or by chance, it may also occur as a result of social cohesion within a species. Suppose that all potentially occupiable spots in an area are not used—that the combined population of two competitors lies below the number supportable by the environment. If either or both species show intraspecific attraction, so that clumping occurs, there will be a negative "codispersion" (a negative correlation in space) of the two species.

The particular cause of spatial separation mentioned immediately above depends on the environment being less than "saturated" with the two competitors. Much more will be said of saturation in a moment. But first, let us examine in somewhat more quantitative terms the impact of spatial separation. The gist of the argument is the same for any population growth model, but to keep things simple and specific, for illustrative purposes, suppose that the Lotka–Volterra model is appropriate. Figuratively subdivide the environment into local areas within which the Lotka–Volterra equations are valid descriptors of population dynamics. Then, for the ith area,

$$\frac{dn_{1i}}{dt} = r_{10}n_{1i}\left(1 - \frac{n_{1i} + \alpha_{12}n_{2i}}{K_i}\right) \tag{5-35}$$

For the population as a whole,

$$\frac{d\bar{n}_1}{dt} = \overline{\frac{dn_1}{dt}} = r_{10}\bar{n}_1 - \frac{r_{10}}{K_1}\overline{n_1^2} - \frac{r_{10}\alpha_{12}}{K_1}\overline{n_1 n_2} \tag{5-36}$$

Recalling that $\text{var}(x) = \overline{(x - \bar{x})^2} = \overline{x^2} - \bar{x}^2$, and that $\text{cov}(x, y) = \overline{(x - \bar{x})(y - \bar{y})} = \overline{xy} - \bar{x}\bar{y}$, (5-36) can be written

$$\frac{\overline{dn_1}}{dt} = r_{10}\bar{n}_1 - \frac{r_{10}}{K_1}[\bar{n}_1^2 + \text{var}(n_1)] - \frac{r_{10}\alpha_{12}}{K_1}[\bar{n}_1\bar{n}_2 + \text{cov}(n_1, n_2)]$$

$$= r_{10}\bar{n}_1\left(1 - \frac{\bar{n}_1 + \alpha_{12}\bar{n}_2}{K_1}\right) - \frac{r_{10}}{K_1}[\text{var}(n_1) + \alpha_{12}\,\text{cov}(n_1, n_2)] \tag{5-37a}$$

Similarly,

$$\frac{\overline{dn_2}}{dt} = r_{20}\bar{n}_2\left(1 - \frac{\bar{n}_1 + \alpha_{12}\bar{n}_2}{K_2}\right) - \frac{r_{20}}{K_2}[\text{var}(n_2) + \alpha_{21}\,\text{cov}(n_1, n_2)] \tag{5-37b}$$

Now both species will coexist (both are persistent) if, for both $i = 1$ and $i = 2$, $\overline{dn_i/dt} > 0$ as $\bar{n}_i$ approaches zero. But when $\bar{n}_i$ approaches zero, $\bar{n}_j$ approaches K_j. Thus the condition for the persistence of species i becomes

$$\lim_{n_i \to 0} \frac{\overline{dn_i}}{dt} = \frac{r_{i0}}{K_i} \lim_{n_i \to 0} [\bar{n}_i(K_i - \alpha_{ij}K_j) - \text{var}(n_i) - \alpha_{ij}\,\text{cov}(n_i, n_j)] > 0 \tag{5-38}$$

In classical Gausian theory, coexistence in this system may occur in any case if $\alpha_{ij}\alpha_{ji} < 1$. So we are interested only in cases where the α values are equal to or exceed 1.0. Suppose that $\alpha_{ij} = \alpha_{ji} = 1$, and $K_i = K_j = K$. Then inequality (5-38) becomes

$$\lim_{n_i \to 0} \text{cov}(n_i, n_j) < \lim_{n_i \to 0} - \text{var}(n_i)$$

But if we have a situation of clumping within species i, $\text{var}(n_i)$ must rise and fall more rapidly than $\bar{n}_i$. So as $\bar{n}_i$ goes to zero, so must $\text{var}(n_i)$, at a faster rate than $\text{cov}(n_i, n_j)$. Hence the expression above becomes simply

$$\lim_{n_i \to 0} \text{cov}(n_i, n_j) < 0 \tag{5-39}$$

Coexistence is assured if, when either species becomes very scarce, there is any negative codispersion at all. If α values become still larger, a higher degree of negative codispersion is required for coexistence.

As discussed in Chapters 2, 3, and 4, it is generally true that environmental fluctuations, of either a density-independent or density-dependent nature, depress average population sizes below their potential equilibria in the absence of such fluctuations. Also, prey–predator instability, competitive exclusion, or chance extinctions may eliminate populations rather regularly on a local scale. And even with high migration rates, time is required to build new populations. Thus it is unlikely that environments are ever truly saturated [i.e., $E(n)$ seldom reaches $E(K)$]. This makes possible spatial separation of competitors by intraspecific social cohesion or interspecific social avoidance. It also gives rise to a high possibility of separation by chance.

Of two competing species, suppose that one always "wins" in local interactions, but the other disperses and colonizes empty areas more readily. It is not difficult, intuitively, to imagine persistence of the "loser" under these circumstances. The problem has been addressed by several authors (Levins and Culver, 1971; Horn and MacArthur, 1972; Culver, 1973; Pielou, 1974), but is discussed most rigorously by Slatkin (1974). Suppose that the environment can be subdivided into patches, of either a discrete or an arbitrary nature. Let p_0, p_1, p_2, and p_3 be the proportion of these patches occupied by neither species, species 1, species 2, and both, respectively. Also let

$$\begin{aligned}
y_1 &= \text{proportion of patches with species 1} \,(= p_1 + p_3) \\
y_2 &= \text{proportion of patches with species 2} \,(= p_2 + p_3) \\
m_1, m_2 &= \text{migration rates in the absence of competition} \\
m_1 - \mu_1, m_2 - \mu_2 &= \text{migration rates in the presence of competition} \\
e_1, e_2 &= \text{extinction rates without competition} \\
e_1 + \epsilon_1, e_2 + \epsilon_2 &= \text{extinction rates with competition}
\end{aligned}$$

Then the invasion rate by species 1, if species 2 is absent, is $m_1 p_1$; if species 2 is present,

$(m_1 - \mu_1)(p_1 + p_2)$. Thus

$$\begin{aligned}
\frac{dp_0}{dt} &= -(m_1 y_1 + m_2 y_2)p_0 + e_1 p_1 + e_2 p_2 \\
\frac{dp_1}{dt} &= m_1 y_1 p_0 - [e_1 + (m_2 - \mu_2)y_2]p_1 + (e_2 + \epsilon_2)p_3 \\
\frac{dp_2}{dt} &= m_2 y_2 p_0 - [e_2 + (m_1 - \mu_1)y_1]p_2 + (e_1 + \epsilon_1)p_3 \\
\frac{dp_3}{dt} &= (m_1 - \mu_1)y_1 p_2 + (m_2 - \mu_2)y_2 p_1 - (e_1 + \epsilon_1 + e_2 + \epsilon_2)p_3
\end{aligned} \tag{5-40}$$

When species 1 is absent, the following equilibrium exists:

$$p_2 = \frac{1 - e_2}{m_2} \qquad p_1 = p_3 = 0$$

It is now possible to perform a neighborhood stability analysis. If the system, at this equilibrium, without species 1, is unstable, it means that species 1 can invade. The analysis will not be presented here, and the final conditions for invasion by species 1 are messy. But in the simple case where $m_1 = m_2$, $\mu_1 = \mu_2$, $e_1 = e_2$, and $\epsilon_1 = \epsilon_2$ (the species have identical population parameters), coexistence is assured if

$$e > \frac{-(m - \mu)(1 - e/m)}{2}$$

which is

$$e > \frac{-m(m - \mu)}{(m + \mu)} \tag{5-41}$$

But this inequality is *always* met. That is (for this special case), any local extinction at all is enough to assure coexistence.

We have seen that, theoretically, spatial structuring of populations, with local extinctions (or at least periodic disturbances) can promote coexistence when competitive exclusion would otherwise occur. What is the evidence? On the United States West Coast, the starfish *Pisaster ochraceus* predates barnacles and mussels by everting its stomach and digesting several prey at a time, in their own shells. Death of these organisms is followed by decay of their adhesive cement or byssal threads and the consequent sloughing into the sea of clumps of dead animals. Rock surface is thereby exposed to colonization by organisms competitively excluded (competition for space) by the mussels and barnacles. Removal of the cause of local disturbance (starfish) permits competition to exclude certain organisms and results in fewer coexisting species (Paine, 1966). A similar situation exists with respect to disturbance by sea urchins (*Strongylocentrotus franciscanus* and *S. purpuratus*) in West Coast tide pools (Paine and Vadas, 1969). Dayton (1971), also working in the West Coast intertidal, found a competitive dominance among sessile organisms, with the mussel *Mytilus californianus* ousting the barnacle *Balanus cariosus*. This barnacle, in turn, is dominant to *B. glandula*, which is dominant to *Chthamalus dalli*, still another barnacle. But exclusion of *Chthamalus* via the last mentioned dominance is effectively broken by limpet predation on newly settled barnacle larvae and predation by the gastropod, *Thais*, on adult *B. glandula*. The mussel fails to eliminate the barnacles by virtue of predation by *Thais* and the battering-ram effects of floating logs. To assess the importance of logs, Dayton drove a series of nails into the rocks and watched for damage. Over a 3-year period, there was a "hit" probability of between 5 and 30%.

A particularly nice example of coexistence due to heterogeneity in time and space is offered by Waser and Case (1981). In the Kibale forest of western Uganda, four large primate species (the redtail monkey, *Cercopithecus ascanius*; the blue monkey, *C. mitis*; the gray-cheeked mangabey, *Cercocebus albigena*; and the chimpanzee, *Pan troglodytes*) coexist primarily on a single food source, fruit. They differ little or not at all in their use of different fruits, so resource separation cannot be claimed. Because it seems likely that food is the critical factor in population limitation for these species, the classic law of competitive exclusion indicates that the observed coexistence is not feasible. What happens is as follows. Fruit appears at very nearly random points in time and space, and must be discovered by its predators. The redtail monkey is the most active of the species, moving about over greater distances in any given period of time, so generally is the first species to discover new sources of fruit. The blue monkey is behaviorally dominant to the redtail and chases the latter away when it comes upon a fruit tree, but its rate of discovery is slow enough to allow the redtails sufficient food to meet their needs. In turn, the blue monkey has a faster rate of discovery of fruit trees than the mangabey, to which it is aggressively subordinate. The champanzees oust the other three species at will, assuring themselves an adequate diet, but are slow enough at discovering new sites that the other species can also satisfy their needs. Hence coexistence is possible.

This chapter is permeated with two basic concepts. First is the idea of competitive exclusion; second, the myriad objections to that idea and the idea that extenuating circumstances are so pervasive that competitive exclusion is a useless hypothesis. In particular, temporal change, including the interactive changes in predator populations and spatial structuring of the environment, seem to vitiate the hypothesis. Indeed, through the maze of pros and cons, various models and endless complications, the reader may emerge wondering whether, in many cases, a change in one competitor's population density really has any impact at all on that of another. This is Wiens's (1977) argument. Competition is a clear force in determining *dispersion* patterns (microhabitat separation, separation to avoid interference—see also Chapter 17), but the concern over *dynamics* may indeed be valid. The literature has saturated us with examples of the differences between ecologically similar species which purport to be responsible for allowing their coexistence, but it has provided few clear-cut cases of competitive effects on population dynamics. Dramatic changes in population levels of one species (e.g., removal) certainly affect the use of space and resources by its competitors and so, ultimately, their dynamics, but documented cases, in the natural environment, where normal population fluctuations in one species have led to significant impacts on population growth of others, other than through predation, are relatively scarce.

Review Questions

1. One problem with the hypothesis of competitive exclusion may be that it is not testable; a given set of observations may be consistent with the hypothesis, or any number of alternatives. Is there any experiment (even imaginary) that could either demonstrate or disprove competitive exclusion? Is competitive exclusion, as some have suggested, a meaningless pseudohypothesis?
2. In Park's carefully controlled experiments with flour beetles there were conditions under which the outcome of competition could not be predicted with certainty; the best Park could do was to assign probabilities. This being the case, under rigidly controlled conditions and in a very simple environment, might we not expect the outcome of competition in nature to be still less deterministic? Should we not then develop stochastic competition models if any sort of realism is to be achieved?
3. Suppose that you grew two species of protozoans *alone* in test tubes, with a constant food input and discover that $K_1 = K_2 = 100$, $r_{10} = r_{20} = 1.0$. You then grow these species together, and rearrange the Lotka–Volterra equations to read

$$\alpha_{ij} = \frac{K_i - n_i - (K_i/r_{i0}n_i)(dn_i/dt)}{n_j}$$

and in this way calculate α_{ij} (and α_{ji}) for various values of n_i and n_j. What you find is that α_{ij} can be written quite accurately as

$$\alpha_{ij} = 1.8 - .01n_j$$

Also,

$$\alpha_{ji} = 2.5 - .02n_i$$

(a) Draw the appropriate isoclines (solve $r_i = 0, r_j = 0$ for either n_i or n_j and stick in some trial values).

(b) Can the two species coexist?

4. Data presentation in many plant and some animal competition studies have made use of DeWitt diagrams: Use the DeWitt diagram given below (Fig. 5-7) to estimate the corresponding prey–predator isocline graphs. Do you expect coexistence? What changes in the DeWitt diagram would reverse this outcome? Will such diagrams always allow prediction of competitive outcome? Why, or why not?

5. Suppose that you were given the following data for a region where species A and B are sympatric and asked to predict expected population densities in regions of allopatry. Suppose you also know that the regions of sympatry and allopatry were essentially identical except in their inclusion of one or both of the species in question. You might use niche overlap to estimate α values and, from these, calculate K values. Or you might look at the density of one species in quadrats where the other is absent and estimate K values in this manner. Both approaches seem reasonable, yet they give very different answers. Why?

Quadrat Number	Number of Individuals of Species *A*	*B*
1	2	0
2	1	2
3	0	4
4	0	3
5	0	3
6	3	0
7	3	0
8	0	3
9	1	1
10	0	4

6. In a small homogeneous area, two very similar species are unlikely to coexist (according to the competitive exclusion hypothesis). If the area is somewhat larger, coexisting mixed populations may occur, with scattered patches of mostly one species or the other. Why does this happen? What determines the minimal area size necessary for coexistence?

7. *Verbally* describe why fluctuating environments may render the competitive process effectively frequency dependent.

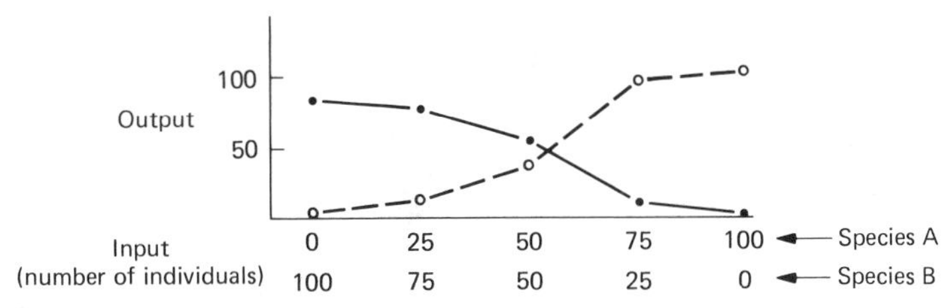

Figure 5-7

Chapter 6

Multispecies Interactions

Biological communities generally consist of a large number of interacting species, not merely isolated prey–predator systems or pairs of competitors. Predators affect the dynamics of competing species, the effects of several competitors on one species' population may or may not be additive, and the complex webs of energy flow through ecosystems have implications for population dynamics which defy the reductionism of autecologists. To what degree must the qualitative predictions of the preceding chapters be tempered by considerations of multispecies interactions? The answer is that, often, we simply do not know. Our theories work quite well in the laboratory, and at least occasionally in the field, but we have as yet not been able to define guidelines for when complex systems can be reasonably broken down into more easily understood, isolated subsystems. As such, we must for the moment, let the conclusions of Chapters 2 through 5 stand as they are, realizing that in at least some instances, they may be considerably less than precise.

A very simple but good, illustrative example of the complexity of multispecies systems is described in the work of Sutherland (1972). On the North American West Coast several species of algae compete for space in the rocky intertidal. They are fed upon by the limpet *Acmaea scabra*, which affects the direction of their competitive outcome. If limpet numbers are sufficiently decreased, the normally devoured, microscopic larvae of superior algal competitors manage to gain a foothold and, in time, oust the limpets' major food species. In adult form, these "winners" are rather poor diet items. Thus the limpets' continued existence depends on their maintaining a critical population density and the persistence of the edible algal species requires a critical level of limpet predation. There exist two quite different equilibria, one with limpets and a diverse algal array, another with no limpets and a depauperate collection of algae. The second equilibrium is clearly more stable at the local level; the first is maintained, in spite of periodic local decays to the other, by occasional disturbances which scour off the inedible algae and immigration by limpets from nearby areas. The existence of more than one stable equilibrium, of several possible configurations of densities for any given collection of species, appears to be a common characteristic of multispecies communities.

6.1 The Nature of Multispecies Interactions

Effects of Predation on Competing Species

One of the first people to investigate the possible implications of predation for competing species was Slobodkin (1961). Slobodkin made use of the Lotka–Volterra equations, tacking on a density-independent predation term:

$$\frac{dn_i}{dt} = r_{i0}n_i\left(1 - \frac{n_i + \alpha_{ij}n_j}{K_i}\right) - Pn_i \qquad (6\text{-}1)$$

where n_i, r_{i0}, α_{ij}, and K_i are as defined in previous chapters, and P describes the predation intensity, which, in turn, is proportional to the predator population density. Equation (6-1) can be rearranged to read

$$\begin{aligned}\frac{dn_i}{dt} &= \frac{r_{i0}n_i}{K_i}\left[K_i\left(1 - \frac{P}{r_{i0}}\right) - (n_i + \alpha_{ij}n_j)\right] \\ &= r^*_{i0}n_i\left(1 - \frac{n_i + \alpha_{ij}n_j}{K^*_i}\right)\end{aligned} \tag{6-2}$$

where $r^*_{i0} = r_{i0}(1 - P/r_{i0})$ and $K^*_i = K_i(1 - P/r_{i0})$. Thus (6-1) is, again, simply a Lotka–Volterra competition equation, but with K_i and r_{i0} modified in such manner as to include the effects of predation. We can now proceed to apply the isocline-graph method of Chapter 5, concluding that the outcome of competition depends on the relationships between

$$\frac{1}{K^*_i} \text{ and } \frac{\alpha_{ji}}{K^*_j} \qquad \frac{1}{K^*_j} \text{ and } \frac{\alpha_{ij}}{K^*_i}$$

Specifically (expression 5-5), coexistence occurs if

$$\frac{1}{K^*_i} > \frac{\alpha_{ji}}{K^*_j} \qquad \frac{1}{K^*_j} > \frac{\alpha_{ij}}{K^*_i}$$

Substituting (from equation 6-2) for K^*_i and K^*_j,

$$\frac{1}{K_i} > \frac{\alpha_{ji}}{K_j}\left(\frac{1 - P/r_{i0}}{1 - P/r_{j0}}\right) \qquad \frac{1}{K_j} > \frac{\alpha_{ij}}{K_i}\left(\frac{1 - P/r_{j0}}{1 - P/r_{i0}}\right) \tag{6-3}$$

If there is no predation, $P = 0$ and the parenthetical terms vanish. Obviously, coexistence is affected by predation. If exclusion is the outcome in the absence of predation, coexistence may be achieved by introducing predation. However, if P is too large, the original outcome may be reversed, leading to exclusion of the former winner.

Slobodkin (1964) attempted to clarify these theoretical results with experiments. Placing brown hydrids (*Hydra littoralis*) and green hydrids (*Chlorohydra viridissima*) in mixed cultures, he found that both species persisted if kept in darkness. In light, the green hydra benefit from the presence of the photosynthesizing green algae living in their tissues (which give them their color) and competitively exclude *H. littoralis*. However, when Slobodkin took on the role of a nondiscriminating "predator," removing fixed fractions of the total hydrid population at regular intervals, something interesting happened. As the predation rate rose from 0 to 25 to 50% per day, the green hydra required longer and longer times to exclude its competitor. At 90% predation per day, Slobodkin achieved coexistence.

Slobodkin's original, theoretical approach has been extended and been made considerably more sophisticated in recent years (Cramer and May, 1972; Comins and Hassell, 1976; Yodzis, 1977). An alternative approach is offered by Roughgarden and Feldman (1975). Consider the following system of three prey and one predator, where (to keep calculations simple) individuals of all three prey species have the same r_{i0}, the same K_i, the same food value for the predator, and the same death rate from predation. Write

$$\begin{aligned}\frac{dP}{dt} &= P\left[\frac{c}{x}(n_1 + n_2 + n_3) - \delta\right] \\ \frac{dn_1}{dt} &= \frac{r_0 n_1}{K}[K - n_1 - \alpha n_2 - \beta n_3] - cn_1P \\ \frac{dn_2}{dt} &= \frac{r_0 n_2}{K}[K - \alpha n_1 - n_2 - \alpha n_3] - cn_2P \\ \frac{dn_3}{dt} &= \frac{r_0 n_3}{K}[K - \beta n_1 - \alpha n_2 - n_3] - cn_3P\end{aligned} \tag{6-4}$$

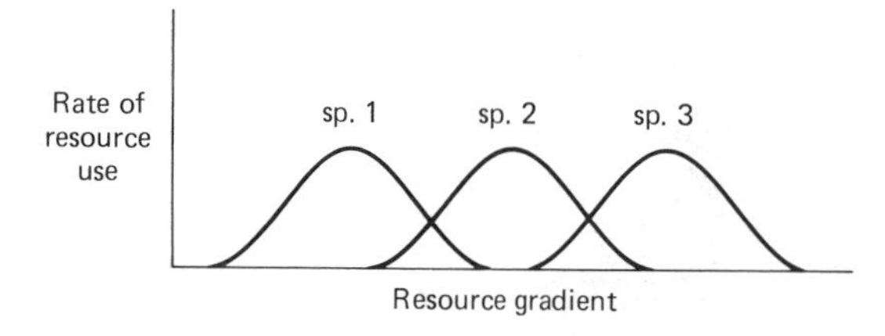

Figure 6-1 Hypothetical distribution of resource use by three species along a resource gradient.

where P is the predator population density, n_i the population density of the ith prey, c the probability an individual prey is caught in some unit time, x the number of prey needed to make one predator, and δ the predator death rate. In Chapter 5 we depicted competition along a resource gradient as in Figure 6-1. In the example above, picture species 2 as occupying the middle niche on the gradient, with species 1 and 3 equidistant on each side and therefore (if all niches are similarly shaped) with equal competitive effect, α. The effect of species 1 on 3, or vice versa, β, is less than α because there is less resource overlap.

To investigate the impact of predation, Roughgarden and Feldman suppose that species 2 is not actually present, and then ask whether invasion by that species is made more likely by the presence of the predator. In other words, if $\lim_{n_2 \to 0} dn_2/dt < 0$, can the presence of the predator make it positive?

In the absence of species 2, and with predation present, equilibrium can be found by setting (6-4) equal to zero for $n_2 = 0$. Because species 1 and 3 are assumed identical (except with respect to their position on the resource gradient), $\hat{n}_1 = \hat{n}_3 = \hat{n}$. If species 2 is introduced under these circumstances,

$$\lim_{n_2 \to 0} \frac{dn_2}{dt} = \frac{r_0 n_2}{K}(K - 2\alpha\hat{n}) - 0 \tag{6-5}$$

Suppose that there were no predation. In the absence of both predator and species 2, let prey equilibrium be given by n^*. Then when species 2 is introduced, its dynamics are

$$\lim_{n_2 \to 0} \frac{dn_2}{dt} = \frac{r_0 n_2}{K}(K - 2\alpha n^*) \tag{6-6}$$

But predation cannot elevate prey equilibrium, only depress it. Thus $\hat{n} < n^*$, so $K - 2\alpha n^* < K - 2\alpha\hat{n}$. Therefore, expression (6-5) (with predator) is always more likely positive than expression (6-6) (without predator). In English, existence of a predator can only make successful invasion of species 2 more likely.

Roughgarden and Feldman's conclusion, straightforward as it seems, is less than the final word—for it represents a rather special case of invasion of a third species, exactly intermediate in niche position between two identical (except for niche position) competitors. Return, for the moment, to Slobodkin's original approach and note that, just as coexistence can be promoted by introducing a predator, an already existing pair of competitors might, just as likely, be destroyed by a predator. Note also that predators do not treat prey species as identical, but usually show preferences. Finally, do not lose sight of the fact that the underlying Lotka–Volterra equations (used also by Roughgarden and Feldman) are somewhat less than precise descriptors. Can Slobodkin's conclusion be saved? As will become increasingly clear, simple and consistent answers do not exist in the realm of multispecies interactions, but as a general rule of thumb it seems probable that predation promotes coexistence unless very intense. First, there is a large and growing body of evidence that food choice by predators can change in time and that a particular food species' relative rate of consumption by a predator rises faster than its relative abundance (see Chapter 16). So predation seems generally to be frequency dependent; as one species comes close to losing a competitive skirmish, its use by predators drops, gaining it a bit more competitive edge. Coexistence is not necessarily assured thereby, but it is certainly made more likely. On the other hand, extremely intense predation will make the demise of favored foods more likely and, in addition, may lower prey densities to levels at which chance extinction is likely (see also Section 6.4).

Evidence is easy to find. On the North American West Coast, the barnacles *Chthamalus* spp. are generally found high in the rocky intertidal, where their resistance to desiccation makes them a superior competitor to their nearest (and partly overlapping) tide-level neighbor, *Balanus glandula. Chthamalus* can also survive on lower rocks, even subtidally, but generally do not because at these tide levels, other barnacles have the competitive edge. But when sea star predation is high, the less preferred *Chthamalus* are able to persist at low tide levels. If predation is sufficiently high, the lowest part of the intertidal may find *Chthamalus* in almost pure stands, as the lone survivor of the predator's decimations. Seasonal suffocation of the competing barnacle populations by shifting sand may have the same effect (Connell, 1961). Similar trends can be found by examining grazed fields. Moderate grazing pressure generally (although not inevitably) increases the number of coexisting competitors. Plant species diversity falls again in the face of overgrazing.

Are Competition Effects Additive?

Where several competitors impinge on a species, it is meaningful to ask whether their competitive effects are additive: to the extent that the Lotka–Volterra model obtains, can we ignore potential, synergistic effects of two species on a third and write, simply,

$$\frac{dn_i}{dt} = r_{io}n_i\left(1 - \frac{n_i + \sum_{j\neq i} \alpha_{ij}n_j}{K_i}\right)? \tag{6-7}$$

In less literal terms (probably more meaningful, since the Lotka–Volterra equations do not incorporate food and predator dynamics), can the outcome of N-way competition be predicted on the basis of pairwise interactions? If (6-7) is correct, it is possible to express the equilibrium community in matrix form.

$$K_i = \hat{n}_i + \sum_{j\neq i} \alpha_{ij}\hat{n}_j = \sum_j \alpha_{ij}\hat{n}_j \tag{6-8}$$

because $1 = \alpha_{ii}$, by definition, and so,

$$\mathbf{K} = \mathbf{A}\cdot\hat{\mathbf{n}} \tag{6-9}$$

where $\mathbf{K}$ is the column vector of K_i's, $\mathbf{A} = \{\alpha_{ij}\}$, *the community matrix*, and $\hat{\mathbf{n}}$ is the column vector of the $\hat{n}_i$'s. This equation can be used to make predictions as follows. Suppose that we know or can estimate the α values. Then, knowing equilibrium n_i values, we can calculate $\mathbf{K}$. Now suppose that one species (species k) is eliminated. What is the effect on the equilibrium densities of the remaining species? We simply solve (6-9) for $\hat{n}$,

$$\hat{\mathbf{n}} = \mathbf{A}^{-1}\cdot\mathbf{K} \tag{6-10}$$

deleting species k (i.e., the kth row and column of $\mathbf{A}$ and the kth element of $\mathbf{K}$). The new vector, $\hat{\mathbf{n}}$, will give the new equilibrium values *provided* that the values of α_{kj}, for all $j \neq k$, are independent of the n_j's values. There is an additional, biological constraint, of course: None of the new $\hat{n}_i$ values realistically can be negative. If (say) n_i is predicted to be negative by (6-10) we know that the system, including species i, cannot persist. However, because we do not know the nonequilibrium behavior of the population it does *not* necessarily follow that species i is ousted. [For a glimpse of dissension in the literature arising from misunderstanding of this point, see Vandermeer (1969, 1981), Brenchley (1979, 1981), and Thomas and Pomerantz, 1981.] To discover the equilibrium configuration of the system, we must now recalculate $\hat{\mathbf{n}}$ several times, deleting one species at a time. If none of these species combinations leads to a strictly positive $\hat{\mathbf{n}}$ vector, we must again recalculate $\hat{\mathbf{n}}$ for all combinations of *two* missing species, and so on. We may discover in the process that there are several, possible species configurations. Such configurations are known as *domains of attraction.* We shall discuss multiple domains of attraction further later in this chapter.

Vandermeer (1969) grew four protozoan species, *Blepharisma* sp., *Paramecium bursaria, P. caudatum,* and *P. aurelia,* in 10-ml test tubes with 5 cc of culture medium. Grown alone, each species revealed a characteristic intrinsic rate of increase r_0 and carrying capacity K. Species pairs were then grown in combination and n_i, n_j, and dn_i/dt monitored. A rearrangement of the Lotka-Volterra equation (6-8) was used to calculate α values:

$$\alpha_{ij} = \frac{K_i\{1 - [(dn_i/dt)/r_{i0}n_i]\} - n_i}{n_j} \tag{6-11}$$

The α and K values obtained predicted coexistence of *P. caudatum* and *P. aurelia,* but suggested that *P. caudatum* would exclude both *Blepharisma* and *P. bursaria,* predictions that accurately reflect the results observed. These results thus also indicate that unless the α values are strongly interactive (i.e., nonadditive), a four-way competition community would decompose into a *P. caudatum–P. aurelia* system. This, in fact, is what Vandermeer observed.

Richmond et al. (1975) raised all possible combinations of *Drosophila willistoni, D. pseudoobscura,* and *D. nebulosa,* at 19 or 22°C, using the serial transfer technique (1 week after flies are introduced to a bottle, with yeast medium substrate, the adults are transferred to a new bottle. The process is repeated every week, adding emerging flies from all previous bottles to the transferred adults). In pairwise competition, *D. nebulosa* and *D. willistoni* coexist at both temperatures; *D. pseudoobscura* outcompetes *D. nebulosa*; and *D. willistoni* and *D. pseudoobscura* coexisted in three of four trials. In three-way competition, *D. nebulosa* was always eliminated. At 19°C, *D. willistoni* was also excluded, leaving only *D. pseudoobscura.* Although quantitative predictions of rates of exclusion were not accurate, the eventual three-way outcomes could be predicted qualitatively from the two-way experiments. The authors thus found no evidence for nonadditivity (although they found little support for additivity either).

On the other hand, Hairston et al. (1968), using *Paramecium caudatum* and *P. aurelia,* varieties 3 and 4, found definite evidence for nonadditivity. And Neill (1974), using microcrustacea, found the same. Neill grew various combinations of 12 microcrustacean species on 23 types of algae in the lab, obtaining equilibrium systems in about 6 weeks. He then removed one or two species and watched the subsequent, equilibrial shifts. He used the Lotka–Volterra model, and wrote

$$\hat{n}_i = K_i - \sum_{j \neq i} \alpha_{ij} n_j$$

If the (say) kth species were removed, then, *given additivity of α values,* it is also true that

$$\hat{n}_i^* = K_i - \sum_{j \neq i,k} \alpha_{ij} \hat{n}_j^*$$

where $\hat{n}_j^*$ is the value of $\hat{n}_j$ obtaining in the absence of species k. Subtracting the second equation above from the first yields

$$(\hat{n}_i - n_i^*) = \sum_{j \neq i,k} \alpha_{ij}(n_j^* - \hat{n}_j) - \alpha_{ik}\hat{n}_k$$

Because, $1 = \alpha_{ii}$, by definition, we can add $(n_i^* - \hat{n}_i) = \alpha_{ii}(n_i^* - \hat{n}_i)$ to both sides to obtain

$$0 = \sum_{j \neq k} \alpha_{ij}(n_j^* - \hat{n}_j) - \alpha_{ik}\hat{n}_k$$

Finally, because $n_k^* = 0$ (species k has been deleted), we can add n_k^* to both sides to get

$$0 = \sum_{j} \alpha_{ij}(n_j^* - \hat{n}_j) \tag{6-12a}$$

Deleting two species, k and p, we can perform similar calculations to obtain

$$0 = \sum_j \alpha_{ij}(n_j^{**} - \hat{n}_j) \tag{6-12b}$$

where $\{n_j^{**}\}$ are the equilibrium values of n_j with both species k and p deleted. If the expressions above are not borne out for all combinations, k and k plus p removed, the underlying assumption of additivity must be in error. Neill, in fact, found discrepancies, and concluded that the α values, at least in his systems, were nonadditive.

Wilbur (1972) constructed small predator-excluding pens along the margin of a pond in southern Michigan. Into these pens he placed young (advanced tail-bud stage) salamanders. Densities of 0, 32, or 64 individuals per pen were used for all combinations of *Ambystoma laterale, A. tremblayi,* and *A. maculatum.* Body weight was checked at intervals up to metamorphosis and mean larval period was noted. A multiple analysis of variance was then performed to assess differences in the dependent variables due to differences among species, density of conspecifics, and density of other species. Wilbur's results are complex; one particular finding is of significance to this discussion. In the three-way system, *A. maculatum* had no direct effect on body-weight growth of the other species, but did influence the effects of these other species on each other's growth. In these experiments, Wilbur was not looking at α values, directly, but inasmuch as α values reflect interspecific density impacts on body growth, it would appear that synergistic effects occurred. Competitive effects were not simply additive.

The evidence above was perhaps best summarized by Vandermeer (1981): "Do such (synergistic) interactions exist? I suspect all biologists presume they exist. The question is how important are they in regard to the use to which the model is to be put?" It also seems useful to add a few, practical observations. Suppose that species A, B, and C compete, and that species C, via interference, keeps B from microhabitats where it most frequently encounters A. It seems likely, then, that removal of C will result in closer contact between A and B and thus higher α_{AB} and α_{BA} values. Suppose that species C is highly attractive as a food to predators that also eat A and B. If C is locally removed, the predators are apt to move elsewhere, relieving predation pressure on A and B. With alleviation of danger from predators, these species may forage more widely, or differently, or at different times than before, all changes that have potential for altering their competitive relationship. As these simple scenarios suggest, it is probably optimistic folly to hope for additivity in competition intensity.

Facilitation

Facilitation has long been recognized and studied in the form of *mutualism* between taxonomically and ecologically divergent species (rhinoceri and tick birds, sea anemones and fish, for example). But as a possible common phenomenon among similar species, it is still far from generally accepted. Part of the resistance comes from the dogmatic position that competition is inevitable among similar species; part comes from the slowness to recognize the importance of indirect species interactions.

Facilitation between similar species is hardly a novel or heretical concept. As long ago as 1961, Mather wrote: "Similarity is not—a sufficient condition for competition even where the individuals actively impinge on one another, for likes that compete under some circumstances can even cooperate under others. Indeed, individuals competing in one respect may be co-operating in another." As a possible example, Mather notes that *Drosophila* larvae compete for yeast, but retard spoilage, to each other's benefit, by also grazing bacteria and fungi. Other examples can be imagined. Suppose that two small mammals in a grassy field share and maintain the same runway system. To the extent that each species can carry out different aspects of maintenance with different efficiencies, each benefits from the other. Of course, in both cases above the *net* interaction need not be facilitative—but at least in the latter case it could be.

Charnov et al. (1976) give a number of possible examples of facilitation. For example, different species in mixed species bird flocks, by foraging in different ways, may stir up insects to each other's advantage or, via different searching techniques, discover food sources at a faster rate. Davidson (1980) provides another. Suppose that species A is in competition with both species B and C, and that B and C are intensely competing with each other. A rise in the density of species C will press B forcefully downward. With no further examination we might expect a decline in species A as well. But though the rise in C may, initially, have a depressing effect on A, the drop in B will considerably *relieve* the competition pressure. The net result, quite possibly, could be a *rise* in A. That is, a rise in C might result in a rise in A—facilitation.

That the influence of one species on another may be mediated via a third species is not hard to see. Consider another example (Fig. 6-2). In this case we have two consumers that compete for food. Their predators are specialists and have no direct influence on each other. However, they are *indirectly* interactive. If the population density of E rises, B is depressed. This, in turn, allows A to rise, providing more food for C whose subsequent increase lets D rise; a change in E incurs the same direction of change in D.

In the general case the full influence of one species on another's dynamics, both direct and indirect, is extremely difficult to assess. But to the extent that the Lotka–Volterra equations accurately describe both prey–predator and competitor relationships, the assessment, at equilibrium, is actually a fairly trivial matter. We define

$$\alpha_{ij} = \frac{\partial r_i/\partial n_j}{\partial r_i/\partial n_i}$$

for all species pairs, *i* and *j*, whether the pair consists of prey and predator, competitors, or direct facilitators. Then the Lotka–Volterra equations tell us that at equilibrium,

$$\mathbf{K} = \mathbf{A}\cdot\hat{\mathbf{n}}$$

so that

$$\hat{\mathbf{n}} = \mathbf{A}^{-1}\cdot\mathbf{K} \tag{6-13}$$

If environmental conditions now change (or there is an outside manipulation) such that K_j increases—the effect on $\hat{n}_i$ can immediately be read off in the form of the *i*, *j*, element of $\mathbf{A}^{-1}$. Levine (1976) argues that the elements of the $\mathbf{A}^{-1}$ matrix should be "normalized" by dividing each element in the *i*th row by (*i*, *i*)th element. In this manner the diagonal elements are made unity, and the other elements describe the *relative* effects of other species. If species n_i and n_j compete, an increase in carrying capacity of *j*, presumably resulting in a rise in $\hat{n}_j$, should lead to a drop in $\hat{n}_i$. Equivalently, if an increase in K_j leads to a *rise* in $\hat{n}_i$, we can conclude that facilitation is occurring. Thus a negative *i*, *j* element in $\mathbf{A}^{-1}$ indicates facilitation.

Davidson (1980) provides us with data and the following analysis. Using a variety of approaches, in combination (see Section 6.5), she determined the **A** matrix for a subcommunity

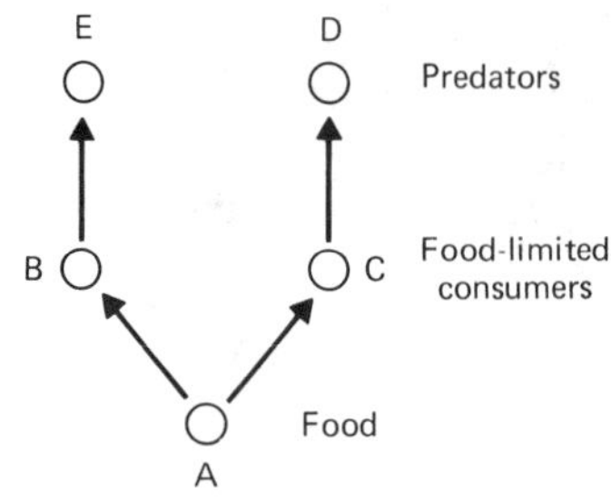

Figure 6-2 Simple food chain.

of six desert ant species:

$$\mathbf{A} = \begin{bmatrix} 1.00 & .06 & .17 & .46 & .03 & .04 \\ .06 & 1.00 & .40 & .17 & .05 & .02 \\ .21 & 1.60 & 1.00 & .18 & .23 & .10 \\ .44 & .24 & .19 & 1.00 & .13 & .22 \\ .05 & .07 & .28 & .15 & 1.00 & .88 \\ .05 & .03 & .11 & .20 & .75 & 1.00 \end{bmatrix}$$

Notice that all *direct* interaction coefficients are positive indicating bona fide competition. However, the normalized inverse of the matrix is

$$\begin{bmatrix} 1.00 & .24 & -.12 & -.52 & .01 & .03 \\ .20 & 1.00 & -.37 & -.33 & .06 & -.01 \\ -.49 & -1.79 & 1.00 & .50 & -.27 & .14 \\ -.40 & -.01 & -.05 & 1.00 & .06 & -.14 \\ .19 & .83 & -.49 & -.14 & 1.00 & -.90 \\ -.06 & -.47 & .29 & -.11 & -.74 & 1.00 \end{bmatrix}$$

indicating a large number of net, facilitative influences. What is happening in this instance can be seen by looking, for example, at the effect of species 3 on species 1. The element, α_{13}, is positive $(= .17)$, indicating that an increase in the density of species 3 depresses the equilibrium density of species 1—if the intermediary influences on and by other species are ignored. The full story is that a population increase in species 3 also depresses species 4 ($\alpha_{43} = .19$). But the effect of species 4 on species 1 ($\alpha_{14} = .46$) is strongly depressing. Thus the fall in species 4 actually alleviates competition pressure on species 1, allowing it to rise. The effect of species 3 on species 1, read from the normalized, inverse matrix, is $-.12$.

Strictly speaking, Levine's formulation tells us only the effect on $\hat{n}_i$ of altering the carrying capacity, not the equilibrium density of species j. More generally, we should like to know the effect of a change in $\hat{n}_j$ on $\hat{n}_i$ when K values are *not* altered. A method for assessing this effect has been provided by Lawlor (1979). The appropriate interaction coefficients,

$$\{\gamma_{ij}\} = \Gamma$$

are found as follows. Write

$$\hat{n}_1 = K_1 - \alpha_{12}\hat{n}_2 - [\alpha_{13} \quad \alpha_{14} \quad \cdots \quad \alpha_{1m}] \begin{bmatrix} \hat{n}_3 \\ \vdots \\ \hat{n}_m \end{bmatrix} \tag{6-14}$$

and

$$\begin{bmatrix} K_3 \\ \vdots \\ K_m \end{bmatrix} = \begin{bmatrix} \alpha_{31} \\ \alpha_{41} \\ \vdots \\ \alpha_{m1} \end{bmatrix} \hat{n}_1 + \begin{bmatrix} \alpha_{32} \\ \alpha_{42} \\ \vdots \\ \alpha_{m2} \end{bmatrix} \hat{n}_2 + \begin{bmatrix} 1\alpha_{34} & \alpha_{35} & \cdots & \alpha_{3m} \\ \alpha_{43}1 & \alpha_{45} & \cdots & \alpha_{4m} \\ \cdot & \cdot & \cdots & \cdot \\ \alpha_{m3} & \alpha_{m4} & \cdots & 1 \end{bmatrix} \begin{bmatrix} \hat{n}_3 \\ \vdots \\ \hat{n}_m \end{bmatrix} \tag{6-15}$$

If we call the $(m - 2) \times (m - 2)$ matrix in (6-15) $\mathbf{B}$, that equation becomes,

$$\begin{bmatrix} K_3 \\ \vdots \\ K_m \end{bmatrix} = \begin{bmatrix} \alpha_{31} \\ \vdots \\ \alpha_{m1} \end{bmatrix} \hat{n}_1 + \begin{bmatrix} \alpha_{32} \\ \vdots \\ \alpha_{m2} \end{bmatrix} \hat{n}_2 + \mathbf{B} \begin{bmatrix} \hat{n}_3 \\ \vdots \\ \hat{n}_m \end{bmatrix}$$

which, rearranged, gives

$$\begin{bmatrix} \hat{n}_3 \\ \vdots \\ \hat{n}_m \end{bmatrix} = \mathbf{B}^{-1} \begin{bmatrix} K_3 \\ \vdots \\ K_m \end{bmatrix} - \mathbf{B}^{-1} \begin{bmatrix} \alpha_{31} \\ \vdots \\ \alpha_{m1} \end{bmatrix} \hat{n}_1 - \mathbf{B}^{-1} \begin{bmatrix} \alpha_{32} \\ \vdots \\ \alpha_{m2} \end{bmatrix} \hat{n}_2 \tag{6-16}$$

Substitution of (6-16) into (6-14) provides us with

$$\hat{n}_1 = \frac{K_1 - [\alpha_{13} \cdots \alpha_{1m}]\mathbf{B}^{-1} \begin{bmatrix} K_3 \\ \vdots \\ K_m \end{bmatrix} - \left\{ \alpha_{12} - [\alpha_{13} \cdots \alpha_{1m}]\mathbf{B}^{-1} \begin{bmatrix} \alpha_{32} \\ \vdots \\ \alpha_{m2} \end{bmatrix} \right\} \hat{n}_2}{1 - [\alpha_{13} \cdots \alpha_{1m}]\mathbf{B}^{-1} \begin{bmatrix} \alpha_{31} \\ \vdots \\ \alpha_{m1} \end{bmatrix}} \tag{6-17}$$

The net per capita effect on $\hat{n}_1$ of altering $\hat{n}_2$ is thus

$$\gamma_{12} = \frac{\alpha_{12} - [\alpha_{13} \cdots \alpha_{1m}]\mathbf{B}^{-1} \begin{bmatrix} \alpha_{32} \\ \vdots \\ \alpha_{m2} \end{bmatrix}}{1 - [\alpha_{13} \cdots \alpha_{1m}]\mathbf{B}^{-1} \begin{bmatrix} \alpha_{31} \\ \vdots \\ \alpha_{m1} \end{bmatrix}} \tag{6-18}$$

This same formulation can be repeated for any species pair i and j to give the matrix of interaction coefficients.

Lawlor has used MacArthur's (1968) α-coefficient matrix for warblers in the northeastern United States (MacArthur, 1958) to calculate the corresponding $\mathbf{\Gamma}$ matrix. The matrices are

$$\mathbf{A} = \begin{bmatrix} 1 & .49 & .48 & .42 \\ .519 & 1 & .959 & .695 \\ .344 & .654 & 1 & .363 \\ .545 & .854 & .654 & 1 \end{bmatrix}$$

$$\mathbf{\Gamma} = \begin{bmatrix} -1 & .053 & .058 & .110 \\ .243 & -1 & 1.22 & .875 \\ .082 & .441 & -1 & -.247 \\ .284 & .537 & -.421 & -1 \end{bmatrix}$$

where the species, in order, are yellow-rumped, black-throated green, Blackburnian, and bay-breasted warbler. Note that the Blackburnian and bay-breasted warblers (if MacArthur's α estimates are valid) are indirectly facilitative.

A particularly nice example of facilitation via indirect interactions is provided by McNaughton (1975), who studied wildebeest (*Connochaetus taurinus albojubatus*) and Thomson's gazelles (*Gazella thomsonii*) in the Serengeti plains of Kenya. In the spring, 50 to 70% of approximately 600,000 migrating wildebeest concentrate along the western border of the Serengeti, and move north. McNaughton erected exclosures along the major migratory route of these animals to measure the impact via grazing and trampling, on the area's grass cover. His findings were that, in the 4 days it took the wildebeest to pass, they ate 84.9% of the green biomass, shortening the grass by 44%. In some areas severe trampling inhibited recovery after the animals had passed. But in most places, for the 28 days following the ravages of the wildebeest, productivity increased markedly. Inside the exclosures productivity was 2.6 g/m^2

per day, outside it was 4.9 g/m^2 per day. Vigorous tillering was stimulated by the grazing and a short, thick mat of green formed. Total biomass was less outside the exclosures, but green, edible biomass was considerably higher. After 32 days, Thomson's gazelles follow the migratory route, showing pronounced preferences, not surprisingly, for the grazed areas. Comparing wildebeest-grazed versus ungrazed areas, McNaughton found ($P < .001$):

	Gazelles Present	Gazelles Absent
Areas grazed by wildebeest	11	4
Areas not grazed by wildebeest	0	8

Clearly, gazelles benefit from the presence of the larger herbivore.

As noted earlier, facilitation among only distantly related species has been long accepted. Here it is usually referred to as mutualism or, sometimes, *symbiosis*. There is not space adequately to describe the variety of well-studied mutualistic interactions here; a brief review cannot justifiably be omitted. The reader interested in this subject is referred to the two-volume set *Symbiosis*, edited by S. M. Henry (1966). A few examples follow:

1. Various homopteran insects, aphids, and scales produce a waste by-product known as honeydew. In some cases this exudate is copious and very rich in carbohydrates. These insects, known collectively as plant lice, may be involved in an intricate, mutualistic relationship with ants. The ants benefit from feeding on the honeydew, whose composition and rate of production has probably evolved to encourage its use, and care for their providers by offering protection from predators. Often, ants will herd the lice into shelters at night; often they will carry individuals with them on their nuptial flights, to seed new colonies. In extreme cases, an ant species may be completely dependent for its survival on its own, specific species of louse. *Acropyga*, a tropical ant genus which keeps lice in its nest, fits this description (Flanders, 1957). A good discussion of the ant–plant louse relationship can be found in a delightful book on the behavior of ants by Sudd (1967).

2. In dry areas of central America—and probably other, climatically similar areas—there is a mutualism involving ants of the genus *Pseudomyrmex* and various species of acacias. The latter, leguminous shrubs, produce thorns with large hollow nodes and thin walls. The ants readily chew through the walls and set up residence inside. In addition, the plants bear foliar nectaries and specialized, modified leaf tips, known as *beltian bodies*, which provide food for their insect tenants. The ants benefit the acacias by repelling insect herbivores, nipping off the invading tendrils of potential plant competitors, and clearing the ground of vegetation for a radius of about 1 m around each acacia. The latter behavior acts to spare acacias from the frequent brush fires that occur in the area.

Some plant structures which benefit insect needs, like galls, for example, are induced by the insects they serve. In the acacias, the large, hollow nodes, the foliar nectaries, and beltian bodies are plant induced; these structures are produced even in the absence of ants, indicating that they evolved into their mutualist role (Janzen, 1966, 1967).

Other, Central American ants, *Azteca* spp., are obligate occupants of *Cecropia* trees. Like the acacias, *Cecropia* produces hollow spaces in which the ants live, this time in swollen internodes of the twigs. Also like the acacia, this tree produces food for its ants, this time in the form of *müllerian bodies*, lipid-, protein-, and carbohydrate-rich structures at the petiole bases. The *Cecropia* tree can survive without its ants, but with their help, in the lowland forests of Costa Rica, it is an "emergent" dominant species; in Puerto Rico, where its ants are absent, *Cecropia* is relegated to a minor position in second-growth forests (Janzen, 1968).

3. Anis are large, black birds (Cuckulids) which are quite common in the New World tropics, and which often have the habit of following cattle. As Smith (1971) has shown, the advantage to the birds of the cattle connection is clear. Cattle stir up insects and in so doing, enable both groove-billed (*Crotophaga sulcirostris*) and smooth-billed (*Crotophaga ani*) anis to feed at almost twice the rate possible in their absence. Presumably, but so far undemonstrably, the cattle benefit from the removal of annoying and parasitic flies. Heatwole (1965) describes the same advantage of cowbirds (*Molothrus ater*) and cattle egrets (*Bulbulcus ibis*). But in North America, where cattle egrets have recently extended their range, grazing animals have been partly deserted for the more lucrative wake of farm machinery. Were Benet's steel-chewing insects a reality, egrets and tractors would be thoroughly preadapted to a beautiful, mutual relationship.

4. Many tropical coral reef communities include species of fish or shrimp that make their living by feeding on the ectoparasites of other fish species. These animals, known as "cleaners", recognizable by a characteristic striped pattern and behavioral display, wait at fixed locations on a reef for the appearance of customers to be cleaned. When the cleanee arrives, it approaches, presents its head or gill region, and the cleaner swims over its body, picking parasites from its skin and, sometimes, even its mouth and gill cavities. To test the effectiveness of the cleaner on its clients, Conrad Limbaugh (described in Feder, 1966) removed cleaners from two small isolated reefs in the Bahamas. The number of fishes plummeted and within a few days those remaining were covered with blotches, sores, and swellings.

The list of mutualistic interactions goes on. Most obviously neglected, perhaps, in this book, are the coevolved relationships of plants and their animal pollinators. There is little point in presenting here a skeletal review of a subject that fills the pages of many papers and several excellent books. For information on animal–plant mutualisms, see Gilbert and Raven (1975) and Harper (1977).

Diffuse Competition

Diffuse competition is a term that refers to the total impact of all competitors on a particular species. We shall not dwell on the subject here—its importance for distribution and resource use by species and the composition of competition communities is discussed at some length in Chapter 17—except to note that it cannot be thought of simply as the sum of the individual effects of all competitors. For example, referring back to Lawlor's (1979) $\mathbf{\Gamma}$ matrix in the preceding section, the addition of bay-breasted warblers to the system of competing warblers actually *decreases* diffuse competitive effects on the Blackburnian warbler. The reader is referred back to the simplistic views presented in "Are Competition Effects Additive?" (p. 159).

6.2 Changing Environments

Stability

In Chapters 2 through 5 we distinguished neighborhood stability, global stability, and persistence. But in Chapter 4 we examined a case where depensatory mortality by predation might give rise to a *second*, neighborhood-stable equilibrium for a species. In the introduction to the present chapter we described a case with two possible stable equilibria of limpets and algae. As a general rule, in fact, when multispecies interactions occur, the systems can be characterized by more than one, and possibly by many, equilibrium states. So it is no longer (necessarily) meaningful to talk about simple limit cycles; perturbations from one equilibrium, if not immediately damping, might move the system from the stability region of one equilibrium (*domain of attraction*) to that of another. So complex communities may periodically, or erratically, change from oscillations about one density configuration of its members to another. This fact suggests another definition for "stability." How far can a system be perturbed before it is forced into another domain of attraction? A system that is not easily affected in this sense can be thought of as "resistant" or, because it returns to its former equilibrium, "resilient." Intuitively, resilient systems are those which are very neighborhood stable—those which return, naturally, to equilibrium quickly. In other words, resilient systems are those whose eigenvalues have strongly negative real parts (continuous models) or very small absolute values (discrete models). The various measures of community stability and their interrelationships are discussed by Holling (1973), Mulholland (1975), Botkin and Sobel (1975), Patten (1975), and Harrison (1979).

Ecosystems can be characterized by the dynamics of their trophic levels or trophic structure as well as by their constituent species. In fact, one can argue that specific species, per se, are of very little consequence in the ecological scheme of things—it is the *role played* that really counts (see Section 6.4). How many levels can we expect to find in trophic chains? Classically, the efficiency with which animals convert food (trophic level $x-1$) into biomass (trophic level x) hovers very crudely about an average of 10%. Thus a productivity of 1000 g/m^2 per year for plants will support about 100 g/m^2 per year of herbivore production. Primary carnivores will produce maybe 10 g/m^2 per year, and so on. A combination of limited productivity at the primary producer level, coupled with the fact that top carnivores must maintain some minimal population density to be reproductively viable, prevents the number of trophic levels from becoming large. This view, however, may be overly simplistic. Pimm and Lawton (1977) review productivity data available in the literature and note that though primary production varies over a range of 10^4 in terrestrial systems and 10^3 in aquatic systems, the number of trophic levels varies very little. In fact, the number of levels is not closely related to primary productivity, and does not appear noticably shorter even in the arctic (very low productivity) than in the tropics. These authors suggest that the more critical determinant is dynamic fragility; as more and more trophic levels are added, there are increasingly great chances of introducing destabilizing subsystems, which then eliminate the higher levels. Their conjecture was qualitatively tested by computer-simulating artificial ecosystems, following the Lotka–Volterra formulation (equation 6-7). As expected, the addition of more trophic levels led to higher rates of random extinction. Allowing *omnivory*—the feeding of individuals of one trophic level on more than one level below them—increased stability, but the trend was unchanged.

Stability: Donor- and Recipient-Controlled Systems

It is not always easy or even possible to define clearly what species are in a given trophic level. Omnivory, cannibalism, and changes in trophic status with age for some species cloud the issue considerably. The complications do not alter the following conclusions in any way. To illustrate the point more effectively, though, suppose that trophic structure is clearly definable; consider the system depicted in Figure 6-3. Here B_i is the biomass in the ith compartment, and the f_{ij}'s denote biomass flow. Individuals in compartment 2 eat individuals in compartment 1, leading to a transfer of biomass, f_{12} units per time; a portion of this, f_{20} units per time, is respired, a portion, f_{23} units per time, goes to decomposers, and the rest goes to growth and reproduction in compartment 2.

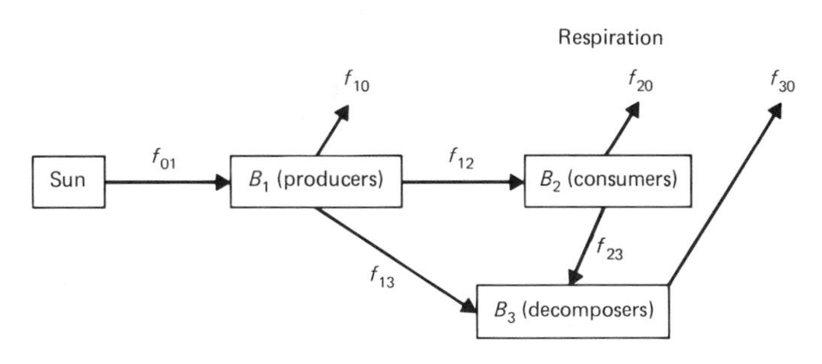

Figure 6-3 Flow diagram of a simple food chain.

The growth equation for compartments 1, 2, and 3 can be written as follows:

$$\frac{dB_1}{dt} = B_1(f_{01} - f_{10} - f_{12} - f_{13})$$
$$\frac{dB_2}{dt} = B_2(f_{12} - f_{20} - f_{23}) \tag{6-19}$$
$$\frac{dB_3}{dt} = B_3(f_{13} + f_{23} - f_{30})$$

Now consider any two-compartment subsystem (say, 1 and 2):

$$\frac{dB_1}{dt} = B_1(f_{01} - f_{10} - f_{12})$$
$$\frac{dB_2}{dt} = B_2(f_{12} - f_{20}) \tag{6-20}$$

Because f_{01} is input directly from sun, soil, and so on—the physical environment—it is reasonable to suppose that its value is independent of B_1 and B_2. f_{10}, on the other hand, and f_{20}, include density self-feedback, and so are functions, respectively, of B_1 and B_2. f_{12} probably depends on both B_1 and B_2. With these considerations in mind we now perform a standard, neighborhood stability analysis. We perturb the equilibrium system by a minute amount, x_1, x_2, and observe the subsequent changes. Using the Taylor expansion yields

$$\frac{dx_1}{dt} = \frac{d(\hat{B}_1 + x_1)}{dt} = \hat{B}_1(f_{01} - \hat{f}_{10} - \hat{f}_{12}) + x_1\hat{B}_1\frac{\partial}{\partial B_1}(f_{01} - \hat{f}_{10} - \hat{f}_{12})$$
$$+ x_2\hat{B}_1\frac{\partial}{\partial B_2}(f_{01} - \hat{f}_{10} - \hat{f}_{12})$$

$$\frac{dx_2}{dt} = \frac{d(\hat{B}_2 + x_2)}{dt} = \hat{B}_2(\hat{f}_{12} - \hat{f}_{20}) + x_1\hat{B}_2\frac{\partial}{\partial B_1}(\hat{f}_{12} - \hat{f}_{20}) + x_2\hat{B}_2\frac{\partial}{\partial B_2}(\hat{f}_{12} - \hat{f}_{20})$$

The first terms on the right side of both equations are merely $d\hat{B}_1/dt$ and $d\hat{B}_2/dt$, and so equal zero. Because f_{01} is independent of B_1 and B_2, its derivatives with respect to these parameters are also zero. The expressions above thus simplify to

$$\frac{dx_1}{dt} = -\hat{B}_1 x_1\frac{\partial}{\partial B_1}(\hat{f}_{10} + \hat{f}_{12}) - \hat{B}_1 x_2\frac{\partial}{\partial B_2}\hat{f}_{12}$$
$$\frac{dx_2}{dt} = \hat{B}_2 x_1\frac{\partial}{\partial B_1}\hat{f}_{12} + \hat{B}_2 x_2\frac{\partial}{\partial B_2}(\hat{f}_{12} - \hat{f}_{20})$$

or, in matrix form,

$$\frac{d}{dt}\begin{bmatrix} x_1 \\ x_2 \end{bmatrix} = \begin{bmatrix} -\hat{B}_1\frac{\partial}{\partial B_1}(\hat{f}_{10} + \hat{f}_{12}) & -\hat{B}_1\frac{\partial}{\partial B_2}\hat{f}_{12} \\ \hat{B}_2\frac{\partial}{\partial B_1}\hat{f}_{12} & \hat{B}_2\frac{\partial}{\partial B_2}(\hat{f}_{12} - \hat{f}_{20}) \end{bmatrix}\begin{bmatrix} x_1 \\ x_2 \end{bmatrix} \tag{6-21}$$

Recall now that stability of this subsystem, in isolation, depends on the signs of the real parts of the eigenvalues of **A**. Solving the characteristic equation of **A**, we find that both eigenvalues have negative real parts (stability) if and only if

$$\hat{B}_1\hat{B}_2\left[\frac{\partial}{\partial B_1}(\hat{f}_{10} + \hat{f}_{12})\left(\frac{\partial\hat{f}_{20}}{\partial B_2}\right) - \left(\frac{\partial\hat{f}_{12}}{\partial B_2}\right)\left(\frac{\partial\hat{f}_{10}}{\partial B_1}\right)\right]_{\hat{B}_1, \hat{B}_2} > 0 \tag{6-22}$$

Rearranged, the above becomes

$$\frac{\partial f_{12}/\partial B_1}{\partial f_{10}/\partial B_1} - \frac{\partial f_{12}/\partial B_2}{\partial f_{20}/\partial B_2} + 1 > 0 \quad \text{(evaluated at } \hat{B}_1, \hat{B}_2\text{)} \tag{6-23}$$

What does expression (6-23) mean? The leftmost term describes the degree to which the biomass flow between compartments is controlled by compartment 1, the compartment from which the flow occurs—the "donor" compartment. The next term describes the degree to which flow is controlled by the "recipient" compartment. When the donor term exceeds the recipient term sufficiently that inequality (6-23) is met, the system is said to be "donor controlled"—and it is stable. This may be put into context (Chapter 4) if we consider a simple prey–predator system. If the predator is very inefficient, much of the biomass flow from prey to predator depends on the prey self-limiting feedback term [i.e., the system is controlled to a large degree by the donor (prey)]. If the predator is highly efficient, prey self-limiting feedback plays a reduced role—the system may be recipient-controlled. Recall that the latter systems are generally unstable.

It is not necessarily true that two stable pairs of interacting trophic compartments (A and B), and (A and C), when linked, will form a stable three-compartment complex, but it is generally true that a system is more apt to be stable if more of its compartment pairs, in isolation, are stable; donor-controlled pathways have a stabilizing influence. Notice that decomposer pathways are donor-controlled (bacteria, fungi, and other detritivores do not hunt down their prey). It is possible, then, that areas with warm, moist climates, with accordingly pronounced decomposer chains, are more stable than colder or dryer areas for this reason? A good discussion of donor and recipient control and the consequences thereof can be found in Lee and Inman (1975).

Stability and Diversity

In 1955, Robert MacArthur published a paper in which he described an intuitively reasonable definition and measure of stability. That paper generated one of the most vociferous and ill-thought-out arguments in the short history of ecology. His measure, derived from an earlier suggestion by Nicholson (1933), stated that if alternative food pathways provided predators flexibility in their consumption patterns, the effects of perturbations at one point in a food web might be buffered by dietary shifts. Look at the graph in Figure 6-4. In the first graph (Fig. 6-4a), perturbation of any species (or trophic level) affects the whole system. The same is true in Figure 6-4b, except that here the effect on higher levels of a change in one producer can, to some extent, be ameliorated by a dietary switch of the consumer. The system in Figure 6-4c is even more buffered from the effects of a local (in the food web) disturbance. MacArthur reasoned that "stability," an intuitive concept of resistance of the system to disturbance, should increase with the number of biomass pathways. Of course, if the relative flow through pathways A → D, B → D, B → C → D in Figure 6-4c were minute, almost all biomass passing by way of A → C → D, then Figure 6-4c would be almost indistinguishable from Figure 6-4a. Under these

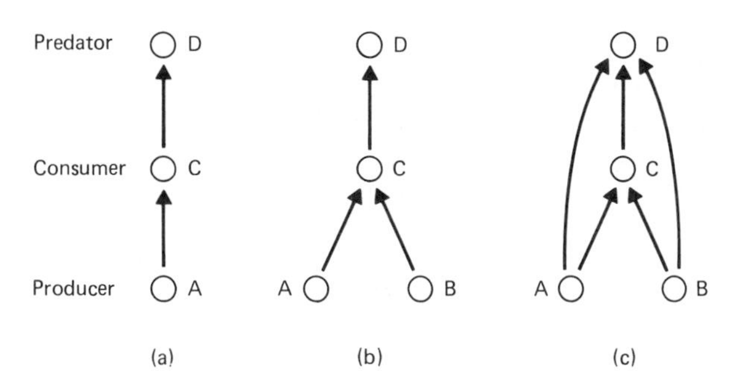

Figure 6-4 Food chains of varying complexity.

circumstances, the resistance of system 6-4c to disturbance might be even less than that of system 6-4b were the flow through the pathways in that system about equal. In other words, it is not only the number of pathways that is important, but also the equitability of the pathways vis-à-vis the proportion of total biomass carried. One measure that faithfully reflects both these considerations is

$$S = -\sum_i p_i \log_2 p_i \tag{6-24}$$

where p_i is the proportion of biomass, flowing from primary producers, that follows pathway i, and S is MacArthur's definition of stability. The form of (6-24) has appeal because of its identity to the "information content" measure of communications theory (the Shannon–Weaver equation).

It is important, to keep the following discussion in perspective, that S does not measure neighborhood stability, nor global stability, nor persistence, nor even resistance as discussed above. It relates, intuitively, to all of them. It is simply a measurable (at least hypothetically) quantity which we *expect* to reflect a purely subjective notion of stability. It states that stability, *by this measure*, is positively related to trophic pathway diversity and thus, generally, species diversity. The purely nonrigorous statement that stability is positively related to diversity thus (a) does not necessarily follow from (6-24) and (b) refers to S, and *not necessarily* any other stability measure. Nevertheless, the stability-diversity hypothesis, even without clear definitions of either term, rapidly became dogma among many ecologists and among an even larger proportion of environmental policymakers. The U.S. Army Corps of Engineers has a well-deserved reputation for trotting out the supposed relationship in their Environmental Impact Statements, and via promised species enrichment (expected under their wildlife mitigation programs) sold their projects; stability, it would seem, is good for us all.

There is considerable evidence that species diversity, in the sense of more species, makes unstable oscillations less likely. But most of the data relate to very simple systems and very simple contrasts—for example, mixed-crop farms seem to have less problem with pest outbreaks than monocultures. Also, there have been some attempts to test the relationship in the laboratory. Hairston et al. (1968), for example, grew various combinations of the protozoans *Paramecium aurelia* (varieties 3 and 4), *P. bursaria* (prey), and *Didinium nasutum* and *Woodruffia metabolica* (predators), starting with equal numbers of all species and defining stability as the reciprocal of the rate of change in population density. They found that an increase in the number of species at the lower trophic level raised the stability of the higher level. But, of course, their measure of stability differs from S and all the other definitions discussed so far. They found no predictable relationship between their stability value and the number of food pathways.

In 1969, R. T. Paine launched an attack on the stability-diversity hypothesis. In this paper he introduced the concept of a "keystone" species, one that plays a major role in shaping the community, but only a minor role as part of the food pathways. The common sea star, *Pisaster*, for example, processes only a very minor proportion of the energy flowing through the West Coast intertidal ecosystem. Yet its removal has drastic effects on persistence of other species (Paine, 1966; see also Section 6.4).

Theoreticians descended on the problem in 1970 (Gardner and Ashby) and 1972 (May). These authors simulated the dynamics of multispecies Lotka–Volterra systems, letting the values of $a_{ij}[= (r_{i0}\alpha_{ij})/K_i]$ vary randomly within bounds (both negative and positive, with mean zero, except for the a_{ii} elements, which were strictly positive). As the proportion of nonzero a values ("connectance," Gardner and Ashby), or the number of species (May) increased, the proportion of systems for which neighborhood stability occurred was found to drop. This result is counter to the stability-diversity hypothesis, at least if stability is taken to be neighborhood stability, and caused quite a furor. The first rejoinder was from Roberts (1974), who pointed out that hypothetical species which were destined to die out should not be considered when assessing neighborhood stability. Therefore, in repeating the earlier simulations, he looked at the neighborhood stability of communities consisting of *persistent* species. His results were intuitively satisfying, the reverse of those of Gardner, Ashby, and May.

Lawlor (1978) dealt a further blow to May's (1972) conclusions. In most biological systems there are no more than seven trophic levels, species do not feed on other species two or more trophic levels above them, and there is at least one plant to be found. Lawlor showed that the proportion of possible species universes that meet these very basic criteria is extremely small—on the order of 10^{-56} for systems of 30 species and connectance one. Thus the chance that May's simulations were of systems representative of real, biological systems is also remote, rendering the outcomes of those simulations meaningless.

In 1975, DeAngelis carried out similar simulations. DeAngelis reasoned that a purely random selection of a values was not biologically realistic. His choice of a values, therefore, was random, but with the following constraints:

1. The rate of decrease in biomass of i due to predation by j must exceed the rate of increase in j due to eating i (i.e., predation cannot be more than 100% efficient).
2. If i eats j, j eats k; then k cannot eat i.
3. If there are no heterotrophs, some autotrophs may *not* be self-damping.
4. At equilibrium for species i, the total effect of j on i is larger for larger n_j.

Simulations were carried out with six primary producers, three consumers, and one carnivore. Predation efficiency (= energy to recipient/energy from donor) was held constant for all prey–predator pairs. Diversity was defined in terms of the number of connections (connectance = proportion of a values that were nonzero). DeAngelis's results show the predictions of the above-mentioned authors to be a bit simplistic. If predation efficiency is on the order of .1 or more, the proportion of randomly designed systems which are neighborhood stable rises with connectance (diversity) at first, and then falls. For predator efficiency = .05, stability rises consistently with connectance. In subsequent runs, DeAngelis allowed for a_{ii} (self-damping) terms to increase at higher trophic levels. On the basis of discussions in Chapters 2 through 5, this seems biologically realistic—species at higher trophic levels are apt to be larger and so less prone to density-independent fluctuations. When this provision is made, stability (fraction of systems which are neighborhood stable) always rises with connectance. Nunney (1980a and b) has elaborated further, pointing out that the Lotka–Volterra equations are less accurate than equations incorporating functional response curves (Chapter 4). Consider the type III curve in Figure 6-5. If prey are sufficiently scarce that predator population growth rate is proportionately faster than prey growth ($\partial r_i/\partial n_j > r_i/n_j$), predation is stabilizing. If (type III curve) $r_i \propto \beta n_j^2/(1 + \gamma_j n_j^2)$ (there is a misprint in the original paper), this inequality can be written

$$\left[\frac{\partial r_i}{\partial n_j} \propto \frac{2\beta n_j}{(1 + \gamma_j n_j^2)^2}\right] > \left[\frac{r_i}{n_j} \propto \frac{\beta n_j^2}{n_j(1 + \gamma_j n_j^2)} = \frac{\beta n_j}{1 + \gamma_j n_j^2}\right]$$

or

$$\gamma_j < \frac{1}{n_j^2} \qquad \text{(6-25)}$$

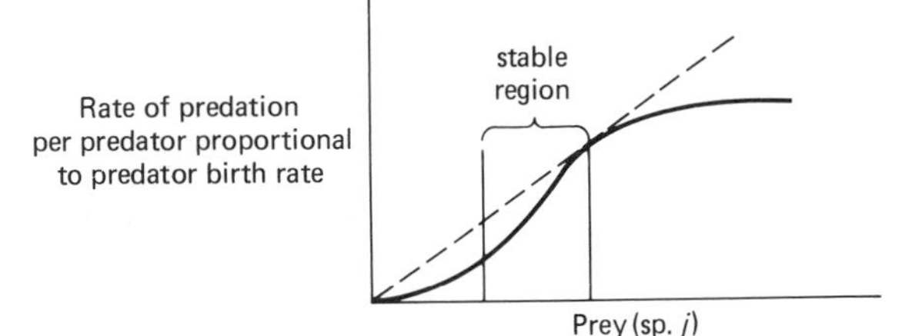

Figure 6-5 Type III functional response curve (see the text for relevance to stability theory).

If, on average, this inequality holds, the addition of more interactions, intuitively, should further buffer the system against perturbation. If C is connectance and s is number of species, each species interacts, on average, with $Cs/2$ others. Thus, when $\gamma < 1/n^2$, on average, stability (in terms of the amplitude of population cycles, and the risk of prey–predator crashes) should rise with Cs, and vice versa. Nunney verified this prediction setting s at 10, simulating system dynamics with varying C values, and defining a system to be unstable if any n fell below a fixed value *or* if any species fluctuated by more than 10%. Nunney's results not only confirm those of DeAngelis, they bring the concept of stability more into line with the intuitive notion envisioned by MacArthur.

It would be satisfying intuitively if we knew for sure that connectance, C, was positively related to species number, s. We owe a debt to Briand (1983) who examined 40 studies of trophic structure and found the empirical relationships,

$$C = 2.2s^{-.65}$$

for fluctuating environments, and

$$C = 2.71s^{-.58}$$

for constant environments. Apparently connectance and species diversity are *inversely* related in real systems!

Qualitative Stability

From Chapter 4, recall that the global stability of Lotka–Volterra systems, for any number of species, is assured by existence of the Liapunov function,

$$V = \sum_{i=1} C_i \left[n_i - \hat{n}_i - \hat{n}_i \ln\left(\frac{n_i}{\hat{n}_i}\right)\right]$$

(Goh, 1977; Case and Casten, 1979). Because multispecies systems far from equilibrium are probably not very well described by the Lotka–Volterra equations, this bit of information may be of less than overwhelming significance. We may also argue that because of environmental perturbations, systems seldom reside at or very near equilibrium and so neighborhood stability analysis also is not very useful. On the other hand, *none* of the population models considered in this book is so nonlinear that neighborhood stability analysis is entirely useless, even if fluctuations are fairly sizable. A system that is neighborhood stable is likely to be more resistant to noninfinitesimal perturbations than one which is not. So eigenvalue solutions may be more meaningful than quick skepticism would suggest. The more likely cause for despair is not the imagined futility of such analyses, but the lack of good quantitative information on the values of α_{ij} at equilibrium (see 6.5; Chapter 17). Fortunately, a number of workers have devised means for assessing neighborhood stability, at least under some circumstances, when only the *signs* of the α values are known. Because it is usually a fairly straightforward matter to figure the signs (positive for competitors, negative for predators, positive for prey, etc.), their schemes may be of some value. The theory is beyond the scope of this book, but may be found in May (1973b), Jeffries (1974), and, particularly, in Levins (1975), under the name *loop analysis*.

Succession

It is of considerable interest, in light of the theoretical difficulties surrounding stability analyses, that ecosystems, even when strongly perturbed, tend to converge to well-defined specific equilibrium configurations characteristic of the local physical environment. That is, most ecosystems are globally stable (as we shall see later, this is true for species *roles*, rather than species per se—in some cases, one species may substituted for another). The process by which these systems return to their equilibrium state, usually called the *climax*, may be rapid or extremely slow. It is known as *succession*, and the pathway followed (which may depend on the nature of the disturbance generating it) is referred to as a *sere*. There may be more than one sere leading to the same climax.

The interspecies interactions which drive succession may be complex, but the process can, to a large extent, be understood by examining competition among the "dominant" species [but see "Succession" (p. 176); Caswell, 1976; Hubbell, 1979]. The dominants are those species which, by virtue of their size, or population density, or important role in the community, set the stage for the other species. In forests it is the trees which most affect the soil, throw the most shade, buffer the wind and the variations in light, temperature, and humidity, and which produce a diversity of structures and textures in which other species can make a living. Trees are the dominants in forest communities. In prairies the dominants are usually the grasses. In rocky intertidal communities the barnacles or mussels form dense beds which determine the nature of the surfaces on which other species settle or forage.

Competition among forest dominants is generally for space, the outward manifestation being a race to avoid shading by other trees. Species that disperse and grow most readily on disturbed sites are usually the "pioneers," the first to colonize after a disturbance, but it is the species that are most tolerant of shade that eventually "win" and become the dominants of the climax community. Thus, in western Washington the successional sequence, beginning with shade-intolerant ponderosa pine (*Pinus ponderosa*), moves through the more shade resistant Douglas fir (*Pseudotsuga canadensis*), Engelmann spruce (*Picea engellmanni*), and mountain hemlock (*Tsuga mertensiana*) to the highly shade tolerant western hemlock (*Tsuga heterophylla*) and western red cedar (*Thuja plicata*). In southern Indiana, sumac (*Rhus* spp.) and "cedar" (*Juniperus virginiana*), which require open sunlight, are displaced by, among others, cherries (*Prunus* spp.), hickories (*Carya* spp.), and finally with the very shade tolerant oaks (*Quercus* spp.) or sugar maple (*Acer saccharum*) and beech (*Fagus grandifolia*). Typically, lower, wetter areas in the Midwest are characterized, at climax, by beech and sugar maple, dryer areas, including south-facing slopes and ridge tops, by oaks and often hickories. On the eastern slopes of the Colorado Rockies, where water is less available then in the Midwest, terrain is even more important in determining community composition. Here, north-facing slopes, cooler and more moist, are covered, below about 9000 ft altitude, with dense stands of Douglas fir. The south-facing slopes, warmer and dryer, have sparse stands of shrubs and small ponderosa pines. An excellent review of the nature of competition and succession, especially in forests, can be found in Connell and Slatyer (1977).

In grasslands, the shortage of water and the frequent fires preclude successful growth of trees. Here shading is clearly secondary to water as an important factor in determining successional seres. The effects of plants on soil chemistry is also important. For example, legumes, which by virtue of their symbiotic nitrogen-fixing bacteria can colonize nitrogen-poor soils and outcompete other species, may soon enrich the soil and seal their own competitive doom as dominants. Legumes are common climax components mostly in areas with very low soil nitrogen content. Grasslands and, even more, desert areas are often characterized by allelopathic species (Chapter 5). Their inhibiting chemicals may play a complicated role in succession (Wilson, 1968). Autoallelopaths may hurry their own demise, while alloallelopaths (such as ragweed) can prolong their existence (e.g., see Neill and Rice, 1971). A typical old field successional sequence (for central Oklahoma and southeastern Kansas) is given by Booth (1941). It begins with a plowed field which, left to nature, first becomes an area of "weeds," fast-growing, fast-dispersing sunflowers (*Helianthus annuus*), daisy fleabane (*Erigeron canadensis*), and other species (*Digitaria*, *Haplopappus*, *Croton*, etc.). This array gives way, after perhaps 2 or 3 years, to annual grasses (mostly *Aristida oligantha*), which dominate for 9 to 13 years, then perennial bunchgrasses (*Andropogon scoparius*) and finally a climax of several perennial grass species (*A. scoparius*, *A. gerardi*, *Panicum virgatum*, and *Sorghastrum nutans*).

In rocky intertidal areas, bare rock surface is usually first invaded by encrusting and film-forming algae and diatoms, followed by limpets and other grazers, and then barnacles and mussels. The species involved and the order in which the groups colonize is, to some degree, determined by the season in which the disturbance takes place, and who happens to have settling larvae at the time. The barnacles and mussels are followed by their predators, starfish and the predatory snail, *Thais* spp. Finally, climax is reached with the addition of the other various animals which live on the barnacle–mussel substrate, such as the chiton *Nuttalina* spp. and the crab *Pachygrapsis* spp. The entire sequence may take $2\frac{1}{2}$ years (Sousa, 1979). See also Connell (1961b) and Dayton (1971).

Succession may occur very slowly, sometimes reaching plateaus, sometimes, because of repeated disturbances, never reaching climax. In the absence of periodic fires, the easternmost extensions of the tall grass prairie of north America would long ago have become forest. Cessation of the fires following settlement by Europeans has resulted in an encroachment by trees. In parts of Wisconsin, the early 1800s saw tall grass prairies with occasional clumps of the fire-resistant burr oak (*Quercus macrocarpa*) scattered about. Today, woodlands cover these sites, and a few clusters of the remnant oaks ("oak openings"), now very old, can still be seen. Prevention of climax by repeated disturbance results in a *dysclimax*. The old prairies of Wisconsin were fire dysclimaxes.

How often do eastern American forests fail to reach climax? It seems pretty clear that most of the few virgin stands still in existence are at climax. But can the younger beech–maple forests of Michigan or the oak–hickory forests of Indiana really be considered climax? Most have fairly common scattered remnants of earlier successional species. Views differ. Henry and Swan (1974) examined, in great detail, a $\frac{1}{4}$-acre area in New Hampshire. Core samples of trees, living and dead, standing and fallen, estimation of sizes and ages of buried, dead wood, and observations on the direction of tree falls and extent of damage to trees hit by falling comrades enabled them to reconstruct the forest history back to before 1665, at which time a huge fire had leveled the area. Trees burned to the core in that fire indicate the presence of much dry wood, suggesting extensive damage by storms not long before. There was evidence of extensive wind damage again in 1898, 1909, 1921, and 1938 (confirmed by weather records). In the 1909 storm all trees with stems more than 9 inches at breast height were flattened. Existence of white pine (*Pinus strobus*) in that area requires fire—fire releases seeds in the cone and allows the tree to invade and gain a quick foothold. In the absence of fire, white pine is competitively excluded. Pines occurred in the area until 1938, suggesting prior, frequent disturbance by fire.

A number of community characteristics seem to change consistently (or, at least are thought to do so) with the course of succession. These changes are listed and discussed by Odum (1969). Some of them are included here. Immediately after a major disturbance, total biomass is apt to be depleted and nutrients may be washed away. But as new growth begins, nutrients are better retained and begin to rebuild. There is little density feedback, and productivity relative to biomass is high. The community is dominated by opportunistic species with high population growth potentials. As biomass builds, density feedback grows and the ratio of productivity to biomass, P/B, drops. Species with lower intrinsic rates of increase, but better competitive ability gradually gain dominance. Since populations of these species are usually comprised of larger individuals, and are age structured, the influence of density-independent fluctuations is diminished and population fluctuations are subdued; the system, vis-à-vis erratic changes in population levels, is stabilized. Population dynamics become more predictable. It is usually true that large individuals, which breed at intervals throughout their lives, are more pliable behaviorally. Behavioral adaptation to changing conditions thus further stabilizes the system. Also, while good colonizers, the early species, are usually successful by virtue of being catholic in their habitat requirements. The later species, living in a more stable and so predictable environment, can afford the luxury of habitat specialization (see also Section 6.4); they have narrower niches. Support for these generalizations is widespread. See, in particular, the field study by Golley and Gentry (1964), and the laboratory microcosm study by Cooke (1967). Alternative approaches, leading to the same conclusions, can be found in the work of Margalev (1958, 1963, 1968).

One consequence of narrowed niches is that more species can fit into the community without extensive overlap of resource utilization. All else being equal—that is, ignoring other changes occurring with succession, such as decline in the amplitudes of population fluctuations—this suggests that the number of species should rise. Whether or not narrowed niches are the *raison d'être*, species diversity does, in fact, usually increase with succession. It is true, for example, for plant species diversity in eastern Canadian forests (Pielou, 1966), true for plants in the ground layer, shrub layer, and in the canopy for old field to forest successional sequences on the Georgia piedmont (Nicholson and Monk, 1974), and it is also true for birds in the latter area (Johnston and Odum, 1956) and on abandoned strip-mined land in Illinois (Karr, 1968). Actually, though, close examination usually reveals that the rise is followed by a slight fall as climax is reached. Occasionally, the usual rise, earlier in succession, fails to materialize. D. L. Taylor (1973), following trends in Yellowstone forests, recorded the number of herb species, shrubs, trees, birds, and mammals in areas which had initiated succession after fires, 1, 3, 7, 13, 25, 57, 111, and about 300 years in the past. Herbs and shrubs rose and then, clearly, dropped in diversity. Trees rose and seemed to reach constancy after only about 25 years. Birds showed no clear trend at all, and mammals, after an initial rise, seemed largely to disappear after about 50 to 100 years. The usual rise in species diversity during succession may well occur simply because first, dispersal to a new area and subsequent colonization takes time and some species take a while to "find" the area, and second, because newly arrived species become established before the old have entirely died out. Whether the apparent trends are the result of deeper-lying more interesting processes, and whether they are even consistent in time and space, seems less than firmly established.

6.3 Spatially Structured Populations

The Effect of Predation on Competitive Coexistence

Earlier in this chapter we noted that predation might affect the outcome of competition between prey species and that, in general, predation promotes competitive coexistence. In Chapter 5 we noted that local disturbances might act to open up areas for colonization by inferior competitors and make possible their persistence. Caswell (1978a) has combined these two kinds of influence into what he calls a *nonequilibrium model*. Consider a habitat with many patches or *cells*. Suppose that, in any one such cell, a species in the presence of a superior competitor is inevitably excluded, and that invasion of a cell by the predator species leads inexorably to extinction of either or both competitors and therefore the predator also. We shall not look at actual population sizes in each cell, only the presence (denoted by 1) or the absence (denoted by 0) of either of the two competitors and the predator. More specifically, suppose the following scenario. Species L, the loser, persists in any cell unless that cell is invaded by a winner, W, in which case it becomes extinct in T_C time units. Species W persists if alone or if its cell is invaded by species L. If the predator, P, invades, it eliminates either competitor, or both, in T_P time units, and then itself dies off. The three species migrate randomly among cells at fixed but arbitrary rates D_L, D_W, D_P. Computer simulation of this system (which, note, has no density feedback and no specified dynamics beyond the simple provisions set forth above) leads us to the following conclusions.

1. Presence of a predator greatly prolongs the time to extinction of the inferior competitor.
2. Median time to extinction of species L rises, with or without predators present, as the number of cells increases.
3. Persistence time increases with a drop in the rate of migration of the winning competitor, and also with a rise in the time required for competitive exclusion in any one cell.
4. When predators are present, persistence time of L increases with a rise in T_P or the migration rate of the predator.

Caswell's results are not surprising; they verify what we might have expected on intuition. Predation opens up cells which otherwise would have remained closed to colonization by the weaker competitor. Spatial structure in the environment stabilizes the species interactions.

Succession

Our earlier discussion of succession implied that the process of ecosystem recovery was without local variation, that after some extensive disturbance had, in effect, destroyed the community, that community would respond uniformly to competitive and other influences and return gradually to an equilibrium state. In fact, the above-cited results of Henry and Swan (1974) notwithstanding, most damage to ecosystems is apt to be local. Whole hectares of beech trees are not flattened every year by catastrophes, but at least this frequently, a large beech falls over or loses a large branch, causing a break, a light gap, in what otherwise might be a fairly uniform, climax canopy. Is it possible, then, that succession, on a highly local scale, is proceeding continuously even in very old communities? Can we think of forests—or any ecosystem for that matter—as a mosaic of local patches each at one or another stage of succession? If so, we should find a close correlation between the size of light gaps in a forest and the size of clumps of earlier successional tree species, which at one time must have grown up in such gaps. To test this concept, Williamson (1975) measured the aerial extent of recent openings caused by tree falls in a 75-by-150-m area in Hoot woods, an old-growth beech–maple forest in Indiana. These gaps varied between 22 and 252 m^2, with a median size of 75 m^2. Williamson then analyzed clump size of various tree species by applying Morisita's (1959) measure of dispersion. Morisita's index is given by

$$I_\delta = q \sum_{i=1}^{q} \frac{n_i(n_i - 1)}{n(n-1)} \tag{6-26}$$

where q is the number of quadrats of some given size, n_i the number of individuals (stems) in the ith quadrat, and n the total number of individuals. Random dispersion is indicated if $I_\delta = 1.0$. Clumping is indicated if $I_\delta > 1.0$. The procedure is to calculate I_δ values *for each of several quadrat sizes* for each tree species, one at a time (or, if desired, groups of species), then plot I_δ against quadrat size. The quadrat size for which I_δ is maximum (>1.0) is the area size at which clumping is most strongly indicated (i.e., the size of the average clump) (Fig. 6-6). Williamson used nested quadrats ranging in size from 1 m^2 to 2500 m^2. I_δ values significantly greater than 1.0 were found for tulip poplar (*Lireodendron tulipifera*) and ash (*Fraxinus* spp.), earlier successional species, and displayed a distinct maximum at 67.5 m^2, very close to the observed (75 m^2) size of the median existing light gaps. Beech and sugar maple, the climax dominants,

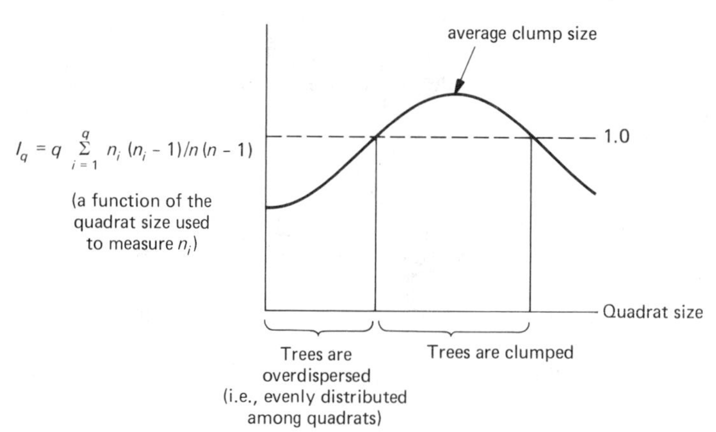

Figure 6-6 Hypothetical example of the use of the Morisita index (I) to determine clump size.

would not, in general, be expected to have colonized light gaps; they would have gradually replaced the tulip and ash which did. Hence beech and maple would not necessarily be expected to show strong clumping. In fact, their I_δ values deviated little from 1.0 over the full range of quadrat sizes.

The concept of biological communities as mosaics of successional stages opens up whole new vistas for the study of succession and ecosystem stability in general. In the simplest view, let us suppose that there is always a chance that, at any point in space, a disturbance will alter the community, leading to a new configuration of species, that over a defined period of time any species configuration will be replaced via local successional processes or disturbance by another with a given probability, and that the probability of configuration i giving way to configuration j is independent of the frequency of local patches of i, j, in the community as a whole. Then we can characterize the community with a vector, $\pi = (\pi_1 \pi_2 \pi_3 \cdots \pi_k)$, describing the fraction of local patches displaying configuration 1, 2, ... , k, and a "transition" matrix, $\mathbf{P} = \{p_{ij}\}$; p_{ij} is the probability that configuration i gives way to configuration j. We write

$$\pi(t + 1) = \pi(t) \cdot \mathbf{P} \tag{6-27}$$

The process described is known as a *Markov process*, and its characteristics are mathematically well understood. A number of interesting features of successional sequences can be predicted if **P** is known. We shall first review some of these features and then look more closely at how **P** can be used to make predictions.

Many forest species, even including the most shade tolerant, may not be able to grow from seedlings in the shade of their parents. Trees are known to alter soil chemistry in their vicinity, changing the pH, or adding various phenolic and other compounds, thereby polluting their own backyard and forcing their own demise. J. F. Fox (1977) examined six forests, from maple–beech climax in Michigan and the Alleghenies, to spruce–fir stands in the Rockies and the Great Smoky Mountains, to oaks in Florida. In all cases, the number of growing saplings beneath the two most dominant tree types were primarily of other species. Fox feels that such behavior may lead to an alternation, a cyclicity of dominants. Gilpin (1975a) presents one possible mechanism for Fox's replacement observations,and shows that, indeed, they *do* lead to cycling. For example, suppose that species A is in exploitation competition with B and is replaced by B ($P_{AB} > P_{AA}$, $P_{BA} = 0$). Then suppose that species C excludes B in the same way ($P_{BC} > P_{BB}$, $P_{BA} = 0$). Finally, let species A oust species C by interference—perhaps allelopathy ($P_{CA} > P_{CC}, P_{CB} = 0$). But Gilpin's argument is valid only for single patches, for environments that do not exhibit spatial patchiness. That cyclic replacement on a local level does not necessarily lead to the same in the community as a whole can be seen readily by application of the Markov model. Suppose that

$$\mathbf{P} = \begin{bmatrix} \alpha & 1-\alpha & 0 \\ 0 & \alpha & 1-\alpha \\ 1-\alpha & 0 & \alpha \end{bmatrix}$$

where $\alpha < .5$. Then B takes over from A, C from B, and A from C, as in Gilpin's example above. The eigenvalues of this matrix are

$$\lambda = 1, \frac{(-1 + 3\alpha) \pm (\alpha - 1)i\sqrt{3}}{2}$$

But cycling will occur only if there is at least one eigenvalue which is negative or complex, and has absolute value equal to or greater than 1.0. Otherwise, oscillations damp out. Absolute value of the second two roots, above, are strictly less than 1.0 unless $\alpha = 1.0$.

Suppose that succession leads directly to dominance by certain species and that these particular species are the ones most likely displaced as a result of outside disturbance (wind storms, fire, insect damage, overgrowing by vines, etc.). Clearly cycling of species may occur in this community (at a local level) also. Forcier (1975) describes what appears to be a very simple system of this sort. In the Hubbard-Brook forest of New Hampshire, pioneer yellow birch (*Betula alleghaniensis*) are replaced in succession by sugar maple, which, in turn, give way to beech. However, beech sprouts are sensitive to disturbance and the weedy yellow birch often replace local patches of beech on the slightest provocation.

Markov processes which are not cyclic lead eventually to a stable state. This does *not* mean a stable community! It means that, over the entire mosaic, the *fraction of patches in various successional states* is constant. The system is constantly changing on a local level, patches of climax dominants are being wiped out, species A is giving way to species B in other patches; the fractions of all patches with climax dominants, or other species, however, remain constant. But the time scale over which this dynamic equilibrium is reached is of considerable importance, for while (for example) northern hardwoods succession in any one local area might require about 300 years, the Markov process may take up to 1000 years. If major widespread disturbances are as frequent as indicated by Henry and Swan (1974), or even as suggested by Borman and Likens (1979), dynamic equilibrium may seldom or never be reached before the entire forest is eliminated and succession must begin *de novo*. On the other hand, 300 years may be enough time for something reasonably close to dynamic equilibrium to be reached.

In very old communities, those at or near dynamic equilibrium, the fractions of patches in various successional stages, π_1, π_2, and so on, represent a steady state, a stable successional structure. These equilibrium π_i values, $\hat{\pi}_i$, can be used in conjunction with the p_{ii} values to calculate various pieces of information. For example, if a local area has just left one successional stage, j, the expected time before it again enters that stage is given by

$$\tau_j = \frac{1 - \hat{\pi}_j}{\pi_j(1 - p_{jj})} \tag{6-28}$$

If j is taken to be the first, pioneer stage in a forest, τ_1 is the average time between formations of light gaps. Once a patch of environment has entered the jth stage, it will remain, on average, in that stage for a period

$$D_j = \frac{1}{1 - p_{jj}} \tag{6-29}$$

Information of this sort requires knowledge of the p_{ii} values and the $\hat{\pi}_i$ values. Observations of the kind described by Fox (1977) or Horn (below) allow us to estimate the p_{ij} values and hence calculate the $\hat{\pi}$ vector. If the community is very old and believed close to equilibrium $\hat{\pi}$ can be observed directly and only the p_{ii} values must be measured.

Horn (1975) attempted to apply the Markov process approach to a forest on the Princeton University campus. Patches ("cells") were defined to encompass a small area, the size of an average, "mature" tree's crown (i.e., basically Williamson's light gap area). Thus only one mature tree could occupy a single cell. To keep calculations manageable, Horn then made the simplifying assumption that establishment of a new tree in a cell occurred simultaneously with death of the old tree. Thus a new tree was treated as if nonexistent if growing under an older, larger, and still living tree, and considered a replacement only when the older tree had died. Hundreds of trees were then examined and the number of saplings growing beneath each tallied. If (say) 9 percent of the saplings growing under the large-toothed aspen sampled were sassafras, it was estimated that aspen was replaced by sassafras 9 percent of the time,

$$p_{\text{aspen,sassafras}} = .09$$

This is, as Horn readily concedes, a very crude way to estimate transition probabilities, and it does not take into consideration the time required for transition, which may vary among species pairs. But a more accurate method is not readily apparent, and the timing problem can be corrected for. So we use what techniques are available to us, make our predictions, and let agreement with or deviation from reality be our guide as to the efficacy of the approach.

Using this technique, Horn formed the following transition matrix (p_{ij} = probability that j is the replacement of i):

	A	B	C	D	E	F	G	H	I	J	K
A	.03	.05	.09	.06	.06	0	.02	.04	.02	.60	.03
B	0	0	.47	.12	.08	.02	.08	0	.03	.17	.03
C	.03	.01	.10	.03	.06	.03	.10	.12	0	.37	.15
D	.01	.01	.03	.20	.09	.01	.07	.06	.10	.25	.17
E	0	0	.16	0	.31	0	.07	.07	.05	.27	.07
F	0	0	.06	.07	.04	.10	.07	.03	.14	.32	.17
G	0	0	.02	.11	.07	.06	.08	.08	.08	.33	.17
H	0	0	.01	.03	.01	.03	.13	.04	.09	.49	.17
I	0	0	.02	.04	.04	0	.11	.07	.09	.29	.34
J	0	0	.13	.10	.09	.02	.08	.19	.03	.13	.23
K	0	0	0	.02	.01	.01	.01	.01	.08	.06	.80

A: large-toothed aspen (*Populus grandidentata*)
B: gray birch (*Betula lutea*)
C: sassafrass (*Sassafras albidum*)
D: black gum (*Nyssa sylvatica*)
E: sweet gum (*Liquidambar styraciflua*)
F: white oak (*Quercus alba*)
G: red oaks (*Quercus* spp.)
H: hickories (*Carya* spp.)
I: tulip poplar (*Lireodendron tulipifera*)
J: red maple (*Acer rubrum*)
K: beech (*Fagus grandifolia*)

Zeros indicate values of less than one-half of 1 percent. Setting

$$\hat{\pi} = \hat{\pi}\cdot\mathbf{P} \tag{6-30}$$

Horn then calculated the equilibrium distribution of cell types. The vector is

$$\hat{\pi} = [0 \quad 0 \quad 4 \quad 5 \quad 5 \quad 2 \quad 5 \quad 6 \quad 7 \quad 16 \quad 50] \tag{6-31}$$

But we have not yet corrected for differences in the time required for the various species to grow up, die, and be replaced. The vector above, therefore, gives us not the fraction of cells with different trees, but rather the relative number of *occurrences* of each species, over time, in a given hypothetical cell. To obtain the first, desired vector, we must weight the relative number of occurrences of each species by its average longevity—if a species occurs twice as often and has twice the longevity of another, it occupies the cell *four* time units for every one the other is there. Longevities of tree species A through K (in years) are 80, 50, 100, 150, 200, 300, 200, 250, 200, 150, 30. Multiplying the corresponding elements in the occurrence and longevity vectors, and normalizing the products, so that they add to 100%, we obtain

$$\begin{matrix}\text{estimated percentage of cells} \\ \text{occupied by species A, B, } \ldots \text{, K}\end{matrix} = [0 \quad 0 \quad 2 \quad 3 \quad 4 \quad 2 \quad 4 \quad 6 \quad 6 \quad 10 \quad 63] \tag{6-32}$$

Because of the one cell–one tree proviso, expression (6-32) is, equivalently, the predicted percent of all individual trees made up by the 11 species. The actual observed vector in a 350-year-old area of the forest presumably fairly close to equilibrium), was

$$\begin{matrix}\text{observed percentage of} \\ \text{tree species A, B, . . . , K}\end{matrix} = [0 \quad 0 \quad 0 \quad 6 \quad 0 \quad 3 \quad 0 \quad 0 \quad 14 \quad 1 \quad 76] \qquad (6\text{-}33)$$

The similarity of vectors (6-32) and (6-33) is less than perfect, but considering the very small amount of information used to calculate (6-32), and the admitted crudeness of the measures used, the agreement is nothing short of astounding. Excellent results, using the same method, have also been obtained for a spruce–fir–birch forest in the Great Smoky Mountains by Culver (1981). This approach gives us a powerful reason for pursuing the Markov process approach in our efforts to predict and to understand how communities change over time in spatially structured environments.

Of course, the conditions required for application of the Markov model—in particular the condition that the transition probabilities are independent of current species densities—may not be met. Horn has discussed this problem (Horn, 1976), and Acevedo (1981) has constructed a more general model which explicitly addresses it. Among Acevedo's conclusions is the assertion that frequency-dependent birth and death rates of the species involved permits possible multiple equilibria.

Before leaving the topic of spatial structure and succession it is important to mention two further studies. Hubbell (1979) has taken an approach to forest dynamics similar to that of the Markov model, but views the colonization of light gaps more literally in terms of the immigration of species. He supposed the probability a gap would be colonized by a species to be proportional to the existing abundance of that species in the forest as a whole. (The Markov model assumes colonization probabilities to be independent of species abundance.) Simulations using this approach led to a lognormal distribution of species abundances, a distribution commonly found in many (most?) organisms (see Preston, 1962; May, 1975a). Caswell (1976) had earlier come to the same conclusion using what he called a "neutral" model. In effect, species abundance distributions and the changes in species diversity with succession can be explained and predicted on the basis of random events, without ever once considering species interactions. These theoretical results do not really mean that we can understand communities as random processes, but they do suggest that most of what we might wish to know regarding general trends can be predicted with little knowledge of the species involved.

6.4 Species Diversity

A number of properties arise from the maze of complex, multispecies struggles and machinations. We have already seen that, at least in most cases, and given sufficient time, communities exhibit a global stability which is exemplified in the successional process. We shall later have to modify this conclusion somewhat, noting that the equilibrium (or system of persistent equilibria and limit cycles) consists of a definable collection, a configuration not of specific species, but rather of ecological *roles*. The paucity of species capable of playing the dominant roles (trees in a forest, grasses in a prairie, etc.) usually assures these species' positions in the climax community. But many other less dominant roles might conceivably be played by any one of a number of alternative species, or alternative configurations of species.

The property that complex communities are, generally, globally stable, in the sense described above, their form determined by the physical environment, suggests a second emergent property: the diversity of ecological roles, or species, in a community may also be determined in part by the physical environment. That is, with some basic data on climate and terrain, we might be able to predict species diversity. As such, species diversity becomes a subject of considerable theoretical interest. Inasmuch as diversity may affect stability (see earlier discussion), diversity may also be a subject of some practical interest.

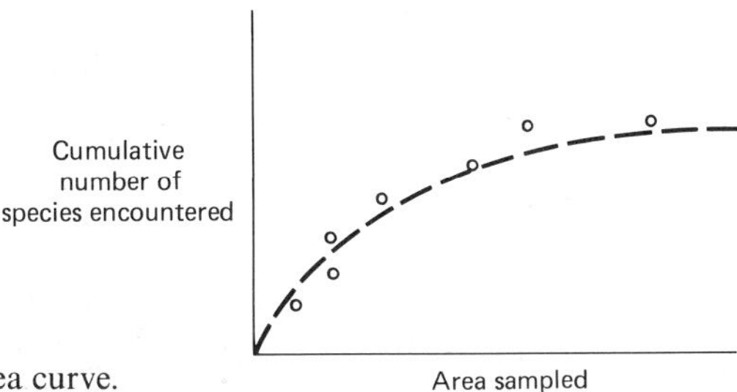

Figure 6-7 Species–area curve.

How Should Species Diversity Be Measured?

The simplest and most obvious index of species diversity is the number of species, *s*, also referred to as "species richness." Unfortunately, even though a species lives in a certain habitat, censusing of that habitat may fail to turn up a representative. Censuses are not exhaustive, and even were they exhaustive, the creature in question might well move from where the census taker had not yet been to a spot already tallied, and so be missed. As a result, species counts must invariably represent lower bounds on the species richness, even in extremely uniform habitats. One way around this difficulty is to examine ever-increasing areas within the habitat of interest and plot cumulative species number against area sampled (Fig. 6-7). The asymptote may be used to estimate *s*. But often these kinds of data are unavailable.

Another shortcoming of *s* as a measure of diversity is that it weights equally both very rare and very common species. Is a community whose four species comprise fractions .99, .006, .003, and .001 of the total population more diverse than one with only three species, each of which is equally abundant? A measure that reflects both number of species and equitability of abundance is the information-theoretic measure (described earlier, in Section 6.2),

$$H' = \text{species diversity} = -\sum_i p_i \log p_i \tag{6-34}$$

Here p_i is the fraction of the total population which is made up of species *i*. Logarithms can be taken to any base, although 2.0 or *e* ($\simeq$ 2.719) is generally used.

Other diversity measures abound. The quantity $2^{H'}$ (or $e^{H'}$ if log base *e* is used to calculate H') gives the number of equally abundant species which would generate the observed H'. $J = H'/H_{\max}$ is referred to as *equitability* and is useful in determining whether differences in species diversity between areas are due primarily to differences in *s* or to different degrees of domination by a few species. Simpson (1949) suggested that diversity could be measured with the expression

$$\frac{1}{\sum_i p_i^2}$$

where p_i is defined as in (6-34). Excellent reviews, criticisms, and further suggestions for measures of diversity may be found in Fisher et al. (1943), Preston (1948), Pielou (1966, 1969), Hurlbert (1971), and Peet (1974). For our purposes in this book we shall stick by *s* and H'. We note at the outset that although these measures can vary in quite different ways among samples, usually they are quite tightly correlated. Thus, in following discussion, unless otherwise noted, "diversity" reflects either measure.

Determinants of Species Diversity

The numbers of species living in a community reflects a balance between three processes: the origin of new species by evolution, the arrival of new species by dispersal and colonization, and the disappearance of species as mediated through species interactions and chance extinctions. The first of these is an immense topic which, out of respect for those who study it, deserves better treatment than could possibly be afforded in a small corner of one chapter. The reader is therefore directed elsewhere. Major references dealing specifically with the role of evolution in determining species diversity include Southwood (1961) and Pianka (1966a).

The importance to species diversity of the dispersal (immigration) of new species to an area is best studied under conditions where immigration events can easily be defined and, ideally, individually detected (i.e., on islands). For a coherent theory of *island biogeography*, including a theory of species diversity on islands (which may be generalizable to spatial areas less physically definable than true islands), we are indebted to MacArthur and Wilson (1963, 1967).

An "island" is an isolated unit of some particular habitat embedded in an inhospitable matrix. Thus small land areas surrounded by water are islands to terrestrial organisms; mountaintops are islands for species which can live on them but cannot survive the challenges of lower altitudes; pitcher plant leaves, with their trapped water, are islands to the protozoans living there, and to a human louse, every man is an island.

MacArthur and Wilson view the dynamics of island biogeography as follows. For every island there is, somewhere, either a mainland or a collection of other islands which serve as the source of immigrants. Each source possesses a species "pool" from which dispersal takes place. If an island is empty of species, or has but a few, most immigrants will be species new to the island. As the island becomes biologically more diverse, however, the rate at which new species arrive must decrease. Obviously, if the island possesses a full complement of the species pool, none of the immigrants will be new forms. Thus the rate at which new species disperse to the island can be depicted as a monotonic decreasing curve when plotted against s. (Fig. 6-8). This relationship, in itself, is rather trivial; but more can be said about it. If the island is small, so far from the source of immigrants, fewer of those immigrants will chance to land on it. So the curve will be lower than one for a larger, or nearer island (Fig. 6-9).

We now need to examine the third component in species diversity determination, extinction. Detailed discussion is deferred until later, but for the moment note that extinction rates are likely to be higher on small islands: (a) small areas support smaller populations which are more likely to experience chance extinction than large populations (Chapter 2), (b) small

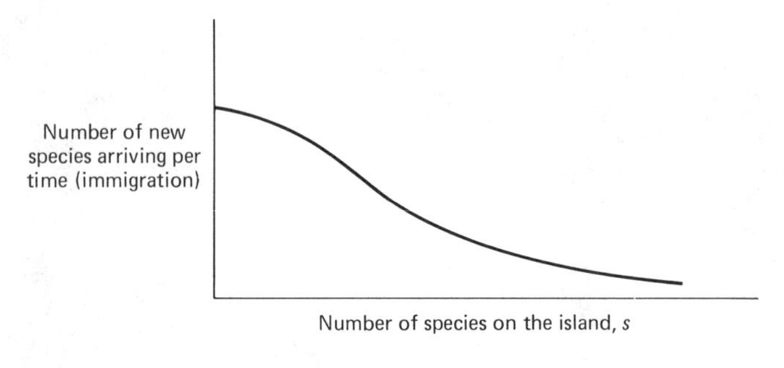

Figure 6-8 Number of new species immigrating to an island as a function of the number of species already present.

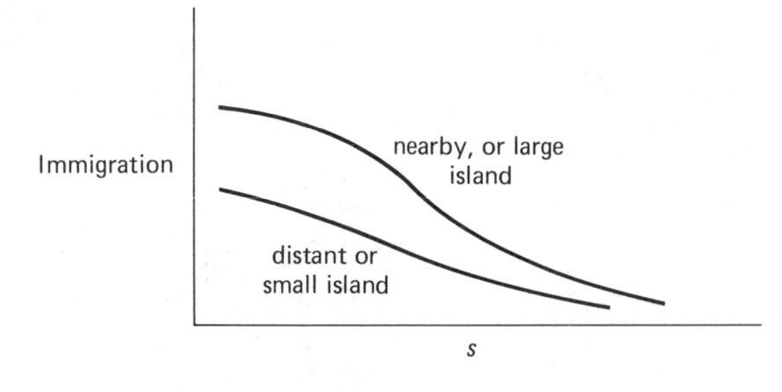

Figure 6-9 Number of new species immigrating to an island as a function of the number of species already present: effects of island size and distance to the mainland.

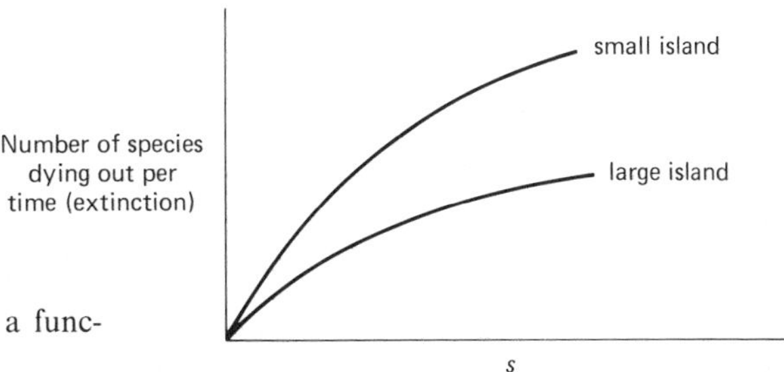

Figure 6-10 Extinction rate as a function of species richness.

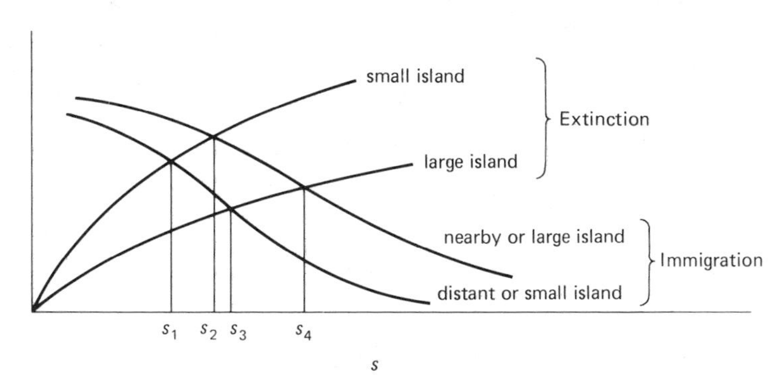

Figure 6-11 Superimposed immigration and extinction curves showing predicted equilibrium species richness.

areas will be less likely to permit competitive coexistence (see reference to Levin, 1974, in Chapter 5), and (c) small areas will less likely possess the variety of habitats which allow for niche separation and coexistence of species. Putting these observations together with the fairly straightforward assertion that the number of species dying out per unit time will generally rise with the number of species available for extinction, we obtain the relationship shown in Figure 6-10.

Figures 6-9 and 6-10 can now be superimposed. Clearly, where the input rate of species (immigration) equals the output (extinction) there must be a dynamic equilibrium. As shown in Figure 6-11, small islands can be predicted to have fewer species than large ones ($s_1 < s_3$, $s_2 < s_4$) and distant islands should be less rich than islands near to the species pool ($s_1 < s_2$, $s_3 < s_4$).

MacArthur and Wilson derive simple equations to describe immigration and extinction curves. But clearly, differences among species and serendipidity are likely to render any quantitative treatment a bit optimistic. Simberloff (1969) puts it this way: The expected errors in observed immigration and extinction are "so large as to render them useless as a measure of true immigration and extinction rates. . . ." One generalization, though, is that smaller and more distant islands should approach their equilibrium species richness more slowly.

A number of experiments have been carried out which relate directly to the theory of MacArthur and Wilson. Cairns et al. (1969) placed two series of sponges in 4 to 5 ft of water in Douglas Lake, Michigan, and periodically checked the number and identity of protozoan species. In this manner they could determine both the rate at which new species arrived and extinction rates. Although the identity of the species differed considerably between the two series, the immigration and extinction curves (Figs. 6-9 and 6-10) were remarkably similar. This study supports the MacArthur–Wilson hypothesis, but suggests that the specific species involved may be fairly unpredictable.

Simberloff and Wilson (1969) and E. O. Wilson (1969), working in the Florida Keys, covered whole mangrove islands, of varying sizes and distances from the shore, with huge plastic tents and killed (almost) all the resident arthropods by fumigation. Careful examination of resident species every few days thereafter enabled them to construct immigration and extinction curves. The scatter was tremendous—indeed the immigration curve appears to *rise* with s on at least one of their islands—but the general predictions of the island biogeography theory are borne out. Large islands and islands nearer shore filled up with species faster and attained higher diversities.

The major qualitative predictions of the theory had been observed well before MacArthur and Wilson's 1963 paper. For example, Darlington, as far back as 1943, had shown, by regression methods, that the number of carabid beetles in the West Indies varied significantly with island size. Hamilton et al. (1961) showed the same relationship for birds in the West Indies, Preston (1962) for the amphibians and reptiles on these same islands. The ponarine ants of Melanesia follow the same trend (MacArthur and Wilson, 1967) as do land plants of the Galápogos (Johnson and Raven, 1973). Working with Darwin's finches on the Galápogos, Hamilton and Rubinoff (1963) also showed the importance of distance to the species pool, finding the following equation to be a reasonably good predictor of species diversity.

$$s = 9.8 - .10D - .02G \tag{6-35}$$

where D is distance to the nearest island and G the distance to the archipelago center (Indefatigable Island) in miles.

Arbitrarily selected areas—the state of Maine; Dane County, Wisconsin; your backyard—are not islands. But they experience immigrations and local extinctions. So might they also, in spite of their lack of discreteness or isolation, follow the same rules? Of two equal-size pieces of habitat, does the one situated in the larger, continuous area of roughly similar habitat possess the greater number of species? Of two equal-size pieces, both in larger, continuous areas of similar habitat, is the one closer to the larger area's center more diverse? Do larger pieces possess more species than smaller pieces, all else being equal? Of course, the last question is biased by the fact that, by virtue of its size, the larger piece might have a greater variety of habitat types. But is the larger area more diverse even if perfectly homogeneous? Johnson et al. (1968) have data which *may* bear on these questions. Regressing the log of the number of native plant species on area (elevation and latitude had significant but less important influences) for arbitrarily selected areas in California, they found an excellent fit with little residual variance (Table 6-1).

Table 6-1

Area	Number of Species	
	Predicted	***Observed***
California coast	2432	2525
Baja California	1450	1480
San Diego County	1316	1450
Monterey County	1473	1400
Marin County	1058	1060
Santa Cruz mountains	1271	1200
Santa Monica mountains	772	640
Santa Barbara area	620	680
San Francisco area	590	640
Tiburon Peninsula	375	370

More About Immigration

A serious flaw in the island biogeography theory is that, unless existing data are used to construct statistical models, it seems incapable, at this point in our knowledge, of *quantitative* prediction. To overcome this drawback, we need to understand much better the processes of immigration and extinction and to construct quantitative models to describe them. MacArthur and Wilson (1967) summarized much of the knowledge available at the time they wrote their book, and contributed extensively to theory. Here we briefly review a few of the traits affecting immigration. In the next section we review the various hypotheses that have been put forward to explain how species avoid extinction.

Species that are at high risk from variations in the physical environment generally possess high intrinsic rates of increase (see Chapter 11) and are poor trackers of the environment (see Chapter 2). High population growth potential means large numbers of offspring from time to time, and a correspondingly large emigration pool. Poor environmental tracking makes efficient dispersal adaptive (see Chapter 13). Such species are usually small, and many encyst readily and can be carried by water or wind currents, or become attached to and carried by mobile animals. Such *opportunistic* species comprise one large class of species with high dispersal rates (and so high immigration rates). Inasmuch as successfully gaining a foothold in an area is a part of dispersal, their high reproductive potential may also be an advantage, allowing propagules to reach large numbers quickly and so avoid chance extinction. Species to which dispersal is important for survival are also, usually, generalists. A specialist may do very well indeed if it arrives on an island with the precise conditions required for its well-being, but a generalist more often finds habitable terrain. The second great class of successful dispersers, quite obviously, is made up of the highly mobile species. Oceanic islands are first colonized and have, at equilibrium, disproportionate numbers of birds and bats as well as the smaller insects, mites, protozoans, and so on, which belong in the first class of dispersers. A prototype of the successful disperser is the Tasmanian white eye, a small bird found throughout the southern hemisphere. It is fecund, mobile, a generalist, and moves in small, social flocks. The latter character makes accidental settling of a new area by one animal, or only males or only females unlikely. Rapid dispersers should make up the bulk of early colonizers to islands, a decreasing proportion as s rises; small and distant islands should have proportionately more of these species. Thus oceanic islands such as the Hawaiian Islands are often rich in avifauna, but poor in mammals.

More About Extinction

Extinction, and its prevention in the face of competition and predation, has been the subject of a considerable number of papers in recent years. In general, those interested in the maintenance of species diversity have concentrated on five broad theories: the productivity theory, the competition theory, the predation theory, the stability theory, and the spatial heterogeneity theory. We include the first four of these in this section, adding an additional, more esoteric, species interaction theory, originated by Levins and Vandermeer (references below). Spatial heterogeneity is discussed in the following section. Excellent reviews can be found in Pianka (1966a, 1974a).

Productivity Theory

One argument has it that areas with high productivity have, ipso facto, more available food than areas with low productivity. As such, food is eliminated as an important force in density feedback and competitive pressures are less. The rate of extinction from competitive exclusion is thereby lower and species diversity is higher. A much more complicated extension of this basic idea is put forth by Connell and Orias (1964). Klopfer and MacArthur (1961) and MacArthur (1972) suggested that higher productivity, meaning more food, would allow organisms to pick and choose their food, to become more specialized, to live in narrower niches. But then narrower niches means that a given range of food resources is more finely subdividable and so can support more species. These arguments are a bit misleading, at least as stated. If food shortages exert significant feedback, then regardless of food production rates, competition pressures are, by definition, high. A region with high productivity will produce commensurate population densities; the food available to each individual will be just the same as in any other region.

Nevertheless, there exist considerable data consistent with the productivity theory. It is a widely accepted generalization that in arid regions, throughout the world, species diversity in almost any taxon increases with precipitation, and precipitation is an excellent indicator and the primary controller of productivity (see, e.g., Pianka, 1966a; Brown, 1973; Davidson, 1977). On the other hand, there is contradictory evidence. Whiteside and Harmsworth (1967) found an inverse relationship between productivity and diversity of chydorids (aquatic crustacea) in Danish and Indiana lakes. A weak logical foundation and conflicting evidence make the productivity theory less than magnetic.

Whether the productivity theory has validity or not, it attempts to describe diversity differences between areas of differing, average productivity, and is not meant to apply to explanations of diversity changes within regions as productivity changes. As noted in Chapter 4, enrichment of a system often leads to destabilization and crashes of populations, and productivity increases associated with widespread disturbance (setting back the clock on succession) often lead to the same (e.g., in the case of some algal blooms).

Competition Theory

Dobzhansky (1950), noting that tropical areas are almost inevitably more diverse than temperate ones, reasoned that with less violent, density-independent variations (the tropics are physically, if not necessarily biologically, less variable than the temperate or arctic zones) there would be more *biological* control of populations (i.e., keener competition). This, according to supporters of the competition theory, leads to narrower niches (avoidance in space, or along a resource gradient). Where niches are narrow, more species can be fit into a community. On the other hand, if niches are maintained narrow, overlapping less because of intense competition, it is not at all clear why the risk of competitive exclusion is less than in another situation where weak competition permits greater overlap. There are other reasons why niches might be less broad in the tropics than elsewhere [see "Productivity Theory" (p. 185) or "Stability Theory" (p. 187). But niche overlap does seem to decline as species diversity rises. Pianka (1974b) considered 28 desert, lizard communities: 10 in North America (with 4 to 10 species), 10 in the Kalahari (11 to 18 species), and 8 in Australia (18 to 40 species). He then calculated indices of niche overlap along three environmental gradients (food, microhabitat, time of activity) for all sympatric species pairs. Overall overlap measures were calculated either by multiplying the values for the three gradients, adding them, or by any one of three other, combinatorial methods. Averaged over all species at a site, all five measures were highly correlated, and among sites, these average niche overlaps were found to diminish as number of species rose.

Niche overlap often drops as competition intensity rises. For example, in England, titmice, tree creepers, and goldcrests are most distinctly segregated in their feeding in winter, when food reserves are lowest (Gibb, 1954) (but see also Chapter 17). But correlation does not necessarily imply cause; the argument for competition as a cause of high diversity remains weak.

Predation Theory

Earlier in this chapter we argued that predation, if not very severe, should more often than not promote competitive coexistence and so increase species diversity. Some evidence has already been mentioned. Lightly grazed fields usually have a richer flora than either ungrazed or heavily grazed fields [although Harper (1969) feels that generalizations should be avoided—see also the data and conclusions of Lubchenco (1978)]. In the preceding chapter we described the predator removal experiments of Paine (1966) and Paine and Vadas (1969), which led to significant drops in species diversity. Glasser (1978) has noted that if predation is heavy, the relative degree of density feedback from food shortages is lightened. This allows prey to become more specialized in diet (see "Productivity Theory," (p. 185)) and live in narrower niches, to display less niche overlap and experience less competitive exclusion. Experiments with a predator (*Palaemonetes vulgaris*, Decapoda), prey (*Artemia* spp., brine shrimp), and food (two algal species) showed that, in fact, the brine shrimp show less niche overlap in the presence of the predator. But the reason for agreement with prediction is obscured by the fact that algal densities *declined* when the predator was present.

The logical foundation for the predation theory is a bit stronger than for those theories already mentioned. Nevertheless, it has its kinks and complications. How intense must predation be to induce significantly greater diversity among its prey? What is the effect of a greater predator species diversity on the diversity of the next lower trophic level? If predation is heavy on trophic level x, how does that affect the intensity of predation on level $x - 1$? At what point does predation lead to destabilization or direct destruction of diversity? Finally, what determines the level of predation intensity in the first place?

Stability Theory

This theory, which is due to Sanders (1968), Sanders and Hessler (1969), and Slobodkin and Sanders (1969), states that regions with only slight density-independent variation may act as species sinks. Few colonizing organisms may be able to survive the physical rigors of a strongly fluctuating environment, but most will be able to tolerate much smaller shocks. Thus immigration rates (taken to include persistence under local, physical conditions) to physically stable environs should exceed those to unstable places, and the expected species diversity should be higher. Where density-independent fluctuations occur, they can more easily be accommodated if they are gradual or predictable. So very constant, or slowly but predictably changing environments should be the most species rich. Sanders (1968) and Sanders and Hessler (1969) cite as evidence the fact that benthic animal species diversity increases in deep waters off the continental shelf of North America, where physical conditions are extremely constant. Along an otherwise increasing trend in diversity of benthonic foraminiferans with depth, drops in species number occur at 45 to 100 m and at 200 to 2500 m. These depth ranges correspond to the relatively unstable regions just above the wave base and where the drop-off beyond the continental shelf takes place (Buzas and Gibson, 1969).

Sanders's stability theory can be supplemented by the observation that density-independent instability (and unpredictability) holds populations to average sizes below what would obtain in stable situations (see Chapter 2). Predator populations are depressed even more because of additional fluctuations affecting them, via their varying food supply. But if predators are less numerous per prey item in unstable areas, predation intensity is less also. So stable areas should have proportionately heavier mortality from predation; the stability theory and the predation theory are linked.

But on the other hand, if population densities average less relative to resource supply in fluctuating environments, resource availability per consumer is up and organisms can afford the luxury of resource specialization and narrower niches. Also (see "Productivity Theory," p. 185), narrower niches mean higher species diversity. Also, as we discussed earlier (Chapter 5), environmental fluctuations, by alternating competition-free and competition-intense periods, may directly encourage coexistence and so diversity. So, not unexpectedly, there are counterexamples to Sanders's data. Abele (1976), for example, studied two coral reef areas off the Pacific coast of Panama. One experienced occasional upwellings, a temperature that varied between 17 and 29°C, and surface salinity changes of 22 to 36 ppt. It supported more decopod crustacean species than the other, more stable area with temperature and salinity ranges of only 28 to 29°C, 31 to 33 ppt. Thus the implications of environmental (density-independent) stability can be argued both ways (see also "Effects of Spatial Heterogeneity," p. 190).

Multispecies Interaction Approaches

At or in the near vicinity of equilibrium, the Lotka–Volterra competition equations are accurate—trivially, we need only to define K and α values such that the expression

$$K_i = \hat{n}_i + \sum_{j \neq i} \alpha_{ij}\hat{n}_j \tag{6-36}$$

is true. Consider any subcommunity of (only) competitors. We can find the average of each side in (6-36), over all species:

$$\begin{aligned} \bar{K} &= \bar{\hat{n}} + (s - 1)\overline{\alpha_{\cdot j}\hat{n}_j} \\ &= \bar{\hat{n}} + (s - 1)[\bar{\alpha}\bar{\hat{n}} + \text{cov}(\alpha_{\cdot j}, \hat{n}_j)] \end{aligned}$$

where $s =$ the number of competitors and $\alpha_{\cdot j} = (1/s)\sum_{i=1}^{s} \alpha_{ij}$, whence

$$s = 1 + \frac{\bar{K} - \bar{\hat{n}}}{\alpha\bar{\hat{n}} + \text{cov}(\alpha_{\cdot j}, \hat{n}_j)} \tag{6-37}$$

The formulation above (Vandermeer, 1970) is not terribly useful as a predictor—by the time data on the means and covariances for all species and species pairs are gathered, we would surely already know s—but is interesting from a theoretical viewpoint. Let us examine each term.

$\bar{K} - \bar{\hat{n}}$ rises (so that s rises) when equilibrium populations are depressed below their carrying capacities. This occurs as the direct result of an increase in the number of competitors sharing the same resources (producing an intuitively trivial component of the equation), but also as a result of density-independent fluctuations or predation. Equation (6-37), therefore, reflects both the predation theory and part of the stability theory.

Species number climbs with a decrease in mean α value, meaning that increased competitive pressures *depress* species number at any point in evolutionary time. Note that $\bar{\alpha}$ is apt to be lowered, just as $\bar{K} - \bar{\hat{n}}$ is raised, by environmental instability and predation.

The covariance term is intriguing. Its value will be small, leading to greater species richness, when $\alpha_{\cdot j}$ varies inversely with the equilibrium density of species j; in other words, when competitive ability is greater in rare, weaker in common species.

Of course, Vandermeer's model holds only for populations at or very near equilibrium. That is, (6-37) is valid only if neighborhood stability occurs. This, in turn, suggests still another approach to species diversity. Levins (1968) performed a neighborhood stability analysis on the Lotka–Volterra competition equations to obtain

$$\frac{d}{dt}\mathbf{x} = (\mathbf{D}\cdot\mathbf{A})\mathbf{x} \tag{6-38}$$

where $\mathbf{x}$ is the column vector of minute perturbations in n_i values from their equilibria, $\mathbf{A} = \{\alpha_{ij}\}$, $\mathbf{D}$ is the diagonal matrix of elements, $\{-r_{i0}\hat{n}_i/K_i\}$ (see also Chapter 5).

Now, neighborhood stability occurs if the real parts of all eigenvalues of $(\mathbf{D}\cdot\mathbf{A})$ are negative. There is a theorem which states that this occurs when the determinants of $(\mathbf{D}\cdot\mathbf{A})$ and all its minors (the matrices obtained by deleting from $(\mathbf{D}\cdot\mathbf{A})$ any corresponding row and column, or combination of corresponding rows and columns) are negative. But because $\mathbf{D}$ is diagonal and negative, this translates to the determinants of $\mathbf{A}$ and all its minors being positive. As long as this is true, the competition community is neighborhood stable. Levins reasoned that one might make use of this fact to predict species diversity in the following way. Construct hypothetical communities by adding random species (randomly chosen α_{ij} values) until the condition of positive determinants is no longer met. At that point the community has become unstable. If limit cycles are assumed not to occur, the highest s for which stability still obtains is the number of species expected in the community. Once s has been determined for a given set of accumulated, random α values, it is possible to go back and calculate the mean and covariances of these α values and relate s to these parameters. Direct calculation of s by this method is difficult, but Vandermeer (1972a) has provided us with Table 6-2.

Table 6-2 **Number of Species Expected in a Competition Subcommunity**

	$\bar{\alpha}$ Value						
$\text{cov}(\alpha_{\cdot j}, \alpha_{j\cdot})$[a]	*.025*	*.120*	*.150*	*.2*	*.4*	*.6*	*.8*
.120	5	4	—	4	3	2	2
.084	6	5	—	5	4	3	2
.030	12	11	—	10	7	4	3
.012	26	22	—	19	13	7	4
.003	65	28	—	22	15	11	6
−.100	—	—	43	35	20	16	12
−.2	—	—	56	43	24	18	14
−.3	—	—	63	48	27	20	16
−.4	—	—	73	57	30	20	18

[a] $\text{cov}(\alpha_{\cdot j}, \alpha_{i\cdot})$ is assumed to be zero for $i \neq j$.

The usefulness of the results in Table 6-2 is this: In a field experiment it is difficult or impossible to measure α values for all species pairs in a competition subcommunity. But a sample of some of the species in that subcommunity will give us *estimates* of $\bar{\alpha}$, $\text{cov}(\alpha_{\cdot j}, \alpha_{j\cdot})$, which can then be applied to predict the total number of species in the subcommunity. Does it work? Culver (1974) has applied the method to ant communities in West Virginia and Puerto Rico (Table 6-3). Clearly, the nuances of variation among communities were not picked up very well by Levins's method, but (except for West Virginia colony number IV) the general magnitude of observed and predicted values were quite similar.

Table 6-3

	Number of Species	
Colony	***Predicted***	***Observed***
Puerto Rico		
A	5	8
B	3	6
C	3	6
F	—	2
G	5	7
H	4	5
I	7	4
J	6	4
West Virginia		
I	—	2
II	4	4
III	—	2
IV	∞	3
V	4	5
VI	—	2
VII	4	3
VIII	4	4
IX	—	2
X	3	4

Effects of Spatial Heterogeneity

We have discussed a variety of reasons for believing competitive exclusion seldom occurs. Based on these discussions the following is fairly trivial: To the degree that competitive exclusion does occur, larger numbers of resources permit coexistence of more species. A more physically heterogeneous environment provides a wider array of microhabitats for plants and sessile animals, and a greater variety of these, in turn, provides a greater variety of potential niches for other organisms. Biological diversity begets still more biological diversity. We shall not pursue this argument in detail, but a few examples will be presented.

In 1961, MacArthur and MacArthur measured the foliage density at litter and ground level, shrub level, and tree (above shrub) level in a number of forests in the eastern United States. Defining

$$p_i = \frac{\text{foliage density in level } i}{\sum_j \text{foliage density in level } j}$$

they then calculated a *foliage height diversity index* (FHD) using the information-theoretic equation

$$\text{FHD} = -\sum_i p_i \log p_i$$

Insectivorous bird species diversity, BSD, was also measured in terms of this equation. When BSD was plotted against FHD, a straight line of slope 2.01 emerged, with almost no scatter; the relationship was excellent. MacArthur and MacArthur reasoned that insectivorous birds have three basic means of subdividing the environment to avoid excessive niche overlap. They might differ with respect to kinds of insects eaten, they might concentrate their efforts on certain tree species, or they might physically subdivide the forest by foraging in vertically stratified zones. The first method, they argued, was inefficient; most insects are rather tediously similar bags of hemolymph—one might avoid certain, noxious types, but among the others there would be little reason for distinguishing and every reason not to specialize. The second method is also somewhat inefficient. Why bypass a given tree species where a perfectly good food supply is waiting, and spend energy flying to the next? Regressions showed some correlation between BSD and tree species diversity, but the relationship was not nearly as neat as for BSD and FHD. Apparently, greater environmental diversity, in the sense of there being more equitable opportunities to pursue food in three environmental compartments, promotes bird species diversity. The BSD–FHD relationship was subsequently tested in Puerto Rico (where division of the forest into two vertical levels gives the best fit), and Panama (four levels) (MacArthur et al., 1966), in Australia (Recher, 1969), Chile (Cody, 1970), and in a variety of New World pine forests (Karr and Roth, 1971). It seems to hold consistently.

Cody (1968) found grassland bird species diversity to increase with structural diversity of the prairie habitat. Pianka (1966b) found the same for arid area lizards. Murdoch et al. (1972) showed a good, linear relationship between insect species diversity and plant species diversity in several midwestern old fields in July and August.

Spatial heterogeneity also affects species diversity in other ways, in conjunction with local disturbances. We have already seen how this works in successional communities. We have also already discussed the locally catastrophic effects of predation by starfish (Paine, 1966) and sea urchins (Paine and Vadas, 1969) on species diversity in the rocky intertidal (see also Sale, 1977), as well as other examples of competitive coexistence spawned by combined temporal and spatial heterogeneity (Chapter 5). Indeed, the existence of spatial heterogeneity would seem to implicate disturbance as a major positive force in maintaining species diversity, in direct contradiction to Sanders's (1968) stability theory. Of course, where communities really are at or near equilibria (true in deep-sea communities?) Sanders's thesis may well be correct. But increasing numbers of workers are insisting that equilibrium communities are rare beasts (see especially Connell, 1978).

It is interesting to note that some species whose continued existence depends on disturbance have evolved traits which promote disturbance. For example, coralline algae are poor competitors which require grazing by other species to clear areas for their settlement and persistence. One such grazer is the lined chiton, *Tonicella lineata*. It seems that the larvae of these chitons settle on the corallines, clearly to the latter's benefit, attracted specifically by a coralline algal extract (Barnes and Gonor, 1973; Paine, 1980).

Multiple Domains of Attraction and Species Roles

In any community, the number of species that can be classified as community dominants is fairly small, and their impact on community structure and dynamics is so sweeping that it is highly unlikely that substitutes could play the identical roles. With respect to less dominant organisms, though, the community impact of one species, or some combination of species, might be matched by some other species or combination of species. Thus, although succession usually leads to a characteristic climax with respect to the dominant species, leading us to view communities as globally stable, the infrastructure of these communities may take on any one of a number of possible species configurations. Perturbation of a system may lead it, without detectable change among the dominants, from the domain of attraction of the original equilibrium species configuration to that of another. The concept of multiple equilibria in multispecies competition systems has been treated by Gilpin and Case (1976). These authors made use the Lotka–Volterra competition equations, scaling density measures and time such that $r_{i0} = 1$, for all i:

$$\frac{dn_i}{dt} = n_i\left(1 - \sum_i a_{ij}n_j\right) \qquad a_{ij} = \frac{\alpha_{ij}}{K_i} \tag{6-39}$$

Several simulations were then carried out, using m (the number of hypothetical, available species) values of 2, 5, 8, 11, 14, 17, and 20, with randomly chosen, initial population densities for each species (initialized so that $\sum_{i=1}^{m} n_i = 1$). The values of a_{ij}, with means of .35, .5, .65, .8, or 1.0, were taken from random number tables. A species was considered extinct and dropped from a simulation if its density dropped below 10^{-5}. Simulations were carried on until a system appeared to reach a stable state. Then the number and identity of the remaining species was noted. For any given initial set of m, $\{n_i\}$, $\{a_{ij}\}$, there were several possible equilibrium configurations of species.

Multivariate analyses of the results showed that 90% of the variance in the number of domains of attraction, D, for each set of m, $\{n_i\}$, $\{a_{ij}\}$ could be explained by the relation

$$D = e^{p(s-1)\bar{a}} \tag{6-40}$$

where s is the number of ultimately surviving species and p is a fitted parameter ($= .062 \pm .001$). Gilpin and Case felt the results might be affected if there were a nonzero covariance between $a_{.j}$ and $a_{j.}$ (see also Levins, 1968; Vandermeer, 1970, 1972a,b). So they redid their simulations, using $a_{ij} = a_{ji}$. The results were identical in form, although p rose to $.182 \pm .025$. Note that the number of possible equilibria rises with the species diversity s and also with the value of $\bar{a}$. Because in Lotka–Volterra competition communities, $a_{ij} = \alpha_{ij}/K_i$ is the intensity of density feedback,

$$\frac{\partial r_i}{\partial n_j} = r_{i0}\frac{\alpha_{ij}}{K_i} = \frac{\alpha_{ij}}{K_i} \qquad \text{(when } r_{i0} \text{ is scaled to 1.0)}$$

a high $\bar{a}$ value indicates a strongly biologically controlled community. That is, $\bar{a}$ and so D, rises as the importance of density-independent influences declines. This has an interesting implication regarding the stability theory and the observed high diversity of species in the tropics. Since the tropics are physically more stable with respect to density-independent factors than other regions, we should expect higher $\bar{a}$ values and so more, potential equilibrium states for any competition subcommunity. If spatial patchiness occurs, an extensive area may experience many or all of these equilibria in various patches. Because each contains a different array of species, the total number of species represented may approach the total number of hypothetical, available species, *m*. In temperate or arctic regions, with fewer equilibrium states, only a small fraction of the hypothetically available species may actually occur.

Gilpin and Case also found that the number of species persisting varied tremendously from one equilibrium state to another, but approached an average asymptotic value of about 2.5. This indicates that, on average, a guild should include two to three species. The fact that observed numbers of guild species are usually greater than this (e.g., there are usually considerably more than two or three species of seed-eating birds in grasslands, many more than two or three grazers on the African plains) indicates any one, or more, of several things: Gilpin and Case's approach is unsatisfactory in that it does not consider systems with limit cycles or the effects of predators; our working definitions of specific guilds are too broad; or most likely, spatial heterogeneity (and simply spatial extent) results in communities being a mosaic of both successional stages and different equilibrium configurations.

The existence of multiple potential equilibria means that a climax community, with well-defined characteristics of stability, trophic structure, productivity, productivity to biomass ratio, nutrient cycling, and respiration can, nevertheless, be constructed, in several ways regarding its species composition. This, in turn, means that from an ecosystem perspective, species per se do not matter; the critical consideration is that the various ecological *roles* be filled. The fact that Cairns et al. (1969) and Simberloff and Wilson (1969), in the island biogeography studies found predictable immigration and extinction rates and equilibrium species diversity, but different species in different experiments, supports this conclusion. But if filling ecological roles is what community organization is all about, we should expect to find islands not only filling up to a reasonably predictable degree of species richness, we should also expect the relative numbers of species in the various guilds to be converging on constant values. Heatwole and Levins (1972) have tested this prediction using Simberloff and Wilson's (1969) Mangrove Island data. Eight crudely defined guilds—herbivores, scavengers, detritus feeders, woodborers, ants, predators, and parasites—were considered. Then, letting $p_i(t)$ represent the proportion of all species in the ith guild at time t after the start of recolonization (they use the term "trophic class"), Heatwole and Levins calculated,

$$D(t) = \sum_i [p_i(t) - \hat{p}_i]^2$$

the mean-square error deviation from the initial (pre-defaunation) equilibrium proportions. The values of $D(t)$ began quite high and fell rapidly over time. Within a year $D(t)$ approached and remained very close to zero.

6.5 Measurement of Species Interaction Coefficients

Much of the theory covered above requires, for testing or application, that we be able to estimate α values. In the laboratory, where most variables can be controlled, this is not particularly difficult. We can simply rearrange the Lotka–Volterra equations describing the type of interaction of interest, and solve for α_{ij}, substituting directly measurable values of n_i, n_j, and dn_i/dt (equation 6-11). If α_{ij} is not constant, we can use this technique to find out how it varies with n_i, and n_j. In the field, life is less simple.

The impact of prey numbers on predator dynamics is the easiest to assess. We can measure food intake rates, and we can estimate the efficiency of biomass conversion into predator growth. If we wish to complicate matters by allowing α_{ij} to vary with population densities, we can quite validly assume the form of a type II or III functional response curve, and from data for any given values of n_{prey}, $n_{predator}$ predict consumption for all combinations of n_{prey}, $n_{predator}$. We can also use the intake rates as a measure of prey removal by predators, and directly calculate prey death rates. Given a functional response curve we can use data for one set of values, n_i, n_j, and write $\alpha_{\text{prey,predator}}$ as a function of these population densities. The more difficult problem is to estimate competition coefficients.

In some rare instances, computation of competition α values may be quite straightforward. If experiments are done on systems which are fairly isolated, and where population changes are rapid enough to generate considerable variation in densities over the experimental period, it may be possible to assess directly the density-feedback quantities,

$$\frac{\partial}{\partial n_j} r_i \qquad \frac{\partial}{\partial n_i} r_i$$

Simply measure $r_i = (1/n_i)(dn_i/dt)$, n_i, and n_j at various times and perform the necessary partial regressions. Alpha, which is merely the ratio of the quantities above (equation 5-3), is then easily found. Siefert and Siefert (1976, 1979) used this technique to find α values among competing (and as it turns out, also facilitative) species of insect species living in *Heliconia* bracts in Costa Rica and Venezuela.

But nature seldom is so kind. Clearly, this approach could not reasonably be used to measure competition between oaks and hickories, or even two species of birds. For situations such as these, ecologists usually use a method originated (at least in rigorous form) by MacArthur (1968, 1972). Write the population dynamics for competitors and their resources as follows:

$$\begin{aligned} \frac{1}{n_i}\frac{dn_i}{dt} &= c_i\left(\sum_m a_{im}w_m R_m - E_i\right) \\ \frac{1}{R_m}\frac{dR_m}{dt} &= r_{m0}\left(1 - \frac{R_m}{K_m}\right) - \sum_j a_{jm}n_j \end{aligned} \tag{6-41}$$

where n_i is population density of the ith competitor, R_m the density of the mth resource, a_{im} the rate of consumption of m by i, w_m the food "value" of m, c_i the biomass conversion efficiency, and τ_i the metabolic cost of living for species i. At equilibrium, the second equation above can be written

$$\hat{R}_m = \frac{\left(r_{m0} - \sum_j a_{jm}\hat{n}_j\right)K_m}{r_{m0}}$$

which, substituted into the first, when $dn_i/dt = 0$, gives us

$$\sum_m \left[a_{im}w_m K_m - \sum_j \left(\frac{K_m}{r_{m0}} w_m a_{im} a_{jm} \hat{n}_j\right)\right] - E_i = 0$$

whence

$$\sum_m \left[a_{im}w_m K_m - \sum_{j\neq i}\left(\frac{K_m}{r_{m0}} w_m a_{im} a_{jm}\right)\hat{n}_j - \left(\frac{K_m}{r_{m0}} w_m a_{im}^2\right)\hat{n}_i\right] - E_i = 0$$

so that

$$\hat{n}_i = \frac{\sum_m (a_{im}w_m K_m) - E_i}{\sum_m [(K_m/r_{m0})w_m a_{im}^2]} - \sum_{j\neq i} \frac{\sum_m [(K_m/r_{m0})w_m a_{im} a_{jm}]}{\sum_m [(K_m/r_{m0})w_m a_{im}^2]}\hat{n}_j \tag{6-42}$$

But the Lotka–Volterra equations, at equilibrium, read

$$\hat{n}_i = K_i - \sum_{j\neq i} \alpha_{ij}\hat{n}_j$$

Thus

$$K_i = \frac{\sum_m (a_{im} w_m K_m) - E_i}{\sum_m [(K_m/r_{m0}) w_m a_{im}^2]} \tag{6-43a}$$

$$\alpha_{ij} = \frac{\sum_m [(K_m/r_{m0}) w_m a_{im} a_{jm}]}{\sum_m [(K_m/r_{m0}) w_m a_{im}^2]} \tag{6-43b}$$

Equation (6-43b), of course, considers only competition for resources and so is a measure of exploitation competition under the assumption of no interference.

It is not usually an easy matter to come up with estimates of K_m, r_{m0}, and w_m, so (6-43b) is seldom used in the form given above. More easily applied expressions are due to Levins (1968) and Culver (1970). Suppose that species use a wide array of resources and that their distribution, over time, in space represents where they need to be to efficiently use those resources (see the discussion of *ideal free distribution* in Chapters 13 and 17). Then

$$\sum_h p_{ih} p_{jh} \tag{6-44}$$

where p_{ih} is the fraction of species i individuals found in location h, gives a fair estimate of $\sum_m w_m a_{im} a_{jm}$. Because the spatial dispersion of the species presumably reflects the relative importances of the different resources to the competitors, the w_m weighting factor is implicit in (6-44). One estimate of α, then, deriving from (6-43b), is

$$\alpha_{ij} = \frac{\sum_h p_{ih} p_{jh}}{\sum_h p_{ih}^2} \tag{6-45}$$

In practice, ecologists can use quadrats or baits (Levins, 1968), and measure population numbers in each to calculate p_{ih} values.

Culver (1970) elaborated on Levins's method to find interference levels. Suppose that we define h not simply as a sample quadrat, but as a specific microhabitat type which might be represented in several sample quadrats. Culver reasoned that interference by species j would drive species i from quadrats in jointly used microhabitats resulting in lowered p_{ih} (p_{ih}^*) values. Considering all quadrats with *only species i* represented, or *only species j*, we can calculate p_{ih}^* and p_{jh}^* values for each microhabitat type. The quantity

$$\frac{\sum_h p_{ih}^* p_{jh}^*}{\sum_h p_{ih}^{*2}}$$

then, is α in the absence of interference. Culver argues that the same quantity, but calculated with p_{ih} (using *all* quadrats) in the place of p_{ih}^*, measures total competition, both interference and exploitation.

There are problems with Levins's (and Culver's) methods. First, possible synergistic effects of third species are not accounted for. Second, there is no provision for frequency- or density-dependent changes in α_{ij} with n_i or n_j. Third, (6-45) cannot give rise to a negative number—that is, the method fails to account for any facilitative effects which may modify the net value of the species interaction. Fourth, by (6-45), the product of α_{ij} and α_{ji} must always be equal to or less than 1.0. This leads to the interesting, but ridiculous consequence that *all* competition subsystems must be neighborhood stable (Neill, 1974; May, 1975b; Abrams, 1980a).

Some workers have used (6-45), or simple modifications of it (see Hurlbert, 1978), to measure niche overlap, defining p_{ih} in terms of the proposition of individuals occupying position h along a resource gradient. Thus along a (say) seed-size gradient, observed sizes of seeds eaten by several granivorous birds can be used to evaluate the degree of their dietary overlap. *If the gradient chosen describes the limiting resource, and if the gradient measure used is that which maximally separates use of that resource by the competitors*, then niche overlap, defined by (6-45), also gives an estimate of (exploitation) α. The drawbacks in applying niche overlap measured in this way to assessing α values are obvious (see also Colwell and Futuyma, 1971; Oksanen et al., 1979). It is nearly impossible to know the critical resources influencing density feedback. There may often be several resources of differing importance [although Cohen's (1978) work suggests that use of multidimensional gradients may seldom be necessary]. Nevertheless, a good many researchers have used niche overlap (as in equation 6-45) to estimate α values (e.g., Cody, 1974a, with birds; Yeaton, 1974, with birds; Davidson, 1980, with ants; Culver, 1974, with ants—see Table 6-3 and the corresponding text). Davidson (1980) used behavioral criteria to approximate interference values also, and combined them with exploitation values to estimate an overall set of competition coefficients (her results have already been discussed in "Facilitation" (p. 161).

Perhaps the most remarkable test of the efficacy of (6-45) to measure α coefficients is given by Yeaton (1974). In his study, similar chapparal areas in southern California and on Santa Cruz Island were censused for birds. Alpha values were estimated between all mainland species pairs, using niche overlap measures along a "horizontal habitat" gradient [$\alpha_{ij}(H)$—Levins's measure, with p_{ih} = fraction of species i found in quadrat i], a "vertical habitat" gradient [$\alpha_{ij}(V)$, with p_{ih} = fraction of species i found at heights 0 to 6 in., 6 in. to 8 ft, above 8 ft,] and a "feeding" gradient [$\alpha_{ij}(F)$—dietary overlap; see original article]. The combinatorial method used to find an overall competition intensity was

$$\alpha_{ij} = [\alpha_{ij}(H)\alpha_{ij}(V)\alpha_{ij}(F)]\left[\frac{1}{\alpha_{ij}(H)} + \frac{1}{\alpha_{ij}(V)} + \frac{1}{\alpha_{ij}(F)} - 2\right]$$

(see the original article for a justification; see Cody, (1974), for further discussion of combinatorial methods). Yeaton then made use of the equilibrium expression

$$\mathbf{K} = \mathbf{A}\cdot\hat{\mathbf{n}}$$

to calculate carrying capacities for the mainland population. On Santa Cruz Island, several important, mainland species are missing. By deleting these species from the matrix **A**, and the vector **K**, whose values should be very similar on the island, Yeaton reasoned that $\hat{\mathbf{n}}$ on the island could now be predicted from

$$\hat{\mathbf{n}} = \mathbf{A}^{-1}\cdot\mathbf{K}$$

For the four species whose population size discrepancies were greatest between island and mainland, the results were as shown in Table 6-4.

Table 6-4

Species	Mainland Population Density	Island Population Density	
		Predicted	*Observed*
Scrub jay	.27	1.11	1.00
Orange-crowned warbler	.07	.55	.50
Hutton's vireo	.14	.55	.60
Bewick's wren	.61	1.27	1.28

With all the simplifications, estimates and assumptions inherent in estimating α values by the methods above, the precision of Yeaton's results is amazing. Perhaps the methods are not as weak as the various objections to them indicate. Or perhaps Yeaton's results are extremely fortuitous.

See Abrams (1980b) for an approach similar to MacArthur's (1972), but more realistic in that it makes use of functional response curves to describe competitors' use of shared resources.

Other methods for estimating competition coefficients exist but are seldom used (see, e.g., Vandermeer, 1972b; Cody, 1974b; Hallett and Pimm, 1979). Still another approach (J. M. Emlen, 1981) is deferred until Chapter 17.

Review Questions

1. Studies of hypothetical bird species along the Maine coast show that three coexisting warblers, A, B, and C, have abundances of 1.0, 2.0, and 2.5 adults per hectare. Ingenious experimental manipulations and observations allow us to calculate the following competition matrix,

$$\mathbf{A} = \begin{bmatrix} 1.0 & .5 & 0 \\ .2 & 1.0 & .2 \\ .25 & 0 & 1.0 \end{bmatrix}$$

   What are the carrying capacities of the three species? On an adjacent island (with identical habitat and so, presumably, identical K and α values) we find only species B and C. What are their equilibrium densities on that island?

2. Suppose that laboratory experiments with three species of flies yielded the following:

   (1) Species 1, alone, reaches an equilibrium density of 100.
   (2) Species 2, alone, reaches an equilibrium density of 200.
   (3) Species 3, alone, reaches an equilibrium density of 150.
   (4) 1000 individuals of species 1 are required to prevent species 2 from gaining a foothold and increasing in numbers when very rare.
   (5) The corresponding figure for species 1 preventing growth of species 3 is 300.
   (6) The corresponding figure for species 2 preventing growth of species 1 is 200.
   (7) The corresponding figure for species 2 preventing growth of species 3 is 300.
   (8) The corresponding figure for species 3 preventing growth of species 1 is 200.
   (9) The corresponding figure for species 3 preventing growth of species 2 is 200.

   Assuming that the competition isoclines are straight lines, find the **A** matrix. What are the equilibrium densities if species 1 and 2 are grown together? If all three are grown together? If n_2 is increased due to experimental manipulation, what happens to n_1 and n_3? How would the isoclines look if you knew competition among all three was frequency dependent?

3. The number of trophic levels in ecosystems rarely exceeds four. Why?

4. Review and contrast the general ecological features of early and late successional species. Why does species diversity usually increase with succession?

5. Studies over a series of hypothetical forests in regions of similar climate and edaphic (soil) characteristics show that the space occupied by a cedar tree at one time is occupied, 100 years later, by another cedar ($p = .5$), a tulip ($p = .3$), or a maple. (Suppose that these are the only species present). Space used by tulips gives way, over 100 years, to maples with probability .4. Wind blows or grape vines kill tulips over this period with probability .2 and the newly opened space is colonized by tulip again half the time (never maples). Maples always replace maples unless disturbance ($p = .2$ every 100 years) kills the maples. If so, they are replaced with cedars 80% of the time, by tulips 20% of the time. What is the transition matrix? In one such forest we find 210 cedars, 170 tulips, and 550 maples. Is this forest in equilibrium?

6. Narrow niche size (implying specialization) is correlated with greater species diversity, and specialized species are more prone to extinction arising from density-independent disturbances. This seems to imply an inverse relationship between diversity and stability. How does this reasoning square with the arguments of MacArthur, May, Roberts, Lawlor, DeAngelis, and so on?
7. Because it is plausible, even likely, that α coefficients are dependent on population densities, the α-matrix approach to competition seems unrealistic. An α matrix whose elements were functions of population densities would probably be mathematically intractable. Might the Markov process approach used for studying succession be usefully applied to competition, and might it not provide more accurate predictions?

Part Three

Evolutionary Mechanisms: The Genetic Basis of Adaptation

In Part Two we discussed at considerable length the structural framework for describing population dynamics. The equations were based, by and large, on broad, sweeping generalities and crudely estimated constraints. There is not much biology in these expressions. Rather they are a kind of compendium of engineering guidelines. To make the predictions of these chapters more applicable to the real world, we must fine-tune them. We should like to be able to know what kinds of species, under what circumstances, will best be described with discrete-time equations, which with continuous-time formulations. Under what conditions will maximum population growth capacity be high or low? We should like to understand why certain modes of density feedback are critical to some populations, other modes are more important for others, while still others seem to encounter intraspecific competition only rarely. We should like to be able to put values on the functional-response-curve parameters, to assess the intensity of competition, and to predict when facilitation might occur and when interaction coefficients are nonadditive. In short, we need to uncover the framework of rules that governs how organisms interact with their environments, physical and biological; we need to be able to predict the manner in which animals and plants adapt to the world that confronts them.

The nature of adaptation has traditionally been approached in rather cavalier fashion, in the following way. First, it is assumed that a trait, be it anatomical, physiological, or behavioral, is to some degree heritable. If the genome does not directly determine the trait—as, for example is the case for most behaviors—then expression of the trait is supposed to be indirectly determined by genes. That is, the genome sets the limits of expression, guides development, and predisposes learning. Learning predispositions might occur via intrinsic feedback mechanisms (genes that lead an animal to interpret the feeling in its gut following ingestion of a toxic food as unpleasant rather than pleasant are clearly adaptive, permitting the animal to learn the correct response in future encounters with food), or simply by making some responses easier to learn than others (via anatomical and physiological constraints). Second, it is assumed that natural selection acts as an efficiency expert, weeding out individuals that survive, mate, or reproduce with less elan than their compatriots. To the extent that these assumptions are true, it follows that organisms, given sufficient numbers of generations, will inherit those combinations of genes that maximize their chances of surviving, mating, and reproducing.

The argument above seems almost trivially obvious. If valid, it allows us an extremely convenient approach to the problem of fine-tuning the descriptive framework of Chapters 2 through 6. We simply consider what particular body size, or mode of density feedback, or pattern of food preference, or social organization, and so on, ought to maximize the survival, mating success, and fecundity of individuals in a particular population and then assume that natural selection has, indeed, generated this constellation of traits. Then we may apply our predictions to the more basic questions of the importance of (for example) body size, mode of density feedback, food preference, and social organization, for predicting population dynamics. Constructing models to explore the optimal "strategies" of organisms, and then using these models to predict what we should expect organisms to do, has become a passion of many ecologists (Maynard Smith, 1978).

But while the optimal strategy approach is convenient, and often seems to work remarkably well in predicting the nature of adaptation, it is fraught with difficulties. At best it is often very misleading. Further discussion of the pitfalls of the approach is deferred until the introduction to Part Four, at which point the reader should be in a better position to appreciate them. In the meantime, because we will make extensive use of optimal strategy models in Chapters 11 through 17, it is necessary to digress into a study of genetics. At the very least, any student of ecological adaptation, and particularly students of the optimal strategy approach, should understand the genetic mechanisms underlying the phenomena they study.

The approach to genetics in this book is fairly straightforward. In Chapter 7 we discuss the genetic structure of populations—the distribution of gene and genotype frequencies—a topic of considerable importance for understanding the process of evolution by chance events and by natural selection. In Chapter 8 the basic rules of evolution by natural selection are set out. Chapter 9 continues the treatment of Chapter 8, discussing the importance of temporal and spatial heterogeneity in the environment. Chapter 10 extends the basic (single-locus) rules to metric (polygenic) traits. Because of the need to treat a very extensive topic in only four chapters, it has been necessary to omit a tremendous amount of detail. I have chosen my deletions with the following rationale: Superficial treatment of a wide variety of genetic systems will not accomplish the goal of instilling in the reader a sense of the strengths and weaknesses of the optimal strategy approach; it is better to understand one system in detail. Therefore, I have limited the following material to diploid organisms. I have not attempted to cover sex linkage, segregation distortion, polyploidy, haplodiploidy, parthenogenesis, or maternal effects—although they may be referred to briefly from time to time. I have also studiously avoided all but a very superficial discussion of such complications as the existence of transposons. I have also simplified much of the mathematics (at minimal cost to precision of the ultimate predictions) by telescoping all allelic forms which might exist at a given locus into two classes of alleles. Finally, because our interests in this book relate to adaptation, and thus microevolution, I ignore entirely the questions of speciation and macroevolution.

Chapter 7

The Genetic Structure of Populations

7.1 Basic Models and Temporal Changes

Panmictic Populations

As in Chapter 2, we begin with the simplest possible case. With respect to genetic structure, this is the situation of a panmictic population—one in which mating is random. In such populations breeding can be thought of as analogous to a process by which individuals toss their gametes into a common *gene pool*, offspring being produced by the random drawing of pairs of gametes from that pool. If we label the two classes of alleles at some, given locus, A and a, where A occurs with frequency p, a with frequency $q = 1 - p$, the probability that two draws from the gene pool result in the acquisition of two A alleles is p^2. Thus the frequency of AA homozygote offspring is p^2. Heterozygotes, Aa individuals, can be produced by first drawing an A and then an a, or vice versa, for a probability value of $pq + qp = 2pq$. Finally, aa homozygotes are drawn with frequency q^2. Notice that the frequencies of the three genotypes, at the time of fertilization (drawing from the gene pool) is always $p^2 : 2pq : q^2$, a function of gene frequency only. Thus differential mortality of adults, and differential fecundity, although they may alter genotype frequencies during life, or affect the makeup of individuals contributing to the gene pool, do *not* affect the algebraic expression describing genotype frequencies at conception. This is one part of the *Hardy–Weinberg law*.

The other part of the Hardy–Weinberg law states that in the absence of differential survival and fecundity among genotypes gene frequency remains unchanged. In such circumstances the $p^2 : 2pq : q^2$ ratio applies not only to the newly conceived portion, but to the population as a whole. A population of this sort is said to be in *Hardy–Weinberg equilibrium*. I will use the term "Hardy–Weinberg equilibrium" here in a somewhat looser fashion: a population in which p is changing may display a Hardy–Weinberg equilibrium among its zygotes.

If the population in question is *semelparous*—that is, individuals breed but once, and then die, so that there is no overlap of generations—then at the beginning of each generation the population should be in Hardy–Weinberg equilibrium. As we shall see in Chapter 8, this fact makes the construction of expressions for evolution by natural selection fairly straightforward. But in age-structured populations Hardy–Weinberg equilibrium can no longer be expected over the full population. If, for any reason, gene frequency has changed between breeding seasons, the population in iteroparous populations (where individuals breed more than once and there is overlap of generations) is composed of both the newly conceived individuals (in Hardy–Weinberg equilibrium), and the surviving adults (not necessarily in Hardy–Weinberg equilibrium). If the change in p has come about via genotypically determined differences in survival ability, the adult population will not be in Hardy–Weinberg equilibrium and the combined adult and offspring cannot be expected to be either.

If the change in p is the result of differential mating success or fecundity, or statistical bias in drawing alleles from the gene pool, the assumption of panmixia is violated and neither the newly conceived nor the adult portion of the population will be in Hardy–Weinberg equilibrium.

Generally, for the sake of simplicity in selection models (Chapter 8), it is assumed that panmixia is realized and that selection is slow—that is, Δp is very small. When this is true, of course, even age-structured populations approach the Hardy–Weinberg ideal. For a detailed discussion of genotype frequencies in age-structured populations, see Charlesworth (1980).

Finite Populations

If we were to toss a fair coin 200 million times, the fraction of tosses coming up heads would be very close to one-half. The chances of coming up heads only one-fourth of the time would be vanishingly small. On the other hand, in four tosses, one would expect, on average, two heads—one-half of the time—but would not be terribly surprised to find heads only once—one-fourth of the time. In very large populations, where there are many random drawings from the gene pool, the frequency with which allele A is drawn should be very close indeed to p. But in small populations there is considerable chance that the frequency may diverge from p by a significant amount. Thus, in most real, finite populations, we must take into account the existence of statistical error in sampling from the gene pool. The classical way to investigate such sampling error is to envision one population and all of its potential (hypothetically possible) daughter populations, and to calculate the variance in p among those daughter populations. Then we look at all potential daughter populations in the following generation and calculate $\text{var}(p)$ in that generation—and so on. We eventually obtain a general expression for $\text{var}(p)$ as a function of time.

Because we wish to keep calculations simple, we will consider only populations within which breeding is random. That is, if the full population is so large that random breeding does not occur (individuals from the north side, say, seldom run into neighbors from the south side simply by virtue of areal extent or density of the population), we must subdivide this population into subunits where breeding *is* random. Such units are known as *demes.*

Within a deme the distribution of alleles in the F_1 generation follows a simple, binomial process, and thus the number of A alleles has an average (expected) value of $2np$, where n is the size of the deme, so $2n$ is the number of alleles. The variance is given by $2npq$. The variance in *frequency*, then, is var (number of A alleles/$2n$) $= 2npq/4n^2 = pq/2n$. But if the variance in p among F_1 daughter populations (demes)—we shall label this $\text{var}(p_1)$—is $pq/2n$, what is the variance among the potential F_2 demes, $\overline{\text{var}(p_2)}$? More generally, what is $\text{var}(p_t)$?

Recall that $\text{var}(x)$ is defined by $\overline{(x - \bar{x})^2}$. Thus $\text{var}(p_t)$ can be written (where Q is the number of potential daughter populations, and superscript i refers to daughter population i, ij to granddaughter population j arising from daughter population i)

$$\begin{aligned}
\text{var}(p_t) &= \frac{1}{Q^2}\sum_{i,j}(p_t^{(ij)} - \bar{p}_t)^2 \\
&= \frac{1}{Q^2}\sum_{i,j}(p_t^{(ij)} - p_t^{(i)} + p_t^{(i)} - \bar{p}_t)^2 \\
&= \frac{1}{Q^2}\sum_{i,j}[(p_t^{(ij)} - p_t^{(i)})^2 + 2(p_t^{(ij)} - p_t^{(i)})(p_t^{(i)} - \bar{p}_t) + (p_t^{(i)} - \bar{p}_t)^2] \\
&= \frac{1}{Q}\sum_i\left[\frac{1}{Q}\sum_j(p_t^{(ij)} - p_t^{(i)})^2\right] + 0 + \frac{1}{Q}\sum_i(p_t^{(i)} - \bar{p}_t)^2 \\
&= \frac{1}{Q}\sum_i \text{var}(p_t^{(i)}) + \text{variance in } p \text{ among daughter populations} \\
&= \overline{\text{var}(p_t)} + \text{var}(p_{t-1}) = \frac{\overline{p_{(t-1)}q_{(t-1)}}}{2n} + \text{var}(p_{t-1})
\end{aligned} \tag{7-1}$$

Also,

$$\frac{\overline{p_{t-1}q_{t-1}}}{2n} = \frac{\overline{p_{t-1}} - \overline{p_{t-1}^2}}{2n} = \frac{\bar{p}_{t-1} - \bar{p}_{t-1}^2}{2n} - \frac{\operatorname{var} p_{t-1}}{2n} \tag{7-2}$$

and

$$\bar{p}_{t-1} = \bar{p}_t = \bar{p} \tag{7-3}$$

Hence,

$$\operatorname{var}(p_t) = \operatorname{var}(p_{t-1})\left(1 - \frac{1}{2n}\right) + \frac{\bar{p}\bar{q}}{2n} \tag{7-4}$$

This recursion equation can also be written

$$\begin{bmatrix} \operatorname{var}(p_t) \\ \dfrac{\bar{p}\bar{q}}{2n} \end{bmatrix} = \begin{bmatrix} 1 - \dfrac{1}{2n} & 1 \\ 0 & 1 \end{bmatrix} \begin{bmatrix} \operatorname{var}(p_{t-1}) \\ \dfrac{\bar{p}\bar{q}}{2n} \end{bmatrix}$$

whence $\lambda = 1, 1 - 1/2n$, and

$$\operatorname{var}(p_t) = C(1)^t + C'\left(1 - \frac{1}{2n}\right)^t$$

when $t = 0$, var $(p) = 0$ (there is only the one, parent deme). Thus $C + C' = 0$. When $t = 1$, we know that $\operatorname{var}(p_1) = \bar{p}\bar{q}/2n$. Thus

$$C + C'\left(1 - \frac{1}{2n}\right) = \frac{\bar{p}\bar{q}}{2n}$$

Solving these expressions for C, C', and substituting,

$$\operatorname{var}(p_t) = \bar{p}\bar{q}\left[1 - \left(1 - \frac{1}{2n}\right)^t\right] \qquad \text{where } \bar{p} = p_0, \quad \bar{q} = q_0 \tag{7-5}$$

The fact that among hypothetical, potential daughter populations, the variance in p grows with time means, simply, that the value of p drifts, randomly, farther and farther away from its original value when traced along any population lineage. Such *genetic drift*, often called *random drift*, eventually results in the chance disappearance of one allele or the other and hence homozygosity. The disappearance of one allele results, of course, in the other going to a frequency of one—that remaining allele is said to be *fixed*. Individuals in populations that have experienced fixation are, by definition, homozygous at that locus.

While genetic drift may lead inexorably to fixation of one allele or another, the rate at which fixation occurs depends critically on the size of the deme, n. Quite obviously, very large demes experience almost no drift and so after many generations may be expected to retain a good deal of genetic diversity. Very small demes, unless the trend is strongly opposed by natural selection, may be expected to be homozygous at many or most loci.

But if the approach described above is applicable to hypothetical, potential daughter populations, it is equally applicable to real daughter populations that satisfy the same assumptions—constant deme size. If a deme generates descendents which persist, or if a large population is subdivided into a series of isolated, "daughter" populations, drift within these demes results just as surely in divergence of p among them as it would among hypothetical descendents. Thus the variance in p among these demes is given by (7-5). Further discussion on this subject is deferred to later in the chapter with the exception of one extremely important point. Note that the expected (or average) homozgosity among demes can be written

$$\overline{p_t^2} = \bar{p}^2 + \mathrm{var}(p) = \bar{p}^2 + \overline{pq}\left[1 - \left(1 - \frac{1}{2n}\right)^t\right]$$

Similarly

$$\overline{q_t^2} = \qquad = \bar{q}^2 + \overline{pq}\left[1 - \left(1 - \frac{1}{2n}\right)^t\right] \tag{7-6}$$

$$2(\overline{pq})_t = \qquad = 2\overline{pq} - 2\overline{pq}\left[1 - \left(1 - \frac{1}{2n}\right)^t\right]$$

The increase in homozygosity predicted by drift can be expressed, quantitatively, then, in the manner shown above.

Inbreeding Coefficients

There is an alternative way to calculate $\mathrm{var}(p_t)$ which makes use of a basic concept of relatedness among individuals and which will be extremely convenient later in this chapter. We define two alleles as being *identical by descent* (ibd) if they are descendents or replicates of a single ancestral allele. Thus if a particular allele in a father is replicated in meiosis and appears, via those replicates in two offspring, the alleles in question in those two offspring are identical by descent. Two alleles may be ibd via common ancestry in a parent, but they may also be ibd via common ancestry in a grandparent, and so on. Thus we must clearly qualify the term "ibd" with a statement as to which generation we consider ancestral. The probability that two genes in a single individual are ibd is denoted by F.

We can write (with respect to parental ancestry)

F = probability that, given one allele, the paired allele in an individual is identical by descent

But the probability that the paired allele is ibd is the probability that of all $2n$ allelic choices in the gene pool, exactly that one allele was drawn. This occurs with probability $1/2n$. Thus with respect to parental ancestry, at time $t - 1$, $F = 1/2n$. In the next generation, the probability that both alleles at a locus in any given individual are ibd is equal to the probability that they are ibd via the parental generation plus the probability that that is not so but that they are ibd via a previous generation. That is,

$$F(t) = \frac{1}{2n} + \left(1 - \frac{1}{2n}\right) F(t - 1) \tag{7-7}$$

This expression is of the same form as (7-4) and may be solved in similar manner to yield

$$F(t) = 1 - \left(1 - \frac{1}{2n}\right)^t [1 - F(0)] \tag{7-8}$$

Because F is defined with respect to some specified, ancestral generation we now let that generation be 0, so that by definition, $F(0) = 0$. Then

$$F(t) = 1 - \left(1 - \frac{1}{2n}\right)^t \tag{7-9}$$

This equation bears an uncanny resemblance to (7-5). Indeed, it is clear that we may now write, for subdivided populations,

$$\mathrm{var}(p_t) = p_0 q_0 F(t) = \mathrm{var}(p_t) = \overline{pq} F(t) \tag{7-10}$$

whence the expected frequencies of genotypes over a subdivided population becomes

$$\overline{p_t^2} = \bar{p}_t^2 + \bar{p}_t\bar{q}_tF(t)$$
$$\overline{q_t^2} = \bar{q}_t^2 + \bar{p}_t\bar{q}_tF(t) \qquad (7\text{-}11)$$
$$2\overline{pq} = 2\bar{p}\bar{q}[1 - F(t)]$$

For later reference, note from (7-7) that the increase in F from one generation to the next is

$$F(t + 1) - F(t) = \frac{1}{2n}[1 - F(t)] \qquad (7\text{-}12)$$

F is known as the *inbreeding coefficient* of a population. As we have seen, it is explicitly related to genetic drift and the expected frequencies of genotypes in subdivided populations. Defined explicitly as the probability that two alleles in an individual are ibd, it is, at least sometimes, easy to measure. As we shall see later, it is an extremely important parameter in population genetics studies.

Nonrandom Mating

The derivation of F and var (p) above assumed random mating. But it is clear, if individuals breed preferentially with relatives or with genotypically similar individuals, that two alleles in the same individual are more likely to be ibd than if breeding were random. Thus inbreeding of this sort increases F. In fact, F, the probability of two alleles being ibd in the same individual, can be thought of as arising from two sources. First, as discussed above, nonzero F values can arise simply by virtue of finite deme size—a nonzero probability that random drawing from the gene pool results in a newborn possessing two alleles ibd. Second, nonzero F values can arise even in infinite demes, where the possibility above approaches zero, if related or similar individuals mate preferentially with each other. In such cases genes are *not* drawn independently from the gene pool; rather, once one allele is drawn, the probability that the next one drawn is ibd is much greater than expected by random chance. It is often convenient to distinguish the amount of inbreeding due to finiteness of deme size as F_{ST}, and inbreeding due to nonrandom events such as preferential mating with kin (*consanguineous mating*), or with genotypically similar individuals (*assortative mating*) as F_{IS}. Note that the probability an individual's genes are *not* ibd is the probability they are ibd neither by virtue of finite population size nor by virtue of mating propensities. That is, $1 - F_{total} = (1 - F_{ST})(1 - F_{IS})$. The total amount of inbreeding is usually referred to as F_{IT}. Hence, from the above,

$$F_{IT} = 1 - (1 - F_{ST})(1 - F_{IS}) \qquad (7\text{-}13)$$

Verbally, these various F values can be understood as follows. F_{IT} is the probability that an individual carries two genes ibd by virtue of ancestral mating of relatives. Some of that probability, F_{ST}, comes from the fact that ancestral mates may have been related purely by chance due to the finite number of mates in a deme. The other contributor to inbreeding, F_{IS}, comes from the additional chance that ancestral mates may have been related due to "intentional" choice of kin as mates.

We have seen how to calculate F_{ST} in theory. In the field its determination is also quite simple, at least if the extent of a deme can be determined (see "Effective Population Size," p. 212) and if $F_{IS} = 0$ and so does not confound observed genotype. Since var $(p) = \bar{p}\bar{q}F_{ST}$ (equation 7-10) when $F_{IS} = 0$, we have only to measure p (or q) in several demes, calculate its variance, and then divide this quantity by the product of the mean values of p and q:

$$F_{ST} = \frac{\text{var}(p)}{\bar{p}\bar{q}} \qquad (\text{assuming that } F_{IS} = 0) \qquad (7\text{-}14)$$

How may we measure F_{IS}?

Consanguineous Mating

We first establish that F_{IS} has the same impact on genotype frequency that F_{ST} does. Consider a single deme, i, which has inbreeding coefficient F_{ISi} by virtue of nonrandom mating. The proportion of AA individuals in the deme is given by the probability that the first allele drawn from the gene pool is of type A, times the probability that the second is also A. The latter probability, in turn, is the probability that the second allele is ibd to the first, plus the probability that it is not ibd but is type (A) in any case:

$$\text{frequency of } AA = p_i(F_{ISi} + (1 - F_{ISi})p_i) = p_i^2 + p_i q_i F_{ISi} \tag{7-15}$$

The heterozygosity, of course, is

$$\text{frequency of } Aa = 2p_i q_i(1 - F_{ISi}) \tag{7-16}$$

Over the entire subdivided population, heterozygosity is depressed not only by inbreeding of this sort, but also by F_{ST}. We write

$$\text{mean heterozygosity} = \overline{2p_i q_i(1 - F_{ISi})} = 2\sum_i \frac{n_i}{n_{tot}} p_i q_i(1 - F_{ISi})$$

$$= 2\left[\frac{\sum_i (n_i/n_{tot})p_i q_i}{\sum_i (n_i/n_{tot})p_i q_i} - \frac{\sum_i (n_i/n_{tot})p_i q_i F_{ISi}}{\sum_i (n_i/n_{tot})p_i q_i}\right]\left(\sum_i \frac{n_i}{n_{tot}} p_i q_i\right)$$

where n_i is the number of individuals in deme i. If we now define F_{IS} for the entire population as

$$F_{IS} = \frac{\sum_i (n_i/n_{tot})p_i q_i F_{ISi}}{\sum_i (n_i/n_{tot})p_i q_i}$$

we obtain

$$\begin{aligned}\text{mean heterozygosity} &= 2(1 - F_{IS})\overline{pq} = 2(1 - F_{IS})[\bar{p}\bar{q} + \text{cov}(p, q)] \\ &= 2\bar{p}\bar{q}(1 - F_{IS})\left[1 - \frac{\text{var}(p)}{\bar{p}\bar{q}}\right] \\ &= 2\bar{p}\bar{q}(1 - F_{IS})(1 - F_{ST}) \qquad \text{(see equation 7-14)} \\ &= 2\bar{p}\bar{q}(1 - F_{IT})\end{aligned}$$

where $\text{cov}(p, q) = \text{cov}(p, 1 - p) = \text{var}(p)$. Similarly,

$$\begin{aligned}\overline{p^2} &= \bar{p}^2 + \bar{p}\bar{q}F_{IT} \\ \overline{q^2} &= \bar{q}^2 + \bar{p}\bar{q}F_{IT}\end{aligned} \tag{7-17}$$

(See also Kirby, 1975.) Notice that if $F_{ST} = 0.0$ (or is not applicable, such as within a deme), equations (7-17) are still valid, but with F_{IS} replacing F_{IT}.

The important consequence of inbreeding is that it distorts the genotype ratio from that expected by the Hardy–Weinberg law. Note, in (7-11) and (7-17), that the frequencies of both homozygotes are increased by an amount pqF, and that the frequency of the heterozygote is reduced by an amount $2pqF$. This distortion has important ramifications for natural selection theory (Chapters 8 through 10). Also, in populations carrying recessive deleterious alleles, inbreeding means an increase in the likelihood of their occurring in homozygous form and thus expressing themselves. This is the cause of *inbreeding depression.*

The values of F_{IS} (actually F_{ISi}, since we calculate inbreeding within a particular deme) can be found by constructing pedigrees. For example, consider Figure 7-1. Picture an allele, at some given locus, which individual I received from its mother, A. Its complement can be ibd only if the

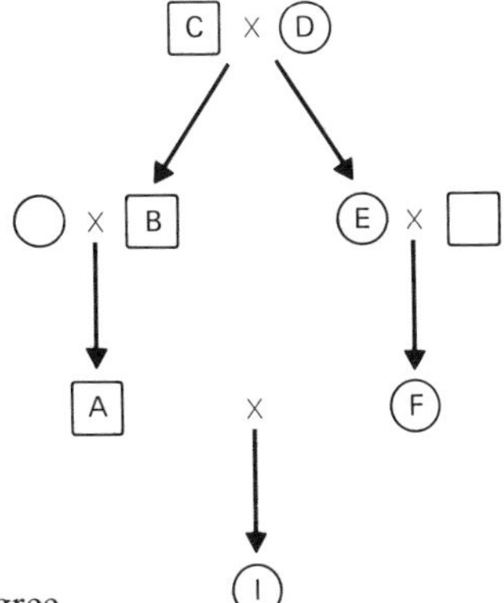

Figure 7-1 Hypothetical pedigree.

two had a common ancestor. Thus identity by descent can occur only if the gene in question was passed to A from B (probability $\frac{1}{2}$), who, in turn, received it from C (with probability $\frac{1}{2}$) *or* D (with probability $\frac{1}{2}$)—for a total of $\frac{1}{2} + \frac{1}{2} = 1$, and if C or D passed it on to E (probability $\frac{1}{2}$), who passed it on to F (probability $\frac{1}{2}$), who finally gave it to I (probability $\frac{1}{2}$). Accounting for all steps in this pathway, we see that the probability of the complementary allele being ibd is $(\frac{1}{2})^4 = \frac{1}{16}$. Thus the value of F_{IS} for individual I, is $\frac{1}{16}$. An easy rule of thumb is that F_{IS} is equal to $(\frac{1}{2})^k$, where k is the number of generational steps between parents (in this case, A and F).

Consider now a slightly more involved situation (Fig. 7.2). Here the two alleles at some locus in I might be ibd by virtue of either of two pathways—A–C–E–F–B, or A–C–E–(H or J)–G–B. In the first case $k = 4$, in the second, $k = 5$. Thus $F_{IS}(I) = (\frac{1}{2})^4 + (\frac{1}{2})^5 = \frac{3}{32}$.

There is actually somewhat more to it than indicated just above. In the first example (Fig. 7-1), either C or D might possess two copies of the allele in question (i.e., they themselves are inbred). If f_C and f_D are the probabilities that individuals C, D are inbred, then the probability that C passed the same allele on to both B and E is not $\frac{1}{2}$, but $\frac{1}{2}(1 + f_C)$. The probability that individual D is the culprit is not $\frac{1}{2}$, but $\frac{1}{2}(1 + f_D)$. Thus, where f_x is the average inbreeding coefficient of individuals C and D, we obtain $F_{IS}(I) = (1 + f_x)/16$.

Although the scheme above allows us to calculate fairly simply the inbreeding coefficients for individuals, it does little for us in finding F_{ISi}, the inbreeding coefficient for the subpopulation as a whole. Of course, if we know the pedigrees of all individuals in a subpopulation, we can calculate f for all members of the subpopulation and take an average. But more generally, we

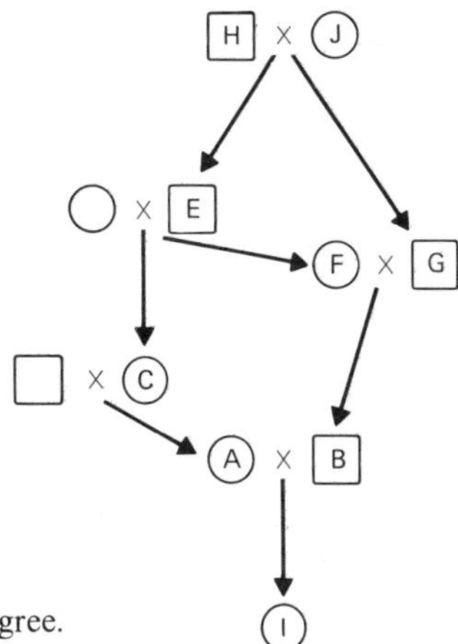

Figure 7-2 Hypothetical pedigree.

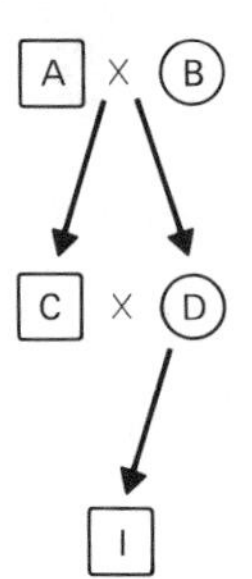

Figure 7-3 Pedigree in which all matings are among siblings.

have information on a mating *pattern* which is characteristic of the population. In such cases F_{IS} is $F_{IS}(I)$ for any I in the population. For example, suppose we knew that all mating was among sibs (Fig. 7-3). Then if individual I appears at generation t, we know that the average inbreeding coefficient of A and B is merely F_{IS} two units past $(t - 2)$. Thus if both genes came from one grandparent or the other, we would write

$$F_{IS}(t) = \tfrac{1}{4}[1 + F_{IS}(t - 2)]$$

But though I might have received its two genes from different grandparents, those genes might nevertheless be ibd. Consider the gene acquired from grandparent A, and note that if I possesses it, then C or D must also have possessed it. Suppose that C possessed it. Then if the gene acquired from grandparent B also passed to C (with probability $\tfrac{1}{2}$), the probability that it is ibd to that from A must be the probability that the two genes held by C are ibd $[= F(t - 1)]$. Hence the probability that the two genes in I are ibd is

$$F_{IS}(t) = \tfrac{1}{4}[1 + F_{IS}(t - 2)] + \tfrac{1}{2}F_{IS}(t - 1) \tag{7-18}$$

In matrix form

$$\begin{bmatrix} F_{IS}(t) \\ F_{IS}(t-1) \\ \tfrac{1}{4} \end{bmatrix} = \begin{bmatrix} \tfrac{1}{2} & \tfrac{1}{4} & 1 \\ 1 & 0 & 0 \\ 0 & 0 & 1 \end{bmatrix} \begin{bmatrix} F_{IS}(t-1) \\ F_{IS}(t-1) \\ \tfrac{1}{4} \end{bmatrix}$$

so that

$$F_{IS}(t) = \sum_j C_j \lambda_j^t \qquad \text{where } \lambda_j = 1, \frac{1 \pm \sqrt{5}}{4}$$

Now, because an individual in this scheme can inherit two genes ibd only from its grandparents (or an earlier generation), $F_{IS}(1) = F_{IS}(0) = 0$. After two units, $F_{IS}(2) = [F_{IS}(0) + 1]/4 = \tfrac{1}{4}$. Thus we can find $C_1 = 1$, $C_j = -1.17$, $C_3 = .17$, and

$$F_{IS}(t) = 1 - (1.17)(.81)^t + (.17)(-.31)^t \tag{7-19}$$

As $t \to \infty$,

$$F_{IS} = 1.0 \tag{7-20}$$

Assortative Mating

Quite often, members of a species mate preferentially either with phenotypically (and so, most likely, genotypically) similar or dissimilar individuals. The first tendency is known as positive, the second as negative assortative mating. The lesser snow goose (*Anser caerulescens*), for example, displays two morphs, a blue form and a white form, determined by genotype at a single locus. Cooke et al. (1976), working with large numbers of banded individuals in a nesting colony at LaPérouse Bay, Manitoba, have shown that offspring of blue × blue parents prefer

Table 7-1[a]

Choice of Offspring	Parents		
	White × White	*Mixed*	*Blue × Blue*
White	52	13	4
Blue	6	7	15

[a] $p < .0001$ by a chi-square contingency table test.

blue mates, those with white × white mates are more likely to breed with white birds. The data for two years (1973, 1974) combined are shown in Table 7-1. A roughly similar situation is found in the arctic skua (*Stercorarius parasiticus*) (O'Donald et al., 1974; Davis and O'Donald, 1976).

When likes mate with like, there is a greater than random possibility that genes ibd will end up in their offspring. There are several, published analyses of the effect of assortative mating on the inbreeding coefficient. The one presented here is taken from Wright (1921). First, let the "value" of an individual be a direct reflection of the number of A alleles carried. Thus the AA homozygote has value 2, the heterozygote 1, and the aa homozygote 0. The critical parameters are given in Table 7-2. Using these values to calculate covariance in value among mates, we write (at time t)

$$\text{cov (value of mates)} = \begin{pmatrix}\text{mean of the products}\\ \text{of mate value}\end{pmatrix} - \begin{pmatrix}\text{product of the means}\\ \text{of mate value}\end{pmatrix}$$

$$= [(4f_1g_{11} + 2f_1g_{12} + 0f_1g_{13}) + (2f_2g_{21} + 1f_2g_{22} + 0f_3g_{23}) + 0] - (2p)(2p)$$

$$= 4(f_1g_{11} + \tfrac{1}{2}f_1g_{12} + \tfrac{1}{2}f_2g_{22} + \tfrac{1}{4}f_2g_{22} - p^2) \qquad (7\text{-}21a)$$

$$\text{var}(x) = (4f_1 + f_2 + 0) - (2p)^2 \qquad (7\text{-}21b)$$

$$\text{correlation (value of mates)} = \frac{\text{cov}}{\text{var}} \equiv \eta \qquad (7\text{-}21c)$$

But also,

$$(\text{Frequency of } AA \text{ at time } t + 1) = p_{t+1}^2 + p_{t+1}q_{t+1}F_{IS}(t + 1)$$

Thus if no selection occurs, so that $p_{t+1} = p_t = p$, then

$$F_{IS}(t + 1) = \frac{(\text{Frequency of } AA \text{ at time } t + 1) - p_{t+1}^2}{p_{t+1}q_{t+1}}$$

$$= \frac{f_1g_{11} + \frac{1}{2}f_1g_{12} + \frac{1}{2}f_2g_{21} + \frac{1}{4}f_2g_{22} - p^2}{pq}$$

Table 7-2

Genotype	Frequency	Value x	Relative Frequency of mating with: AA	Aa	aa
AA	$f_1 = p^2 + pqF$	2	g_{11}	g_{12}	g_{13}
Aa	$f_2 = 2pq(1 - F)$	1	g_{21}	g_{22}	g_{23}
aa	$f_3 = q^2 + pqF$	0	g_{31}	g_{32}	g_{33}

An examination of the numerator above shows it to be equal to ($\frac{1}{4}$cov). Thus

$$F_{IS}(t+1) = \frac{(\frac{1}{4})\,\text{cov}_t}{pq} = \frac{(\frac{1}{4})\eta\,\text{var}\,(x)_t}{pq} = \frac{\frac{1}{4}\eta[4f_1 + f_2 - 4p^2]}{pq} \qquad \text{(from 17-21b)}$$

$$= \frac{\frac{1}{4}\eta[2pq + 2pqF_{IS}(t)]}{pq}$$

$$= \tfrac{1}{2}\eta[1 + F_{IS}(t)]$$

Solution of this recursion equation gives

$$F_{IS}(t) = \frac{\eta}{2-\eta} + \left[F_{IS}(0) - \frac{\eta}{2-\eta_i}\right]\left(\frac{\eta}{2}\right)^t \tag{7-22}$$

At equilibrium,

$$F_{IS} \rightarrow \frac{\eta}{2-\eta} \tag{7-23}$$

and over one generation,

$$\Delta F_{IS} = F_{IS}(t+1) - F_{IS}(t) = \frac{\eta}{2} - \frac{2-\eta}{2}F_{IS}(t) \tag{7-24}$$

Genotype frequencies within a subpopulation practicing assortative mating are thus given by

$$\begin{aligned} \overline{p^2} &= p^2 + pq\frac{\eta}{2-\eta} \\ 2\overline{pq} &= 2pq\left(1 - \frac{\eta}{2-\eta}\right) \\ \overline{q^2} &= q^2 + pq\frac{\eta}{2-\eta} \end{aligned} \tag{7-25}$$

(See Stark, 1976.)

Inasmuch as the effects of assortative mating can be expressed in the form of an inbreeding coefficient, (7-22) to (7-25) are valid. However, note that F_{IS} as used here is no longer, as it was originally used, a probability. This fact will be driven home if we consider negative assortative mating—individuals preferentially mating with genetically unlike individuals. In this case η, and so F_{IS}, become *negative*. Returning to the midpoint of the derivation of (7-17), we see that nothing in the calculations need be changed on account of a possibly negative F_{IS} value. However, one possible outcome now is a negative F_{IT} value. Therefore, because negative assortative mating may be occurring in our study populations, it may often be appropriate to cease viewing F values as probabilities of identity by descent, and rather view them merely as measures of distortion in the genotype frequencies (equation 7-25).

Sexuality

We return now to inbreeding resulting from random processes in finite demes and discuss some processes that require modification of our original equations. Suppose that the population consists of gonochorists (separate sexes). Because males cannot mate with males, females with females, such a population is not truly random in its breeding pattern. However, we can easily compute F_{ST}, although its value will now differ slightly from that in (7-8) and (7-9).

At some locus consider the allele arising from one parent. Its complementary allele can be ibd in any of three ways:

1. The two parents (generation $t - 1$) had no common parent (probability $1 - 1/n$), in which case the contribution to $F_{ST}(t)$ is $(1 - 1/n)F_{ST}(t - 1)$.
2. The two parents *had* a parent in common (probability $1/n$) and got the same gene (ibd) from that parent (probability $\frac{1}{2}$).
3. The two parents had a parent in common, but did not get the same gene. However, the genes they got were ibd anyway [probability $F_{ST}(t - 2)$].

Putting the above together, we obtain

$$F_{ST}(t) = \left(1 - \frac{1}{n}\right) F_{ST}(t - 1) + \frac{1}{n}\frac{1}{2} + \frac{1}{n}\frac{1}{2} F(t - 2)$$

or

$$1 - F_{ST}(t) = \left(1 - \frac{1}{n}\right) [1 - F_{ST}(t - 1)] + \frac{1}{2n} [1 - F_{ST}(t - 2)]$$

so that

$$\begin{bmatrix} 1 - F_{ST}(t) \\ 1 - F_{ST}(t - 1) \end{bmatrix} = \begin{bmatrix} 1 - \dfrac{1}{n} & \dfrac{1}{2n} \\ 1 & 0 \end{bmatrix} \begin{bmatrix} 1 - F_{ST}(t - 1) \\ 1 - F_{ST}(t - 2) \end{bmatrix}$$

The characteristic equation is

$$\lambda^2 - \left(1 - \frac{1}{n}\right)\lambda - \frac{1}{2n} = 0 \qquad \text{so that } \lambda = \frac{(1 - 1/n) \pm \sqrt{1 + 1/n^2}}{2}$$

If n is not very small ($\geqslant 10$), we may write, to a good approximation,

$$\lambda = 1 - \frac{1}{2n}, -\frac{1}{2n}$$

Thus

$$1 - F_{ST}(t) = C_1 \left(1 - \frac{1}{2n}\right)^t + C_2 \left(-\frac{1}{2n}\right)^t$$

Because genes ibd cannot be inherited from the parental generation, $F_{ST}(1) = F_{ST}(0)$. Thus

$$1 - F_{ST}(0) = C_1 + C_2, \qquad 1 - F_{ST}(1) = 1 - F_{ST}(0) = C_1 \left(1 - \frac{1}{2n}\right) - C_2 \left(\frac{1}{2n}\right)$$

and

$$C_1 = \left(1 + \frac{1}{2n}\right) [1 - F_{ST}(0)]$$

$$C_2 = -\frac{1}{2n} [1 - F_{ST}(0)]$$

Finally,

$$\begin{aligned} 1 - F_{ST}(t) &= \left(1 + \frac{1}{2n}\right) [1 - F_{ST}(0)] \left(1 - \frac{1}{2n}\right)^t - \frac{1}{2n} [1 - F_{ST}(0)] \left(-\frac{1}{2n}\right)^t \\ &= \left\{\left[1 - \left(\frac{1}{2n}\right)^2\right]\left(1 - \frac{1}{2n}\right)^{t-1} + \left(\frac{1}{2n}\right)^2 \left(-\frac{1}{2n}\right)^{t-1}\right\} [1 - F_{ST}(0)] \end{aligned}$$

and for n not very small ($n \gtrsim 10$),

$$1 - F_{ST}(t) \rightarrow \left(1 - \frac{1}{2n}\right)^{t-1} [1 - F_{ST}(0)]$$

so that

$$F_{ST}(t) = \left\{1 - \left(1 - \frac{1}{2n}\right)^{t-1} [1 - F_{ST}(0)]\right\} \tag{7-26}$$

Comparing this expression with (7-8), it is clear that, to an approximation, the effect of sexual reproduction is to delay the increase in F_{ST} by a single generation.

Effective Population Size

The amount of genetic drift can be measured by the change in F_{ST} from one generation to the next,

$$\Delta F_{ST} = \frac{1}{2n}(1 - F_{ST}) \qquad \text{from (7-12)} \tag{7-27}$$

where n is the size of a deme. Of course, this expression was derived under situations of no consanguineous or assortative mating (i.e., for an "idealized" population), and were these tendencies present, would be inaccurate. But in any case, (7-27) can be made accurate by replacing n with an appropriately altered *effective population size* defined, simply, as that value of n such that the ΔF_{ST} predicted agrees numerically with the observed value of the full inbreeding coefficient, F_{IT}. Put another way, the effective population size, n_e, is that value of n which, if substituted into (7-27), describes the amount of genetic drift in a real population. Because that drift is described with the use of var(p), this measure is also referred to as the *variance effective number. Inbreeding effective number*, also denoted by n_e, is that n which, when used in place of true n in equations describing the equilibrium value of F in an idealized population, gives the inbreeding value for a real population. Generally, these two effective values are numerically very similar or even identical, but this need not be true in all cases (for a discussion of this problem, see Kimura and Crow, 1963). For our present purposes we are interested primarily in changes in inbreeding and how they effect natural selection. Thus we will deal with variance effective numbers.

Over very long periods of time, the value of F_{IT} will approach some equilibrium value and further change in F_{IT} becomes, by definition, zero. Thus if we use (7-27) to describe the change in F_{IT} for an idealized population, it is clear that eventually either $F_{IT} \rightarrow 1.0$ or n_e must approach infinity. The variance effective population size of completely inbred populations is undefined. The concept of variance effective population size, then, is of practical interest only for populations not at inbreeding equilibrium. We accordingly constrain our previous definition somewhat and write: n_e is that value of n which, if substituted into the equation describing ΔF_{ST} for an idealized *non-inbred* population, describes the amount of genetic drift actually occurring in a population.

Suppose that the population in question is characterized by consanguineous mating. Without knowledge of the general mating scheme, we cannot specify n_e precisely. But we can write a general expression. If the population were an idealized one, then

$$\Delta F_{IT} = \Delta F_{ST} = \frac{1}{2n} \qquad [\text{for } F_{ST}(0) = 0]$$

If F_{ST} remains zero, then

$$\Delta F_{IT} = \Delta F_{IS} = \frac{1}{2n_e} \tag{7-28}$$

In the case of sib matings, (equation 7-18), there is no change in F_{IS} between $t = 0$ and $t = 1$, so we look at the change from $t = 1$ to $t = 2$, obtaining $\Delta F_{IS} = \frac{1}{4}$. Therefore,

$$\frac{1}{4} = \frac{1}{2n_e}$$

so that $n_e = 2$, regardless of true population size. If F_{ST} does *not* remain zero, then

$$\begin{aligned} \Delta F_{IT} &= \Delta[1 - (1 - F_{IS})(1 - F_{ST})] \\ &= \Delta F_{IS} + \Delta F_{ST} - \Delta(F_{IS}F_{ST}) \end{aligned}$$

The latter term can be written

$$\begin{aligned} \Delta(F_{IS}F_{ST}) &= F'_{IS}F'_{ST} - F_{IS}F_{ST} \\ &= (F'_{IS} - F_{IS})F'_{ST} + F_{IS}(F'_{ST} - F_{ST}) \\ &= (F'_{IS} - F_{IS})(F'_{ST} - F_{ST}) + (F'_{IS} - F_{IS})F_{ST} + F_{IS}(F'_{ST} - F_{ST}) \\ &= \Delta F_{IS}\,\Delta F_{ST} + F_{ST}\,\Delta F_{IS} + F_{IS}\,\Delta F_{ST} \end{aligned}$$

Taken from generation zero, where $F_{ST} = F_{IS} = 0$, then

$$\begin{aligned} \Delta F_{IT} &= \Delta F_{IS} + \Delta F_{ST} - \Delta F_{IS}\,\Delta F_{ST} \\ &= \frac{1}{4} + \frac{1}{2n} - \frac{1}{4}\frac{1}{2n} = \frac{2n + 3}{8n} \end{aligned}$$

But also $\Delta F_{IT} = 1/2n_e$ so that

$$n_e = \frac{4n}{2n + 3}$$

where n is deme size.

Suppose that the population were characterized by assortative mating so that (between $t = 0$ and $t = 1$) $\Delta F_{IS} = \eta/2$ (see equation 7-24). If F_{ST} remains zero, then

$$\Delta F_{IT} = \Delta F_{IS} = \frac{1}{2n_e} = \frac{\eta}{2}$$

$$n_e = \frac{1}{\eta} \tag{7-29}$$

Note that in a randomly breeding population, F_{IS}, the probability that two genes drawn sequentially from the gene pool are ibd is given by the probability that, given one gene, the next one drawn is ibd, which is simply $1/2n$. Thus in such a population,

$$\Delta F_{IS} = \frac{1}{2n} = \frac{\eta}{2}$$

so that $\eta = 1/n$ and

$$n_e = \frac{1}{1/n} = n$$

That is, effective population size is identically deme size.

It was noted previously that sexually reproducing populations are not truly random in their breeding patterns and hence not idealized populations. To investigate the impact of sex on the effective population size, consider that the probability that two genes in an individual are ibd is given by the probability that both genes came from a *male* grandparent (probability $= 1/n_m$), that it was the *same* male grandparent (probability $= 1/2$), and that both alleles were replicates of the same grandparental gene (probability $= \frac{1}{2}\frac{1}{2} = \frac{1}{4}$), or that they came from a grand-*mother* (probability $= 1/n_f$), that it was the same grandmother ($\frac{1}{2}$), and that both were replicates of the same gene ($\frac{1}{4}$). That is,

$$F_{IT} = \frac{1}{8n_m} + \frac{1}{8n_f}$$

where n_m and n_f are the numbers of breeding males and females in the population. This is accurate if the grandparents were not themselves inbred (i.e., if F_{IT} in the preceding two generations was zero). Thus

$$F_{IT} - 0 = \Delta F_{IT} = \frac{1}{8n_m} + \frac{1}{8n_f}$$

But also,

$$\Delta F_{IT} = \frac{1}{2n_e}$$

Thus

$$\frac{1}{n_e} = \frac{1}{4n_m} + \frac{1}{4n_f} \tag{7-30}$$

If P_m and P_f are, respectively, the proportion of the population consisting of males and females, then (7-30) can be written

$$n_e = 4nP_mP_f \tag{7-31}$$

If there is a balanced sex ratio (1 : 1), then $n_e = n$. Skewed sex ratios effectively lower the population size. For example, in polygynous systems, where the *breeding* sex ratio is skewed in favor of females, the effective population size is lower for a given n than for monogamous species.

All of the derivations above implicitly assume a constant deme size. But suppose that n varies. This situation can be analyzed as follows. The probability that two genes in a randomly chosen individual are *not* ibd after one generation of breeding is $1 - F_{IT}(1) = 1 - 1/2n_e(0)$. After two generations, $1 - F(2) = [1 - 1/n_e(0)][1 - 1/n_e(1)]$. After t generations,

$$1 - F(t) = \prod_{i=0}^{t-1}\left[1 - \frac{1}{2n_e(i)}\right]$$

But from (7-9), this must also be

$$1 - F(t) = \left[1 - \frac{1}{2n_e}\right]^t$$

where n_e is the effective population size over time. If no values of $n_e(i)$ are very small, and if t is large, the expressions above can be approximated with

$$\prod_{i=0}^{t-1}\left[1 - \frac{1}{2n_e(i)}\right] \simeq 1 - \sum_i \frac{1}{2n_e(i)} = \left(1 - \frac{1}{2n_e}\right)^t \simeq 1 - \frac{t}{2n_e}$$

Hence

$$\frac{1}{t}\sum_i \frac{1}{n_e(i)} = \frac{1}{n_e} \tag{7-32}$$

That is, n_e is the harmonic mean of the $n_e(t)$ values.

An interesting consequence of population fluctuations is that the decrease in heterozygosity due to inbreeding varies from generation to generation. If a population has had, in its recent past, a very low population size, it is said to have gone through a "genetic bottleneck" a period during which inbreeding was severe. The result is that we should expect high homozygosity in the present, even though the population size may now be quite large. An example of the bottleneck effect is provided by Bonnell and Selander (1974). During an earlier part of the present century the northern elephant seal (*Mirounga angustirostris*) was hunted nearly to extinction. The southern elephant seal (*M. leonina*) has experienced no equivalent depression in numbers. Bonnell and Selander used electrophoretic techniques to examine gene frequencies at 21 protein loci from 159 northern elephant seals and 18 protein loci from the southern species (42 individuals). In the former they found absolutely no polymorphism whatsoever, indicating extreme inbreeding. In the latter, five of the loci (28%) showed more than one allelic type, about standard for mammals.

Social organization in animals often may have a profound effect on the degree of inbreeding. Most animal populations are at least somewhat "viscous." That is, they are less than freely dispersing. As a result, neighboring individuals are generally more related to one another than would be expected on the basis of random mixing. Many species, in addition, form fairly discrete social groups, generally of related individuals, and occasionally the bulk of breeding activity occurs within such groups. The result of such organization, of course, is to cause animals to mate consanguineously, raising the inbreeding coefficient, depressing heterozygosity, and increasing the rate of genetic drift. One example of such a social system is found in the house mouse (*Mus musculus*), where the basic breeding unit is an extended family group. Genetic subdivision along family lines is often so well marked that several essentially isolated subpopulations can be found within the population inhabiting a single barn. For a detailed discussion of inbreeding in this species, see Anderson (1964), Selander (1970), and Selander and Yang (1970).

The impact of age structure on effective population size has been examined by several authors. The most thorough is probably that of Felsenstein (1971). His treatment is beyond the scope of this book, but his final expression is given below:

$$n_e = \frac{n_0 \tau}{1 + \sum_{i \geq 1} \left(\frac{1}{l_i} - \frac{1}{l_{i-1}} \right) Q_{i+1}^2} \tag{7-33}$$

where

$$\tau = \sum_{i=0}^{\infty} i l_i b_i = \text{generation time}$$

$$Q_i = \sum_{j \geq i} l_j b_j$$

$$n_0 = \text{number of individuals in the zero age class}$$

This equation is not entirely general; it assumes constant age structure and constant population size, but its form is instructive. As an illustration of its use, consider a population with the following life table:

x	l_x	b_x
0	1	0
1	.5	1
2	.2	2
3	.1	1

so that $n_0 = n/1.8$ and $\tau = 1.6$, where n = total population size (see Chapter 3). Then, from (7-33),

$$n_e = \frac{1.6/1.8n}{1 + (.5 + .3)} = .494n$$

The Effect of Mutations on the Inbreeding Coefficient

The effect of mutations is to alter the allelic form and thus to destroy identity by *type*. But the concept of inbreeding coefficient is useful to us in later applications only if it describes identity both by descent *and* by type. We therefore refine our definitions of F (F_{IS}, F_{ST}, F_{IT}) to mean exactly what they have meant to this point, but conditional upon there being no alteration of type due to mutation. If mutations occur, the values of F can still be described by the equations above, but the values obtained must be discounted for the loss in identity by mutation. We write (from equation 7-7)

$$F_{IT}(t) = \left[\frac{1}{2n_e} + \left(1 - \frac{1}{2n_e}\right) F_{IT}(t-1)\right](1-\mu)^2$$

where μ is the mutation rate and reverse mutation is assumed negligible. $(1 - \mu)^2$ is therefore the probability that both alleles have been unaltered and so remain identical by type. At equilibrium, $F_{IT}(t) = F_{IT}(t-1)$, so that

$$F_{IT} = \frac{(1/2n_e)(1-\mu)^2}{1 - (1 - 1/2n_e)(1-\mu)^2}$$

which, because μ is generally extremely small, approaches

$$F_{IT} \simeq \frac{(1/2n_e)(1-2\mu)}{1 - (1 - 1/2n_e)(1-2\mu)} \simeq \frac{1}{1 + 4\mu n_e} \tag{7-34}$$

Partly because of mutation the inexorable drift in F_{ST}, and so also F_{IT}, toward 1.0, with corresponding fixation at all loci, may never be consummated.

7.2 Spatially Structured Populations

Earlier in this chapter we used the inbreeding coefficient, F_{ST}, to describe variance in the frequency of a gene among subpopulations (demes). We now note that demes cannot necessarily be treated as if there were no mixing of individuals, for indeed there is almost inevitably some migration among them. To deal with the calculation and prediction of F_{IT} values for populations whose demes experience some degree of migration, we examine three models. The first is the *island model*, in which the population is pictured as subdivided into a series of randomly distributed demes among which migration is random; that is, an individual leaving one deme is equally likely to reappear in any other. In such circumstances,

$$F_{IT}(t) = \frac{1}{2n_e} + \left[1 - \frac{1}{2n_e} F_{IT}(t-1)\right](1-m)^2 \tag{7-35}$$

where m is migration rate, and $(1 - m)^2$ is the probability that neither parent came from outside the deme and thus carried genes not ibd (and type) to those of the other parent. At equilibrium, we obtain

$$F_{IT} = \frac{(1-m)^2}{2n_e - (2n_e - 1)(1-m)^2}$$

But also, from (7-10 and 7-17) we have

$$F_{IT} = \frac{\sigma_p^2}{\bar{p}\bar{q}}$$

Hence

$$\sigma_p^2 = \frac{\bar{p}\bar{q}(1-m)^2}{2n_e - (2n_e - 1)(1-m)^2} \tag{7-36}$$

Thus, using (7-6) in conjunction with the variance given above, the expected genotype frequencies can be obtained.

The island model has been used to estimate the among deme variance in gene frequency in *Drosophila nigrospiracula*, a desert adapted fruit fly, by Johnson and Heed (1976). This particular species of fly oviposites and feeds on necrotic tissue of saguaro cacti in the American Southwest. Thus each cactus supports a subpopulation. Inasmuch as breeding takes place in the vicinity of the cactus on which eggs are laid, each of these subpopulations constitutes a deme. Johnson and Heed marked an enormous number of flies on individual cacti, using fluorescent powders, and then tried to recapture them over the next 24 days to measure dispersal. Fifty-three percent of the subsequently captured flies had been marked, providing them with a good sample for estimating the desired parameter; 391 of 5721 marked, recaptured flies had moved to new cacti, indicating a migration rate of 6.83%. The marked, recaptured flies also provided an estimate of the population sizes on the various cacti, which ranged from 343 to 14,725. Assuming that these n values are good estimates of n_e (i.e., $F_{IS} = 0$), equation 7-36 can be used to estimate var(p). Using the two extreme values of n_e, we have

$$\sigma_p^2 = \frac{.868\bar{p}\bar{q}}{2n_e - (2n_e - 1)(.868)} \leqslant \begin{cases} .0024 & \text{where } n_e = 343 \\ .0001 & \text{where } n_e = 14{,}725 \end{cases}$$

(The values above are maxima, occurring when $\bar{p} = \bar{q} = .5$.) Johnson and Heed concluded that spatial structure of the *D. nigrospiracula* population had little impact on genetic drift or genotype frequencies.

There is a major problem with the island model; it presumes that migration is random. In reality, individuals are far more likely to disperse to nearby than to far-off demes, resulting in an increase in gene frequency difference with distance. Two alternative models have been derived to deal with this difficulty, the stepping-stone model (Kimura and Weiss, 1964), and the isolation by distance model (Wright, 1940, 1969). The *stepping-stone model* deals with discrete demes, separated into distinct subpopulations, while the *isolation by distance model* treats the case of continuous distribution where effective demes are isolated by virtue of finite home ranges (neighborhoods) of its members. The former is quite complicated and so will be bypassed here in favor of the latter.

Occasionally, populations are naturally subdivided into discrete units within which mating is random: isolated pockets of individuals occupying small areas, for example, or populations on some islands. House mice (*Mus musculus*) form tightly knit, small groups within which virtually all mating takes place. These groups are discrete entities even when not separated in space. In such cases the local subpopulation constitutes a deme. But in other cases demes may be definable only in more statistical terms, as when reproductive isolation occurs not through separation into subpopulations, but only by distance between individuals in a continuously distributed population. Any given individual cannot be expected to wander freely throughout the entire population, encountering potential mates without regard to their position. Indeed, an individual must possess a center of activity and the likelihood of encountering and mating with other individuals must fall off in some fashion with their distance from this center. In order to make use of the equations of Section 7.1 in such populations, we must be able to define an effective population size centered around any point in space, and a corresponding "neighborhood," relating roughly to the area over which an individual confines its wanderings. Models attempting to define these neighborhoods are known as isolation by distance models.

We will derive both the one- and two-dimensional isolation by distance models. River- and streambank habitats or cliff faces are examples of situations best characterized by the one-dimensional model, open fields or forests by the two-dimensional model. Let the position of a parent at the time it gives birth relative to that of its offspring when the latter reproduces be normally distributed with mean zero. Thus if x is this position, its probability density is

$$\phi(x) = \frac{1}{\sqrt{2\pi}\sigma} e^{-x^2/2\sigma^2}$$

If n is the number of individuals in an ideal deme (so that $F_{IT} = F_{ST}$), evenly distributed in space, in a habitat strip of length 2σ, then ρ, the population density, is $n/2\sigma$. The chance that a particular gamete comes from a member of the parental generation, at distance x, is $\phi(x)/\rho$ and the chance that the offspring possesses two genes ibd is the probability that *both* parents are the same individual times one-half:

$$\frac{1}{2}\phi(x)\frac{\phi(x)}{\rho} = \frac{1}{2\rho}\phi(x)^2$$

provided that the parents are not inbred (i.e., F_{IT} is initially zero). Then, in an ideal deme,

$$\Delta F_{ST} = \Delta F_{IT} = F_{IT}(1) = \frac{1}{2\rho}\int_{-\infty}^{\infty}\frac{1}{2\pi\sigma^2}e^{-2x^2/2\sigma^2}\,dx = \frac{1}{2n\sqrt{\pi}} = \frac{1}{4\sigma\rho\sqrt{\pi}}$$

But, also, $\Delta F_{IT} = 1/2n_e$. Therefore,

$$n_e = (2\sigma\sqrt{\pi})\rho \tag{7-37}$$

If we define home range (neighborhood) size by the 95% confidence interval on an individual's wanderings, then home range (HR) $\simeq 2\sigma$, and

$$n_e = (2\sigma\sqrt{\pi})\rho = (\sqrt{\pi}\,\text{HR})\rho = (1.77\text{HR})\rho \tag{7-38}$$

The degree of genetic drift in such a population is given by (equation 7-10)

$$\text{var}(p) = \bar{p}\bar{q}\left[1 - \left(1 - \frac{1}{2n_e}\right)^t\right] = \bar{p}\bar{q}\left[1 - \left(1 - \frac{1}{4\sigma\rho\sqrt{\pi}}\right)^t\right] \tag{7-39}$$

As home range size (2σ) increases, and as population density increases, n_e rises and drift slows.

In two dimensions the derivation is essentially the same. Here we let n be the number of individuals in a square 2σ units on a side, so that $\rho = n/4\sigma^2$. Then

$$\Delta F_{IT} = \frac{1}{2\rho}\int_{-\infty}^{\infty}\int_{-\infty}^{\infty}\left(\frac{1}{2\pi\sigma^2}e^{-2x^2/2\sigma^2}\right)\left(\frac{1}{2\pi\sigma^2}e^{-2y^2/2\sigma^2}\right)dx\,dy = \frac{1}{2\pi n}$$

But also, $\Delta F_{IT} = 1/2n_e$. Thus

$$n_e = \pi n = 4\pi\sigma^2\rho \tag{7-40}$$

so that

$$\text{var}(p) = \bar{p}\bar{q}\left[1 - \left(1 - \frac{1}{8\pi\sigma^2\rho}\right)^t\right] \tag{7-41}$$

These models, and similar models (Malécot, 1969), suffer from the fact that they assume uniform distribution of individuals. This may, at first glance, appear to be only a minor source of distortion of the real world; we know that most populations are not really uniformly dispersed. However, Felsenstein (1975) has demonstrated that any variation at all in reproduction among individuals leads, unavoidably, to clumping of individuals. Felsenstein's conclusion that the models are thus "biologically irrelevant" may be a bit strong, but his paper stands as a warning not to interpret (7-37) through (7-41) too literally.

Consider the pattern of distribution in gene frequency p over space. If migration is very free, then we should expect large, fairly uniform areas of similar p; there will be little local variation in p. If migration is very limited, we should find a random mix of gene frequencies with much patchiness (very small patches). For intermediate levels of migration, we might expect to find cohesive groups of neighboring demes with high or low frequencies. Patterns of this sort have been simulated by Endler (1977), who used a computer to follow the genetic drift patterns of 2500 demes, arranged in a hexagonal grid. Each deme was held fixed at 100 individuals.

However, gene frequencies were permitted to change, draws from the local gene pools being determined by a random number generator. Migration was without bias to any one of the six, adjacent demes. His results are extremely interesting; not only do they support the conjectures above, but they indicate that patterns of gene frequency distribution, the peaks and troughs in p, may be extremely persistent in time. This means that differences in gene frequency within a species, from place to place, even if quite stable over time, do *not* necessarily implicate natural selection. Endler's results may be summarized briefly as follows:

1. Over the entire system of 2500 demes there was virtually no genetic drift at all. This is entirely expected for a population totaling $2500 \times 100 = 250{,}000$ individuals.
2. At migration levels of .20 (per generation probability of an individual emigrating), F_{ST} rose to about .15 in the first 100 generations, and appeared to level off at about .20.
3. Apparently stable and steep clines in gene frequency arose and persisted over many generations. Boundaries between high- and low-frequency areas shifted randomly in space over the distance of a few deme diameters, but the general position of the clines remained quite stable.
4. The size of "patches"—number of adjacent demes with $p > .55$ or $p < .45$—rose over time and asymptoted at about 10 (mean gene frequency over the entire area $= p = .5$).

These patterns are of interest in regard to the role of statistical variations in the spatial dispersion patterns of gene frequencies. They are not necessarily representative of the real world, but are interesting as possibilities which might arise from migration in the absence of selection. Similar results were obtained earlier by Rohlf and Schnell (1971).

Two final, but by no means unimportant aspects of the genetic structure of populations over space are the founder effect and the Wahlund effect. Occasionally, a few individuals from a population migrate and found a new subpopulation. Because the number of individuals involved in setting up the new subpopulation may be quite small, it is very likely that gene frequencies in this new unit will differ significantly from the frequency of the parental gene pool. What we see is a genetic drift, a deviation from expected gene frequencies arising not from sampling error during reproduction but rather during colonization of a new subpopulation. The importance of this *founder effect* is considerable, as we shall see in Chapter 9.

The *Wahlund effect* is important to researchers who wish to measure genotype frequencies in the field. It states simply that if the area over which sampling is carried out exceeds the neighborhood size, the frequencies obtained may be biased—heterozygote frequency is underestimated. The reason is quite simple, and is directly related to the earlier equations used to measure genetic drift. If gene frequencies p differ among demes and sampling occurs over more than one deme, the observed heterozygosity is

$$2\overline{pq} = 2[\bar{p}\bar{q} + \text{cov}(p, q)] = 2\bar{p}\bar{q} + 2\,\text{cov}(p, 1 - p) < 2\bar{p}\bar{q}$$

But the expected heterozygosity, predicted from gene frequencies obtained over the area, $\bar{p}$ and $\bar{q}$, is $2\bar{p}\bar{q}$. Thus there would appear to be a shortage of heterozygotes. The problem is eliminated if sampling is done on an areal scale commensurate with or smaller than the deme neighborhood.

Review Questions

1. If the Hardy–Weinberg conditions are met, we can expect a $p^2 : 2pq : q^2$ equilibrium distribution of genotypes. To what extent is the inverse true—that is, if a population is observed to maintain Hardy–Weinberg equilibrium at a given locus, can it be assumed that Hardy–Weinberg conditions are being met?
2. If we go back in time far enough, is F_{ST} really zero? Or is F_{ST} merely arbitrarily set to zero to give a base value against which later inbreeding values can be compared?

3. Explain why the inbreeding coefficient might differ from locus to locus in the same population.
4. Effective population size, n_e, usually deviates from true population size, n, as a function of sexuality, sex ratio, changing population size, age structure, and so on. How should these various factors be amalgamated when more than one applies to a given population? Imagine any real population. Design a series of observations or experiments that could be used to tease out these separate contributions to the inbreeding coefficient.
5. In the spatial structure models described in (7-37) to (7-40), dispersal is assumed to be randomly distributed with a mean of zero (i.e., symmetric in direction). What effect might nonzero means (directional dispersal) have on the results? How might kurtosis affect the value of neighborhood size n_e? If males and females had different-size home ranges, which home range size would be critical in determining n_e for the population? Should we simply use a mean?

Chapter 8

Natural Selection: I. Single-Locus Models

It should be obvious that genetic drift is not the only means by which changes in gene frequency can be effected. Between genotypes there may be differences in survival, mating propensity or, once mated, the number of offspring produced. Individuals that mate more successfully than others or produce larger numbers of offspring effectively cause more of their genes to be drawn from the gene pool during reproduction. Whatever allele is responsible for the greater reproductive success is thus disproportionately represented and shows an increase in frequency in the next generation. If genotypes endowing their owners with better powers of thermoregulation under high temperatures share a hot environment with other genotypes, they are likely, on average, to constitute a larger, surviving proportion of the population at breeding time than they did at birth. Consequently, they contribute a disproportionate number of genes to the gene pool and so generate a higher frequency of the responsible gene in the next generation. This would not necessarily happen were the population inhabiting a cold environment.

The brief argument above points out two basic facts. First, a gene that conveys on its owner a higher probability of survival, a greater chance of mating, or a greater fecundity than its complementary allele will increase in frequency over time. Second, the action and thus the "selective advantage" of a gene depends on the environment in which its bearer lives. The second fact should not be interpreted without caution. Part of an organism's environment is, in fact, its own internal milieu. Thus the effect of a gene depends on the anatomy, physiology, and behavior of the individual it inhabits, some of which is determined by other genes, some by chance. Geneticists often speak of the "genetic background" against which the action of a particular gene must be assessed. It is important to consider the fact that differences among individuals do *not* necessarily lead to genetic change; the differences must be due specifically to genetic as opposed to experiential or chance differences. In fact, as we shall see shortly, the differences must belong to a rather specific class of genetic differences.

It is not only gene frequencies that are apt to change due to differential characteristics of genotypes. Chromosomal aberrations such as inversions of portions of a chromosome, or translocations—where a part of one chromosome becomes attached to another chromosome—may also affect the survival and so on of the individual carrying them and so be selectively passed on to future generations. In both cases what happens is the same. Alternative forms of a gene at some locus, or alternative forms of a chromosome segment, are contributed to or drawn from the gene pool with unequal probability. This process is the modus operandi of natural selection and the changes generated are what allow organisms to become adapted to their environment. The outcome of natural selection, together with the random changes described in Chapter 7, are the stuff of microevolution.

As described above, the process of natural selection seems so straightforward that it should hardly appear necessary to support the basic prediction of adaptation with specific data. But recall that a gene's environment includes the genetic background of the individual that contains it. If a beneficial allele (in the sense of increased survival, mating success, or fecundity) should somehow be found always in conjunction with an allele, at another locus, which is harmful to its owner, do we still expect that allele to change in frequency in such manner as to increase survival, mating success, and fecundity in the population? There are other cautions as well, which we shall investigate later—and, finally, there are surprisingly few examples where a particular gene can be shown to be adaptive (convey greater survival, mating success, or fecundity than would a different gene). It is, on the other hand, fairly simple to show correlations between gene frequencies and environmental conditions which *suggest* adaptation. For example, Schopf and Gooch (1971), using electrophoretic techniques, investigated the frequencies of two alleles coding for a "fast"- and a "slow"-migrating form of leucine aminopeptidase (Lap) in the marine ectoproct *Schizoporella unicornis.* Among five areas characterized by different temperatures, they found a fairly regular increase in the observed frequency of the slow allele from .35 at the area of lowest temperature to .76 at the warmest area. McKechnie et al. (1975), also using electrophoresis, examined six enzyme loci in the butterfly *Euphydras edithea* from each of 21 localities in California. Two different measures of genetic distance (difference) (Nei, 1972; Rogers, 1972) were calculated for all pairs of localities and these distances were correlated with a variety of environmental parameters [altitude, latitude, soil type, larval food plant species, food plant genus, food plant family, whether the food plant was annual or perennial, 10-year averages for annual minimum and maximum and daily minimum and maximum (by month) temperature]. Collectively, these variables accounted for 91, 89, 93, 87, 82, and 81 percent of the variances in genetic distance at the six loci, respectively.

But while the results of the studies reported above are suggestive, none shows that the genetic differences observed are adaptive. In Schopf and Gooch's system, perhaps the "slow" genotype preferentially seeks out warmer areas. In the case of McKechnie et al.'s system, perhaps the various populations were different for historical reasons. There are a few more-convincing studies. Morgan (1975) raised fruit flies (*Drosophila melanogaster*) on media containing various types of alcohol (toxic) and no sucrose, and used the number of F_1 generation flies emerging per parental adult for two genotypes—electrophoretic fast- and slow-migrating forms (isozymes) of alcohol dehydrogenase (Adh)—as an indicator of selection for one or the other allele. He then took crude extracts of the fast and slow isozymes and tested their relative activity levels in breaking down (detoxifying) the various alcohols. If the natural selection response is indeed adaptive, there should be a positive correlation between relative activity of the fast isozyme form and the relative emergence success of the fast genotype. Morgan's results are given in Table 8-1. If data for isopropanol are deleted, there is a significant ($p < .01$) negative correlation between relative activity and an *inverse* measure of success, just as expected (see Table 8-1 for more details). Morgan argues that the deletion of isopropanol is appropriate;

Table 8-1

Alcohol Type	Activity Ratio Fast : Slow	β^a
N-Propanol	1.21	.4192
Isobutyl	1.31	.3719
N-Butanol	1.51	.3569
Ethanol	1.82	.3324
Cyclohexanol	2.08	.2556
Isopropanol	2.52	.4315

$$^a\ \beta = \frac{\ln\left(\dfrac{\text{fast genotype parental input}}{\text{fast genotype } F_1 \text{ emergence}}\right)}{\ln\left(\dfrac{\text{slow genotype parental input}}{\text{slow genotype } F_1 \text{ emergence}}\right)}$$

which is an inverse measure of the relative success of the fast genotype ($\beta < .5$ indicates selective advantage for the fast genotype).

for this alcohol, the breakdown product, acetone, is itself toxic. Thus enzyme activity cannot, in this case, be considered detoxification.

Later work on the Adh locus, and on the glycerol-3-phosphate dehydrogenase locus, is suggestive of selective advantages of different allelomorphs in quite another way. Oakeshott et al. (1982) examined distribution patterns for two electrophoretic variants at these loci over Australia, North America, and Asia and found widespread polymorphisms (more than one form maintained in the populations) and consistent latitudinal clines in allelomorph frequencies. These findings suggest selection gradients.

The classic case of adaptation via genetic change at a single locus is that of the peppered moth (*Biston betularia*) in Great Britain (Kettlewell, 1955a,b, 1956, 1958, 1961; Clarke and Sheppard, 1966). Prior to the start of the Industrial Revolution in the mid-nineteenth century, this species was characteristically a dirty white color with gray markings. It spent its hours perched on tree trunks among the similarly patterned lichens, against which it was difficult to see. When coal soot began spreading in England, many of the lichens died off and the color of the trunks darkened. As a result, the light-colored moth probably became much easier to spot by its predators. Thus a mutant gene causing dark coloration, until now quite rare, suddenly found itself at a selective advantage with respect to avian predation, and spread rapidly in the population. The correlation of dark moth frequency and degree of industrialization was marked. Also, direct observations on moths of both morphs, released in equal numbers into both unpolluted and highly industrialized areas, showed dark moths being disproportionately predated by birds in the unpolluted areas and less predated in the dirty areas (see the references cited above). The cause of the selection pressure being convincingly demonstrated, *industrial melanism* became, almost overnight, a classic example of ongoing evolution. Genetic experiments confirmed that the melanism was caused by an allele at a single locus. Subsequently, in other areas of England, melanic forms produced by mutations at different loci were discovered (see Sheppard, 1960, for a review), illustrating the fact that similar effects may be caused by the genetic machinery in more than one way. Industrial melanism has also been found in other lepidopterans and different orders of insects (Creed, 1971). Recently, a gradual decline in the use of coal has resulted in the cleansing of the air, a decline in the commonness of blackened tree trunks, and the corresponding decrease in frequency of melanic forms of both the peppered moth and other insect species.

Use of a clearly discernible, morphological trait to illustrate natural selection and adaptation at a single gene locus is a risky proposition. In the case of the peppered moth, strong correlations of gene frequency and environmental conditions could be complemented by experiments on the susceptability of the different morphs to predation. Thus the story that emerged was unambiguous. But this is not generally the case. Studies using specific enzymes, such as that of Morgan (1975), leave no doubt as to exactly what phenotypic character is being investigated. But morphological traits may well be associated with physiological traits not readily apparent to the investigator, which may be controlled by genes at the same or at other, strongly associated loci. Thus selection for one physiological type or another may drag along with it a particular morph. This "piggyback" effect can be quite misleading; in such cases a correlation between morph and environmental conditions will not be due to selection acting on the morphological trait per se, and any conclusions we might draw to that effect would be entirely erroneous.

A particularly well studied case, where conclusions are considerably less than satisfying, is that of the European garden snail (*Cepaea nemoralis*). This species, extravagantly abundant in scattered localities of western Europe, Great Britain, and where it has been introduced, North America, can be characterized by the banding patterns and background color of its shell. At one locus, a recessive allele provides the snail with a set of five dark bands. Other loci carry genes which modify expression of these bands, making them translucent, or thinner and broken, or fusing them, or suppressing all but the middle band. At still another locus a series of perhaps seven alleles codes for the background shell color. Alleles coding for brown are dominant to those promoting fawn, or "pink," which, in turn are dominant to those rendering the shell a pale yellow (for a review of the genetics, see Cain et al., 1960). Such clear-cut morphological differences, each determined by a single locus, cry out for simple explanations. In addition, some populations are monomorphic (one or another of the alleles at any particular locus is fixed), others polymorphic, and the gene frequency patterns appear to have remained quite constant over the species' geographic range for thousands of years (Cain and Sheppard, 1954). This certainly suggests a strong influence of natural selection. Moreover, we know that snails with different shell characteristics possess different physiological characteristics. For example, after warm, dry weather, *Cepaea* tends to lay larger egg clutches, and to lay more frequently. This behavior is particularly marked in the yellow morph. In the laboratory, unbanded yellow snails survive better than other morphs at 20°C, but not at 12°C. Yellow banded snails collected in one locality produced clutches of eggs at 6.2 times the rate at 20°C than at 12°C, while the corresponding multiple for pink banded snails was 3.1. The same ratios for these morphs collected at another locality was 2.1 and 1.1. Yellow banded snails from this latter site oviposited at 57% the rate of pink banded snails at 12°C, at 110% the rate when raised at 20°C. Oviposition by unbanded color morphs apparently is not differentially affected by temperature. Yellow unbanded snails always produce more eggs than yellow banded snails; there is no difference between pink banded and pink unbanded (Wolda, 1967; but see Wolda, 1963). We are thus left with the question: Are the shell pattern and color frequencies observed in nature due to natural selection acting on shell pattern and color, or on physiological traits that are somehow correlated, not necessarily causally, with shell characteristics?

Forest habitats contain *Cepaea* populations with high frequencies of dark morphs, while open habitats have more of the light-colored morphs. In addition, banding against a yellow background is characteristic of grassy habitats (Sheppard, 1952; Curry et al., 1964; Jones, 1973; Harvey, 1976). These observations suggest that shell color and pattern is selected for its cryptic qualities—that is, the ability of an animal to blend into its surroundings. Sheppard (1951) noted that near Oxford, England, *Cepaea* was predated by thrushes which carry the shells off to "anvils," rocks against which they break the shells, and that the frequency of broken yellow shells fell off during the period of April to June. It is during this period that the appearance of foliage would offer relatively greater protection to the yellow shell morph (which appears somewhat greenish when containing a living organism). Sheppard concluded that predation and protective coloration might well be important causes of the observed shell morph patterns in this species.

But the commonly observed patterns of dark morphs in woods and light morphs in fields occurs also in areas without thrush predation. Furthermore, there are areas where thrush predation occurs, but the usual morph dispersion patterns do *not* occur. Thus whereas predation may be a partial explanation for observed morph frequencies, and perhaps the entire explanation in some places, it is not the whole story. Jones (1973a) reviews the data showing quite clear latitudinal and altitudinal trends in color morph. Toward the north, and with higher altitudes, the frequency of the darker morphs increases. Jones found a regression coefficient of $b = 9.06$ ($p < .001$) for the percent of brown morphs on mean summer temperature (see also Jones, 1973b,c; Bantock, 1975a). At the northernmost limits of its range, *Cepaea nemoralis* populations contain no yellow shelled individuals (Jones, 1973c). The notion that springs to mind is that dark morphs may more readily absorb sunlight and so maintain biochemically efficient body temperatures in cold climates more reliably than do light morphs (or that light morphs can better withstand heat via greater albido). Indeed, laboratory experiments show that yellow unbanded snails survive higher temperatures than do dark and banded snails (Richardson, 1974; see also Heath, 1975). Richardson also collected dead and dying snails during hot days in the field and found a significant difference among the frequencies of yellow and brown snails represented; there were fewer dying yellow snails than expected by chance. Bantock (1975b) found higher survival of brown morphs under cold conditions in population cage experiments. Finally, examining subfossils in Great Britain, Cain (1971) showed that climatic changes over thousands of years, from cool to warm and back again, resulted in corresponding increases and then decreases in frequencies of the yellow morph and the unbanded form. On the other hand, Arnason and Grant (1976) found that higher temperatures were correlated, over Europe, with the frequency of the *brown* morph in a closely related species, *C. hortensis*. This last observation throws some doubt on the simple heat absorption–albido hypothesis and suggests that the underlying cause may be something quite different—that the relation with shell color and banding may be spurious.

8.1 Basic Models

Gene Frequency Changes

To this point we have referred, somewhat vaguely, to the success of an individual with respect to survival, mating success, and fecundity, as the driving force in natural selection. We need a more rigorous base on which to erect quantitative expressions for rate of gene frequency change. Thus we define *fitness* for genotype i as the average number of offspring produced per individual of that genotype born. Genotypes that differ in survival, mating success, or fecundity will clearly possess different fitnesses. Fitness is also sometimes called *selective value*, and differences in fitness among genotypes are referred to as *selective advantages* (or disadvantages) or *selection differentials*. Fitness of a genotype may depend on a variety of factors, anatomical and morphological, physiological and behavioral. The latter two may vary with the nature of the environment and with the types of interindividual interactions in the population. Thus these components of fitness are especially prone to gene–environmental interactions and to genotype frequency dependence. (The impact of frequency-dependent selection differentials will be discussed below.) That fitness has a behavioral component also leaves open the possibility that one genotype may have high relative fitness not only by virtue of possessing the ability to survive, mate, and reproduce well, but by actively interfering with the success of individuals of other genotype. Indeed, an individual which sufficiently reduces the ultimate reproductive success of its fellows may do so at personal expense and *still* come out with higher relative fitness. Such behavior has been dubbed "spiteful." Because selfishness (generally *not* spiteful) may enhance the relative fitness of the genotype whose members practice it, selfish behavior is generally selected for. On the other hand, selfishness that feeds back negatively on the perpetrator, via detrimental impact on the deme, would clearly be disadvantageous. Thus we must be careful in supposing that any act that increases an individual's survival, mating success, or fecundity is necessarily selectively advantageous. In addition, we must recognize that even behaviors deleterious to the individual performing them may under some circumstances have high fitness. For example, an individual that protects a conspecific by diverting the attack of a predator onto itself surely does not benefit from the act. But if the act is directed at saving others of the same genotype, and is successful, the individual loss may be more than compensated for by the overall gain to the genotype. Thus the fitness of the genotype is enhanced. In this manner genes predisposing individuals to act altruistically (aiding others at their own expense) may evolve. Clearly, such behavior is most apt to evolve when individuals affected by the act are genetically similar to the actor—which is most likely when they are related. For this reason altruism is most likely to occur in species characterized by viscous populations—where dispersal is limited. Clearly, in such cases the fitness of a genotype is determined not only by individual characteristics, but also by the summed impact of its actions on other individuals' success. Workers dealing with such phenomena often speak of *inclusive fitness* to indicate extension of the more limited concept of individual fitness to include these behavioral impacts.

Before proceeding, it will be useful to make a distinction between fitness, usually designated with the symbol W, and the population growth parameter R. Averaged over all genotypes i in a population, the mean fitness must be

$$\bar{W} = \sum_i \begin{pmatrix}\text{number of offspring}\\ \text{per individual of}\\ \text{genotype } i \text{ born}\end{pmatrix} \begin{pmatrix}\text{number of individuals}\\ \text{of genotype } i \text{ born}\end{pmatrix} \Bigg/ \sum_i \begin{pmatrix}\text{number of individuals}\\ \text{of genotype } i \text{ born}\end{pmatrix}$$

$$= \frac{n(t+1)}{n(t)} \quad \text{(if generation time is scaled to 1.0)}$$

But this (Chapter 2) is the definition of R. Hence

$$\bar{W} = R$$

But $R_i = n_i(t+1)/n_i(t)$ is *not* generally equal to W_i, for the contributed genes will be reassorted during the reproductive process; even if all homozygotes die or fail to reproduce and the entire contribution to the gene pool is by heterozygotes, only one-half of the offspring will be heterozygous. A specific example may further clarify this distinction. If we denote by A and a the two allelic forms at some locus, there will be three possible genotypes, AA, Aa, and aa. If these genotypes have equal R values, then, by definition, their relative frequencies must remain the same. However, their fitness values may differ considerably. Suppose that we have $W_{AA} = 1.0$, $W_{Aa} = 1.5$, and $W_{aa} = .5$. Then the number of A genes contributed to the next generation by AA individuals $= n_{AA}W_{AA}$ times 2, because each such individual carries two A genes. The number contributed by Aa individuals is $n_{Aa}W_{Aa}$ times 1, and the aa individuals contribute none. The total number of A genes put into the next generation is thus $2n_{AA}W_{AA} + n_{Aa}W_{Aa}$. Similarly, the input of a alleles is $n_{Aa}W_{Aa} + 2n_{aa}W_{aa}$, and the frequency of A alleles in the gene pool, and hence the next generation, is

$$p' = \frac{2n_{AA}W_{AA} + n_{Aa}W_{Aa}}{(2n_{AA}W_{AA} + n_{Aa}W_{Aa}) + (n_{Aa}W_{Aa} + 2n_{aa}W_{aa})} = \frac{2n_{AA}W_{AA} + n_{Aa}W_{Aa}}{2n_{\text{tot}}\bar{W}}$$

If we assume that mating is random, so that Hardy–Weinberg equilibrium obtains at birth, then $n_{AA} = p^2 n_{\text{tot}}$, $n_{Aa} = 2pqn_{\text{tot}}$, and the frequency above becomes

$$p' = \frac{2(p^2 n_{\text{tot}} W_{AA}) + 1(2pqn_{\text{tot}} W_{Aa})}{2n_{\text{tot}}\bar{W}} = p\frac{pW_{AA} + qW_{Aa}}{\bar{W}}$$

$$= p\frac{1.0p + 1.5q}{p^2(1.0) + 2pq(1.5) + q^2(0.5)}$$

If p is equal to $\frac{2}{3}$, then $p' = p$, and the genotype frequencies in the next generation are unchanged. That is, $R_{AA} = R_{Aa} = R_{aa}$.

Some geneticists may find the foregoing a bit confusing. The usual approach used by population geneticists is to avoid the (most obvious) complications posed by density feedback on W by using *relative* fitness values, w:

$$w_{AA} : w_{Aa} : w_{aa} = W_{AA} : W_{Aa} : W_{aa}$$

where one of the values of w (usually w_{AA}) is set to 1.0. Very often the notation used is of the general form

$$w_{AA} = 1 \qquad w_{Aa} = 1 - s \qquad w_{aa} = 1 - t$$

The conclusions in this and other chapters are in no way affected by the difference in notation, and because this book's primary thrust is ecological the fact that W must be a function of population density, is not considered a liability. Accordingly, the notation already introduced will be continued (for a discussion of density feedback and fitness, see "Frequency-Dependent and Density-Dependent Selection" (p. 236).

The general formulation of the expression for change in gene frequency with natural selection follows the calculations above. Because of the convenience of being able to assume Hardy–Weinberg equilibrium at birth, geneticists usually make the simplifying assumption that mating success and fecundity differ little among genotypes and that the major contribution to selection differentials comes from differential mortalities. We write:

	Genotype		
	AA	Aa	aa
Number at birth in parental generation	np^2	$2pqn$	q^2n
Fitness	W_1	W_2	W_3
Number at time of reproduction	p^2W_1n	$2pqW_2n$	q^2W_3n

Therefore, the number of A and a genes in the new gene pool is

$$2(p^2W_1n) + 1(2pqW_2n) \qquad 1(2pqW_2n) + 2(q^2W_3n)$$

and the new frequency of A, after one generation of selection, is

$$p' = \frac{\text{number of } A \text{ genes}}{\text{number of } A \text{ plus } a \text{ genes}} = \frac{p(pW_1 + qW_2)}{\bar{W}} \tag{8-1}$$

The change in p from one generation to the next is thus

$$\Delta p = (p' - p) = \frac{p(pW_1 + qW_2)}{\bar{W}} - p = \frac{p(pW_1 + qW_2) - p(p^2W_1 + 2pqW_2 + q^2W_3)}{\bar{W}}$$

$$= \frac{pq[p(W_1 - W_2) + q(W_2 - W_3)]}{\bar{W}} \tag{8-2}$$

Alternatively,

$$\Delta p = \frac{p(pW_1 + qW_1)}{\bar{W}} - p = \frac{p\bar{W}_A}{\bar{W}} - p = \frac{p(\bar{W}_A - \bar{W})}{\bar{W}} \tag{8-3}$$

If inbreeding occurs in the population, we must replace the genotype frequencies p^2, $2pq$, and q^2 with the values $p^2 + pqF$, $2pq(1 - F)$, and $q^2 + pqF$. If this is done, we obtain the more general expression

$$\Delta p = \frac{pq[p(W_1 - W_2) + q(W_2 - W_3)] + pqF[q(W_1 - W_2) + p(W_2 - W_3)]}{\bar{W}_0 - pqF(2W_2 - W_1 - W_3)} \tag{8-4}$$

where $\bar{W}_0$ is the population mean fitness in the absence of inbreeding. The first bracketed expression in (8-4) is commonly referred to as α. Also, for convenience in this book, I shall refer to the second bracketed expression as β, and let $2W_2 - W_1 - W_3$ be k. Thus (8-4) can be more compactly written

$$\Delta p = \frac{pq\alpha + pqF\beta}{\bar{W}_0 - pqFk} \tag{8-5}$$

where $\alpha = p(W_1 - W_2) + q(W_2 - W_3)$
$\beta = q(W_1 - W_2) + p(W_2 - W_3)$
$k = 2W_2 - W_1 - W_2$

and if $F = 0$,

$$\Delta p = \frac{pq\alpha}{\bar{W}_0} = \frac{pq\alpha}{\bar{W}} \tag{8-6}$$

It is important to note three things. First, the maximum impact of inbreeding occurs when $F = 1$. This means that $\bar{W} \geqslant \bar{W}_0 - pq(2W_2 - W_1 - W_3) = pW_1 + qW_3 > 0$, unless both homozygotes are lethal. Second, inbreeding *can* change the direction in which natural selection alters gene frequency. For example, suppose that $W_1 - W_2 = -.5$, $W_2 - W_3 = .1$, and $p = .1$, so that $q = .9$. Then, when $F = 0$,

$$\Delta p = \frac{(.1)(.9)[.1(-.5) + .9(.1)]}{\bar{W}_0} = \frac{.0036}{\bar{W}_0} > 0$$

but when $F = 1$,

$$\Delta p = \frac{(.1)(.9)[.1(-.5) + .9(.1)] + (.1)(.9)[.9(-.5) + .1(.1)]}{\bar{W}_0 - (.1)(.9)(.6)} = \frac{.036}{\bar{W}_0 - .054} < 0$$

Finally, recall from Chapter 7 that the impact of assortative mating on genotype frequencies is similar in form (although possibly opposite in direction) to that of inbreeding. Hence there is no need to formulate additional equations to investigate assortatively mating populations.

Equilibrium gene frequencies under natural selection occur when $p = 0, q = 0$ (fixation of one or the other allele), or when $\alpha + F\beta = 0$ (see equation 8-5). The first two situations are trivial: When there is only one allele, there can be no change in its frequency barring mutation. The third case is more interesting. Write

$$0 = \alpha + F\beta = [p(W_1 - W_2) + q(W_2 - W_3)] + F[q(W_1 - W_2) + p(W_2 - W_3)]$$

Then, remembering that $q = 1 - p$, and collecting terms, we obtain

$$p = \frac{(W_2 - W_3) - F(W_2 - W_1)}{(2W_2 - W_1 - W_3)(1 - F)} \tag{8-7}$$

If this value lies between 0 and 1, both alleles are maintained in the population: We say that the population is *polymorphic*.

To discover under what conditions polymorphism occurs for a given population size and environmental state, we write

$$\begin{aligned}\frac{\partial \bar{W}}{\partial p} &= \frac{\partial}{\partial p}[(p^2W_1 + 2pqW_2 + q^2W_3) - pqF(2W_2 - W_1 - W_3)] \\ &= 2[p(W_1 - W_2) + q(W_2 - W_3)] + (p - q)(2W_2 - W_1 - W_3)F \\ &\quad + \overline{\frac{\partial W}{\partial p}} - pqF\left(2\frac{\partial W_2}{\partial p} - \frac{\partial W_1}{\partial p} - \frac{\partial W_3}{\partial p}\right) \\ &= 2\alpha + \overline{\frac{\partial W}{\partial p}} + \left[(p - q)k - pq\frac{\partial k}{\partial p}\right]F \end{aligned} \tag{8-8}$$

If we ignore inbreeding, this becomes

$$2\alpha + \overline{\frac{\partial W}{\partial p}}$$

so that

$$\alpha = \frac{1}{2}\left(\frac{\partial \bar{W}}{\partial p} - \overline{\frac{\partial W}{\partial p}}\right)$$

Substituting into (8-6) gives us

$$\Delta p = \frac{pq}{2\bar{W}}\left(\frac{\partial \bar{W}}{\partial p} - \overline{\frac{\partial W}{\partial p}}\right) \tag{8-9}$$

The second term in parentheses describes the average change in genotype-specific fitness values with gene frequency. That is, it describes the influence on selection of (gene) frequency-dependent changes in fitness.

Inclusion of inbreeding and the frequency-dependent component of selection makes our equations rather messy. Thus we shall ignore these complications for the moment, and write (8-9) as

$$\Delta p = \frac{pq}{2\bar{W}} \frac{\partial \bar{W}}{\partial p}$$

It is now clear that if an increase in p raises $\bar{W}$, so that $\partial \bar{W}/\partial p > 0$, then Δp is positive. On the other hand, if an increase in p decreases $\bar{W}$, then $\partial \bar{W}/\partial p < 0$, and Δp is negative. Therefore, natural selection always changes the value of p in such manner as to increase $\bar{W}$. Refer now to Figure 8-1. If $W_1 > W_2, W_3$, mean fitness is maximized when $p = 1.0$. If $W_3 > W_2, W_1$, mean fitness is greatest when $p = 0$. Thus both of these circumstances lead to fixation. If $W_2 < W_1, W_3$, mean fitness is increased either by fixation of A, or fixation of a, depending on the initial values of p. Only when $W_2 > W_1, W_3$ will selection lead to polymorphism. In other words, the polymorphic frequency given by (8-7) (ignoring inbreeding) is valid only if $W_2 > W_1, W_3$. The condition where fitness of the heterozygote is greater than that of either homozygote is known as *heterosis*. A locus displaying heterosis is referred to as "heterotic."

The conclusion that natural selection increases mean fitness has been used by many biologists, ecologists in particular, to justify the optimization approach to prediction: If a particular physiology or behavior will increase an organism's survival, mating success, or fecundity (i.e., fitness) it is assumed to evolve (or to have evolved). This is an excellent time to

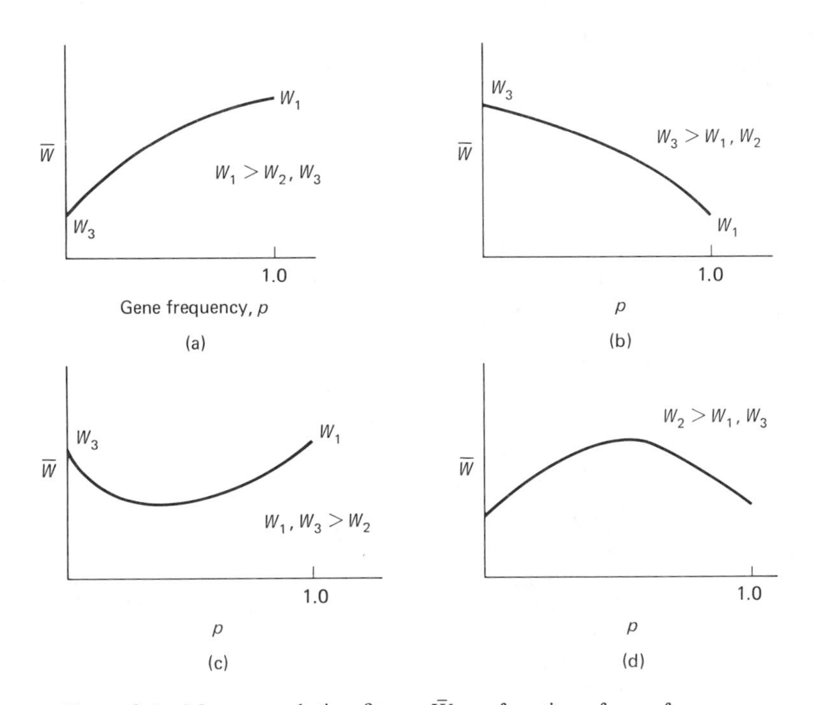

Figure 8-1 Mean population fitness $\bar{W}$ as a function of gene frequency.

sound a warning: Natural selection does *not* necessarily maximize mean fitness! Both frequency-dependent selection and inbreeding can alter the direction and thus ultimate effect on fitness of the selection process. Furthermore, "spiteful" behavior may act to *minimize* mean fitness (Chapter 15).

In spite of this warning, to keep matters uncluttered, let us for the moment ignore the impact of frequency-dependent selection and inbreeding. We look first at a classic case of heterosis.

In Old World tropical areas, Falciparum malaria is a significant cause of mortality in human populations. In these same areas there is found a polymorphism at a gene locus affecting the nature of the red blood cells. The less common "sickle-cell" allele, found at a frequency approaching zero in people without tropical ancestry, causes distortion and rigidity of erythrocytes and thereby makes it impossible for the erythrocytes to squeeze through the capillaries. The result is blood blockage and anoxia of tissues. Presence of the malaria parasite in blood cells lowers the pH of those cells and hastens the process of sickling. But the parasite cannot live in sickled cells. Hence the sickling tendency kills off a sizable portion of the parasite population. In heterozygous individuals, those with a partial complement of normal erythrocytes, this mutual destruction of sickle cells and parasites is advantageous. But individuals homozygous for the sickle cell trait are at an extreme disadvantage. In parts of east Africa, examination of blood samples showed that varying proportions, up to about 40%, of the adult inhabitants carried the sickle-cell allele, and of these, about 2.9% were homozygous (Allison, 1955, 1961). These numbers allow us to calculate the frequency of the allele among adults:

$$\frac{2(.029)(.40) + 1(1 - .029)(.40)}{2} = .206$$

Therefore, at birth (given random mating), the frequencies of the sickle-cell homozygote, the heterozygote, and the normal homozygote genotypes should be

$$= .042 : .327 : .630$$

But the observed frequencies were (for the data presented above)

$$.012 : .388 : .600$$

The ratio of fitnesses of these genotypes is thus

$$\frac{.012}{.042} : \frac{.388}{.327} : \frac{.600}{.630} = .30 : 1.25 : 1.00$$

The heterozygote displays a (roughly) 25% selective advantage over the normal homozygote. For a much more complete review on the genetics of the sickle cell trait see Templeton (1982).

There are no other cases of heterosis that are so clear cut. More generally, data take the form shown below for the plant *Liatris cylindracea*, studied by Schaal and Levin (1976). *L. cylindracea* produces corms that display a series of rings, the number of which is at least crudely correlated with age of the plant. Schaal and Levin looked at 14 enzyme loci, measuring the deviation in observed heterozygote frequency from the expected Hardy–Weinberg value in local neighborhoods:

$$F_{IS} = 1 - \frac{\text{observed heterozygosity}}{2pq}$$

(see equation 7-17 and the comments immediately following). In all cases, F_{IS} was positive, indicating inbreeding. But for 12 of the 14, F_{IS} declined with age. Alcohol dehydrogenase, for example, dropped from .916 in plants between 1 and 5 years of age to .294 for individuals more than 26 years old. The drop in F_{IS}—that is, the rise in heterozygosity—with age implies heterozygote advantage (although not necessarily at these loci; see Chapter 10).

As an aside, it appears that F_{ST}, the *among* deme component of inbreeding, also declines with age, meaning a convergence in p values. This suggests that natural selection is acting to oppose genetic drift and push the local subpopulations toward a common genetic type.

The Evolution of Traits

Whereas the geneticist is often happy to deal with genes per se, the ecologist is apt to be more concerned with morphological, physiological, or behavioral traits whose expression depends on those genes. Consider any trait, X, whose value is given by x_1, x_2, and x_3, respectively, for the genotypes AA, Aa, and aa. We may construct the following table:

	Genotype		
	AA	Aa	aa
Frequency	$p^2 + pqF$	$2pq(1 - F)$	$q^2 + pqF$
Fitness	W_1	W_2	W_3
Trait value	x_1	x_2	x_3

Where a prime denotes the next generation, we may write

$$\bar{x} = p^2x_1 + 2pqx_2 + q^2x_3 \tag{8-10}$$

$$\bar{x}' = p'^2x_1 + 2p'q'x_2 + q'^2x_3 \tag{8-11}$$

But from (8-5), we have

$$p' = \frac{pW_0 + pq\alpha + pqF(W_1 - W_2)}{\bar{W}} \tag{8-12}$$

Substituting (8-12) into (8-11), we find, after considerable algebraic duress,

$$\Delta\bar{x} = \bar{x}' - \bar{x} = \frac{2pq\alpha^{(x)}(\alpha + \beta F)}{\bar{W}_0 - pqFk} + \frac{p^2q^2(\alpha + \beta F)^2}{(\bar{W}_0 - pqFk)^2}k^{(x)} \tag{8-13}$$

where
$$\alpha = p(W_1 - W_2) + q(W_2 - W_3)$$
$$\alpha^{(x)} = p(x_1 - x_2) + q(x_2 - x_3)$$
$$\beta = q(W_1 - W_2) + p(W_2 - W_3)$$
$$\beta^{(x)} = q(x_1 - x_2) + p(x_2 - x_3)$$
$$k = 2W_2 - W_1 - W_3$$
$$k^{(x)} = 2x_2 - x_1 - x_3$$

If selection is slow (i.e., Δp is small), then unless $k^{(x)}$ is quite large, the second term in (8-13) must be much smaller than the first. Because of this fact it may generally be ignored. Therefore, as a good approximation, we write

$$\Delta\bar{x} \simeq \frac{2pq\alpha^{(x)}(\alpha + \beta F)}{\bar{W}_0 - pqFk} \tag{8-14}$$

Alternatively, when Δp is small,

$$\Delta\bar{x} \simeq \frac{d\bar{x}}{dp}\Delta p = \frac{pq(\alpha + \beta F)}{\bar{W}_0 - pqFk}\cdot 2\alpha^{(x)} = \frac{2pq\alpha^{(x)}(\alpha + \beta F)}{\bar{W}_0 - pqFk}$$

It is convenient, at this point, to introduce some new notation. Let x_{ij} be the trait value of the jth individual belonging to the ith genotype. Then

$$x_{ij} = \bar{x} + (x_{ij} - \bar{x}_i) + (\bar{x}_i - \bar{x})$$

The first term in parentheses, the deviation of x_{ij} from the mean value for genotype i, $\bar{x}_i$, must be due to events unique to individual j–finding itself at a particular point in space, having had particular experiences, and so on. Part of this term reflects environmental factors whose net influences on x_{ij} we label E_{ij}. Part, labelled ϵ_{ij} is random error. The means of both terms over all individuals of genotype i (being the averages of deviations from that mean) must be zero. The next term, $(\bar{x}_i - \bar{x})$, is the deviation due to genetic differences between the ith genotype and the hypothetical average individual. We shall call this deviation due to genetic differences G_i. Finally, because different genotypes may respond to different environments in different ways, $x_{ij} - \bar{x}$ cannot be written simply as the sum of deviations due to genetic and environmental differences and random error. We define

$$I_{GEi} = (x_{ij} - \bar{x}) - (E_{ij} + \epsilon_{ij} + G_i)$$

the error by which $G_i + E_i + \epsilon_{ij}$ misses the value of $x_{ij} - \bar{x}$, as the deviation due to gene–environment interactions. Thus, finally,

$$x_{ij} = \bar{x} + E_{ij} + \epsilon_{ij} + I_{GEi} + G_i \quad \text{(8-15a)}$$

The value G_i, known as the *genetic value*, can be further dissected. The genetic value is comprised of the joint effects of two alleles. But if the effect of one allele is dependent on the form of the other, G_i is not simply the sum of the allelic effects; there is a deviation due to allelic interaction. The sum of allelic effects we shall call the *additive genetic value*, A_i. The deviation in G_i from A_i, the interaction term, we call the *dominance genetic value*, D_i. Thus

$$x_{ij} = \bar{x} + A_i + D_i + I_{GEi} + E_{ij} + \epsilon_{ij} \quad \text{(8-15b)}$$

We lose nothing at this point by simplifying notation; we can drop the i, j subscripts and write

$$x = \bar{x} + A + D + I_{GE} + E + \epsilon \quad \text{(8-16)}$$

where the genotype-specific A, D, and so on, values are understood to be that of the individual in question.

Although the concept of additive genetic value may seem a bit esoteric at first, it will turn out to be extremely useful. We therefore proceed to show how to calculate it. First, we denote by $\alpha_A^{(x)}$ the effect of a single A allele on x, the effect of an a allele by $\alpha_a^{(x)}$. Then the additive genetic values are

$$\begin{aligned} A_1 &= 2\alpha_A^{(x)} \\ A_2 &= \alpha_A^{(x)} + \alpha_a^{(x)} \\ A_3 &= 2\alpha_a^{(x)} \end{aligned} \quad \text{(8-17)}$$

The effect of a single allele A can be found simply by calculating the difference in x between the average individual containing an A and the average individual in the population as a whole. If an individual is known to possess an A, the probability that it contains another is $F + (1 - F)p$—the probability that the other allele is ibd plus the probability that it is not but is identical by type anyway. This individual is homozygous. The probability that the allelic complement is a is given by $1 - [F + (1 - F)p]$, and the individual is heterozygous. Therefore, the average value of x for individuals possessing an A allele is

$$[F + (1 - F)p]x_1 + \{1 - [F + (1 - F)p]\}x_2$$

The effect of the A allele, $\alpha_A^{(x)}$ is then

$$\alpha_A^{(x)} = [F + (1 - F)p]x_1 + \{1 - [F + (1 - F)p]\}x_2 - \bar{x}$$

After some algebraic manipulation this reduces to

$$\alpha_A^{(x)} = q\alpha^{(x)} + qF\beta^{(x)} = q(\alpha^{(x)} + F\beta^{(x)}) \quad \text{(8-18a)}$$

(see the definitions of $\alpha^{(x)}$ and $\beta^{(x)}$ in equation 8-13). Similarly,

$$\alpha_a^{(x)} = -p(\alpha^{(x)} + F\beta^{(x)}) \tag{8-18b}$$

Thus

$$\begin{aligned} A_1 &= 2q(\alpha^{(x)} + F\beta^{(x)}) \\ A_2 &= (q - p)(\alpha^{(x)} + F\beta^{(x)}) \\ A_3 &= -2p(\alpha^{(x)} + F\beta^{(x)}) \end{aligned} \tag{8-19a}$$

and

$$\begin{aligned} G_1 &= 2q(\alpha^{(x)} + F\beta^{(x)}) + D_1 \\ G_2 &= (q - p)(\alpha^{(x)} + F\beta^{(x)}) + D_2 \\ G_3 &= -2p(\alpha^{(x)} + F\beta^{(x)}) + D_3 \end{aligned} \tag{8-19b}$$

For future reference note that

$$\bar{A} = p^2 A_1 + 2pqA_2 + q^2 A_3 = 0$$

Because the A values are deviations from the mean, the last equation has to be true.

We can now calculate the variance in additive genetic values, known as the *additive genetic variance*, $V_A(x)$.

$$\begin{aligned} V_A(x) &= (p^2 + pqF)[2q(\alpha^{(x)} + F\beta^{(x)})]^2 \\ &\quad + (2pq - 2pqF)[(q - p)(\alpha^{(x)} + F\beta^{(x)})]^2 \\ &\quad + (q^2 + pqF)[-2p(\alpha^{(x)} + F\beta^{(x)})]^2 \\ &= 2pq(1 + F)(\alpha^{(x)} + F\beta^{(x)})^2 \end{aligned} \tag{8-20}$$

Thus (equation 8-14)

$$\begin{aligned} \Delta\bar{x} &= \frac{2pq\alpha^{(x)}(\alpha + F\beta)}{\bar{W}_0 - pqFk} \\ &= \frac{V_A(x)}{\bar{W}}\left[\frac{\alpha^{(x)}}{\alpha^{(x)} + F\beta^{(x)}} \frac{\alpha + F\beta}{\alpha^{(x)} + F\beta^{(x)}}\right] \end{aligned} \tag{8-21}$$

If there is no inbreeding, then

$$\Delta\bar{x} = \frac{V_A(x)}{\bar{W}}\left(\frac{\alpha}{\alpha^{(x)}}\right) = \frac{2pq\alpha\alpha^{(x)}}{\bar{W}} \tag{8-22}$$

Thus evolution by natural selection is directly proportional to the *additive* genetic variance in the population. All of the calculations above, of course, assume a lack of frequency-dependent selection.

It should now be clear that additive genetic variance is a less esoteric quantity than it might have appeared. But it remains to put some biological interpretation on the quantities α and $\alpha^{(x)}$. Again ignoring inbreeding, recall that

$$\frac{\partial\bar{W}}{\partial p} = 2\alpha$$

Similarly,

$$\frac{\partial\bar{x}}{\partial p} = 2\alpha^{(x)}$$

Thus for constant population size and environment,

$$\alpha = \alpha^{(x)} \frac{\partial\bar{W}/\partial p}{\partial\bar{x}/\partial p} = \alpha^{(x)} \frac{\partial\bar{W}}{\partial\bar{x}}$$

Therefore,

$$\Delta\bar{x} = \frac{2pq\alpha\alpha^{(x)}}{\bar{W}} = \frac{V_A(x)}{\bar{W}}\frac{\partial\bar{W}}{\partial\bar{x}} \tag{8-23}$$

Also,

$$\operatorname{cov}_A(x, W)$$

the covariance of the additive values of x and W, known as the "additive covariance," is

$$\begin{aligned}
\overline{(A_x - \bar{A}_x)(A_w - \bar{A}_w)} &= \overline{A_x A_w} - \overline{A_x}\,\overline{A_w} \\
&= [p^2(2q\alpha^{(x)})(2q\alpha) + 2pq(q-p)\alpha^{(x)}(q-p)\alpha + q^2(-2p\alpha^{(x)})(-2p\alpha)] \\
&\quad - (\bar{0})(\bar{0}) \\
&= 2pq\alpha^{(x)}\alpha
\end{aligned} \tag{8-24}$$

Substituting into (8-22) give us

$$\Delta\bar{x} \simeq \frac{V_A(x)\,(\partial\bar{W}/\partial\bar{x})}{\bar{W}}$$

$$\Delta\bar{x} \simeq \frac{\operatorname{cov}_A(x, W)}{\bar{W}} \tag{8-25}$$

if there is no frequency-dependent selection, and $F = 0$. All of the forms above are useful, as we shall see in Chapter 10.

Additive genetic variance (and the equations above that use it) must remain rather abstract unless there is some realistic means of measuring it. In fact, there are several methods for assessing additive genetic variance in the laboratory, methods that are at least crudely applicable also in the real world. Suppose that parents and offspring experience the same environment, so that the environmental component of phenotypic value can be eliminated as a cause of variance between parents and offspring. Then phenotypic value is equal to the genetic value, which equals $A + D$, the additive plus dominance values. Suppose that the parent in question is AA. Then if that parent's mate was chosen randomly from within the population (or if we look at many AA parents, so that, on average, this is true), its offspring are random, unbiased representatives of individuals carrying at least one A gene. The deviation of such offspring from the population mean is thus, by definition, $\alpha_A^{(x)}$. This is exactly one-half of the additive genetic value of the parent in question. Suppose that the parent is Aa. Then half of its offspring carry at least one A gene and have genetic value $\alpha_A^{(x)}$, and half carry at least one a gene and have genetic value $\alpha_a^{(x)}$. On average, the genetic value of the offspring is $(\alpha_A^{(x)} + \alpha_a^{(x)})/2$, which, again, is exactly one-half the parent's additive genetic value. Finally, if the parent is aa, all its offspring possess at least one a gene and therefore have a genetic value of $\alpha_a^{(x)}$, which is one-half the parent's additive genetic value. Letting subscripts p and $\bar{o}$ denote the parent and the mean offspring, then

$$\begin{aligned}
\operatorname{cov}(x_p, x_{\bar{o}}) &= \operatorname{cov}(G_p, G_{\bar{o}}) = \operatorname{cov}(A_p + D_p, G_{\bar{o}}) \\
&= \operatorname{cov}(A_p + D_p, \tfrac{1}{2}A_p)
\end{aligned}$$

But since the additive and dominance values are, by definition, uncorrelated, this leads to

$$\tfrac{1}{2}\operatorname{cov}(A_p, A_p) = \tfrac{1}{2}V_A(x)$$

Summarizing, when parent and offspring live under identical environmental conditions, the covariance of a parent's trait value (observed, phenotypic value) with that of its average offspring gives one-half the additive genetic variance.

$$\operatorname{cov}(x_p, x_{\bar{o}}) = \frac{V_A(x)}{2} \tag{8-26}$$

If average offspring phenotype is *regressed* on parent phenotype, the slope of the relationship is given by

$$\frac{\operatorname{cov}(x_p, \bar{x}_o)}{\operatorname{var}(x_p)} = \frac{1}{2}\frac{V_A(x)}{\operatorname{var}(x)}$$

The ratio of variances is the proportion of the total phenotype variance that is additive. At such, it is also the proportion of the total variance that is available for natural selection to work on or, as said in the genetics vernacular, "heritable." $V_A(x)/\operatorname{var}(x)$ is called the *heritability* and is designated by the symbol h_x^2. Equations (8-25) can now be rewritten

$$\begin{aligned}\Delta\bar{x} &\simeq V_A(x)\frac{(\partial\bar{w}/\partial\bar{x})}{\bar{W}} = \frac{1}{\bar{W}}h_x^2 \operatorname{var}(x)\frac{\partial\bar{W}}{\partial\bar{x}} \\ &\simeq \frac{1}{\bar{W}}h_x h \operatorname{cov}(x, W)\end{aligned} \tag{8-27}$$

if there is no frequency-dependent selection, and $F = 0$, where h^2 is the heritability of fitness itself.

If either allele is fixed, $V_A(W) = 0$, $V_A(x) = 0$, and if heterosis occurs, equilibrium is reached when $\partial\bar{w}/\partial p = 2\alpha = 0$, that is, when $V_A(W) = 2pq\alpha^2 = 0$. But if $\partial\bar{W}/\partial p = 2\alpha = 0$, then $\Delta\bar{x} = 0$: Natural selection comes to a halt when additive genetic variance in *fitness* is zero.

Note that, because $\frac{1}{2}h_x^2$ is the regression coefficient of $x_{\bar{o}}$ on x_p,

$$(x_{\bar{o}} - \bar{x}) \simeq \tfrac{1}{2}h_x^2(x_p - \bar{x}) \tag{8-28}$$

if $x_{\bar{o}} \simeq x_{\bar{p}} = \bar{x}$ (i.e., if selection is very slow or nonexistent). We shall make use of this relationship later.

Continuous Time Models

Before leaving basic models we shall look very briefly at the continuous-time analog to the expressions above. Just as W designated fitness in the discrete case, m will denote fitness in the continuous case. The parameter m bears the same analogous relationship to the population growth parameter r that W does to R. If there are no mating and fecundity differences among genotypes, then, for the AA genotype, $m_1 = (1/n_{AA})(dn_{AA}/dt)$, and so on. Constructing the following table:

	Genotype		
	AA	Aa	aa
Frequency at birth	$p^2 + pqF$	$2pq(1 - F)$	$q^2 + pqF$
Fitness	m_1	m_2	m_3

we see that

$$\begin{aligned}\frac{dp}{dt} &= \frac{d(n_{AA}/n)}{dt} + \frac{1}{2}\frac{d(n_{Aa}/n)}{dt} \\ &= \frac{1}{n^2}[n(n_{AA}m_1) - n_{AA}(n\bar{m}) + \tfrac{1}{2}n(n_{Aa}m_2) - \tfrac{1}{2}n_{Aa}(n\bar{m})] \\ &= (p^2 + pqF)m_1 - (p^2 + pqF)\bar{m} + \tfrac{1}{2}2pq(1 - F)m_2 - \tfrac{1}{2}2pq(1 - F)\end{aligned}$$

which eventually reduces to

$$\frac{dp}{dt} = pq(\alpha + \beta F) \qquad (8\text{-}29)$$

where $\alpha = p(m_1 - m_2) + q(m_2 - m_3)$
$\beta = q(m_1 - m_2) + p(m_2 - m_3)$

(Alpha and beta are not quite the same as before because they now refer to continuous fitness measures.) Finally,

$$\frac{d\bar{x}}{dt} = \frac{d\bar{x}}{dp}\frac{dp}{dt} = [2\alpha^{(x)} - F(q - p)k^{(x)}]pq(\alpha + \beta F)$$

$$= 2pq\alpha^{(x)}\alpha + [2pq\alpha^{(x)}\beta - pq(q - p)\alpha k^{(x)}]F - pq(q - p)\beta k^{(x)}F$$

if there is no frequency-dependent selection. If, in addition, we let $F = 0$, then

$$\frac{d\bar{x}}{dt} = 2pq\alpha^{(x)}\alpha = \begin{cases} h_x^2 \operatorname{var}(x)\dfrac{\partial\bar{m}}{\partial\bar{x}} \\ h_x h \operatorname{cov}(x, m) \end{cases} \qquad (8\text{-}30)$$

(See equations 8-27.)

8.2 Some Extensions of the Basic Models

Frequency-Dependent and Density-Dependent Selection

Both frequency-dependent selection and inbreeding complicate the expressions for evolution of a single-locus trait under natural selection. We have discussed inbreeding at some length in Chapter 7. It is now time to examine frequency-dependent selection more closely.

Frequency-dependent selection arises because of interactions among individuals within the population. Because interactions can also lead to density feedback, it is unrealistic to look at frequency dependence without also considering density dependence. A number of approaches have been used (see, e.g., Anderson, 1971; Charlesworth, 1971; Roughgarden, 1971; Clarke, 1972; Smouse, 1976; Slatkin, 1978). One of the most straightforward approaches, which is also in keeping with the methods used in Part Two of this book, views genotype interactions as following the Lotka–Volterra equations (Smouse, 1976). As before (continuous case), we can use (8-29), but now

$$m_i = M_i\left(1 - \frac{\sum_j \alpha_{ij}n_j}{K_i}\right) \qquad (8\text{-}31)$$

Also,

$$\frac{dn}{dt} = \bar{m}n = (p^2m_1 + 2pqm_2 + q^2m_3)n \qquad (8\text{-}32)$$

Because dp/dt depends on the value of n (equation 8-31), and because n is changing as a function of gene frequency (equation 8-32), it is not possible to reach genetic equilibrium without also reaching population equilibrium. Inasmuch as populations fluctuate due to density-independent influences almost incessantly, we may conclude that genetic equilibrium is never (quite) reached. Even in the case where population variation is minute, the equilibrium solution (for n as well as p) is fairly complicated. Inasmuch as the Lotka–Volterra equations are probably not really accurate (as was also true in Chapters 4, 5, and 6), there is little value in deriving this solution. However, a few general conclusions are warranted.

1. Selection, because it chooses genotypes with the highest m values, should favor high M, high K, and low α values ($\partial m/\partial M > 0$, $\partial m/\partial K > 0$, $\partial m/\partial \alpha_{ij} < 0$).
2. There is no a priori reason to suspect that selection will increase n.
3. If the environment fluctuates, selection may lead to a polymorphism in which one (or two) genotypes is at a selective advantage during periods of low-density feedback and the other(s) is favored when density feedback is strong (see Chapter 5).
4. If α values are less than 1.0, the density depression in fitness (m) of a genotype is maximum when that genotype is *relatively* as well as absolutely abundant. In such cases, fitness varies inversely with relative abundance. If the relationship between abundance and fitness is sufficiently marked, the genotypes are protected from extinction and a polymorphism ensues.

A more general model of frequency-dependent selection is that of Cockerham et al. (1972). Suppose that mating is random in a population without inbreeding. We may depict fitness values as shown in Table 8-2. In this case

$$\Delta p = \frac{pq}{\bar{W}} [p(W_1 - W_2) + q(W_2 - W_3)] \tag{8-2}$$

which, after considerable algebra, reduces to

$$\Delta p = \frac{pq}{\bar{W}} [p(d_1 - d_2) + q(d_2 - d_3)] \tag{8-33}$$

where $d_{ij} = W_{ij} - W_{jj}$
$d_i = \text{average of } d_{ij}\text{'s over all } j\cdot = W_i - W^*$

W^* is the fitness, averaged over all genotypes, when individuals compete only with those of their own genotype. To make biological sense of (8-33), suppose (for example) that in competition with individuals of like genotype, all genotypes are equally fit, but that in competition with other genotypes, their fitness increases. Then $W_i - W^*$ is positive, measuring the ability of the ith genotype to survive, mate, and reproduce in the presence of density feedback from genetically different individuals. The genotype that is most successful under such competitive situations is favored by natural selection. Notice, in this case, that fixation leads to competition with one's own genotype only and thus lower, mean population fitness!

It seems obvious that different genotypes should often respond to density feedback in different ways and that, therefore, density- and frequency-dependent selection might be common. What evidence do we have that this is, indeed, true? Most of the classic examples come from the work of Allard and his colleagues. For example, Workman and Allard (1964), growing pure strains of wild oats (*Avena barbata*) in various combinations, found that fitness of a strain often increased when it was relatively rare. In a later paper Allard and Adams (1969) reported

Table 8-2

Genotype	**In Competition with:** *AA*	*Aa*	*aa*	**Mean =** $p^2W_{i1} + 2pqW_{i2} + q^2W_{i3}$
AA	W_{11}	W_{12}	W_{13}	W_1
Aa	W_{21}	W_{22}	W_{23}	W_2
aa	W_{31}	W_{32}	W_{33}	W_3

similar results from experiments on wheat and barley. Harding et al. (1966), working with the plant *Phaseolus lunatus*, looked at the genetics of the *S* locus. They found that the average fitness of genotypes carrying the *S* allele and those carrying its complement, *s*, under their laboratory conditions, were very nearly the same, $\bar{W}_S \simeq \bar{W}_s$, but that the relative fitness of the heterozygote, W_{Ss}, dropped from a value of about 3.0 when the frequency of *Ss* was .02 to a value of 1.0 when *Ss* reached a frequency of .14. As relative abundance of the heterozygote continued to rise above .16, W_{Ss} drops still further. Recall that the equilibrium frequency of an allele (let p be the frequency of S) is given by (equation 8-7, with $F = 0$)

$$p = \frac{W_{Ss} - W_{ss}}{2W_{Ss} - W_{ss} - W_{SS}}$$

If, in the case above, $p = .02$, then W_{Ss} (relative to $W_{SS} = W_{ss}$) is 3.0. Substituting into the expression above yields

$$p = \frac{3 - 1}{6 - 1 - 1} = .5$$

Thus p should increase above .02. But as it does so, the relative fitness of the heterozygote drops and, with it, the equilibrium value of p. In this case it is clear that an equilibrium should be reached somewhere between $p = .02$ and $p = .5$.

The mating success component is also known to show frequency dependence on occasion: Ehrman (1970), also working with *D. pseudoobscura*, this time using orange versus purple eye genotypes, found a consistent rare male mating advantage leading to a stable polymorphism (see also Petit, 1958; Spiess, 1970; Spiess and Kruckeberg, 1980).

Nassar et al. (1973), studying the "Payne" inversion in *D. melanogaster*, generated a strain of flies carrying the inversion and another not carrying it which, in all other respects, were (nearly) identical genetically. Thirty-four generations after extraction of the two strains, they performed +/+ by +/+ crosses, +/+ by I/I crosses, and I/I by I/I crosses (where I designates the Payne inversion). The results of these crosses would be, of course, homozygote +/+ flies, heterozygote +/I, and homozygote I/I flies, respectively. Once inseminated, the females were placed in vials and allowed to oviposit. The frequency p of the inversion, in the parental gene pool was manipulated by regulating the relative numbers of mothers chosen from each of the mating schemes above, and choosing them in Hardy–Weinberg ratios. In one set of runs, the high–density experiment, the mothers were allowed 3 days to lay their eggs. In the low-density experiment, only 12 hours was available for ovipositing. Fitness estimates (flies emerging/flies introduced) are given in Table 8–3. At high densities fitness of both homozygotes decreased as the corresponding allele became more common. Frequency-dependent selection pressures should result in an equilibrium p of about .28. At low densities, no such frequency dependence is apparent.

Nassar (1979) repeated essentially the same experiments examining the leucine amino peptidase locus (Lap). His results are shown in Table 8-4. Again a frequency dependence of

Table 8-3

	Fitness Estimate					
	High Density			*Low Density*		
Initial Frequency of Payne Inversion	+/+	+/*I*	*I*/*I*	+/+	+/*I*	*I*/*I*
.1	.986	1.025	1.677	1.121	.469	.685
.3	1.114	.854	1.048	1.353	.564	1.11
.7	1.630	1.095	.799	—	—	—
.9	3.508	1.311	.898	2.072	.777	1.03

Table 8-4

Initial Frequency of Fast Migration Lap Allele (*F*)	Fitness Estimate					
	High Density			*Low Density*		
	FF	*FS*	*SS*	*FF*	*FS*	*SS*
.2	6.10	1.325	.511			
.2	5.87	1.072	.658	5.02	1.184	.654
5	2.00	.686	.628			
.5	1.74	.684	.89	2.447	.510	.53
.7	1.066	.972	.981	1.035	1.034	.61
.8	1.037	.846	1.624	1.096	.856	.60

fitness was found, this time in both the high- and low-density populations. In the crowded populations this was such as to provide for a stable polymorphism. In the uncrowded populations, the fast allele was invariably favored. Finally, some particularly illustrative experiments of the same sort have been performed by Tošić and Ayala (1981) on the Mdh-2 locus in *D. melanogaster*. As can be seen in Figure 8-2, there is positive frequency-dependent fitness for all three genotypes at low population density (*SS* designates the genotype

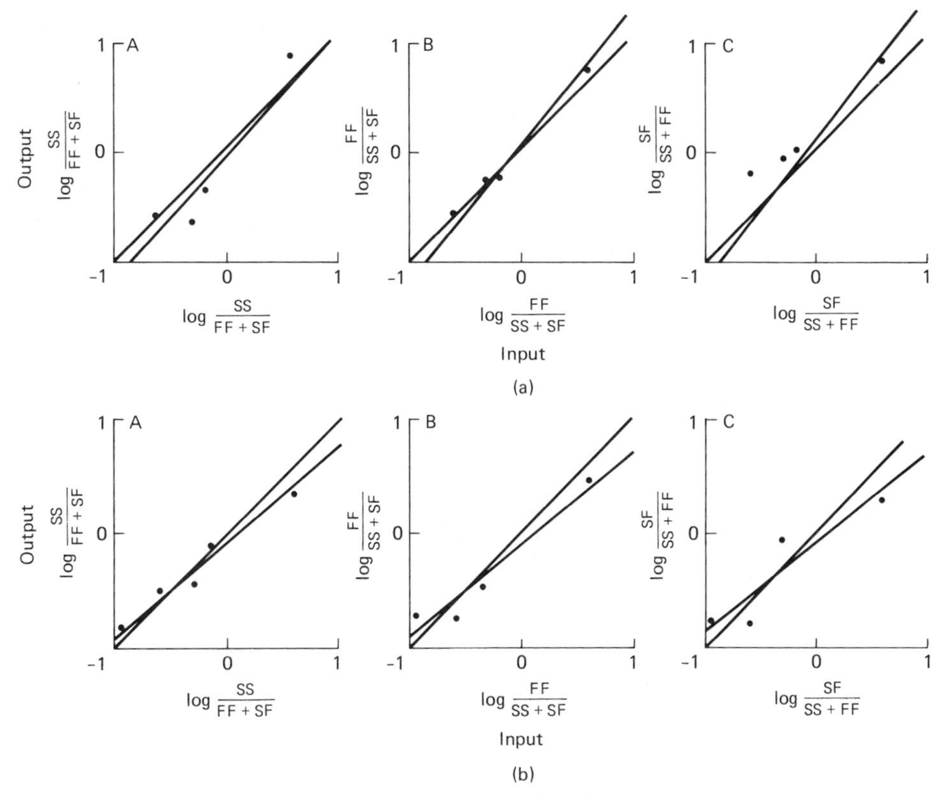

Figure 8-2 Logarithms of the relative frequency of various genotypes in one generation plotted against the corresponding logarithms of relative frequency in the previous generation: (a) low population densities; (b) high population densities. Locus depicted codes for malic dehydrogenase (Mdh). SS, electrophoretically slow-migrating homozygote; FF, fast homozygote; SF, heterozygote. (From Tošić and Ayala, 1981.)

homozygous for the slow-migrating allele under electrophoresis, *SF* the heterozygote, and *FF* the "fast" homozygote). However, at high population densities, all three genotypes display negative frequency dependence.

Complications Due to Age Structure

All of the models above pertain to the simple situation where there is no overlapping of generations or, in the continuous case, continuous reproduction without age-specific differences in fecundity or survival. Does age structure alter the nature of our conclusions so far?

In 1971, King and Anderson (see also Anderson and King, 1970) ran a series of computer simulations of natural selection acting on a population of three genotypes (one locus, two alleles), each of which was characterized by its own Leslie matrix. Every generation an age vector of each genotype was premultiplied by the appropriate matrix to generate a new age vector, and the zero-age class was then reassorted according to the expected distribution of genotypes under random mating. In one set of runs the elements of the Leslie matrices were constant (no density feedback). Viabilities did not differ among genotypes, and fecundity was allowed to display heterosis. The results showed cycling of $n_0(t + 1)/n_0(t)$, as might be expected, until both stable age distribution and genetic equilibrium were established. Except for the apparent cycling change in fitness due to the shifting frequencies of the age classes, no surprises were recorded. Equilibrium gene frequency was as expected in the absence of age structure. In another set of runs, density feedback was introduced by multiplying fecundity values by a logistic correction factor, $1 - n/K_{ij}$, where n was the total population and K_{ij} the carrying capacity of the A_iA_j genotype. Fecundity was truncated to prevent negative values when n exceeded K_{ij}. In these runs, survival did not vary among genotypes; selection was directional for fecundity (favoring one homozygote) and heterotic for K. The results obtained are shown in a very general way in Figure 8-3. Over time, one genotype, then another, appeared to be favored by selection; the total population size grew, reached a plateau, then rose again to a new plateau. Different genotypes were favored at different stages in the population cycle. Similar simulations by Charlesworth and Giesel (1972) gave similar results. The latter authors stressed the fact that changes in gene frequency, by changing population fitness ($= R$) would perpetuate population cycles which, by causing changes in the relative fitnesses of different genotypes would promote continuing gene frequency change. Thus equilibrium gene frequencies could come about only once the population reached stable age distribution, and vice versa.

The most extensive studies of selection in age-structured populations have been carried out by Charlesworth (1970, 1974, 1980). He suggests two approaches. First, suppose that the Leslie matrix elements are unchanging—that is, the population is growing in an unlimited environment or is in the neighborhood of equilibrium and experiences only minute fluctuations, In either case we should expect stable age distribution to be reached fairly quickly, at least when selection is slow. The number of A alleles passed on via parents of age x in this case, under

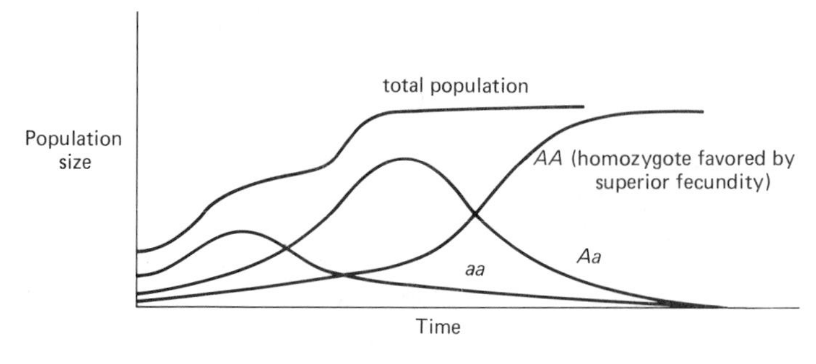

Figure 8-3 Changes in population density of various genotypes over time in an age-structured population (see King and Anderson, 1971).

random mating, is given by

$$2(n_x p^2 b_{xAA}) + 1(n_{x'} 2pqb_{xAa})$$

where b_{xij} is the fecundity of an average ij genotype individual of age x. Thus

$$2n_0(t)p(t) = \sum_x 2n_x(t)[p(t)^2 b_{xAA} + p(t)q(t)b_{xAa}]$$

But (Chapter 3)

$$n_x(t)p(t)^2 = n_0(t-x)p(t-x)^2 l_x = n_0(t)p(t-x)^2 R^{-x} l_x$$

at stable age distribution. Thus, dividing by $n_0(t)$, we get

$$p(t) = \sum_x R^{-x}[p(t-x)^2 k_{xAA} + p(t-x)q(t-x)k_{xAa}] \tag{8-34}$$

where: $k_x = l_x b_x$. Having assumed selection to be slow, we may now apply the Taylor expansion and write

$$p(t-x) \simeq p(t) - x\Delta p$$

Substituting into (8-34) gives us

$$p = \sum_x R^{-x}[(p^2 - 2px\Delta p)k_{xAA} + (p - x\Delta p)(q - x\Delta q)k_{xAa}]$$

so that

$$\Delta p = \frac{\sum_x R^{-x}(p^2 R_{xAA} + pqk_{xAa}) - p}{\sum_x xR^{-x}(2pk_{xAA} + (1-2p)k_{xAa})} \tag{8-35}$$

But because selection is slow, $k_{AA} \simeq k_{Aa} (= \bar{k})$, and (8-35) approaches

$$\Delta p = \frac{p\left[\left(\sum_x R^{-x}\bar{k}_{xA}\right) - 1\right]}{\sum_x xR^{-x}\bar{k}_x} \tag{8-36}$$

Defining $\sum xR^{-x}\bar{k}_x$ = generation time (τ), recalling that $1 = \sum R^{-x}\bar{k}_x$, and noting that

$$1 = \sum_x R^{-x}\bar{k}_x = \sum_x R^{-x}(p^2 k_{xAA} + 2pqk_{xAa} + q^2 k_{xaa})$$

$$= p^2\left(\sum_x R^{-x}k_{xAA}\right) + 2pq\left(\sum_x R^{-x}k_{xAa}\right) + q^2\left(\sum_x R^{-x}k_{xaa}\right)$$

we have, finally,

$$\Delta p = \frac{p(\bar{W}_A - \bar{W})}{\tau\bar{W}} \qquad \text{where } W_{ij} = \sum_x R^{-x}k_{xij} \tag{8-37}$$

Charlesworth's alternative approach is to ignore the problem of reassortment bias due to differential mating and fecundity among genotypes (as was done in the derivation of equations 8-2 and 8-3), so that fitness for a genotype is equivalent to its population growth rate, and to write

$$W_{ij} = R_{ij}$$

where R_{ij} is defined by

$$1 = \sum R_{ij}^{-x}\bar{k}_{xij}$$

While the second method fails to account for mating and fecundity differences, the first requires slow selection and stable age distribution. Charlesworth (1974) confirmed these advantages and disadvantages by comparing the predictions of these two approaches with changes in gene frequency arising from a computer simulation of the sort used by King and Anderson (1971).

We now define the *reproductive value* of an individual age x (Fisher, 1958):

$$v_x = \frac{v_0 \lambda^x}{l_x} \sum_{y>x} \lambda^{-y} l_y b_y \tag{8-38}$$

where λ is understood to mean the real positive root of the Leslie matrix characteristic equation. The meaning of this quantity can be gleaned by supposing that the population is at stable age distribution, so that $\lambda = R$. When this is so,

$$v_x = v_0 \frac{R^x}{l_x} \sum_{y \geqslant x} R^{-y} l_y b_y$$

$$= v_0 \sum_{y \geqslant x} R^{-(y-x)} \frac{l_y}{l_x} b_y$$

Now, because in a population growing at a rate R, young born g years after age x form R^{-g} as large a proportion of the total population as young born 0 years after age x, we can say that, in a sense, young born at age $x + g$ are "worth" only R^{-g} as much as those born at age x. With this in mind, and noting that l_y/l_x is the probability that the individual now age x lives to age y, the above can be seen to be

$v_x = v_0($ expected number of future offspring produced by an individual now age x, weighted by the worth of these offspring)

That is, the reproductive value at age x (when stable age distribution holds), v_x, is the expectation of further contribution to the gene pool of future generations by an individual now age x. The concept of reproductive value has been used extensively to study the evolution of life histories. For the moment, however, we shall apply it to the problem of selection in age-structured populations. Equation (8-38) can be rewritten

$$v_x = \frac{\lambda^x v_0}{l_x}\left(\lambda^{-x} l_x b'_x + \sum_{y \geqslant x+1} \lambda^{-y} l_y b'_y\right)$$

$$= v_0 b'_x + \frac{1}{\lambda}\frac{l_{x+1}}{l_x}\left(\frac{\lambda^{x+1} v_0}{l_{x+1}} \sum_{y \geqslant x+1} \lambda^{-y} l_y b'_y\right) \tag{8-39}$$

$$= v_0 b'_x + \frac{1}{\lambda} p_x v_{x+1}$$

Suppose that the fecundity and survival values are constant. Then define

$$V(t) = \sum_{x>0} n_x(t) v_x(t) = \sum_{x>0}\left[n_x(t) v_0 b'_x + n_x(t) \frac{1}{\lambda} p_x v_{x+1}\right] \tag{8-40}$$

The first bracketed expression is

$$v_0 \sum_{x>0} n_x(t) b'_x = n_0(t) v_0 = n_0(t) \frac{1}{\lambda} p_0 v_1 = \frac{1}{\lambda} n_1(t+1) v_1 \qquad \text{(from 8-38)}$$

The second is

$$\sum_{x>0} n_x(t) \frac{1}{\lambda} p_x v_{x+1} = \frac{1}{\lambda} \sum_{x>0} n_{x+1}(t+1) v_{x+1}$$

Hence (8-40) becomes

$$V(t) = \frac{1}{\lambda} \sum_{x>0} n_x(t+1)v_x = \frac{1}{\lambda} V(t+1)$$

so that

$$V(t+1) = \lambda V(t) \tag{8-41}$$

Crow (1979) has recently shown that the derivation above works also for $V_i(t)$, where V_i is the total reproductive value for the average individual carrying the ith allele calculated from genotype-specific p_x and b'_x values. But if this is true we can write, where

$$\tilde{q}_i = \frac{V_i}{V}$$

$$\Delta\tilde{q}_i = \frac{V_i(t+1)}{V(t+1)} - \frac{V_i(t)}{V(t)} = \frac{\lambda_i V_i(t)}{\lambda V(t)} - \frac{V_i(t)}{V(t)}$$

$$= \frac{V_i(t)}{V(t)} \frac{\lambda_i - \lambda}{\lambda} = \tilde{q}_i \frac{\lambda_i - \lambda}{\lambda} \qquad \text{where } \tilde{q}_1 = \frac{V_i(t)}{V(t)} \tag{8-42}$$

Thus natural selection increases the proportion of the total population reproductive value that is contributed by individuals with higher (stable age distribution) growth rates. In the nontrivial case, equilibrium occurs when $\lambda_i = \lambda$. The last approach does not yield an explicit expression for change in gene frequency, but contains this information hidden in the value $\tilde{q}_i$, and is probably accurate over a broader set of circumstances than those given by Charlesworth.

For an entirely different approach which does not provide an expression for change in gene frequency but which gives precise equilibrium values, the reader is referred to Mode (1969).

Review Questions

1. The notion of fitness is perhaps the most difficult and controversial in all biology; any attempt to use it or define it is at best tenuous. But if we assume for a minute that we in fact know what it is, of what could we say it is predicated? Do *individuals* exhibit fitness, or do they merely share in an average fitness for their genotype? If the former, then an enormous proportion of the individual's fitness must be a function of chance events: a good year or a bad year, good or bad mate, good or bad location, and so on, and this seems to rob the term of what we want it to mean. But what can it mean for a *genotype* to have fitness? It clearly cannot mean any kind of average fitness for a given genotype, since presumably all genotypes are unique, the chances of two identical ones vanishingly small. If it is to mean that particular genotypes have particular fitnesses, we need a way to characterize genotypes much more specifically than is now possible. Some biologists speak of fitness at a particular locus within the genotype. To speak in this way seems to ignore that, in fact, not loci but rather individuals are reproductive "units," each of which contains many loci, all interacting in unknown ways. It is possible that fitness is a pseudoscientific metaphysical term that upon analysis would be found to be meaningless?
2. Can we get at "fitness" *before* the fact—before we observe its effects on gene frequency? Can we measure it in the present? If not, then is our definition of fitness not circular?
3. In this chapter it is shown mathematically how inbreeding might alter the direction of selection. Can you explain why this is so without recourse to equations?
4. Construct a simple model along the lines of (8-2) for a locus with *three* alleles. Under what conditions will all three alleles be maintained at equilibrium?

5. At a locus displaying complete dominance, $W_{AA} = W_{Aa} > W_{aa}$, maintenance of both alleles (i.e., polymorphism) under frequency-dependent selection requires that, at equilibrium,

$$W_{AA} = W_{aa}$$

   Why?
6. Suppose that there is *not* complete dominance. Now what are the conditions, under frequency-dependent selection, for maintenance of polymorphism? (The interested reader is referred to Slatkin, 1978.)
7. How much of this chapter, with all of its equations describing genetic processes in populations with discrete generations, is applicable to populations with overlapping generations? To what extent can be merely substitute $\tilde{q}_i = V_i/V$, where V is reproductive value, for q_i, the frequency of the ith allele and obtain the same results?
8. Derive an expression for Δp when fitness is frequency-dependent, say,

$$W_1 = c(1 - n_{AA}),\ W_2 = W_3 = c(1 - n_{Aa} - n_{aa})$$

   Calculate $\hat{p}$, $\hat{n}$ for $c = 10.0$. Can the population reach equilibrium if $p \neq \hat{p}$?
9. In question 8, what is V_A when $p = 0.5$? What is V_A at genetic equilibrium?

Chapter 9

Single-Locus Models in Time and Space

9.1 Changing Environments

Basic Models

Simple models of gene frequency change, like those describing population density change, are thrown into disarray by temporal fluctuations in the environment. Indeed, when the nature of change is capricious, it is impossible to predict the moment-by-moment behavior of gene frequencies. And even when environmental change is entirely deterministic, the equations governing gene frequency dynamics may be sufficiently complex as to compromise their utility seriously. Consequently, we shall not attempt detailed descriptions, but rather look at the qualitative implications of temporal change for natural selection. Spatial heterogeneity, both in the physical and biological environment and in the genetic structure of populations, may promote stability in gene frequencies just as it does for population densities (Chapters 2 and 3). It also provides for an array of natural selective processes that can substantially alter the course of evolution predicted in Chapter 8. We begin with a discussion of the consequences of environmental change.

Because we are interested in predicting gene frequency change due to natural selection, and because short-term predictions in changing environments are both impossible, due to random aspects of the change, and involve enormous mathematical complexity, it would seem reasonable to compromise, settling for descriptions of average gene frequency change over long time periods. In the continuous case, we write

$$\frac{dp}{dt} = pq[p(m_1 - m_2) + q(m_2 - m_3)]$$

for a specific instant in time. Over some defined time interval, then,

$$E\left(\frac{dp}{dt}\right) = E\{pq[p(m_1 - m_2) + q(m_2 - m_3)]\}$$

If selection is weak, or if environmental fluctuations result in frequent reversals of the m_i so that Δp is small over small time intervals, then to a reasonable approximation the expression above becomes

$$E\left(\frac{dp}{dt}\right) \simeq pq[p(E(m_1)E(m_2)) + q(E(m_2) - E(m_3))] \tag{9-1}$$

That is, the average long-term rate of gene frequency change can be expressed in the same simple manner as that used when the environment is constant if mean values of fitness are used in place of instantaneous values. Unfortunately, the discrete time case is not so simple. Let us suppose that the environmental changes occur strictly within generations and the pattern is similar each generation, so that the differences in environment experienced by individuals in different generations are negligible. Then the value of W does not vary among generations and, effectively, from the perspective of the evolutionary process, there is no environmental change. The equations appropriate for constant environments are still valid. Short-term fluctuations of this sort have been termed "fine-grained" (Levins, 1968). Of course, short-term fluctuations will not necessarily be so repeatable that they generate only insignificant fitness differences among generations. Furthermore, there are also long-term changes in the environment that will be felt minimally within, but strongly among, generations ("coarse-grained" fluctuations). To deal with the latter case we simplify the concept of fitness again, as we did quite consistently in Chapter 8, to apply only to viability differences between the genotypes, assume random mating, and write

$$W_i(t) = \frac{n_i(t+1)}{n_i(t)}$$

for the ith genotype. But then

$$\frac{n_i(t+T)}{n_i(t)} = \prod_{k=t}^{t+T-1} W_i(k)$$

and the change in p over the long-term interval t, $t + T$, is

$$\Delta p_{t\ \mathrm{tot}+T} = \frac{pq[p(W_1^{*T} - W_2^{*T}) + q(W_2^{*T} - W_3^{*T})]}{2\bar{W}_g} \tag{9-2}$$

where

$$W^{*T} = \prod_{k=t}^{t+T-1} W(k) = \exp\left[\sum_{k=t}^{t+T-1} \ln W(k)\right] = \exp[TE(\ln W)]$$

But using the Taylor expansion, we can write

$$W^{*T} = \exp\left\{TE\left[\sum_{i=0}^{\infty} \frac{[W - E(W)]^i}{i!} \frac{d^i}{dW^i}(\ln W)\right]_{W=E(W)}\right\}$$

which, if fitness differences are small, is approximately

$$W^{*T} = \left[E(W) - \frac{1}{2}\frac{\mathrm{var}(W)}{E(W)}\right]^T \tag{9-3}$$

Equation (9-3) is, basically, the expression derived by Real (1980, and references therein). Because the general equation describing Δp is not altered in form, we can still conclude that natural selection will (in general) maximize population mean fitness. But now, that mean fitness is given by W^*. Thus natural selection will act to maximize, for a given T, the bracketed value in (9-3). Of course, maximization of W^* is predicted only over the long run; there is no reason to expect it to increase over any particular series of years.

Real (1980) has discussed at some length the implications of maximizing W^*; not only should natural selection be acting to increase $E(\bar{W})$, it should also promote a decrease in var(W). How is this to be accomplished? Real notes that behavioral or physiological tendencies which maximize $\bar{W}$ in one generation may not be the same ones that do so in another. Furthermore, it seems likely that, in general, accomplishing the behavioral or physiological tasks appropriate to one generation may require sacrificing the tasks appropriate to another. Thus if a genotype is to minimize var(W), it must be able to retain a degree of flexibility, switching from one set of behaviors and physiological reactions to another as the situation demands. For organisms that live only one generation, this argument is not pertinent, but for species whose lives span several breeding seasons, there should be strong selection for plasticity. The plasticity might come in either of two forms: First, we usually define "genotype" in terms of genetic constitution at a particular locus. Thus a "genotype" is really a composite of genotypes, each of which could be adapted to a different set of environmental conditions. Alternatively, somatic plasticity might evolve at the individual level. In the first case, adaptation to the changing environment will involve polymorphisms. In the second, polymorphisms are not necessary (although we have no firm notion as to the mechanism by which the genome codes for plasticity).

Let us briefly review the foregoing conclusions. Fine-grained temporal change will make somatic plasticity adaptive but will not promote the natural selection of polymorphisms (see also Strobeck, 1975). In coarse-grained environments temporal change will make somatic plasticity beneficial and may also render polymorphisms adaptive. Great flexibility in behavioral and physiological responses, however, *if they are appropriate to the environmental circumstances*, will make polymorphisms less important to adaptation.

Two important questions have now been raised. Will natural selection indeed promote polymorphisms where they are appropriately adaptive? Will somatic flexibility or tolerance be appropriate (i.e., adaptive) to the changing environmental circumstances?

Let us suppose that the environment fluctuates in a coarse-grained manner between two possible states. In the first, which occurs with frequency Q, the AA genotype is most fit; in the second, the aa genotype is at a selective advantage. We suppose that there is no heterosis, so that neither environmental state will, of itself, promote polymorphism. In fact, suppose that there is no dominance whatsoever, so that the heterozygote is exactly halfway between the homozygotes in fitness.

	Relative Fitness of Genotype		
Environment	AA	Aa	aa
Type I	1	$1 - \frac{s}{2}$	$1 - s$
Type II	$1 - t$	$1 - \frac{t}{2}$	1

Were the environment to remain in state I, the A allele would come to fixation; in state II, the a allele would eventually become fixed. But in a fluctuating coarse-grained environment, we have

$$\begin{aligned} W_1^* &= (1)^Q(1 - t)^{1-Q} \\ W_2^* &= \left(1 - \frac{s}{2}\right)^Q \left(1 - \frac{t}{2}\right)^{1-Q} \\ W_3^* &= (1 - s)^Q(1)^{1-Q} \end{aligned} \tag{9-4}$$

If W_2^* is larger than W_1^* and W_3^*, there is an effective heterosis and polymorphism should prevail. If $Q = 1/2$, this condition becomes, after some algebraic rearrangements,

$$\frac{2s}{2 + s} < t < \frac{s}{2 - s}$$

Thus if s were, for example, .75, t would have to lie in the interval $\frac{30}{55} < t < \frac{33}{55}$ in order for polymorphism to occur. It is fairly easy to see that polymorphism is most likely if the environmental states are at least roughly the same in frequency. Also, effective heterosis (and so polymorphism) is enhanced by any degree of dominance in favor of the more fit genotype. Of course, if there is already heterosis in both environmental states, fluctuations will have no qualitative impact on maintenance of polymorphism (see also Haldane and Jayakar, 1962). For more sophisticated treatments of this general approach, see Gillespie (1973a,b, 1974).

The question of adaptation to fine-grained variation has been addressed by Hartl and Cook (1973). Let S be a random variable which reflects fine-grained changes in the environment, and suppose that we are given the following fitness table:

	Genotype		
	AA	Aa	aa
Fitness	$W + k_1 S$	$W + k_2 S$	$W + k_3 S$

where $E(S) = 0$ and $\text{var}(S) = \sigma^2$. In this case the parameters k denote the degree to which environmental fluctuations affect fitnesses. A large k indicates very little somatic plasticity on the part of the genotype in question. The evolution of $\bar{k}$, then, describes the evolution of somatic plasticity or tolerance in response to changing conditions. Using the standard equation (Equation 8-4), and ignoring inbreeding, gives us

$$\Delta p = \frac{pq\{p[(W + k_1 S) - (W + k_2 S)] + q[(W + k_2 S) - (W + k_3 S)]\}}{\bar{W}(1 + \bar{k}S)}$$

$$= \theta \frac{S}{1 + \bar{k}S} \qquad \text{where } \theta = \frac{pq[p(k_1 - k_2) + q(k_2 - k_3)]}{\bar{W}} \tag{9-5}$$

But θ is always positive if $k_1 > k_2 > k_3$, and negative if $k_3 > k_2 > k_1$. Because S is changing, we must now look at the *expected* change in p. Using the Taylor expansion, this becomes

$$E(\Delta p) = \theta E \frac{S}{1 + \bar{k}S} = \theta E \left[\sum_{i=0}^{\infty} \frac{(\Delta S)^i}{i!} \left(\frac{d^i}{dS^i} \right) \frac{S}{1 + \bar{k}S} \right]$$

$$\simeq -\theta \bar{k} \sigma^2 \tag{9-6}$$

But θ is equal to $(pq/2\bar{W})(d\bar{k}/dp)$. Therefore,

$$E(\Delta p) \simeq -\frac{pq\bar{k}\sigma^2}{2\bar{W}} \frac{d\bar{k}}{dp} \tag{9-7}$$

Thus selection always decreases $\bar{k}$ or, equivalently, favors increased somatic plasticity or tolerance (which is appropriate to reducing the impact on fitness of environmental variability.)

The relative importance of individual somatic plasticity and population plasticity via polymorphism as mechanisms of adaptation is a matter that will be taken up later, in Chapter 13. But it is important to deal with one aspect of individual plasticity at this point in the text. Consider a particular genotype, defined by the genic configuration at some particular locus (say AA). This genotype's realized phenotype may vary, depending either on the genetic makeup at *other* loci, or on the basis of individual plasticity. Changes in this realized phenotype may occur either over generations via natural selection (with respect to the genetically coded variation), or much more quickly (with respect to behavioral and physiological adjustments within individuals). And regardless of the mechanism, what is selected for genetically is that which increases the genotype's fitness, and what is changed somatically is that which is adaptive (genotypes whose responses were maladaptive would be quickly eliminated). That is, natural selection should predispose individuals to respond to their environment in the same direction as the slower, gene change process would lead them. Thus if we can make a case for the spread of a gene coding for a particular response to an environmental change, we may also argue for an existing predilection to the same response. One example is the following. We shall argue later in this chapter that certain conditions should lead to the spread of genes which predispose their owners to behave altruistically. We should recognize, however, that with respect to behavior traits, genes seldom mandate certain behaviors; they merely predispose. Thus if the "altruist" genes have already spread, individuals nevertheless will not *necessarily* behave altruistically. However, if the conditions for spread of the altruist gene occur momentarily, we may predict expression of the gene, the performance of altruism, at that moment.

Table 9-1

Temperature Flies Housed at:	Average Percent Heterozygosity (over 22 Loci, Several Replicates)
25°C	7.77 ± .26
19°C	7.98 ± .84
Alternating	9.09 ± .42

We return now to coarse-grained environments and ask for evidence in support of the prediction that environmental fluctuation may lead to polymorphism. In laboratory experiments, Powell (1971) kept *Drosophila willistoni* under conditions of constant temperature—19°C or 25°C—or in environments with biweekly switches from one temperature regime to the other (two weeks is roughly one generation). Twenty-two protein loci were examined and the fraction of individuals heterozygous at each was determined. The averages of these fractions are given in Table 9-1. The average heterozygosity of the flies in the alternating environment was significantly ($p < .05$) greater than that of the flies raised at 25°C. The heterozygosity at 19°C is not significantly different, but deviates from that in the alternating temperature population in the expected direction.

Levinton (1973), looking at the genetic structure of bivalve populations, reasoned that the deeper under water and mud a population was found, the more buffered it should be from environmental fluctuations. Thus subtidal infaunal species should, under the hypothesis that temporal variability favors polymorphism, be more monomorphic than intertidal epifaunal species. Levinton, using electrophoretic techniques, estimated the number of alleles at each of two loci as a measure of polymorphism. His results, given in Table 9-2, are as predicted.

A warning is in order regarding the interpretability of the foregoing experiments and observations. Individuals of a species living in different environments are likely to have evolved to be somewhat different, each adapted to its particular living conditions. Thus what appears variable to individuals adapted to a dramatically changing environment may be quite different from what appears variable to individuals living in a relatively stable environment. This must be particularly true when different species are compared. Daily temperature ranges (say) are a human-made measure which may or may not be interpreted by the organism in question in a humanlike manner. What is a dramatic range to one species may be barely detectable by another. To the extent that observed polymorphisms reflect local adaptation, the experiments above are meaningful. This seems likely within species. But where the enzyme polymorphisms observed are the secondary biochemical consequences of more holistic (behavioral or physiological) adaptations, they are not: If a bivalve species living in the high intertidal alters its physiology such as to make temperature fluctuations unimportant to its survival or reproduction, it no longer (necessarily) can be considered to be living in a more changeable environment than its deep-water infaunal cousin. This warning is given impetus by

Table 9-2

Species	Habitat	Number of Alleles	
		Phosphohexose Isomerase	Leucine Aminopeptidase
Mytilus edulis	Epifaunal, intertidal	7	5
Modiolus demissus	Semi-infaunal	6	4
Mercenaria mercenaria	Shallow, infaunal	6	4
Macoma balthica	Medium, infaunal	3	4
Mya arenaria	Deep, infaunal	3	3
Nucula annulata	Infaunal, subtidal	2	—

the following two studies. Ayala et al. (1975) looked at heterozygosity in deep-sea asteroids, expecting to find monomorphism because of the constancy of the deep-sea environment. They found no differences from asteroids found living elsewhere. Gooch and Schopf (1972) checked eight other deep-sea animals, with similar results. The findings were not attributable to sampling only a few loci; Ayala et al. looked at 24 loci, and Gooch and Schopf looked at 74. More likely, deep-sea animals have never been under selection pressure to adapt to uncertainty, and thus, to them, the surroundings are just as variable as are those of any organism. The rationale behind this warning may serve as a possible explanation for the fact that large animals (vertebrates), which as a result of their high mobility experience wider varieties of habitats, are generally less polymorphic than are small, less mobile invertebrate species (Selander and Kaufman, 1973).

Genetic Flushes

Populations grow because environmental conditions permit a rise in mean fitness (or, equivalently, R) above 1.0. At times when mean fitness exceeds 1.0, even genotypes normally in the process of dying out may have fitnesses in excess of 1.0. Deleterious genes will continue to drop in relative frequency, but the *numbers* of individuals carrying them in expressed form will decline less rapidly and may even increase. When the environment returns to normal, or becomes unusually harsh, signaling a subsequent population decline, on the other hand, the disappearance of genetically inferior types will be accelerated. It is reasonable, then, to think of periods of population growth as times of genetic innovation—when new, normally deleterious genes are able to survive long enough to form new combinations, with new phenotypic consequences, and of population declines as times of weeding out those new innovations that did not work well. Perhaps fluctuations in the environment, by inducing periods of genetic experimentation and elimination, promote more rapid evolution and make available to the evolving population a broader array of potential solutions to adaptive challenges. This suggestion was made by Ford (1964) and again by Carson (1968). Ford notes the case of the marsh fritillary butterfly, which between 1920 and 1925 experienced a period of very rapid population growth. The result was that one butterfly variety became many, and even grossly deformed individuals managed to live (and perhaps, reproduce). After 1925 the population declined and the variants disappeared.

Age Structure

In Chapter 8 we gained a glimpse of the genetic complexity that age structure has the power to promote. Even in constant environments evolutionary progress of a gene is intimately entangled in the mechanics of fluctuating age class frequencies. The task of adequately describing evolution at the genetic level in continuously disturbed environments is truly horrible to contemplate. Even long-term predictions may be difficult: Charlesworth (1970) has shown that R (or r) is a good measure of fitness in sexual populations with overlapping generations *unless* population density fluctuates. Perhaps Crow's (1979) approach using reproductive values (see Chapter 8), allied with the work of Boyce (1977) and Cohen (1977a,b) on the use of Leslie matrices in fluctuating environments (Chapter 3), would provide a profitable springboard for future work.

9.2 Spatially Structured Populations

In Chapter 7 we saw how isolation by grouping or by distance could profoundly affect inbreeding and genetic drift. We shall see now that it is also extremely important to the process of natural selection. First, note that there are two qualitatively different ways in which organisms can group themselves. The first is by forming groups (or ranges in space) which effectively isolate them reproductively from distant individuals. Groups (or what are effectively groups by way of isolation by distance) of this sort are referred to as *demes*. But within demes individuals will form temporary subgroups in which they interact for a period of time and then separate. Such *trait groups* (D. S. Wilson, 1980, and references therein) include fish schools, family groups, agonistic encounter groups, and pairs in copulo. The behavioral interactions among individuals in trait groups are the stuff of inclusive fitness (Chapter 8), the source of intra- and intergenotypic effects on fitness. We shall begin by discussing the importance of trait group formation on natural selection. A massive review of material covered in the first five parts of this section can be found in Uyenoyama and Feldman (1980).

Evolution Among Trait Groups

Although the term "trait group" was coined by D. S. Wilson in 1975, a great deal of work on special cases of trait group selection had already revolutionized certain aspects of ecology. We begin with an illustrative example, very loosely interpreted from Boorman and Levitt (1973).

Suppose that the allele A, with frequency p, is dominant to its complement, a, but that aa individuals, when they meet, interact in some hypothetical, cooperative manner that increases the fitness of their genotype. We suppose that encounters are random, proportional to the genotype's frequency, that W_3 falls short of $W_1 = W_2$ by an amount c, and that cooperation among individuals of the inferior genotype raises that genotype's fitness by an amount b. If the rate of encounter of any two individuals is $1/T$, the following fitness table is appropriate:

	Genotype		
	AA	***Aa***	***aa***
Fitness	W_1	W_1	$W_1 - c + \dfrac{bq^2}{T}$

Using the standard equation (8-4) for rate of change in gene frequency, letting p represent the frequency of the A allele,

$$\Delta p = \frac{pq^2}{\bar{W}}\left(c - \frac{q^2 b}{T}\right) \tag{9-8}$$

Were it not for the cooperation among aa individuals, the a gene would disappear. But because of the cooperation, a will come to equilibrium at a frequency of

$$q = \sqrt{\frac{cT}{b}} \tag{9-9}$$

Suppose that all genotypes have identical fitness except that aa homozygotes behave "altruistically" toward each other. That is, given the appropriate stimulus, individuals act in such manner as to lower their own viability, but aid others of their kind. Let the cost in lowered viability to the altruist be c, the gain to the recipient of the behavior be b. Then if individuals find themselves interacting (in trait groups) with others in proportion to genotypic frequency (i.e., random mixing), trait groups consist of but two individuals, and only one individual acts at a time, we have

	Genotype		
	AA	***Aa***	***aa***
Fitness	W_1	W_1	$W_1 + q^2(b - c)$

whence

$$\Delta p = -\frac{pq^4}{\bar{W}}(b - c) \tag{9-10}$$

Hence the altruist-promoting gene spreads as long as $b > c$; that is, the benefit to the recipient exceeds the loss to the altruist. This kind of mechanism is one possible explanation for the existence of apparently altruistic traits in nature.

Of course, it seems unlikely that genotypes will behave altruistically (or in any other way) only toward others of the same genotype. The genotype of a potential recipient of altruism may not be readily identifiable prior to the performance of the act in question. How does an individual recognize another carrier of the gene? One way around this possible dilemma is to suppose that population viscosity—the lack of free, random dispersal (Gadgil, 1975)—results in a high likelihood that members of the trait group are kin. Then altruism might still evolve not by virtue of its being specifically directed at individuals of like genotype, but rather because the relatedness of individuals in the trait group increases the probability that they are of similar genotype. Thus although recognition of and direction of the behavior toward genetically similar individuals will speed the process, a random dispensation of favors may still lead to evolution of altruistic behavior.

The concept of inclusive fitness and its importance for the evolution of social behaviors, and altruism in particular, goes back to Haldane (1932, 1955; see also Williams and Williams, 1957). Its most recent appearance, and its most in-depth examination, has grown from the work of Hamilton (1963, 1964, and later works cited below). Because of the reliance of Hamilton's (and many others') models on genetic relationships among actors and recipients, natural selection based on inclusive fitness has become known as *kin selection* (Maynard Smith, 1964).

Selection based on inclusive fitness—"indirect selection" is probably more appropriate than "kin selection" (Brown and Brown, 1981)—can be approached in three quite distinct ways. First, it is possible to construct models of natural selection mediated via interindividual interactions within families (usually within litters). Second, both intra- and interfamilial interactions may be examined for cases where frequency of the gene responsible for the behavior in question is very low, so that for all practical purposes a recipient will be carrying the gene only if it is related to the actor with which it finds itself. This is classical kin selection of the sort investigated by Hamilton. Finally, where frequency of the gene is not low, and quite unrelated individuals are both apt to be carriers, the (early) models of Hamilton are no longer applicable. We must turn to other approaches in order to predict the evolutionary course of the gene. We shall cover these three approaches one at a time.

Family Selection Models

Several authors have used the family selection approach to kin selection (Cavalli-Sforza and Feldman, 1978; Templeton, 1979; Wade, 1978a, 1979). The following is from the papers by Wade. We suppose that selection for the behavioral trait in question occurs via effects on viability among interacting offspring in a family. The *aa* genotype is the actor, although we suppose that *Aa* acts a fraction, $1 - h$ of the time. *AA* is passive. As before, c is the cost to the actor, b the impact on the recipient. If mating in the population as a whole is random, then fitnesses are as given in Table 9-3. In the absence of the behavior, fitness is equal to W for all genotypes, and family size is assumed constant at n.

An understanding of Table 9-3 is perhaps facilitated by a more detailed explanation of a couple of its entries. Consider, for example, the fitnesses of the two genotypes in a family with one *AA* parent and one *Aa* parent (row 2). First, we must realize that in this case there are no *aa* offspring. Next, recall that *AA* individuals do not act, and that *Aa* individuals act with probability $1 - h$. Finally, in such a family, *on the average*, half of the young will be *AA*, half *Aa*. Thus the fitness of the *AA* genotype is W plus the expected number of acts, $\frac{1}{2}n(1 - h)$, times the

Table 9-3

Mating Scheme	Frequency	Fitness of AA	Fitness of Aa	Fitness of aa
$AA \times AA$	p^4	W	—	—
$AA \times Aa$	$4p^3q$	$W + b(1-h)\frac{n}{2}$	$W + (1-h)(-c-b) + (1-h)b\frac{n}{2}$	—
$AA \times aa$	$2p^2q^2$	—	$W + (1-h)(-c-b) + (1-h)bn$	—
$Aa \times Aa$	$4p^2q^2$	$W + \left[n(1-h) + \frac{n}{2}\right]\frac{b}{2}$	$W + (1-h)(-c-b) + \left[n(1-h) + \frac{n}{2}\right]\frac{b}{2}$	$W + (-c-b) + \left[n(1-h) + \frac{n}{2}\right]$
$Aa \times aa$	$4pq^3$	—	$W + (1-h)(-c-b) + [(1-h)+1]\frac{bn}{2}$	$W + (-c-b) + [(1-h)+1]\frac{b}{2}$
$aa \times aa$	q^4	—	—	$W + (-c-b) + bn$

$$\bar{W}_a = p\bar{W}_2 + q\bar{W}_3 = \left(-c - b + \frac{bn}{2}\right)[pq + (1-h)p(p-q)] + W$$

$$\bar{W} = (-c - b + bn)[q^2 + 2pq(1-h)] + W$$

impact of being the recipient of an act, b. Fitness of the Aa genotype is W plus the net impact of acting, $-c$, times the probability of acting, $(1 - h)$, plus the expected number of acts affecting that individual, $(\frac{1}{2}n - 1)(1 - h)$, times the effect of receiving those acts, b. Other entries in the table are calculated in a similar manner. Multiplying the fitness in each row by its corresponding frequency of occurrence—the frequency with which each mating combination takes place—we can find the population (over all families) mean fitnesses for each genotype. Substituting these into the basic equation (8-3),

$$\Delta q = \frac{q(\bar{W}_a - \bar{W})}{\bar{W}}$$

and performing some algebraic rearrangements, we find,

$$\Delta q = \frac{npq\left(\frac{-c - b}{n} + \frac{b}{2}\right)[qh + p(1 - h)]}{\bar{W}} \qquad (9\text{-}11)$$

Therefore, the a allele spreads if

$$\frac{-c - b}{n} + \frac{b}{2} > 0 \qquad (9\text{-}12)$$

If the behavior in question is altruistic, so that c is positive, expression (9-12) becomes

$$\frac{b}{c} > \frac{2}{n - 2} \qquad (9\text{-}13)$$

An altruistic trait that expressed itself only among sibs would be unlikely to evolve in a species with small families (or in one which dispersed so that sibs were separated). Indeed, if the number of sibs were two, altruism could not evolve at all. As family size increases, the condition for spread of an altruism-inducing gene is more likely to be satisfied.

Clearly, in the case above, the a allele spreads because the group of sibs containing an altruist is more fit, as a whole, than sib groups without an altruist. Within a sib group, the altruist himself must necessarily be at a selective disadvantage compared to his brothers and sisters. Thus selection acts against the trait within, but in favor of the trait among families. Wade, in a later paper (1980), illustrates this quite elegantly, proving that family selection can be broken down into two components, one reflecting negative pressure within the family, the other describing the positive, interfamily pressure.

Classical Kin Selection

Originating with its resurrection by Hamilton in 1963 and 1964, kin selection has gone through a flurry of sophisticated developments in theory, and has given a tremendous boost to the entire field of behavioral ecology. Some of the more important papers dealing with the subject are those of Hamilton (1970, 1971b, 1972), Gadgil (1975), Orlove (1975), Harpending (1979), Michod (1972), Michod and Hamilton (1980) and Hughes (1983). In very simple terms the argument is as follows. Consider the effect of a newly introduced gene that is still very rare. This gene is one that predilects its owner to behave in some manner affecting the well-being of other individuals. As before, let c be the cost of the behavior to the actor and let b be the effect on the other individuals, the recipients. Finally, let the probability that a given recipient also carries the gene in question be r.

Label the gene a; then the following reasoning applies. With each act performed, the fitness of the actor is depressed by an amount c. Because a is very rare, the actor will, almost invariably, be carrying only one such gene. Thus the effect of an act on the number of a genes via the cost to the actor is $-(1)(c) = -c$. Now, because each of (say) $n - 1$ recipients carries, on average, r of the genes, the total effect of an act on the number of a genes in recipients is $b(n - 1)r$. The total effect on the number of a genes, then, is $-c + br(n - 1)$. Gains exceed losses if this quantity is positive, that is, if

$$\frac{b}{c} > \frac{1}{r(n - 1)} \tag{9-14}$$

where the value to the recipients times r must exceed the cost to the altruist. If the behavior is spiteful (hurts the actor as well as the recipients), then b is negative, the a allele gains when $-c - |b|r(n - 1) > 0$, and inequality becomes

$$\frac{|b|}{c} < \frac{1}{r(n - 1)} \tag{9-15}$$

where the harm to the recipients times r must exceed the harm to the actor. Cooperative behavior, where b is positive and cost is negative, will always be favored by natural selection, because then $c + br(n - 1)$ is positive and the inequality (9-14) is always met.

The value of r is of considerable importance in (9-14). For example, an increase in the number of altruist genes becomes increasingly likely as r rises. r is easily calculated from pedigrees (see Chapter 7). With respect to sibs it is not difficult to see that r, the probability that one sib contains the gene given its occurrence in another, is on average $\frac{1}{2}$; for half-sibs, r averages $\frac{1}{4}$; for parents–offspring, r is $\frac{1}{2}$; and so on. In fact, referring to Figures 7-1 through 7-3, the value of r for any two individuals is exactly twice the inbreeding coefficient (F_{IS}) of their hypothetical offspring if inbreeding of the common ancestor(s) is ignored. For an illustration, see Figure 9-1. The more closely related the members of the trait group are, the more effective is the selection pressure on the indirect component of inclusive fitness. r is known as the *coefficient of relationship.*

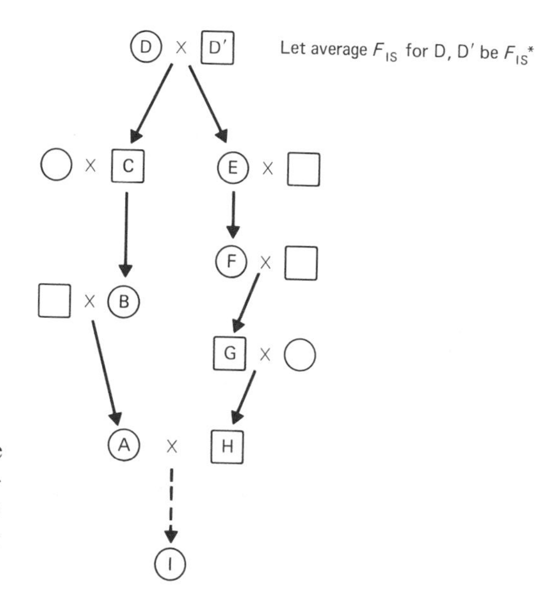

Figure 9-1 Hypothetical pedigree: Let average F_{IS} for D, D' be F_{IS}^*. Then F_{IS} of hypothetical offspring of A and H equals $(\frac{1}{2})^7(1 + F_{IS}^*)$. If F_{IS}^* is set to zero, then r (for A, H) is twice the value above, or $2(\frac{1}{2})^7 = \frac{1}{64}$.

A little thought will elucidate the kin selection process further. Notice that while the number of altruist genes, a, may rise within a trait group as a result of altruism, its complement, A, benefits equally but without the cost. Thus A should rise in number even more than a. We have already shown that the increase in the number of a genes is $br(n - 1) - c$. But the expected number of A genes in the group is 1.0 (in the altruist—recall that the gene is assumed very rare so appears in heterozygous form) plus $(n - 1)(1 - r)$ in the other individuals. The first gene is affected by an amount $-c$, the others by b. Thus the total change in number of A genes is $b(n - 1)(1 - r) - c$. Comparing the gains in the two alleles,

$$\text{gain in } A = b(n - 1)(1 - r) - c = br(n - 1) + b(1 - 2r)(n - 1) - c > br(n - 1) - c$$

unless $r > \frac{1}{2}$. Thus within a trait group, the number of a alleles may rise, but the frequency of a generally declines. Gene a will rise in frequency in the population as a whole only if the advantage to individuals in the altruist-containing trait groups is large enough to offset this relative decline within each group. Like the family selection process (Wade, 1980), the general kin selection process consists of a within- and an among-group component. Quantitatively, let $\Delta a'$, $\Delta a''$ be the changes in number of a genes within an altruist-containing group due to the altruistic interchange, and the change in number of a genes due to overall population growth, respectively ($\Delta a'' = 2q\Delta n$). Because of the assumption that q is very small, the proportion of all potential trait groups of size n ($2n$ alleles) containing an altruist is, on average, $1 - (1 - q)^{2n} \simeq 2qn$. Therefore, averaged over the entire population,

$$\Delta\bar{a} = (2qn)(\Delta a' + \Delta a'') + (1 - 2qn)(\Delta a'') = 2qn\Delta a' + \Delta a'' = 2qn\Delta a' + 2q\Delta n$$

$$\Delta\bar{A} = (2qn)(\Delta A' + \Delta A'') + (1 - 2qn)(\Delta A'') = 2qn\Delta A' + \Delta A'' = 2qn\Delta A' + 2p\Delta n \tag{9-16}$$

Thus

$$\begin{aligned} \bar{W}_a &= 1 + \frac{\Delta\bar{a}}{\bar{a}} = 1 + \frac{\Delta\bar{a}}{2qn} \\ &= 1 + \frac{2qn\Delta a' + 2q\Delta n}{2qn} = 1 + \frac{\Delta n}{n} + \Delta a' = \bar{W} + \Delta a' \end{aligned} \tag{9-17}$$

Thus $\bar{W}_a > \bar{W}$, so that the frequency of a rises if $\Delta a' > 0$, that is, if inequality (9-14) is met.

The derivations above assumed not only that the a allele is very rare, but also that all trait groups are of the same size, that the relatedness of individuals in trait groups, r, is constant, and that only one altruist acts in a given situation. Let us relax the assumptions on group size and invariant r. If n_i is the size of altruist-containing trait group i, m_i is the number of nonaltruists in the group, and r_i the appropriate coefficient of relationship, the gain in number of a alleles due to an altruistic act in the ith group is

$$b(n_i - m_i - 1) - c$$

But because a is very rare, the proportion of recipients which are themselves potential altruists is $(n_i - 1)r_i$. Thus $m_i = n_i - 1 - (n_i - 1)r_i = (n_i - 1)(1 - r_i)$, and the quantity above can be written

$$b(n_i - m_i - 1) - c = br_i(n_i - 1) - c$$

Averaged over all trait groups containing an altruist, the gain in number of A genes is then

$$b\overline{r(n - 1)} - c$$

In the case of altruism, where $c > 0$, this becomes

$$\frac{b}{c} > \frac{1}{\bar{r}(\bar{n} - 1) + \operatorname{cov}(r, n)} \tag{9-18}$$

Evolution of an altruism-inducing gene is enhanced if the coefficient of relationship, r, is positively correlated with group size.

There is a wide variety of scenarios possible describing the reaction of altruists to an altruism-provoking stimulus. In the preceding models it was assumed that only one altruist acted at a time. This might occur if the probability of any individual's acting were very low; then the likelihood of two or more showing the behavior would be vanishingly small. Or perhaps once a single individual acts, the benefits deriving from another act disappear while the cost to the actor remains. In such a case, natural selection should weed out individuals that give expression to their altruistic urges once another individual has acted. This might be the case for warning calls by prey individuals. If producing such a call while warning others serves to direct the predator's attention to the caller, the behavior qualifies as altruism. But after a single call has alerted other members of the trait group, another caller accomplishes nothing but to put itself in danger. But what happens to the natural selection process if more than one potential altruist may act? Suppose that each individual carrying the a gene demonstrates a probability α of responding in a given situation, and that altruists act, or fail to act, independently of each other. Then the probability of x altruistic acts being performed in group i is

$$p_i(x) = \binom{n_i - m_i}{x} \alpha_i^x (1 - \alpha_i)^{n_i - m_i - x}$$

and the number of a genes gained per incident is

$$b(n_i - m_i - x) - cx$$

given that $x \geqslant 1$. The gene will be selected for if

$$b(\bar{n} - \bar{m} - \bar{x}) - c\bar{x} = b(\bar{n} - \bar{m}) - (b + c)\bar{x} > 0 \tag{9-19}$$

The value of $\bar{x}$, given that $x \geqslant 1$, is

$$\begin{aligned} \bar{x} &= \overline{1 + \sum_{x-1=0}^{n-m-1} \binom{n - m - 1}{x - 1} \alpha^{x-1}(1 - \alpha)^{(n-m-1)-(x-1)}} \\ &= \overline{1 + \sum_{y=0}^{n-m-1} y \binom{n - m - 1}{y} \alpha^{y}(1 - \alpha)^{(n-m-1)-y}} \\ &= \overline{1 + (n - m - 1)\alpha} \qquad \text{where } y = x - 1 \end{aligned} \tag{9-20}$$

Substituting (9-20) into (9-19), we see that a will spread in the population if

$$\frac{b}{c} > \frac{1 + \overline{(n - m - 1)\alpha}}{(\bar{n} - \bar{m} - 1) - \overline{(n - m - 1)\alpha}} \tag{9-21}$$

But (see the derivation of equation 9-18)

$$n_i - m_i = 1 + r_i(n_i - 1)$$

Thus (9-21) becomes

$$\frac{b}{c} > \frac{1 + \overline{rn\alpha} - \overline{r\alpha}}{(\overline{rn} - \bar{r}) - (\overline{rn\alpha} - \overline{r\alpha})} \tag{9-22}$$

A general bibliography on the use of inclusive fitness in kin selection is given by Aoki (1981). For an alternative approach, see Griffing (1981a–c). For the effects of inbreeding on kin selection, see Wade and Breden (1981).

General Models of Trait Group Selection

The classic kin-selection models assume that q is very small. Thus they provide us with the conditions under which a particular gene will spread in the population *when it is very rare*. But as the gene becomes more common, the models break down. It becomes necessary to build models in which either r is redefined or the value q appears explicitly. The difficulty lies in the fact that as a becomes common, it is increasingly likely to appear in nonrelatives of the actor. Thus the conceptually simple measure r no longer provides us with a useful means of assessing which individual is likely to possess the gene and which is not.

Probably the earliest and simplest model fitting our new criterion is that of Charnov (1977). Charnov's approach is to suppose a random mix of individuals, that given fixed proportions of all altruistic acts are performed by aa as opposed to Aa individuals, and that given fixed proportions of altruistic acts are directed at each of the three genotypes. Inasmuch as the assumption of fixed values seems highly unrealistic biologically, we shall view this model as primarily of historic interest and will not pursue it further here. A less restrictive (although less easily applicable) approach was used Abugov and Michod (1981). As before, let c and b be the effects of an altruistic act on the actor and recipient, respectively. Let $P(uv)$ be the frequency of the uv genotype, $P(ij \mid uv)$ be the probability that a recipient is of genotype ij given that the act was performed by an uv individual, and let H_{uv} be the number of acts per individual of genotype uv per unit time. Then inclusive fitness of the ij genotype is given by

$$W_{ij} = W - cH_{ij} + b(n - 1)\sum_{uv} H_{uv}\frac{P(uv)P(ij \mid uv)}{P(ij)} \tag{9-24}$$

provided that one individual acts at a time and that fitness in the absence of altruistic interactions is W.

We shall now apply Abugov and Michod's model to a hypothetical case in order to illustrate two points: (a) natural selection favors the preferential treatment of individuals of like genotype (or, equivalently, close kin), and (b) conditions for the continued spread of a change as the frequency of the gene changes. Suppose that the frequencies with which acts are performed by aa, Aa, and AA individuals are $H_{aa} = \theta$, $H_{Aa} = h\theta$, and $H_{AA} = 0$. Then (9-24) yields

$$W_{aa} = W - c\theta + b(n - 1)\frac{\theta P(aa)P(aa \mid aa) + h\theta P_{Aa}P(aa \mid Aa)}{P(aa)}$$

$$W_{Aa} = W - ch\theta + b(n - 1)\frac{\theta P(aa)P(Aa \mid aa) + h\theta P(Aa)P(Aa \mid Aa)}{P(Aa)}$$

$$W_{AA} = W + b(n - 1)\frac{\theta P(aa)P(AA \mid aa) + h\theta P(Aa)P(AA \mid Aa)}{P(AA)}$$

If mating is random and we ignore inbreeding, so that $P(AA) = p^2$, $P(Aa) = 2pq$, $P(aa) = q^2$, then

$$\bar{W}_a = W - c\theta(q + ph) + b(n-1)\theta\left\{\left[qP(aa\,|\,aa) + \frac{q}{2}P(Aa\,|\,aa)\right]\right.$$
$$\left. + h[2pP(aa\,|\,Aa) + pP(Aa\,|\,Aa)]\right\}$$
$$\bar{W}_A = W - c\theta qh + b(n-1)\theta\left\{\left[\frac{q^2}{2p}P(Aa\,|\,aa) + \frac{q^2}{p}P(AA\,|\,aa)\right]\right.$$
$$\left. + h[qP(Aa\,|\,Aa) + 2qP(AA\,|\,Aa)]\right\}$$

We now address the first point above. I claim that natural selection favors altruists that preferentially direct their acts toward genetically similar individuals. Because we are interested in the altruists' behavior, we need to look at $\bar{W}_a$. Whatever genetic changes at other loci collectively act to increase this mean fitness value should evolve. We thus look to see what values of $P(ij\,|\,uv)$ will maximize $\bar{W}$. When q is small, the dominant bracketed term is the second. That term is maximized by maximizing $P(Aa\,|\,Aa)$ at the expense, if necessary, of $P(aa\,|\,Aa)$. When q is large, the dominant bracketed term is the first. This term is maximized by making $P(aa\,|\,aa)$ as large as possible, at the expense, if necessary, of $P(Aa\,|\,aa)$. In either case, mean fitness is maximized when the most important altruist genotype (with respect to number of acts) directs its altruism toward individuals of like genotype. How can the altruist bias its acts toward recipients of identical genotype? Perhaps by limiting its acts to situations when it is among kin.

The second point, that conditions for further spread of the gene depend on its current frequency, can be proven with a simple example. Suppose that $q \to 0$. Then *aa* genotypes virtually never appear and

$$\bar{W}_a \simeq W_{Aa} \to W - ch\theta + b(n-1)\theta h p(Aa\,|\,Aa)$$
$$\bar{W}_A \simeq W_{AA} \to W + b(n-1)\frac{2pqh\theta p(AA\,|\,Aa)}{p^2} \to W$$

a will spread if $\bar{W}_a > \bar{W}_A$, that is, if

$$\frac{b}{c} > \frac{1}{(n-1)P(Aa\,|\,Aa)} \tag{9-25}$$

Of course, in a trait group within which altruistic acts fall equally on all members, the proportion of acts effectively directed at other altruists, $P(Aa\,|\,Aa)$, is r. Hence (9-25) is equivalent to (9-14), as indeed it must be.

Now consider the situation as *a* approaches fixation. In this case,

$$\bar{W}_a \simeq \bar{W}_{aa} \to W - c\theta + b(n-1)\theta P(aa\,|\,aa)$$
$$\bar{W}_A \simeq \bar{W}_{Aa} \to W - ch\theta + b(n-1)\left[\theta\frac{P(aa)}{P(Aa)}P(Aa\,|\,aa) + h\theta P(Aa\,|\,Aa)\right]$$

Again, *a* continues toward fixation if $\bar{W}_a > \bar{W}_A$.

To evaluate $\bar{W}_a$ relative to $\bar{W}_A$, we now need to know how $P(ij\,|\,uv)$ varies with $P(ij)$ and $P(uv)$. In the simplest case we suppose that $P(ij\,|\,uv) = p(ij)$—that is, favors are spread randomly— and that selection is slow so that $P(aa) = q^2$, and so on. Then

$$\bar{W}_a \simeq W - c\theta + b(n-1)\theta$$
$$\bar{W}_A \simeq W - ch\theta + b(n-1)\theta$$

and $\bar{W}_a > \bar{W}_A$ when $h > 1$. Thus heterotic enhancement of altruism promotes fixation of the altruist gene.

When an altruism-inducing gene approaches fixation in a population, the probability that a potential recipient carries the gene is (almost) independent of whether that individual and the potential altruist are related. When this is so, it may be argued that there is no longer a selective advantage to directing favors at kin. Yet as we shall see in Chapter 15, there is an uncanny tendency for social cooperation and (what appears to be) altruism to occur primarily, in fact almost exclusively among kin. Why? As pointed out by Alexander and Borgia (1978), there are invariably genes, new mutations not yet fixed, that act to *enhance* already evolved tendencies toward altruism, or that code for a disposition toward some other kind of altruistic act, or which expand the circumstances under which altruism may occur. Natural selection for social behaviors is a continuing process, and the presence of altruism-promoting genes not yet at fixation generates selection pressures for preferential dispositions toward kin.

One of the most extensive treatments of trait group selection is by D. S. Wilson (1975, 1977, 1980) (see also, Bell, 1978). Let n_A be the number of potential actors in a trait group (the number carrying the gene that predisposes for the behavior of interest), n the total number of individuals in the group, and $n_B = n - n_A$ the number of nonactors. Define a time unit large enough that the probability that it contains an act is small. Then the probability of an act occurring in that time unit is proportional to n_A, say kn_A. Suppose that only one individual acts at a time. Then the fitness increment of an actor over this time unit is

$$\Delta W_A = \frac{kn_A[-c + b(n_A - 1)]}{n_A} = k[-c + b(n_A - 1)]$$

and that for a recipient is

$$\Delta W_B = \frac{kn_A(bn_B)}{n_B} = kbn_A$$

Over the entire population, each increment weighted according to the number of individuals in a group,

$$\Delta \bar{W}_A = k\frac{-c\bar{n}_A + b\overline{n_A^2} - b\bar{n}}{\bar{n}_A}$$

$$= -ck + bk\left[\bar{n}_A + \frac{\operatorname{var}(n_A)}{\bar{n}_A} - 1\right] \tag{9-26a}$$

$$\Delta \bar{W}_B = bk\left[\bar{n}_A + \frac{\operatorname{cov}(n_A, n_B)}{\bar{n}_B}\right] \tag{9-26b}$$

The trait in question will spread if $\Delta\bar{W}_A > \Delta\bar{W}_B$, that is, when

$$\frac{b}{c} > \frac{\bar{n}_A\bar{n}_B}{\bar{n}_B \operatorname{var}(n_A) - \bar{n}_B\bar{n}_A - \bar{n}_A \operatorname{cov}(n_A, n_B)}$$

when the denominator is positive,

$$\frac{b}{c} > -\frac{\bar{n}_A\bar{n}_B}{\bar{n}_B \operatorname{var}(n_A) - \bar{n}_B\bar{n}_A - \bar{n}_A \operatorname{cov}(n_A, n_B)} \tag{9-27}$$

if the denominator is negative.

There is an implicit assumption here. The trait in question must be heritable. Inasmuch as nothing has been said in this derivation about the genetics of the trait in question, the treatment above is surely a bit cavalier. But if the traits are, in fact, genetically based, and if the behavior of interest effects only viability, so that recombination (differential mating ability and fecundity) is not at issue, then Wilson's W values are indeed fitnesses, and the treatment above is legitimate. Let us examine some specific situations.

1. If the number of actors is the same in all trait groups, then $\mathrm{var}(n_A) = 0$, the denominator is negative, and inequality (9-27) is

$$\frac{b}{c} > 1 \quad \text{or} \quad b - c > 0 \tag{9-28}$$

which is the condition for spread of the trait under standard individual selection (indirect fitness component $= 0$); recall that we are using c to denote a *cost*. If the act is beneficial to the actor, c is negative.

2. Suppose that n_A is random (binomial) among groups and independent of n_B. Then $\mathrm{var}(n_A) = \bar{n}(\bar{n}_A/\bar{n})(1 - \bar{n}_A/\bar{n})$, $\mathrm{cov}(n_A, n_B) = 0$, and inequality (9-27) is

$$b > \frac{c\bar{n}}{\bar{n}_A} \tag{9-29}$$

In this case, the condition for further spread of the gene becomes less stringent as the gene increases in frequency.

3. Suppose that group size is constant. Then $n_A = (\text{constant} - n_B)$, so that $\mathrm{cov}(n_A, n_B) = -\mathrm{cov}(n_A, n_A) = -\mathrm{var}(n_A)$, and the appropriate inequality, where P_A and P_B are the proportions of A and B-type individuals in the population as a whole,

$$\frac{b}{c} > \frac{\bar{n}P_A P_B}{\mathrm{var}(n_A) - \bar{n}P_A P_B} \tag{9-30}$$

If altruists are randomly scattered throughout the population, $\mathrm{var}(n_A) = \bar{n}P_A P_B$, and the denominator in (9-30) becomes zero: altruism cannot evolve under such circumstances. However, if altruists are sufficiently clumped in their distribution among trait groups [i.e., $\mathrm{var}(n_A) > \bar{n}P_A P_B$], altruism can evolve.

We shall return to Wilson's model in later chapters.

Evolution Among Demes

Demes periodically become extinct, and their places in space are taken over by new demes arising from immigrants. It seems reasonable, then, if probability of extinction is a function of gene frequency within a deme, that immigrants are likely to carry a disproportionate number of those genes that slow extinction rate. Alternatively, if emigration or survival during migration are genetically affected, immigrants are apt to carry a greater proportion of genes promoting emigration and migratory survival than that carried by a random sample of individuals from the whole population. In short, extinction and replacement of demes can occur in a way that results in changes in gene frequency. Any gene that promotes population growth (if growth results in demes more quickly reaching a point where emigration occurs), propensity to migrate, survival during the migration process, or maintenance of a deme will be favored. The spread of genes due to selection acting among demes was discussed by Wright in 1945.

Group selection, or more properly, *interdeme selection* as envisioned by Wright has its share of problems, and thus its share of detractors. First, as Wright himself showed, even very small amounts of migration among existing demes (as opposed to the founding of a new deme in the place of one that has died out), effectively destroy the demes as discrete units of selection.

Migration, then, must be almost exclusively of a very specific kind if interdeme selection is to work as a nonnegligible force. Second, natural selection relies on variation among the units of selection. But if individuals are randomly mixed among demes with respect to genotype, the variance in fitness among demes must be only $1/k$ times the variance in fitness within demes, where k is the size of a deme. Therefore, the rate of evolutionary change arising from interdeme selection is apt to be miniscule unless deme size is very small. Then, because of the alleviation of density feedback as deme size becomes very small, the extinction process may be prolonged, again, resulting, in slow genetic process. For these reasons, group selection has, at least until recently, generally been thought to be of minor significance in comparison to the classic form of individual selection. Also, because most behaviors that would benefit a deme might be expected, via the deme's well-being, to feed back beneficially on the perpetrator, it was not at all clear how group and individual selection might, analytically, be separated. Clear support for the efficacy of group selection would have to await evidence in the form of the successful evolution of an altruist trait. But this form of evidence is not easy to obtain. First, as we have seen, altruism can arise by trait group selection. Since members of a deme are invariable somewhat related, it would be extremely difficult to separate the influences of trait group and interdemic selection. Second, there has been much recent speculation as to whether there really are many (any) traits which are truly altruistic as opposed to manipulative or indirectly beneficial (see, e.g., Charnov and Krebs, 1975).

For the reasons stated above, interdemic selection languished quietly until in 1962, Wynne-Edwards published a book under the title *Animal Dispersion in Relation to Social Behavior*. This was a very large book, and it raised an even larger controversy. Central to Wynne-Edwards's thinking was the notion that social behavior had evolved primarily as a mechanism to restrain reproductive effort voluntarily. Inasmuch as voluntary restraint of reproduction is flagrantly antithetical to the basic precepts of individual selection, it was clear that such behavior must have arisen via group selection. Wynne-Edwards, although he defended no effective mechanism for its operation, actively promoted this reasoning. The reaction to this book was swift and almost certainly overdone. More than one worker trotted out the view that if a trait could be explained via classic individual selection, the law of parsimony demanded that the group selection explanation be dropped. A flurry of activity ensued in the attempt to show that virtually any behavior could, in some way or other, have arisen from simple selection processes acting at the level of the individual.

Although there are still objections from a variety of quarters (see, e.g., Charlesworth, 1979), the furor of the 1960s and 1970s had spawned several serious attempts at elucidating the evolutionary role of interdemic selection. Because the existence of this mode of selection as a significant force in evolution would be illustrated most clearly in the case of the evolution of altruism, most models are expressed in such terms. This does not at all imply, of course, that group selection, like trait group selection, acts only on such traits. Some of the increasingly numerous workers in the field, many with quite divergent approaches, are Levins (1970), Eshel (1972), Levin and Kilmer (1974), Matessi and Jayakar (1976), Boorman (1978), and Wade (see references below). In this text we shall concentrate on the work of Levins and Wade.

Levins (1970) reasoned as follows. Designate by $f(p, t)$ the frequency (proportion) of demes with gene frequency p at time t. Let the probability of demic extinction be a function of gene frequency in the deme, $e(p)$, and suppose that the gene frequency in colonizing propagules follows the distribution function $g(p)$. Finally, suppose that these founding propagules are the only migrants. Then

$$f(p, t + dt) = f(p, t)\frac{\text{fitness of demes with gene frequency } p}{\text{mean demic fitness}}$$

$$= f(p, t)\frac{1 - e(p)\,dt + mg(p)\,dt}{1 - \bar{e}\,dt + mf(p, t)} \qquad \text{where } m = \text{migration rate} \tag{9-31}$$

From this it follows that

$$df = f(p, d + dt) - f(p, t) = f(p, t)\frac{[1 - e(p)\,dt + mg(p)\,dt] - [1 - \bar{e}dt + mf(p, t)\,dt]}{1 - \bar{e}\,dt + mf(p, t)\,dt}$$

$$\rightarrow f(p, t)[\bar{e} - e(p) + mg(p) - mf(p, t)]\,dt \tag{9-32}$$

This expression can be manipulated to become [calling $f(p, t)$ simply f]

$$\frac{df}{dt} \rightarrow f[\bar{e} - e(p) + mg(p) - mf]$$

$$= [1 - g(p)](\bar{e} - m) + (f - 1)(e(p) - m) + [\bar{e}g(p) - e(p)f]$$

But if the number of demes remains constant, then $\bar{e}$ must equal m. Also, if group selection is slow, $e(p) \simeq \bar{e}$, so $e(p) \simeq m$. Thus the first two terms above vanish and we are left with

$$\frac{df}{dt} \rightarrow \bar{e}g(p) - e(p)f \tag{9-33}$$

To relate *within*-group changes in gene frequency to f, we reason as follows. Demes are finite in size, so the change in gene frequency is a discrete process which, in a given period of time, may or may not occur. The expected change in p over one time unit is given by dp/dt. Thus over the interval t to $t + dt$, the probability that a change in gene frequency will take place is $|dp|$. If we now define $M(p) = dp/dt$, and suppose M to be positive, we see that

$$\begin{pmatrix}\text{probability a change in gene frequency occurs}\\ \text{and that the change is from } p - dp \text{ to } p\end{pmatrix} = M(p - dp)$$

Similarly,

$$\begin{pmatrix}\text{probability a change in gene frequency occurs}\\ \text{and that the change is from } p \text{ to } p + dp\end{pmatrix} = M(p)$$

Finally, the change per infinitesimal period of time dt in the number of demes with gene frequency p is (where N is the total number of demes),

$$\frac{d}{dt}(Nf(p)) = \begin{pmatrix}\text{probability that a deme with gene}\\ \text{frequency } p - dp \text{ becomes a deme}\\ \text{with gene frequency } p\text{, given}\\ \text{that a change takes place at}\\ \text{all, summed over all such demes}\end{pmatrix} - \begin{pmatrix}\text{probability that a deme with gene}\\ \text{frequency } p \text{ becomes a deme}\\ \text{with gene frequency } p + dp\text{, given}\\ \text{that a change takes place at}\\ \text{all, summed over all such demes}\end{pmatrix}$$

(If M is positive, no other changes are possible.) Multiplying by the probability that a change takes place, we obtain

$$|dp|\,\frac{dNf(p)}{dt} = \begin{pmatrix}\text{probability that a change in gene}\\ \text{frequency occurs, \textit{and} the}\\ \text{change is from } p - dp \text{ to } p\text{,}\\ \text{summed over all demes with}\\ \text{gene frequency } p - dp\end{pmatrix} - \begin{pmatrix}\text{probability that a change in gene}\\ \text{frequency occurs, \textit{and} the}\\ \text{change is from } p \text{ to } p + dp\text{,}\\ \text{summed over all demes with}\\ \text{gene frequency } p\end{pmatrix}$$

$$= M(p - dp)Nf(p - dp) - M(p)Nf(p)$$

so that, dividing by $N\,|dp|$,

$$\frac{df}{dt} = \frac{M(p - dp)f(p - dp) - M(p)f(p)}{dp} = -\frac{d}{dp}(Mf) \tag{9-34}$$

The total effect on the frequency of demes with gene frequency p, both from among- and within-demic forces, is thus

$$\frac{df}{dt} = \bar{e}g(p) - e(p)f - \frac{d}{dp}(Mf) \tag{9-35}$$

If M is negative, the derivation is essentially the same, leading to the same expression. The expression we really want is one that will give us the change in mean gene frequency. To obtain this, substitute from (9-35) and write

$$\begin{aligned} \frac{d\bar{p}}{dt} &= \frac{d}{dt}\int pf(p)\,dp = \int p\frac{df}{dt}\,dp = \int (p-\bar{p})\frac{df}{dt}\,dp + \int \bar{p}\frac{df}{dt}\,dp \\ &= \bar{e}\int (p-\bar{p})g(p)dp - \int (p-\bar{p})e(p)f\,dp - \int (p-\bar{p})d(Mf) + \bar{p}\int \frac{df}{dt}\,dp \end{aligned} \tag{9-36}$$

The last term, of course, is zero.

To simplify this rather unpleasant equation, we now examine each of the first three terms. If the mean gene frequency among colonizing propagules is p^*, then

$$\int pg(p)\,dp = p^* \qquad \text{and} \qquad \int \bar{p}g(p)\,dp = \bar{p}$$

Thus the first term is

$$\bar{e}(p^* - \bar{p}) \tag{9-37}$$

To evaluate the second term, expand $e(p)$ using the Taylor series. Then

$$\begin{aligned} \int (p-\bar{p})e(p)f\,dp &= \int \Delta p e(\bar{p} + \Delta p)f\,dp \\ &= \int \sum_{i=0}^{\infty} \Delta p \frac{\Delta p^i}{i!}\frac{d^i}{dp^i}(e(p))_{\bar{p}} f\,d\bar{p} \\ &= \sum_{i \geqslant 1} \frac{d^i e(\bar{p})}{d\bar{p}^i}\mu'_{i+1} \end{aligned} \tag{9-38}$$

where μ'_i is the ith moment about the mean, $\overline{\Delta p^i}$.

Suppose that deviations Δp are quite small, so that moments of order higher than 2 almost disappear. Thus (9-38) becomes

$$\int (p-\bar{p})e(p)f\,dp = \frac{de(\bar{p})}{d\bar{p}}\sigma^2 \tag{9-39}$$

The third term is

$$\int (p-\bar{p})\,d(Mf) = \int \Delta p\,M\,df + \int \Delta p\,f\,dM$$

But if $f(p)$ is normal,

$$f(p) = \frac{1}{\sqrt{2n}\,\sigma}e^{-p^2/2\sigma^2}$$

it follows that

$$\frac{df(p)}{dp} = -\frac{\Delta p}{\sigma^2}f(p)$$

Hence the expression above becomes

$$\int -\frac{\Delta p^2}{\sigma^2} Mf(p)\,dp + \int \Delta p\, f(p)\,dM$$

Expanding $f(p)\,dp$ and dM with the Taylor expansion, this approaches

$$-\frac{1}{\sigma^2} M(\bar{p})\,\overline{\Delta p^2} + \frac{d^2M(\bar{p})}{d\bar{p}^2}\overline{\Delta p^2} = -M(\bar{p}) + \frac{d^2M(\bar{p})}{d\bar{p}^2}\sigma^2 \tag{9-40}$$

Finally, substituting (9-37), (9-39), and (9-40) into (9-36), and noting (Chapter 7) that $\sigma^2 = \bar{p}\bar{q}f = pq/2n_e$, where n_e is the effective size of a deme, we have

$$\frac{d\bar{p}}{dt} = \bar{e}(p^* - \bar{p}) - \left[\frac{de(\bar{p})}{d\bar{p}} + \frac{d^2M(\bar{p})}{d\bar{p}^2}\right]\frac{\bar{p}\bar{q}}{2n_e} + M(\bar{p}) \tag{9-41}$$

where $\bar{p}$ is the population mean gene frequency, p^* is the mean gene frequency in colonizing propagules, $e(\bar{p})$ is the extinction rate of demes with gene frequency $\bar{p}$, and $M(p) = dp/dt$ within demes. We now illustrate this equation's use with a simple, hypothetical example.

Suppose that there is no genetically related propensity to migrate or survive migration, so that $p^* = \bar{p}$. Suppose further that extinction is very much a function of gene frequency, $e(p) = p^2$. Finally, suppose that $\bar{W}_A = 2\bar{W}_a$ is a constant. Then $M(p) = p((\bar{W}_A - \bar{W})/\bar{W})$ (see equation 8-3). The system described experiences strong individual selection for A, and group selection, via differential extinction, against A. Equation (9-41) becomes

$$\frac{d\bar{p}}{dt} = \bar{e}(0) - \left[\frac{d}{d\bar{p}}\bar{p}^2 + \frac{d^2}{d\bar{p}^2}\bar{p}\left(\frac{\bar{W}_A - \bar{W}}{\bar{W}}\right)\right]\frac{\bar{p}\bar{q}}{2n_e} + \bar{p}\left(\frac{\bar{W}_A - \bar{W}}{\bar{W}}\right)$$

where

$$\frac{\bar{W}_A - \bar{W}}{\bar{W}} = \frac{\bar{W}_A - (\bar{p}\bar{W}_A + \bar{q}\frac{1}{2}\bar{W}_A)}{\bar{p}\bar{W}_A + \bar{q}(\frac{1}{2}\bar{W}_A)} = \frac{\bar{q}}{1 + \bar{p}}$$

Performing the necessary differentiations yields

$$\frac{d\bar{p}}{dt} = \frac{\bar{p}\bar{q}}{(1 + \bar{p})^3 n_e}[(1 + \bar{p})^2 n_e + 2 - \bar{p}(1 + \bar{p})^3] \tag{9-42}$$

When $n_e = 1.3$, $\bar{p}$ is .942, and because $d\bar{p}/dt$ is positive when $\bar{p}$ is less than .942 and negative when $\bar{p}$ is above .942, that value represents a stable equilibrium. Group selection successfully impedes the fixation of A by individual selection. Notice though that as n_e becomes larger, the equilibrium value of $\bar{p}$ will rise. Thus, at least in this example, the existence of group selection as a significant force in the evolution of $\bar{p}$ depends very much on the deme's being effectively very small. Group selection is enhanced by positive assortative mating, consanguineous mating, and small social groupings, and is weakened by any factor tending to raise n_e—large groupings, for example.

Wade's approach (Wade, 1976, 1977, 1978b; Wade and McCauley, 1980) is experimental rather then theoretical. Using a highly cross-bred (and thus genetically variable) strain of the flour beetle *Tribolium castaneum*, Wade (1976; see also 1977) started four sets of 48 laboratory populations with 16 randomly drawn individuals each. Each of the four sets were subjected to a different treatment. In treatment I the 48 populations were allowed to grow without interference until day 37. At that time 48 new populations were begun with 16 individuals each, the founders coming first from the largest of the 48 previous generation populations and then, when this group was exhausted, the next largest, and so on. This procedure was continued for nine generations. It was meant to simulate group selection for rapid population growth (to day 37) in which extinction claims most of the demes each generation and sees them replaced by propagules from only the fastest-growing demes. Treatment II was similar to the above except

Table 9-4 **Population Size (Mean) per Deme at Day 37**[a]

	Treatment[b]		
Generation	*I*	*II*	*III*
1	256	271	278
2	223	211	215
3	196	122	152
4	178	88	137
5	279	108	170
6	228	76	155
7	219	48	106
8	129	26	68
9	178	20	49

[a] 48 replicates for each treatment.

[b] I, high-selected line; II, low-selected line; III, control.

that individuals forming the new populations were drawn from the *smallest* populations of the previous generation. This, then, was a line selected for slow growth. Treatments III and IV were different sorts of controls. In III new populations were formed with 16 individuals drawn together from a randomly chosen deme. In IV each set of 16 was randomly chosen from the pooled members of all demes. Wade's results are shown in Table 9-4. Treatment IV results were almost identical to those of treatment III through generation seven and are not given here.

The data show a general decline in population size at day 37. Why this should be so is not clear. However, a comparison of the experimental with the control lines shows clear trends. The high-selected line demonstrated a relative increase, and by the ninth generation the difference between it and the controls was significant at the .0001 level. The low-selected line responded less dramatically but over the nine generations showed a significant decline relative to the controls ($p < .0005$) (see Figure 9-2). Treatments III and IV diverged at generation seven, the former showing a relative rise. This is not unexpected. By generation seven the smaller demes were really quite minute in size, and only a very few randomly chosen propagules could be taken from them before they were used up as colonizer sources. Thus, in the later generations, an increasing proportion of the next generation's demes were founded by individuals taken with bias from larger demes.

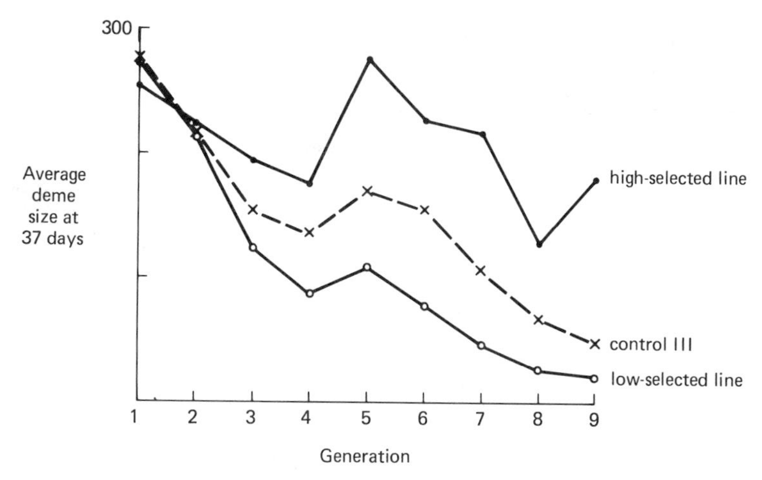

Figure 9-2 Results of group selection for deme size at day 37.

In another set of experiments Wade repeated the procedures, splitting each treatment into three subtreatments. In the first, each new deme was begun with 16 individuals chosen as before. In the second, 12 individuals were chosen from the appropriate donor deme, and the other four from the next largest (high line), next lowest (low line), or next randomly chosen (treatment III) deme. This was meant to mimic a group selection situation with 25% migration. The third subset involved eight individuals chosen from the appropriate donor deme, rather than 12, and eight from the next, rather than four, simulating a 50% migration scheme. Results were qualitatively the same for all three subtreatments, equivalent to those shown in Table 9-4 and Figure 9-2, but progress was slowed, especially with high migration in the high-selected line. Before concluding, contrary to the comments made earlier in this section, that group selection can still operate as a significant force even in the presence of high migration, it should be noted that Wade's migration patterns were *not* random. Had the additional four or eight founders been drawn not from the next largest, or next smallest deme, but without bias, haphazardly, the results might have been quite different.

Perhaps Wade's major contribution so far has been to point out that whereas natural selection acting within demes has available to it only the additive component of genetic variance in fitness, interdeme selection, not complicated by genetic recombination due to sexual reproduction of individuals, can act on the total genetic variance. Because strong directional selection often results fairly quickly in a proponderance of the dominance variance component (see Chapter 10), it may be that after a unidirectional, individual selection process has been slowed by the waning of $V_A(x)$, interdemic selection becomes relatively powerful. Wade and McCauley (1980) have, in fact, measured a "realized" demic heritability,

$$h_x^{*2} = \Delta\bar{x}\frac{\bar{W}}{\partial\bar{W}/\partial\bar{x}\ \mathrm{var}(\bar{x})} \qquad \text{(see equation 8-27)}$$

where $\Delta\bar{x}$ is the observed change in trait value among all demes, $\bar{W}$ is the population mean fitness value, and var $(\bar{x})$ is the among deme variance in the trait's value. For population growth to day 37 this value was large and fairly consistent over a range in deme size of 6 to 48, indicating that deme size may, after all, be less critical than suggested by Levins' model.

The lessons derived from Wade's results have been corroborated by Craig (1982) who individually selected for fast or slow migrators by providing a string up which flour beetles (*Tribolium confusum*) could climb to escape their vials. Group selection was accomplished by choosing vials with the most or the fewest migraters. The effects of the group selection were very strong, one extra source of genetic variability for its workings being a tendency for density-dependent migration. Craig notes that much of the drop in variance among groups, as opposed to within groups, is due to the averaging out of phenotypic (nongenetic) noise; a much higher proportion of var (x) than var $(\bar{x})$ is nongenetic.

Spatial Heterogeneity and Polymorphism

Analogous to the situation in which the environment varies with time, an individual that wanders over a variety of spatial backgrounds during its life views the environment in a fine-grained manner. Fitness reflects the mean environment to which a genotype is subjected. But if generational differences occur, the environment is coarse-grained and the effective (geometric mean) fitness is

$$W^* = \prod_i W(i)^{Q_i}$$

where $W(i)$ is the fitness in the ith type of environment and Q_i is the proportion of generations spent living in the ith-type environment. This being the case, coarse-grained spatial heterogeneity, like coarse-grained temporal heterogeneity, may be expected to give rise to polymorphism.

Steiner (1977) searched the literature for data on genic variance in Hawaiian *Drosophila*. He then regressed the values obtained on niche width, measured as the number of plant species used as oviposition sites. He found the expected relationship (regression coefficient $= 1.45, p < .05$). Lacy (1982) found that genic variation among mycophagous Drosophilidae in the eastern United States was positively correlated with diversity of host mushroom species. On the other hand, Sabath (1974) looked much more carefully at an average of seven loci over 11 *Drosophila* species in southern Indiana and measured niche width on the basis of measurements of habitat types over quadrats. He found no relationship whatsoever. Sabath's results do not necessarily contradict the hypothesis, however. Without extensive information it is difficult to tell whether heterogeneity will result in effective heterosis. So neither the finding of polymorphism nor its absence necessarily supports or refutes the expectation. For example, Mitter and Futuyma (1979) looked at 10 tree (foliage)-feeding moths, each polyphagous but sufficiently immobile that migration of larvae among food plants is rare. The adults are short-lived, nectar feeding, and can disperse their eggs among different tree types. Here, then, is a clear-cut case of coarse-grained spatial heterogeneity. Mitter and Futuyma, accordingly, expected to find polymorphism with respect to genes adapting the larvae to their particular host tree. However, although some polymorphism occurred at various enzyme loci, differences in gene frequency among populations on different tree types were observed only twice. This suggests that there is no differential selection at those loci and, therefore, that the polymorphisms observed are independent of the spatial heterogeneity. Perhaps Mitter and Futuyma were simply looking at inappropriate gene loci. Perhaps fitness differences are so slight as to be undetectable: that is, perhaps the individuals are sufficiently adaptable (generalized) that genetic differences contribute little to their foraging efficiencies on different substrates.

Laboratory studies can be better controlled and so, perhaps, are better capable of evaluating predictions about polymorphism in spatially heterogeneous environments. Powell (1971) used Brazilian *Drosophila willistoni*, allowing them to oviposite in population cages on either Carolina instant *Drosophila* medium (f_1), Spassky's medium (f_2), or their choice of either. Food was available in the form of baker's yeast (y_1), brewer's yeast (y_2), or both. Because eggs can be laid only on one medium or the other, and are not mobile, f_1 and f_2 constitute alternatives in a coarse-grained environment. Limited larval mobility also makes coarse-grained resources out of y_1 and y_2. After 45 weeks in the population cages, Powell examined 22 gene loci over 50 individuals in each cage. The results (Table 9-5) show quite clearly that the artificially induced spatial heterogeneity promoted (or prevented the disappearance of) polymorphism.

The analytical difficulty that spatial heterogeneity presents, which temporal heterogeneity does not, is the blunting of "coarse-grainedness" due to migration. Thus we must temper any

Table 9-5

Environment	Average Percent of Individuals Heterozygous per Locus
y_1, f_1	$7.28 \pm .83$
	$7.80 \pm .82$
	$7.98 \pm .84$
y_2, f_1	$8.04 \pm .81$
y_2, f_2	$7.98 \pm .84$
y_1 and y_2, f_1	$11.09 \pm .89$
	$9.06 \pm .84$
y_1, f_1, and f_2	$10.16 \pm .88$
	$9.33 \pm .88$
y_1 and y_2, f_1 and f_2	$13.90 \pm .98$
	$12.81 \pm .92$

predictions of polymorphism with considerations of the amount of migration in a system. But not only might migration render the environment sufficiently fine-grained that polymorphism may not arise, it might alternatively *produce* polymorphism by enabling the mixing of two populations, each of which is being selected in a different manner. To untangle the role of migration, a number of models have been built, beginning with that of Levene (1953) (see also Levins and MacArthur, 1966; Bulmer, 1972; Pollak, 1974; and Gillespie, 1976). The treatment below is from Pollak.

Consider a single locus with two alleles, A with frequency p, and a with frequency $q = 1 - p$. If natural selection causes an increase in p when A is rare and an increase in q when a is rare, both alleles are protected from extinction and we are left with a polymorphism. Now picture the population as being divided into patches. Let n_{ij} be the expected number of Aa individuals in patch j arising from parents in patch i, m_{ij} the probability that an offspring ends up in patch j given that its parents lived in patch i (migration rate), W_i the fitness in patch i and $\boldsymbol{\rho}$ the vector describing the number of parents in patches $1, 2, \ldots, k$. If migration precedes selection, we have

$$n_{ij} = W_i m_{ij}$$

or, in matrix form, where

$$N = \{n_{ij}\} \qquad M = \{m_{ij}\} \qquad W = \begin{bmatrix} W_1 & 0 & 0 & \cdot \\ 0 & W_2 & 0 & \cdot \\ 0 & 0 & W_3 & \cdot \\ \cdot & \cdot & \cdot & \cdot \end{bmatrix}$$

$$N = WM$$

But then

$$\boldsymbol{\rho}(t+1) = N\boldsymbol{\rho}(t) = WM\boldsymbol{\rho}(t)$$

whose general solution is

$$\rho_i(t) = \sum_k C_{ik}\, \lambda_k^t$$

where λ_k are the eigenvalues of WM. If migration *follows* selection,

$$n_{ij} = m_{ij} W_j$$

or $N = MW$, and

$$\boldsymbol{\rho}(t+1) = MW\boldsymbol{\rho}(t)$$

whence

$$\rho_i(t) = \sum_k C_{ik} \lambda_k^t \tag{9-43}$$

where λ are the eigenvalues of MW. Because the eigenvalues of WM and MW are the same, there is no difference in the general solution and the general conclusion is reached that ρ_i will not vanish (there will exist at least one patch in which the heterozygote is retained—that is, polymorphism) if and only if the largest eigenvalue is equal to or greater than 1.0. Consider the simplest case of just two patches.

$$M = \begin{bmatrix} 1 - m_{12} & m_{12} \\ m_{21} & 1 - m_{21} \end{bmatrix}$$

$$W = \begin{bmatrix} W_1 & 0 \\ 0 & W_2 \end{bmatrix}$$

Then

$$MW = \begin{bmatrix} W_1(1 - m_{12}) & W_2 m_{12} \\ W_1 m_{21} & W_2(1 - m_{21}) \end{bmatrix}$$

and

$$\lambda^2 - [W_1(1 - m_{12}) + W_2(1 - m_{21})]\lambda + W_1 W_2(1 - m_{12} - m_{21}) = 0$$

The eigenvalues are thus

$$\lambda = \frac{W_1(1-m_{12})+W_2(1-m_{21}) \pm \sqrt{[W_1(1-m_{12})+W_2(1-m_{21})]^2 - 4W_1W_2(1-m_{12}-m_{21})}}{2}$$

The largest one will equal or exceed unity only if

$$W_1(1 - m_{12}) + W_2(1 - m_{21}) - W_1W_2(1 - m_{12} - m_{21}) > 1 \qquad (9\text{-}44)$$

Suppose that the population is near fixation for one allele or the other. Then an increase in the rare gene in either patch requires that one or the other value of W exceed 1.0. But condition (9-44) cannot be met unless one of the W exceeds 1.0. Thus the condition for maintenance of polymorphism is equivalent to protection of one allele in one patch, the other in the other patch. That is, we require that the heterozygote be more fit than one homozygote (when that homozygote is at very high frequency) in one patch, more fit than the other homozygote (when that homozygote is at very high frequency) in the other patch. As an example, suppose that $W_1 = 1.1$, $W_2 = .8$, and $m_{12} = m_{21} = m$. Then the left side of (9-44) is

$$1.1(1 - m) + .8(1 - m) - (1.1)(.8)(1 - 2m) = 1.02 - .14m$$

which exceeds 1.0 only if $m < \frac{1}{7}$. High rates of migration destroy the role of patchiness in maintaining polymorphisms.

An important fact demonstrated in the approach above, is that as more patch types are added, the chances that any two of them will promote polymorphism increases. That is, with increasing numbers of patches, the probability of polymorphism (and thus, presumably, also the number of polymorphic loci) should rise.

Another prediction of Pollak's model is that high levels of migration destroy the conditions for polymorphism. But of course Pollak's model made the implicit assumption that migration, as in Wright's island model, is undirected; and there is good reason to believe that migration is far from random. Fretwell and Lucas (1969), C. E. Taylor (1976), and J. M. Emlen (1981) have all made strong cases for organisms' choosing their surroundings in such a way as to maximize their viability, mating success, or fecundity—in short, their contribution to their genotype's fitness. Thus AA individuals (say) will preferentially migrate to areas in space where W_{AA} is highest. When this is so (and note in passing that this raises among-group genetic variance to the benefit of group selection processes), the conditions for polymorphism are considerably relaxed. As C. E. Taylor (1976) states, all that is now required for the maintenance of polymorphism is that

$W_{Aa} > W_{aa}$ somewhere, when A is rare, and

$W_{Aa} > W_{AA}$ somewhere, when a is rare

or

$W_{AA} > W_{aa}$ somewhere, and

$W_{aa} > W_{AA}$ somewhere else

When viewed from this perspective it is difficult to see why any populations in the real world should be anything but polymorphic. The question of interest seems no longer to be what maintains genic variability, but rather why it is not almost universal. The problem with the latter attitude lies in the question of whether organisms really do migrate preferentially to the most salubrious environs and in just how many patch types they are apt to be found. It certainly seems reasonable that they should show appropriate migratory behavior, at least on a microhabitat scale, and there is considerable evidence that they do (see Chapters 11 through 17). Among the more classical tests of adaptive habitat selection are those of Kettlewell (Kettlewell, 1955a; Kettlewell and Conn, 1977), who demonstrated that melanic forms of the peppered moth preferred dark substrates, whereas the light morphs preferred light surfaces. Powell and Taylor (1979) captured *Drosophila* from moist woods and open, warm woods, marked them with fluorescent dusts, and then released them at various points in space. They were able to show that flies returned not necessarily to the area of capture, but to the area *type* in which they were originally found.

Review Questions

1. How should the population genetic response to environmental fluctuations differ with the time scale of those fluctuations? Why? What role does age structure play in the nature of the response?
2. It seems plausible that in many cases in which kin selection favors a trait when it is rare, individual selection will take over and favor the same trait as it becomes more common. Thus there is generally no need to appeal to group selection models. For example, suppose that a certain insect is toxic to predators because it contains some poisonous substance derived from its food plant. If an insect of this species possesses a gene for bright coloration and is killed by a predator, the gene will spread if the predator learns to avoid the brightly colored kin of the first insect. When the brightly colored type becomes more common, bright color is no longer altruistic, but individually advantageous. The same sort of altruistic to selfish switch might occur in many other cases. In the light of these comments is it not a fundamental flaw in all kin and interdemic selection models to suppose that the cost of the altruistic behavior remains constant (and nonzero) over time?
3. Is it possible that Wade's group selection experiments were so well designed that the outcomes were tautological? If so, what do these experiments tell us? Is this problem also to be found in individual selection experiments?
4. What is the relationship between the inbreeding coefficient and the coefficient of relationship r used in the kin-selection models?
5. Evaluate the following statement: There are a number of serious problems with Wade's group selection experiments. The variance in reproductive success among male insects is often great enough that a single male could have fathered a high percentage of the offspring born in a particular group. Thus Wade would have been showing no more than individual selection for some traits influencing male fertility. He may have been selecting for genes in the male (the heterogametic sex) biasing the sex ratio. Large groups would result when sex ratio was skewed in one direction, small populations when the skew was reversed.

Chapter 10

Natural Selection: II. Multilocus and Quantitative Models

10.1 Interacting and Interconnected Loci

Epistasis and Pleiotropy

In the preceding chapters we developed some basic expressions for predicting the change in frequency of a gene, or the change in mean value of a trait controlled by a single gene locus due to natural selection. But most traits are affected by more than one, and often by many, loci. Can we think of such "polygenic" traits as being merely the sum of the contributions of the various gene loci affecting them? Unfortunately, no. We have already commented in Chapter 8 that the effect of a gene on its owner may depend on the owner's environment, and that the environment includes, in addition to the external and organismal components, the genetic milieu; the impact of an allele at one locus may depend very much on the genotypic configurations at a number of other loci. In other words, the genes at one locus may modify the effects of genes at other loci. This phenomenon is known as *epistasis*.

Where epistasis occurs, we should predict that similar natural selection pressures, acting on a particular gene locus, may have somewhat different effects, depending on the genetic background against which selection takes place. Of course, such seemingly innocuous processes as genetic drift may change the genetic background. Thus it is not surprising that Selander et al. (1969), looking at protein polymorphisms in several Danish populations of house mice, found a variety of different gene frequencies for each locus, depending on the location in which the mice were caught, even though the ecological situations were almost identical from place to place.

Dobzjansky and Pavlovsky (1957) studied the equilibrium frequencies of the third chromosome inversions Pikes Peak and Arrowhead in laboratory populations of *Drosophila pseudoobscura*. In identical laboratory environments the ultimate eventual frequencies differed with the locality from which the stock flies were collected. If populations were started with a single pair of flies, each individual taken from a different geographic area, the equilibrium frequencies eventually reached were less variable among the various experimental replicates than if populations began with a series of pairs originating from 10 geographic areas. In the former case there was a more consistent genetic background (Dobzjansky and Spassky, 1962).

More recently, Sokal and Taylor (1976) measured the fitnesses of genotypes in two housefly (*Musca domestica*) strains and their crosses and tried unsuccessfully to use these measures to predict the evolutionary progress of a gene. During the selection process the fitnesses changed significantly; properties of hybrids were in no way predictable from those of their parents, and the basic equations of Chapter 8 were useless.

The existence and degree of genetic dominance may be the result of modifier (epistatic) genes. Fisher (1958) argued that if genotypes at a second locus could act in such manner as to make *Aa* phenotypically more like *AA*, when *A* was a beneficial allele, such genotypes would be selectively advantageous. This theory for the existence of dominance has been hotly debated. Recently, Kacser and Burns (1981) have argued simply but pursuasively that while Fisher's evolutionary mechanism might be incorrect, it is probably true that dominance is a result of epistatic interactions, and that it is almost unavoidable that established beneficial genes should be dominant and that new mutations be recessive. It is certainly a widespread phenomenon that beneficial alleles are dominant and that detrimental alleles and new mutant forms are recessive.

Kojima (1971) in a discussion of epistasis asked the question: Is there a constant fitness value for a given genotype? His answer was a resounding no.

Not only are the fitnesses and therefore rates and directions of allelic changes at one locus generally dependent on genes elsewhere in the genome, but it is now a well-established fact that many, and probably most, genes have effects on a variety of morphological, physiological, or behavioral traits, a phenomenon known as *pleiotropy*. In keeping with the theories of Fisher (1958) and Kacser and Burns (1981) mentioned above, there is a distinct tendency for an allele to be dominant with respect to those traits for which its impact on fitness is positive, and to be recessive with respect to those traits for which its impact on fitness is negative. This fact has an interesting implication: If *Aa* is phenotypically like *AA* for those traits for which *AA* is most fit, and like *aa* for those traits where *aa* is superior, then *Aa* demonstrates the superior phenotype with respect to all traits affected by the locus in question, and thus must have higher fitness than either homozygote. That is, if dominance is a usual characteristic of alleles that contribute to a beneficial character, pleiotropic genes will usually be heterotic.

The complications noted above do not prevent natural selection from promoting the genotypic configurations of the most viable, successfully mating, or fecund individuals in a population. But the existence of epistasis and pleiotropy mean that at any particular locus there will no longer necessarily be a single value of p that maximizes fitness; perhaps $p = 0$ maximizes fitness for one particular configuration of gene frequencies at other loci, $p = .3$ maximizes fitness for another configuration, and $p = .8$ for still another. In the simple case where only two loci are involved, picture a simple coordinate system where frequency of the *A* allele at one locus (p_A) is plotted on the x-axis, the frequency of the *B* allele at another locus (p_B) on the y-axis. Each hypothetical point on this coordinate system is associated with some population mean fitness $\bar{W}$, and if we connect the points of equal fitness we can generate a kind of topographic map of fitness. Wright (1932, 1965; see also his 1980 review) refers to such topographic maps as *adaptive surfaces*, and of course they needn't be limited to the case of just two gene loci (two dimensions). Where adaptive surfaces have more than one peak, these are known as *adaptive peaks*. Natural selection, because (in the absence of frequency-dependent interactions) it acts to increase fitness, "pushes" gene frequencies up the slope of the adaptive hill on which the population finds itself.

A very nice example of a two-dimensional adaptive surface is given by Lewontin and White (1960) for inversions on two chromosomes in the locust, *Moraba scurra*. The fitness values for the various inversion frequency combinations were generated in laboratory populations (see Fig. 10-1). The arrows point "uphill," the direction followed by the evolving system. Note that there are alternative adaptive peaks; which one is climbed depends on the initial gene frequencies of the population.

But while we may still expect natural selection to result in an increase in frequency of the most fit genotype, the adaptive peak climbed by the population may not be the highest peak available. Thus, as a result of epistasis, pleiotropy, and their encouragement of multiple adaptive peaks, we must make a distinction between "increasing fitness" and "maximizing fitness." Natural selection may accomplish the former, but not necessarily the latter. Note also that the relative heights of the peaks may be frequency dependent. To wit, "climbing a peak" may have the effect of lowering that peak's relative height.

Is there any way that populations can jump from one adaptive peak to another and thus eventually reach the top of the highest peak? Because selection acts to increase fitness it is hard to see how evolution could occur *down* one slope and then up another; there is no obvious way in which the adaptive valleys can be crossed. But suppose that a propagule from one population, inhabiting one adaptive peak, broke away and formed a new, geographically isolated popu-

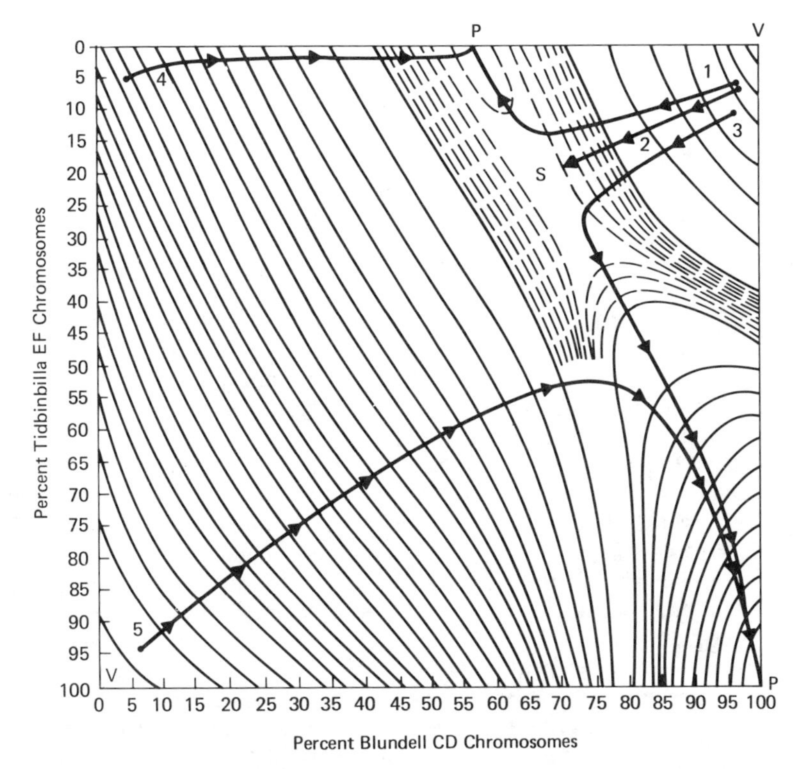

Figure 10-1 Topographic map (adaptive surface) of population mean fitness for various frequencies (percents) of two chromosomal inversions in the locust *Moraba scurra*. (After Lewontin and White, 1960.)

lation in a similar (external) environment with identical selection pressures? If that propagule were small enough, its genetic constitution might differ considerably from that of the parent population by chance alone—maybe the genetic makeup of the new, founder population occurs on the slope of another adaptive peak. If so, selection will subsequently act to steer its evolutionary course up that new peak, and the parent and daughter populations will diverge genetically and, therefore, probably also phenotypically. One possible example of an event of this sort is that of the Kaibab and Abert's squirrels of the western United States ponderosa pine forests. Abert's squirrels (*Sciurus aberti*), are a diagnostic feature of montane forests of Colorado, but their extensive population is almost totally isolated from the north rim (Kaibab plateau) of the Grand Canyon, which possesses an ecologically similar habitat. Sometime in the not too distant past it appears that a contingent of Abert's squirrels somehow reached the plateau and successfully colonized it.Evolving in similar conditions but with different genetic backgrounds, the two populations diverged. Abert's squirrels are dark gray to black with tufted ears. Kaibab squirrels are very similar, but sport a white tail.

In addition to adaptive peak jumping by founder populations, populations that are sufficiently small to experience strong genetic drift may suddenly find themselves, quite by accident, on the slopes of a new peak, which may be higher or lower. Of course, under such circumstances, selection must be a relatively weak force for genetic change and we might expect to find the population, instead of climbing steadily up the new peak, wallowing randomly about on the adaptive surface.

Interdemic selection, too, may affect the choice of peaks climbed. If the population is subdivided into demes that are sufficiently different genetically, the individual selection pressures pushing the population as a whole up one peak may possibly be overcompensated by group selection favoring a different path. But analysis is difficult here, for the traits that benefit or harm a group need bear no simple relationship to those affecting individual success. Thus the individual and group selection adaptive surfaces may not be the same—in fact, they are likely to be quite different.

With spatial (and genetic background) variation in the environment, the adaptive landscape describing the fitness effects of any chosen set of gene loci will undoubtedly change. Thus subpopulations at one locale may be climbing one adaptive peak, to a particular configuration of gene frequencies, while an adjacent subpopulation may be climbing another peak, to another set of frequencies. Migration among such subpopulations may hopelessly muddle the prospects of adaptation of subpopulations to local conditions. On the other hand, it may produce the jolting changes in gene frequencies needed by the local subpopulation to shift its evolutionary course to another, higher adaptive peak. Temporal variation, by continually changing the heights and even positions of the adaptive peaks, presents further complications.

There is a glimmer of hope in this morass of complexities: Whether the adaptive peak climbed is the highest one, whether fitness is really maximized, is probably unimportant to the ecologist who wishes to make predictions about the nature of adaptive characteristics. Because population mean fitness in a persistent population may always tend toward a value of $m = 0$ (geometric mean of $\bar{W} = 1$ in the discrete case), all peaks, ultimately, are of the same height. The fact that for a current, existing situation one peak is higher is unimportant, for if that peak were actually climbed, the eventual fitness would still be 1.0. In fact, by virtue of the population having climbed the "higher" peak, the old, existing situation no longer obtains; now the old peak from which that population jumped may be the higher one. The geneticist will be interested in gene frequencies per se, so which peak has been climbed is important. The ecologist wishing to predict the current direction of evolution or to discover whether, given existing circumstances, a particular trait is evolutionarily stable, need be concerned only with the fact that natural selection increases fitness for those existing circumstances. (See Templeton, 1982, for an excellent recent discussion of adaptive surfaces.)

Linkage

The glimmer of hope mentioned above is consoling only if we can assume that natural selection really results in the climbing of adaptive peaks. Suppose that we look at two loci, with alleles *A* and *a*, *B* and *b*. Suppose that *A*, and *B* are the more fit alleles (individuals carrying *A* or *B* average higher fitness than the average individual). Then the most fit hypothetical genotype is *AABB*. But suppose that the loci occur very close to each other on the same chromosome and that chromosomes carrying *A* also carry *b*, that chromosomes with *B* also have *a*. When the beneficial form of one gene is "linked" in such a manner to the less beneficial form of the other, natural selection will still pass on a higher proportion of the more fit gene *combination*, but the hypothetically most fit genotype never materializes, and so cannot spread. If the linkage is extremely "tight," so that, as above, *Ab* and *aB* combinations never split up, the two loci may be treated quite simply as a single locus. If linkage is extremely "loose," so that chromosome breaks result commonly in crossing over—the *Ab*, *aB* forms break up and *AB*, *ab* gametes are formed—then *A*, *a*, *B*, and *b* assort (almost) independently and can be treated as alleles at separate loci. But if linkage occurs at an intermediate level, the situation is messier; the alleles at the two loci do not assort independently and the equations of Chapter 8 are inadequate to deal with dynamics at *either* locus. What is worse is that if both linkage and epistasis occur concurrently between two (or more) loci, neither locus is free to evolve, either functionally or physically independently of the other, and there is no longer any guarantee that selection will push populations up adaptive peaks. We are apt to find evolutionary equilibria which represent complex, utterly unanalyzable compromises between selection pressures and constraints imposed by nonindependent assortment at many loci (see Allard et al., 1968; Feldman et al., 1974).

We shall begin hopefully, by looking at linkage in a two-locus system. There will be four gametic types:

	Gametic Type			
	AB	***Ab***	***aB***	***ab***
Frequency	r	s	t	u

We shall let c be the crossover rate, the proportion of meiotic pairs of homologous chromosomes in which breakage results in a subsequent recombination of allelic forms. Then, in the absence of selection, the passing of one generation finds

$$r' = r - c(ru - st)$$

The change in r is due to the fact that whenever AB and ab come together in meiosis, with frequency ru, they recombine to form Ab and aB a fraction c of the time, resulting in the *loss* of an AB gamete. Whenever Ab and aB join, with frequency st, crossing over results in the *gain* of an AB. If we refer to the quantity $ru - st$ as d, then

$$\begin{aligned} r' &= r - cd \qquad & s' &= s + cd \\ t' &= t + cd \qquad & u' &= u - cd \end{aligned} \tag{10-1}$$

More generally, if there is inbreeding, (and throughout this chapter $F = F_{IT}$),

$$r' = \frac{[2(r^2 + r(1 - r)F) + 1(2rs)(1 - F) + 1(2rt)(1 - F) + 1(2ru)(1 - F)]}{2}$$

$$+ \frac{c[1(2st)(1 - F) - 1(2ru)(1 - F)]}{2}$$

$$= [r(r + s + t + u)(1 - F) + rF] - c(ru - st)(1 - F)$$

which reduces to

$$r' = r - cd(1 - F)$$

Similarly,

$$\begin{aligned} s' &= s + cd(1 - F) \\ t' &= t + cd(1 - F) \\ u' &= u - cd(1 - F) \end{aligned} \tag{10-2}$$

The value of d in the new generation is

$$d' = (r'u' - s't') = \cdots = d[1 - c(1 - F)] \tag{10-3}$$

Therefore, d after n generations, d_n, is

$$d_n = d_0[1 - c(1 - F)]^n \tag{10-4}$$

showing that d declines asymptotically to zero over time. What does this tell us? If we are to assume that the two loci assort independently, this means that the probability of drawing an A, as opposed to an a, from the gene pool is independent of whether the A and a in question are connected to B or b. That is, independent assortment occurs when

$$\frac{r}{t} = \text{the relative probabilities of drawing } A, a \text{ when } B \text{ is attached}$$

$$= \frac{s}{u} = \text{the relative probabilities of drawing } A, a \text{ when } b \text{ is attached}$$

Equivalently, independent assortment occurs when

$$ru - st = 0$$

But $ru - st$ is d, which we have shown above approaches zero over time, quickly if the crossing over rate is high, more slowly in the presence of heavy inbreeding. Thus in populations not experiencing natural selection, the alleles at gene loci, even quite tightly linked loci (c very small), will eventually combine in frequencies such that their assortment in the gene pool is independent. Once this occurs the equations in Chapters 8 and 9 are applicable— the genetic system is said to be in *linkage equilibrium.*

But suppose that the A and B loci experience differential fitnesses. We may set up the following table, giving fitness values for each possible combination of gametes (Table 10-1). We may calculate the new values of r and r', as for (10-2), but now each term must be properly weighted by the appropriate fitness term:

Table 10-1

		Maternal Gamete			
		AB	***Ab***	***aB***	***ab***
	AB	W_{11}	W_{12}	W_{13}	W_{14}
Paternal	***Ab***	W_{21}	W_{22}	W_{23}	W_{24}
Gamete	***aB***	W_{31}	W_{32}	W_{33}	W_{34}
	ab	W_{41}	W_{42}	W_{43}	W_{44}

$$r' = \frac{2(r^2 + r(1 - r)F)W_{11} + 1(2rs)(1 - F)W_{12} + 1(2rt)(1 - F)W_{13} + 1(2ru)(1 - F)W_{14}}{2\bar{W}}$$

$$+ c\frac{1(2st)(1 - F)W_{23} + 1(2ru)(1 - F)W_{14}}{2\bar{W}}$$

$$= [r\bar{W}_1(1 - F) + rFW_{11}] - cQ \tag{10-5}$$

where $\bar{W}_1 = \bar{W}_{AB}$ and $Q = (ruW_{14} - stW_{23})(1 - F)$. Further analysis becomes quite complex for the general case, so we shall now simplify the situation by ignoring inbreeding and supposing that there are no maternal effects. The latter assumption is equivalent to saying that $W_{ij} = W_{ji}$, $W_{23} = W_{14} = W_H$, the fitness of the double heterozygote. Equation (10-5) and the corresponding equations for the other gametic frequencies now become

$$r' = r\frac{\bar{W}_1}{\bar{W}} - cd\frac{W_H}{\bar{W}}$$

$$s' = s\frac{\bar{W}_2}{\bar{W}} + cd\frac{W_H}{\bar{W}} \tag{10-6}$$

$$t' = t\frac{\bar{W}_3}{\bar{W}} + cd\frac{W_H}{\bar{W}}$$

$$u' = u\frac{\bar{W}}{\bar{W}_4} - cd\frac{W_H}{\bar{W}}$$

where $\bar{W}_1 = \bar{W}_{AB}$, $\bar{W}_2 = \bar{W}_{Ab}$, $\bar{W}_3 = \bar{W}_{aB}$, and $\bar{W}_4 = \bar{W}_{ab}$, which leads us to

$$\Delta r = r\frac{\bar{W}_1 - \bar{W}}{\bar{W}} - cd\frac{W_H}{\bar{W}} \qquad \Delta s = s\frac{\bar{W}_2 - \bar{W}}{\bar{W}} + cd\frac{W_H}{\bar{W}}$$

$$\Delta t = t\frac{\bar{W}_3 - \bar{W}}{\bar{W}} + cd\frac{W_H}{\bar{W}} \qquad \Delta u = u\frac{\bar{W}_4 - \bar{W}}{\bar{W}} - cd\frac{W_H}{\bar{W}} \tag{10-7}$$

and

$$d' = \frac{ru\bar{W}_1\bar{W}_4 - st\bar{W}_2\bar{W}_3}{\bar{W}} - \frac{cdW_H}{\bar{W}} \tag{10-8}$$

Solving for equilibrium d by setting $d' = d$, we obtain

$$d = \frac{\bar{W}_1\bar{W}_4 st(1 - \bar{W}_2\bar{W}_3/\bar{W}_1\bar{W}_4)}{\bar{W} - \bar{W}_1\bar{W}_4 + cW_H} \tag{10-9}$$

Except for the trivial case where $s = 0$ or $t = 0$, d will not tend to zero unless $\bar{W}_1\bar{W}_4$ is exactly equal to $\bar{W}_2\bar{W}_3$. As long as the genetic system is not at fixation, it is (almost) impossible for selection to act on two linked gene loci as if they were assorting independently; the system is said to be in *linkage disequilibrium*. Even at a given point in time (to avoid changes in W values due to epistasis), the equations in Chapters 8 and 9 would seem inapplicable.

We shall not give up hope yet. Just how large are linkage disequilibrium values in nature, and how inaccurate are the basic equations in Chapters 8 and 9 as a result? Cook and O'Donald (1971) studied the ecological genetics of the European garden snail (*Cepaea nemoralis*) in Ben Whiskin, Ireland, and observed frequencies of the pink (AA or Aa) and yellow (aa) shell color, and the unbanded (BB or Bb) and banded (bb) morphs. The observed values, given in Table 10-2, allow us to calculate r, s, t, and u:

$$u = \sqrt{z} = .559$$

$$t = \frac{-2u + \sqrt{4u^2 + 4w}}{2} = -u + \sqrt{z + w} = .084$$

$$s = -u + \sqrt{z + y} = .287$$

$$r = 1 - u - s - t = .070$$

so that $d = ru - st = +.015$. A second site in the Pyrenees, reported in the same article, yielded values of $x = .026$, $y = .380$, $w = .050$, and $z = .545$, whence $u = .738$, $t = .033$, $s = .224$, $r = .005$, and $d = -.004$. The positive value found in Ireland suggests an "excess" of AB or ab gametic types in the population, suggesting selective advantage of one or both of these forms. In the Pyrenees, the Ab or aB form may be at a selective advantage.

Table 10-2

	Shell Type			
	Pink Unbanded	***Pink Banded***	***Yellow Unbanded***	***Yellow Banded***
Morph frequency	.184	.403	.101	.313
Frequency in algebraic terms	$r^2 + 2rs + 2rt + 2ru + 2st = x$	$s^2 + 2su = y$	$t^2 + 2tu = w$	$u^2 = z$

The disequilibrium values above seem awfully small. But let us pursue a bit further the matter of their importance to the evolutionary process. Substituting d from Ireland into equations (10-7), we obtain

$$\Delta r = r\left(\frac{\bar{W}_1 - \bar{W}}{\bar{W}}\right) - .015c\frac{W_H}{\bar{W}} \quad \text{(for example)}$$

If, as Cain et al. (1960) have calculated, c between these loci, in this species, is about 2.25%, the expression above becomes, upon substitution of r,

$$\Delta r = (.070)\left(\frac{\bar{W}_1 - \bar{W}}{\bar{W}}\right) - (.0003375)\frac{W_H}{\bar{W}}$$

If the population is at genetic equilibrium (and observations of this species over time indicates extremely stable morph frequencies), we obtain

$$\bar{W}_1 = \bar{W} + .00482W_H$$

In similar manner, we can calculate

$$\begin{aligned} \bar{W}_2 &= \bar{W} - .00118W_H, \\ \bar{W}_3 &= \bar{W} - .00402W_H, \\ \bar{W}_4 &= \bar{W} + .00060W_H \end{aligned}$$

Therefore, in the presence of linkage disequilibrium we would have (from equation 10-7)

$$\begin{aligned} 0 = \Delta r &= \frac{r(.00482W_H)}{\bar{W}} - \frac{cdW_H}{\bar{W}} \\ 0 = \Delta s &= \frac{s(-.00118W_H)}{\bar{W}} + \frac{cdW_H}{\bar{W}} \\ 0 = \Delta t &= \frac{t(-.00402W_H)}{\bar{W}} + \frac{cdW_H}{\bar{W}} \\ 0 = \Delta u &= \frac{u(.00060W_H)}{\bar{W}} - \frac{cdW_H}{\bar{W}} \end{aligned} \qquad (10\text{-}10)$$

But if we designate by p_A, p_a, p_B, and p_b the frequencies of the A, a, B, and b alleles, then, were there *no* linkage disequilibrium, $r = p_A p_B$, $s = p_A p_b$, $t = p_a p_B$, $u = p_a p_b$, and $d = 0$, and substituting from (10-10), we have

$$\Delta p_A = p_B\,\Delta r + p_b\,\Delta s = p_A[.00482p_B^2 - .00118p_b^2]\frac{W_H}{\bar{W}} \qquad (10\text{-}11)$$

Similarly,

$$\Delta p_B = p_B[.00482p_A^2 - .00402p_a^2]\frac{W_H}{\bar{W}}$$

In addition to the trivial solutions (fixation), this system has an equilibrium at $p_A = .475$, $p_B = .331$. But in this case, a perturbation in p_A above .475 leads to a rise in p_B which, in turn, results in a further increase in p_A. If p_A once drops below .475, p_A falls and gives rise to a further drop in p_A. Perturbations in p_B have the same effect; this equilibrium is unstable. The only stable situation is fixation. If d were zero, given the present frequencies and fitnesses in the Irish population, we would have $p_A = r + s = .357$, $p_B = r + t = .154$, and

$$\Delta p_A = -(.357)(.000256)\frac{W_H}{\bar{W}} < 0$$

$$\Delta p_B = -(.154)(.000864)\frac{W_H}{\bar{W}} < 0$$

Given sufficient time, only the *ab* gametic type should remain—a far cry from reality!

Not only may linkage disequilibrium significantly alter the course of natural selection, but change in the course of natural selection may affect d. Nevo et al. (1977) looked at allozyme polymorphisms in the barnacle *Balanus amphitrite* in cool water, power plant intake canals, and warm outfalls. Barnacle larvae were allowed to settle on glass plates and then brought into the laboratory for analysis. A depression of heterozygote frequencies in the intake canals probably reflected a diverse source of larvae (Wahlund effect). But the further drop in heterozygotes ($p < .001$) between the intake and outflow canals and the change in frequency of some of the alleles examined could only have been due to selection differences between the two habitats. Linkage disequilibrium was small in the intake canal population, very large in the outflow population.

In spite of the comments and calculations above there are reasons for downplaying the significance of linkage disequilibrium as a distorter of the evolutionary process. First, in many cases, natural populations seem to be very close to linkage equilibrium. Muki et al. (1977), for example, checked the second and third chromosomes of individual *Drosophila melanogaster* in a large population and found no evidence for linkage disequilibrium at all. Then Nei and Li (1973) showed that d is affected in an encouraging manner by subdivision of populations. Noting that

$$d = ru - st$$

$$p_A = r + s \qquad p_B = r + t$$

$$p_a = t + u \qquad p_b = s + u$$

it follows that, over subpopulations,

$$\bar{d} = \overline{ru} - \overline{st} = \bar{r}\bar{u} + \text{cov}(r, u) - \bar{s}\bar{t} - \text{cov}(s, t)$$

and that

$$\text{cov}(p_A, p_B) = \text{cov}(r + s, r + t) = -\text{cov}(r, u) + \text{cov}(s, t)$$

Thus

$$\bar{d} = (\bar{r}\bar{u} - \bar{s}\bar{t}) - \text{cov}(p_A, p_B) \tag{10-12}$$

Most field measurements will be based on population-wide gene frequencies, and so will actually describe only the $\bar{r}\bar{u} - \bar{s}\bar{t}$ term. But it is $\bar{d}$ that is the important quantity. Note now that if $\bar{r}\bar{u} - \bar{s}\bar{t} > 0$, then A and B tend to be associated (and also a and b) more than expected if the alleles assorted independently. Therefore, $\text{cov}(p_A, p_B)$ will probably be positive. If $\bar{r}\bar{u} - \bar{s}\bar{t} < 0$, the opposite is apt to be true. Therefore, spatial structuring of populations at least partly counteracts local disequilibria. Finally, Kimura (1965a) (see also Crow and Kimura, 1970) discovered when two loci are linked and nonindependent in their effects on fitness, d rapidly approaches a value *close to zero* before its further progress is drastically slowed or stopped. Kimura termed this state of almost being at linkage equilibrium, *quasi-linkage equilibrium.*

To demonstrate Kimura's argument, we shall use a continuous model:

$$\frac{dr}{dt} = r(\bar{m}_1 - \bar{m}) - cdQ \tag{10-13}$$

where $\bar{m}_1 = rm_{11} + sm_{12} + tm_{13} + um_{14}$ and $Q = rum_{14} - stm_{23}$ (see equations 10-5 and 10-7), and define $z = ru/st$. When $z = 1$, d must equal zero and the population is in linkage equilibrium. Adopting the notation, $dx/dt = \dot{x}$, we have

$$\begin{aligned} \dot{z} &= \frac{st(r\dot{u} + \dot{r}u) - ru(s\dot{t} + t\dot{s})}{(st)^2} \\ &= z\left(\frac{\dot{r}}{r} + \frac{\dot{u}}{u} - \frac{\dot{s}}{s} - \frac{\dot{t}}{t}\right) \\ &= z(\bar{m}_1 + \bar{m}_4 - \bar{m}_2 - \bar{m}_3) - cdQz\left(\frac{1}{r} + \frac{1}{s} + \frac{1}{t} + \frac{1}{u}\right) \end{aligned}$$

Writing this

$$\dot{z} = zM - cdQzP$$

and noting that

$$d = (ru - st) = st\left(\frac{ru}{st} - 1\right) = st(z - 1)$$

we have

$$\dot{z} = zM - cQ(z - 1)ruP$$

But

$$ruP = \frac{rust}{st}P = \frac{ust}{st} + \frac{rut}{st} + \frac{rus}{st} + \frac{rst}{st}$$

$$= r + (u + s + t) + \frac{ru}{st}(s + t) + u - (u + s + t)$$

$$= 1 + (z - 1)(s + t)$$

so that, finally,

$$\dot{z} = zM - cQ(z - 1)[1 + (z - 1)(s + t)] \tag{10-14}$$

Now suppose that M is at its maximum value. This occurs when $\bar{m}_2$ and $\bar{m}_3$ are small and hence leads to $s, t \to 0$. Under this circumstance, z is also maximal. Equation (10-14) now becomes

$$\dot{z} \to zM_{\text{max}} - cQ(z - 1) = z(M_{\text{max}} - cQ) + cQ$$

whose solution is

$$z = \frac{cQ}{cQ - M_{\text{max}}}\left\{1 + \left[z_0\frac{cQ - M_{\text{max}}}{cQ} - 1\right]e^{-(cQ - M_{\text{max}})t}\right\}$$

Because z is maximal, and certainly positive, cQ must exceed M_{max}. Therefore, as time increases, z drops and approaches

$$z^* = \frac{cQ}{cQ - M_{\text{max}}} \tag{10-15}$$

Because z is very large when s and t are very small, the drop to this value is rapid. Now suppose that M is at its minimum value ($s + t \to 1.0$, $z \to 0$). Then (10-14) is

$$\dot{z} \to zM_{\text{min}} - cQ(z - 1)(1 + z - 1) = zM_{\text{min}} - cQz(z - 1)$$

whose solution, as time increases, approaches

$$z^{**} = \frac{M_{\text{min}} + cQ}{cQ} \tag{10-16}$$

The gist of the argument is that, regardless of its initial size, z rapidly approaches a value $\hat{z}$, such that

$$\left(z^{**} = \frac{M_{\text{min}} + cQ}{cQ}\right) \leqslant \hat{z} \leqslant \left(z^* = \frac{cQ}{cQ - M_{\text{max}}}\right)$$

or

$$1 + \frac{M_{\text{min}}}{cQ} \leqslant \hat{z} \leqslant \frac{1}{1 - M_{\text{max}}/cQ} \tag{10-17}$$

If selection is fairly slow, then $M = \bar{m}_1 + \bar{m}_4 - \bar{m}_2 - \bar{m}_3$ has only a small range about zero as gene frequencies change. Thus $M_{\min} \simeq M_{\max} \simeq 0$, and z lies very close to 1.0. Although quasi-linkage equilibrium, defined by (10-17), is not really a state of $d = 0$ (or $z = 1$), it is generally very close. Therefore, because quasi-linkage equilibrium is usually approached quite rapidly, the assumption that linked loci assort *almost* independently seems warranted.

The Evolution of Gene Complexes

Selection is more efficient at altering gene frequencies at several individual loci if those loci are not linked. However, suppose that individuals carrying an A gene or a B gene are, collectively, more fit than the population as a whole, but that a and b are the better alleles *when in combination with each other*. In this case absence of linkage would lead to the fixation of the A and B alleles, even though the ab genotype was the most fit. Here, the presence of an epistatic interaction would make tight linkage between a and b advantageous. Can natural selection modify the degree of linkage among loci? The answer is yes, by at least two mechanisms, although it is not obvious that the modifications wrought by selection are necessarily advantageous. First, chromosomes may occasionally break into three pieces and then rejoin. Often, the middle piece gets turned around and its ends reattach to the wrong end segments (Fig. 10-2). The reversed midsegment is called an *inversion*. Consider two genes within an inversion and examine what happens during meiosis if crossing over takes place. To achieve synapsis, a normal chromosome and one containing the inversion must align themselves as shown in Figure 10-3a. Now if crossing over occurs between locus B and C, the daughter chromosomes (Fig. 10-3b and c) are incomplete and, therefore, usually lethal. Gametes formed from such chromosomes never find their way into the genomes of the next generation. If two normal or two inversion chromosomes come together, crossing over is not lethal, but it also results in no change in the resultant gametes. Effectively, then, the B and C loci are tightly linked. In fact, all genes within an inversion act as if they assorted as a single locus. We may assign the inversion a name (x, say), the normal chromosome another (Q), and examine natural selection acting on the QQ, Qq, and qq "genotypes" in the manner described in Chapters 8 and 9. If chromosomes are inclined to break at specific points, then inversions involving the same, or almost the same, chromosome segment may originate more than once. If the allelic configurations carried are different, we may find several forms of the same inversion—analogous to multiple alleles at a

Figure 10-2 Schematic representation of a (part of a) chromosome showing four gene loci.

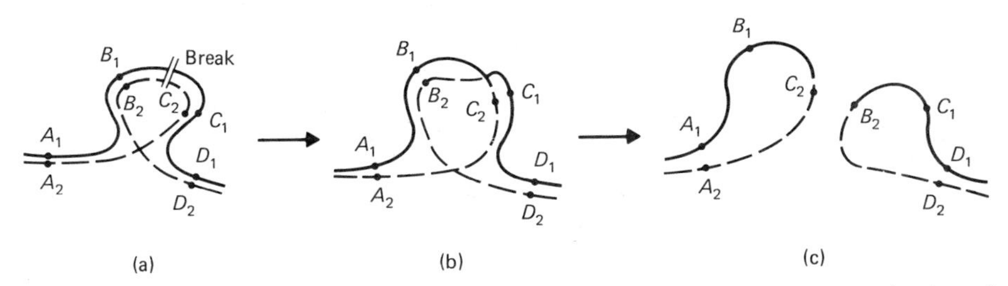

Figure 10-3 Effect of a chromosomal inversion, with crossing over, on the production of gametes.

single locus. The appearance of inversions permits selection to act on gene combinations rather than individual loci.

Linkage may also be affected by the presence of epistatic genes which encourage chromosome breakage and *deletions*. If pieces of the chromosome lying between two loci, B and C, are deleted without lethal effect the two genes will be closer together. Therefore, the chance that another break will occur between them, leading to the possibility of crossing over, is diminished. Linkage is tightened. Conversely, an epistatic gene may promote breakage specifically within a chromosomal region between the B and C loci and thereby enhance the possibility of crossing over. Whether and how such genes might evolve, though, is problematical. Wills and Miller (1976) made a computer model of a system of multiple linked loci with weak heterosis (so that independent selection at each locus would produce polymorphism). Perturbing the system from equilibrium, they simulated the subsequent changes in genotype frequency. Tight linkage resulted in the loss of heterozygosity at a faster rate than loose linkage, whereas no linkage at all led to the original stable state of polymorphism at all loci. Tight linkage, by allowing the loss of beneficial heterozygosity, was clearly disadvantageous. This observation, however, does not guarantee the spread of genes encouraging higher crossing-over rates. Charlesworth and Charlesworth (1979) simulated a system essentially the same as that of Wills and Miller, except that they added a locus, unlinked to the others, whose alleles coded for different degrees of linkage. Almost invariably the allele promoting *decreased* crossing over increased in frequency. This result is in agreement with the predictions of a number of earlier workers, dating all the way back to the first edition, in 1930, of Fisher's *The Genetical Theory of Natural Selection* (Fisher, 1958). On the other hand, if the environment is not constant, it appears that increased recombination *can* be selected for. This arises from the likelihood that the combinations of alleles favored by natural selection is changing in time (Charlesworth, 1976).

What evidence is there for gene loci mediating recombination rates by crossing over? Moriwaki (1940) demonstrated the existence of a dominant gene on the second chromosome of *Drosophila ananassae* that promotes recombination between almost every pair of loci on the same chromosome. Several workers, selecting artificially for high or low crossing-over rates, have had their efforts meet with success (see Nei, 1967, for a review and references). A particularly nice study of this sort is that of Chinnici (1971a), who selected for increased or decreased crossing over between the gene loci "scute" (sc) and "crossveinless" (cv) in *Drosophila melanogaster*. After 33 generations, beginning with a value of $c = .154$, he obtained populations with $c = .221$ (positive-selected line) and $c = .085$ (negative-selected line). Unlike the case described by Moriwaki, the selection response here was specific to the chromosomal region containing the sc and cv genes. Chinnici then examined the polytene chromosomes to check if his results might be due to the spread of an inversion or a deletion. Not so. Finally, Chinnici (1971b) took the experimental populations and backcrossed them to stock flies, keeping only the progeny that carried the (say) kth chromosome in identical form to that of the experimental line. In this way he produced a line identical to the experimental flies with respect to the kth chromosome, but with the remaining chromosomes a mix of those of the original stock population. This was done, one at a time, for all chromosomes. Testing the derived lines for crossing-over frequency showed that the experimental results, cited above, were due to influences eminating from *all* chromosomes. In other words, the modification of crossing over which he had achieved was polychromosomal and thus polygenic in nature.

In the field it is difficult to find clear evidence for the evolution of tendencies toward greater or less crossing over. However, different populations of the same species may exhibit striking differences in crossing-over rates, as detected by counting chiasma frequencies in germ cells. Price and Bantock (1975), for example, looking at the snail, *Cepaea nemoralis*, in marginal habitats in England, found much higher chiasma frequencies in low- than in high-density populations. They speculated that sparse populations experience more inbreeding and that the selective advantage of rapid approach to linkage equilibrium therefore puts greater importance on a high value of c (see equation 10-4; but see also the discussion above pertaining to the evolution of tighter linkage).

If the looseness or tightness of linkage can really evolve in an adaptive fashion, and if genes even on different chromosomes may be mutually epistatic, it seems reasonable to suppose that any population will gradually evolve interactive clusters of genes, some of whose members assort together as a unit, and others whose epistatic influence on this unit is adaptive, appropriately turned on or off, according to the genetic nature of this unit. We speak of *coadapted gene complexes* and it is not difficult to find examples. In a variety of insect species particularly the Lepidoptera, some species, quite palatable to their predators, appear to gain some protection by mimicking noxious-tasting species. Often the mimetic species is polymorphic, mimicking one species, or another, or none at all. The quality of the mimetic resemblance depends on attention to a variety of details, morphological and sometimes behavioral, which must differ from one mimetic morph to another. Thus there must exist whole constellations of coadapted genes one of which is turned on if the individual mimics one particular species, another when another species is the model. In nonmimetic morphs, these coadapted complexes presumably are turned off and another gene complex, appropriate to the nonmimetic life-style takes over. Sheppard (1960) and others hypothesize a *switch gene*, which determines the morph by turning on or off the appropriate gene complex.

Suppose that two populations, isolated by structure or distance, have evolved up different adaptive peaks. Each population will be characterized by an appropriate set of coadapted gene complexes. If these two populations were to be brought together and allowed to interbreed, either of two things might happen. First, the closely linked portions of the coadapted complexes would breed true, but the mixing of epistatic genes might well result in genetic chaos and a drop in population fitness. The consequence might be the evolution of reproductive barriers: Those individuals that bred with like individuals—with the same genetic origins—would still occupy their adaptive peak, whereas those with more catholic tastes would produce inferior offspring, positioned at some intermediate position between the peaks on the adaptive surface. Any gene predisposing individuals to avoid breeding with unlike individuals would be at a selective advantage. An alternative consequence of interbreeding would be the evolution of switch genes which turned on the gene complex contributed by one parent, and turned off the other. If this occurred, there would not be a selective disadvantage to cross breeding and a gradual redesigning of both gene complexes would ensue, utilizing the additional genetic material contributed by both parents. Our population would be in a position to occupy either adaptive peak, which one depending on the genotype at one (or perhaps a few) switch gene loci. Either consequence would lead us to expect polymorphism or, alternatively, sudden shifts from one adaptive peak to another in space. The latter, known as *area effects*, might also arise as a result of isolation by distance in a continuously distributed population where ancestral subpopulations at different points in space found themselves on the slopes of different adaptive peaks quite by accident (genetic drift) or due to subtle changes along an environmental gradient and subsequently climbed these respective peaks (for a discussion of area effects, see Wright, 1965). What appears to be an example of a very simple area effect involving two loci is reported by Wolda (1969) for the garden snail, *Cepaea nemoralis*. Censusing the snail population along a roadside in Holland in 1956, Wolda noticed a point at which an abrupt change occurred in the frequencies of the color and banded morphs. This change occurred in the conspicuous absence of any measurable, correlated change in soil, vegetation, or microclimate. Returning 12 years later, in 1968, Wolda found the same sudden transition, shifted in space a few meters.

In the face of the complexities discussed so far in this chapter, what is the role of the rather simple models of gene frequency change presented in Chapters 8 and 9? First, in spite of the existence of alternative adaptive peaks, we have seen that natural selection still increases fitness, causing the population to climb the peak defining *optimal adaptation* for the present set of environmental conditions and genome composition. Thus selection, in a more restrictive sense, still maximizes fitness locally, even though individual gene frequency changes may be difficult to predict. Second, we have seen that epistasis and linkage effectively cancel each other's complicating effects as a population rapidly approaches quasi-linkage equilibrium. Thus in many cases, at least where selection is slow, even the future course of individual gene loci may be closely approximated with the simple expressions of the preceding chapters. Finally, even where the above reassurances will not work, where the genes at a locus have minute effects on a character it is valid to use single-locus models to check whether under present circumstances a new mutant altering the status quo will be favored by selection. In other words, the one-locus models are useful for testing the neighborhood stability of existing genetic equilibria (Karlin and MacGregor, 1974).

10.2 The Evolution of Polygenic Traits

We have stated a general conclusion that even in the presence of epistasis, pleiotropy, and linkage, natural selection should cause populations to climb adaptive peaks. But how rapidly is this done; and how can we predict, quantitatively, the rate of change in a given character comprising a part of the overall adaptive complex of traits?

We begin as follows. First let x_{ij} be the value of the trait in question in the jth individual with genotype i at some locus. Next, we write

$$x_{ij} = \bar{x} + (x_{ij} - \bar{x}_i) + (\bar{x}_i - \bar{x})$$

As in Chapter 8, the first term, the deviation of x_{ij} from the mean value for genotype i, can be broken down into an environmental component E, a gene-environment interaction component I_{GE}, and a random component, ϵ. Next, as in Chapter 8, we note that $(\bar{x}_i - \bar{x})$ is comprised of a genetic value, G so that

$$x = \bar{x} + G + I_{GE} + E + \epsilon \tag{10-18}$$

Recall also that the genetic value at any one locus can be divided into an additive, A, and dominance, D, parts. But as used here to explore polygenic traits, genotype i is a composite of many loci, at least some of which are likely to be epistatic. That is, the genetic value must be the sum of the additive values over all loci, plus the sum of the correction factors accounting for allelic interactions (dominance) at these loci, *plus still another correction factor* accounting for epistatic interactions among the loci. We write

$$G = A + D + I_G$$

where I_G is the epistatic correction term. Finally, then

$$x = x + A + D + I_G + I_{GE} + E + \epsilon \tag{10-19}$$

where each term is a sum of effects over all loci defining genotype i.

Now lump terms in (10-19) to read

$$x = \bar{x} + A_m + (D + I_G + I_{GE} + E) + \left(\sum_{k \neq m} A_k + \epsilon\right) \tag{10-20}$$

where A_m is the additive value at the mth locus. Because all the terms in parentheses have means of zero, this can also be written

$$x = \bar{x} + A_m + \epsilon' \tag{10-21}$$

where ϵ' is a new "error" term also with mean zero. If A_m is independent of the other terms in 10-20, and so ϵ', and if, given this independence, A_m is identical in form to the A of (8-19a),

$$\alpha_A^{(x)} = q\alpha^{(x)} \qquad \alpha_a^{(x)} = -p\alpha^{(x)} \tag{10-22}$$

then (10-21) is equivalent to the single-locus case (equation 8-16) and we may use (8-23) to obtain

$$\Delta\bar{x}(m) = \frac{1}{\bar{W}}\frac{d\bar{W}}{d\bar{x}} V_A(X(m)),$$

where $\Delta\bar{x}(m)$ is the change in x due to genetic change at the mth locus and $V_A(x(m))$ is the additive genetic variance in x due to genotypic variation at the mth locus. Furthermore, because the A_m values are independent, by definition, the total additive variance is simply the sum of the additive variances at all loci, thus

$$\Delta\bar{x} = \sum_m \Delta\bar{x}(m) = \frac{1}{\bar{W}}\frac{d\bar{W}}{dx}\sum_m V_A(x(m)) \tag{10-23}$$

But how can we be assured that the A_m values are as in (10-22)? To show that they are, use (10-21) (i.e. *define* A_m to be independent of ϵ') and write, for any locus,

$$x_{AA}(= x_1) = \bar{x} + 2\alpha_A^{(x)} + \epsilon'$$
$$x_{Aa}(= x_2) = \bar{x} + \alpha_A^{(x)} + \alpha_a^{(x)} + \epsilon'$$
$$x_{aa}(= x_3) = \bar{x} + 2\alpha_a^{(x)} + \epsilon'$$

Then use the least-squares technique to find the values of $\alpha_A^{(x)}$ and $\alpha_a^{(x)}$ which best fit (10-21). That is, we define $\alpha_A^{(x)}$ and $\alpha_a^{(x)}$ such that independence is realized. If the resultant definitions are those in (10-22), we may proceed to use (10-23) to predict evolutionary changes in polygenic traits. The sum of errors squared is

$$\psi = p^2(x_1 - \bar{x} - 2\alpha_A^{(x)})^2 + 2pq(x_2 - \bar{x} - \alpha_A^{(x)} - \alpha_a^{(x)})^2 + q^2(x_3 - \bar{x} - 2\alpha_a^{(x)})^2$$

Differentiating with respect to $\alpha_A^{(x)}$ and $\alpha_a^{(x)}$, setting the derivatives equal to zero, and solving, we obtain

$$\alpha_A^{(x)} = q\alpha^{(x)} \qquad \text{where } \alpha^{(x)} = p(x_1 - x_2) + q(x_2 - x_3)$$
$$\alpha_a^{(x)} = -p\alpha^{(x)}$$

Equation (10-23) is therefore valid. We shall now write the sum of the additive variances,

$$V_g(x) = \sum_i V_A(x_i)$$

so that

$$\Delta\bar{x} = \frac{1}{\bar{W}}\frac{d\bar{W}}{d\bar{x}} V_g(x)$$

Equivalently,

$$\Delta\bar{x} = \frac{1}{\bar{W}}\frac{d\bar{W}}{d\bar{x}} h_x^2 \operatorname{var}(x) \tag{10-24}$$

The Determination of Additive Variance

A_m has been defined so as to be independent of $\sum_{k \neq m} A_K + C'$, where C' includes the dominance, epistatic, environmental, and genotype–environmental terms. Therefore, *if the environment is controlled to be the same for all genotypes and for parents and their offspring*, and *if the genetic changes from one generation to the next do not alter the epistatic interaction term*, we may find additive variance in x, at locus i, by comparing offspring and parent as we did in Chapter 8:

genetic value of average offspring $(G_{\bar{o}i}) = \frac{1}{2}$ additive value of one parent (A_{pi})

so that

$$\operatorname{cov}(x_{\bar{o}i}, x_{pi}) = \operatorname{cov}(G_{\bar{o}i}, A_{pi}) = \operatorname{cov}(\tfrac{1}{2}A_{pi}, A_{pi}) = \tfrac{1}{2}V_A(x_i)$$

But then

$$V_g(x) = \sum_i V_A(x_i) = 2\sum_i \operatorname{cov}(G_{\bar{o}i}, A_{pi})$$
$$= 2\operatorname{cov}(x_{\bar{o}}, x_p) \qquad (10\text{-}25)$$

Additive and dominance variance values can also be gleaned from various other kin relationships. The following is a partial list from Hartl (1980).

$$\begin{aligned} \operatorname{cov}(x_{\bar{o}}, x_p) &= \tfrac{1}{2}V_g(x) \\ \operatorname{cov}(x \text{ of Half sibs}) &= \tfrac{1}{4}V_g(x) \\ \operatorname{cov}(x \text{ of Full sibs}) &= \tfrac{1}{2}V_g(x) + \tfrac{1}{4}V_d(x) \\ \operatorname{cov}(\text{Monozygotic twins}) &= V_g(x) + V_d(x) \end{aligned} \qquad (10\text{-}26)$$

The dominance variance, $V_d(x)$, is defined as follows. Writing the variance in total genetic value, we sum over all loci to obtain

$$\begin{aligned} \operatorname{Var}(G) &= \sum p^2(\bar{x}_1 - \bar{x})^2 + 2pq(\bar{x}_2 - \bar{x})^2 + q^2(\bar{x}_3 - \bar{x})^2 \\ &= \sum [2pq\alpha_x^2 + p^2q^2k_x^2] \\ &= \sum 2pq\alpha_x^2 + \sum p^2q^2k_x^2 = \sum V_A(x) + \sum V_D(x) \\ &= V_g(x) + V_d(x) \end{aligned} \qquad (10\text{-}27)$$

where $\alpha_x = p(\bar{x}_1 - \bar{x}_2) + q(\bar{x}_2 - \bar{x}_3)$ and $k_x = 2\bar{x}_2 - \bar{x}_1 - \bar{x}_3$.

By using expressions (10-26) it is possible to estimate both additive and dominance genetic variance. Subtracted from the total genotypic variance, the sum of these two values gives the epistatic term. Gathering of such information under a variety of environmental conditions also allows us to calculate the E and I_{GE} terms. However, we must proceed with caution. Because we cannot be sure that I_G is constant over generations, and because we have ignored linkage and possible frequency-dependent selection, the relationships above cannot be relied upon as accurate. It is best to use several such covariance calculations and figure the true variances to be some compromise of the results. A *realized heritability* can be measured directly from actual selection experiments, making use of (10-24).

The Effect of Selection on Additive Variance

Write

$$\Delta V_g(x) \simeq \sum \frac{dV_g(x)}{dp}\Delta p \qquad \text{(summation is over all loci)}$$

$$= \sum \frac{d}{dp} 2pq\alpha_x^2 \frac{dp}{dt} = 2\sum [(q - p)\alpha_x^2 + 2pq\alpha_x k_x] \frac{dp}{dt}$$

where

$$k_x = 2\left(x_2 - \frac{x_1 + x_3}{2}\right)$$

as before. Let us designate A as the favored allele. Then if there is no dominance, $k_x = 0$, $dp/dt > 0$, and as p increases above .5, $V_g(x)$ drops and continues to drop until it reaches zero. If A is dominant, $k_x > 0$, and $V_g(x)$ drops sooner and faster. If A is recessive, $k_x < 0$, and $V_g(x)$ may rise until p is considerably above .5. Eventually, in this case also, $V_g(x)$ falls to zero. Additive variance never quite reaches zero in the real world only because of temporal or spatial structuring of populations, or the appearance of new genes by mutation or new genetic recombinations.

Relationships of Selection History and Genetic Architecture

We are now in a position to show how genetic architecture, the relative sizes of the components of x (A, D, I_G, etc.), and their variances and covariances may tell us something of the evolutionary history of a population.

1. We have seen that natural selection, although it may temporarily increase additive variance, eventually acts to eliminate it. If $d\bar{W}/d\bar{x}$ is large—that is, if the trait X is important to fitness—selection acts more rapidly, both to change $\bar{x}$ and to decrease $V_g(x)$. For a given rate of mutation and genetic recombination, then, and for a given degree of temporal and spatial heterogeneity, $V_g(x)$ should come to a lower equilibrium with respect to traits directly impacting on fitness (see Robertson, 1955). Falconer (1960a) lists several examples gleaned from the literature. For example, the heritability of butterfat content in cow milk is .6, while milk yield (important for feeding calves) has an h^2 of .3 and conception rate (directly connected to fitness) shows $h^2 = .01$. In pigs h^2 is .55 for back-fat thickness, .5 for body length, but only .15 for litter size.

2. If selection, rather than being stabilizing (selecting against extremes), is directional, genes whose recessive forms are favored should increase in frequency faster than those that are dominant for the favored extreme. The result is that dominant alleles reach fixation at a lower rate than recessive alleles. But if loci with dominance in the direction of selection do not go to fixation as rapidly, there should be proportionately more unfixed loci dominant for the favored extreme than for the disfavored extreme. This *directional dominance* can be seen with the use of simple crossing experiments. If (say) large x has been favored by selection, then individuals with large x values are more apt to be heterozygous than individuals with small x values, and those with small x are more likely to be homozygous for the recessive alleles. Crosses among the larger x individuals, then, should result in a wide variety of genotypes and so a large variance in x. This will not be so for crosses between small x individuals. Also, the similarity of homozygous and heterozygous individuals and the expected high frequency of the selected, dominant alleles mean that there will be more individuals occupying the larger half of the range in x than the lower half. This results in a skewed distribution of x in the natural population (Fig. 10-4). The discovery of a strong skew in x, or greater variance in x among the offspring of crosses of large x as opposed to small x individuals (or vice versa), may be evidence for past directional selection. Clearly directional dominance might affect the success with which a trait can be pushed in one direction or the other by selection (see Falconer, 1954; Bohren et al., 1966).

3. In the absence of dominance, $V_d(x)$ is zero. Thus the appearance of a large dominance variance term indicates strong dominance effects. If dominance is *not* directional, we should suspect *overdominance* (that is, $x_2 > x_1, x_3$ or $x_2 < x_1, x_3$), and a selective advantage for the x_2 genotype (i.e., heterosis). Kearsey and Kojima (1967) found a large additive, small dominance variance for body weight in *Drosophila melanogaster*, but the opposite for egg hatchability. This suggests (see also point 1) that body weight is not terribly important to fitness, that an intermediate value is favored, and that egg hatchability is important and has a strong heterotic component.

4. Directional selection, in addition to generating directional dominance, might be expected to favor modifier genes which make heterozygotes as similar as possible to the favored homozygote. This is Fisher's (1958) proposed mechanism for the evolution of dominance,

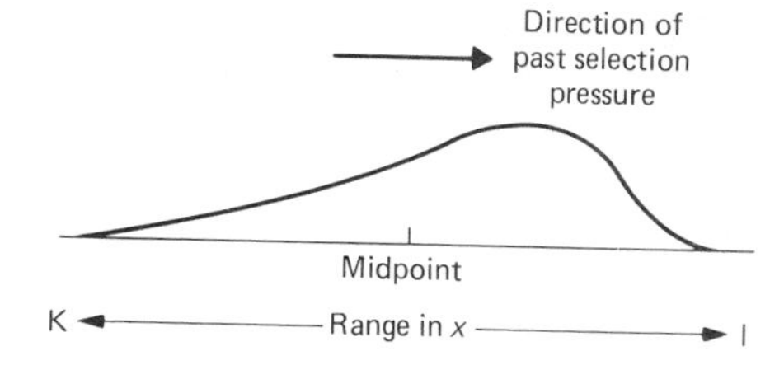

Figure 10-4 Skewed distribution in trait value over a population suggesting directional selection.

mentioned earlier. The concept is controversial, but if true should lead in the present circumstance to increased epistasis. Thus the I_G term might be expected to be larger in populations undergoing directional as opposed to stabilizing selection. A demonstration that genetically identical populations might realize a higher I_G value under conditions of directional selection is provided by the study of Jinks et al. (1973) on the genetics of the plant *Nicotiana rustica* in California. Jinks and his coworkers performed crosses and discovered evidence that epistatic effects were weakest in the plants' normal habitat and strongest in marginal habitats (where we might expect directional selection toward new adaptive strategies).

In the case of Jinks's plants, the *same strain* of organism appears to have had a different genetic architecture, depending on the environmental conditions. Differences in genetic architecture may also occur between the sexes. Holmes et al. (1974) studied the inheritance of avoidance behavior in laboratory mice (*Mus musculus*). The mice were given an unconditioned stimulus (electric shock) and a conditioned stimulus (sound); the behavioral response whose inheritance was studied was a tendency to flee across a line without returning for at least 3 seconds. In the males a strong $V_d(x)$ component was noted, indicating overdominance at most loci; in the females the $V_d(x)$ term was small. Hay (1973) investigated the genetics of preening behavior in *Drosophila melanogaster*. A strong, directional dominance was observed, but the direction was one way for flies raised in tubes, the other way, *for the same strain of flies*, when raised in dishes. These results raise new questions about the relationship of the genetic system to the environment, questions whose implications for evolution by natural selection is not at all clear.

Before leaving the topics of components of genetic value and genetic architecture, it should be pointed out that the work of Jinks et al., Holmes et al., and Hay as well as many other genetic studies have not involved the used of equations of the sort given in (10-26). Those expressions were designed for use with noninbred individuals, randomly mated. There is another school of quantitative genetics which makes use of *diallel crosses*, carefully designed crosses among individuals of a large variety of inbred lines (see Hayman, 1954, and Griffing, 1956, for a discussion of the techniques). The practitioners of this method are attracted by its power to dissect quite accurately and in minute detail the genetic components of a trait; the followers of the other approach, described in this text, are inclined to believe that results obtained with highly inbred strains are unlikely to be meaningful in the real world. To my knowledge the only studies designed to compare results of the two techniques are those of Lynch and Sulzbach (in press). Using diallel crosses of inbred strains as well as parent–offspring regressions from a genetically mixed random-bred population to assess the genetic architecture of thermoregulatory traits in house mice, they obtained the results (given in Table 10-3). The numbers are highly encouraging. They suggest that the critical, heritability values obtained in random-bred populations and values obtained from inbred descendant lines of such populations may be very similar. Regression methods have the advantage of being applicable, at least, crudely, in the field (e.g., Perrins and Jones, 1974; Boag and Grant, 1978, 1981; Smith and Zach, 1979; Smith and Dhondt, 1980; Garnett, 1981). The diallel technique is more difficult, but more powerful.

The Influence of Inbreeding and Assortative Mating

Recall from Chapter 7 that both inbreeding, arising either from finiteness of demes or from consanguineous mating, and assortative mating affect genotype frequencies in similar manner. The effect of such genotype frequency distortion on the process of natural selection can be seen by writing (from equation 8-21)

$$\Delta\bar{x} = \sum_m \Delta\bar{x}_m = \sum_m \frac{2p_m q_m \alpha^{x(m)}}{\bar{W}} (\alpha^{(W(m))} + F^{(m)}\beta^{(W(m))})$$

where p_m and q_m are gene frequencies, $F^{(m)}$ the inbreeding coefficient at the mth locus, and $\alpha^{(x(m))}$ and $\alpha^{(W(m))}$ are as defined in (8-13). The value of $\bar{W}$ is also affected by the F values, but is far

Table 10-3

Trait	Heritability Estimate Using:	
	Parent–Offspring Regression	*Diallel Crosses*
Female body weight at 100 days, following 50 days of cold acclimation	.43	.48
Male body weight at 50 days	.19	.20
Amount of cotton used in nest		
Males	.32	.23
Females	.31	.32
Amount of brown fat	.08	.09

more responsive to density feedback and adjusts to 1.0 at population equilibrium. Thus in the following analysis $\bar{W}$ will be treated as if independent of $F^{(m)}$. We now differentiate the expression above with respect to $F^{(m)}$, obtaining

$$\frac{\partial \Delta \bar{x}}{\partial F^{(m)}} = \frac{2p_m q_m \alpha^{(x(m))} \beta^{(W(m))}}{\bar{W}} \tag{10-28}$$

Now consider a case of complete dominance,

$$x_1 = x_2 \neq x_3$$

If selection favors the dominant phenotype, then

$$W_1 = W_2 > W_3$$

and the right side of (10-28) becomes (subscript m deleted)

$$\frac{2p^2q^2}{\bar{W}}(W_2 - W_3)(x_2 - x_3) \tag{10-29}$$

If $x_2 > x_3$, then $\Delta\bar{x}$ is positive and (10-29) is positive. If $x_2 < x_3$, then $\Delta\bar{x}$ is negative and (10-29) is negative. Thus inbreeding and positive assortative mating speed the process of selection for or against dominant traits. If selection favors the recessive, the results are reversed—inbreeding and positive assortative mating slow the spread or elimination of recessive traits by natural selection. With respect to polygenic traits, the effect of inbreeding and positive assortative mating on the evolution of a trait will depend on the distribution of F values across all the loci involved, and on the degree and direction of dominance for all those loci.

The mean value of a polygenic trait, for any given gene frequency configuration, is also affected by inbreeding and assortative mating. For some locus, write

$$\begin{aligned}(\bar{x} - \bar{x}_0) &= (p^2 + pqF)(x_1 - \bar{x}_0) + 2pq(1 - F)(x_2 - \bar{x}_0) + (q^2 + pqF)(x_3 - \bar{x}_0) \\ &= pqF[(x_1 - \bar{x}_0) - 2(x_2 - \bar{x}_0) + (x_3 - \bar{x}_0)]\end{aligned} \tag{10-30}$$

where $\bar{x}_0$ is the trait mean in the absence of inbreeding. If the A gene is dominant, so that $x_1 = x_2 \neq x_3$, the right side of (10-30) becomes

$$pqF[(x_3 - \bar{x}_0) - (x_2 - \bar{x}_0)]$$

If the dominant gene is the more fit (usually true), this quantity is negative when $(x_2 - x_0) > (x_3 - x_0)$, positive if $(x_2 - x_0) < (x_3 - x_0)$, and in either case $\bar{x}$ is displaced away from $\bar{x}_0$ in the direction that decreases fitness. For further information on the importance of inbreeding and assortative mating in population genetic processes, see Kempthorne (1973).

Alternative Approaches to the Study of Quantitative Traits

In addition to the method described above, there are several approaches to the study of evolution of quantitative traits. Suppose we assume that rather than one or two allelic types at every locus, there are a very large number—effectively an infinite number (see Lande, 1975, and references cited therein for a defense of this thesis). Where this is the case, the distribution of a character determined by the genotype at several, or even just one, locus will be continuous. Furthermore, if the allelic effects are additive both within and among loci, and numerous, then by the central limit theorem (a statistical cure all), the trait, or some transform of it, should be approximately normally distributed. Characters with normally distributed values fit certain assumptions that make their evolution fairly easy to describe. This particular approach has recently been championed by Lande (see references below). Another approach is to suppose the existence of a given distribution in the trait of interest and to ask whether a mutation at a modifier locus affecting that distribution will be selected for. Equilibrium distribution of the trait can be defined as that which admits no new mutant forms (at the hypothetical modifier locus) with fitness higher than $\bar{W}$ in the absence of the form. This approach differs from that of Lande in specifically admitting an epistatic gene. We treat the first of these approaches in this section. We shall return to the second in Section 10.3.

Suppose that the probability density of a trait, X, given by $\phi(x)$, is normal:

$$\phi(x) = \frac{1}{\sqrt{2\pi}\,\sigma_x} \exp\left[-\frac{(x - \bar{x})^2}{2\sigma_x^2}\right]$$

Suppose, in addition, that selection is stabilizing and that fitness is a *Gaussian* function of x:

$$W(x) = k \exp\left[-\frac{(x - x^*)^2}{2w^2}\right] \tag{5-31}$$

where x^* is the "optimal" value of x—that value which maximizes fitness for the given environmental situation, and w^2 is a measure of eurytopy. A large w^2 means the population has reasonably high fitness over a broad range of x values. Thus $1/w^2$ is a measure of the intensity of stabilizing selection. The value k is where density feedback is reflected. In a persistent, nongrowing population, k must take on whatever value makes $W = 1.0$. Because $\phi(x)$ is normal about a mean of $\bar{x}$ and with variance σ_x^2, we have

$$\bar{W} = \int_{-\infty}^{\infty} W(x)\phi(x)\,dx = \int_{-\infty}^{\infty} \frac{k}{\sqrt{2\pi}\,\sigma_x} \exp\left[-\frac{(x - \bar{x})^2}{2\sigma_x^2}\right] \exp\left[-\frac{(x - x^*)^2}{2w^2dx}\right]$$

$$= \frac{kw \exp\left[-\dfrac{(\bar{x} - x^*)^2}{2(w^2 + \sigma_x^2)}\right]}{\sqrt{w^2 + \sigma_x^2}} \tag{10-32}$$

To apply (10-32) for finding $\Delta\bar{x}$, we first calculate

$$\frac{1}{\bar{W}}\frac{d\bar{W}}{d\bar{x}} = -\frac{(\bar{x} - x^*)}{\sigma_x^2 + w^2}$$

Then

$$\Delta\bar{x} = \frac{1}{\bar{W}}\frac{d\bar{W}}{d\bar{x}} h_x^2\,\sigma_x^2 = -(\bar{x} - x^*)h_x^2 \frac{\sigma^2}{\sigma_x^2 + w^2} \tag{10-33}$$

The expression is independent of density feedback; it shows that selection always pushes $\bar{x}$ toward the optimal value (x^*), and that evolutionary progress is fastest if h_x^2 is large and if fitness falls off rapidly on either side of x^*. This expression was independently derived by Latter (1970) and by Lande (1975). Both authors have argued that the Gaussian fitness function is reasonable and that traits affected by many independently contributing loci should be normally distributed. Even if the allelic effects are *not* independent, because of dominance and epistasis, there is no reason, in traits under stabilizing selection, for the direction of the interactions to be biased in one way or another. Therefore, the nonindependent (nonadditive) effects may be expected to at least approximately cancel out, and (10-33) should be a reasonably accurate depiction of evolution in the real world. We shall return to this approach in Sections 10.3 and 10.4.

Phenotypic Selection Models

In 1978, Deniston published a paper entitled "An incorrect definition of fitness revisited." In this article Deniston reiterated the distinction between fitness (W or m) and the population growth parameter (R or r). We can write

$$\text{fitness of genotype } ij = m_{ij} = B_{ij} - d_{ij} = \text{number of births } by \text{ genotype } ij - \text{death rate of genotype } ij$$

$$\text{population growth rate of genotype } ij = r_{ij} = b_{ij} - d_{ij} = \text{number of births } of \text{ genotype } ij - \text{death rate of genotype } ij$$

B_{ij} will not generally equal b_{ij} because not all births *by* ij will result in ij offspring (because of recombination), while some of the births *by* other genotypes *will* be ij offspring. There have been attempts by various authors to link m and r mathematically; the expressions are cumbersome and will not be discussed here (see, e.g., Slatkin, 1970). A much simpler, and perhaps more useful way to relate them is as follows (after J. M. Emlen, 1980). Write

$$\bar{x} = \int x\phi(x)\,dx \tag{10-34}$$

where $\phi(x)$ is again the probability density of x over the population. If we let $n(x)\,dx$ be the number of individuals in the population with trait value in the range $x, x + dx$, and let $r(x)$ be the population growth rate of such individuals, then, by definition,

$$\phi(x) = \frac{n(x)\,dx}{n} \tag{10-35}$$

and

$$\begin{aligned}\frac{d\bar{x}}{dt} &= \frac{d}{dt}\int x\frac{n(x)}{n}\,dx = \int x\frac{n[dn(x)/dt] - n(x)\,(dn/dt)}{n^2}\,dx \\ &= \int x\left[\frac{r(x)n(x)}{n} - \frac{n(x)}{n^2}\bar{r}n\right]dx \\ &= \int x[r(x) - \bar{r}]\phi(x)\,dx \\ &= \operatorname{cov}(x, r)\end{aligned} \tag{10-36}$$

The rate of change in the mean trait value is exactly equal to the covariance of trait value and r. The discrete-time case, derived in similar manner, is

$$\Delta\bar{x} = \frac{1}{R}\operatorname{cov}(R(x), x) \tag{10-37}$$

Deniston (1978) derived the same expression but defined r (or R) to be the growth rate of *genotypes* with trait value x. He goes on to say that the "Mendelian shuffle is simply hidden in (the r) and no particular assumptions concerning number of loci, linkage or the mating scheme need be made...." It is "truly a general theorem about evolution, although one may pay a bitter price for such generality." The bitter price he had in mind was the disadvantage that if r were used instead of m, (10-36) would have to be recalculated continuously because r values are, by nature, not constant. We may challenge his phrase "bitter price", for although his approach and its extension to phenotypes (equations 10-36 and 10-37) does indeed make use of r (or R) values which change over time, the classic genetic models use m (or W) which are equally nonconstant. The bitter price would seem to apply equally to both kinds of model. In fact, the genetic models are, strictly speaking, *less* accurate, both over time and at any instant in time, because they fail to account, except in very complex additions and amendments to the equations of Chapters 8 and 9, as well as the present chapter, for differential mating success and fecundity among genotypes, and frequency-dependent selection. On the other hand, the phenotypic formulation must be used with extreme caution inasmuch as all of these nasty details are subsumed in not at all transparent fashion in the single variable r (or R). Equations (10-36) and (10-37) do have the advantage, though, of being precise (deterministically). To show the relationship between this phenotypic approach and the genetic approach derived earlier, and to address the question of the relationship between fitness and the population growth parameter, write

$$\Delta\bar{x} = \frac{1}{R}\operatorname{cov}(x, R) \simeq \frac{1}{R}\left(\frac{dR}{dx}\right)_{\bar{x}} \operatorname{var}(x) \tag{10-38}$$

This compares to (equation 10-24)

$$\Delta\bar{x} = \frac{h_x^2}{\bar{W}}\frac{d\bar{W}}{d\bar{x}} \operatorname{var}(x) \tag{10-39}$$

Therefore, R and W are related in the manner

$$h^2\left(\frac{1}{\bar{W}}\frac{d\bar{W}}{d\bar{x}}\right) = \left(\frac{1}{R}\frac{dR(\bar{x})}{d\bar{x}}\right) \qquad \text{where } R \equiv \bar{R} \tag{10-40}$$

Genetic Covariance

The equations derived so far implicitly assume that $\bar{x}$ is affected only by selection acting on the trait, X. But we must recognize that a change in the mean value of one trait, X_i, can be influenced, due to pleiotropic effects, by selection acting on other traits, X_j. If selection is slow, this piggyback effect is

$$\Delta\bar{x}_i \simeq \sum_j \frac{dA_i}{dA_j}\Delta\bar{x}_j$$

where A_i, and A_j are the heritable (additive) values of x_i and x_j. But the regression of the additive value of x_i on the additive value of x_j is

$$\frac{dA_i}{dA_j} = \frac{\operatorname{cov}_g(x_i, x_j)}{V_g(x_j)}$$

Thus

$$\Delta\bar{x}_i = \sum_j \frac{\operatorname{cov}_g(x_i, x_j)}{V_g(x_j)}\Delta\bar{x}_j \tag{10-41}$$

Substituting for $\Delta\bar{x}_j$ from (10-24) and collecting terms, we have

$$\Delta\bar{x}_i = \sum_j \frac{d \ln \bar{W}}{d\bar{x}_j} \operatorname{cov}_g(x_i, x_j) \tag{10-42}$$

Lande (1979) has put this into matrix form:

$$\Delta\mathbf{x} = \mathbf{G}\nabla \ln \bar{W} \tag{10-43}$$

where G is the additive genetic covariance matrix

$$\mathbf{G} = \begin{pmatrix} V_g(x_1) & \operatorname{cov}_g(x_1, x_2) & \cdot & \cdot \\ \operatorname{cov}_g(x_1, x_2) & V_g(x_2) & \cdot & \cdot \\ \cdot & \cdot & \cdot & \cdot \end{pmatrix}$$

and ∇ is the operator vector

$$\begin{pmatrix} \dfrac{\partial}{\partial x_1} \\ \dfrac{\partial}{\partial x_2} \\ \dfrac{\partial}{\partial x_3} \\ \vdots \end{pmatrix}$$

The conclusions above mean that we cannot look at the evolution of one phenotypic trait independently of others. Does natural selection still cause populations to climb adaptive peaks? Yes, if selection is slow:

$$\Delta \ln \bar{W} \simeq \sum_i \frac{d \ln \bar{W}}{d\bar{x}_i} \Delta\bar{x}_i$$

and substituting from (10-42) gives us

$$\Delta \ln \bar{W} \simeq \sum_{i,j} \frac{d \ln \bar{W}}{d\bar{x}_i} \operatorname{cov}_g(x_i, x_j) \frac{d \ln \bar{W}}{d\bar{x}_j}$$

$$= \sum_{i,j} \left(\frac{d \ln \bar{W}}{d\bar{x}_i}\right)^2 \frac{\operatorname{cov}(x_i, x_j)}{\operatorname{var}(x_j)} \operatorname{cov}_g(x_i, x_j)$$

But

$$\operatorname{cov}_g(x_i, x_j) \simeq \frac{dA_i}{dA_j} V_g(A_j)$$

or

$$\operatorname{cov}_g(x_i, x_j) \simeq \frac{dA_j}{dA_i} V_g(A_i)$$

Multiplying and taking the square root, we have

$$\operatorname{cov}_g(x_i, x_j) \simeq \sqrt{\frac{dA_i}{dA_j} V_g(x_j) \frac{dA_j}{dA_i} V_g(x_i)}$$

$$= \sqrt{V_g(x_j)V_g(x_i)}$$

$$= h_i h_j \sqrt{\operatorname{var}(x_j) \operatorname{var}(x_i)}$$

and reversing the calculation process,

$$h_i h_j \sqrt{\frac{dx_i}{dx_j} \operatorname{var}(x_j) \frac{dx_j}{dx_i} \operatorname{var}(x_i)} = h_i h_j \operatorname{cov}(x_i, x_j)$$

Substituting, we arrive at

$$\Delta \ln \bar{W} = \sum_{i,j} \left(\frac{d \ln \bar{W}}{d\bar{x}_i} \right)^2 \frac{h_i h_j \operatorname{cov}(x_i, x_j)^2}{\operatorname{var}(x_j)} \tag{10-44}$$

All terms on the right side of (10-44) are positive, so $\Delta \ln \bar{W}$ must be positive; natural selection increases $\bar{W}$. But if natural selection changes the mean values of all traits such that the composite effect is to raise fitness, we may picture the adaptive surface in a new way. No longer will we define the axes to measure gene frequency at the various loci; instead, we will define them to measure trait value, $\bar{x}_i$, for all traits, $i = 1, 2, \ldots$. This new adaptive surface is more intuitive and more heuristic, at least for the ecologist, and will be the one used henceforth in this book. Before blithely moving on, however, note that although populations climb adaptive peaks on a k-dimensional adaptive surface defined by k traits, they do *not* necessarily climb peaks defined by *subsets* of the k traits. That is, natural selection acts on survivorship, fecundity, mating ability, foraging and feeding rates, sex ratio, and so on, in such manner as to increase $\bar{W}$, but this does not necessarily mean that (say) foraging and feeding rates considered in isolation are optimized! Perhaps a feeding stategy that is optimum with respect to nutrient intake somehow impacts negatively on mating ability, and that an optimal mating strategy detrimentally affects the ability to eat a good diet. Maximum fitness, then, will be a compromise in which neither feeding nor mating, by themselves, are optimal. This fact is a serious obstacle to the indiscriminate application of optimization models for purposes of ecological prediction.

Were pleiotopic effects only of rare occurrence, the discussion above would have little meaning. But many experiments have demonstrated their near ubiquity. Hanrahan et al. (1973) and Eisen et al. (1973), for example, selected, in separate lines of laboratory mice, for body weight at age 3, 6, or 8 weeks, and for litter size, and watched the resultant changes in the nonselected traits. It is intriguing that the indirect responses varied in intensity with population size and were occasionally opposite in direction to that predicted. These results may give us pause concerning the completeness of our theoretical formulations. Nevertheless, the indirect responses to selection were loud and clear. Pleiotropy, and the additive covariance among traits that it generates, cannot be ignored in any predictive model of evolutionary change.

Measuring Additive Covariances

If we are to incorporate indirect selection into our models, we must know how to calculate additive covariances. The technique is basically the same as that for finding additive variances. Because (ideally) the additive genetic values of traits are independent of the interactive and environmental values, and because the genetic values of a trait in the average offspring of a randomly mated parent is one-half the additive value of that parent, we may write

$$\begin{aligned} \operatorname{cov}(x_{1\bar{o}}, x_{2p}) &= \operatorname{cov}(G_{1\bar{o}}, A_{2p}) = \operatorname{cov}(\tfrac{1}{2}A_{1p}, A_{2p}) \\ &= \tfrac{1}{2} \operatorname{cov}_g(x_1, x_2) \end{aligned} \tag{10-45}$$

Realized covariance can be directly calculated by applying experimental data to (10-41).

10.3 The Maintenance of Variance

The continuing action of natural selection depends on the existence of additive genetic variance. What maintains this variance? It seems clear that mutation, by preventing fixation, is one force for preventing loss of variance. Heterosis and frequency-dependent selection are others, although the former does not contribute to additive variance. Finally, in Chapter 9, we noted that environmental heterogeneity in the form of temporal change or the dispersal of individuals over patches in space (amounting, in essence, to the same thing) might promote a kind of effective heterosis, resulting in the coexistence of more than one allele at a locus. Do the same forces also act to provide variance for quantitative traits such as height, color, tail length, or age of peak reproductive value? We shall look at traits under the influence of stabilizing selection.

In the case of a trait determined at a single locus, the mean value and variance are interdependent. The reason is simply that both are functions of gene frequency p:

$$\bar{x} = p^2x_1 + 2pqx_2 + q^2x_3$$
$$\mathrm{var}(x) = x^2 - \bar{x}^2 = (p^2x_1^2 + 2pqx_2^2 + q^2x_3^2) - \bar{x}^2$$

If selection changes $\bar{x}$, it does so by altering p and that, in turn, affects $\mathrm{var}(x)$. But if $\bar{x}$ is determined by a very large number of alleles at one or more loci, so that x and its additive genetic component are normally distributed under stabilizing selection, that is no longer true; recall that the normal distribution is characterized by the independence of its mean and variance. Kimura (1965a), Latter (1970), and Lande (1975, 1977) have used this fact to generate models for the maintenance of genetic variance in quantitative traits.

Lande (1975) viewed additive genetic variance in a trait, $V_g(x)$ ($= \sigma_g^2$) as reflecting a balance between stabilizing selection, recombination, and mutation. His mathematical arguments are rather complex, so we reproduce only his final result here. For derivations, see the original paper. Lande showed that, at equilibrium,

$$\sigma_g^2 \; [= V_g(x)] = 2\left(\sum_i \sqrt{\mu_i m_i^2}\right)\sqrt{w^2 + \sigma_e^2 + \left(\sum_i \sqrt{\mu_i m_i^2}\right)^2} + 2\sum_i \sqrt{\mu_i m_i^2} \qquad (10\text{-}46)$$

gives the additive genetic variance in a trait, X,

where w^2 = degree of eurytopy ($1/w^2$ is a measure of selection intensity)
σ_e^2 = environmental variance in x
μ_i = mutation rate at locus i
m_i^2 = increment in additive variance due to the appearance of a new mutant allele at locus i

In a later paper, Lande (1977) showed that, contrary to intuition, the effects of inbreeding and assortative mating on expression (10-46) are vanishingly small. Note that in Lande's idealistic system (all allelic effects are additive both within and among loci) there is no dominance or interaction variance. Hence phenotypic variance, σ_x^2, is simply $\sigma_g^2 + \sigma_e^2$.

Where mean and variance of a trait are genetically independent as above, selection for $\bar{x}$ and σ_x^2 are uncoupled. This has important consequences (when it is true). Specifically, it means that fluctuations in the direction of natural selection cannot affect the genetic variance in a trait. In the single-locus case, we saw (Chapter 9) that coarse-grained changes in selection over time might lead to polymorphism and so greater variance in x. This, as shown by Lande's calculations, should not occur for normally distributed traits. But it is important to point out the drawbacks of Lande's approach. Allelic effects are *not* invariably additive within and among loci. Nonadditive genetic contributions are often sizable. Thus fluctuating selection may be expected to result in small changes in directional dominance that periodically reverse. Selected changes in x *will* affect σ_x^2. Second, when variance is uncoupled from the mean, there may nevertheless be fitness interactions between them. For example, at small $\bar{x}$ values a small variance may be adaptive; at large $\bar{x}$ values, a large variance. Perhaps a mixed strategy is better than monomorphism when $\bar{x}$ is large, and vice versa. We can imagine selective advantages for, say, positive or negative skew. Finally, interdemic or trait group selection may act on σ_x^2 in a manner closely dependent on $\bar{x}$ (see equation 10-67 below).

In contrast to Lande's approach, let us again assume that x is normally distributed, but this time suppose that there exists a *modifier* gene that specifically alters σ_g^2. This approach was used by Slatkin and Lande (1976), who, for simplicity, also assumed the heritability to be constant, the within-family distribution of x at birth to be independent of parental phenotype and, as in the preceding argument, the selection function to be Gaussian. Again the calculations are exceedingly messy, so we go straight to the conclusion. Slatkin and Lande predict that in a constant environment σ_g should always tend to zero. Slight fluctuations in the environment *may* act to maintain a finite additive genetic variance in x, in the absence of mutation, but only if

$$\operatorname{var}(x^*) \geqslant \frac{w^2}{1 + h_x^2\left(1 + 2\sum_i \rho_i\right)} \tag{10-47}$$

where x^* = optimal phenotype value which fluctuates in time

$1/w^2$ = measures the strength of stabilizing selection

h_x^2 = heritability of the trait X

ρ_i = ith autocorrelation coefficient (see Chapter 2)

This expression says that fluctuating selection pressures will prevent loss of all genetic variance only when the amplitude of change in optimal phenotypic value exceeds some critical threshold. Nonzero variance is most likely to be maintained when w^2 is small (strong selection), h_x^2 is large, and $2\sum_i \rho_i$ is large. This last term translates to: Nonzero additive variance is most likely to be maintained when directional changes in selection are slow (i.e., coarse-grained) and regular.

If environmental fluctuations are sufficiently strong, and also rapid, a population will experience, in effect, selection pulling it in opposite directions simultaneously. Such *disruptive selection*, as opposed to stabilizing selection, will not favor zero variance. Felsenstein (1979) has analyzed a situation in which fitness is given by

$$W(x) = k\left\{\exp\left[\frac{-(x - x^*)^2}{2w^2}\right] + \exp\left[\frac{-(x + x^*)^2}{2w^2}\right]\right\}$$

the sum of two Gaussian functions. He finds that selection equilibrium is reached when the population either specializes at one adaptive peak ($\bar{x} \to x^*$ or $\bar{x} \to -x^*$) or generalizes with $\bar{x}$ halfway between the two optima. The condition for specialization is that $W(g)$, the fitness value as a function of additive genetic value, be bimodal. In other words, specialization is more likely with respect to a trait with high heritability, or if the adaptive peaks are widely separated.

We can summarize as follows: Disruptive selection can act to lead a population to one or another adaptive peak if h^2 is high enough or if the peaks are widely separated, but favors an intermediate generalist strategy otherwise. In the latter case, the opposing forces should maintain a fairly high level of phenotypic variance. Under stabilizing selection, temporal changes at modifier loci acting on variance may maintain some variance if the environmental fluctuations are sufficiently large in magnitude. But for an ideal genetic system that determines trait values in a continuous additive fashion, environmental fluctuations have no influence on those traits' variances.

The primary shortcoming of the discussion just above is that the models ignore density feedback. It is not difficult to see if all individuals in a population displayed $\bar{x} \to x^*$, and if the trait x were somehow related to a population-limiting factor, that density feedback would be very severe indeed. Effectively, all individuals would be occupying an extremely narrow niche. For example, suppose that x represented the average food item size eaten by a particular genotype. Then if all genotypes were nearly identical, all would be competing for the same narrow range of food sizes. A mutant or recombinant form that displayed x different from x^*, but not by a great amount, might, by being different, avoid a degree of competition and therefore density feedback, but without sacrificing greatly the quality of its diet. Such an individual would be at a selective advantage. In this manner a certain amount of genetic variance might be maintained. Of course, if individuals of the population in question exhibited strong phenotypic plasticity, this might preclude the need to diverge genotypically. Hence we might expect to find a negative correlation between plasticity and genetic variance [see "Habitat Preference" (p. 366)]. The question of maintaining variance in x (niche width) has been explored by Slatkin (1979). (See also Bulmer, 1971, Fenchel and Christiansen, 1977, and the discussions of work by Roughgarden, 1976, and Slatkin, 1980, in Chapter 17.)

Slatkin (1979) pointed out that if some genotypes were more sensitive, some more important at generating density feedback, both frequency- and density-dependent selection would occur. In this context he considered two genetic models. The first was Lande's ideal—x normally distributed and determined by additive genetic contributions within and among loci. The second also made use of an assumption that x was normally distributed, but that $\bar{x}$ and σ_g^2 were altered by a rare mutant modifier (nonadditive effect). In the case of the purely additive model, note that at equilibrium

$$\bar{x} = \bar{x}_w = \int x\, \phi(x) W(x)\, dx \tag{10-48}$$

where $\bar{x}_w$ is mean trait value after selection, $\bar{x}$ is value at birth, $\phi(x)$ is the distribution of the trait, and $W(x)$ is the fitness function. Also,

$$\sigma_x^2 = \sigma_{xw}^2 = \int (x - \bar{x})^2 \phi(x) W(x)\, dx \tag{10-49}$$

But if $\phi(x)$ is normal, then

$$\int \frac{\partial \phi(x)}{\partial \bar{x}} W(x)\, dx = \int \frac{x - \bar{x}}{\sigma_x^2} \phi(x) W(x)\, dx = \frac{\bar{x}}{\sigma_x^2}(1 - \bar{W})$$

But density-dependent selection assures that $\bar{W} = 1$ at equilibrium. Hence the parenthetical expression is zero, and

$$\int \frac{\partial \phi(x)}{\partial \bar{x}} W(x)\, dx = 0 \tag{10-50a}$$

Similarly,

$$\int \frac{\partial \phi(x)}{\partial \sigma_x^2} W(x)\, dx = \int \frac{1}{2\sigma_x} \frac{\partial}{\partial \sigma_x} \left\{ \frac{1}{\sqrt{2\pi}\, \sigma_x} \exp\left[\frac{-(x - \bar{x})^2}{2\sigma_x^2} \right] W(x)\, dx \right\}$$

$$= \frac{1}{2\sigma_x}\left(-\frac{1}{\sigma_x} + \frac{\sigma_x^2}{\sigma_x^3} \right) = 0 \tag{10-50b}$$

We shall return to equations (10-50) in a moment. But first consider the modifier gene approach. Picture a rare mutant modifier gene that affects $\bar{x}$ and σ_x^2 by minute amounts $\delta\bar{x}$ and $\delta\sigma_x^2$. Because these effects are tiny, we can write

$$\delta\sigma_x^2 \simeq \frac{d\sigma_x^2}{d\delta_x} \delta\bar{x} = g(\bar{x}, \sigma_x^2)\, \delta\bar{x} \tag{10-51}$$

Now if $\phi(x; \bar{x}, \sigma_x^2)$ is the distribution of x before appearance of the mutation, then after its appearance we must have

$$\phi(x; \bar{x} + \delta\bar{x}, \sigma_x^2 + \delta\sigma_x^2)$$

Therefore, if $W(x)$ is the fitness function, the selection coefficient of the mutant form is

$$S = W_{\text{mutant}} - \bar{W} = \int \phi(x; \bar{x} + \delta\bar{x}, \sigma_x^2 + \delta\sigma_x^2) W(x)\, dx - \bar{W}$$

This can be expanded, using the Taylor series, to give

$$\begin{aligned} s &= \int \left[\phi(x; \bar{x}, \sigma_x^2) + \frac{d\phi}{d\bar{x}} \delta\bar{x} \right] W(x)\, dx - \bar{W} \\ &= \int \left[\phi(x; \bar{x}, \sigma_x^2) + \left(\frac{\partial\phi}{\partial\bar{x}} + g(\bar{x}, \sigma_x^2) \frac{\partial\phi}{\partial\sigma_x^2} \right) \delta\bar{x} \right] W(x)\, dx - \bar{W} \\ &= \delta\bar{x} \int \left[\frac{\partial\phi}{\partial\bar{x}} + g(\bar{x}, \sigma_x^2) \frac{\partial\phi}{\partial\sigma_x^2} \right] W(x)\, dx \end{aligned} \tag{10-52}$$

At equilibrium, under frequency-dependent selection, this new mutant must have fitness equal to the hypothetical average individual in the population. Hence s must be zero and

$$\int \left[\frac{\partial\phi}{\partial\bar{x}} + g\,(\bar{x}, \sigma_x^2) \frac{\partial\phi}{\partial\sigma_x^2} \right] W(x)\, dx = 0 \tag{10-53}$$

But there must surely be more than one modifier gene acting on this system. Let us consider only one such gene, with selective value

$$s = \int \left[\frac{\partial\phi}{\partial\bar{x}} + g'(\bar{x}, \sigma_x^2) \frac{\partial\phi}{\partial\sigma_x^2} \right] W(x)\, dx$$

This too must equal zero at equilibrium:

$$\int \left[\frac{\partial\phi}{\partial\bar{x}} + g'(\bar{x}, \sigma_x^2) \frac{\partial\phi}{\partial\sigma_x^2} \right] W(x)\, dx = 0 \tag{10-54}$$

But there is no reason to expect $g = g'$ (and surely the corresponding functions cannot be identical over *all* modifier loci). Hence (10-53) and (10-54) require that

$$\begin{aligned} \int \frac{\partial\phi}{\partial\bar{x}} W(x)\, dx &= 0 \\ \int \frac{\partial\phi}{\partial\sigma_x^2} W(x)\, dx &= 0 \end{aligned} \tag{10-55}$$

which is identical to the conditions for the purely additive effects case.

Equations (10-55) or (10-50) describe the distribution of a quantitative trait, X, at equilibrium under frequency- and density-dependent selection. A more concrete description must depend on a knowledge of the fitness function, $W(x)$. Let us consider a logistic analog,

$$W(x) = 1 + r_0 - \frac{r_0 n}{K(x)} \int \alpha(x - x')\phi(x')\, dx' \tag{10-56a}$$

with

$$K(x) = \frac{K_0}{\sqrt{2\pi}\, \sigma_k} \exp\left[\frac{-(x - x_0)^2}{2\sigma_k^2} \right] \tag{10-56b}$$

$$\alpha(x - x') = \exp\left[\frac{-(x - x')^2}{2\sigma_\alpha^2} \right] \tag{10-56c}$$

Slatkin (1979) uses this form, with equations (10-55), to show that at equilibrium,

$$\begin{aligned} \sigma_x^2 &= \sigma_k^2 - \tfrac{1}{2}\sigma_\alpha^2 \\ \hat{n} &= \frac{K_0}{\sqrt{2\pi}} \sigma_\alpha \end{aligned}$$

An important conclusion to be drawn from Slatkin's calculations is that when x is normally distributed and mean and variance are unconstrained, the equilibrium mean, variance, and population size under frequency- and density-dependent selection are independent of the underlying genetic architecture. Under these conditions we are justified in using the phenotypic selection models presented earlier (p. 291) if not with impunity, then at least with less trepidation.

Attempts to explore the impact of environment fluctuations on σ_x^2 for a system experiencing frequency- and density-dependent selection have been made but, to the author's knowledge, never published

10.4 Spatially Structured Populations

Kin- and Interdemic Selection on Polygenic Traits

Let $\bar{C}_s$ be the mean value of a propensity to act altruistically in trait group s, V_s be the variance in C_s. Let b, c be the benefit and the cost to individuals involved in such an altruistic act. Then the fitness of individual i in trait group s is

$$W_s = W_0 + B_i - cC_{si}$$

and

$$\bar{W}_s = W_0 + \bar{B} - c\bar{C}_s$$

where B_i is the benefit to individual i of the acts of others and $\bar{B}$ is *numerically* $b(n_s - 1)\bar{C}_s$. Also

$$\Delta\bar{C} = \sum_s n'_s\bar{C}'_s - \sum_s n_s\bar{C}_s = \sum_s n'_s\,\Delta\bar{C}_s + \sum_s \bar{C}_s\,\Delta n_s$$

where n_s *is the size of the* s-th group and a prime denotes the subsequent time unit. But within the s-th group (equation 10-24),

$$\Delta\bar{C}_s = \frac{1}{\bar{W}_s}\frac{d\bar{W}_s}{d\bar{C}_s}\,V_g(C_s) = -c\,\frac{V_g(C_s)}{\bar{W}_s}.$$

Also,

$$n'_s = n_s\frac{\bar{W}_s}{\bar{W}},\ \bar{W} = \sum_s n_s\bar{W}_s/n$$

Putting the above expressions together we obtain,

$$\begin{aligned}\bar{W}\Delta\bar{C} &= -c\sum_s \frac{n_s}{n}\,V_g(C_s) + [b(n_s - 1) - c]\sum_s \frac{n_s}{n}(\bar{C}_s - \bar{C})^2 \\ &= -cV_W(C) + [b(n_s - 1) - c]\,V_B(C)\end{aligned}$$

where $V_W(C)$, $V_B(C)$ are within and between group (additive genetic) variances in C. Defining

$$t = \frac{V_B(C)}{V_B(C) + V_W(C)}$$

we see that the trait, C, spreads by kin selection if

$$t > \frac{c}{b(n_s - 1)}$$

But t is an intraclass correlation–the (additive genetic) correlation between two randomly chosen individuals within a trait group. For a one-locus situation then, t is the coefficient of relationship r (see equation 9-14). The above derivation is from Aoki (1982).

The phenomenon of group selection acting on polygenic traits is not well studied. We present first a formulation due to Slatkin and Wade (1978). We consider first the individual selection component, and then the group components of selection, finally combining them into compound formulae. As in our earlier discussion, we suppose the fitness function to be Gaussian,

$$W(x) = k \exp\left[\frac{-(x - x^*)^2}{2w^2}\right]$$

If, for simplicity, we scale x such that x^* is zero, then

$$W(x) = ke^{-x^2/2w^2} \tag{10-57}$$

Suppose that k, w^2, h_x^2, and the within-deme variance in x, σ_x^2, are the same for all demes, and that the total number of demes in the population is constant. Then, within any deme, (10-33) tells us that

$$\Delta\bar{x} = -\bar{x}\frac{h_x^2\sigma_x^2}{\sigma_x^2 + w^2} \tag{10-58}$$

The change in the grand population mean $\bar{\bar{x}}$ is

$$\Delta\bar{\bar{x}} = -\bar{\bar{x}}\frac{h_x^2\sigma_x^2}{\sigma_x^2 + w^2} \tag{10-59}$$

We will need an expression also for the change in among-deme variance in $\bar{x}$, $\sigma_{\bar{x}}^2$. To obtain this, note that

$$\begin{aligned}\sigma_{\bar{x}}^2(t+1) &= \overline{\bar{x}(t+1)^2} - \overline{\bar{x}(t+1)}^2 = \overline{[\bar{x}(t) + \Delta\bar{x}]^2} - \overline{[\bar{x}(t) + \Delta\bar{x}]}^2\\ &= \overline{[(1-z)\bar{x}]^2} - \overline{[(1-z)\bar{x}]}^2\\ &= (1-z)^2\sigma_{\bar{x}}^2(t) \qquad \text{where } z = \frac{h_x^2\sigma_x^2}{\sigma_x^2 + w^2}\end{aligned}$$

Thus

$$\Delta\sigma_{\bar{x}}^2 = [(1-z)^2 - 1]\sigma_{\bar{x}}^2 \rightarrow -2\frac{h_x^2\sigma_x^2}{\sigma_x^2 + w^2}\sigma_{\bar{x}}^2 \tag{10-60}$$

for slow selection.

There are several group-selection components. With respect to random migration among the demes, let g be the fraction of a deme replaced each generation by individuals randomly drawn from all other demes. Then, considering only migration within any one deme,

$$\bar{x}(t+1) = (1-g)\bar{x}(t) + g\bar{\bar{x}}(t)$$

so that

$$\Delta\bar{x} = -g(\bar{x} - \bar{\bar{x}})$$

The change in $\bar{\bar{x}}$, then, is

$$\Delta\bar{\bar{x}} = -g\overline{(\bar{x} - \bar{\bar{x}})} = 0 \tag{10-61}$$

and the change in $\sigma_{\bar{x}}^2$ is

$$\begin{aligned}\Delta\sigma_{\bar{x}}^2 &= \sigma_{\bar{x}}(t+1)^2 - \sigma_{\bar{x}}(t)^2 = \left[\overline{\bar{x}(t+1)^2} - \overline{\bar{x}(t+1)}^{\,2}\right] - \sigma_{\bar{x}}(t)^2 \\ &= \overline{[(1-g)\bar{x} + g\bar{\bar{x}}]^2} - \overline{[(1-g)\bar{x} + g\bar{\bar{x}}]}^{\,2} - \sigma_{\bar{x}}^2 \\ &= [(1-g)^2 - 1]\sigma_{\bar{x}}^2 \to -2g\sigma_{\bar{x}}^2 \end{aligned} \qquad (10\text{-}62)$$

for low migration rates.

Via genetic drift the change in $\bar{\bar{x}}$ should be zero; the change in $\sigma_{\bar{x}}^2$ is calculated by first writing $\bar{x}$ as a function of gene frequencies at all loci, $\{p\}$. Using the Taylor expansion, it can be shown that

$$\sigma_{\bar{x}}^2 \simeq \sum \left(\frac{d\bar{x}}{dp}\right)^2 \sigma_p^2$$

where summation is over all loci. Therefore (Chapters 7 and 8),

$$\Delta\sigma_{\bar{x}}^2 \simeq \sum \left(\frac{d\bar{x}}{dp}\right)^2_{\bar{p},\bar{q}} \Delta\sigma_p^2 = \sum (2\alpha_x)^2_{\bar{p},\bar{q}} \left(\frac{\bar{p}\bar{q}}{n_e}\right) = \sum \left(\frac{4\bar{p}\bar{q}\alpha_x^2}{n_e}\right)_{\bar{p},\bar{q}}$$

But $V_g(x)$ over the total population $[= h_x^2 \operatorname{var}(x)_{\text{total}}]$ is equal to

$$\sum (2pq\alpha_x^2)_{\bar{p},\bar{q}}(1 + F)$$

so

$$\Delta\sigma_{\bar{x}}^2 = \frac{2h_x^2 \operatorname{var}(x)_{\text{total}}}{n_e(1 + F)} = \frac{2h_x^2(\sigma_x^2 + \sigma_{\bar{x}}^2)}{n_e(1 + F)}$$

But suppose that the distribution of individuals among groups is random. Then $\sigma_{\bar{x}}^2 = \sigma_x^2/n_e$. Also, $F = 1/2n_e$. Thus

$$\Delta\sigma_{\bar{x}}^2 = \frac{2h_x^2(1 + 1/n_e)\sigma_x^2}{n_e(1 + 1/2n_e)} \simeq \frac{h_x^2\sigma_x^2}{n_e} \qquad (10\text{-}63)$$

The founder effect, the replacement of extinct demes by propagules drawn from a single parent deme, is calculated as follows. Let $e(\bar{x})$ be the extinction rate of a deme with average trait value $\bar{x}$. Then the change in $\bar{x}$ must be

$$\begin{aligned}\Delta\bar{x} &= \bar{x}(t+1) - \bar{x}(t) \\ &= [(1 - e(\bar{x}))\bar{x} + e(\bar{x})\bar{\bar{x}}] - \bar{x} = (\bar{\bar{x}} - \bar{x})e(\bar{x})\end{aligned} \qquad (10\text{-}64)$$

whence

$$\begin{aligned}\Delta\bar{\bar{x}} &= \overline{(\bar{\bar{x}} - \bar{x})e(\bar{x})} = -\overline{\bar{x}e(\bar{x})} + \bar{\bar{x}}\overline{e(\bar{x})} \\ &= -\operatorname{cov}(e, \bar{x}) \simeq -\left(\frac{de(\bar{x})}{d\bar{x}}\right)_{\bar{\bar{x}}} \sigma_{\bar{x}}^2\end{aligned} \qquad (10\text{-}65)$$

The variance at time $t + 1$ will be a weighted average of the variance after extinction ($\simeq \sigma_{\bar{x}}^2$ if $\Delta\bar{\bar{x}}$ is very small), and the variance among the new founder populations (i.e., among the propagules). If propagules of size ρ are drawn randomly from the extant demes, this variance is (see argument for drift)

$$\sigma_{\bar{x}}^2 + \sum \left(\frac{d\bar{x}}{dp}\right)^2 \Delta\sigma_p^2 = \sigma_{\bar{x}}^2 + \sum (2\alpha_x)^2_{\bar{p},\bar{q}} \frac{\bar{p}\bar{q}}{\rho} \simeq \sigma_{\bar{x}}^2 + \frac{h_x^2\sigma_x^2}{\rho}$$

so

$$\Delta\sigma_{\bar{x}}^2 \simeq \left[(1 - \bar{e})\sigma_{\bar{x}}^2 + \bar{e}\left(\sigma_{\bar{x}}^2 + \frac{h_x^2\sigma_x^2}{\rho}\right)\right] - \sigma_{\bar{x}}^2 = \frac{\bar{e}h_x^2\sigma_x^2}{\rho} \qquad (10\text{-}66)$$

Putting all these terms together, we have, finally,

$$\begin{cases} \Delta\bar{\bar{x}} \simeq -\dfrac{h_x^2\sigma_x^2}{\sigma_x^2 + w^2}\,\bar{\bar{x}} - \left(\dfrac{de(\bar{x})}{d\bar{x}}\right)_{\bar{\bar{x}}} \sigma_{\bar{x}}^2 \\ \Delta\sigma_{\bar{x}}^2 \simeq -2\dfrac{h_x^2\sigma_x^2}{\sigma_x^2 + w^2}\,\sigma_{\bar{x}}^2 - 2g\sigma_{\bar{x}}^2 + \dfrac{h_x^2\sigma_x^2}{n_e} + \dfrac{\bar{e}h_x^2\sigma_x^2}{\rho}, \end{cases}$$

or

$$\begin{cases} \Delta\bar{\bar{x}} = -\dfrac{h_x^2\sigma_x^2}{\sigma_x^2 + w^2}\,\bar{\bar{x}} - \left(\dfrac{de(\bar{x})}{d\bar{x}}\right)_{\bar{\bar{x}}} \sigma_{\bar{x}}^2 \\ \Delta\sigma_{\bar{x}}^2 = \left[-2g + h_x^2\left(1 + \dfrac{\bar{e}n_e}{\rho} - 2\dfrac{n_e\sigma_{\bar{x}}^2}{n_e\sigma_{\bar{x}}^2 + w^2}\right)\right]\sigma_{\bar{x}}^2 \end{cases} \quad (10\text{-}67)$$

We can illustrate the use of equations (10-67) with an example. Suppose that $g = 0$, $h_x^2 = .5$ (and essentially constant over time and space), extinction rises proportionately with x [$e(\bar{x}) = (1 + \bar{x})/2$, where $-1 \leqslant x \leqslant +1$], effective population size is 10, propagule size is 2, and w, describing the intensity of within-deme selection, is 1. Then $\sigma_x^2 = 10\sigma_{\bar{x}}^2$, so $\sigma_x^2/(\sigma_x^2 + w^2) = 10\sigma_{\bar{x}}^2(10\sigma_{\bar{x}} + 1)$, and

$$\Delta\bar{\bar{x}} = -\left(\frac{5\bar{\bar{x}}}{10\sigma_{\bar{x}}^2 + 1} + .5\right)\sigma_{\bar{x}}^2$$

$$\Delta\sigma_{\bar{x}}^2 = \left(1.75 + 1.25\bar{\bar{x}} - \frac{10\sigma_{\bar{x}}^2}{10\sigma_{\bar{x}}^2 + 1}\right)\sigma_{\bar{x}}^2$$

As long as $\sigma_{\bar{x}}^2 > 0$, the following is true. If $\bar{\bar{x}}$ is initially below x^* ($= 0$ in this formulation), and less than $-0.1(10\sigma_{\bar{x}}^2 + 1)$, then $\Delta\bar{\bar{x}}$ is positive. Now $\Delta\sigma_{\bar{x}}^2 \propto 1.75 + 1.25\,\bar{\bar{x}} - [10\sigma_{\bar{x}}^2/(10\sigma_{\bar{x}}^2 + 1)]$. Thus $\Delta\sigma_{\bar{x}}^2$ is positive when

$$\bar{\bar{x}} \geqslant -\frac{7.5\sigma_{\bar{x}}^2 + 1.75}{12.5\sigma_{\bar{x}}^2 + 1.25}$$

Call the value on the right side of the equation above z. If $\bar{\bar{x}}$ falls below z, $\sigma_{\bar{x}}^2$ falls. $\bar{\bar{x}}$ may continue to fall, depending on the value of $\sigma_{\bar{x}}^2$, but will eventually bottom out and begin to rise. $\sigma_{\bar{x}}^2$ will continue to fall while $\bar{\bar{x}}$ increases until $\bar{\bar{x}}$ reaches z. At this point $\sigma_{\bar{x}}^2$ begins to increase. By this time $\sigma_{\bar{x}}^2$ may have become sufficiently small that $\bar{\bar{x}}$ continues to increase. But if so, its further rise is braked and eventually reversed by the rise in $\sigma_{\bar{x}}^2$. Therefore, as long as among deme variance does not actually fall to zero (and this is reasonably assured by the existence of mutations), $\bar{\bar{x}} = z$ is the (not necessarily neighborhood-stable) equilibrium. But also $\bar{\bar{x}}$ rises or falls depending on whether $\bar{\bar{x}} \lesseqgtr -.1(10\sigma_{\bar{x}}^2 + 1)$. So $-.1(10\sigma_{\bar{x}}^2 + 1)$ is also the equilibrium. We write

$$-.1(10\sigma_{\bar{x}}^2 + 1) = \bar{\bar{x}} = -\frac{7.5\sigma_{\bar{x}}^2 + 1.75}{12.5\sigma_{\bar{x}}^2 + 1.25}$$

whence

$$\sigma_{\bar{x}}^2 \simeq .61 \qquad \text{and} \qquad \bar{\bar{x}} \simeq -.71$$

A somewhat different approach was taken by Hamilton (1975). Taking advantage of the fact that the covariance of two variables can be partitioned into within- and among-group components, he wrote

$$\Delta\bar{\bar{x}} = \frac{\mathrm{cov}_g(x, w)}{\bar{W}} = \frac{\mathrm{cov}_{g_A}(\bar{x}, \bar{W})}{\bar{W}} + \frac{\mathrm{cov}_{g_w}(x, W)}{\bar{W}}$$

$$= \frac{1}{\bar{\bar{W}}}\left(\frac{d\bar{W}(\bar{x})}{d\bar{x}}\right)_{\bar{\bar{x}}} h_{x_A}^2\,\mathrm{var}_A(\bar{x}) + \frac{1}{\bar{\bar{W}}}\left(\frac{d\bar{W}(x)}{dx}\right)_{\bar{x}} h_{x_g}^2\,\mathrm{var}_w(x)$$

where the subscripts A and w indicate among- and within-group components. Hamilton wrote this

$$\Delta\bar{\bar{x}} = \beta_1 \operatorname{var}_A(\bar{x}) + \beta_0 \operatorname{var}_w(x) \tag{10-68}$$

A similar form follows directly from the phenotypic selection model:

$$\begin{aligned}\Delta\bar{\bar{x}} &= \frac{1}{R} \operatorname{cov}\big(R(x), x\big) \\ &= \frac{1}{R} \operatorname{cov}_A(R, \bar{x}) + \frac{1}{R} \operatorname{cov}_w(R, x) \\ &\simeq \frac{1}{R}\left(\frac{dR(\bar{x})}{d\bar{x}}\right)_{\bar{\bar{x}}} \operatorname{var}_A(\bar{x}) + \frac{1}{R}\left(\frac{dR(x)}{dx}\right)_{\bar{\bar{x}}} \operatorname{var}_w(x)\end{aligned} \tag{10-69}$$

If we know R as a function of x within groups, and group mean R as a function of $\bar{x}$ among groups, (10-68) and (10-69) can easily be applied. But the relative extinction rates and the propagule compositions are subsumed in the derivative terms and their effects on these terms are not at all obvious.

Environmental Gradients and Peripheral Populations

We mentioned earlier that populations might be structured in such a manner than, even in homogeneous environments, one subpopulation occupies one adaptive peak, a second inhabits another. If selection pressures shift gradually over an environmental gradient, this is particularly likely. In a simple case we may picture a founder population to lie at some given point in the adaptive space, representing its mean phenotype, and simulate changes in optimal x_1 and x_2 over an environmental gradient by sliding the adaptive surface (viewed as a topographic fitness map) past the point (see Figure 10-5). In Figure 10-5a, the population lies in the sphere of influence of peak A. Selection can be expected to push the population to that peak. Somewhere else on the gradient (Fig. 10-5b) the population lies in the sphere of influence of peak B. The switch from selection pressures forcing the population toward A or B may occur over very short distances in space. The *area effect* in the Dutch roadside population of *Cepaea nemoralis* has already been cited as a possible example. Kettlewell and Berry (1969) provide a less dramatic example, involving a very simple trait. The melanic form of the moth *Amathes glareosa* on one of the Shetland islands, varies from 97% frequency at the north end of the island

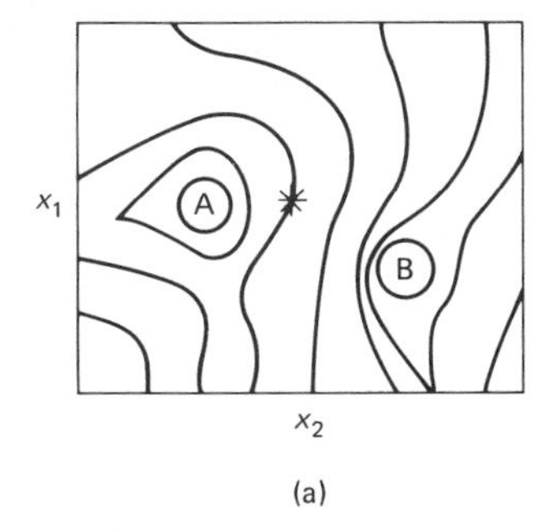

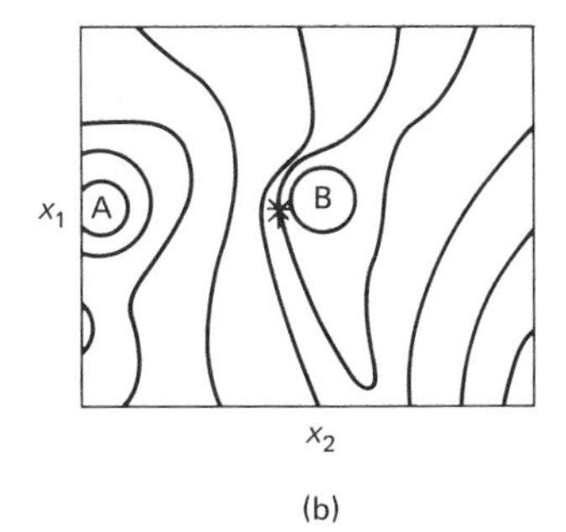

Figure 10-5 Depiction of a hypothetical adaptive surface at two points in space: (a) point 1 on the gradient; (b) point 2 on the gradient. x_1 and x_2 measure the population mean values of two traits, X_1 and X_2.

to only 1% at the south end, showing a drop over one 8-mile length of 22%. Antonovics (1968), Cooke et al. (1972), and Wu and Antonovics (1976), for a variety of plant species, demonstrated sharp clines in phenotype that are adaptive to soil gradients in heavy metal toxicity. But the clines exhibited by *Cepaea* and *Amathes* occur in the absence of any obvious environmental change, Large differences among subpopulations are generated, apparently, by only minute environmental trends.

We can get a more explicit idea of the mechanism of sudden shifts in population characters over space by examining the following model of Clarke (1966) (see also Dodson, 1976). Define the distance along a gradient from one end to be D, where $0 \leqslant D \leqslant 1$. Suppose that genotypic fitnesses at one locus take the form

Genotype	AA	Aa	aa
Fitness	xD	xk	$x(1 - D)$

$0 \leqslant k \leqslant 1$

If there is no heterosis, one or the other allele becomes fixed, which one depending on the point in space (D). The reason for a sudden switch in gene frequency from zero to 1.0 is obvious. But suppose that heterosis occurs for some intermediate range in D. Then

$$\hat{p} = \frac{W_2 - W_3}{2W_2 - W_1 - W_3} = \frac{xk - x + xD}{2xk - x} = \frac{k + D - 1}{2k - 1} \tag{10-70}$$

and the equilibrium value of p varies with D as shown in Figure 10-6. The slope, in the region of heterosis, is $1/(2k - 1)$. Now let us introduce a modifier gene that effects fitness as shown below:

		Genotype		
		AA	Aa	aa
Modifier Gene	BB or Bb	$xD + xr$	$xk + xs$	$x(1 - D) + xt$
	bb	xD	xk	$x(1 - D)$

If genotype at the modifier locus is bb, then $\hat{p}$ is as given in Figure 10-7 and (10-70). But if the modifier displays BB or Bb, then

$$\hat{p} = \frac{k + D - 1 + (s - t)}{2k - 1 + (2s - r - t)} \tag{10-71}$$

which has a slope of $1/(2k - 1 + 2s - r - t)$. The two lines defined by (10-70) and (10-71) intersect if there exists a value of D where $0 \leqslant D \leqslant 1$, for which

$$\hat{p} = \frac{k + D - 1}{2k - 1} = \frac{k + D - 1 + (s - t)}{2k - 1 + (2s - r - t)}$$

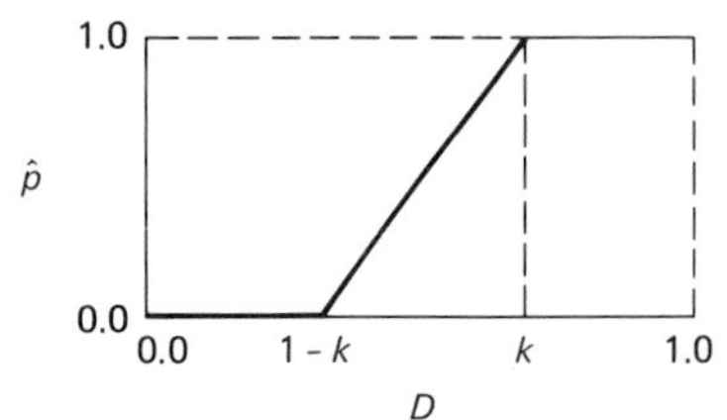

Figure 10-6 Hypothetical gene frequency cline over an environmental gradient.

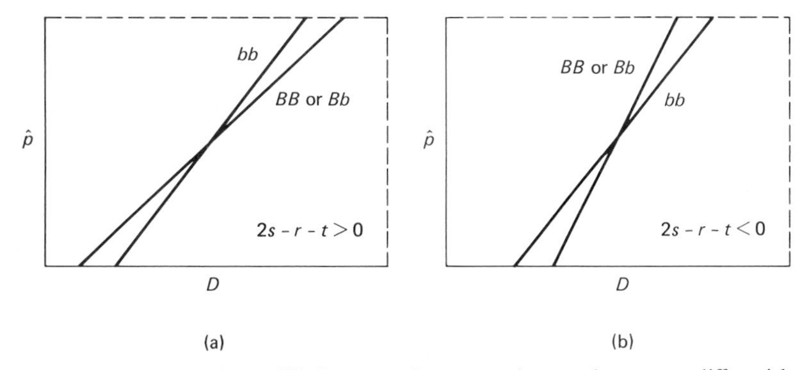

Figure 10-7 Clines in equilibrium gene frequency $\hat{p}$ at one locus may differ with the genotype at another (B, b) locus.

Such a value of D always exists, so that intersection is guaranteed at

$$D = \frac{k(t - r) + (r - s)}{r + t - 2s}$$

$$\hat{p} = \frac{t - s}{r + t - 2s} = \frac{s - t}{2s - r - t} \tag{10-72}$$

If $2s - r - t > 0$, the slope of (10-71) is less than that when genotype *bb* is found. The opposite is true when $2s - r - t < 0$. Therefore, we may plot $\hat{p}$ versus D as in Figure 10-7. Now consider that the B allele will be selected (so we should find *BB* or *Bb* individuals predominating) when

$$\bar{W}_B > \bar{W}_b$$

After some algebraic manipulation this becomes

$$\hat{p} < \frac{(t - s) + \sqrt{s^2 - rt}}{r + t - 2s} = \frac{(s - t) - \sqrt{s^2 - rt}}{2s - r - t} \qquad \text{if } 2s - r - t > 0$$

$$\hat{p} > \frac{(t - s) + \sqrt{s^2 - rt}}{r + t - 2s} \qquad \text{if } 2s - r - t < 0 \tag{10-73}$$

But when $2s - r - t > 0$, this critical value lies to the *left* of the intersection (see equation 10-72), and if $2s - r - t < 0$, to the right. Thus the population switches from *bb* to *BB* (or *Bb*) to the left of the intersection in Figure 10-7a, to the right in Figure 10-7b (Figure 10-8). The

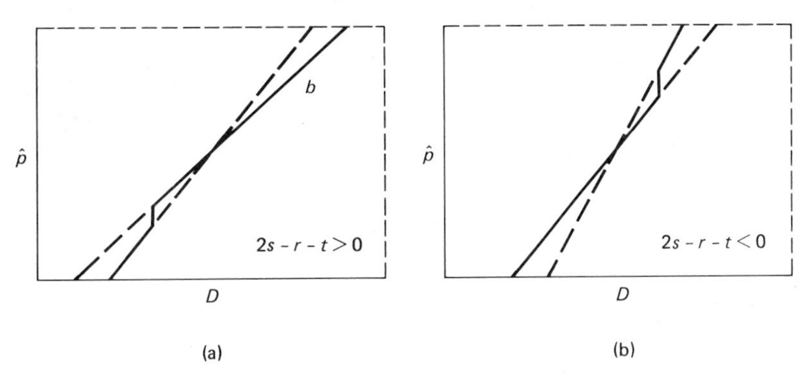

Figure 10-8 Step clines in gene frequency at one locus resulting from selected changes in gene frequency at another (B, b) locus.

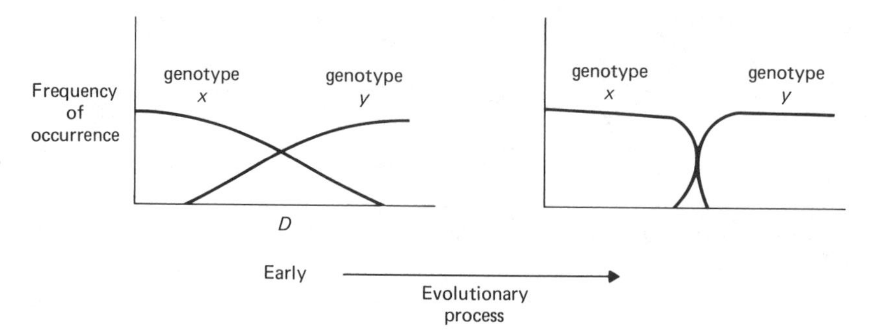

Figure 10-9 Genotype distribution along a gradient might change in this manner when each genotype is most fit (for a given population density) over different sections of the gradient.

addition of a modifier locus has resulted in the appearance of a step cline. If additional modifiers are added, they will show the same behavior. If the jump in $\hat{p}$ is not small, there is a reasonably good chance that the critical $\hat{p}$ values for a third modifier fall in the range of that jump, making it larger. This increases the chance that a fourth modifier will possess a critical p in that range, and so on. With large numbers of epistatic loci, the step could become extremely large.

Sharp clines might also come about as a result of genotype interaction. If one genotype (or class of genotypes) is more fit along part of an environmental gradient, another along another segment, and the two compete in regions of overlap along the gradient, natural selection will favor behavior that keeps each genotype in the areas where it is competitively most successful (most fit). We might expect to see something on the order of that shown in Figure 10-9 (see, e.g., Endler, 1973). The frequency of genotype x drops suddenly at some point along D.

Examine the evolution of heavy metal tolerance in plants over intense gradients, from heavily contaminated soil to clean, Antonovics (1968) found that the small, localized populations of tolerant individuals displayed an unusual degree of self-pollination. Self-fertilization also reaches high proportions in the cobwebby gilia (*Gilia* pp.) in marginal habitat, and incompatibility barriers to outcrossing appear (Wallace, 1957; Grant, 1966). Both of the plants above, where they occur in stressful habitats, one to which they have not yet reached a reasonable evolutionary accomodation, have become inbreeding. One possible reason is that in stressful habitats populations are, by and large, sparse. Therefore, successful propogation requires that outbreeding become merely facultative, inbreeding more common. The evolution of incompatability barriers, however, suggests a second possibility. Gene flow, via pollen or seed dispersal in plants, migration in animals, will generally be from a high-density region (see Chapter 2). New adaptive gene combinations which arise in the peripheries of a species' range may be swamped by such inflow. Those subpopulations which will ultimately be successful in adapting to local conditions, then, are the ones that inbreed, thereby blocking gene flow from elsewhere. Individuals that possess new, adaptive genetic innovations will more successfully pass those innovations on to their offspring if they also self-fertilize (or "accept" pollen only from genetically similar individuals). Parthenogenesis, certainly a successful mechanism for blocking gene flow, has also evolved in a number of animals, where, at least for lizards, it often seems correlated with life under marginal conditions (Darevsky, 1966; Wright and Lowe, 1968).

Review Questions

1. Might we expect the evolution of switch genes or "rheostat" genes for recombination (or similar shuffling events that generate variability) to evolve in response to various ecological conditions that merit more or less genetic diversity? Would this occur by individual, trait group, or interdemic selection? What might the ecological conditions be?

2. It seems likely that similar genotypes are more likely to compete than dissimilar ones. How might this affect trait group selection? How might it affect genetic structure of populations in space? (see Chapter 5.) How might it affect interdemic selection?
3. Recall the inequality used in Chapter 9 to explore the evolution of altruism by kin selection:

$$\frac{b}{c} > \frac{1}{r(n-1)}$$

where r is the probability that when the altruist carries one altruist gene, any given (potential) recipient also carries the gene. When only one rare gene is involved, r is simply the coefficient of relationship between altruist and recipient. If an altruistic trait is polygenic, how is r to be defined? Measured? Remember that it is possible for some of the genes involved to be rare, others common; and that different gene loci may express different inbreeding values; perhaps epistasis occurs, or linkage (For a discussion of this problem, see Yokoyama and Felsenstein, 1978.)
4. Suppose that you discovered a very sharp cline in body color pattern in a species of beetle. You can find no corresponding environmental change and so hypothesize the cline to be an area effect due to occupancy of alternate adaptive peaks. But, of course, it may be that you simply neglected to detect an environmental gradient because you failed to measure the critical parameter to which the beetle population, unbeknown to you, is responding. How might you test your hypothesis?
5. All phenotypes are more or less arbitrarily scaled by us as observers. This is particularly true of ecological and behavioral phenotypes, for which the scoring of even a single individual is generally the result of some sampling procedure. Thus phenotype and genotype are uncoupled in a fundamental way that is not merely based on lack of mechanistic knowledge. Even if we knew the entire genome of an organism, all biochemical mechanisms of gene expression and interaction, and had a complete knowledge at the molecular level of all environmental effects on development, this would not necessarily translate into a deterministic knowledge of the phenotype. Is this a good argument for a phenotypic approach in modeling selection?
6. Refer to (10-67). Explain *verbally* the meaning of each term in the equations. What is the biological meaning of an increase or decrease in g, ρ, and so on? Under what conditions would you expect to find large or small g, ρ, and so on? What is the impact of large versus small values of these parameters for the impact of group selection?

Part Four

Adaptation: Fine-Tuning the Structural Framework

Part Three began with the proposal that population genetics theory could be used to predict ecological adaptations. The rationale was to use this theory as justification for the hypothesis that natural selection acts as an efficiency expert, weeding out the less fit and so choosing that (those) phenotype(s) with maximum fitness under the circumstances in which they live. This "optimization" approach seems a quite reasonable one (Maynard Smith, 1978), and certainly does not violate any tenet of population genetics. But for many reasons raised in Chapters 7 through 10, it should not be considered either consistent or accurate in its predictions (see Lewontin, 1977). More appropriately, it should be used to make educated guesses as to what might be expected, given incomplete knowledge of genetical, ecological, and physical constraints. It should be used not to make predictions but to guide investigation, to suggest what might be interesting to look for. This last point has been made forcefully by Gould and Lewontin (1979) in a paper that should rightfully be made mandatory reading for all students of evolutionary biology. Their presentation begins with a description of the spandrels of the Cathedral of San Marco, those spaces between the shoulders of two adjoining arches which are an inevitable by-product of any architectural construction in which arches are used. Were the cathedral a biological organism, a biologist following the optimization school might marvel at those spandrels and their elaborate ornamentation, and accord to them a great evolutionary purpose. But, indeed, they have no purpose at all; they are merely necessary incidentals of design which the architect and artists have made the most of. It is easy to "explain" the existence or form of a biological trait using optimization arguments. If one argument fails, it is a simple matter to find another one. But there is no way to test these "explanations," and there is always the possibility that the trait is not even an adaptation, but only another spandrel, a consequence of other traits—or even purely serendipitous. Evolutionary biologists should not attempt to explain the evolutionary significance of what they observe; rather they should use their knowledge of evolutionary genetics to suggest how natural selection might be expected to fine-tune gross features which, themselves, did not necessarily evolve as adaptations.

Consider a typical ecological character, such as interspecific competitive viability. Such a character consists of a large number of behaviorally, physiologically, and morphologically interdependent traits, such as feeding efficiency, development rate, fecundity, and aggressive predilections. Not only does the extragenetic interdependence of these component traits make predictions of evolutionary change problematic, but the ubiquity of pleiotropy also binds these traits at a genetic level, compounding the problem. Consider a locus for which one allele, A, affects all subsidiary traits in a manner that promotes increased competitive viability. Such genes, under selection for competitive viability, will presumably reach fixation rather quickly. But those loci where A affects some subsidiary traits in the favored direction, others in the opposite direction, will be torn between two opposing directions of selection. They will reach equilibrium much less quickly. The result is that as time progresses, an increasing proportion of the alleles still under selection affect component traits in a confounding manner—additive genetic covariance among the traits becomes more and more negative (Falconer, 1960a). As a result, evolutionary progress is greatly slowed, or grinds to a halt short of the adaptive summit. In fact, strong negative additive covariance among traits may indicate strong joint selection pressures. It is possible that alternating selection pressures at a few loci, or periodic infiltration of the gene pool, via migration, of new genes and gene combinations may serve to knock loose the bound-up additive variances, but this is purely conjecture.

Can epistatic interactions evolve such as to minimize additive covariance, thus allowing independent evolution of traits? The answer is vague. Lande (1980) produces calculations indicating that it may be so. On the other hand, laboratory studies are less encouraging. Reeve and Robertson (1952) selected for 50 generations for long wings in Drosophila melanogaster, expecting to achieve homozygosity at the pertinent loci. Instead, $V_g(x)$ *increased from 30% to 50%. Upon relaxation of selection it subsequently dropped to 40%. Part of this unexpected result may have been due to an initial additive genetic correlation of about 75% between wing length and thorax length. We might have expected to find this correlation dwindling with time as natural selection acted to prevent thorax length from following wing length to maladaptive extremes. But, in fact, the correlation increased to 80%. Another result of the artificial selection for wing length noticed by Reeve and Robertson was the development of a negative correlation between body size and viability. Perhaps their harsh selection pressures, even without breaking up the wing–thorax correlation, somehow disrupted various coadapted gene complexes. Such a consequence should be fully expected. Indeed, there is abundant evidence indicating that progress, under artificial selection, generally comes at the price of sacrificing viability or fecundity via deterioration of other, adaptive traits (see, e.g., Lerner and Dempster, 1951; Mather, 1953). Bell and Burris (1973) selected simultaneously for larval weight at 13 days, and pupal weight in the flour beetle, Tribolium castaneum. The changes expected on the basis of selection pressures imposed and knowledge of additive variances and covariances were realized with fair accuracy. As expected,* $V_g(x)$ *dropped for both traits during the course of selection (by about 20%). But even though the selection scheme strongly favored a decrease, the additive covariance showed no detectable change at all (see also Cheung and Parker, 1974).*

Another sort of problem encountered in the genetics of polygenic traits is that a change selected for under one environmental situation may or may not be expressed under another. For example, Falconer (1960b) selected for growth rate in mice under conditions of plentiful food and sparse food. Animals selected for fast growth while being well fed exhibited a marked evolutionary response. But the same selected strain did negligibly better than unselected individuals when subsequently placed on a sparse diet. If selection for fast growth was carried out on poorly fed animals, the response was poor. But if these evolutionarily nonresponsive animals were then put on a high diet, they grew much more rapidly than unselected individuals. Similar, strange results occurred when selection was for slow growth rate.

In a more empirical vein, it is not at all unusual to see disfigured organisms surviving and reproducing quite successfully. One-legged birds scrapping aggressively at a feeder are not rare. Is that to say that were such anomalies to arise from a genetic rather than accidental basis, the individuals would be any less apt to survive? Are even gross aberrations only slightly less fit than the norm? Do such misfits display lower survival only during a short, critical period—perhaps when very young, or briefly, right after the accident? If so, what sense does it make to conjecture that natural selection fine-tunes organisms to their environment? Yet as we shall see in later chapters, fine-tuning often, and perhaps regularly, occurs. Why? Perhaps the answer lies in the existence of somatic plasticity. It seems eminently reasonable to suppose that (J. M. Emlen, 1968a), "evolution has modified the nervous and endocrine systems in such a way that:

1. Stimulus situations "motivate" animals into certain physiological states;
2. In response to any one such physiological state, and depending in part on previous experience, certain motor patterns and associations are more easily learned than others; and
3. The positive or negative reinforcing aspect of a given behavior pattern under a given stimulus is genetically as well as environmentally determined."

If an animal eats a food that is toxic, there follows a chemical reaction in the body. Those individuals which are genetically predilected to interpret that reaction as a positive reinforcement will be quickly eliminated in the evolutionary game. Those who interpret it as negative and possess the capacity to learn will survive. A tremendous array of complex, interactive, adaptive responses can be directed by a much smaller collection of genetically coded guidelines. Given the capacity for plasticity and the existence of genetically based directives, it does not really matter whether constraints prohibit the spread of a hypothetical gene coding for some behavior (or physiological) modification. The conditions that favor the gene's spread, whether consummating or not, also predispose the same behavior modification by somatic means. Adaptive peaks can be climbed, partly aided by changes in the genetic complex, partly in spite of them.

I end this introduction with a very brief look at what J. H. Campbell (1982) calls autonomy in evolution. Throughout Chapters 7 to 10, genes have been thought of as rather passive chunks of DNA. If they happen to convey superior benefits to their owners or to other individuals carrying genes derived from the same ancestral DNA, they will see their likenesses passed on disproportionately to future generations. But recent research, and recent rediscovery of some old research by McClintock (1951, 1956a,b), has shown that genes can be much more than passive participants in the natural selection lottery. For example, some gene sequences, known as transposons or insertion sequences, are able to move about, to leave one part of a chromosome (perhaps leaving a replicate in place, perhaps not), and move *to another position. In doing so they may alter their effect or that of other genes nearby upon their owner's development. In effect, transposition may act as a mutation inducer and, as such, serve to break up genetic correlations and free additive variance for selection to act upon (for a recent collection of papers on this subject, see the 1980 volume of the Cold Spring Harbor Symposium on Quantitative Biology). Campbell refers to genes that can alter their own expression, such as transposons, as "profane." He lists four categories of profane genes for which there is good-evidence, and speculates about a fifth and even a sixth.*

Campbell's first level of profane genes includes those which are structurally violable in the sense that they include "special target sites" for gene-processing enzymes to recognize and act on. For example, insertion sequences with such target sites might respond to these enzymes by moving, thereby altering their expression. In such a way the gene can change its expression in response to changes wrought by other genes or perhaps long-term environmental alterations. A second, "self-dynamic" level contains more sophisticated genes, those that behave like level I genes but, in addition, code for replicases that multiply their numbers and so affect their intensity of expression. Level III genes, "contingently dynamic," are like those of level II, but also can specify the conditions under which they will change or enhance their expression. That is, while the expression as well as the frequency of level I and II genes can be acted upon by selection, level III genes can evolve the ability to direct further the manner in which natural selection acts on them. "Automodulation" genes of level IV, like level III genes, include information that programs changes in their own structure, but now in response to self-determined environmental or phenotypic condition rather than the conditions in which they just happen to find themselves. That is, these genes can even alter the conditions under which natural selection acts on them. All of these levels are well documented in both procaryotes and eucaryotes. There is some evidence for a fifth level of "experiential genes" that can be passed on to offspring in modified form. Future geneticists may find themselves resurrecting some Larmarckian principles. Finally, Campbell speculates on a sixth level of "anticipatory/cognitive genes" that could utilize chemical or electrical cues in their environments to "anticipate" future changes in their milieu and alter their expression appropriately in advance.

Whether the fifth and sixth levels occur is not known. But the existence of levels I through IV genes introduces to population genetics a "complicated form of causality" in which genes can evolve so as to direct their own evolution. They do not stand peacefully by and simply increase or decrease in frequency according to the whims of the external environment. The hierarchical framework they provide for genetic response to selection vastly confounds the rather simplistic notions we have portrayed in the preceding chapters. Complex multiple options of response are furnished, and it is tempting to say that the pathways for adaptive fine-tuning are multiplied to an extent that linkage and epistasis can be relegated to a trash heap of outmoded irritations. Unfortunately, we have no idea at this time just how common profane genes are (although at least 25 families of such genes appear to exist in Drosophila melanogaster). The future of population genetics may be quite different from the field we recognize today.

Chapter 11

The Evolution of Population Parameters

We discuss this subject in three parts. First we ask whether the effort put into reproduction by an organism should be expended in one massive effort or partitioned gradually. Species that breed all at once generally die soon thereafter and are often referred to as *big-bang strategists.* A more awkward but genteel term is *semelparous.* Species that hold back their effort, use the saved energy for survival, and live to breed again are termed *iteroparous.* Next we raise the question of when populations behave in such a way that their dynamics is best described with discrete as opposed to continuous models. Finally, we ask how the effort spent in reproduction should be allocated, either into a small number of large well-cared-for young, or a large number of smaller less-well-provisioned young.

11.1 The Allocation of "Effort" into Reproduction and Survival

Before beginning we need a workable definition of "effort." That offered by Trivers (1972)—"any investment by the parent in an individual offspring that increases the offspring's chance of surviving (and hence reproductive success) at the cost of the parents' ability to invest in other offspring"—is, at best, difficult to measure and so is probably unusable except in theoretical constructs. In practice, *reproductive effort* has been used pretty loosely. In this book we will define it as the proportion of a parent's available energy supply that is allocated to egg production and care of young (see Williams, 1966). We begin with a consideration of survivorship curves (Fig. 11-1).

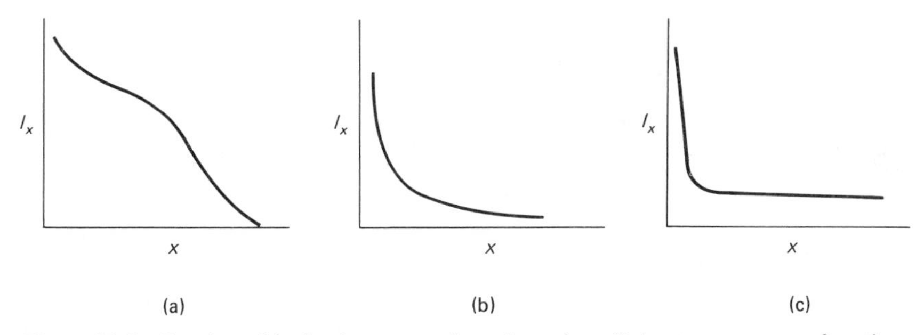

Figure 11-1 Survivorship l_x, the proportion of a cohort living to age x, as a function of age: (a) type I curve, characteristic of birds and mammals; (b) type II curve, characteristic of opportunistic species; (c) type III curve, characteristic of species with high larval mortality and low adult mortality, such as large marine fish.

Iteroparity and Opportunism

At a very superficial level we can argue that individuals in populations following a curve such as that shown in Figure 11-1b cannot, at any age, expect to live to the next year with much certainty and should therefore reproduce as soon as possible, while they are still around to do so. In fact, a partial reproductive effort is likely merely to waste the rest of a full expenditure. Species of this sort should be semelparous. Of course, if full effort is put into one massive reproductive binge, the individual will subsequently be drained, and will increase still more its chances of dying. We should expect to find survivorship curves which begin by dropping rapidly and smoothly to be truncated at the point where reproduction occurs. The only apparent reason for discrediting this argument is that in some cases a species may display indeterminate growth. If growth is indeterminate and if larger females produce more eggs, we are faced with opposing advantages: breed now and assure yourself a contribution to future generations, or wait and if you are lucky enough to live, endow the future with a great many more descendents. But large size often brings with it a superior buffering against the physical elements and so a smaller mortality rate. In species with indeterminate growth, then, we might generally expect to see the survivorship curve progressively leveling off after a juvenile or larval period of high death rate (Fig. 11-1c). If the leveling off is very marked, the risk of dying before the next breeding season is greatly alleviated and there is less need to breed now. On the other hand, there is also less advantage in waiting. Even if some "effort" is expended on present reproduction, the organism can probably expect to live and reproduce again. We should expect iteroparity. Species displaying semelparity usually exhibit survivorship curves of the sort shown in Figure 11-1b (called type II curves). Most insects are examples, as are annual plants. Species that show type III survivorship curves (Figure 11-1c) are usually iteroparous: large marine fish, for example, or perennial plants. Species demonstrating type I survivorship curves (Figure 11-1a) can, until they get quite old, expect to live to the next breeding season with reasonable assurance. Like species of the survivorship type III variety, they too can be expected to be iteroparous. Type I curves are characteristic of birds and mammals.

The arguments above are a bit primitive, and they do not account for additional kinds of life histories—deferred big bang strategies, for example—but they set the stage for the kind of reasoning that we shall employ in this chapter. We now look a bit more quantitatively at this question of semelparity versus iteroparity.

The classic approach, and the first attempt to make such arguments, comes from Lamont Cole (1954). Later, the same line of reasoning, both criticisms and extensions of Cole's work, appeared in papers by Charnov and Schaffer (1973), Goodman (1974), and others. The various ideas were synthesized and reviewed by Stearns in 1976. All of the formulations above, however, are superseded by a simpler, more general, verbal argument presented by Horn (1978). Horn's approach is this. Suppose that once maturity is reached, mortality falls to zero, and that breeding begins at age 1 (in whatever units are appropriate) and continues at unit intervals thereafter, forever. Such individuals would be the ultimate immortal exemplifiers of iteroparity. In such a population the reproductive value, the expectation of future offspring produced, would be the same for all individuals once they reached maturity, regardless of age. But this means that a young individual, at first reproductive age, can expect to contribute the same number of descendents as its parent. Thus the parent can do equally well in terms of offspring contribution to the future by continuing in the old way, living and reproducing forever, or dying and producing a replacement in the form of one extra surviving young. If the survivorship of newborns to maturity is p_0, the second alternative calls for increasing litter size by exactly $1/p_0$. If we now relax the assumption of immortality and suppose that survival from one breeding season to the next is p_a, a dying parent can make up for its death by producing only p_a extra offspring—that is, increasing its litter size by p_a/p_0. If we wish to couch contribution in terms of *genetic* contribution, so that the argument pertains to selective advantage, we simply note that if mates are genetically unrelated, the offspring will carry half of a parent's genome; offspring of an offspring will carry only one fourth of the parent's genome. Thus to equal the genetic consignment of continuing to live and reproduce, a dying parent must produce not p_a/p_0 but $2p_a/p_0$ extra young. Where inbreeding occurs and mates may have more than zero relatedness, the appropriate number is $\alpha p_a/p_0$, where $1 < \alpha < 2$.

The argument can be stated more formally. For each parent, the genetic contribution via descendents, including the parent if it lives, in the next time unit, is

$$W = p_a + \tfrac{1}{2}b'p_0 \tag{11-1}$$

where b' is the birth rate. (As in Chapter 2, we use a prime to denote discrete-time processes.) If the parent dies and uses the released effort otherwise required for survival in making extra offspring, this becomes

$$W = 0 + \tfrac{1}{2}b^*p_0 \tag{11-2}$$

where b^* is the litter size of a semelparous parent. Semelparity is favored when

$$\tfrac{1}{2}b^*p_0 > \tfrac{1}{2}b'p_0 + p_a$$

or when

$$b^* > \frac{2p_a}{p_0} + b' \tag{11-3}$$

Therefore, in species where for reasons of energetics, physical size, and so on, individuals find it a relatively trivial task to increase litter size by a fraction $(2p_a/p_0)/b'$ by dying, semelparity should be favored. In species whose individuals cannot do so, iteroparity should prevail.

A test of this approach is almost impossible. To some degree we can estimate the relative demands of survival and of producing young, but this is difficult or impossible in many cases. Therefore, because we see the *faits accomplis* of natural selection and seldom the process, this appealing argument for predicting semelparity or iteroparity is, to a large degree, merely an academic titbit. But like the logistic equation, it possesses some heuristic value. To wit: Populations that suffer high juvenile mortality relative to adult mortality have high p_a/p_0 values. Such populations are less likely to be able to meet the inequality in (11-3), so are more likely to be iteroparous than populations with equal b', suffering relatively high *adult* mortality. Trees and large marine fish suffer high mortality when young, very little when adult, and should be (and are) iteroparous. Mayflies, which suffer very high larval mortality, but also very high adult mortality (even before reproducing), are much more likely to meet the inequality (and are, plainly, semelparous). One may argue that species which are already semelparous, and so suffer high adult mortality by definition, automatically present low p_a/p_0 values; that iteroparous species, by definition, must display high p_a/p_0 values; and that the argument is thus circular. If we attempt to "explain" the fact that some species are semelparous, others iteroparous, this is a valid objection. But if we simply suggest that once one of these life history strategies is present, the resulting p_a/p_0 value encourages positive feedback and the maintenance of that strategy, then the approach is valid. If we were to bring an iteroparous species into the laboratory and subject it to artificially high adult mortality, greatly lowering p_a/p_0, we could predict an evolutionary change predisposing an earlier concentration of reproductive effort and, perhaps eventually, a full shift to semelparity.

Suppose that a population lives in a temporally uncertain environment. This complication has been addressed by Schaffer (1974a) and Schaffer and Gadgil (1975). If uncertainty affects primarily the birth rate, b', or juvenile survival, p_0, a parent cannot count on a successful litter. In fact, there is always a chance that $\tfrac{1}{2}b'p_0$ will drop to zero, leaving the parent with no progeny at all. The parent that, so to speak, puts all her eggs in one basket, stands to lose everything. Under such circumstances iteroparity gains an added advantage. Stearns (1976) has referred to species that practice iteroparity, even when the average values of b' and p_a/p_0 otherwise suggest an advantage to semelparity, as "bet hedgers." If uncertainty, on the other hand, acts primarily on adult survival, then p_a is effectively lowered, (11-3) is more easily met, and we might expect semelparity.

Refer for a moment to the logistic equation with density-independent fluctuations:

$$r = r_0\left(1 - \frac{n}{K}\right) + S$$

where S is a random variable changing in time. Species that live in a very uncertain environment (unpredictable changes in S), or one that may be predictable but nevertheless fluctuates, experience population fluctuations. In such populations the expected value of n falls below, perhaps much below K. In terms of selective advantage, then, an individual's tolerance for those around it, its ability to live and glean resources under crowded conditions, is seldom at issue. Fitness depends on the ability to produce offspring rapidly when conditions are amenable. Natural selection can be expected to act primarily to increase r_0. Recalling the phenotypic selection model of Chapter 10, we can write

$$\text{rate of change in } r_0 \text{ under natural selection} = \lim_{n/K \to 0} \frac{dr}{dr_0} = 1$$

$$\text{rate of change in } K \text{ under natural selection} = \lim_{n/K \to 0} \frac{dr}{dK} = 0$$

Therefore, species that suffer great population fluctuations tend to evolve large r_0 values which enable them to capitalize on periods of environmental benevolence to grow. Such *opportunistic* species are often referred to as *r-selected. Equilibrium species*, though, those for which n/K is close to 1.0, can be characterized as follows:

$$\text{rate of change in } r_0 \text{ under natural selection} = \lim_{n/K \to 1} \frac{dr}{dr_0} = 0$$

$$\text{rate of change in } K \text{ under natural selection} = \lim_{n/K \to 1} \frac{dr}{dK} = \frac{r_0}{K}$$

Equilibrium species, which by definition undergo only slight population fluctuations, are often called *K-selected*. The first thorough discussion of r and K selection is found in MacArthur and Wilson (1967); see also Pianka (1970, 1972). A more recent and formal presentation of the role of population fluctuation in the evolution of growth parameters is given by Armstrong and Gilpin (1977).

Inasmuch as r_0 and K are terms used with the logistic equation, an equation that is biologically not very accurate, and are otherwise not rigorously defined, there has been a recent trend to downplay the terms r and K selection. Horn (1978) calls the terms "barbaric." We can be a little more precise in our objections. First, r_0 in the logistic equation is not only a measure of maximum growth potential. It is also the value of r when $n \to 0$ (the two, in general, need not be the same). Also, r_0 is ostensibly independent of K, an assertion that is patently untrue. Recall that $r = b - d$, where b and d are birth and death rate. Now suppose that either one or both of these more biologically definable parameters varies (as in the logistic equation) in linear fashion with population density:

$$b = b_0(1 - \xi n)$$
$$d = d_0(1 + \nu n) \qquad \text{where } \xi \text{ and } \nu \text{ are constants}$$

Birthrate declines from a maximum of b_0 when the population is very sparse; death rate increases from a minimum of d_0. Combining these expressions gives

$$r = b_0(1 - \xi n) - d_0(1 + \nu n) = (b_0 - d_0) - (b_0\xi + d_0\nu)n$$

Hence

$$r = r_0\left(1 - \frac{n}{K}\right) \tag{11-4}$$

where
$$r_0 = b_0 - d_0$$
$$K = \frac{b_0 - d_0}{b_0\xi + d_0 v}$$

(see the derivation of equation 2-13 in Chapter 2). Thus it is unrealistic to view natural selection as acting to increase r_0 or to increase K, for the two are strongly interdependent. Indeed, from (11-4) we can see that an evolutionary change in r_0 may affect K in the *same* direction. Write

$$\frac{dK}{dr_0} = \frac{\partial K}{\partial b_0}\frac{db_0}{dr_0} + \frac{\partial K}{\partial d_0}\frac{dd_0}{dr_0} = d_0(v + \xi)\frac{1}{d(b_0 - d_0)/db_0} - b_0(v + \xi)\frac{1}{d(b_0 - d_0)/dd_0}$$
$$= \left(d_0\frac{db_0}{dd_0} - b_0\right)(\xi + v)\left(\frac{db_0}{dd_0} - 1\right)$$

This quantity is positive, meaning that evolutionary changes move r_0 and K in the same direction, if $db_0/dd_0 < 0$ or if db_0/dd_0 exceeds b_0/d_0. If r_0 and K evolve in parallel direction, it is nonsense to talk about r and K selection as a dichotomy. Other classification schemes have been proposed. For example, Grime (1977) suggests that plants might be categorized as C (competitive), S (stress tolerant), or R (ruderal). The first group is characterized by species showing maximum vegetative growth in productive, reasonably undisturbed areas. The second displays less vegetative and reproductive vigor but is adapted to endure low nutrient or otherwise stressed conditions. Ruderal plants have a short life span and high seed productivity. They grow in severely disturbed but potentially productive areas. The reader is referred to Chapter 12 for further discussion of r-selected, K-selected, and other possible categories of species. The general argument that strong population fluctuations serve to raise reproductive potential and that populations mostly near their equilibrium value evolve higher equilibrium values is still valid, but it is probably best to stick to the less loaded terms, "opportunistic" and "equilibrium" species.

The immediately preceding line of reasoning can be wedded to that regarding the evolution of semelparity and iteroparity in the following way. Populations with small p_a are prone to large fluctuations and so evolve large population growth potential. Large growth potential, by definition, means large numbers of offspring (high b' value). But as b' rises, for a given reservoir of parental "effort" available, the provisioning per individual offspring must go down. This must, almost certainly, result in lowered juvenile survival (p_0 drops). Because p_a and p_0 are *both* low, it is not possible to draw any general conclusions about their ratio. However, because b' is large, p_a/p_0 is likely by comparison to be very small. Thus inequality (11-3) is apt to be met and semelparity expected. If environmental fluctuations cause a high variance in $b'p_0$, and environmental variations are not perfectly synchronized in space, the opportunistic, semelparous species can also alleviate the uncertainty to some degree by actively dispersing. Finally, species which to some degree rely on dispersal in order that at least some of their offspring end up in suitable habitat, will do best if their offspring are habitat generalists. Generalists can be expected to make the best of the largest variety of situations in which they may find themselves. The emerging pattern is summarized in Figure 11-2. It would be impossible to include all conceivable pathways in this figure, the cause–effect relationships are only qualitative, and the patterns only general rules of thumb with, undoubtedly, many exceptions. But the relationships are, nevertheless, not without interest. Species with type III survivorship curves can generally be expected to follow pathway III, those with type II curves, pathway II, and type I species, such as birds and mammals, will usually follow pathway I.

A few kinks, as well as additions need to be pointed out.

1. Parental care should be expected only when the associated effort on the parent's part is likely to pay off in increased p_0. Species with parental care, then, will be those that first, have a reasonably high rate of adult survival, and second have the wherewithal, through size, mobility, or teeth, actually to have an impact on the survivorship of their young. On the other hand, some

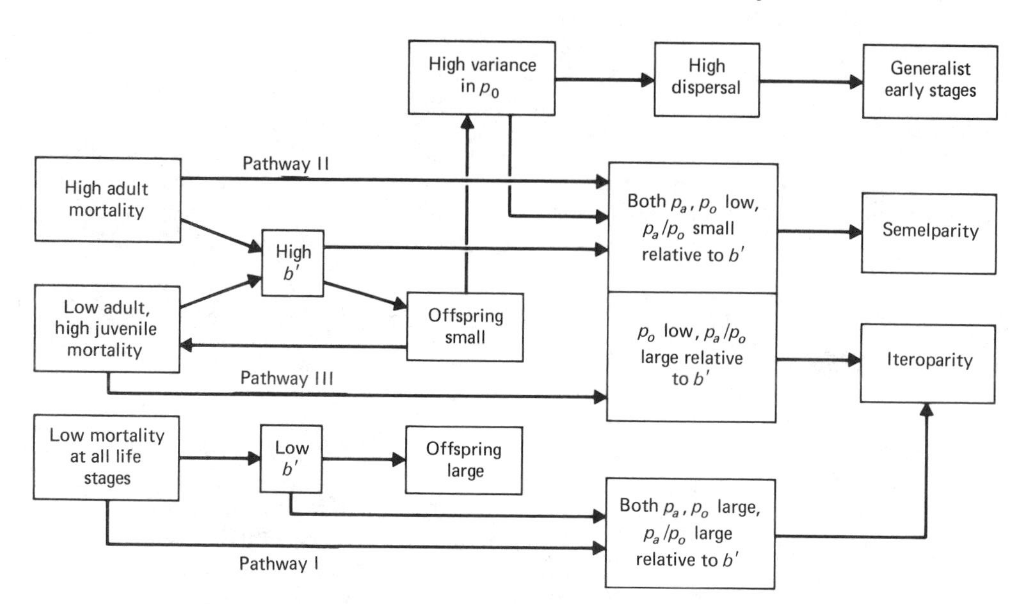

Figure 11-2 Summary diagram of some of the relationships between mortality pattern, fecundity, body size, dispersal tendencies, and reproductive life history. The arrows indicate positive causal pathways.

small, passive species may be preadapted to carry their young with them. For example, a number of microcrustaceans have "marsupia" in which the young develop. In such cases p_a may be low but p_0 will be scarcely lower. Also, there is no reason to expect b' to be particularly large in such cases. As long as parental care can extend until the young reach the neighborhood of adult size, in fact, we can expect b' to be small if for no other reason than limited capacity of the marsupium. Thus it would seem possible to find semelparous species with parental care and small numbers of young. This initially unlikely seeming strategy is indeed found in some microcrustacea (Childress and Price, 1978). The intuitive notion that very small size is necessarily associated with opportunism is therefore in error.

2. Species that to some degree rely on dispersal of their offspring would do well to ensure that those offspring maximize their likelihood of ending their dispersal phase in suitable habitat. We have already mentioned that being a generalist might help in this respect. But in addition, any rudimentary form of behavioral disposition to end one's wanderings when the conditions at hand are appropriate would also help. For example, barnacle larvae may be quite choosy about where they settle to metamorphose. Both substrate and the presence of conspecifics have been shown to be important in this group of animals (Knight-Jones, 1953; Crisp, 1961). Even plant seeds in a sense choose their final germination place by waiting until wind, water, or animal vector carry them to appropriate surroundings. Harper and co-workers (Harper et al., 1965; Harper and Benton, 1966) have experimented with a variety of species, finding that each has its own characteristic tastes as to what circumstances call for germination. Of course, if a barnacle larva or a seed spend too much time drifting before finding suitable surroundings, they will die and even any minute chance of success will be lost. Both barnacles and seeds will eventually reach a point where "desperation" overtakes them; they give up the search and progress, conditions willing, to their next life stage.

3. One founding principle of the lines of reasoning above is that by dying, a semelparous species is able to divert what energy would otherwise go into homeostasis into reproduction. This principle need not be all or none; some species may put their all into reproduction and then die, some will hold back a bit, and live with some acceptable probability, and some will dribble out their reproductive energies and live long lives. Both of the latter two categories are iteroparous, but the second is less removed from the big bang strategist than is the last; we can discern degrees of iteroparity. In the purely semelparous region are such adult-wasted species as Pacific salmon. Preying mantid males may give their all as the female with whom they copulate munches on their hindbrain. In spiders the female often eats the male, the added energy from the meal ideally going toward the production of a bigger brood. Carpenter bees, in their fanatic guarding of a nest, burn themselves out with frightening regularity and are replaced by other zealots, which also die in short order. Small mammals fall in the middle category, turning out litters that exact a considerable energy drain on the mother (McClure et al., 1977). Extreme iteroparous species include most primates, carnivorous mammals, and most trees. Let us see if, in fact, the reproductive effort really varies inversely with survival. Browne and Russell-Hunter (1978), in an exhaustive scouring of the literature on molluscs, found that (averaged over all species for which data were available) 29.90% of all nonrespired energy expended by semelparous species went into production of young. Among iteroparous species the corresponding figure was 19.21%. Tinkle (1969) looked at weight per clutch per parental body weight, presence of conspicuous sexual coloring, use of elaborate courtship, aggressiveness, and existence of multiple clutches per season as measures of "risk" in reproduction. As such they represent measures of parental expenditure. Each of 25 lizard species was scored 0 or 1 on each of these variables and the mean used as the amount of total parental expenditure. Tinkle found that this measure was considerably higher among short-lived species (average of .8 for 12 species) than among long-lived species (average of .37 for 13 species). Primack (1979) used three estimates of reproductive effort in a study of annual and perennial plants of the genus *Plantago*. With respect to field-collected plants, the number of seeds per 10 cm^2 of leaf surface averaged 2.12 for nine species of annual (semelparous). For perennials (iteroparous) the average was 2.24 for six species. With respect to milligrams of seed per 10 cm of leaf surface, the means were 2.05 for annuals and 1.79 for perennials. Finally, the number of capsules per plant averaged 1.84 for annuals, 1.56 for perennials. Primack also took data from herbarium specimens. For annuals the number of seeds, milligrams of seeds per 10 cm^2 of leaf surface and number of capsules were 2.56, 2.28, and 2.08. For perennials the corresponding figures were 1.36, 1.82, and 1.64. All comparisons are in the direction expected. Of course, we are dealing here with different species, with whole constellations of differently evolved characteristics, so the observations are only soft confirmation of the theory.

4. The theory predicts that species (or, more reliably, variants within a species) whose young live in more uncertain environments than the adults should be iteroparous. Additional theoretical impetus for this view comes from the work of Murphy (1968), who ran computer simulations of competition between two populations, one of which was strongly, the other weakly iteroparous. In both cases R, calculated from the Euler equation,

$$1 = \sum R^{-x} l_x b'_x$$

before density feedback, was 4.0. Density effects were modeled using a variant of the Ricker equation,

$$n_i(t + 1) = n_i(t) \exp\left[T_i - \frac{n_i(t) + \alpha_{ij} n_j(t)}{K_i} - S_i(t)\right]$$

where T_i is a constant, K_i is carrying capacity, α_{ij} the competition coefficient, and S_i a fluctuating number with mean 1.0 and variance σ^2. Parameter values were set so that coexistence occurred in the absence of variation in S. Note that variations in S describe density-independent mortality over the first interval of life. Thus as σ^2 increases we should expect to find increasingly strong competitive advantage for the more strongly iteroparous population. This is just what Murphy found (see also Holgate, 1967). With respect to support from real world data, Murphy found that a plot of reproductive span against variation in spawning success showed a linearly increasing trend for certain fish.

The converse of the assertion above is that species whose adults live in uncertain, fluctuating environments are more likely to be semelparous or marginally iteroparous than are

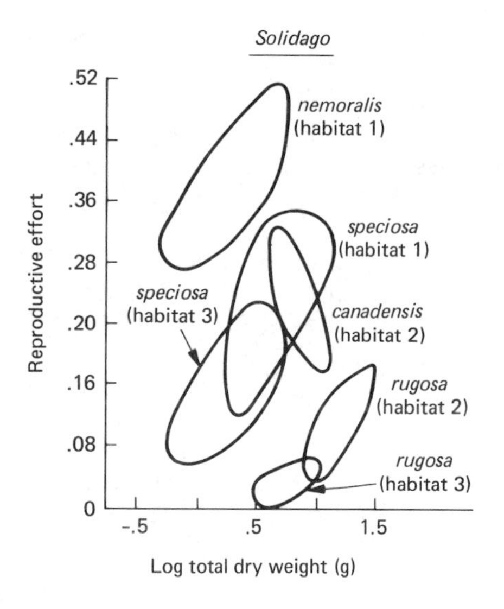

Figure 11-3 Reproductive effort (i.e., ratio of dry weight of reproductive to total aboveground tissue), as a function of total aboveground dry weight for various goldenrod populations. Each closed curve embraces the points representing individuals of a population. 1, dry site; 2, wet site; 3, hardwood site. (From Gadgil and Solbrig, 1972, Fig. 5.)

forms living under more constant conditions. Gadgil and Solbrig (1972) examined four species of goldenrod (genus *Solidago*) growing on three areas ranging from highly disturbed (uncertain) to undisturbed. The most disturbed area was dry, early successional, highly trampled, with mostly herbaceous vegetation. The intermediate area was wetter, later succession, less trampled, and contained a much larger proportion of woody shrubs. The last area was mesic, late successional hardwood, with very little trampling. According to theory we should expect higher reproductive effort among those plants growing in the first, lower effort among those growing in the last area. Gadgil and Solbrig estimated reproductive effort with the ratio of dry weight of reproductive tissues (flowers, flower buds) to total, aboveground biomass. Reproductive effort so estimated varied with time, but the three areas fell into a consistent, relative pattern by about halfway through the growing season. Plotted against log total dry weight the reproductive effort for the four species in the three areas is shown in Figure 11-3. As was pointed out above, it might be misleading to compare different species. But within species, *S. speciosa* displays greater reproductive effort, as expected, in habitat 1 than in habitat 3; *S. rugosa* puts more into reproduction in area 2 than area 3, as expected (see also Abrahamson and Gadgil, 1973). Gadgil and Solbrig, in the same paper, report on dandelion (*Taraxacum*) populations from three (other) areas of high, medium, and low disturbance. In this species, where asexual vegetative reproduction is important, the concept of reproductive effort is a bit unclear. However, we shall consider sexual reproductive effort, estimated relatively by the number of flower heads, and forge on. Using electrophoretic techniques, Gadgil and Solbrig identified four genetic "biotypes," which they labeled A, B, C, and D. Observations on phenotype and on distribution of the four biotypes showed that type A produced the most flower heads, with B second, C third, and D last (Table 11-1).

Table 11-1 **Number of Flow Heads per Plant**

	Biotype			
Population	*A*	*B*	*C*	*D*
1	3.6	2.3	1.5	—
2	2.6	2.1	1.9	—
3	3.8	2.3	0.5	1.2

This observation would support the theory if A were found to be most common in area 1, D in area 3. As can be seen in Table 11-2, this is clearly the case.

Table 11-2 **Frequency (Number Found in Samples)**

	Biotype			
Population	*A*	*B*	*C*	*D*
1	73	13	14	—
2	53	32	14	1
3	17	8	11	64

That there is a genetic basis for the observed life history variation in dandelions is strongly suggested by Gadgil and Solbrig's data. That is hardly surprising, but the evidence is reassuring.

In a further set of investigations, Solbrig and Simpson (1977) raised biotype D and other dandelions in undisturbed and disturbed (defoliation or removal of the entire adult plant at midsummer for the first 2 years) sites. After 4 years, type D, the low-reproductive-effort form, made up 85% of the total number of stems and 91% of the biomass in the undisturbed area, but only 7 to 9% and 3 to 7% of stems and biomass, respectively, in the disturbed site.

On the other hand, while genetic differences for life history traits within a population indicate the efficacy of natural selection, the evolution of life history traits does not necessarily require that we observe such differences. It is possible that selection has molded an ability to respond appropriately via developmental plasticity. Thus Hickman (1975), working with the annual plant, *Polygonum cascadense*, in the Oregon Cascades, noted that

1. Individuals apportion up to 71% more energy to reproduction in harsh habitats than in moderates ones (as expected if the harshness affects the adult plant).
2. This difference is developmental, not genetic.

A number of genetic studies of the evolution of life history traits have been performed. In some cases the results are supportive of the theory above; in others, not at all. C. E. Taylor et al. (1981) raised *Drosophila pseudoobscura* in one of two ways. In the first, flies were simply allowed to grow and reproduce undisturbed in population cages. The presence of density feedback was forcefully indicated by the presence of adults with obviously stunted growth. Presumably the average population size was close to the maximum obtainable under completely constant conditions, so that selection should be for increased survival rather than high birthrate. The second set of populations were raised under "*r*-selected" conditions. Individual females were placed into a bottle with unlimited food. As their offspring emerged, they were pooled and 10 individuals, half of each sex, were randomly assigned to vials. Subsequently, emerging flies were removed and put into new, clean vials every 2 to 4 days, so that crowding of adults never occurred and competition with offspring never occurred. As expected, the first group of populations gradually evolved greater longevity, but only among females. In the second set, longevity dropped as expected, but again only for females. Why should the theory be supported for females only? The reason given by the authors is that females demonstrate a strong preference for virgin males. Thus, even though males may live longer, this fact is of no consequence and cannot be selected for. In the second set of populations, male longevity is not selected against for the same reason. Another possibility is that since females store sperm, a male's best strategy is always to fertilize as many females as possible in a short time, then die so as not to compete with its own offspring. Quirks of this sort are grist for bewilderment and frustration for the theoretician. They can seldom be anticipated and are often of critical importance. The reader should keep the possibility of such unexpected wonders in mind while reading the rest of this chapter and those to follow.

Other genetic problems exist, problems that raise questions about the entire line of reasoning we have pursued so far. Throughout the preceding pages we have woven a theory of trade-offs: Energy that goes into growth or survival is unavailable for reproduction, greater reproductive effort early in life endangers subsequent survival, larger numbers of young must come at the expense of energy expended per young. Available data confirm the almost trivial certainty of these trade-offs within individuals. But natural selection molds life histories not from *phenotypic* correlations *within* individuals, but rather from *additive genetic* correlations *among* individuals. Is there evidence that individuals that put large allocations of energy into growth put less into reproduction? Or do individuals with the capability of growing faster also possess the ability to produce more offspring? Are they simply all-round superior individuals? A glance at some data should serve to raise a small red flag. Temin (1966), working with *Drosophila melanogaster*, found a positive correlation between fertility and viability. Mukai and Yamazaki (1971) found a positive correlation between viability and development rate in the same species. Also investigating *D. melanogaster*, Giesel and Zettler (1980) measured genetic correlations between fecundity in the first third of life, the last third of life, and peak reproductive output and found all three to be positive. Individuals with high peak reproduction also generally were reproductively viable later in life than individuals with low peak reproduction (except for extremely fecund flies, for whom a trade-off between life span and reproductive output *did* appear). Giesel et al. (1982) extend the list to include positive genetic correlations between development rate and age at death. Among inbred strains still other correlations were found between life history traits contributing to fitness. Finally (Giesel and Zettler, 1980), among the various strains, fitness (measured by the population growth parameter, r) under extremely marginal ($n \ll K$) conditions was positively correlated with fitness under equilibrium ($n \rightarrow K$) conditions. These results all indicate that, among individuals (or strains), those flies that are superior with respect to one trait are also superior with respect to others. That is, there are *not* trade-offs of the sort needed to fuel the energetic approach to the evolution of life histories. Of course, these observations are on phenotypic or genetic values, not additive genetic values. Also, were the subjects not thoroughly adapted to the laboratory conditions under which the measurements were made, the positive correlations might have been due to a general evolutionary improvement of the stock. Finally, some trade-offs *are* apparent. Giesel and Zettler (1980) found that strains with an early age of peak reproduction (contributing to higher fitness—see Chapters 3 and 12) also had lower fecundity at all ages, and were less long-lived. Snell and King (1977) report that in rotifers, life span and number of offspring produced per day are negatively correlated, as are life span and the number of offspring produced on the first day of maturity. The trade-off in this case, between fecundity and longevity, is such that very long and very short lived strains have almost identical contributions to population growth rates.

Additive genetic correlations are more meaningful to evolutionary theory than phenotypic or total genetic correlations. Here again we find a mixed picture. Rose and Charlesworth (1981a) report a negative additive genetic correlation between longevity and early-life fecundity, and between early- and late-life fecundity in *D. melanogaster* (but see Giesel, 1979). They also (1981b) found a negative additive genetic correlation between late fertility and both early fertility and longevity. On the other hand, Eisen and Johnson (1981) found that when laboratory mice were selected for larger litter size, they also evolved larger body size, larger testes per unit body weight, and larger epididymus and seminal vesical weights. Selection for larger body weight resulted not only in the intended gain, but also in larger litter sizes. The partial additive genetic correlation of testis weight and litter size, corrected for body weight, was $.60 \pm .04$. Islam et al. (1976) selected for testis weight in laboratory mice and achieved, inadvertently, higher ovulation rates. Land et al. (1980), cited in Eisen and Johnson, (1981), report the same for sheep. Clearly, we cannot simply assume the existence of trade-offs in a theory of life history evolution. Finally, the evolutionary models are vastly complicated by the observation that genetic architecture for life history traits may change with environmental conditions. For example, Giesel et al. (1982) report that significant changes in genetic correlations occur with temperature shifts in *D.*

melanogaster At 22°C, the genetic correlation between development rate and age of peak reproduction was −.721. At 25°C and 28°C it is very nearly +1.0. The genetic correlation between development rate and age of death is −.50 at 22°C and +.708 at 28°C. For peak reproduction and age of death the corresponding figures are about −1.0 at 22°C and +1.0 at 28°C.

For more information on heritabilities and genetic correlations among life history traits, see Istock et al. (1976), Derr (1980), and references given in Giesel et al. (1982). An interesting discussion of the parallel changes in additive genetic and environmental correlations among traits can be found in Hegman and DeFries (1970).

Age of Reproduction

If a species is semelparous, at what age should it reproduce? (The same question, as it applies to iteroparous species, is the subject of the next chapter.) Three variables are critical. First, the longer an organism waits to breed, the more likely it will be to die without doing so. Second, the longer it waits, the larger, or stronger, or more experienced it is apt to be and so the more fecund. Third, an individual that breeds early, and whose offspring inherit the tendency to breed early, compounds its contribution to future generations at a faster rate. A comparison of early and late breeding is rather like a comparison of compounding principal quarterly or annually. All else being equal, it is selectively advantageous to reproduce as early in life as possible. These three considerations are conveniently coordinated in the Euler equation,

$$1 = \sum_x R^{-x} l_x b'_x$$

In the case of a semelparous species, of course, there is only one x for which b_x is nonzero. Therefore, if we call the age of reproduction T, we have

$$1 = R^{-T} l_T b'_T \tag{11-5}$$

or

$$\ln R = \frac{1}{T} \ln (l_T b'_T) \tag{11-6}$$

Recall that, in theory, natural selection acts to increase population mean fitness or, equivalently, ln R. Thus if we know the manner in which l_T and b_T vary with T, we can solve (11-6) explicitly for $\hat{T}$; simply find that value of T which maximizes ln R. But before surrendering to such temptations, consider some complications. In many species survivorship and fecundity are not functions of age per se, but rather of size. In such cases it must still be possible to write the two life history parameters as functions of age, but the form of the functions now depends on such esoteric quantities as food availability, how much time individuals can afford to spend out of hiding to forage, to what extent population crowding stunts growth, and the population density. The reader interested in dwelling on such matters is referred to Lynch (1980) and the discussion of his work at the end of Chapter 12. But it gets even more complicated. If we consider our species in conjunction with other interacting species, it is no longer even clear that l_T necessarily drops as age of reproduction, T, rises! The latter statement deserves clarification.

Consider a situation where the offspring of our species are highly palatable and, being in great numbers, attract predators. If the development period is long, there may be little that natural selection can do about it. But if reproductive timing of the various individuals in an area is variable, and heritable, evolution may take interesting courses. If individuals are well dispersed and each produces few enough young that predators gain little by congregating, it may prove advantageous to increase the variance in timing. But if individuals congregate, spreading offspring production over time will attract and hold predators, allowing them to eat a prolonged feast. In such a circumstance it would seem best to synchronize. But there is still a problem for the prey species. One good, reproductive year provides a large quantity of food for predators. If the prey species is very important in the diet of one or more predators, that predator too must have a good year. This means that next year there will be more predators. A good year now for the prey species seems unavoidably to lead to a bad year following. But this problem, too, can be overcome. Suppose that the prey species not only breeds synchronously within a given year, but also skips a number of years before again breeding synchronously. If the periodicity of reproduction is longer than the generation time of the predator, the effects of the predator's good reproductive year will have run their course and the prey will be able to breed safely. Thus predation may favor long delays between birth and reproduction, and so may increase the age of reproduction, T.

How such periodic reproduction, a deferred big-bang strategy, might actually have evolved is a matter of some controversy (see, e.g., Bulmer, 1977), but it is clear that it has happened many times. It is also quite clear that the explanation given above, a "predator saturation" ploy, is the reason for the maintenance of the strategy in species where it has been studied. It is not at all difficult to see how synchrony might be maintained. Individuals that breed early will not be able to saturate the predator and may have difficulty finding mates; individuals breeding late will reap the legacy of their synchronized, earlier-breeding colleagues—hordes of needy predators. Some examples of species with periodic reproduction are the following.

There exist two North American species of periodic cicada, *Magicicada septendecim* and *M. cassius*, which emerge to breed in a period of only a few days once every 17 years. The larvae live beneath the soil, among roots of trees, and emerge only briefly as adults before dying. Where they co-occur they are synchronized in their breeding (Dybas and Lloyd, 1962). Two 13-year periodic cicadas live in the same area; they, too, are synchronized where they cooccur. Karban (1982) confirmed the predator saturation hypothesis for these creatures by catching first instar nymphs of *M. septendecim* in nets as they fell to the ground to start their life cycle. In this way he estimated population density at each of 16 sites. He also estimated predation by collecting and counting bird-discarded wings in mesh traps. Relative abundance of birds at the 16 locations was also recorded. The number of wings recovered (rate of predation) was directly proportional to the number of birds but independent of the density of adult cicadas. Thus the danger of predation (probability any given individual is eaten) *does* drop with a rise in cicada density. Also, the number of nymphs recovered per adult *rose* with the local number of adults. This suggests, perhaps, an intraspecific facilitation—an Allee effect—or possibly a tendency for the cicadas to move toward population centers, thus exaggerating the predator saturation strategy.

Bamboos also are commonly periodic. Janzen (1976) lists a large number of such species, with their generation times, and discusses the problem of the evolution of periodicity in great detail. Among these bamboo the seeds vary greatly in size, from rice kernel dimensions to between 100 and 350 g. But in all cases they are nontoxic and highly palatable (as required by the evolutionary hypothesis discussed above). Indeed, practically every kind of animal imaginable eats them, from insects to birds and mammals. Also, when seed is set, it is set in immense quantities. *Phylostachys bambusoides*, for example, breeds in highly synchronized fashion once every 120 years or so, producing pear-sized seeds that cover the ground to as much as 5 or 6 inches. Following seed set of this and other periodic species, there is evidence (recent and historical) of huge migrations, and massive population increases followed, when the seed source is gone, by mass starvation. Rodents, whose populations had burgeoned, are suddenly left with nothing to eat and so invade agricultural fields. Crop disasters and famine often follow. There is no doubt whatsoever about the predator saturation abilities of these bamboo.

But let us take the simple course, assuming that (11-6) suffers no complications due to density feedback or species interactions. We offer this analysis, then, in the spirit that life may generally, but not necessarily always, be complicated. In many fish species fecundity (b' = number of eggs produced) is approximately proportional to body length (see, e.g., Hubbs, 1958; Nikolskii, 1963). If we know suppose that length varies approximately with age, T, we have $L = 1 + \beta T$, and

$$b'_T = \alpha(1 + \beta T) \tag{11-7}$$

where α is a measure of fecundity and β a measure of growth rate. Consider now (say) any one of the species of Pacific salmon. Let survival to the end of its first year of life be p_0 and its survival thereafter be p_a. Then

$$l_T = p_0 p_a^{T-1}$$

We have

$$\begin{aligned}\ln R &= \frac{1}{T}\ln(l_T b'_T)\\ &= \frac{\ln p_0 + (T-1)\ln p_a}{T} + \frac{\ln[\alpha(1+\beta T)]}{T}\end{aligned} \tag{11-8}$$

To find the breeding age, T, that maximizes $\ln R$, differentiate (11-8) and set the derivative equal to zero:

$$0 = \frac{d(\ln R)}{dT} \propto \ln\frac{P_a}{P_0} + \frac{\beta T}{1+\beta T} - \ln[\alpha(1+\beta T)] \tag{11-9}$$

Although not easily, T can be calculated explicitly from this function. That is, if we know how fecundity varies with age (we know α and β) and also the first-year and subsequent adult survivorships (p_0, p_a), we can predict the age at which spawning should occur in order to maximize population mean fitness (i.e., the optimal strategy). Note that we can use (11-9) to calculate

$$\frac{\partial T}{\partial p_a} = \frac{(1+\beta T)^2}{p_a \beta^2 T} > 0, \qquad \frac{\partial T}{\partial p_0} = -\frac{(1+\beta T)^2}{p_a p_0 \beta^2 T} < 0$$

$$\frac{\partial T}{\partial \alpha} = -\frac{(1+\beta T)^2}{\alpha \beta^2 T} < 0, \qquad \frac{\partial T}{\partial \beta} = -1 < 0$$

Thus an increase in juvenile survivorship, p_0, selects for a drop in reproductive age, while an increase in adult survivorship favors a rise in that age. Populations displaying high α or β values (high fecundity or high growth rate) should ideally reproduce at younger ages than populations with correspondingly low values. For a roughly analogous treatment of seed production in plants, see Paltridge and Denholm (1974).

Suppose that environmental fluctuations resulted in changes in (say) p_a. The result would be to force variations in R over time. Recall from Chapter 9 that when $\bar{W}(\simeq R)$ varies, natural selection theoretically acts in such a manner as to increase the geometric mean fitness over time—equivalently, to increase $E(\ln R)$. To investigate the impact of such fluctuations (assuming them to be of small magnitude), we therefore first calculate $E(\ln R)$, using the Taylor expansion, and then, as with (11-9), differentiate and set the derivative equal to zero. Doing so, we obtain

$$\begin{aligned}E(\ln R) &= \frac{\ln p_0 + (T-1)\ln E(p_a) - \frac{1}{2}(T-1)\operatorname{var}(p_a)/p_a}{T} + \frac{\ln \alpha(1+\beta T)}{T}\\ 0 &= \frac{\partial E(\ln R)}{\partial T} \propto \ln\frac{E(p_a)}{p_0} + \frac{\beta T}{1+\beta T} - \ln\alpha(1+\beta T) - \frac{\operatorname{var}(p_a)}{2E(p_a)^2}\end{aligned} \tag{11-10}$$

Now use (11-10) in place of (11-9) to calculate

$$\frac{\partial T}{\partial \operatorname{var}(p_a)} = -\frac{(1+\beta T)^2}{2\beta^2 T E(p_a)^2} < 0$$

Thus a slight increase in variability of adult survival (in the absence of variability in p_0, α, β) selects for earlier reproduction. As long as variation in p_a, p_0, α, and β are uncorrelated (!), this and similar calculations can be used to predict the impact of environmental fluctuations on optimal age of reproduction. Calculations become considerably more complex when these variables fluctuate in a correlated manner.

Implications for Population Dynamics

What are the consequences of semelparity and iteroparity for population dynamics? Consider semelparity. Most important, it means a likely separation of life stages in time. Then if density feedback acts on the young, it must act while they are maturing. It is felt as a continuous process and so promotes stability. This is true, for example, in the periodic breeders in which a very short, active stage may result in temporal separation between prey and predator. This, to a large degree, eliminates predators as effective feedback monitors and forces density feedback to occur on the long-lived larval stage. Periodic cicadas can be predicted to have very tightly governed stable populations. If density feedback acts only on adults, on the other hand, its effects may be discrete, for the adult stage is often very short. In this instance, especially with species possessing huge reproductive potentials, as most semelparous species do, there is ample opportunity for density overadjustment and consequent instability (see Chapter 2). Of course, this need not always be true. Where an extended period of growth is required, as in many annual plants, intense competition among adults may act in continuous fashion. Thus seed populations might very well fluctuate dramatically, but adult population densities should be reasonably well controlled. Stability is also enhanced in some plant species by the production of seeds that can lie fallow for long periods of time, thus effectively creating an overlap of generations (Templeton and Levin, 1979). Also, plants may evolve seeds whose germination times are staggered so as to maximize the likelihood that at least some will begin to grow under benign conditions. This strategy both cuts the realized reproductive potential and buffers the population against environmental fluctuations. Semelparous species, then, display either highly consistent, density-regulated or quite unpredictable, widely fluctuating populations, depending on the relative lengths of their life stages and the manner in which density feedback operates on them. Because there is no overlapping of generations (in most cases), the simple population dynamics and population genetics equations are applicable to those semelparous species whose numbers are at least reasonably well density regulated.

Iteroparous species may be extremely population stable, and their numbers are seldom as erratic as those of the more opportunistic, semelparous breeders. Iteroparous species benefit from the buffering effect of overlapping generations, and each age class may respond to environmental influences differently, with a consequent averaging effect. Extreme equilibrium species are almost inevitably iteroparous. Iteroparity promotes longer life (see also Chapter 12) and is more often than semelparity a characteristic of large species. These are the species whose individuals are most apt to display parental care, with consequent, longer-term associations with their young and mates. Both parental care and prolonged contact with kin set the stage for the evolution of social organization (Chapters 14 and 15). While overlap of generations stabilizes populations of iteroparous species, the buffers and time lags inherent in such life histories also causes an inertia and so an inefficiency in tracking environmental changes, both numerically and genetically. The latter puts a premium on the evolution of physiological and behavioral plasticity that may be critical to the development of "higher" social organization.

11.2 Discrete or Continuous Population Growth?

If organisms live for a period w, and bouts of density feedback occur at intervals of less than period w, individuals will, on average, feel the crunch more than once in their lifetimes. This will almost invariably be true of iteroparous species. But for a population to behave in a discrete-time manner, there cannot be a hanging on of individuals over more than one time interval;

there must be a complete turnover. Thus iteroparous populations cannot ever experience the kind of discrete-time phenomena described in Chapter 2—instability due to overcompensation and chaos. Our discussion, therefore, must be restricted to semelparous species. Another preliminary point can be made. Just above we argued that if density feedback acted only on the young, it was likely to be of a continuous nature. We can thus limit our discussion even further; discrete population equations may be used *descriptively* on *any* population, but *predictively*, to ascertain whether density overcompensation or chaos are expected, only on semelparous species where feedback acts on the adult stage. We shall limit our discussion, therefore, to such populations and ask under what circumstances the adult stage or the breeding season (if feedback on the adult occurs via reproduction) is apt to be very short and synchronized.

1. We have already pointed out the advantage to synchronizing reproduction when conditions are such that predator saturation is desirable and possible. But if predator saturation occurs, can we claim that density feedback occurs primarily at this time.

2. In many marine species, fertilization is external. If there is no advantage to a pair bond and if social groups do not exist, sperm can be transferred efficiently to eggs only if many individuals congregate in swarms. Obviously, if this strategem is to work, shedding of gametes must be highly synchronized. A classic example of this sort of behavior is the palolo worm (*Eunice viridis*), which swarms in immense numbers in various parts of the South Pacific primarily on one day of the year. Whether these swarms experience density feedback and whether the density feedback they experience at such times, via attraction of predators, perhaps, is the primary limiting force on their populations is not known.

3. Species that breed colonially or in close proximity may exhibit *social stimulation*. This is particularly apt to be so in polygynous species in which (for example) a precopulatory display by one female may arouse males in the vicinity and prompt other females to display similarly to keep their mates from straying. Darling (1938) reports social stimulation in a variety of birds, and Svardson (1949) describes it in fish. Social stimulation acts to bring individuals in a group into reproductive activity at the same time. But there is no evidence that social stimulation occurs in semelparous species with short adult stages.

In conclusion, it may well be premature to conclude that populations which really behave in discrete-time fashion are rare. But the remarks above are convincing to this author, at least, that they are certainly not common.

11.3 Brood Size

General Considerations

Once a semelparous or an iteroparous strategy is established, so that the general magnitude of parental effort in a given season is determined, we may ask just how that effort is to be allocated. Should a parent produce tremendous numbers of young, each small, poorly provisioned, and without parental care? Or would genetic contribution be improved were a small number of young born each well provisioned? A few generalities are in order.

1. If the parents are incapable of effective action with respect to feeding or protecting their young without debilitating effort, parental care is likely to greatly constrain the size of b' (fecundity) while making little difference in p_0 (juvenile survival). Parental care will certainly not be selectively advantageous. If there is no parental care, natural selection should favor larger b' values than if parental care occurs. The expected inverse relationship between degree of parental care and egg number can be seen quite nicely in parasitoid, ichneumonid wasps. As the solicitude with which eggs are lain—deeply within a host (which requires a longer ovipositor and higher risk), directly on the host, or merely on foliage near the host's eggs—declines, the number of eggs increases (Price, 1973).

Another side of this argument involves the ability of the young to care for themselves. Ito and Iwasa (1981) conjecture that if food procurement ability of the young is poor, parents must, of necessity, aid them somehow. The extra parental energy expended per young, in turn, forces production of a smaller clutch. Ito and Iwasa's argument is elegant in its simplicity, and may explain why certain species which might, by other criteria, be expected to be opportunistic, actually lay small numbers of eggs. In particular, deep-sea fishes, terrestrial amphibians, freshwater clams, and small stream snails come to mind.

2. If dispersal is an important component of the life history, even size and yolk content of the eggs is likely to be without significant benefit. Species that fit this criterion should lay very large numbers of relatively poorly provisioned eggs.

3. Large eggs usually make large young, even within a species (Svardson, 1948). If large young have a competitive edge over their fellows, parents would be well advised to weigh this advantage of increased egg size against the advantage of increased egg number (see Pope et al., 1961; Parrish and Saville, 1965; Bagenal, 1969). McLaren (1966) found that females of the arrowworm (*Sagitta elegans*) may postpone breeding, in spite of the risk of death while they bide their time. He suggests that this occurs so that they can lay larger eggs, thereby generating more advantaged young without suffering the consequences of reduced brood size.

4. Uncertain conditions for the adults, we have seen, seem to lead to larger reproductive effort. We should also expect them to lead to a greater value of b' per unit effort. If we accept the notion that open, field habitat is more prone to disturbance of plants (less buffering from sun, wind, etc.) than closed, forested habitat, we should expect the number of seeds produced per average seed weight to be greatest in open areas and smallest in closed areas. The data in Table 11-3, taken from Salisbury (1942), crudely support this expectation. The species involved are a sampling of the seed plant species in each of the type areas.

Table 11-3

Habitat	Number of Seeds per Average Seed Weight (g)
Open	74,524
Semienclosed (grasses)	9,937
Semienclosed (forbs)	1,610
Woodland	7,211
Shade species on woodland floors	1,973
Woodland shrubs	234
Woodland trees	31

The discussion above was presented mainly in the interest of reviewing some ideas and speculations. The arguments are too vague, too nonrigorous, and too prone to unforeseen complications to be terribly meaningful. For example, although it may be advisable to provision eggs well to encourage the young to be large and competitively viable, this may also place the young squarely in a highly predated size range. Plant seeds are likely to be large not only as a consequence of environmental uncertainty, but because shaded conditions require seedlings to be of some critical size before they can safely become obligate photosynthesizers. Plants may also benefit from a hefty endosperm boost if germinating in dry or unreliably mesic areas. Baker (1972) presents abundant evidence that xeric species create larger seeds than do mesic species or wet-habitat species.

Lack's Hypothesis

We shall now address the question of family size in considerably more detail for iteroparous species. What is the optimal number of young per brood? Do the data support the contention that organisms really evolve to optimal strategies? If the reader detects an unmistakable avian orientation, that is because the bulk of research has involved birds.

To some degree, the brood (clutch) size is characteristic for a given family of birds. For example, among shorebirds, most of the Jacanidae lay four eggs, as do the Rostratulidae, the Vanellinae, and the Thinocoridae. The Burhinids lay two (MacLean, 1972). In general, smaller bird species lay larger numbers of eggs. In many groups the characteristic number is sufficiently rigid that environmental changes have no detectable effect on b'. But in most species, egg number, though fixed within species-characteristic limits, may vary considerably with (say) the food supply. Let us look at some examples of such species.

In 1954, Lack reviewed a large body of evidence in support of a hypothesis he had been proposing for a number of years. The hypothesis is simply that a breeding pair should lay that number of eggs that can be expected to result in the largest number of offspring surviving to maturity. More specifically, look at Figure 11-4. As shown in Figure 11-4a, the survivorship of young may rise when b' is very small because of the thermoregulatory advantage of huddling. But as clutch size increases, parents may find it increasingly difficult to find adequate food for the young, or to competently herd the young and fend off predators during the fledgling stage. Thus the p_0 curve drops. (In reality the "curve" should be depicted as a series of points, each corresponding to an integer number of eggs; it is drawn in continuous form simply for convenience.) But reproductive success is really measured in number of surviving young, $b'p_0$. Multiplying the p_0 values in Figure 11-4a by b', and plotting the products against b', generates the curve in Figure 11-4b. With such an illustrative tool it is easy to see that an optimal clutch size exists at $\hat{b}'$ (actually the next largest or smallest integer) (see also Royama, 1966). That young in smaller clutches are better fed than their cohorts from larger clutches is abundantly supported with data. Royama (1966) found that in the great tit (*Parus major*), food consumption rates were highest (by direct observation), at all nestling ages, in smaller clutches. Also, the young from smaller clutches ate, on the average, larger food items. As we shall see in Chapter 16, this is an indication of better diet. Furthermore, the average weight of the fledglings was approximately the same for all clutches up to 8 or 10 young, but then dropped as clutch size increased. In the kittiwake (*Rissa tridactyla*), pairs breeding for the first time produced offspring which gained weight at a rate of 15.81 g per day in nests with one offspring, but only 14.75 g per day in nests with two. For older more experienced birds, the corresponding figures were 16.15 and 15.57 (Coulson and White, 1958). Hogsted (1980) manipulated clutch size in the American magpie (*Pica pica*). He found no significant differences in nestling weight as a result of his adding eggs, but he found that the addition of eggs caused a significant drop in survivorship. Among nonpredated nests, survivorship in unmanipulated clutches was 58%. With one egg added, this figure fell to 49%, and with two additional eggs, to 26%. In a mammalian species, the white-footed mouse, (*Peromyscus leucopus*), body growth rate of the young was enhanced by removal

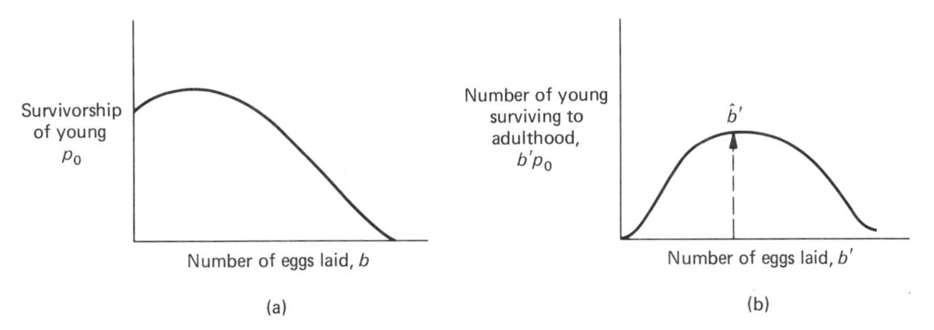

Figure 11-4 (a) Hypothetical relationship between survival of young and the number of young produced by a (bird) species displaying parental care. (b) Hypothetical relationship between reproductive success and the number of young produced in a (bird) species displaying parental care.

of litter mates, depressed by the addition of extra young. When the young were tested for toughness by dumping them in ice water and measuring time to exhaustion, individuals from reduced litters lasted significantly longer than individuals from natural and enlarged litters (Fleming and Rauscher, 1978). Cameron (1973) kept desert wood rats, *Neotoma lepida*, in the laboratory with unlimited food, cuttings of their natural diet, supplemented with sunflower seeds, oats, lettuce, celery, carrots, and rat chow, and found that offspring from small litters not only were more apt to survive, but were weaned earlier. Perrins (1963, 1965) First provided the necessary link between body size and subsequent survival (see also Jarvis, 1974). For the great tit in England he found, over 6 years, a 10% recovery rate of fledglings that had left the nest weighing 14 to 19 g. For 15- to 21-g birds, the recovery rate jumped to 20%, and for all birds over 18 g, an average of 30%. Perrins also found that larger clutches, perhaps as a result of extra noise or a larger number of parental trips to and from the nest, suffered increased predation.

Clearly, the expected decline in p_0 with increasing b' is well documented. We can therefore move on to the task of making predictions based on Lack's model. We first move one more conceivable complication out of the way. Can clutch size evolve? That is, is clutch size heritable? Perrins and Jones (1974) performed parent–offspring regressions of clutch size (statistically corrected insofar as possible for environmental factors) in the great tit (*Parus major*) and calculated a heritability of .51. This answers the question above in the affirmative, but is a bit surprising. Traits that are important components of fitness are expected to have low h^2 values (Chapter 10). Leamy (1981), estimating heritability of litter size in *P. leucopus*, obtains more equivocal but more reasonable results. Among males he found values of $.028 \pm .131$, and $.091 \pm .090$ (two generations of tests); among females the figures were $-.086 \pm .097$ and $-.105 \pm .089$. It appears in this case that $h^2 \simeq 0$. If clutch size varies with environmental conditions as a result of adaptive somatic flexibility on the part of the parent, we can reasonably expect to make predictions based on what the breeding pair *ought* to be doing to maximize reproductive success. If clutch size changes as a response to genetic change—then conditions favoring large clutch size in one year will result in increased clutch sizes the next. Environmental tracking becomes a problem, precluding optimal behavior. As we shall see shortly, the data indicate a remarkable coincidence between observed and expected optimal response, with no hint of time lag.

Referring to Figure 11-4b, we can make the following conjectures. First, if food supply increases or, if predation is the clutch-limiting factor and the number of predators drops, the p_0 curve is higher and falls off less rapidly with b'. The resulting peak in Figure 11-4b responds by shifting to the right. Because the data reviewed immediately above, and the bulk of other studies not reported here, nearly all implicate food supply as the critical factor (at least for birds), we shall accordingly conclude that increased food supplies promote larger clutches. Some corollaries are in order. For a given food supply, a larger breeding population density means less food per foraging parent and so should result in smaller clutches. It will certainly be highly beneficial, because of the considerable cost of egg laying itself (a small, passerine bird may lay roughly its own weight in eggs over a period of perhaps a week) to be able to read environmental cues in advance and base clutch size on a best estimate of conditions to come. But when this cannot be done, it may pay to produce an extra egg and, if conditions later prove not to warrant the extra insurance, get rid of it. Koskimies (1950) reports that the European swift (*Micropus apus*), whose flying-insect food source is highly dependent on poorly predictable future weather conditions, lays three eggs in years that bode especially well, two or three in less certain years, but if it becomes necessary, subsequently rolls one out of the nest.

Lack himself has amassed impressive data sets for a number of species at a number of sites, relating brood size to breeding success. He has also brought together similar information from other workers for both birds and other kinds of animals. Two examples are provided in Table 11-4. The data for Swiss starlings (*Sternus vulgaris*) (Table 11-4a) (Lack, 1948a), although the differences in the recovery column are not statistically significant, suggest that the birds really do adjust clutch size toward an optimum. The swift data (Table 11-4b) (Lack and Arn, 1947; Lack and Lack, 1951) tell the same story. Lack also presents abundant evidence for

Table 11-4 **Brood Size Data**

Brood Size	Number of Broods Observed	Number of Recoveries (3 Months Later) per 100 Broods
(a) Swiss Starlings (Lack, 1948)		
Early broods		
1	65	—
2	164	3.7
3	426	6.1
4	989	8.3
5	1235	10.4
6	526	10.1
7	93	10.2 (7–8 combined)
8	15	
Late broods		
1	44	5.8 (1–3 combined)
2	96	
3	381	
4	391	8.9
5	285	8.8
6	73	8.2
7	7	—
(b) Swifts (Lack and Arn, 1974; Lack and Lack, 1951)		**Number Raised per Brood**
Alpine swift (*Apus melba*)		
1	58	1.0
2	281	1.8
3	541	2.4
4	5	2.4
Common swift (*Apus apus*)		
1	36	.8
2	102	1.7
3	32	1.7

declines in clutch size with increasing population density for a number of species. A plethora of information on these topics is available in his book *Population Studies of Birds* (1966), and in his truly remarkable, earlier book, *The Natural Regulation of Animal Numbers* (Lack, 1954).

A slight variant on Lack's hypothesis is provided by Smith and Fretwell (1974). These authors argue, for a given set of environmental conditions and species facies, that offspring fitness (read "contribution to the genetic ancestry of future generations") rises in a characteristic manner with the amount of parental energy expended on them (Fig. 11-5). If W_y represents "fitness" of an average offspring, and I_y the expended parental effort on raising that offspring, then

$$\text{parental "fitness" } W_p = \tfrac{1}{2}W_y \times \text{number of young} \propto W_y \frac{1}{I_y}$$

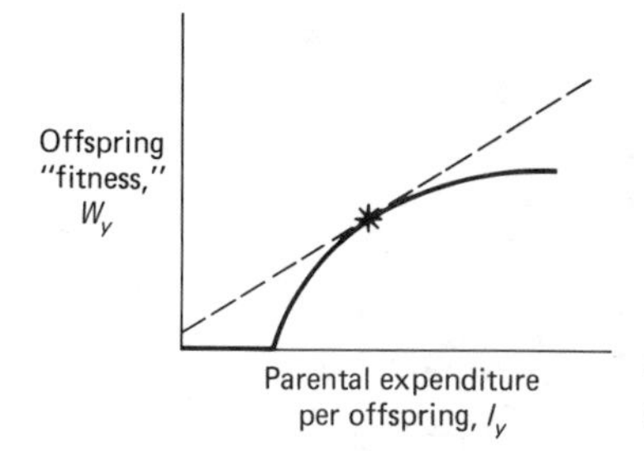

Figure 11-5 Hypothetical relationship between "fitness" of an average offspring and parental expenditure.

Parental "fitness" is maximized when

$$0 = \frac{dW_p}{dI_y} \propto \frac{1}{I_y}\frac{dW_y}{dI_y} - \frac{W_y}{I_y^2}$$

or when

$$\frac{dW_y}{dI_y} = \frac{W_y}{I_y}$$

That is, the optimal parental strategy is to realize that value of I_y such that W_y/I_y equals the slope of the W_y versus I_y curve (the point of tangency marked in Fig. 11-5). Thus, for a given set of environmental conditions and species facies, the value of I_y, and hence the number of young, is determined. Note that as conditions in the environment improve, W_y rises, W_y/I_y increases, so I_y drops. Thus for a given offspring size, more young should be produced.

Clutch (or litter) size varies with food supply not only among birds. Darling (1937) reports that in red deer (*Cervus elephas*), 92% of all females are pregnant in "good food years," 33% with one young, 60% with twins, and 7% with triplets. In "bad food years" the pregnancy rate falls to 78%, with the percent bearing one, two, and three young being 81, 18, and 1, respectively.

A variety of observations on clutch size have been invoked as evidence for Lack's hypothesis, or have spawned elaborations of the hypothesis. For example, it seems that the number of young produced in a variety of animal species declines toward the tropics. Of a variety of nonpasserine bird species common to south and equatorial Africa, 56 species had clutches of essentially the same size in both areas, but 17 had larger clutches and only 3 had smaller clutches in South Africa. Among passerines 84 species showed no detectable difference, 54 had larger clutches, and 4 had smaller clutches (Moreau, 1944). The same trend is supported in a more recent compendium of clutch sizes by Ricklefs (1980). The same trend appears in North American mammals (Lord, 1960); at higher latitudes, a given species produces larger litters. Ashmole (1963) felt that the likely explanation for this phenomenon was that temperate-climate animals suffered heavy mortality due to winter conditions or, if they left the winter conditions, due to the stresses of migration. In particular, in temperate-zone breeders, winter food supply would be low relative to summer food supply. This should result in large amounts of food per breeding pair and permit, by Lack's hypothesis, large clutch sizes. By contrast, tropical species would experience higher winter productivity and so less mortality, therefore experiencing lower food supply per individual during the breeding season. Ricklefs (1980) has followed up this suggestion by collecting vast amounts of data on clutch size and regressing log clutch size, C, on log winter productivity, W, and log summer productivity, S. He finds, for open-breeding (as opposed to hole-nesting) passerines,

$$\log C = a - .09 \log S - .90 \log W \qquad (11\text{-}11)$$

The $-.09$ is significant at the .05 level, the $-.90$ at the .001 level. He also reports that the ratio of summer to winter population density is negatively correlated with clutch size for North and Central American nesters. Finally, in parts of the world where winter migrants are funneled from large land areas into very small land areas, the winter food supply must also be greatly depressed for the resident tropical birds. Therefore, equatorial birds in such places as Java and Borneo should have clutches larger than otherwise expected. This, in fact, happens. Ricklefs' analysis makes the interesting point that food availability in one season may have profound effects on fecundity in another.

There are other reasons why equatorial species might have smaller clutches than their temperate-zone relatives, most consistent with Lack's hypothesis. Annual productivity at cooler latitudes may be lower than productivity in the tropics, but the tremendous flush of growth that occurs in the temperate zone in early summer might often exceed the productivity at any one time of year in warmer climates. This is a variant, really, of Ashmole's thesis. Also, in the tropics, daylight is shorter than in early summer farther north or farther south, providing diurnal foragers with more time to gather food. (This approach would not account for the trend found in nocturnal mammals.) Higher observed rates of predation in the tropics might slow the rate of parental coming and going from the nest, calling for fewer, more cautious approaches. Further discussion and data are provided by Skutch (1949, 1954, 1960). Of course, if any of these arguments are valid, we should not expect to find the same, latitudinal trend for precocial birds. The fact is that in general, we do not.

Cody (1966) has pushed Lack's hypothesis a bit farther, arguing in familiar r- versus K-selection terms. If a breeding pair has only so much energy to expend, it should opt for more, less-well-provided-for young in environments that are uncertain. Because tropical areas are more climatically predictable than temperate areas, clutch size should therefore be larger in the latter (if it is climate that affects food supply or otherwise controls foraging). By the same token, coastal areas are more stable than inland areas of comparable latitude, so populations of a given species nesting in both habitats should have larger clutches inland. There is some evidence that this prediction is borne out. Johnston (1954) found that song sparrows (*Melospiza melodia*) laid fewer eggs where they nested along the Pacific coast than where they bred farther inland, and Nice (1937) presents data showing that the same species increases its clutch size both progressively northward from southern California to Alaska, and progressively eastward from the West Coast.

Uncertainty also arises from the existence of predators. If predation on young is high (low p_0) or uncertain, the basic theory developed in Section 11.1 provides us with an argument for expecting a larger clutch size—less well developed iteroparity. But here, at least for birds, there is an opposing consideration. Where predation is low, parents can afford to prolong the development period of parental care without undue risk. This allows them to provide food for the young at a slower rate and so makes larger clutches possible. Lack made this argument in 1948 (1948b) and provided the data in Table 11-5 for European passerines. Nice (1957) reprinted Lack's data and added her own for some North American passerines (Table 11-6).

Of course, clutch size can also be expected to vary with the experience a parent can muster in finding food for its offspring and protecting them from predators. More experienced individuals might also be better equipped to know how to choose the best microhabitat for breeding, and to know how to acquire it in the face of competition from others; and experience comes with age, among other things. The literature is full of studies showing the impact of age and experience on clutch size; we mention only three here. Coulson and White (1958) found that

Table 11-5

Nest Location	**Predation**	**Incubation and Nestling Period (days)**	**Clutch Size**	**Number of Species**
In holes	Low	31.1	6.9	18
Roofed	↓	29.9	6.8	7
In a niche		28.4	5.5	5
Open	High	26.3	5.1	54

Table 11-6

Nest Location	Incubation and Nestling Period (days)	Incubation, Nestling, and Fledgling Period (days)	Average Clutch Size	Number of Species
In holes	32.6	38.0	5.4	10
Open	23.0	27.0	4.0	11

kittiwakes nesting for the first time had greater difficulty feeding their young than did birds with at least one year's practice in raising young. The evidence was in the form of lower growth rates of the offspring and lower fledgling weight. Lack's hypothesis would predict, given this information, that first-year breeders should lay smaller clutches. And indeed they do. In their first reproductive attempt, kittiwakes lay an average of 1.83 eggs. In their second year this number is raised to 2.06, and birds with at least 2 years' experience lay 2.33 eggs. In spite of this discrepancy, inexperienced birds fledge a much lower percent of their young than do older birds; 29 first-year nests fledged only 19 birds, 16 second-year nests fledged 19, and 10 nests of older birds yielded 19 young.

Perrins (1965), in his study of the great tit, provides us with the data, given in Table 11-7. Note that the patterns are reminiscent of those discussed in Chapter 3. (A theoretical construction for the evolution of patterns of this sort is presented in the next chapter.) There is no way of knowing exactly what proximate factors are at work here, but the relative ability of individuals of different ages to compete successfully for suitable nest sites, and their relative proficiency at foraging, undoubtedly play a role.

DeSteven (1978) studied the influence of age on the breeding biology of the tree swallow (*Iridoprocne bicolor*). He found that over each of two years, clutch sizes were smaller for yearling females than for older birds. He also found that egg size was smaller ($p < .01$ in one year, not significant the other) and that the percent of hatchlings fledged was smaller ($p < .01$ one year, not significant the other). Fledgling success might be expected to be less for birds laying smaller eggs because there is a significant, positive correlation between size of egg and size of nestling (see also Nolan and Thompson, 1978). Although these data are rather convincing with respect to the heightened success rate and larger clutch sizes of older birds, DeSteven also found an additional, confounding advantage of older age. Probably because of the cost of producing eggs, these birds almost invariably lay their eggs at such a time that the young will hatch after the optimal time for food finding. But the benefits of size or experience somehow allow the older females to realize hatching closer to the optimal time than young females. Thus *some* of the greater success, and (under Lack's hypothesis) *one* of the reasons for larger clutch sizes among older birds, is not age per se, but an indirect by-product of age. Better timing of the nesting cycle and, as we shall see in Chapter 14, better positioning of the nest, is a pretty universal advantage enjoyed by older birds of many species.

Table 11-7

Age of Female (yr)	Mean Clutch Size	Sample Size
1	9.4	54
2	10.0	33
3	9.7	29
4	9.7	9
5	9.5	2
6	9.0	1

In spite of all the data above, the hypothetical link between experience and reproductive success should not be accepted unquestioningly. It is almost certainly true, but the possibility remains that the improved performance reflects not greater experience but increased reproductive effort.

Brood Reduction

Lack's hypothesis as it has been applied so far contains an implicit assumption that all young in a nest are essentially alike. That is, all are fed about equally, all grow at about the same rate, and all experience the same risk of starvation or predation. Individual differences among the offspring and runts of the litter aside, this assumption is not far afield if all young are born or hatched simultaneously. In mammals the members of a single litter certainly meet this criterion. In birds, where eggs are laid at 1-day or greater intervals, the assumption will be met only if earlier-laid eggs develop at a slower rate or delay development until the entire clutch is laid. Many passerines, at least with their first clutch of a season, begin incubation only when the last or penultimate egg has been produced (for a compendium of data, see Clark and Wilson, 1981). In these cases the implicit criterion for Lack's hypothesis are at least loosely met. But there are many bird species that do not delay incubation—raptors, for example. In these cases the nestlings will be gradated in size. Usually, the first to be fed will be the largest (the most difficult to ignore); the last to receive food will be the smallest, the last to hatch. Where environmental conditions are easy to predict and thus birds can assess their optimal clutch size, the former synchronous hatching scheme is probably the best for maximizing reproductive contribution. But where divination of future food prospects is not so easy, in more uncertain environments, it probably cannot work. Here it is best to stagger the hatching of young. If food becomes scarce, the last to hatch, the weakest individual, is sacrificed. It is that individual which has accrued the least parental investment, *sensu* Trivers (1972), and so is the most dispensible. The only cost to the parent is the initial price of the wasted egg, plust a very small amount of feeding time. There is even some speculation that in some species females may lay an extra egg that they are almost certain not to raise to independence, partly as insurance, perhaps, against hatching failure, partly as an opportunistic gesture in the event of unexpected good fortune. The regulation of clutch size by death of offspring is known as *brood reduction.*

The study of brood reduction has become increasingly popular in recent years, and some rather interesting insights have been gained through these studies. Howe (1978), working with the common grackle (*Quiscalus quiscala*), discovered that the first two to four eggs layed hatch synchronously, but that the fifth and sixth, if such occur, hatch later. Birds laying four eggs or fewer usually fledge all their young; birds laying five or six eggs often lose one or two. The most successful clutch size, and the size usually observed, is five, suggesting that the fifth egg is laid "hopefully," the sixth as insurance, or opportunistically. As a partial aid to the fifth and sixth hatchlings, these last two eggs are more heavily provisioned with yolk and albumen.

The laughing gull (*Larus atricilla*) usually lays three eggs. Time between laying of the first and second is about .92 day, between the second and third, 1.36 days. Hahn (1981), attempting to evaluate the significance of staggered hatching in this species, switched second and third eggs among nests in such a way that 13 experimental nests had three synchronously hatching young. His results are shown in Table 11-8. Quite clearly, the practice of staggering the hatching of young has its advantages. The brood reduction hypothesis states that when death of offspring occurs, it should be the last offspring that is most likely to die. In the control nests, 62.5% of third chicks survived to fledge compared to 83.3% of first chicks; third chicks were only $62.5/83.3 = .75$ as likely to survive as first chicks. In the experimental nests, the numbers are 53.3% and 54.2%, and $53.3/54.2 = 1.02$. Staggering offspring production in time is apparently advantageous, but its advantage does not lie in the opportunity for brood reduction. In this case it probably works by spreading out the parental care period by 2 or 3 days, thereby alleviating some of the strain of parental foraging.

Table 11-8

	Control Nests (Natural, Asynchronous Hatching)		Experimental Nests (Synchronous Hatching)	
Number of nests	48		13	
Number of young fledged per pair	2.13		1.62	
Number of nests fledging:				
Three young	26	(54.2%)	3	(23.1%)
Two young	9	(18.8%)	5	(38.5%)
One young	8	(16.7%)	2	(15.4%)
No young	5	(10.4%)	3	(23.2%)
Number of first chicks fledging	40 of 48	(83.3%)	8 of 15	(53.3%)
Number of third chicks fledging	30 of 48	(62.5%)	13 of 24	(54.2%)

The concept of brood reduction raises a number of rather interesting subsidiary questions. Will starvation be an event actively encouraged by a parent, or by siblings? Might it even take the form of suicide—one member of the clutch voluntarily giving up? As a first step in answering these questions, let us suppose that the probability of an individual young surviving the food-competitive period of parental care is the same for all individuals. This is not a reasonable assumption, since staggering, for example, clearly designates the first one to go where brood reduction is practiced. But it makes calculations simple, and serves as a conservative end point in the argument to follow; the conditions for infanticide, fratricide, and suicide derived below are all relaxed when there is variance in p_0. We follow the treatment of O'Connor (1978). Let p'_0 be the probability that a young individual survives the environmental stresses whose effects occur independently of brood size, and p''_{0n} be the survivorship of young after competition within a clutch of size n. The probability of living to adulthood is $p_0 = p'_0 p''_{0n}$. We set the brood size to n, let W be the genetic contribution of a young over its entire life subsequent to fledging, and let $W_A(n)$ represent the contribution of all offspring to the parent's inclusive fitness. Consider first the case of possible infanticide. We suppose that a new gene has appeared, still very rare, that predilects the parent to kill one offspring. If the parent happens *not* to carry the gene, its inclusive fitness will be

$$W_A(n) = p'_0 p''_{0n} n(\tfrac{1}{2}W)$$

the number of offspring successfully raised, times $W/2$ because the offspring will carry half that parent's genome. If the parent carries the infanticide gene, the clutch size is reduced to $n - 1$, and

$$W_A(n-1) = p'_0 p''_{0n-1}(n-1)(\tfrac{1}{2}W)$$

The new gene will be favored if the parent's inclusive fitness is higher for carrying it [i.e., if $W_A(n-1) > W_A(n)$]. Some simple calculations show that this condition is equivalent to

$$\frac{p''_{0n-1}}{p''_{0n}} > \frac{n}{(n-1)} \qquad (11\text{-}12)$$

Infanticide is advantageous if the relative chance of survival to adulthood in clutches of $n - 1$ as opposed to n exceeds $n/(n-1)$.

In the case of a gene promoting fratricide, the inclusive fitness of the sib carrying the gene is

$$p'_0 p''_{0n-1} W + p'_0 p''_{0n-1}(n-2)(\tfrac{1}{2}W)$$

and for a sib without the gene,

$$p'_0 p''_{0n} W + p'_0 p''_{0n}(n-1)(\tfrac{1}{2}W)$$

The gene is favored if the former quantity exceeds the latter, or if the relative chance of survival to adulthood in clutches of $n - 1$ as opposed to n exceeds $(n + 1)/n$:

$$\frac{p''_{0n-1}}{p''_{0n}} > \frac{n+1}{n}$$

With respect to suicide, an individual carrying such a self-destructive gene has an inclusive fitness of

$$p'_0 p''_{0n-1}(n - 1)(\tfrac{1}{2}W)$$

while, without the gene, the value would be

$$p'_0 p''_{0n} W + (n - 1)p'_0 p''_{0n}(\tfrac{1}{2}W)$$

The suicidal gene is favored when the relative chance of survival to adulthood in clutches of $n - 1$ as opposed to n exceeds $(n + 1)/(n - 1)$:

$$\frac{p''_{0n-1}}{p''_{0n}} > \frac{n+1}{n-1}$$

Any gene that indiscriminately predisposes its owner toward infanticide, fratricide, or suicide will very quickly reduce clutches to nothing at all (or, actually, one in the case of fratricide), and could not possibly be advantageous (Howe, 1978). But coupled with an ability on the part of the bearer to act on that gene's instructions only under appropriate circumstances, such genes could presumably spread. Referring to these inequalities it can be seen that conditions for the advantageous disposition of fratricide are more easily met than those for infanticide and these, in turn, more easily than for suicide.

In addition, we can expect a parent whose conditions for infanticide have not been met to resist any attempts at fratricide within her clutch. The fact that the female is larger and stronger than her young also make her the ultimate arbiter of brood reduction. Young will not likely die unless the parent herself witholds food (we exclude other sources of mortality in this discussion). These conclusions, taken together, lead to an interesting conclusion. The gulf between the conditions for infanticide and fratricide are greatest when n is small. Thus we are more likely to find situations where fratricide is advantageous, but prevented by the parent, in small clutches. Put differently, sibling conflict can be expected to be more common in species with small clutches. On the other hand, the condition for infanticide is most easily met when n is large. Therefore, we should expect brood reduction most commonly in species with large clutches. O'Connor (1978) gathered data on birds from the literature and, where possible, estimated p''_{0n-1}, p''_{0n}, and n. He then determined which species met the criterion for infanticide. A contingency table test of the data supported the prediction that starvation occurred more often in species meeting the criterion than in those which did not ($p = .009$). Several other points are worth mentioning. Because of the lower cumulative parental investment, brood reduction should occur primarily during the early stages of nesting. O'Connor reports that this seems to be true. Because the conditions for fratricide occur before it becomes advantageous for a parent to kill an offspring, fighting may occur in which an individual is injured or weakened, or deprived of food and so rendered smaller than its peers. At this point the weak individual is the most dispensible, genetically, and it becomes advantageous to pick up the pace of picking on it. That young animals often take malicious advantage of an underdog, then, is not unexpected. The presence of a weak youngster also makes the condition for infanticide more likely to be met. In such a way offspring may force their selfishness on a parent. Of course, staggering the young in time provides a built-in potential victim.

The Trade-off Between p_a and $b'p_0$

Referring to Table 11-4, it can be observed that although the fit between predicted optimal clutch size and observed clutch size is reasonably good, the number of young actually found falls

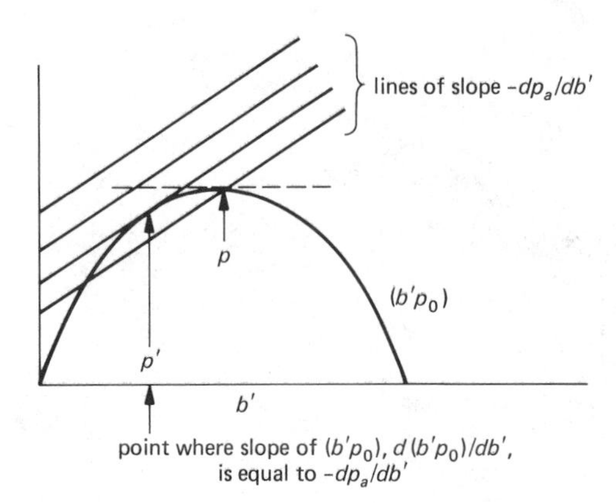

Figure 11-6 Graphical model of optimal clutch size for an iteroparous species.

below the expected number. This bias is highly consistent across virtually all data sets that are clear enough to suggest any trends at all. The reason is really quite straightforward. Lack's hypothesis does not take into consideration the need for parents of iteroparous species to hold back some effort in the interests of survival. If the parent *can* spend the effort to successfully raise (say) five offspring, then Lack's hypothesis predicts a clutch size of five. But in reality, it is not advantageous for the parent to do so. Rather, it should spend a little less, maybe enough to raise four young, and assure itself a reasonable chance to live and breed again. A discussion of this matter and an illustrative, graphical model have been presented by Charnov and Krebs (1974). Using the terms p_0, p_a, and b' as before, we rewrite (11-1):

$$W = p_a + \tfrac{1}{2}b'p_0 \tag{11-1}$$

Fitness is maximized when b takes on a value such that

$$0 = \frac{dW}{db'}$$

so that

$$-\frac{dp_a}{db'} = \frac{d(b'p_0)}{db'} \tag{11-13}$$

Because p_a can be expected always to fall with a rise in b', $-dp_a/db'$ will be positive. The situation can be graphed as shown in Figure 11-6. The point P is where the number of surviving young, $b'p_0$, is maximized, and corresponds to Lack's predicted clutch size. The point P' is where equality (11-13) obtains, and is the optimal clutch size corrected for considerations of parental survival.

Review Questions

1. Is it valid to argue that species experiencing heavy mortality have evolved high birthrates as compensation? Why or why not?
2. Review the advantages and disadvantages of iteroparity and the relationship of iteroparity to density-independent fluctuations in the environment.
3. Suppose that experimental manipulation showed the following. Individuals of a given bird species are able to lay two clutches of N eggs each in a season or a single clutch of $2N + 2$ eggs, and that the latter practice invariably leaves more surviving offspring. N varies from individual to individual. Suppose we also know that some birds were predisposed to laying one clutch, some to laying two, and that this predisposition had a fairly high heritability. On the basis of this information, would we be justified in concluding that a pattern of two

clutches per season should be evolving? Why or why not? [Refer to the last part of "Iteroparity and Opportunism" (p. 314).]

4. What conditions encourage evolution of an early onset of reproduction? Why do some species mature slowly? Do these arguments apply equally to both males and females?
5. To some degree plants may be able to avoid density-independent rigors by producing seeds with the ability to lie dormant until appropriate conditions occur. What implications does this have for the evolution of semelparity versus iteroparity? For "litter" size? Might the answers to these questions depend on whether the environmental cues triggering germination were indicative of short- or long-term environmental changes (i.e., affecting success of the plant throughout much of its life, or success of the seedlings only)?
6. Suppose that we wished to perform a stability analysis on a synchronized semelparous population with delayed maturity. What determines whether we should view population growth as discrete or continuous? What difference would it make?
7. Do you suppose that brood reduction is an adaptive strategy or simply unavoidable loss of young? Are those who promote brood reduction as an adaptation merely misinterpreting the spandrels of San Marco?

Chapter 12

The Evolution of Age Structure

12.1 A General Theory of Life Histories

The question to be addressed in this chapter is, given the existence of iteroparity, what age-specific patterns of p_x and b'_x should we expect to find for a given population? For general reviews of this topic, the reader has ample sources (Giesel, 1976; Stearns, 1976, 1977).

Following the lead of the preceding chapter, note that p_x must be held within certain bounds by susceptibility of the organism to the physical conditions of the environment in which it lives. Small organisms inevitably fall victim to such (to us) innocuous threats as being trapped by surface tension in a drop of water, or being smashed by a waving twig; and there is really nothing they can do about it. Also, recall that $b'p_0$ and p_a evolve in response to the death rates at various ages and at least according to one school of thought, also reflect trade-offs between the advantages and risks of parental care. Let us for the moment join the ranks of the "trade-off" school of life history evolution, suppose that the general magnitude of the birth and death rates has been determined, and try to fine-tune our predictions a bit. For a given amount of *potential* parental investment, how much should be spent in current reproduction and how much withheld in the interests of parental survival?

An Energetics Approach: Trade-Off Theories

We must first establish that the assertion of a trade-off between $b'p_0$ and p_a is true; that is, we must show evidence that parents produce offspring only at risk to their own survival. Most evidence to this effect comes in indirect form, showing that breeding individuals suffer considerable energy drain and weight loss as a result of their reproductive activities. In redwing blackbirds both sexes lose weight over the breeding season. Between March and July, in California, males dropped from 67.9 g, on average, to about 62.0 g. Females lost about 8% of their weight, from 42.1 g to 38.9 g. (Orians, 1961a, see also Brenner, 1967). Figures compiled for other bird species are 10 to 20% loss for tree sparrows (*Spizella arborea*) (Heydweiler, 1935), 9% in song sparrows (*Melospiza melodia*) (Nice, 1937), and 7 to 14% in the bullfinch (Newton, 1966). Many birds, particularly cold temperate or arctic breeders, have only a short period in which to breed, and then to molt and build up fat reserves for migration. Here the reproductive season loss in body weight must surely affect subsequent survival. Stickel (1973) looked at individuals comprising the large spawning masses of the intertidal snail (*Thais lamellosa*), and found that females lost 41% (one year) and 55% (a second year) of their weight while in the spawning act. It seems very unlikely that a loss of about half one's body weight will be beneficial to subsequent survival. But these data, while showing within-individual trade-offs, do not necessarily indicate *among*-individual trade-offs. Recall that it is the latter that are critical to the evolution of life histories.

Fisher (1975) took a somewhat different and more direct approach. Studying the Laysan albatross (*Diomedea immutabilis*), which often breeds for the first time at quite an advanced age, he related the age of first reproduction to longevity (Table 12-1). The survival percentages for all

Table 12-1

First Age of breeding (yr)	Number of Birds	Percent Surviving Through Age:			
		9	*10*	*11*	*12*
6	64	75.0	71.1	68.0	50.1
7	271	91.7	83.3	76.4	55.0
8	483	86.0	81.6	75.0	65.0
9	217	—	80.3	75.0	54.0
10	27	—	—	85.0	71.3
11	6	—	—	—	83.4

Table 12-2

Number of Females	Number of Eggs per Female	Life Span (days)
18	89.0	10.3
20	82.7	9.9
16	78.2	13.0
13	72.9	10.7
12	67.9	11.0
17	67.8	11.9
19	64.8	11.9
19	21.5	16.5
	Nonovipositors	19.4 to 23.0

birds breeding first at other ages exceed those for birds first breeding at age 6. Birds that wait to breed clearly live far longer than those breeding early.

Loschiavo (1968) gathered the same sort of data, but for a dermestid beetle, *Trogoderma parabile* (Table 12-2). The same declining trend in survivorship with reproductive effort is clearly visible. These data all suggest that increased allocation of available energy reserves to reproduction lowers subsequent survival and hence future reproductive contribution (but see below).

Making use of the foregoing relationship between reproductive effort and lowered survivorship the following year, Gadgil and Bossert (1970) proposed the following model. First define a *profit function* that measures the increase in an individual's fitness with increased parental, reproductive effort. This function clearly varies with the number of offspring the individual produces and also, in some way, the quality of the offspring (regarding their mating potential, viability, etc.). The reproductive effort we shall call *E*. Thus profit function relates fitness to *E*. A plot of fitness against *E* will show a concave curve (Fig. 12-1a) if the first young

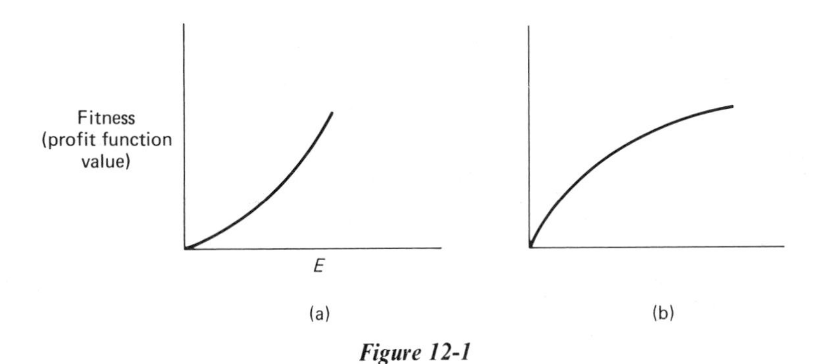

Figure 12-1

exacts a large cost, and once it is produced subsequent young are relatively less costly. Most species, from the various Pacific salmon to birds to human beings are likely to show a curve of this sort. Salmon put a huge effort into the first egg by way of swimming all the way to their natal streams. After that, the next eggs come cheap. Birds exert a great deal of time and energy courting, defending territories, and building nests, all of which is necessary for the laying of the first egg. Once those tasks are done, the increased cost of additional eggs is relatively small. The degree of concavity, obviously, will vary from population to population. It is somewhat hard to envision a species where the first young is cheap and subsequent young are increasingly expensive [although R. M. Fagen (personal communication) suggests that in human beings, the 3:00 AM last straw factor may provide an example], but for what it is worth, the corresponding curve is convex (Fig. 12-1b).

Next we define a *cost function*, also a function of E, that describes the loss in fitness (via lost future reproduction) due to present reproductive efforts. We might, as a first approximation, suppose that this cost rises approximately linearly with E. But it seems likely that when E reaches some critical level, the reproducing individual might well be in a state of health that puts it beyond redemption. If so, cost will, at that point, rise dramatically with E. Again, we expect a concave function (Fig. 12-2a), although we must concede that convex curves might, under some circumstances, be possible. We can now put the various forms of profit functions and cost functions together on the same graphs to make predictions about a population's life history strategy (Fig. 12-3).

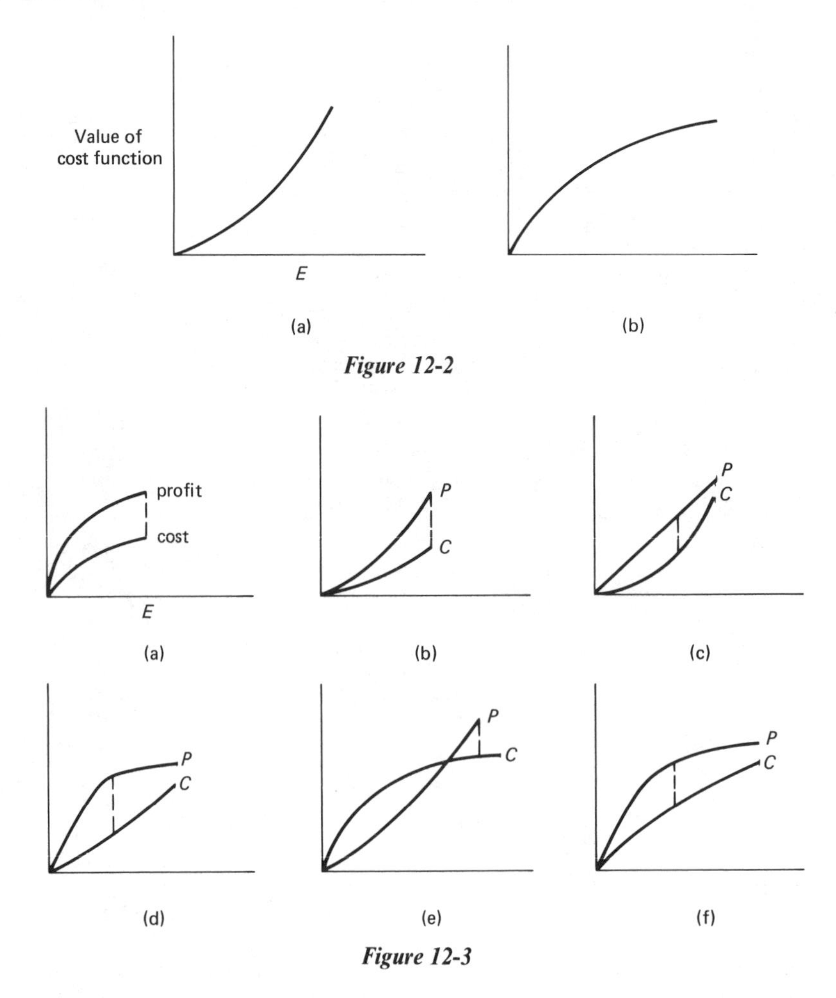

Figure 12-2

Figure 12-3

Natural selection should adjust E so as to maximize P, the profit function, relative to C, the cost function. In Figure 12-3a, b, and e, this difference is always maximum when E is as large as possible. These are the species that put their all into reproduction; they are semelparous. In Figure 12-3c, d, and f, $P-C$ is largest when E takes an intermediate value. These species should hold back and save energy for survival to the next breeding period. With respect to the latter, iteroparous species, we can, hypothetically, draw the corresponding curves for all age groups and find the E_{opt} for each. These values can then be related back to find the corresponding $P(= b'p_0)$ and p_a values. In this way age-specific survivorship and fecundity patterns can hypothetically be predicted.

Gadgil and Bossert proposed (and see also Charlesworth and Leon, 1976) that as an organism approachs the end of its life, the reproductive potential remaining drops and so, therefore, must C. Thus (see, e.g., Fig. 12-3c, d, or f) E should be expected to rise with age. There is considerable evidence that E does generally increase with age in a great many species of animals and plants (little of it well documented). But as Fagen (1972) has pointed out, it is quite easy to invent perfectly reasonable life histories in which reproductive effort does *not* increase with age. Thus the conclusion that E increase with age should not be taken to be necessarily or perhaps even generally true.

Schaffer (1974b) used the model of Charnov and Schaffer (1973; see Chapter 11), which is really just a less general and conceptually clearer version of Gadgil and Bossert's approach, to assess the effects of environmental fluctuations on E_{opt}. Writing

$$W = b'p_0 + p_a$$

let $b'p_0$ vary in time due to environmental fluctuations between values $b'p_0(1 + S)$ and $b'p_0(1 - S)$. Then

$$W^* = \sqrt{[b'p_0(1 + S) + p_a][b'p_0(1 - S) + p_a]} = \sqrt{(b'p_0 + p_a)^2 - S^2(b'p_0)}$$

W^* can be increased by emphasizing the p_a component. That is, variation in $b'p_0$ favors a lower reproductive effort (see also Murphy, 1968). On the other hand, if fluctuations in the environment cause variations in p_a between values of $p_a(1 - S)$ and $p_a(1 + S)$, then

$$W^* = \sqrt{(b'p_0 + p_a)^2 - p_a^2S^2}$$

Fitness is now maximized by playing down the relative contribution of p_a; that is, variation in p_a favors greater reproductive effort. These arguments parallel those in Chapter 11 dealing with the more coarse-focus problem of semelparity or iteroparity.

A number of authors have attempted to address the question of age-specific birth and death rate evolution from a population dynamic rather than an energetic viewpoint. Lewontin (1965) used simple mathematical expressions reflecting general observed trends in l_x and b'_x with x, and showed that the population growth parameter was affected by a number of factors, particularly early first reproductive age. In 1966, Williams showed that reproductive value could be broken down into constituent parts as follows:

$$\frac{v_x}{v_0} = \frac{\lambda^x}{l_x}\left(\sum_{y \geqslant x} \lambda^{-y} l_y b_y\right) = b'_x + \lambda^{-1} p_x \frac{v_{x+1}}{v_0} \qquad \text{(see eq. 8-39)} \qquad (12\text{-}1)$$

This led H. M. Taylor et al. (1974) to produce an elaborate proof that maximizing λ (or, equivalently, $\bar{W}$ at stable age distribution) by natural selection was the same as maximizing v_x/v_0 for all x, and to use their proof to generate extremely general conclusions (in agreement with those of the simpler models above).

Schaffer (1974b) took a somewhat similar approach, a kind of hybrid between that of Taylor et al. and that of Gadgil and Bossert. Noting that at population equilibrium $\lambda = 1$, so that (12-1) becomes

$$\frac{v_x}{v_0} = b'_x + v_x^* \qquad \text{where} \qquad v_x^* = p_x \frac{v_{x+1}}{v_0} \qquad (12\text{-}2)$$

he argued that fitness would be maximized when

$$0 = \frac{d(v_x/v_0)}{dE} = \frac{db'_x}{dE} + \frac{dv_x^*}{dE}$$

or

$$\frac{db'_x}{dE} = -\frac{dv_x^*}{dE} \qquad (12\text{-}3)$$

He envisioned b'_x rising asymptotically with E_x, the reproductive effort at age x, and v_x^* falling at an increasing rate with E_x (Fig. 12-4), although he acknowledged that both curves might also be concave. If, as expected, Figure 12-4 accurately describes real-life situations, it makes sense to proceed and plot the derivatives of these parameters with respect to E_x against E_x, as shown in Figure 12-5. The point where the derivatives are equal (equation 12-3; point of intersection in Fig. 12-5) defines optimal E_x and hence also the values of b'_x and v_x^*.

A similar procedure was followed by Pianka and Parker (1975). Referring to v_x^* as the "residual" reproductive value, they plot v_x^* against b'_x for various values of v_x/v_0. The result is a set of fitness (actually v_x/v_0) isoclines (an "adaptive function," Levins, 1968) (Fig. 12-6). Natural selection can be expected to "choose" from the set of all possible combinations, that combination of v_x^* and b'_x that maximizes v_x/v_0. To make predictions, we obviously must have some idea of what those possible combinations are. If changes in reproductive effort make a large difference in the number of offspring that can be turned out, but have little effect on subsequent survival, an organism's options are included within a set of points outlined by the

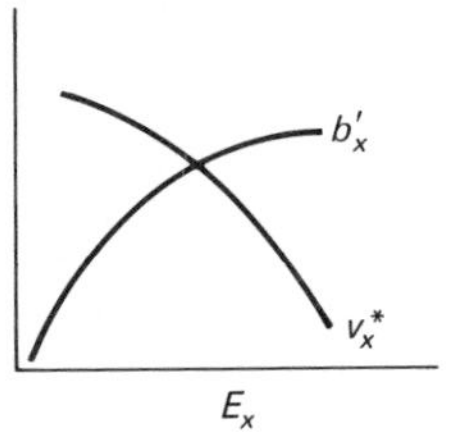

Figure 12-4 Hypothetical relationship of fecundity (b'_x) and residual reproductive value (v_x^*) to reproductive effort (E_x) at age x.

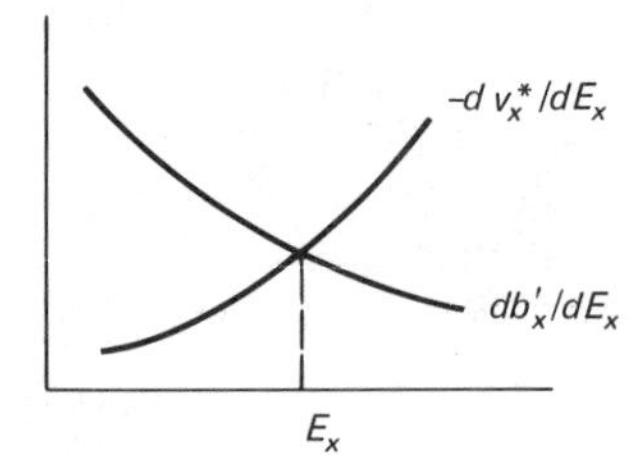

Figure 12-5 $-dv_x^*/dE_x$ and db'_x/dE_x as functions of E_x (from Fig. 12-4). The value of E_x where the lines cross gives the optimal reproductive effort.

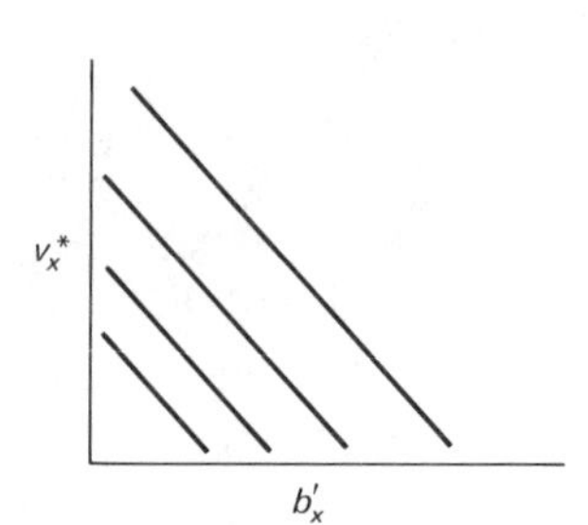

Figure 12-6 Adaptive function for v_x^* and b'_x.

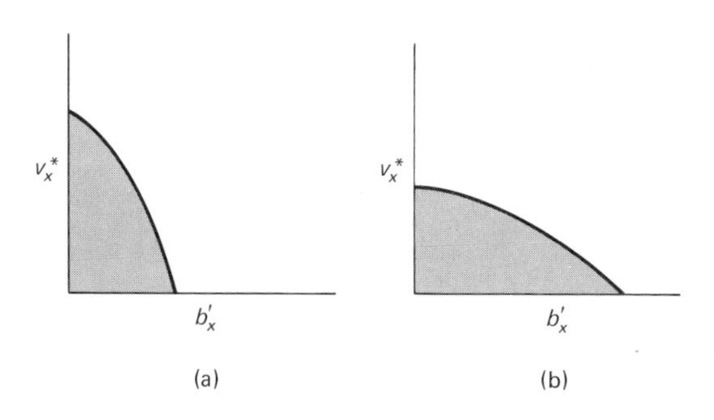

Figure 12-7 Possible fitness sets for v_x^* and b'_x.

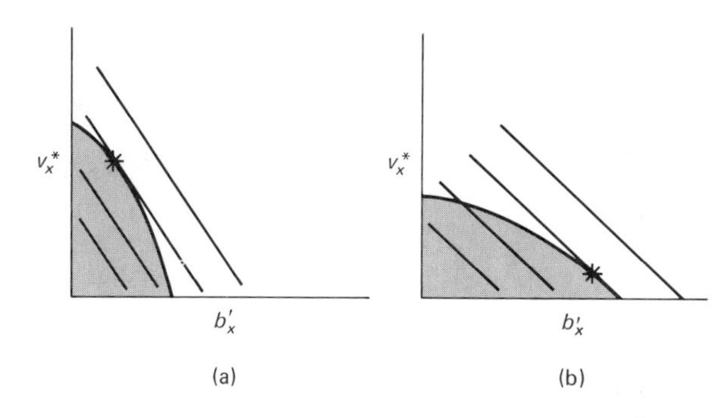

Figure 12-8 Superimposed fitness sets and adaptive function from Figures 12-6 and 12-7. The point of tangency gives the optimal b'_x, v_x^* combination.

curve in Figure 12-7b. If increased reproductive effort makes little difference to the number of young produced but has drastic implications for survival, Figure 12-7a is appropriate. Clearly, these represent the extremes of a whole continuum of possibilities. Superimposing the curves of Figures 12-6 and 12-7, we can find quite simply the realizable option that maximizes v_x/v_0 (Fig. 12-8).

The models above suffer from a bad case of redundancy. There are other problems. First, it is likely that by the time the information necessary to make quantitative predictions has been gathered, the researcher will already know everything that might subsequently be predicted. Second, the models all rely directly or indirectly on the existence of stable age distribution. That is, they assume either that the populations are at equilibrium, with unchanging density feedback, or have been growing freely for some time in an environment with no density feedback at all. The models, then, have some value, but only in a very general qualitative sense, and are testable only for specific cases with the garnering of large amounts of data. But for more useful kinds of predictions it might be more profitable to turn to another set of models, based on a slightly different view of the world.

Some Difficulties with the Use of Trade-off Theories

While the trade-off school of life history evolution presents a rather tidy series of intuitively satisfying hypotheses, there are some problems with its lines of reasoning. As discussed in Chapter 11, most evidence for energetic trade-offs are intraindividual. Here they must be trivially true. Many fewer data exist showing negative correlations between (say) reproductive effort (or clutch size) and subsequent survival *among* individuals. Indeed, even were such data commonplace, they still would not necessarily support the assumptions behind the trade-off models: What is necessary for these models to work is not negative phenotypic covariance, but negative additive genetic covariance among the traits that are, hypothetically, being traded off.

Recall from Chapter 10 that the evolutionary progress made by a trait under natural selection was the sum of the effects of direct selection and indirect effects via selection on other traits genetically tied to the first via pleiotropy. Where the trait of interest is X_i, we have

$$\bar{W}\Delta\bar{x}_i = V_g(x_i)\frac{\partial\bar{W}}{\partial\bar{x}_i} + \sum_{j\neq i}\text{cov}_g(x_i, x_j)\frac{\partial\bar{W}}{\partial\bar{x}_j}$$

Thus if $V_g(x_i)$ is small, but one or more of the additive covariance terms is large, the character, X_i, might even evolve in opposite direction to the force of selection acting directly on it. If selection acted strongly on two characters, we might expect them to rather rapidly evolve negative covariance (Falconer, 1960a; see Introduction to Part Four), but this argument in itself does not make it so. These problems are discussed more fully in Chapter 11.

"Fitness" Maximization Models

The approach presented here was first suggested by Bidder (1932), and again by Haldane (1941) and Comfort (1956). It was formalized verbally by Medawar (1952, 1957) and put into more rigorous terms by Williams (1957), Hamilton (1966) and J. M. Emlen (1970). The basic argument is as follows. We begin with a population of hypothetical ancestral organisms that are potentially immortal, experiencing accidental deaths at a rate invariant with age, and with an age-invariant birth rate. Now introduce a new mutant gene that endows its carrier with increased survivorship or fecundity, but whose effects become felt only after some particular age, say age x. Clearly, such a gene will be selected for. But suppose that at a given locus *AA* individuals experience onset of the advantage at age x_1, *aa* individuals at age x_2, where $x_1 < x_2$. The advantage of the trait will be expressed for a longer proportion of an *AA* individual's life than an *aa* individual's life. Therefore, *AA* individuals will be more fit and the *A* allele will spread faster than, or at the expense of, *a*. Alternatively, suppose that there exist two alleles at an epistatic locus, one of which (call it *X*) encourages earlier expression of the advantageous trait. Then allele *X* is selected for. In either case the result of natural selection is to encourage the evolution of beneficial traits, but to encourage them most strongly at young ages (at least for young ages beyond the age of first reproduction). If conflict situations arise where the development of a beneficial trait at one age can only be realized at the expense of its expression at another age, the trait can be expected to appear at a young rather than an old age. In such ways beneficial traits are pushed toward early ages. By the same kind of argument, detrimental traits are selected against, but again most strongly at young ages—onset of such a character at an early age has a much greater period of time in which to wreak its influence, on average, than had it appeared in a very old individual. As beneficial traits accumulate at young ages, deleterious ones at old, the changing life table milieu accelerates the process. An old individual in a population that has already felt the influence of such selection pressures is likely to be feeling enough accumulated genetic problems that it can expect little reproductive life to remain. Then the selection pressures to eliminate further deleterious gene effects will be almost nonexistent.

The reasoning above does not necessarily require the existence of negative genetic covariance among "trade-off" traits (although the argument would be greatly strengthened if negative covariances occurred, and in fact this assumption formed the basis for Williams (1957) argument.)

Hamilton's and Emlen's mathematical formulation of this argument—which we shall call the Medawar hypothesis—also applies only to equilibrium populations (stable age distribution) or those growing in an unlimited environment (no density feedback). It is given as follows. We make use of the Euler equation,

$$1 = \sum_x \lambda^{-x} l_x b'_x$$

and recalling that $l_x = \prod_{y=0}^{x-1} p_y$, differentiate both sides to obtain

$$0 = \frac{d(1)}{dp_k} = \sum_x \left(-x\lambda^{-x-1} \frac{\partial \lambda}{\partial p_k} l_x b'_x \right) + \sum_{x>k} \left(\lambda^{-x} \frac{l_x}{p_k} b'_x \right) \tag{12-4}$$

$$0 = \frac{d(1)}{db'_k} = \sum_x \left(-x\lambda^{-x-1} \frac{\partial \lambda}{\partial b'_k} l_x b'_x \right) + \lambda^{-k} l_k \tag{12-5}$$

so that

$$\frac{1}{\lambda}\frac{\partial \lambda}{\partial p_k} = -\frac{\sum_{x>k} \lambda^{-x} l_x b'_x}{p_k \sum_x x\lambda^{-x} l_x b'_x} \tag{12-6}$$

$$\frac{1}{\lambda}\frac{\partial \lambda}{\partial b'_k} = \frac{\lambda^{-k} l_k}{\sum_x x\lambda^{-x} l_x b'_x} \tag{12-7}$$

Because $\lambda = R$ at stable age distribution, (12-6) and (12-7) describe the rate of evolution of the age-specific survivorship p_k and the age-specific birth rate b'_k [refer to "Phenotypic Selection Models" (p. 291)]. But because $(1/\lambda)(\partial\lambda/\partial x) = (1/R)(\partial R/\partial x)$ is *not* identical to $(1/R)(dR/dx)$, we cannot procede to make quantitative predictions without further knowledge of genetic covariances (and physiological relationships) among the p_k and b'_k values and the nature of other constraints. The general predictions of the verbal argument above emerge in qualitative form, however, if we suppose p_k and b'_k to evolve independently. Figures 12-9 and 12-10 are plots of (12-6) and (12-7).

From an examination of these graphs we can predict that the age-specific birthrate should rise shortly after the age of first reproduction is reached, peak in early adulthood, and then fall. Age-specific survivorship is harder to deal with. After age of first reproduction, it is clear that p_x

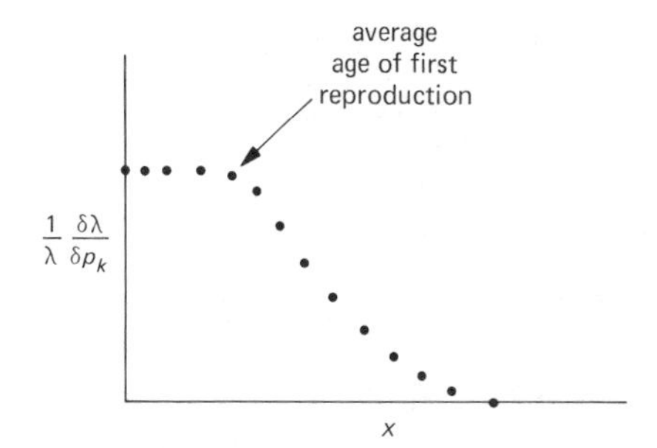

Figure 12-9 Selection intensity on age-specific survival p_x as a function of x.

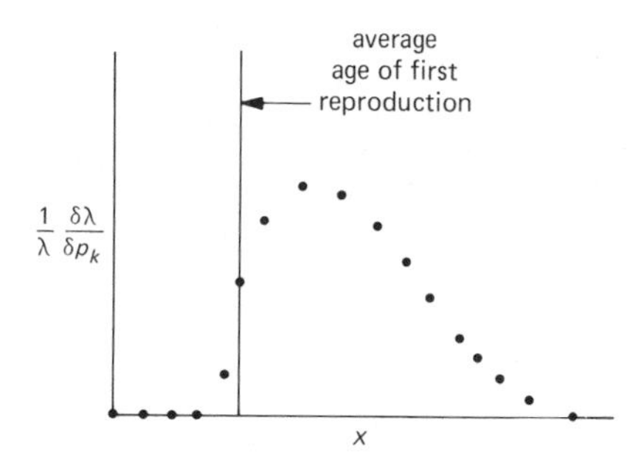

Figure 12-10 Selection intensity on age-specific fecundity b'_x as a function of x.

should fall with increasing x, but what of the early years? J. M. Emlen (1970) argued that the values of p_x reflected the interaction of many physiological, morphological, and behavioral traits, all of which were simultaneously changing in the developing organism, and that the particular configuration of these traits that promoted minimum mortality should be expected to occur, on the basis of (12-6) (Fig. 12-9), somewhere between birth and first reproduction. Exactly where this optimal age for survivorship falls we cannot say, but we can predict that p_x should rise, reach a peak *before* first reproduction, and then fall throughout life.

Both Hamilton's and Emlen's discussion ignored complications due to density feedback acting on the population parameters. It can be argued that at stable age distribution the population must either be rising exponentially, thus in all likelihood experiencing no feedback to speak of, or at equilibrium, in which case density factors do not enter the argument because they are not changing. Ricklefs (1981) has shown this to be true; density feedback does not alter the argument above.

Data in support of the above predictions come in many forms. The survivorship pattern for human beings is shown in Table 12-3. Data are from Pearl (1940). The same general pattern is found in mammals and birds in general (Taber and Dasmann, 1957; Caughley, 1966; Tanner, 1975; see also Dittus, 1977). It also occurs in plants (Watson, 1979). To the extent that onset of disease is a measure of genetic predisposition to dying, data on incidence of diabetes also support the prediction (Table 12-4, after Simpson, 1969).

Sokal (1969), working with two strains of *Tribolium castaneum*, a flour beetle, kept control lines and selected lines. The selected lines were allowed to oviposite for 3 days and were then killed. According to Medawar's hypothesis, failure to survive to older ages (which clearly precludes reproduction at those ages) means no selection to maintain the genetic integrity of the system at those ages. Hence we might expect the selected populations eventually to become more short-lived. Sokal tested for longevity in his two sets of lines by taking 30 female and 30 male pupae, at random, from both his control and his selected lines, and raising them singly in vials. The selected line, as expected, displayed significantly shorter life spans. Similar results have been obtained with *D. melanogaster* (Luckinbill, personal communication.)

Table 12-3

Age (yr)	q_x
0–1	.062
1–2	.010
3–4	.004
5–6	.003
7–8	.002
9–10	.00155
14–15	.002
19–20	.003
24–25	.004
29–30	.004
34–35	.005
39–40	.006
44–45	.009
49–50	.012
54–55	.017
59–60	.025
64–65	.036
69–70	.054
74–75	.079
79–80	.120

Table 12-4

Age (yr)	Percent Increase in Incidence of Diabetes	
	Male	*Female*
0–9	.05	.03
10–19	.08	.12
20–29	.14	.05
30–39	.05	.30
40–49	.18	.25
50–59	1.44	1.73
60–69	.68	2.12
70–79	1.19	1.44

It can be argued that the age-specific death rate rises almost perfectly exponentially in many species, thus supporting an alternative model which holds q_x to increase due to the accumulation of errors of transcription or injuries. But although the latter view must surely be part of the story, it does not account for the *drop* in q_x in early life. Perhaps that drop is due to the incompleteness of ontogeny at those early ages—maybe the various body repair or elimination mechanisms have not yet fully developed, allowing an accumulation of deleterious somatic mutations that are only later culled. If so, why do those repair mechanisms fail to continue such an efficient culling process later on? Maybe the damages accumulate at an accelerating rate, or the repair mechanisms themselves are subject to cumulative damage and gradually lose their effectiveness. The degree to which Medawar's hypothesis "explains" the aging process, and the degree to which it embellishes the spandrels of San Marco, is not clear.

With respect to fecundity there is a considerable literature on birds (see some of the data presented in Chapter 11). Mostly these data compare first year versus older birds, showing that clutch sizes are larger in the latter. The few data sets in which "older" is broken down show a decline among the oldest birds. In mammals the pattern is true of mice (e.g., *Peromyscus leucopus*, Drickamer and Vestal, 1973) and also big game. For example, Simpson (1968) offers the data in Table 12-5 for the greater kudu (*Tragelaphus strepsiceros*) in Rhodesia (Zimbabwe). Human data are abundant; those in Table 12-6 are from Keyfitz (1968).

One measure of fecundity, F_x, includes p_0 as a component. That is, the health status of the newborn may be included in the fecundity measure if we use the slightly altered form of the Euler equation (with F and L in place of b' and l) to calculate the corresponding forms for (12-6) and (12-7). That being the case, the following data on incidence of birth defects as a function of parent's age is pertinent to the predictions we have made about declining fecundity with age (Table 12-7). Data are from Polednak (1976). The predicted trend is obvious for females, less

Table 12-5

Age of Female (months)	Percent of Females Pregnant or Lactating
1–10	About 25
11–20	70
21–30	80
31–40	100
41–50	85
51–60	50
60	0

Table 12-6

Age (yr)	Number of Births per Female per Year, F_x
0–5	.00000
5–10	.00000
10–15	.00042
15–20	.03556
20–25	.10742
25–30	.08765
30–35	.05074
35–40	.02444
40–45	.00675
45–50	.00039

Table 12-7 **Number of Common Birth Defects per 10,000 Live Births (Northern New York State)**

Mother's Age (yr)	Father's Age (yr)										
	15–19	20–24	25–29	30–34	35–39	40–44	45–49	50–54	55–59	60+	Mean
15–19	1.03	1.19	1.39	.98	.92	1.33	1.67	—	—	—	1.16
20–24	.90	1.06	.97	1.04	1.36	1.23	2.01		(1.00)		1.03
25–29	.93	1.11	.97	1.08	1.03	1.07	1.32	.66	.73	—	1.02
30–34	—	.89	1.25	1.14	1.14	1.06	.80	1.29	1.75	1.71	1.14
35–40	—	3.31	1.25	.99	1.34	1.25	1.50	1.69	2.31	—	1.31
41–49	—	—	—	.59	2.29	1.96	1.98	2.44	2.20	1.96	2.02
Mean	1.01	1.09	.99	1.09	1.20	1.31	1.51	1.68	1.77	1.70	

obvious but still clearly present for males (see also Milham and Gittelsohn, 1965). Note the *drop* in birth defect incidence for mothers between the 15–19 age class and the 25–29 age class. This suggests the subsequent rise is not simply the result of accumulated somatic mutations.

In species with determinate growth, where size promotes increased egg-laying capacity and perhaps escape from predation, both b' and p continue to increase until late in life. However, although peak years for fecundity and minimum mortality are considerably delayed from the time of pre- or early adulthood, there is nevertheless still a rise and then fall in reproductive output in age, a rise and then fall in survival probability. For example, the data in Table 12-8

Table 12-8

Snail Size (mm)	Number of Snails	Estimated Number of Capsules per Snail	Estimated Number of Eggs per Capsule	Estimated Number of Eggs per Snail
40–50	7	202	39	7,900
51–60	1.0	214	57	12,200
61–70	3	344	75	25,800
71–80	2	243	94	22,800
81–90	3	52	112	5,800
91–100	1	16	130	2,400

describe the relationship between length (an indicator, although not necessarily linear, of age) and estimated number of eggs layed in the intertidal gastropod *Thais lamellosa* (J. M. Emlen, 1966a). For details, see the original reference. See also Spight and Emlen, 1976.

Other, somewhat indirect forms of support of particular interest to the behaviorist can be found in the research done with Japanese macaques (*Macaca fuscata*). Presuming that learning ability exerts a positive effect on fitness, it is worth noting the age patterns of acquisition of new skills and the cultural transmission of these skills. Monkeys in the Koshima colony had for some time been fed, among other things, grains of wheat. These grains often got mixed up with sand on the shores of the island where the population was maintained, making their ingestion a bit gritty. The animals eventually learned to placer-mine the wheat, taking a handful of grains plus sand, carrying it to the water, and holding it under water until the sand settled out. Forty percent of 2- to 3-year-olds learned to do this, 85% of the 6- to 7-year-olds, 40% of 8- to 11-year-olds, and only 10% of animals over 12 years of age. The skill was first learned by 6- and 7-year-old monkeys and then passed on to the younger members of the group. The older monkeys were considerably slower in learning the task and refused to copy their youngers (Kawai, 1965). Other tasks, such as learning to dig for buried food items, or washing off sweet potatoes, followed the same pattern. The animals at the peak of reproductive value learned fastest. The value of these data in supporting the Medawar hypothesis is complicated by the fact the subordinate individuals learned from dominant individuals, but not vice versa, and that, at least among males, adults are dominant over younger animals.

In hypothetically culture free tests involving the learning of nonsense syllables, Inglis et al. (1968) report that in human subjects, learning is most rapid in children about 11 years old. Ability to learn the nonsense syllables increased from age 6 (their youngest subjects) to age 10 (statistically significant at the .05 level), then showed a nonsignificant decrease to age 20, a further decrease from 20 to 30 (the difference between 10 and 30 *is* statistically significant), and thereafter an increasingly rapid deterioration with age. Short-term memory of the syllables also increased with age at first, and then fell.

We end this discussion with a warning that the predictions *cannot* be taken to have *quantitative* meaning. The model does not in any way take constraints into consideration. But as long as we view the predictions in purely qualitative fashion—a rise and then a fall in p_x and b'_x—it works quite well.

An ingenious and unfortunately not very well received effort to make more quantitative inroads into the question of age-specific mortality and fecundity—really a question of aging—has been made by Demetrius (1975a,b, 1976). Demetrius's entire line of reasoning will not be replicated here; briefly, he defines two quantities which he calls population "entropy," H, and "reproductive potential," Φ. These are defined by

$$H = -\frac{\sum_x (\lambda^{-x} l_x b'_x) \ln (\lambda^{-x} l_x b'_x)}{\sum_x x\lambda^{-x} l_x b'_x} \tag{12-8}$$

$$\Phi = -\frac{\sum_x (\lambda^{-x} l_x b'_x) \ln (l_x b'_x)}{\sum_x x\lambda^{-x} l_x b'_x} \tag{12-9}$$

To appreciate the meaning of H, the critical function in Demetrius's analysis, note that $\lambda^{-x} l_x b'_x$ is one measure of the contribution of age x to the rate of population growth. Then

$$\frac{\lambda^{-x} l_x b'_x}{\sum_x \lambda^{-x} l_x b'_x} = \lambda^{-x} l_x b'_x$$

is the *fraction* of total contribution by all ages that is attributable to individuals of age x. If we call this quantity q_x, then H becomes

$$H = -\frac{\sum_x \tau_x \ln \tau_x}{\sum_x x\tau_x} \qquad \text{where} \qquad \sum_x \tau_x = 1.0 \tag{12-10}$$

The numerator is the standard form of the Shannon–Wiener information-theoretic equation, which can be used as a measure of *equitability* of age-specific contributions. The denominator is one possible measure of generation time. Note for future reference, that after some collecting of terms,

$$H - \Phi = \ln \lambda \tag{12-11}$$

Demetrius next argued that because the survivorship and fecundity values of genotypes differed, each would demonstrate its own value of H. Let H_{ij} be the entropy of the ijth genotype, where p_i and p_j give the frequencies of the two allelic types. Then we can write

$$\begin{aligned} H &= \sum_{i,j} H_{ij} f_{ij} \qquad (f_{ij} \text{ is the frequency of genotype } ij) \\ &= \sum_{i,j} H_{ij} p_i p_j = \sum_i \bar{H}_i p_i \end{aligned}$$

Therefore, under the influence of natural selection,

$$\Delta H = \sum_i \bar{H}_i \Delta p_i = \sum_i \bar{H}_i p_i \left(\frac{\lambda_i - \lambda}{\lambda} \right)$$

(see Chapter 8). If $\lambda_i \simeq \lambda \simeq 1$ (slow selection), then

$$\begin{aligned} \Delta H &= \sum_i \bar{H}_i p_i \left(\frac{\lambda_i - 1}{\lambda} - \frac{\lambda - 1}{\lambda} \right) \\ &\simeq \sum_i \bar{H}_i p_i (\ln \lambda_i - \ln \lambda) \\ &= \sum_i \bar{H}_i p_i [(\bar{H}_i - \bar{\Phi}_i) - (H - \Phi)] \qquad \text{(from 12-11)} \\ &= \sum_i p_i \bar{H}_i (\bar{H}_i - H) - \sum_i p_i \bar{H}_i (\bar{\Phi}_i - \Phi) \\ &= \text{var}(H) - \text{cov}(H, \Phi) \end{aligned} \tag{12-12}$$

Before pressing on, we shall include two derivations, not considered by Demetrius, that considerably strengthen the eventual conclusion he reached. Recalling that $\ln \lambda = H - \Phi$, so that $\Phi = H - \ln \lambda$, we can write

$$\begin{aligned} \text{cov}(H, \Phi) &= \text{cov}[(H - \ln \lambda), H] \\ &= \text{var}(H) - \text{cov}(H, \ln \lambda) \end{aligned}$$

Thus

$$\begin{aligned} \Delta H &= \text{var}(H) - \text{cov}(H, \Phi) \\ &= \text{cov}(H, \ln \lambda) \end{aligned} \tag{12-13}$$

This is a somewhat satisfying conclusion, because it states that natural selection indeed increases H, but only when H is positively correlated with $\ln \lambda$. At stable age distribution $\lambda = R = \bar{W}$, so this seems very reasonable.

Next we note that λ and $\ln \lambda$ vary in the same direction, so that $\operatorname{cov}(H, \ln \lambda)$ is of the same sign as $\operatorname{cov}(H, \lambda)$. But if $\operatorname{var}(\lambda)$ is not too large, $\operatorname{cov}(H, \lambda)$ can be closely approximated with $(dH/d\lambda)\operatorname{var}(\lambda)$. Thus $\operatorname{cov}(H, \ln \lambda)$ is the same sign as

$$\frac{dH}{d\lambda} = \sum_x \frac{\partial H}{\partial \tau_x}\frac{d\tau_x}{d\lambda} \qquad \text{where } \tau_x = \lambda^{-x} l_x b'_x$$

The expression above is not easily amenable to analysis, but a very large number of calculations with trial combinations of hypothetical τ_x values consistently led to $dH/d\lambda > 0$. It seems, then at least for most imaginable situations, that $\operatorname{cov}(H, \ln \lambda)$ is positive, so that ΔH is also positive.

In Chapter 11 the r- versus K-selection dichotomy was attacked as being overly simplistic. A better approach would be to include H selection as a third dimension. Because of the genetic covariation problems in viewing different life history selection schemes as trade-offs, and the interdependence of r, K, and H, such an approach would still be simplistic, but it would represent at least a small advance in thinking. The major drawback of Demetrius's model is that, like the preceding models (at least the formal, quantitative analyses of them), it equates λ with R. That is, it is valid under circumstances where the effects of all but the dominant (Perron) root of the Euler equation have damped out—it works for populations at stable age distribution. Thus it states that H should increase under natural selection in strongly equilibrium species; it need not apply to opportunistic species.

If natural selection maximizes H, as the analysis indicates above and as Demetrius states, a very interesting prediction becomes possible. To find the values of τ_x that maximize H, under the constraint that the τ_x must sum to 1.0, we use the method of Lagrange multipliers. This constitutes defining a function, $f(\{\tau_x\}) = H + \mu(\sum_x \tau_x - 1)$, where μ is a constant of unknown value, and then differentiating this new function, setting the derivatives equal to zero:

$$0 = \frac{\partial f}{\partial \tau_x} = -\frac{(1 + \ln \tau_x)\sum_x x\tau_x - x\left(\sum_x \tau_x \ln \tau_x\right)}{\left(\sum_x x\tau_x\right)^2} + \mu \qquad (12\text{-}14)$$

for all τ_x. Multiplying by τ_x, summing over all ages, and remembering that the τ_x sum to 1.0, we then obtain

$$\mu = \sum_x \mu\tau_x = \frac{\left(\sum \tau_x + \sum \tau_x \ln \tau_x\right)\left(\sum_x x\tau_x\right) - \left(\sum_x x\tau_x\right)\left(\sum_x \tau_x \ln \tau_x\right)}{\left(\sum_x x\tau_x\right)^2} = \frac{1}{\sum_x x\tau_x} \qquad (12\text{-}15)$$

Now that μ has been found, it can be substituted back into (12-14), yielding

$$\frac{1 + \ln \tau_x}{\sum x\tau_x} + \frac{xH}{\sum x\tau_x} = \mu \qquad \left(= +\frac{1}{\sum x\tau_x}\right)$$

so that

$$\tau_x = e^{-xH} \qquad (= \lambda^{-x} l_x b'_x) \qquad (12\text{-}16)$$

Equation (12-16) is really full of biological information. First, because the τ_x must sum to 1.0, it ties together the first age of reproduction, the last age of reproduction, and the value of H. If we know two of these values, the last is calculable. Second, it gives us an explicit, quantitative prediction about age-specific survival and fecundity (we shall return to this point in a moment). Finally, the entire analysis rests on the conclusion that natural selection acts to maximize H. But because H is a measure of the equitability of contribution to population growth by all ages, an increase in H means that all ages move toward equal representation in terms of offspring production. This has a profound effect on population stability, spreading the risk of mortality and loss of brood over all ages more or less equally. Iteroparous populations, Demetrius is saying, evolve greater and greater stability.

Referring specifically to the prediction of τ_x by (12-16), we ask the obvious question: Does it work? A quick referral to the data in Tables 12-1 and 12-2 and an examination of the literature in general shows that it does not. Why? There are at least four possibilities. First, as remarked above, the model assumes stable age distribution and so applies only loosely to a limited number of species. Second, owing to environmental uncertainty or genetic or physical constraints, most populations may simply have failed to converge very accurately on the ideal, expected end point of evolution. Third, consideration of the constraints previously mentioned in regard to Hamilton's and Emlen's treatments of Medawar's hypotheses, and dealt with specifically in the energetics models, is lacking. It would be interesting to try and expand Demetrius's model to include those additional, energetic constraints. Fourth, the additive covariance structure of populations may, in general, preclude or make unlikely the successful climbing of the highest of perhaps many adaptive peaks, that peak defined by Demetrius.

The question of adaptive peaks in the evolution of life histories has been rather nicely touched on by Lande (1982). His model is genetic and so not prone to the weak assumptions of the trade-off models. The following version is abbreviated, but gives the flavor. Recall from Chapter 10 (continuous analog to equation 10-43) that we can write

$$\frac{d}{dt}\mathbf{z} = G\,\nabla \ln \bar{W} \tag{12-17}$$

where $\mathbf{z}$ is a vector describing the population mean values of a constellation of traits, $\bar{z}_i$, G is the additive genetic variance–covariance matrix for those traits, $\bar{W}$ is population mean fitness, and

$$\nabla = \begin{bmatrix} \frac{\partial}{\partial \bar{z}_1} & \frac{\partial}{\partial \bar{z}_2} & \cdots \end{bmatrix}$$

We can think of each $\bar{z}_i$ element of the vector describing some life history feature. Because (12-17) is valid only when there is no change in age structure, we shall limit further analysis, as before, to populations at or very near stable age distribution. Because the z_i will be functions of population density, we also limit discussion to populations near equilibrium (these two constraints are usually more or less effectively equivalent). Next note that we can write

$$\frac{d \ln \bar{W}}{dt} \simeq \sum_i \frac{\partial \ln \bar{W}}{\partial z_i}\frac{dz_i}{dt} = (\nabla \ln \bar{W})^T G (\nabla \ln \bar{W}) \tag{12-18}$$

But then selection proceeds until $d \ln \bar{W}/dt \to 0$, at which point

$$(\nabla \ln \bar{W})^T G (\nabla \ln \bar{W}) \to 0 \tag{12-19}$$

Remember that $\ln \bar{W}$ and G are both functions of the z's. The vector $\mathbf{z}$, for which (12-19) holds, gives the expected evolved life history. But notice that there may be several solutions. Because each solution describes a combination of the z values such that $(d/dt)\,\bar{\mathbf{z}} \to 0$, each represents an adaptive peak. A very simple illustration is in order. Consider a population with discrete breeding seasons and two reproducing age classes. Then one way of describing the life history of this population's members is with the vector

$$[b'_1 \quad b'_2 \quad p_0 \quad p_1]$$

But suppose that we were interested in the evolution of reproductive effort at ages 1 and 2. Then, denoting reproductive effort by E, we might alternatively write

$$\mathbf{z} = [E_1 \quad E_2]$$

Substituting into (12-19), we then have

$$0 = \begin{bmatrix} \dfrac{\partial \ln \bar{W}}{\partial E_1} & \dfrac{\partial \ln \bar{W}}{\partial E_2} \end{bmatrix} \begin{pmatrix} V_g(E_1) & \text{cov}_g(E_1, E_2) \\ \text{cov}_g(E_1, E_2) & V_g(E_2) \end{pmatrix} \begin{bmatrix} \dfrac{\partial \ln \bar{W}}{\partial E_1} \\ \dfrac{\partial \ln \bar{W}}{\partial E_2} \end{bmatrix}$$

By writing this is the form

$$0 = [\text{A} \quad B] \begin{bmatrix} g_{11} & g_{12} \\ g_{12} & g_{22} \end{bmatrix} \begin{bmatrix} A \\ B \end{bmatrix} = A^2 g_{11} + 2AB g_{12} + B^2 g_{22}$$

we can find

$$A = -B \frac{g_{12} \pm \sqrt{g_{12}^2 - g_{11} g_{22}}}{g_{11}} \tag{12-20}$$

We now need to find the E_1 and E_2 for which the above is true.

To proceed further, we must consider a particular situation. Suppose we knew from physiological experiments that

$$b'_1 = .6E_1 \qquad b'_2 = .6E_2 \qquad p_1 = .8 - .2E$$

and from genetical experiments that

$$V_g(E_1) = .5 \qquad V_g(E_2) = .2 \qquad \text{cov}_g(E_1, E_2) = -.4$$

From the Euler equation ($1 = R^{-1} p_0 b'_i + R^{-2} p_0 p_i b'_2$) we also know that

$$\begin{aligned} \frac{\partial \ln R}{\partial b'_1} &= \frac{p_0}{cR} = \frac{p_0}{c} \\ \frac{\partial \ln R}{\partial b'_2} &= \frac{p_0 p_1}{cR^2} = \frac{p_0 p_1}{c} \\ \frac{\partial \ln R}{\partial p_1} &= \frac{p_0 b'_2}{cR^2} \end{aligned} \tag{12-21}$$

where $c = R^{-1} p_0 b'_1 + 2R^{-2} p_0 p_1 b'_2$ and $R = 1$ at population equilibrium. Finally

$$\begin{aligned} \frac{\partial \ln R}{\partial E_1} &= \frac{\partial \ln R}{\partial b'_1} \frac{\partial b'_1}{\partial E_1} + \frac{\partial \ln R}{\partial p_1} \frac{\partial p_1}{\partial E_1} \\ \frac{\partial \ln R}{\partial E_2} &= \frac{\partial \ln R}{\partial b'_2} \frac{\partial b'_2}{\partial E_2} \end{aligned} \tag{12-22}$$

Noting that $\partial R / \partial x = \partial \bar{W} / \partial x$ and putting (12-21) and (12-22) together, we obtain

$$\frac{\partial \ln \bar{W}}{\partial E_1} = \frac{p_0}{c}(.6) + \frac{p_0 b'_2}{c}(-.2) = \frac{p_0}{c}(.6 - .12E_2) = A$$

$$\frac{\partial \ln \bar{W}}{\partial E_2} = \frac{p_0 p_1}{c}(.6) = \frac{p_0}{c}(.48 - .12E_1) = B$$

Substitution into (12-21) then leads to

$$.6 - .12E_2 = -(.48 - .12E_1)(-1.29) \tag{12-23a}$$

$$.6 - .12E_2 = -(.48 - .12E_1)(-.31) \tag{12-23b}$$

In the first case (12-23a),

$$E_2 = -0.16 + 1.29E_1$$

so that

$$b'_1 = .3E_1 \qquad b'_2 = -.10 + .77E_1 \qquad p_1 = .8 - .2E_1 \tag{12-24}$$

In the second case,

$$E_2 = 3.76 + .31E_1$$

so that

$$b'_1 = .3E_1 \qquad b'_2 = 2.256 + .186E_1 \qquad p_1 = .8 - .2E_1 \tag{12-25}$$

If there are no nongenetic constraints on E_1 or E_2, then

$$1 = R^{-1}p_0b'_1 + R^{-2}p_0p_1b'_2 = p_0b'_1 + p_0p_1b'_2$$

$$= p_0[.3E_1 + (.8 - .2E_1)(-0.10 + .77E_1)] \qquad \text{for case 1} \tag{12-26a}$$

and

$$1 = p_0[.3E_1 + (.8 - .2E_1)(2.256 + .186E_1)] \qquad \text{for case 2} \tag{12-26b}$$

If p_0 is .80, case 1 yields $E_1 = 3.81$ and 2.26, and case 2 yields E_1 = complex. Case 2 is clearly not biologically realistic. Therefore, we turn our attention to case 1. When $E_1 = 3.81$, we find, from (12-24),

$$\begin{aligned} p_0 &= .80 \\ p_1 &= .8 - .2E_1 = .038 \\ b'_1 &= .3E_1 = 1.143 \\ b'_2 &= .6E_2 = -.10 + .77E_1 = 2.83 \end{aligned}$$

When $E_1 = 2.26$, we obtain

$$\begin{aligned} p_0 &= .80 \\ p_1 &= .8 - .2E_1 = .348 \\ b'_1 &= .3E_1 = .678 \\ b'_2 &= -.10 + .77E_1 = 1.64 \end{aligned}$$

These configurations represent the two adaptive peaks where the energetic trade-offs and the genetic variances and covariances come to a balance. Without some change in the genetic system or some shift in the response of the reproductive system to parental effort, no further selection can occur.

12.2 The Real World

By this point any self-respecting field biologist will justifiably be fidgeting. We have covered a lot of theory, much of it rather superficial, most of it so general as to be almost unusable except as intellectual exercise, and much of that which is quantitative leading to predictions at variance with observation. Is life history theory, then, worth pursuing, or should it be discarded as a waste of time? Beyond the observation that b'_x and p_x values rise and then fall with age, as predicted, do even the very superficial predictions work with real consistency? Stearns (1980) has reviewed the data looking for correlations between life history traits expected on the basis of theory, such as late breeding and increased reproductive effort or even iteroparity and parental care, and sounds a very discouraging note. To some degree the expected correlations are there, but they are by no means consistent. The tone of this 1980 paper is completely different from that of his 1976 and 1977 review papers, downplaying the basic theory and expanding at length on the problems posed by unexpected genetic correlations and allometric constraints in development. Still other problems exist. As noted in Chapters 10 and 11, genetic architecture may change from one set of environmental conditions to another. Thus adaptive peaks shift over space. Might we find one set of area effects giving way to another over environmental gradients? Where environmental fluctuations occur so that selection regimes change, the existence of multiple adaptive peaks could mean that any evolutionary trajectory might appear utterly random. It would be comforting to believe that such possibly chaotic trajectories would be bounded into small regions of the adaptive space—then straightforward predictions might still be possible. But at this point in time we do not know this to be the case.

Then there are the problems raised by density feedback. Each age group will exert its own feedback on the rest of the population, and each age group will be uniquely susceptible to feedback from others. Barclay and Gregory (1981) raised *Drosophila melanogaster* in bottles. Density was controlled in their various, selected lines by (a) allowing the populations simply to build up on their own without disturbance, (b) removing all adults once a week, (c) removing all adults twice a week, (d) removing 80% of the larvae each week, or (e) removing both adults and 80% of all larvae each week. These various treatments had no apparent effect on egg production, but survival to adulthood was lowest when adults were removed, highest when larvae were removed, and intermediate in the unmolested control lines. Recalling that low p_0 is expected to encourage iteroparity, which ought to be correlated positively with longevity and that depressed p_a should have the opposite impact, these results make very good sense. The other results obtained by these authors, though, are confusing. For example, there is no apparent pattern of relationship between nature of treatment and time to adult emergence. But although there is no pattern to the relationships, there were, nevertheless, statistically significant changes in development rate with removal. Boonstra (1978) found that selective removal of male and female adults of the Townsend vole (*Microtus townsendii*) had a profound impact on the mortality and growth rates of juveniles. Survivorship of the young was highest in the area where all adults were removed. Removal of male adults leads to faster maturation and body growth of young males; adult males have an apparently adverse affect on body growth of young females as well, and adult females appear to shorten the life expectancy of young females. Mertz (1971), working with the flour beetle *Tribolium castaneum* discovered that the distribution of fecundity with age depended on population density. In sparse populations most reproduction was early in life; in dense populations it was more evenly spread throughout life. Interage class competition could certainly be a major force in molding life histories, and as we noted in Chapter 2, could possibly lead to population cycling, which, in turn, could be expected to affect the pattern of evolution of mortality and fecundity. Also, the additive genetic covariance matrix may change with population density.

Schaffer and Rosenzweig (1977) have demonstrated a completely different sort of complicating factor. Even the simple, trade-off theory of life history evolution can generate nightmares for would-be predictors of evolutionary trends. Referring to Schaffer's (1974a) arguments presented previously, and to Figures 12-4 and 12-5, suppose that the b'_x and v^*_x curves, instead of being convex as shown in Figure 12-4, were concave (Fig. 12-11). Then the

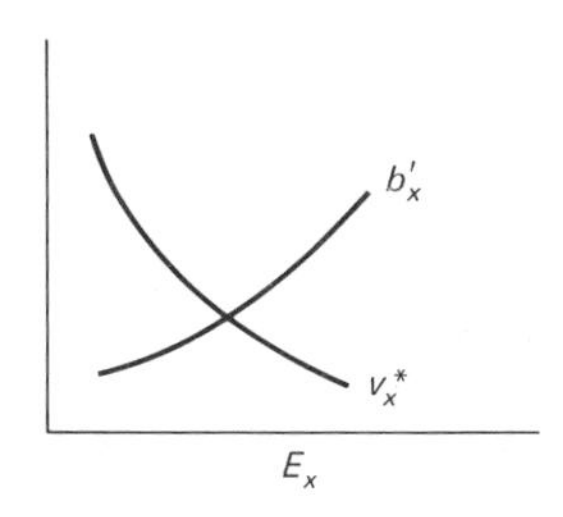

Figure 12-11 Hypothetical relationship of fecundity (b'_x) and residual reproductive value (v^*_x) to reproductive effort (E_x) at age x.

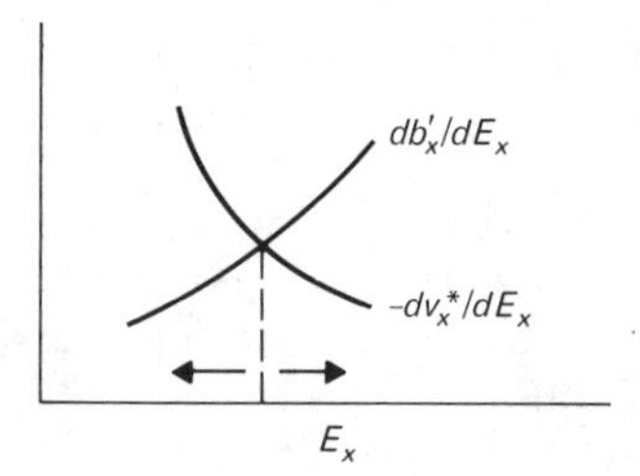

Figure 12-12 $-dv_x^*/dE_x$ and db'_x/dE_x as functions of E_x (from Fig. 12-11).

curves of the derivatives (corresponding to Fig. 12-5) would appear as in Figure 12-12. Consider the equilibrium point where the two curves cross in Figure 12-12. If E_x is displaced to the right, then db'_x/dE_x exceeds $-dv_x^*/dE_x$, so that

$$\frac{d(v_x/v_0)}{dE_x} = \frac{db'_x}{dE_x} + \frac{dv_x^*}{dE_x} > 0$$

and E_x increases still more under natural selection. If E_x is pushed to the left, the inequalities are reversed and E_x continues to decline. The equilibrium is unstable, and the value of E_x favored by selection is either 100% or nothing. That is, there are now two possible equilibria predicted by theory. Of course, there is no particular reason to expect the b'_x versus E_x, v_x^* versus E_x curves to be uniformly convex or concave. Consider a highly plausible organism in which fecundity rises at an increasing rate with reproductive effort at first and then shows diminishing returns, in which residual reproductive value drops rapidly with increased E_x at first, but then levels off until a critical expenditure is reached, beyond which death is very likely. The appropriate curves are shown in Figure 12-13. Plotting the derivative curves, in this case, we obtain something like that shown in Figure 12-14. Again there are two possible equilibria, one at zero (wait until next year to reproduce) and one at an intermediate value.

Polymorphisms in life history might also come about in a different way. Giesel (1974) ran computer simulations of populations, monomorphic or mixed with respect to age-specific fecundity distributions, in fluctuating environments, and found that the time-average population growth rate and density, $E(r)$ and $E(n)$, were often higher in the latter cases. That is, mixed populations maintained higher average population sizes and displayed better ability to bounce back from disturbances. The reason is not difficult to discern. Some life schedules are better

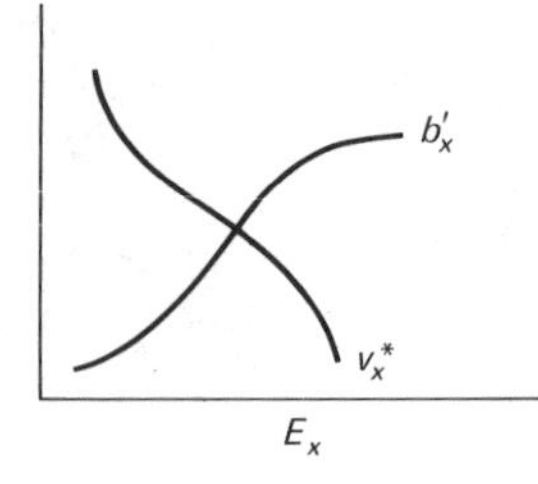

Figure 12-13 Hypothetical relationship of fecundity (b'_x) and residual reproductive value (v_x^*) to reproductive effort (E_x) at age x.

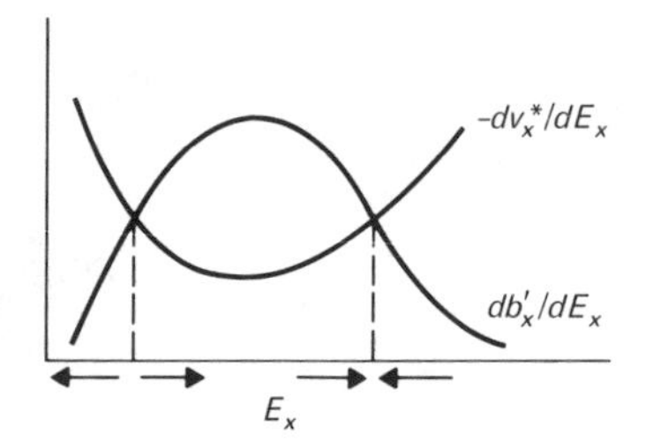

Figure 12-14 $-dv_x^*/dE_x$ and db'_x/dE_x as functions of E_x (from Fig. 12-13).

equipped to keep population growth rate high under some circumstances (e.g., early reproduction following a crash), others serve better under other conditions. A mixed population is more likely than a monomorphic one to possess, at any point in time, an appropriate life schedule. Interdemic selection might well favor such polymorphism.

Cases of apparent polymorphism in life history strategy do occur. Recall the dandelion biotypes of Gadgil and Solbrig (1972), for example. Gross and Charnov (1980) describe an example in the bluegill sunfish (*Lepomis machrochirus*). In this species there exist two types of male. One, which the authors describe as a "sneaker," is nonterritorial and nonaggressive, with the habit of dashing after a female who has just dropped her eggs and releasing sperm before the territory owner has a chance to do so. Sneakers survive on quick daring, and retreat before threats from a larger male. They grow up into "satellite" males, larger, still living mostly by the sneaker's code, but now moving more deliberately and showing more willingness to fight back a challenge. The cost of this life-style, both in energy and risk, makes it unlikely that any sneaker–satellite males get very large or live past an age of perhaps 6 years; the advantage to this strategy is the early passage of sperm. The other morph consists of the territorial males, which, without the energetic drains and dangers of the former way of life, grow large and begin to defend territories at about the age of 7. They lose the advantage of early breeding, but once mature can generally exclude the sneakers and satellites, and experience considerable reproductive success. Note that it is not clear (yet) whether this polymorphism is determined directly by genotype or is only indirectly genetic—a case of developmental plasticity. While the different life history forms of the bluegill are obviously lived by different individuals, there are also cases in which alternative life history strategies are quite flexible, any one individual shifting back and forth between them. As in many frogs, the green tree frog (*Hyla cinerea*) is characterized by male choruses. Females are attracted to the chorus and move about among the males looking for a mate. In a study of these animals Perrill et al. (1978) discovered that 16 percent of calling males on a pond in Georgia had associated with them a silent male, lurking less than half a meter away and in 30 cases of amplexus recorded at the stations of these shadowed males, 13 involved not the caller but the "satellite." Sexual parasitism of this sort is made possible by the indirect route taken by the female to reach the caller of her choice. This gives the satellite an opportunity to chase and grab the female before she reaches her goal. Unlike the case of the bluegill, the parasitic habit in this frog is definitely not genetically fixed. A male who calls one night may be a satellite the next.

A somewhat similar situation is reported by Howard (1981) for the bull frog, *Rana catesbiana*. In this case behavior is determined by size. The largest males are territorial. A second group, usually midsized frogs, will call, but if challenged, leave the area without fuss. Individuals of the third group, almost always the smallest, wait unobtrusively on the large males' territories, and actively attempt to waylay and clasp females en route to their chosen male. The females are quite obvious about attempting to avoid the ambush, swimming more quickly toward their favored mate when pursued.

12.3 An Alternative Approach and Some Implications for Population Dynamics

Different types of organisms are often best described with different sorts of models. With the exception of Gadgil and Bossert's approach, which is extremely general, most of the models above are best suited to populations in which the critical life history parameter is *age*. But in many species mortality and fecundity are less functions of age per se than they are dependent on *size*. This is true of many fish and a wide variety of invertebrates and plants. A rather elegant approach to investigating life histories of this kind of organism has been introduced by Lynch (1980). We shall begin with three simple, and probably not very restrictive assumptions: (1) egg volume is invariant with age of the female, (2) clutch size is invariant with age, and (3) after adulthood, p_x is constant. These assumptions will not be universally true, quite obviously, but for certain kinds of organisms—the model was designed for use with Cladocerans—they are not far from accurate and, even if so, their violation will not greatly affect the predicted results. Write

$$1 = \sum_x e^{-rx} l_x b'_x$$

Then, if reproduction occurs every other day, as in the case of Lynch's Cladocerans, summation is over $x = 2, 4, 6, \ldots$, and this equation becomes

$$\begin{aligned} 1 &= e^{-rk} l_k b'(1 + e^{-2r} p_a^2 + e^{-4r} p_a^4 + \cdots) \\ &= \frac{l_k b' e^{-rk}}{1 - e^{-2r} p^2} \end{aligned}$$

where k is age of first reproduction, whence

$$e^{2r} - p_a^2 - l_k b' e^{-r(k-2)} = 0 \qquad (12\text{-}27)$$

Now let F be the per individual rate of accumulation of biomass for growth and reproduction, where B is biomass. Lynch finds that the form

$$F = \frac{dB}{dt} = \frac{F_{max} + 1}{1 + F_{max} e^{-aB}} - 1 \qquad \text{where } a = \text{constant} \qquad (12\text{-}28)$$

describes this quite accurately. Integrating from zero to k, the age of first reproduction

$$\begin{aligned} B_k - B_0 &= \int_0^k \frac{F_{max} + 1}{1 + F_{max} e^{-aB}} \, dt - t = \int_0^k \frac{F_{max} + 1}{1 + F_{max} e^{-aB}} \frac{1}{dB/dt} \, dB - t \\ &= \frac{1 + F_{max}}{F_{max}} \left[(B_k - B_0) + \frac{1}{a} \ln \frac{1 - e^{-aB_k}}{1 - e^{-aB_0}} \right] - k \end{aligned}$$

so that

$$k = \frac{(B_k - B_0) + \dfrac{1 + F_{max}}{a} \ln \left(\dfrac{1 - e^{-aB_k}}{1 - e^{-aB_0}} \right)}{F_{max}} + C \qquad (12\text{-}29)$$

Lynch added the C to account for an extra period of time required for accumulation of energy for the first clutch and for egg development. He estimated that C should take a value of 4, but found excellent fits of equation (12-29) to his data for $C = 1$. This expression is not only useful for predicting k, but when k and F_{max} are measured, can be used to find a. Once a is known, $F(B_k)$ can be calculated. But from age k, all biomass accumulation is in the form of offspring. Hence $b' = F(B_k)/B_0$, the rate at which biomass is produced divided by the biomass of one young. All values can finally be reinserted into (12-27) to predict r.

Using this technique simply to predict r, which is undoubtedly already known by the time other parameters have been measured, would not be very interesting. What makes this model useful are the relationships, (12-27) through (12-29) themselves. Plugging in a variety of different hypothetical values for F_{max}, p_a, a, B_0, and B_k, it becomes possible to predict optimal values for one parameter as a function of the others. Lynch supposed that $l_k = 1.0$, not unreasonable under ideal laboratory conditions, and then generated the graphs in Figure 12-15 (crudely redrawn from some of his figures). In Figure 12-15a, note that when p_a is constant, the population might be expected to evolve large B_k values and intermediate B_0 values (these are the values that maximize r). When adult survival rate increases with body size (Fig. 12-15b), the same trends are apparent, although B_0 might be expected to be somewhat larger for a given value of B_k. When survival *drops* with increased size (as might occur for cladocerans in lakes

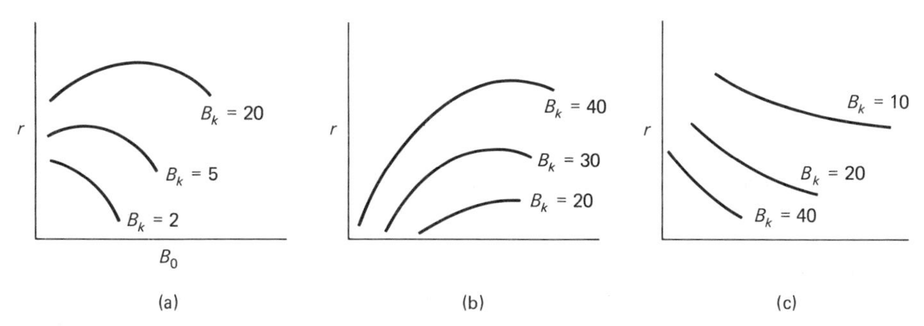

Figure 12-15 Population growth rate (r) as a function of body weight at birth (B_0): (a) p_a constant; (b) $dp_a/dB > 0$; (c) $dp_a/dB < 0$.

with planktivorous fish; see below), Figure 12-15c intimates that natural selection should favor small birth weight and small size at reproductive maturity. These predictions are in line with our previous discussions. When survival increases with size (usually positively related to age), iteroparity is selected for—birth size is generally larger than in semelparous species and reproductive maturity is not rushed. When survival drops with size (and thus age), the parents are less likely to be able to risk waiting to reproduce—semelparity, with small birth size (related to large b'), and early maturity are favored. The advantage to models of this sort is that they combine in simple form the kinds of measurements of growth and energetics that can be easily obtained, and a *quantitative* set of predictions about the likelihood of alternative life history patterns. They are also easily extendable to include the effects of predation. They therefore can be fitted directly into expressions describing population dynamics. Brooks and Dodson (1965) describe the drop in average size of planktonic organisms in lakes with large populations of planktivorous fish, which preferentially predate the larger prey items. In the case of their study the size decrease was at least largely due to change in species and perhaps age composition. Lynch (1979) pushes such observations farther. In ponds where both fish and *Cyclops* (a copepod predator of the cladoceran *Ceriodaphnia*) are present, the *Ceriodaphnia* reach maturity at a smaller size—B_k is not as large as in ponds where these predators are less abundant. This is in support of his prediction that when p_a rises with B, B_k should decline.

We end this chapter with a brief review of the effects of alternative life history strategies on the dynamics of populations. For the most part, these comments merely belabor the obvious. Populations that evolve semelparous life patterns generally produce many more young of smaller size, and are more prone to fluctuations in the physical environment that keep their mean density below the level that would obtain in the absence of such fluctuations. The same is true for species that are only slightly iteroparous—that is, those that invest large reproductive effort and have high mortalities. These species are the ones that are almost impossible to describe with any accuracy using deterministic equations. The species that are reasonably amenable to analysis as to their dynamics, their competitive interactions, and their behavioral adaptations are those which are strongly iteroparous, the long-lived species whose age structure buffers them against environmental fluctuations. The lesson in this last conclusion is that because they have age structure, the nature of competition, both intra- and interspecific, in these species must include consideration of interage class interactions. Very few researchers have spent time worrying about such topics.

Strongly iteroparous species are, by nature of their low birthrates, not good colonizers; they simply are not very opportunistic. Furthermore, in addition to their low birthrates, they have a much greater average age of reproduction. Because population growth is strongly affected, in the manner of accrual of compound interest, by generation time, this further slows their population growth potential. Such species respond only ponderously to changes in the environment. More detailed discussion of the impact on the population growth parameter of changes in various life history parameters can be found in Lewontin (1965) and Caswell (1978b).

Review Questions

1. In human beings the age of menarche has decreased over the past few decades. During this same period survivorship has increased for the general age range where menarche occurs. What implications does this fact have for the energetics (trade-off) approach to life history evolution? What does it say about the genetic architecture of age at first reproduction and survivorship? What implications does it have for human population dynamics? Are these changes best characterized as r-selected, K-selected, or H-selected changes? Is the r-, K-, H-selection framework appropriate here?
2. In Chapter 10 we presented an expression for phenotypic selection that stated that a trait, X, would evolve if its presence in the population raised R, the population growth parameter. In the present chapter we presented a number of models that presumed a trait would evolve if $\lambda(= \lambda_1$, the Perron root) was increased by its presence. But λ_1 is not necessarily equivalent to R. What implications does this have for the validity and the generality of the models? Is the problem alleviated if we base our models on reproductive value instead of λ_1? How do R, λ_1, and V_x relate to the basic genetic concept of fitness?
3. In the discussion of Demetrius's assertion that population entropy, H, should be increased by natural selection, it was mentioned that larger H values would have a stabilizing effect on population numbers. Why should this be so?
4. Refer to (12-17) through (12-26) (the presentation of Lande's formulation). If p_0 had taken on the value .5 instead of .8, both case 1 and case 2 would have yielded complex solutions for E_1. How would you interpret such a result biologically?
5. What general conclusions about the distributions of survivorship and fecundity over age can be made, if any? What effect do environmental fluctuations have on the evolution of these distributions?

Chapter 13

Genetic Feedback and the Evolution of Dispersal

Density feedback acts on a population in two ways. First, it depresses birth rates and survival probabilities and influences rates of dispersal. Second, it exerts selection pressures on those parameters and molds them into the patterns we observe in nature. Chapters 11 and 12 discussed some of the expected life history effects of strong as opposed to weak density feedback pressures. Less direct effects on the evolution of birth, death, and dispersal rates come via behavioral adaptation, the subject of the next two chapters, and through the adaptive responses to other, interacting species, covered in Chapters 16 and 17. In this chapter we remain, for the moment, with direct influences of density feedback on the basic, population parameters. First we shall talk about possible short-term reversible changes in the genome that might occur as a result of changes in feedback intensity and which, in turn, might influence population growth. We shall then turn to a discussion of the evolution of dispersal; to what degree can we expect it to be density mediated?

13.1 Genetic Feedback

General Considerations

The term *genetic feedback* was originally coined by David Pimentel (see Chapter 16) in the context of prey–predator interactions, but is quite applicable to the subject to be covered here. Even at very low crossover frequencies the myriad number of genes in an organism permits a hypothetically tremendous amount of genetic recombination every generation. This recombination, plus a much lower rate of mutation, permits the maintenance in most populations of a considerable amount of genetic variance, much of it additive (see Lande, 1975; Chapter 10). What prevents this variance from continually growing is selection, primarily stabilizing selection. Suppose now that a population experiences environmental conditions more benign than usual—in the sense that there is either less density-independent suppression of growth rate or less density-dependent feedback (more food, fewer predators, etc.). When this is the case, the population mean fitness $\bar{W}$, which is numerically R, rises and the population grows. Those genotypes, defined over as many loci as we wish, that represent one or both tails of genetic distribution for characters now experiencing relaxed selection would, under normal circumstances, have had fitnesses well below 1.0. But now their fitnesses are much less depressed and may even rise above 1.0. Although the gene combinations normally low in fitness will still generally decline in frequency, the selection pressures operating against them are weaker relative to the variance-generating forces of recombination and mutation; genetic variance in the population will rise. Thus population increase is a time when new genetic innovations are retained, and less rare, and thereby are given the opportunity to combine in single individuals when two novel genotypes mate and produce offspring. Occasionally, this combination of normally unfit genotypes will give rise to a new genotype with superior fitness. The vast bulk of the new genetic variation is apt to be deleterious, and when the eventual population decline occurs, will be rapidly weeded out. But those innovations of superior fitness that might never have come about without the population increase phase will persist and spread. This is a somewhat more rigorous elaboration of the suggestion in Chapter 9 that population boom-and-bust cycles might promote rapid selection.

It has been suggested that the dynamics of genome and population might interact to cause population cycling. Wellington (1964), studying the western tent caterpillar (*Malacosoma pluviale*) found that the larvae, living in "tent" nests, come in two varieties. One variety is active, and independent, moving about readily and feeding rapidly. The other can be further divided into three subcategories: (a) active, but not traveling very far unless in the company of members of the active, independent class; (b) less active; and (c) positively sluggish. The latter hardly bother to move, sometimes failing to free themselves from the eggs they hatch from and occasionally dying of starvation from lack of feeding effort. The active and inactive caterpillars give rise to accordingly active and inactive adult moths. Moths of the first class tend to disperse readily and often oviposit far from their place of birth. If environmental conditions are particularly benign, these dispersers are quite successful and their phenotype survives in large numbers to lay eggs which give rise to disproportionate numbers of active larvae. The less active forms oviposit near their birthplace. Under normal conditions, the active form invites disaster with its long dispersal, and the stay-at-home, less active adults spread their genes in the local population. As their "slow" genes accumulate locally, though, the population as a whole becomes more and more slow, until failure to even bother to feed begins to result in a marked decline in numbers. Generally, before extinction occurs, however, there is a very good year, the independent, active morph is highly successful, and its numbers are replenished. In this way, two morphs, one adapted to ideal environmental conditions but effectively suicidal under normal conditions, and one better but not well adapted to everyday conditions, fluctuate in relative frequency, affecting population dynamics in a complex manner, depending on changes in the environment.

That genetic feedback of the sort described here might be (one of) the causes of the observed cycling in microtines (see Chapter 2) was first suggested by Dennis Chitty (1957) and Chitty and Phipps (1966). Krebs and his associates (Tamarin and Krebs, 1969; Gaines, 1971; Gaines et al., 1971; Myers and Krebs, 1971; Krebs et al., 1973) have spent a considerable amount of time and effort attempting to test Chitty's ideas with suggestive but discouragingly indecisive results. In all these studies, the workers found indications that behavioral and/or physiological changes occurred in the population as density climbed, peaked, and then fell. (But see Hofmann et al., 1982). Also, they found that the biological changes and the demographic changes were rather nicely correlated with changes in gene frequencies at one or more loci. Garten (1976) has found similar phenomena working with *Peromyscus*. The patterns found are complex, differing among species and even among sexes of the same species, but they are at least consistent with the hypothesis of genetic feedback. Cycling might also arise if different genotypes displayed different ages of maturity and survival. Then changes in age structure could lead to population cycling. But the patterns observed cannot be adequately explained in this manner (Charlesworth and Giesel, 1972).

There are two major difficulties with the studies cited above. For all their sophistication, breadth, and sheer magnitude, they have never quite led to a clear demonstration of either cause or effect of density and gene frequency changes, much less a mutual interaction leading to cycles. In fact, the evidence for the whole idea is terribly weak. Second, a rather nasty problem has just recently surfaced (McGovern and Tracy, 1981). The loci described as changing in all of the foregoing studies are the transferrin (Tf) locus and the leucine aminopeptidase (Lap) locus. McGovern and Tracy monitored fluctuations in genotype for these two loci over an 8-year period for a population of *Microtus ochrogaster*. What they found was a bit disturbing; they noticed that genotypic change often occurred *within individuals*. To check these rather astounding results, they captured a number of animals in February, checked their blood for Tf and Lap gene frequency immediately, and then put them into a controlled laboratory situation for various periods of time, after which the genomes were again checked. One group was kept at 3°C, under a 10 hours of light/14 hours of darkness (10L : 14D) lighting regime (winter simulation); a second was held at 10°C, with 12L : 12D (spring conditions); and the third was maintained at 16°C, 15L : 9D (summer). In the winter treatment, 5 of 11 animals changed genotype, one with respect to the Tf locus, two at the Lap locus, and two at both loci. In the summer populations, six of nine changed, two at the Tf locus, three at the Lap locus, and one at both. Three of the eight spring animals changed, all at the Lap locus. The changes were generally in the direction toward the genotype of unchanging individuals. Field animals were also captured in the summer, and subsequently put into winter-simulated conditions for one month before return to a summer environment. Out of 13 individuals, nine changed at the transferrin locus, and two of these also switched genotype at the Lap locus. All ended up homozygous. These results make good sense biochemically. The electrophoretic mobility of the isozymes, which is used to distinguish allelic types, is affected by binding of metals other than just iron, and are also affected (both of these loci) by the concentration, in vivo, of sialic acid. It is known that sialic acid concentrations are affected by pregnancy and disease which thereby influence the electrophoretic mobility of Lap enzymes. Perhaps the observed genotype frequency changes observed in field populations of cycling voles are not caused by changes in gene frequency, but by changes in the physiological state of the population's members. Recall also the existence of "profane" genes mentioned in the introduction to Part Four. Chitty's hypothesis is still intruiguing, but is far from proven.

Spatial Heterogeneity and Ideal Free Distribution

Feedback, both population density feedback and genetic feedback, may be affected by the manner in which organisms adapt to heterogeneity in their environments. If organisms insist on remaining in areas where the physical habitat is benign, they may grow to such numbers that feedback has an extraordinary impact not only on their population growth parameter, but also on their general state of health. At least under some circumstances genotypes that predispose their owners to disperse from high-density situations to physically less ideal but socially more amenable surroundings will be favored, at least to a point. Ideally, an organism will move to new places if there is reason to expect that the move will increment its genotype's fitness. In J. M. Emlen's (1981) words, we might expect "that natural selection has predilected organisms to show a degree of appropriate microhabitat selection such that beneficent areas (higher W for individuals occupying these areas) are used preferentially to those with less cover, less food, lurking predators, clues suggesting some pending disaster, etc. This being so, populations will, at any time, tend to aggregate in the 'better' areas. Indeed, the aggregations should grow in density until the disadvantages born of crowding feed back negatively on the area's quality to the point that relative beneficence disappears." Individuals should continuously test their immediate surroundings and respond by moving or not. The result will be a tendency toward a dynamic dispersion equilibrium, where the contribution of every individual to its genotype's fitness is equal. This idea of *ideal free distribution* was suggested by Fretwell and Lucas (1969), Fretwell (1971), and later developed by C. E. Taylor (1975, 1976). It has already been briefly discussed in Chapter 9, where some of its genetic implications were noted.

The impact of an ideal free distribution pattern on population dynamics can be seen as follows. First, $\operatorname{var}(n/\hat{n})$ is minimized. If $f(n, \hat{n})$ is the growth descriptor,

$$\frac{d\bar{n}}{dt} = \overline{f\left(\frac{n}{\hat{n}}\right)} \simeq f\left(\frac{\bar{n}}{\bar{\hat{n}}}\right) + \frac{1}{2}\left(\frac{d^2f(n, \hat{n})}{d\hat{n}^2}\right)_{\bar{n},\bar{\hat{n}}} \operatorname{var}(n/\hat{n}) \tag{13-1}$$

Where $f(n, \hat{n})$ is the logistic equation and r_0K is constant,

$$\frac{d\bar{n}}{dt} = \overline{(r_0K)\frac{n}{K}\left(1 - \frac{n}{K}\right)} = r_0K\left[\overline{\left(\frac{n}{K}\right)} - \overline{\left(\frac{n}{K}\right)}^2 - \operatorname{var}\left(\frac{n}{K}\right)\right]$$

Equilibrium occurs when

$$\overline{\left(\frac{n}{K}\right)}_{\text{equil}} = \frac{1}{2} + \sqrt{\frac{1}{4} - \text{var}\left(\frac{n}{K}\right)} \tag{13-2}$$

Ideal free distribution allows the variance term to be very small, thus maximizing $\overline{n/K}$. Ideal free distribution also requires that as population density rises, more individuals spread into what were before relatively inferior microhabitats; as the quality of the already occupied habitat deteriorates due to density feedback, new types of areas are used. Thus although the average quality of habitat declines with increased population density, the area over which the population lives is enlarged, buffering the impact of the feedback. If we were to draw a curve of $(dn/dt)/n$ against n, depicting density feedback, it would appear linear under logistic growth; with ideal free distribution it would fall off more gradually at first, and then more precipitously as even marginal habitat was filled up (see Section 13.4).

Dispersal of the sort discussed just above reflects adjustive movements from one microhabitat to another, generally over fairly short distances. Clearly, if the distances required for such adjustment are too large to permit reasonably free movement, it will be impossible for an organism to assess its well-being at various points in space and to act such as to maximize the contribution to its genotype's fitness. Now consider dispersal, the movement of organisms over large distances. We are talking of distances such that a whole lifetime may be spent in one area so defined or another, distances that preclude the organism, once it arrives, from readily moving on to still another area. Dispersal among such "coarse-grained" patches is critical to the operation of the simpler sorts of intergroup selection (where local extinctions and the founding of new populations by colonizers is the driving mechanism), and contributes to the stability of large populations (Chapters 2 through 6). If we wish to understand the nature of dispersal—for example, the degree to which it is density dependent—we must first ask (a) whether organisms show habitat preferences and whether these preferences are adaptive (this question clearly is also important to the assertion that ideal free distribution occurs), and (b) what the risks are, the costs to fitness, of moving between patches. These questions are dealt with in the following section on habitat preferences, and are discussed further in the context of the evolution of dispersal in Section 13.2.

Habitat Preference

That field sparrows prefer open to heavily wooded habitat, and that pileated woodpeckers prefer mature forest, are such obvious, observed facts that the naturalist rarely even questions them. But it is worth asking to what extent they are under genetic control. Inasmuch as this topic is discussed under the broad heading of genetic feedback and will lead us shortly to the next topic of the evolution of dispersal tendencies, we should also ask the degree to which these habitat preferences may change over time and to what extent they are affected by population density. It seems fairly clear that much of the habitat preference we observe has at least some genetic basis. Wecker (1963) tested the preferences of prairie deer mice for field versus woods in a outdoor pen containing both. Field-caught mice, offspring of field-caught mice raised in the laboratory, offspring of lab mice raised in field habitat, and even offspring of field-caught mice raised in woods, all preferred the field habitat. Dark- and light-colored individuals of the peppered moth (*Biston betularia*) preferentially rest on dark- and light-colored substrates, respectively (Kettlewell, 1955a). On the other hand, experience also seems to play a part. Laboratory-born and raised, and laboratory-born, woods-raised deer mice showed no detectable preference for fields or woods. Wiens (1970) raised tadpoles (*Rana aurora*) in aquaria with either striped or checkerboard patterns of tape on the walls, or in aquaria with no pattern, or first in striped and then in checkerboard (or vice versa) aquaria. Just before metamorphosis, individuals were placed in test chambers with stripes at one end and checkers at the other. Animals with stripe experience preferred the striped habitat, as did, to a lesser degree, the controls. Only one of the groups raised half the time with checkers showed preference for the checkered end of the test aquaria, and this tendency was not at all marked. Wiens explains these results by noting that this species is native to a striped habitat of cattails and reeds, and uses a stripe gestalt to orient to the habitat in which it is best adapted. More to the point for our discussion, perhaps, is the work of Grant (1971). In experiments with young voles (*Microtus pennsylvanicus*), Grant took animals without any grassland experience, held them in woodland for 2 weeks, and then released them in 1-acre enclosures about half grassland and half forest. The animals almost immediately entered and remained in grassland habitat. Movement into the wooded area occurred *only at very high population densities.* This observation supports the notion of ideal free distribution. The fact that strong habitat preference occurs, but that it may be, in effect, reversed at high densities, is also of obvious importance to the theory of the evolution of dispersal behavior.

Dispersal can be thought of in at least two ways. First, it may be a means by which individuals overcrowded in an area of otherwise appropriate habitat risk the dangers of travel across less beneficent terrain in order to find less hectic surroundings of, again, their preferred habitat. Or, on the other hand, dispersal may be one way that an individual can react to deteriorating environmental conditions from any one of a number of causes and seek better living conditions. The former directly implicates population pressures, the latter alludes to temporal changes in the physical environment. Consider the latter first. Species that experience strong temporal fluctuations in their environment are those most likely to benefit by occasional, or perhaps even regular, dispersal. In such circumstances the grass really is apt to be greener on the other side of the fence. This should be particularly true of species characterized by behaviorally rigid individuals. Species whose individual members are very flexible in their responses to change, on the other hand, are those least likely to feel the need to disperse. Animals in the latter group, we should expect, will make use of their somatic plasticity to traverse hostile terrain successfully only when conditions are particularly severe, or when high population densities significantly overburden their ability to adjust to resource shortages or social pressures.

Consider those species whose individuals are not very flexible in their behavioral and physiological responses to change, but whose *populations* are flexible by virtue of phenotypic variability *among* individuals. Populations of these species are bufferred to some degree against environmental disturbances, but those particular phenotypes affected in adverse ways at any time are still benefited by dispersal. Thus populations with high phenotypic variability are likely to be reasonably stable and yet also be characterized by high dispersal. We should, as a corollary, find dispersing individuals to differ genetically from their stay-at-home brethren, and to find the genetics of dispersers to be a bit different depending on the nature of the environmental stress perpetrating the dispersal. Genetic heterogeneity also benefits a population not only by permitting it to adjust to greater spatial variability in its range, but also by providing its dispersing members, on average, a wider range of suitable destinations. We have already talked a bit about the genetic mechanisms by which spatial and temporal heterogeneity act to promote genetic variance in a population. We bring up the topic again here because of the ecological correlate. Populations in heterogeneous environments can be expected to be, if not somatically flexible, then phenotypically variable. In fact, it is reasonable to conclude that the degree of phenotypic variability should parallel the environmental variability. A corollary is that phenotypic variability should also parallel breadth of diet or use of other resources.

The proposal that niche breadth and phenotypic variability should be positively correlated is known as the *niche-variation hypothesis.* Although it may seem a bit contrived to have to even state such a trivially obvious-sounding idea, the niche-variation hypothesis is far from universally accepted, and data to the contrary occasionally rear their heads. As a general rule of thumb, it seems beyond serious doubt, but the compulsion of a number of workers to keep questioning it and digging up more data requires that the niche-variation hypothesis be stated with the proper tone of suspicion. We shall very quickly review some high points from the literature. Soulé and Stewart (1970) calculated the male/female ratios of bill depth and bill length for a number of bird species, and used the values obtained as a measure of phenotypic variability. These ratios, by a U-test, were significantly higher among species with broad diets, for both measures ($p < .02$). Rothstein (1973) argued that wider niches should enable species to inhabit a wider array of habitats (the converse of our line of reasoning just above). But if a wider array of habitats can be used, the population has more room for growth and should be larger than those of species with narrow niches. He claims that available data from the biological literature support this contention. Grant et al. (1976) looked at a medium-size bill finch (*Geospiza fortis*) on Daphne and Santa Cruz islands in the Galápagos. On the latter island, bill size was more variable among individuals and the species occupied a wider range of habitats. An excellent review of the niche-variation hypothesis, with a good listing of recent literature, can be found in Grant and Price (1981).

Before leaving the niche-variation hypothesis, it is worthwhile to recall the arguments of Lande (1975, 1977), Slatkin and Lande (1976), and Felsenstein (1979), discussed in Chapter 10. Recall that, respecting traits whose fitness values are unaffected by density feedback, those determined by additive allelic effects should *not* be more variable in fluctuating environments, regardless of the short- or long-term nature of those fluctuations (Lande, 1975, 1977). The introduction of epistasis, via selection at modifier loci (Slatkin and Lande, 1976), does permit increasing genetic variance in traits (i.e., niche width, genetic component) in response to environmental variation, but only if the magnitude of that variation exceeds a critical threshold. Selection for nonzero niche width was also predicted to be more likely when environmental variation was coarse-grained (slow, generation to generation). Wandering from place to place, provided that the environment is spatially heterogeneous, furnishes the equivalent of fine-grained temporal change. Long-range dispersal results in new surroundings for offspring, generation to generation, and thus provides effectively coarse-grained fluctuation for a population. We should expect, then, that traits not affected by density feedback (presumably such things as color, fur thickness, and length) should indeed demonstrate wider genetic variability in response to coarse-grained heterogeneity, temporal or spatial, if selection pressures are sufficiently variable. Of course, if selection pressures are extremely divergent, the traits should track slow fluctuations and respond to quicker variation, as if experiencing disruptive selection. Whether the trait will demonstrate genetic variability or not now depends on the manner in which fitness varies as a function of additive genetic value (see Section 10.3; Felsenstein, 1979). Of course, the variability to which we refer above is *genetic* variability. Phenotypic variability (realized niche width) can clearly be augmented by individual differences due to environmental factors—an organism's experiences and the conditions under which its development took place. Species whose members exhibit strong plasticity adjust somatically to fine-grained changing conditions and so can be expected to experience weaker selection. As pointed out in Chapter 10, this fact suggests that we might find a negative correlation between genetic variance σ_G^2 and phenotypic plasticity.

Now consider those traits whose fitness values *are* affected by density feedback. These must include almost all traits of ecological interest, including those affecting birthrate, death rate, emigration rate, age structure, food preferences, and social behavior. These traits are molded by density-dependent selection. Inasmuch as individuals with certain trait values are likely to exert stronger feedback than others, or to be more sensitive than their fellows to feedback, these traits are also affected by frequency-dependent selection. Thus the appropriate method for investigating evolution of genetic variance (niche width) is to use frequency- and density-dependent selection models. Recall from Chapter 10 that Slatkin (1979) provided a general formula describing the relationships between $\bar{x}$, σ_x^2, and $\hat{n}$ for such models:

$$0 = \int \frac{\partial \phi(x)}{\partial \bar{x}} W(x)\, dx \qquad 0 = \int \frac{\partial \phi(x)}{\partial \sigma_x^2} W(x)\, dx \tag{13-3}$$

where $\phi(x)$ is the distribution in value of the trait X (assumed normal), and $W(x)$ is the fitness as a function of trait value. Knowledge of $W(x)$ permits calculation of $\bar{x}$, σ_x^2, and equilibrium population density (see the example in Section 10.3).

An alternative approach follows in an expansion of a paper by J. M. Emlen (1975). In Chapter 6 (see also Chapter 17) we noted that under certain circumstances the competition coefficient α might be estimated as a function of niche overlap. Here we make use of this fact to estimate the impact of competition by genotype j on genotype i:

$$\alpha_{ij} = \frac{\int f_i(x) f_j(x)\, dx}{\int [f_i(x)]^2\, dx} \qquad \text{(see equation 6-45)} \tag{13-4}$$

where $f_i(x)$ is the distribution of resource use along a gradient, X, by genotype i. We suppose that $f_i(x)$ and $f_j(x)$ are Gaussian,

$$f_i(x) = c' \exp\left[\frac{-(x - I)^2}{2\sigma_e^2}\right] \tag{13-5}$$

where c' represents the maximum rate of resource utilization, I the mean value of x for genotype i, and σ_e^2 the environmental variance in x, assumed identical for all genotypes. Performing the integrations in (13-4), we find

$$\alpha_{ij} = \exp\left[\frac{-(I - J)^2}{4\sigma_e^2}\right] \tag{13-6}$$

In the spirit of Lande (1975) and Slatkin and Lande (1976), suppose next that the mean positions of the various genotypes along the gradients are normally distributed, $\phi(\bar{x}; \bar{\bar{x}}; \sigma_G^2)$, with mean $\bar{\bar{x}}$ $(= 0)$ and total genetic variance σ_G^2. Then the average competitive effect, per individual, on one organism by its fellows is

$$\alpha = \iint \alpha_{ij} \phi(I; \bar{\bar{x}}, \sigma_G^2)\, \phi(J; \bar{\bar{x}}, \sigma_G^2)\, dI\, dJ$$

$$= \frac{\sigma_e}{\sqrt{\sigma_G^2 + \sigma_e^2}} \tag{13-7}$$

Now suppose that the population obeys the logistic equation with density-independent fluctuations. Then

$$r(t) = r_0 \left[1 - \frac{n(t)}{K}\right] + \gamma S(t) \qquad \text{where } n = n_{\text{total}} \tag{13-8}$$

But this can also be written

$$r(t) = r_0 \left[1 - \frac{\alpha n(t)}{K^*}\right] + \gamma S(t) \tag{13-9}$$

where K^* is the value of K in the absence of any among-genotype variance. Thus

$$K = \frac{K^*}{\alpha} \tag{13-10}$$

(We write α, K, and K^* as constants because we can almost surely count on their changes under natural selection being of a vastly slower nature than variations in n.)

Finally, we make use of the ideal free distribution hypothesis [asserting that individuals distribute themselves in space in such manner that their expected contributions to their genotype's fitness is everywhere equal—see "Spatial Heterogeneity and Ideal Free Distribution" (p. 365)]. Then in respect to each genotype, each point along the resource gradient contributes equally to its inhabitants. (Evidence for the validity of this assertion comes from the work of M'Closkey, 1982, on desert rodents.) Now we should expect the K for each individual genotype (i.e., K in the absence of genetic variance, or K^*) to equal the sum of resource values obtained over the entire gradient. But because the value acquired at any one point is equal to that at any other under the ideal free distribution pattern, this total value is simply proportional to the area under the resource utilization curve for that genotype,

$$K^* = \xi \int c' \exp\left[\frac{-(x - I)^2}{2\sigma_e^2}\right] dI = \xi c' \sqrt{2\pi}\, \sigma_e \equiv c\sigma_e \tag{13-11}$$

where ξ is a factor converting amount of resource used to standing crop. We therefore have, from (13-7), (13-9), (13-10), and (13-11),

$$K = \frac{K^*}{\alpha} = \frac{c\sigma_e}{\sigma_e/\sqrt{\sigma_G^2 + \sigma_e^2}} = c\sqrt{\sigma_G^2 + \sigma_e^2} = c\sigma_x \tag{13-12}$$

So far we have dealt only with the effects of density feedback on fitness (actually the population growth rate); we have not considered the impact of stabilizing selection. If we suppose there to be no effect of n/K on the intensity of stabilizing selection, we may write (analogous to the Gaussian function used previously)

$$r_0(x) = \hat{r} - \frac{(x - x^*)^2}{2w^2}$$

where x^* = the optimal phenotype, $1/w^2$ measures the intensity of stabilizing selection, and $\hat{r}$ is the intrinsic rate of increase exhibited by the optimal phenotype. Let the bar notation, $\bar{x}$, indicate means over phenotypes, $E(\cdot)$ the mean over time. It follows that, because selection leads to $\bar{x} \to x^*$,

$$\overline{r_0\left(1 - \frac{n}{K}\right)} = \overline{\left[\hat{r} - \frac{(x - \bar{x})^2}{2w^2}\right]\left(1 - \frac{n}{c\sigma_x}\right)} = \left(\hat{r} - \frac{\sigma_x^2}{2w^2}\right)\left(1 - \frac{n}{c\sigma_x}\right)$$

$$E(r) = \left(\hat{r} - \frac{\sigma_x^2}{2w^2}\right)\left[1 - \frac{E(n)}{c\sigma_x}\right] \tag{13-13}$$

We can now picture a new genetic form with different σ_x^2 and a correspondingly different expected fitness, r', arising in the population. If $E(r') > E(r)$, both individual and group selection will favor the spread of the new form. At selection equilibrium we should expect to find the value σ_x^2 that maximizes $E(r)$. Accordingly, differentiate (13-13) to obtain

$$0 = \frac{dE(r)}{d\sigma_x} = -\frac{\sigma_x}{w^2}\left[1 - \frac{E(n)}{c\sigma_x}\right] + \left(\hat{r} - \frac{\sigma_x^2}{2w^2}\right)\frac{E(n)}{c\sigma_x^2}$$

But at population equilibrium, $n = K = c\sigma_x$ (13-12), so that $E(n)/c\sigma_x = E(n)/K$, and the expression above becomes

$$0 = -\frac{\sigma_x}{w^2}\left[1 - \frac{E(n)}{K}\right] + \left(\hat{r} - \frac{\sigma_x^2}{2w^2}\right)\frac{E(n)}{K}$$

from which

$$\sigma_x^2 = \frac{\hat{r}w^2[E(n)/K]}{1 - E(n)/2K} \tag{13-14}$$

where σ_x^2 = phenotypic variance
$1/w^2$ = stabilizing selection intensity
K = carrying capacity
$\hat{r}$ = maximum intrinsic rate of increase
$E(n)$ = mean population size over time

Thus high maximum intrinsic rates of increase encourage broader niches (larger phenotypic variance), as does weak stabilizing selection (large w^2). For a given value of $\hat{r}$, equilibrium species should be more generalized (larger σ_x^2) in their habits than opportunistic species, for whom $E(n) \ll K$. The second prediction is intuitively obvious; the first and third are discussed in more general contexts in Chapters 11 and 16, respectively.

13.2 The Evolution of Dispersal

Dispersal can be classified into three types. First is the commonly observed dispersal by individuals of a new generation when still quite young. Most birds, mammals, and the larvae of innumerable fish and invertebrates appear to leave home grounds either as passive floaters or in response to social pressures. Once settled as adults, further movement on anything approaching this scale is far less likely. Some species, in fact a great many species, have larval forms whose sole "purpose" seems to be dispersal. As we noted in Chapter 11, this should not be unexpected for opportunistic species. Second, most populations are patchy in nature, and the occasional wandering of individuals from one patch to another is a commonly observed phenomenon. Such wandering is most pronounced, and sometimes highly organized, in species with tremendous growth potential. Locust swarms or bee swarms are examples. Third, there is the very long distance dispersal, generally seasonal, involving most or all of a population simultaneously, and often including a return trip with the return of appropriate weather. Migration is quite a different beast from the first two kinds of dispersal; it will not be considered further here. The reader with more than a passing interest in the evolutionary ecology of long-distance migration should look at Lack (1954), MacArthur (1959), Cox (1968), or Berthold and Querner (1981) for information and ideas concerning birds, and Brower (1961) or Dingle (1972, 1978) for migration studies of insects (see also Baker, 1978).

We shall deal together with the first two classes of dispersal. Some advantages and disadvantages of dispersal have already been mentioned. Advantages include escape from deteriorating environmental conditions (recall the ideas of Lidicker, 1962, reviewed in Chapter 2), and in opportunistic species, spread of genes and enhanced chances of at least one offspring finding itself in suitable surroundings for survival and propagation. At the group selection level, a possible advantage to dispersal is the avoidance of intense density feedback, with the instability that might result, especially when density-response time lags occur (see Section 13.4). There are other advantages not necessarily dependent on growing, local densities or environmental changes. For example, inbreeding will be alleviated by dispersal; if a significant number of loci display heterosis, or if there exist a significant number of recessive deleterious alleles, it is beneficial to avoid inbreeding.

Bulmer (1973) and Greenwood et al. (1978) have documented the role of dispersal in preventing inbreeding depression in the great tit (*parus major*). Between 1964 and 1975, of 885 pairings where the identity of both parents was known, 194 were between resident males and females, 239 between resident males and immigrant females, 158 between resident females and immigrant males, and 294 involved immigrants of both sexes. Of the total, 13 (1.5%) were closely inbred pairs (five between mother and son, seven between brother and sister, one between aunt and nephew). Two of the mother–son pairs bred twice in this period, and one of the brother–sister pairs, for a total of 16 matings. There was no significant difference in clutch size between these close inbreeders and the other pairs, but mortality during the egg plus nestling period was higher among offspring of the inbred couples (27.7% versus 16.2%, $p < .05$). Of 86 inbred offspring successfully fledged, six subsequently were found breeding (6/94 = 6.4%). The corresponding figure for offspring of outbred parents was 10 percent. The difference is not statistically significant but is suggestive. Finally, none of the offspring of mother–son or sib matings lived more than one breeding season. Van Noordwijk and Scharloo (1981) report, on the basis of a long-term study of the same species in Holland, that number of eggs failing to hatch rises significantly with inbreeding.

A number of workers have found that populations of mixed genotype often have higher yield ($\bar{W}$) than monomorphic populations (e.g., Allard and Adams, 1969; see also Chapter 8). Where this is so, there is an advantage to parents either outbreeding or passing to their offspring genes which induce them to disperse to areas where they will mingle with genetically different individuals (see Real, 1980). In species that display a rare male mating advantage (Chapter 8), dispersal should be particularly beneficial for young males. Is this one reason why young animals, particularly young males, generally disperse?

On the other hand, inbreeding does not have to be disadvantageous. Once forced inbreeding has resulted in the elimination of most deleterious recessives, inbreeding depression will be much less severe, and individuals sporting particularly high fitness genotypes may pass more genes by mating with a genetically similar individual, thus maximizing the likelihood of producing similar offspring. This may be particularly true in behaviorally and physiologically rigid species. In fact, such species do tend to be more homozygous (Selander and Kaufman, 1973). Further advantages and disadvantages of dispersal behavior are given below.

In 1981, Gillespie devised a model, applicable to haploid species, defining the conditions under which dispersal would be selectively advantageous. Because dispersal generally occurs once or at most only a few times in a generation, natural selection can be expected to maximize the geometric mean fitness over generations, $\bar{W}_g$. That being so, Gillespie was able to show that dispersal should evolve when spatial heterogeneity was low and temporal heterogeneity high. This makes intuitive sense—for diploid species as well. If temporal variability is low, there is no advantage to disperse, and if spatial heterogeneity is great, the chances of landing in a bad location after dispersal are correspondingly raised.

Gadgil (1971) used a discrete, logistic analog to construct a phenotypic selection model of dispersal. He wrote

$$n_{ij}(t + 1) = (1 - p_{ij}(t))\,[n_{ij}(t)(1 + r_{ij}(t))] + \sum_{s \neq j} p_{is}(t) z_{sj} [n_{is}(t)(1 + r_{is}(t))] \qquad (13\text{-}15)$$

where $n_{ij}(t)$ = number of individuals of phenotype i in subpopulation j at time t

$p_{ij}(t)$ = fraction of phenotype i individuals dispersing from site j at time t

z_{jk} = fraction leaving site j which colonize site k

$r_{ij}(t) = r_0[1 - \sum_{\text{all phenotypes, } u} n_{uj}(t)/K_j(t)]$

r_0 = intrinsic rate of increase of phenotype i (all phenotypes are assumed to have the same r_0)

$K_j(t)$ = carrying capacity at site j (same for all phenotypes) at time t

Gadgil allowed

$$K_j(t) = K_j^* \left[1 + \gamma \sin \left(\frac{t - \phi_j}{\theta} \pi \right) \right]$$

to vary:

$1 \geqslant \gamma_j \geqslant 0$ = amplitude of the variations

K_j^* = mean value of K_j over time

θ = frequency of fluctuation in K_j

ϕ_j = phase-angle difference among sites

He then set

$$p_{ij}(t) = a_i \left[\frac{\sum_u n_{uj}(t)}{K_j(t)} \right]^{x_i}$$

where a_i and x_i described the ith phenotype and simulated the dynamics of the system. The value of a_i describes the average tendency of the ith phenotype to disperse; x_i describes the density response of dispersal. If $x_i \to 0$, dispersal is (nearly) density independent. If x_i is large, the probability of an individual leaving its natal colony rises rapidly with population density. Negative values of x_i are possible but biologically unlikely. Gadgil considered only positive values. The n_{ij} values were constrained to be positive, and the initial population sizes at all sites were set equal. Simulations involved various values of z, and mixtures of various a, x combinations. The a, x combination (phenotype) that fared the best in mixed company, showing the greatest relative increase in abundance, was declared the winner. Gadgil's results are shown in Table 13-1. As in Gillespie's model, low spatial heterogeneity and high temporal variability led to high dispersal rates, a. In addition, Gadgil's approach indicates *sensitivity* of dispersal response to population density, x, to be high when a is low, and vice versa. Perhaps we can interpret these results as follows. If a population shows high sensitivity (x high) to density feedback, there is less need for migration (lower a) to achieve escape from overcrowding. If both x and a are high, the population overreacts, with consequent lowered fitness.

Table 13-1

Variability in Time (in γ_j)	**Variability in Space (in K_j^*)**	
	Low	***High***
Low	a intermediate x intermediate	a low x high
High	a high x low	a intermediate x intermediate

If all fluctuations in K are in phase, there is little advantage in dispersal—an emigrating individual is likely to find itself right back in the same situation that induced it to leave its original subpopulation. Gadgil found that when the phase angles were small, a large value of a was never favored. When phase angles were large, so that the destinations of the dispersers were more likely to differ significantly from their place of origin (so that leaving a bad site was more likely to mean arriving at a better site), higher a values were advantageous. In either case the *relative* trends in Table 13-1 hold.

Hamilton and May (1977), although not refuting the general predictions of Gillespie's and Gadgil's models, take a somewhat different tack, arguing that dispersal can be advantageous even in utterly constant homogeneous environments. Consider the case of an environment with no heterogeneity in K in either time or space. Suppose that there exists a stay-at-home (nondispersing) genotype, producing k offspring all of which compete for their parents' site, only one successfully. Suppose, also, that there exists a new mutant form that sends off all but one young into this system of, say, Q sites. Then this mutant lands $(k - 1)/(Q - 1)$ young, on average, on each of the other sites, and so has a chance

$$\frac{k-1}{Q-1}\bigg/\left(\frac{k-1}{Q-1} + k\right)$$

of winning in a *new* site, occupied by stay-at-home individuals, as well as a 100 percent chance of retaining the home site. Because established stay-at-homes have *no* chance of reclaiming a site occupied by the mutant form and a less-than-perfect chance of retaining home ground, the mutant can successfully invade the system—its expected fitness is higher. That is, dispersal is selected for over no dispersal at all, even in constant homogeneous environments.

But what probability of dispersal is optimal in such habitats? Consider an established strategy with migration (dispersal) probability $= m$ and a new mutant with corresponding value m'. Let p be the proportion of dispersers surviving the act of dispersal. Then the fitness of the established strategy is

$W(m) = [(Q - 1)$(probability retain established sites) + (1)(probability regain mutant site)]$/Q$

$$= \frac{1}{Q}\left\{(Q-1)\left[\frac{\text{number of established young from given home site remaining there plus number arriving from elsewhere}}{\text{the above, plus number of mutant young arriving at that site = total number of young on the site}}\right]\right.$$

$$\left.+ (1)\left[\frac{\text{number of established young migrating to mutant site as opposed to any one of } Q-1 \text{ possible sites other than home}}{\text{the above, plus the number of mutant young remaining on the mutant site = total number of young on the mutant site}}\right]\right\}$$

$$= \frac{1}{Q}\left\{(Q-1)\frac{k(1-m)+\frac{kmp}{Q-1}(Q-1)}{k(1-m)+\frac{kmp}{Q-1}(Q-1)+\frac{km'p}{Q-1}} + (1)\frac{\frac{kmp}{Q-1}(Q-1)}{\frac{kmp}{Q-1}(Q-1)+k(1-m')}\right\}$$

(13-16)

When Q is large this approaches

$$\frac{Q-1}{Q} \rightarrow 1.0 \tag{13-17a}$$

Similarly, for Q large,

$$W(m') = \frac{1-m'}{1-m'+mp} + m'p \tag{13-17b}$$

The mutant strategy, m', can invade if $W(m') > W(m) = 1$. The established strategy, m, is uninvadable only if there exists no m' such that $W(m') > 1$. To find this uninvadable strategy, suppose that $W(m') < 1$:

$$W(m') = \frac{1-m'}{1-m'+mp} + m'p < 1$$

Then, after some manipulation,

$$m'^2 - m'(1+mp) + m > 0 \tag{13-18}$$

This is *least* likely to happen [meaning that $W(m')$ is most likely to exceed 1.0]. When m' takes the value that minimizes the left side of the inequality. This particular m' is the strategy presenting the worst threat to m. We find this m' by differentiating:

$$0 = \frac{d}{dm'}[m'^2 - m'(1+mp) + m] = 2m' - (1+mp)$$

so that

$$m' = \frac{1+mp}{2} \tag{13-19}$$

The second derivative is 2, which is positive. Hence (13-19) represents a minimum. But if there exists a unique, stable (i.e., uninvadable) strategy, m, the worst threat is the strategy,

m', that approaches infinitesimally close to m. That is $m' = (1 + mp)/2 \to m$, so that $m = 1/(2 - p)$. This strategy is indeed uninvadable if inequality (13-18) is met. Substituting, we find

$$\left(\frac{1}{2 - p}\right)^2 - \frac{1}{2 - p}\left(1 + \frac{1}{2 - p} p\right) + \frac{1}{2 - p} = \frac{1}{2 - p}\frac{1 - p}{2 - p} > 0$$

Thus the stable, and presumably evolved optimal strategy is

$$m = \frac{1}{2 - p} \tag{13-20}$$

The most interesting, and somewhat surprising outcome is that even in the total absence of environmental variation, and even if virtually no migrants survive ($p \to 0$), the optimal strategy is still to send off half of all offspring as dispersers each generation.

This model has been extended (Comins et al., 1980; Comins, 1982) to cover the more general cases of colonization and recolonization, where sites can be vacated by virtue of all newborn dispersing or because of local catastrophes, and where dispersion is continuous rather than patchy. General conclusions are essentially identical. This model of Hamilton and May, incidentally, can be made still more realistic and powerful with the addition of possible trade-offs between dispersal ability (or tendency) and competitive ability. If these two attributes are negatively correlated among individuals (at the additive genetic level), the conditions for evolution of dispersal are more easily met. Furthermore, those individuals with lower competitive ability should evolve higher dispersal tendencies under conditions of high population density. Dispersal might be expected to be density dependent and to involve individuals generally different from stay-at-homes.

Other sorts of evolutionary models have also been devised. For example, Van Valen (1971) notes that if local populations go extinct from time to time, a gene that promotes dispersal and thus the founding of new populations can be selected for. The equilibrium frequency of such a gene would reflect a balance between the group selection advantage of avoiding extinction of the group's genome, and the individual selective disadvantage from migration mortality. A more detailed, quantitative analysis of this model might be possible using Levins's (1970) approach to group selection (Chapter 9). Gillespie's and Gadgil's models of dispersal rest on advantages born of escaping deteriorating local conditions in a heterogeneous environment. Hamilton and May's model shows that dispersal can evolve even in homogeneous environments. Dispersal may also evolve via its beneficial effects on overcrowded subpopulations if individuals within subpopulations are genetically similar.

D. S. Wilson (1980) provides the following trait group selection model for the evolution of dispersal. Consider two phenotypes: A, which occurs with frequency p, and among whose individuals a fraction m emigrate each generation, and B, with frequency $q = 1 - p$, none of whose members emigrate. Picture the worst scenario for dispersal, one in which *all* dispersers die. Then if we use the expression

$$n(t + 1) = r_0 n(t)\left[1 - \frac{n(t)}{K}\right]$$

for the population dynamics in any local area, we have

$$\begin{aligned} n(t)_{\text{after dispersal}} &= n(t)_{\text{before dispersal}}[q + p(1 - m)] \\ &= n(t)_{\text{before dispersal}}(1 - pm) \end{aligned}$$

and

$$\frac{n_A(t + 1)_{\text{after dispersal}}}{n_A(t)_{\text{after dispersal}}} = (1 - m)\frac{r_0}{K}\left[K - n(t)_{\text{before}}(1 - pm)\right]$$

$$\frac{n_B(t + 1)_{\text{after dispersal}}}{n_B(t)_{\text{after dispersal}}} = \frac{r_0}{K}[K - n(t)_{\text{before}}(1 - pm)] \tag{13-21}$$

Letting $R = n(t + 1)/n(t)$,

$$R_A = (1 - m)\frac{r_0}{K}[K - n(1 - pm)] \tag{13-22a}$$

$$R_B = \frac{r_0}{K}[K - n(1 - pm)] \tag{13-22b}$$

Within any single trait group, R_A clearly is less than R_B. However, we must take into consideration the fact that the population as a whole is composed of many trait groups of different composition. This is done (recall from Chapter 9) by setting p equal to p_A^*, the value experienced by individuals of type A, in (13-22a), and p_B^*, the value experienced by individuals of type B, in (13-22b). The values of p_A^* and p_B^* are

$$p_A^* = \frac{\sum_i n_{A_i}(n_{A_i}/n_A)}{n} = \left[\bar{n}_A + \frac{\text{var}(n_A)}{\bar{n}_A}\right] \Big/ n \tag{13-23a}$$

$$p_B^* = \frac{\sum_i n_{B_i}(n_{A_i}/n_A)}{n} = \left[\bar{n}_A + \frac{\text{cov}(n_A, n_B)}{\bar{n}_B}\right] \Big/ n \tag{13-23b}$$

where the subscript i denotes the ith trait group. The overall values of $\bar{R}_A$ and $\bar{R}_B$, then, the means over all trait groups, are

$$\bar{R}_A = (1 - m)\frac{r_0}{K}\left[K - \bar{n} + m\bar{n}_A + m\frac{\text{var}(n_A)}{\bar{n}_A}\right] \tag{13-24a}$$

$$\bar{R}_B = \frac{r_0}{K}\left[K - \bar{n} + m\bar{n}_A + m\frac{\text{cov}(n_A, n_B)}{\bar{n}_B}\right] \tag{13-24b}$$

The A-type individuals, the dispersers, will evolve if $\bar{R}_A > \bar{R}_B$, which occurs when (after some simplification)

$$m\left(\bar{n}_A + \frac{\text{var}(n_A)}{\bar{n}_A}\right) < \left[\frac{\text{var}(n_A)}{\bar{n}_A} - \frac{\text{cov}(n_A, n_B)}{\bar{n}_B}\right] - (K - \bar{n}) \tag{13-25}$$

A new phenotype that disperses, although only rarely ($m \to 0$), can thus invade a population of nondispersers when

$$\frac{\text{var}(n_A)}{\bar{n}_A} - \frac{\text{cov}(n_A, n_B)}{\bar{n}_B} > K - \bar{n}$$

Let us take a specific example and suppose types A and B to be randomly (Poisson) and independently distributed. Then $\text{var}(n_A) = \bar{n}_A = p\bar{n}$, and $\text{cov}(n_A, n_B) = 0$, and the dispersal trait can invade when

$$m(p\bar{n} + 1) < (1 - 0) - (K - \bar{n})$$

or

$$(1 - m) > (K - \bar{n})$$

when A is very rare ($p \to 0$). Clearly, the conditions for evolution of dispersal under these extremely stringent circumstances are not easily met—the population must be extremely close to carrying capacity. But evolution of dispersal *is* possible. Where variance in the commonness of the new genotype among trait groups is higher (i.e., individuals are clumped, particularly if the established and the mutant types are negatively correlated in space), dispersal might become established quite easily. In the case above, were dispersal to get a foothold, equilibrium would be reached when p reached

$$\hat{p} = \frac{(1 - m) - (K - \bar{n})}{m\bar{n}}$$

Note that p drops as trait group size, $\bar{n}$, increases. This is true in the general case, meaning that where the maintenance of dispersal behavior is due to trait group selection, it will be most common in species that live in small subpopulations.

Refer again to (13-25). Suppose that the tendency for dispersal has approached fixation ($p \rightarrow 1.0$) and that the few nondispersing genotypes mingle randomly with other individuals [so $\operatorname{cov}(n_A, n_B) = 0$]. Noting that $\bar{n}_A = p\bar{n}, \bar{n}_B = (1 - p)\bar{n}$, a little rearrangement then yields

$$m < \frac{\operatorname{var}(n_A) - p\bar{n}(K - \bar{n})}{p^2\bar{n}^2 + \operatorname{var}(n_A)} \rightarrow \frac{\operatorname{var}(n) - \bar{n}(K - \bar{n})}{\bar{n}^2 + \operatorname{var}(n)} \tag{13-26}$$

Now all individuals have a predilection to disperse. But if they possess some phenotypic plasticity, we can argue that dispersal will be realized only under those conditions that would favor its evolution [i.e., only when inequality (13-26) is met]. Thus dispersal should occur when $\operatorname{var}(n)$ is high, and when mean population density is high compared with carrying capacity. A local clue to the value of $\bar{n}$ for the potential disperser is n, the local density. Thus we might expect to find dispersal to be density-dependent.

We have, in the preceding section, made one very important implicit assumption—that dispersal is affected by the genome. What is the evidence? Prus (1966; cited in King and Dawson, 1972) explored the genetics of emigration tendencies in the flour beetles *Tribolium confusum* and *T. castaneum.* The procedure was to dangle a string into the top of one vial, containing beetles, then loop it over the top and down into another vial. Prus watched the frequency with which beetles climbed the string out of their home vial and dropped into the vial next door. In both species there were marked interstrain differences. Krebs et al. (1973), studying the dispersal from vole enclosures (*Microtus pennsylvanicus* and *M. ochrogaster*), found not only variation in timing of dispersal with population cycles stage (Chapter 2), but clear evidence of genetic differences among those dispersing and those remaining in the enclosure. For example, in *M. pennsylvanicus*, homozygotes for the Tf locus were less prone to emigrate, particularly among females. Greenwood et al. (1979) have actually managed to estimate heritability of dispersal distance in first-year individuals of the great tit (*Parus major*). Dispersal is sufficiently short range in this species that young can be banded in the nest and found breeding the following year. Parent–offspring regressions for father–son, father–daughter, mother–son, and mother–daughter showed no significant differences, and a mean of .28, indicating a heritability of $h^2 = (2)(.28) = .56$.

We should note before leaving evolutionary models of dispersal that dispersal is not always a simple matter of occasional wandering among patches or of a one-shot emigration of young or subadults. In *Mus musculus*, for example, in eastern Europe and central Asia, each litter disperses in the standard fashion as its members reach adulthood—until the last litter comes along. The last litter remains and ultimately replaces the parents (Naumov, 1940, cited in Anderson, 1970). Such complications make evolutionary modeling a bit less simple and the results gained thereby a little less reliable than intimated in the discussion above.

13.3 Population Consequences of Dispersal

We have already discussed the role of dispersal in keeping local populations from reaching extremely high densities and in buffering population changes over areas comprised of several subpopulations (Chapter 2). Dispersal is also of importance to the genetic structure of populations and has implications for group selection. There is still another consequence of some importance. Brown (1957) noted that the expansion of a population into new areas, resulting from dispersal responses to high density, followed by contraction when density dropped, would result in the formation of periodically isolated population pockets. If isolation were of sufficiently long duration, genetic drift or differential selection might lead to divergence. Recontact during the next period of population expansion could then lead to the evolution of genetic isolating barriers, resulting in speciation. We can carry Brown's argument to a less extreme point with interesting results. First, in order to remain isolated during periods of population expansion, occasionally isolated pockets would need to have evolved ways of slowing or stopping the influx of outside genes into their local patch. Where adaptation to local conditions makes blockage of gene flow advantageous, there exist various means to accomplish this end (see Chapter 9). Also, we have noted that individuals that disperse may be genetically somewhat different from other individuals in their source population. What this adds up to is that peripheral populations with limited gene flow between them and their more central parental population may differ considerably in their genetic composition. These differences, however, may fail to appear or to be maintained if there have not evolved the gene-flow blocking devices necessary to maintain genetic isolation.

An ecological correlate of the foregoing argument is that peripheral populations that have not successfully been able to stem the gene flow from denser populations elsewhere will likely have failed to fine-tune their adaptations to local conditions. Thus we might expect lower reproductive potential and lower equilibrium population densities. The work of J. T. Emlen (1978) bears directly on this question. Emlen examined avian biomass in pine forest at a site in southern Georgia, nine sites along the Florida peninsula, one site in the Florida Keys, and two in the Bahamas. All bird species were partitioned by tenths into one of several guilds [see "Fluctuations in the External Environment" (p. 145)] defined according to nine foraging categories and eight foraging techniques. Examples of guilds defined are ground seed gleaners, fruit and bud eaters, ground insect gleaners, and pine foliage insect gleaners. Thus he looked not at bird species per se, but at guilds—ecospecies, in a sense. The various study sites were physiographically quite similar, although they differed somewhat in canopy height and cover, shrub cover, density of palmetto, degree of hardwood intrusion and nature of the surrounding area, and the particular pine species present. The most marked difference he found was in the total bird biomass per unit area. Relative to the biomass density in southern Florida, migrant bird species showed biomasses of 2.2 in Georgia, 1.4 in central Florida, and 3.4 in the Bahamas. For permanent resident species, the ratios for these areas were 3.7 : 3.8 : 18.1. The decline in biomass from north to south was found to hold individually for each guild except one seed-eating guild. The marked increase between the mainland and the Bahamas was primarily due to the insect eaters. Emlen made a detailed analysis of these observations, concentrating mainly on the insect-eating guilds, particularly the pine foliage insect gleaners. A plot of estimated insect biomass, based on exhaustive sampling and counting, against bird biomass for this guild is shown in Figure 13-1. The correlation for Florida birds is weak ($p > .05$), and disappears if the two southernmost sites are dropped from the analysis. Hence the change in bird density with latitude cannot be accounted for by changes in food availability. It appears that the Georgia birds maintain lower population densities per unit food than Florida birds and that Bahamian birds do very much better than Georgia birds. Why? Emlen hypothesized that the peninsular shape of Florida, connected to the much more extensive landmass to the north, would result in the funneling of southward dispersal into Florida, thereby exacerbating gene flow into these southern populations. Because the northern and southern habitats *do* differ somewhat, it is reasonable to suppose that different adaptations might be appropriate. Gene flow thus might interrupt adjustment to local conditions in the south, with a resulting lower equilibrium bird biomass density. The Bahamian populations, on the other hand, experiencing very little gene flow, may have been able to adapt without interference to the local conditions. Why Georgian populations should be less dense than the island populations is less clear, although the former occur in a mosaic of habitat types, and pine forest birds are commonly found in other habitats.

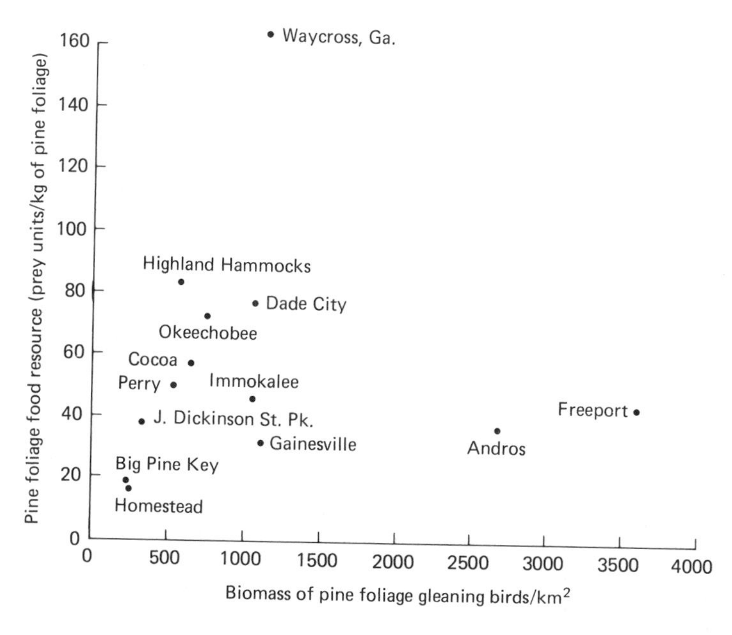

Figure 13-1 Plot of food availability against bird biomass for the pine foliage gleaner guild at several sites in the southeastern United States and Bahamas. (From J. T. Emlen, 1978.)

Thus, to some extent, the members of mainland populations might tend to be jacks of all trades, making them less finely tuned to the pine forest habitat specifically, while the island populations have nothing but pine forest to adapt to and become masters of one habitat.

Emlen tested his hypothesis by gathering information on degree of subspeciation of the various birds in his study. Where subspeciation occurred, he reasoned, some isolation must exist among local populations. Where gene flow was not inhibited, subspeciation could not occur. Thus if the bird biomass was reduced to the south because of gene flow preventing local adaptation, birds that showed subspeciation should show less density attenuation than those showing no subspeciation. His findings of birds of Georgia and northern Florida are shown in Table 13-2. The expected trend is statistically significant whether the table is taken as shown, or the "reduced" category is lumped with the "absent" category, or "reduced" is lumped with "not reduced."

Table 13-2

Species	Populations in the South		
	Absent	***Reduced***	***Not Reduced***
With subspeciation	3	8	12
Without subspeciation	24	10	4

In a later paper J. T. Emlen (1979) made similar observations and measurements on the birds of the California mainland and the Baja peninsula, with similar findings. In this later paper he also elaborated on why island populations might be so dense. Like the Bahama–Florida comparison, the coastal islands off Baja California are also very heavily populated. Perhaps this could be due to lack of gene flow, or perhaps simply to a "fence effect." Krebs et al. (1969), in their studies of microtines, found that populations in enclosures, where dispersal was greatly inhibited, built up to huge levels, with consequent habitat destruction. Dispersal may allow animals to leave local areas before tremendous density-feedback pressures occur, and so keep local population densities below levels obtaining in situations where dispersal cannot occur. Certainly, on islands dispersal into secondary habitat is limited. Evidence that this explanation may be appropriate for the Baja California islands is found in the population trends of the black-throated sparrow, *Amphispiza bilineata.* On the mainland this species maintains fairly low numbers and disperses readily from its preferred scrub habitat into extensive areas of woodland (the density ratio—in birds/km^2—in scrub to woodland areas is 7.6 :26.9). On the islands, such spillover is restricted and the corresponding ratio is 137.1 : 147.4. This suggests that available alternate habitat on the mainland allows social pressures to keep population density at low levels, while on islands the lack of escape opportunities prevent this. In passing, note that if this explanation of black-throated sparrow population differences is valid, it provides evidence for an overwhelming role by interference, as opposed to exploitation competition as the usual arbiter of density feedback.

13.4 Models for the Evolution of Density-Feedback Patterns

A plot of fecundity, or the growth parameter r, against population density would, according to the logistic equation, be a straight, monotone-decreasing line. A plot of $\ln R$ versus n, as predicted by the Ricker curve, also gives a straight line. But in reality, at least in the laboratory, where the considerable noise and distortion of these curves due to density-independent factors and species interactions can be controlled, birth rate and r usually rise to a peak at some critical n, and *then* fall, not necessarily linearly (see, e.g., Chapman, 1928; MacLagan, 1932; Park, 1932; for a more recent review, see King and Dawson, 1972). The initial rise is often termed the *Allee effect*, after W. C. Allee, who first speculated as to its *raison d'être.* Most modern workers have ignored the Allee effect, appearing to be interested primarily in the shape of the r versus n curve at higher densities. Let us look more closely at the curve. Should r drop linearly with n, exponentially, or in some other way? Can we make any generalizations from theoretical considerations?

A very simplistic discussion was presented by J. M. Emlen (1973). If we begin with a hypothetical population demonstrating the relationship in Figure 13-2, and note that a population which fluctuates spends a decreasing amount of time at values of n progressively farther from $\hat{n}$, we can argue that selection pressures increasing r act most commonly in the vicinity of $\hat{n}$. Furthermore, to the extent that adaptive arguments are valid, behavioral, physiological, and morphological changes that raise r at values of n near $\hat{n}$ may come at the expense of r at values of n farther from $\hat{n}$. Hence we might view progressive changes in evolutionary time altering the curve as shown in Figure 13-3. The appearance of an Allee effect, then, is not unexpected.

Concentrating on the right-hand portion of the curve—so that such complications as the Allee effect can be ignored, we can manage to describe a very wide array of curve shapes with the expression

$$r = r_0\left[1 - \left(\frac{n}{K}\right)^{\theta}\right] \qquad \text{where } \theta > .0 \qquad (13\text{-}27)$$

This particular form was used by Gilpin et al. (1976), who speculated in somewhat more sophisticated fashion on the evolution of the shape of the density-feedback curve. Let r_0 be fixed and ask what value of θ we should expect to evolve. A phenotype that suffers the least density

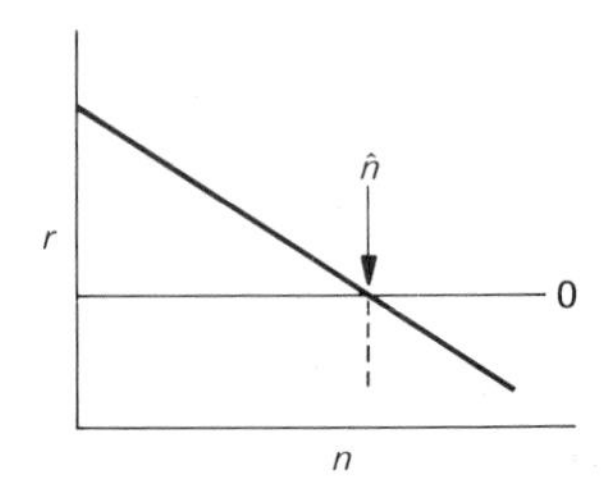

Figure 13-2 Plot of the population growth rate (r) against population density (n) as described in the logistic equation.

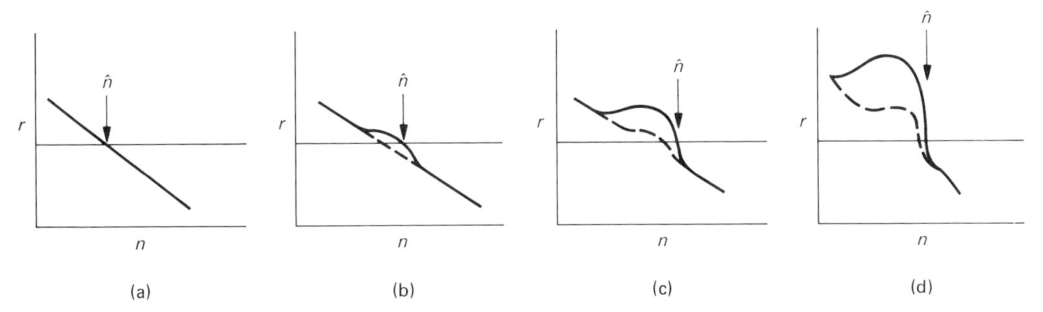

Figure 13-3 Hypothetical evolutionary changes in the relationship between r and n.

feedback is always at an advantage over its less resistant partners. But given that density feedback cannot be escaped, we next address the question of how best to distribute the unwelcome pressures over the various values of n/K. Gilpin and his coworkers reasoned that opportunistic species, because they operated mostly well below K, should experience the strongest selection for resistance to feedback when n/K is low, at the expense of greater sensitivity when n/K is high. The feedback pattern which should therefore emerge is realized by making θ large. The magnitude of feedback, $(n/K)^\theta$, approaches zero when θ is large and n/K is small. The trade-off is overwhelming pressures when n is close to K. Equilibrium species, on the other hand, benefit from low feedback pressures when n/K is about 1.0, but can absorb the trade-off of strong feedback when n/K is small. Such species ought to evolve low θ values. Expected curves are shown in Figure 13-4. Similar reasoning led to the same conclusion for Fowler (1981). Fowler showed that the expected pattern (Fig. 13-4b) occurred in 23 of 27 cases of large mammals (equilibrium species) for which data were sufficient to determine an unambiguous curve. Analysis of density-feedback curves by Stubbs (1977) for 46 opportunistic species, 37 of them insects, mostly demonstrated patterns like that in Figure 13-4a.

Simulation models employing (13-27) and introducing density-independent noise in the form

$$r = r_{\text{determinate}}[1 + S(t)]$$

and density-dependent noise in the form

$$K = K_{\text{determinate}}[1 + S'(t)]$$

were explored by Turelli and Petry (1980). A rare new genotype with altered r_0, K, or θ values was introduced and its fate determined. The evolutionary progress of r_0 was quite model-specific—in the linear case ($\theta = 1$), density-dependent variation had no effect on the evolution of r_0. Large K values were always selected for. Of primary interest to the present discussion, density-dependent noise depressed θ, and density-independent noise raised θ, in agreement with findings of the authors noted above.

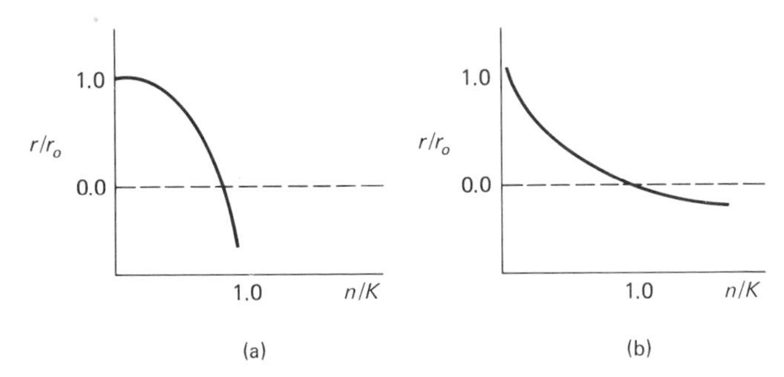

Figure 13-4 General predicted forms of the r versus n curve for (a) an opportunistic species, and (b) an equilibrium species.

Thomas et al. (1980) combined the best of theory and experiment by examining 58 populations of *Drosophila*—27 species at different temperatures—introducing various numbers, n, of adults, allowing them to oviposit for 7 days, removing them, and then counting the emerging progeny. Using nonlinear techniques, they then tried fitting r_0 and θ parameters to either

$$n(t+1) = n(t) + r_0 n(t)\left[1 - \left(\frac{n(t)}{K}\right)^{\theta}\right] \tag{13-28}$$

a discrete analog of (13-27), or

$$n(t+1) = n(t)\exp\left\{r_0\left[1 - \left(\frac{n(t)}{K}\right)^{\theta}\right]\right\} \tag{13-29}$$

a generalized form of the Ricker model. The second, Ricker group of equations fit the data extremely well when appropriate values of r_0 and θ were chosen. What is particularly interesting about the results, though, is not so much that they were able to get good fits, but that in *all* 58 cases, $r_0\theta$ was less than 2.0. A stability analysis of (13-29) shows that populations should be neighborhood stable when $r_0\theta < 2.0$, limit cycle when $r_0\theta > 2.0$, and when $r_0\theta > 2.69$, go into chaotic fluctuations. The authors reasoned that local populations that evolved $r\theta$ values above two were in danger of chance extinction due to their instability. Natural fly populations should, presumably, retain the genetic consequences of selection for increased reproductive potential, but also the genetic impact of group selection for stability. Because species experiencing considerable density-independent noise are those that we generally expect to have high r_0 values (Chapter 11), this hypothesis suggests an upper limit to the value of θ that should evolve in opportunistic species.

Review Questions

1. Might genetic feedback, as it is presented in this chapter, act to limit population size? Why or why not?
2. Considering the findings of McGovern and Tracy and the possible existence of "profane" genes (see the introduction to Part Four), is there any way in which population fluctuations arising from genetic feedback can clearly be demonstrated or refuted—even hypothetically?
3. Ideal free distribution is a very convenient concept for a number of reasons. Why? For what sorts of creatures, and under what circumstances, does such dispersion seem most likely or least likely to be realized? What data can you find that bear on this question?
4. What implications does ideal free distribution have for the spatial-genetic structure of populations of phenotypically rigid as opposed to phenotypically plastic species?
5. On the basis of the discussions in Chapters 11, 12, and 13, what do *you* think about the concept of K selection? What does it mean to select for higher K (or, more properly, $\hat{n}$)? How might selection affect dispersal, dispersion, and social behavior when $n \to K$? $n \ll K$?
6. What affect might dispersal and ideal free distribution have on the shape of the density-feedback curve?
7. Discuss the implications for neighborhood and global stability of the density feedback curves expected (and apparently found, in general) for opportunistic as opposed to equilibrium species.

Chapter 14

Sociality: I. Spacing and Mating Systems

Chapters 11 through 13 dealt with adaptations of populations to their physical and biotic environment only in a very general way. Interindividual interactions were largely ignored. However, the manner in which one organism acts toward or responds to another is basic to the dynamics of the population in which those two individuals live. In this chapter we discuss first (Section 14.1) the factors influencing the way individuals arrange themselves relative to one another in space. Spatial arrangement influences rather directly the reproductive ecology of the organisms. This constitutes Section 14.2. Finally, both the spacing and the mating system of a species set the evolutionary mileu for social organization. This topic will be covered in Chapter 15.

14.1 Spacing Systems

A number of species, particularly smaller invertebrate species, appear to have little or no social organization beyond the simple act of copulation. Such species have little reason to overdisperse or clump in response to intraspecific stimuli. Dispersion is apt to be dictated largely by the whims of prey, predators, or physical processes. Aggregations of these species are, in a sense, passive; they are not considered to be social.

Among individuals belonging to species where interindividual interactions are important to fitness, we shall recognize four general, but not mutually exclusive categories of spacing system. Individuals may congregate into loosely organized clusters of reproducing pairs (and sometimes their extended families). We shall call these *colonies*. The cells constituting a volvox, or the collection of hydranths forming a stony coral or a Portuguese man-of-war are also colonies, but of a somewhat different sort. The colonies considered in this chapter are of the kind demonstrated by gulls nesting en masse on oceanic rocks, or herons in a rookery.

The second category of spacing system is *territoriality*. A number of definitions of "territory" exist, including "defended area" and "exclusive area" (for a criticism, see J. T. Emlen, 1957), but perhaps the most useful is that of Wolf (1970): an area "within which the resident controls or restricts use of one or more environmental resources." Territories are usually rather discrete entities with well-defined boundaries. They come in a variety of sizes, sometimes as small, contiguous areas within a colony, sometimes as more or less contiguous spaces of large enough size that "colony" seems inappropriate to describe a collection of them, often as nonadjacent yet exclusive areas. Territories may be "defended" by individuals of both sexes, as with pine squirrels (Smith, 1968), or by pairs (the male generally taking the major, aggressive role), or by whole groups (as, for example, in the vicuña, Koford, 1957).

Home ranges constitute the third spacing system we consider. Burt (1943) defined home range simply as the area to which an animal normally confines its movements. That is, when a nondispersing individual settles into an area and stays there, the extent of its wanderings defines its home range. Home range, then, is not an exclusive area (although many species possess a "core" area somewhere in the home range that meets the definition of territory); rather, it is just a convenient label, reflecting the fact that many animals seem attached to an area in space rather than wandering freely (see Jewell, 1966). Both colonial and territorial animals, unless they are totally vagrant, have home ranges, but not all creatures with home ranges are colonial or territorial. There are a large number of difficulties involved in measuring home ranges. The interested reader is referred to Blair (1940a,b), Hayne (1949), Young et al. (1950), Harrison (1958), and Jorgenson and Tanner (1963).

Often individuals come together to form *groups.* [We avoid worrying about the nuances between "group," "flock," "troop," and so on, or transcending the mundaneness of biological scientism with venereal terms such as prides of lions or gaggles of geese—more elevated and relaxed reading can be found in Tipton (1977).] Groups may space themselves out as already described for individuals, by being group territorial or possessing group home ranges. What is of particular interest to us here, though, is the role of population dynamics and evolution in the molding of social interactions within groups, and in turn, the effect of such social organization on population genetic structure and population dynamics.

Territoriality or Coloniality?

The distinction between territoriality and coloniality is largely one of scale. At one extreme, large exclusive areas act to overdisperse—to spread out—individuals or breeding pairs. This is *territoriality.* At the other extreme, individuals within groups, even tightly organized groups, may space themselves at distances which human beings, in their own social organization, would refer to as *personal space.* The swallows congregating on a telephone wire manage to preserve about 4 inches between themselves and the next bird, probably reflecting the individual distance needed to extend the wings for takeoff (J. T. Emlen, 1952). *Coloniality* is the middle ground between these extremes.

Horn (1968) was one of the first to address the question of relative adaptive value of territoriality versus coloniality. Suppose that the adaptive significance of the spacing system reflects the promise of sufficient food. Then during the breeding season, we can picture the distribution of individuals and their food as in Figure 14-1. In both Figure 14-1a and b, the dots represent potential food sites that may or may not actually contain food at any one time, and the crosses represent nest sites. In the left figure four breeding pairs are evenly dispersed over the breeding area (territorial), each with access to the immediately surrounding food sites. In Figure 14-1b all pairs are congregated at one point and share all food sites. The latter case represents colonial breeding. Horn reasoned as follows: If resources are uniformly distributed and predictable, each dot corresponds to a *reliable* food source. If we call the distance between dots k, then individuals (or pairs) distributed as in Figure 14-1a will travel an average distance of

$$\sqrt{(\tfrac{1}{2}k)^2 + (\tfrac{1}{2}k)^2} = .71k \tag{14-1}$$

to get to their food. If colonial (Figure 14-1b), on the other hand, the distance traveled is

$$\frac{4\sqrt{(\tfrac{1}{2}k)^2 + (\tfrac{1}{2}k)^2} + 8\sqrt{(\tfrac{1}{2}k)^2 + (\tfrac{3}{2}k)^2} + 4\sqrt{(\tfrac{3}{2}k)^2 + (\tfrac{3}{2}k)^2}}{4 + 8 + 4} = 1.50k \tag{14-2}$$

If, as seems obvious, spending less time and energy on meeting food needs is advantageous, a predictable uniform distribution of resources favors the evolution of territoriality.

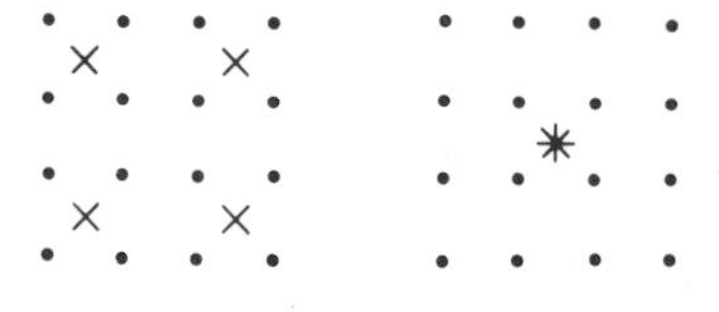

Figure 14-1 Schematic depiction of food sources (dots) with (a) four territorial breeding pairs and (b) four colonially nesting pairs. (After Horn, 1968.)

If resources are predictable and highly clumped at one point rather than uniformly distributed, it will be advantageous for all pairs to breed as close as possible to the food source. Coloniality, or perhaps communal, group breeding might be expected.

If resources are unpredictable and clumped, we may picture the foraging individuals as having to search all points in space. A colonial species will have to travel an average distance of $1.50k$ to its food (as in equation 14-2). Individuals of a territorial species would have to move a distance of $1.93k$. Where food is clumped and unpredictable, coloniality is favored.

There is considerable evidence to back up this theory's predictions. Most marine birds, which forage for food in moving fish schools, are colonial. The herring gull (*Larus argentatus*) nests solitarily where it breeds inland near small bodies of water, whereas seacoast or offshore island-nesting individuals are colonial. Common terns (*Sterna hirundo*) are solitary on river islands, colonial where marine and often on the great lakes. Osprey (*Pandion haliaetus*) which are usually solitary nesters inland, are colonial along the East Coast. Birds foraging in terrestrial habitat, as opposed to aquatic or marine habitat, utilize areas of relatively uniform food distribution and are, by and large, territorial (sources from Wynne-Edwards, 1962). Erwin (1977) studied two terns and the black skimmer (*Rynchops nigra*) in some detail in Virginia. Royal terns (*Thalasseus maximus*), which feed far offshore, where there is great spatial variation in fish abundance from place to place, forage socially. Social feeding is likely to be advantageous when food sources are unpredictable and must be searched for, and opportunities for social foraging would certainly be maximized by colonial habit. These terns are colonial. Black skimmers feed close to shore where spatial heterogeneity of food is far less, and they are solitary foragers. Nevertheless, they are colonial. The common tern (*Sterna hirundo*) feeds both inshore and far offshore, forages socially, and is colonial. Is the coloniality of the black skimmer an exception to Horn's hypothesis? Not really. In this species foraging cannot occur on the breeding grounds, so social feeding or not, it can be argued that areas of predictable food availability cannot be "defended." Thus Horn's thesis is a bit misleading in its simple elegance. When reliable uniform food sources occur on the breeding grounds, territoriality is expected and found, when food is unpredictable in time and space, coloniality is predicted and found. But where food is fairly predictable but to be found off the breeding ground, the spacing system cannot be easily foreseen. As Brown (1964) has forcefully pointed out, territoriality does not automatically follow from the existence of foraging advantages of the sort described above; a territory cannot be "defended" unless it is defendable. If maintenance of an exclusive area becomes costly in agonistic energy expenditure or risk, owing perhaps to very high population density or lack of perches from which to survey the domain or aggressively display, it may no longer be profitable. There is also some awkward evidence to the effect that many territorial bird species forage largely *off* their territories. Of course, food is not the only resource for which defending exclusive rights is beneficial. And there are other, non-resource-related advantages to territoriality. Often territories perform the function of courting arenas. Or perhaps, as suggested by Crook (1961), territories have evolved—or at least are retained by natural selection—because the overdispersion of individuals promulgates crypsis (concentrations of individuals are often more noticeable to predators). Overdispersion might also act to restrain the spread of disease.

Coloniality

The practice of coloniality affects population dynamics in a number of ways. It affects foraging efficiency, defense against predators, and the timing of reproduction. Where individuals cluster small territories into a colony, they have ready visual access to the activities of their fellows. If one individual is foraging successfully, it is possible to surmise that fact from the steady arrival of food items and to follow that individual on its rounds. If food is evenly dispersed, there will be little advantage in doing so (generally, we should not expect to find coloniality under these circumstances anyway), but if food is clumped and unpredictable, the discoverer of the food source should suffer little loss from others tagging along, and the followers may benefit considerably. Thus the close proximity of colonial breeders may enhance the rate of food finding, with beneficial effects on nesting success. To use a term of Ward and Zahavi (1973), the importance of certain assemblages of birds such as colonies, may be to act as "information centers" for food finding.

Krebs (1974), studying the great blue heron (*Ardea herodias*), which nests in colonial "rookeries," noted that the birds seemed to forage in different places on different days and that they often left in groups. Indeed, it was not uncommon for individuals from one colony to follow those of another, close-by colony to particular feeding spots. To check the hypothesis that such social foraging enhanced food-gathering efficiency, Krebs regressed food intake per individual on size of the feeding group and found (y is the food intake per individual, x the number of herons in a group)

$$y = .25x - .004x^2 + .99 \tag{14-3}$$

According to this empirical expression, the optimal group size is 31.75. The usual group size, in fact, varied between 20 and 30. At a group size of 30, an individual gathers approximately four times as much food per time as when foraging alone.

Horn (1968), who speculated on the advantages of coloniality for foraging efficiency in his study of Brewer's blackbirds (*Euphagus cyanocephalus*), reasoned that the information center function of a colony would be most felt by birds with a maximum number of nearby neighbors (i.e., birds nesting in the middle of the colony). Accordingly, he identified the colony center (center of gravity) and classified all nests as above or below median distance from that center. He then compared weight gain of young per nest per day, clutch size (which should rise with foraging success—Chapter 11), and nesting success (number of young hatched per nest) for these outer and inner nests. The results are shown in Table 14-1. The birds from the colony interior seem to be feeding their young more efficiently, have larger clutches, and fledge more young.

Table 14-1

Nesting Area	Number of Nests	Clutch Size	Weight Gain per Nest per Day	Number of Young Hatched per Nest
Outer	21	5.5	23.8	5.1
Inner	20	5.7	27.5	5.4

With respect to defense, it is clear that close breeding means more predator lookouts and a better chance of hearing an alarm call. Also, many species of birds "mob" predators; that is, they respond to the presence of predators unlikely to catch them on the wing by flying up and harassing them. If colonial birds mob only when a predator approaches within a given, critical distance, colony interior birds will benefit from larger numbers of mobbers and so be better protected (Fig. 14-2). Examining predation for outer and inner sites, Horn found the results

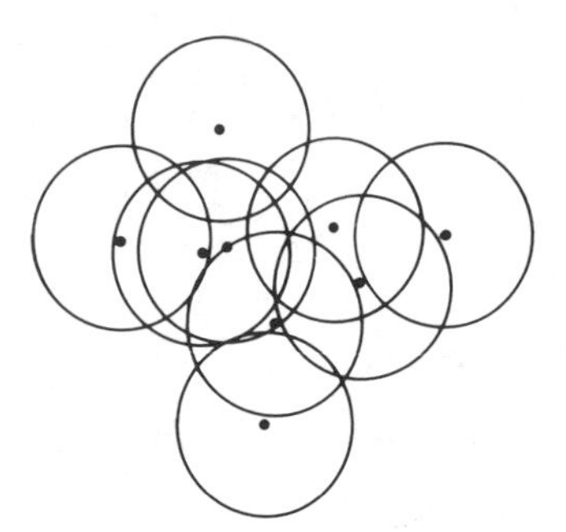

Figure 14-2 Schematic diagram of a nesting colony. Dots represent nest sites, circles about the dots denote the area over which the presence of an aerial predator elicits mobbing behavior. (After Horn, 1968.)

given in Table 14-2. The figures support the conjecture for the egg-laying and incubation period. During the nestling period there is virtually no mobbing, so equal values should be expected. A second test of this mobbing hypothesis was managed as follows. Where colonies followed a river bank or cliff face ("linear" as opposed to two-dimensional or "open" colonies), Horn reasoned that mobbing might backfire, discouraging the predators to some degree but also clearly indicating to them the positions of the nests. In eight linear colonies, during the egg-laying and incubation period, the inner nests suffered heavier destruction by predators than the outer nests (20 versus 9 nests predated), as expected.

Table 14-2

	Number of Nests Predated During:	
Nesting Area	***Egg-laying and Incubation Stages***	***Nestling Period***
Outer	35	20
Inner	16	21

Similar data but with opposite implications have been gathered for the great tit (*Parus major*) by Krebs (1971). This species is not really colonial, but the same argument that nesting density can deter predators should be the same. Krebs recorded number of nests predated as a function of distance to nearest-neighboring nest. Results are shown in Table 14-3. A plot of percent predation against nearest-neighbor distance yields an almost linear, inverse relationship with an average of roughly 50% predation at 10 to 29 m, about 20% at 70 to 89 m. For further information on the relationship between nest spacing and predation, see Post (1974).

Table 14-3[a]

Nearest-Neighbor Distance (m)	**Number of Nests** *Predated*	**Number of Nests** *Not Predated*	**Percent Predated**
<45	43	147	23
>45	34	273	11

[a] $X^2 = 11.1, p < .01$.

The idea that proximity leads to reproductive stimulation and breeding synchrony is an old one. Among many marine invertebrates and a few fish, reproduction is often an event in which large masses of individuals stimulate each other to shed gametes in great spawning storms. Darling (1938) suggested that "social stimulation" induced synchrony and hence greater reproductive success in birds. Lehrman and his associates (e.g., Lott et al., 1967) studied this phenomenon in great detail in the ring dove and found that merely the presence of a male, even behind a glass partition, induces ovarian development in the female. Horn (1966) noticed that one female Brewer's blackbird giving a precopulatory display triggered other females in the colony to do likewise. Sex is contagious. He also found that the expected synchronizing effect of such behavior was (as expected regarding frequency of visual contacts) more pronounced among inner than outer nests (Horn, 1970).

To test the assertion that synchrony is beneficial, that it improves reproductive success, S. T. Emlen and Demong (1975) measured the standard deviations in hatching data for 15 colonies of bank swallows (*Riparia riparia*). The percent of hatchlings that fledged was then regressed on this standard deviation, yielding a coefficient of $-.72$, $p < .001$. Percent of eggs laid that fledged young, regressed on this standard deviation in hatching date, gave a value of $-.82$, $p < .001$.

Findlay and Cooke (1982), studying snow geese (*Anser caerulescens*) in a large colony at La Perouse Bay, Manitoba, found that at all nesting stages—laying, hatching, fledging—those birds whose efforts lay closest in time to the colony mean timing were the most successful. Obviously, synchrony is beneficial to reproductive success (barring reversals of fortune during the post fledging period). But why should this be so? One possible (but to this author, unlikely) hypothesis is predator saturation (Chapter 11). Another is that social stimulation among individuals all of whom are (presumably) attempting to optimize their hatching date for most efficacious use of food for their young leads to a kind of group decision. A group decision, by incorporating input from all breeders, might be expected to be more accurate than any one individual's decision. Although this argument may smack a bit of black magic, it can be tested for human beings quite simply. S. T. Emlen (personal communication) experimented by asking students in his animal behavior class to close their eyes and give a sign when they thought exactly 7 minutes had elapsed. If the sign given was silent, so that other class members were unaware of their colleagues' guesses, the estimates were highly variable. If the sign was more audible (such as raising a hand), the variance in response was considerably smaller and the mean quite amazingly close to 7 minutes.

One difficulty with the foregoing arguments and data are that, generally speaking, older birds are more experienced and so expected to produce larger clutches, have healthier young, and suffer lower clutch mortality (Chapter 11). But it is precisely these older birds that inhabit the center area of colonies. Is reproductive success higher for inner nests because of their position or because of the age of the breeding pairs? One rather shaky way around this dilemma is to suppose that the more experienced birds prefer the colony center because it is superior for nesting purposes, and that their greater experience (and often size) enables them to displace younger birds and realize this preference. Two papers pertinent to this matter are by Patterson (1965) and Coulson and Wooller (1976).

Territoriality

Territoriality may serve any one of many functions for its owner, but the two adaptive advantages most often considered are reproductive and food gathering. The small territories of colonial birds often encompass little more than the nest and its immediate surroundings. The advantage to keeping nonfamily members out may be related to protection of the eggs and young and minimizing theft of nest materials, but is surely not the protection of a food supply. In some species these territories may serve as courtship centers. Other species may use one small territory for purposes of attracting a mate and then abandon the area to move into the colony (e.g., the black-headed gull (*Larus ridibundus*); Moynihan, 1955, Tinbergen, 1956). In many noncolonial species the ultimate nesting area serves first as a reproductive territory. In others, groups of males come together and occupy very small territories known as *leks*. Females wander about among the males on these leks, choose a mate and copulate, and then leave to breed on their own. Among birds this is particularly common in cryptic, ground-nesting, precocial species, such as grouse, where the male has little to contribute to the offsprings' welfare and might serve only to attract predators were he to remain with the female (Brown, 1964).

Leks are also found among mammals—especially ungulates (e.g., the Uganda kob, *Adenota kob*; Buechner et al., 1966). Male bullfrogs (*Rana catesbiana*) position themselves along shorelines in the spring, where they remain in exclusive areas for up to 2 or 3 weeks, driving off intruders. At dusk, most leave these areas and take up "calling" stations. Later in the season the males aggregate into groups (choruses) of 14 to 37, in deeper water, each at a particular spot. Here they inflate their lungs, float high in the water, and expose the yellow color of the gular area, excluding all other males from their immediate surroundings (about 2 to 6 m between males). When a female arrives, attracted by the chorus, she moves among the males. When she selects her mate, she swims to him and is clasped. Males not possessing a calling station do not mate (S. T. Emlen, 1968) [although recall Howard's (1981) satellite males described in Chapter 12].

Dragonflies hold reproductive territories that are more than just leks. Here the males occupy exclusive contested lengths of shoreline around ponds and intercept females as they fly to and from the water (Moore, 1964; Campanella and Wolf, 1974; Campanella, 1975). Females oviposit on these territories. The last two papers add the observation that in cases where female arrivals at a pond are not predictable in time and space, territoriality does not occur.

In many species reproduction may be only one function of territoriality. Indeed, much of the behavioral and ecological theory surrounding territoriality presumes the function of guaranteeing an adequate food supply. Data are, by and large, indirect, and will discussed below in the context of territory size. It is worth noting in passing, though, that territories of the Costa Rican hummingbird (*Panterpe insignis*) shift spatially with the seasonal flowering peaks of the major nectar-bearing plant species (Wolf, 1969; Wolf and Stiles, 1970).

Where territories serve as courtship arenas and in those species in which female choice of mate seems largely based on the "quality" of the lucky male's territory (see Section 14.2), it is not clear how territoriality might have evolved. Are the females choosing merely a symbol, perhaps reflecting a male's prowess in intimidating other males? If so, her male offspring will likely inherit the same prowess, thereby becoming successful in their own attraction of females. In this way sexual selection can lead to a steady evolution of apparently meaningless behavior and otherwise useless traits. But we are justified in asking whether this process is really occurring. Just how does a female assess territory quality, and are her criteria ecologically meaningful? Perhaps the preferred territories are rich in food potential, protective cover, or superior nest sites. The fact that food availability is so critical to clutch size and the successful fledging of young (Chapter 11), and that females mating polygynously on one territory have roughly equal reproductive success to monogamous females on other territories (Section 14.2), certainly suggests that females choose their mates on the basis of territorial food potential. But this is by no means certain. In the greater prairie chicken (*Tympanuchus cupido*), a lek species, Robel (1966) found that number of copulations per male increased with lek size. Why size is important is not clear. Verner (1964) discovered that in long-billed marsh wrens (*Telmatodytes palustris*), in Washington State, bigamous males had larger territories than monogamous males, who, in turn, had territories bigger than those of bachelors. In this species it *appears* that females choose their mates primarily on the basis of incomplete, male-built nests, so the relationship of mating success and territory size is, again, obscure.

More sophisticated attempts to evaluate territory "quality" have recently become plentiful. C. C. Smith (1968) noted that territory size in tree squirrels was inversely proportional to food density on the territory. This strongly suggests that tree squirrels "defend" a rather well-defined food supply, presumably one that matches their anticipated needs. In this spirit Schoener (1968b), following a suggestion by McNab (1963), fitted a regression of the form

$$\log_{10}(\text{Territory size}) = b \log_{10}(\text{body weight})$$

and found a slope of $b = 1.09 \pm .11$ for birds. For predatory birds alone the value was $b = 1.31$. This is consistent with the hypothesis that territoriality is related to energy needs; however the b values are somewhat different from the value expected from "existence energy" studies:

$$\log_{10}(\text{existence energy}) = .1965 + .6210 \log_{10}(\text{body weight}) \quad \text{for passerines}$$
$$\log_{10}(\text{existence energy}) = -.2673 + .7545 \log_{10}(\text{body weight}) \quad \text{for nonpasserines} \quad (14\text{-}4)$$

(Lesiewski and Dawson, 1967; Kendeigh, 1970). Perhaps the energetic costs of capturing food increase more rapidly with body weight than does existence energy.

Gass (1979) studied the rufous hummingbird (*Selasphorus rufus*) and found that a plot of log territory size versus log resource density (nectar supply) yielded a curve of slope (almost precisely) -1.0. If, in fact, individuals in this species hold territories that exactly meet their energetic needs, this is what should be expected. Kodric-Brown and Brown (1978), working on the same hummingbird species, made plots of log territory size versus number of flowers per square meter and found a rather clean linear-decreasing relationship: $A = 738D^{-1.0}$, $A = 649D^{-.82}$ for 2 years, where A is territory size and D is flower density. If we suppose the exponent to be -1.0, with error in the second year, then for both years, AD, the number of flowers per territory, was constant across territories. Again, this is what we should expect if territories are maintained as protected food reservoirs. An inverse relationship between territory size and food abundance has also been reported for a lizard, *Sceloporus jarovi*, by Simon (1975).

Seastedt and MacLean (1979), looking at the lapland longspur (*Calcarius lapponicus*) in Alaska, an entirely different sort of bird in an entirely different sort of habitat, used sticky boards to estimate insect (food) biomass in wet, mesic, and dry areas. Then if b_{ij} represents the biomass of insect type j in type area i, and if a_i is the proportion of an individual bird's territory that is of type i, a biomass density index for that territory is given by

$$\frac{\sum_j \sum_i a_i b_{ij}}{\sum_i a_i}$$

An analysis showed Spearman rank correlation coefficients between this index and territory size of $-.43$ (data from 1967–1972 and 1975 combined). When the biomass density index was adjusted to account for specific insect preferences by the birds, this coefficient dropped to $-.52$ ($p < .01$).

Carpenter and MacMillen (1976) constructed a quantitative model based on the food resource hypothesis for territorial maintenance and outlined the conditions under which territoriality should occur in the iiwi (*Vestiaria coccinea*), a Hawaiian honeycreeper. Suppose that a is the fraction of food (nectar) productivity, P, used by a nonterritorial bird, and let e be the fraction of potential intruders successfully excluded by a territorial bird. Then the gain to a nonterritorial individual is aP, while the gain to a territorial bird is $[a + e(1 - a)]P$. The cost to the first individual is simply the basic cost of living E, while the territorial bird expends an additional amount T, accumulated through the agonistic acts needed to expell intruders. Territoriality should be advantageous when its resulting gains exceed its costs:

$$aP < E$$

and

$$E + T < P[a + e(1 - a)]$$

Putting the two inequalities together, we should expect to find territorial behavior when

$$\frac{E + T}{a + e(1 - a)} < P < \frac{E}{a} \tag{14-5}$$

Carpenter and MacMillen kept time and energy budget data on both territorial and nonterritorial individuals over two summers. Feeding areas were mapped, inflorescences counted (the birds feed from the red-flowered tree *Metrosideros collina*) and it was assumed that the amount of nectar removed from a nonterritorial area was proportional to the number of visits to the area. Intruders took 75% of the nectar on nonterritorial bird foraging areas, and two strongly territorial birds excluded 92% and 100% of intruders from their own areas. On the basis of these data, the relevant parameters were estimated to be

$$\begin{aligned} E &= 13.4 \text{ kcal/day} \\ T &= 2.3 \text{ kcal/day} \\ P &= \text{nectar flow} = .259 \pm .010 \text{ kcal/day} \\ a &= .25 \\ e_{max} &= 1.0 \end{aligned}$$

(For all but the two birds mentioned above, e was calculated according to the proportion of time devoted to territorial advertisement.) In all cases territorial birds in the study area met inequality (14-5). All but one nonterritorial bird fell outside the range for which territoriality was predicted.

Armed with data of the sort presented so far, it seems not too ambitious to conclude that in at least some species, territoriality serves a food reservoir function, and that the area maintained is exactly (or nearly so) sufficient to meet the owner's needs. One further study should give us pause, however. Myers et. al. (1981) observed the linear territories of sanderlings (*Calidris alba*) along a central California beach where a negative relationship between territory size and food density is readily apparent. Suppose that we designate the territory length by TL, the prey (*Excirolana linguifrons*, an isopod) density, sampled from cores bored in the sand, by PD, and the number of territorial intrusions per 50-minute period by other sanderlings by ID. The authors performed a partial correlation analysis (designed to test relationships between two variables, independent of the effects of other variables) on the transformed variables, 1/TL, ln (1 + ID), PD (for an explanation of these transformations see the original paper). Although territory size appears at first to be related to food density, after effects of intruders are accounted for, there is no correlation at all between these variables (partial $r = -.03$, $p > .40$). On the other hand, 1/TL is strongly correlated with ln (1 + ID): $r = .64$, $p < .001$. In other words, territory size is strongly and negatively related to the number of intruders (i.e., the sanderling population density), but is apparently independent of food supply. Path analysis, a technique used to try gleaning cause-and-effect relationships, was then applied to the data. The results suggest that high food densities attract conspecifics, which then exert increased pressure on territory holders, forcing diminution of territory size. Changes in food density *do* affect territory size, but not so much in respect of changing the area needed to meet foraging needs as by changing the numbers of intruders and so the cost of maintenance.

One more observation on territory size is pertinent. In the black-capped chickadee (*Parus atricapillus*), territory size is quite large—nearly 2 ha—just before and at the time of nest building. Thereafter it drops precipitously to about .6 ha at the time of egg laying, .1 ha during incubation, and then gradually rises again to .8 ha at the time of fledging (Stefanski, 1967). Thus territory size is near its minimum at just the time when resources to feed the young are needed most. To some degree this may reflect a pulling in of the area vigorously "defended" because familiarity with neighbors by that time has reached a point where aggression is no longer needed to demarcate and maintain territorial boundaries. The matter is worth pursuing.

The implications of territoriality for population dynamics are the subject of some debate. Many authors have assumed that the act of staking claim to areas of less than freely compressible size (if area really reflects energetic needs) by breeding individuals limits the number of animals that can breed. This, then, holds the population below levels that might obtain were the species nonterritorial. The idea can be traced back as far as 1953, when Kluyver and Tinbergen remarked on the limited compressibility of titmouse territories with increased population density and noted that in high-population years, the birds spilled over from their preferred mixed forest habitat into inferior pine woods. In support, it has been shown quite clearly that territoriality prevents some individuals, sometimes a very large portion of the population, from mating. For example, Orians (1961b) found that if all territorial male redwings on a breeding marsh were shot, their places were taken immediately, within a day, by others. The same replacement by floating males occurs with removal of territory owners in the red grouse (*Lagopus lagopus*) (Watson and Jenkins, 1968). The gecko (*Gehyra variegata*) seems to be population limited by its territorial habit. In this species males are territorial all year beneath the loose bark of dead trees. There is only one male to a tree. Females will tolerate a bigamous or sometimes a trigamous relationship, but no more, chasing off additional would-be mates. Thus regardless of the abundance of food or the number of predators, the reproductive potential of this species seems locked inexorably to the number of dead trees with loose bark and the territorial behavior of its individuals (Bustard, 1970).

But there are a number of problems with the territory-population limitation concept. First it is (perhaps) applicable only to species with basically incompressible territories. As we have seen, not all territories share this attribute. Second, during breeding season it is usually the male whose territoriality limits the number of breeding males. Females can mate at will, unless they too exclude one another, as in the case of *Gehyra.* In polygynous birds, though, only one female (the first mated, or "primary" female) receives a significant amount of paternal aid in raising her young. Hence, unless most feeding is done on the territory, there is no advantage to mated females' expending the energy and suffering the risks of expelling newcomers. Indeed, even if all feeding *were* on the territory, the organized distribution of individuals over space induced by territoriality might well be expected to maximize food-gathering efficiency. Therefore, even though not all individuals reproduce, the total resource base may be used in the maximally efficient manner, resulting in a maximizing of reproductive output: Territoriality rather than limiting population size may actually raise the birthrate, leading to increased population density (see Brown, 1969). Brown also argues that the spilling over of titmice into pine woods described by Kluyver and Tinbergen should ideally not occur until such time as fitness in the preferred mixed forest falls from density feedback to a point where pine woods nesters in fact do *not* suffer a relative fitness loss (recall the ideal free distribution argument of Fretwell, Lucas, C. E. Taylor, and J. M. Emlen in Chapter 13). He then shows, using Kluyver and Tinbergen's data, that in keeping with this hypothesis, the exchange ratio (number of young hatched in habitat i settling in j/number hatched in j settling in i) is almost exactly 1 : 1. In this case, at least, territoriality may buffer fluctuations in population size, but it does not act as a population-limiting factor. Based on considerations of exploitation competition within a species, Brown's argument would seem to be a very general one. But it may, nevertheless, not always apply. Recall that the proximate factor mediating territory size may be aggression. And note that while we might expect the level of aggression to reflect reasonably accurately the ultimate cause (say food shortage) of the aggression, this expectation could be a bit naive. It is rather hard to believe that the eternal bickering among birds at a feeder leads to a net benefit to anyone; and there is some evidence to suggest that interference competition may be more important than exploitation competition within some species, and that "misplaced" aggression may be the cause of interspecific territoriality (see Chapter 17). Animals cannot be expected always to behave as they "should." For a review of territoriality as a population-limiting factor, see Klomp (1972).

Home Ranges

In its purest form a home range should be thought of not as a discrete area, like a territory, but rather as a statistical consequence of site attachment. Individuals tend to concentrate their movements in familiar surroundings and so, over a given period of time, appear to be wedded to a particular area. In fact, given enough time, home ranges seem to wander in space (Jewell, 1966). But home ranges, like Truth and ethanol, seldom occur in pure form. In pronghorns (*Antilocapra americana*), males in rutting season move about over loosely defined, overlapping home ranges, but actively "defend" a moving core area against all male comers (Bromley, 1969). Woodmice (*Peromyscus leucopus*) in southern Michigan have home ranges that overlap only very slightly (are these home ranges or territories?) for males or for females, but which allow slightly freer mixing of individuals of opposite sex (Metzgar, 1971). (Metzgar has suggested that "territoriality" acts to limit woodmouse populations.) The burrow entrance dirt mounds of the kangaroo rat (*Dipodomys spectabilis*) were found to be quite evenly dispersed in space in a New Mexico study (Schroder and Geluso, 1975), suggesting individual avoidance. Using radioactive gold implants to track harvest mice (*Reithrodontomys humulis*), Kaye (1961) discovered that his subjects did not use the whole home range area, but traversed paths between a small number of places, usually on the periphery, where most time was spent. Eastern chipmunks (*Tamias striatus*) have broadly overlapping home ranges, but manage to avoid encounters with each other by a system of spatial time sharing. Such a system could arise only through familiarity of individuals with each other's activity patterns. The social pattern, then, is admirably suited to avoiding fights with the neighbors and ensuring joint action to deter immigrants (Getty, 1981).

The advantages and disadvantages of the territorial aspects of home range are no different from those discussed above with respect to territoriality. It remains to ask why individuals are so often site-attached, why homing behavior is so well developed in some species. [As just one example, *Microtus californicus*, the California vole, has been shown capable of returning from a 600-ft distance, six times the diameter of its home range (Fisler, 1962).] One possible reason is that familiarity with the terrain best enables individuals to avoid predation (Davis and Emlen, 1948). Blair (1940a) noted that individuals of the prairie deermouse (*Peromyscus maniculatus*) released in their home range after capture appeared to know the terrain well, running directly to holes, often (several releases) by exactly the same path. Sometimes a released animal would initially appear lost, eventually orient itself, and dive for a hole. In 93% of all observations the hole chosen for refuge was in the animal's home range. Metzgar (1967) put woodmice, one at a time, into a 2.8 m by 1.4 m by 3.1 m, windowless test room with litter on the floor, and then introduced a screech owl. Mice with one day's exploration experience in the room were taken less often by the owl than those with no prior experience.

Another factor encouraging site attachment is social conflict. The argument can be illustrated with the use of voles (*Microtus*), which, as anyone who has worked with them can attest, treat strangers in rather savage fashion. An individual that confines its wanderings to a small area encounters only a few individuals. A small number of individuals can be handled without undue risk once familiarity leads to spatial time sharing, an established dominance hierarchy, or simply avoidance by subordinate animals. Larger areas involve the establishment of more extensive interindividual arrangements, with the increased concomitant risks, and also increases the rate at which strange newcomers are intercepted. It is patently dangerous to wander very far from home. On the other hand, there is a distinct advantage to expanding one's horizons, and the balance between these conflicting forces may play a major role in determining home range size. The advantage, which accrues mostly to males, is that a large home range will, in general, encompass more female home ranges. The picture that emerges from this simple conjecture is that (adult) males will have larger home ranges than females, that more aggressive dominant individuals will have larger home ranges than their meeker competitors, and that home ranges will expand during peak breeding periods. In fact, all of these predictions seem borne out, at least for small mammals. Male home ranges exceed female home ranges in size in *Peromyscus maniculatus*, both in woodland and prairie populations (Blair, 1940a, 1942), the eastern chipmunk (Blair, 1942), the European vole (*Apodemus sylvaticus*) (Chitty, 1937), montane and long-tailed voles (*Microtus montanus* and *M. longicauda*) (Jenkins, 1948), and the cotton rat (*Sigmodon hispidus*) (Erickson, 1949), to mention only a few (see also Brown, 1966). Home range is larger in the male than in the female for the lizard *Uta stansburiana*, as well (Tinkle, 1965). In small mammals, dominant males also are known to have the largest home ranges and to restrict the movements of subordinates (Brown, 1966). Finally, home ranges do, in fact, often increase in size during the breeding season (Brown, 1966; Jewell, 1966; Dunaway, 1968—but see Blair, 1951).

An alternative explanation for home range size is that proposed by McNab (1963) and presented above in the context of territoriality. McNab obtained data on home range sizes and body weights for a wide variety of mammals and found

$$\log_{10}(\text{home range size}) = 6.76 + .63 \log_{10}(\text{body weight})$$

The slope (which we should expect to be roughly 2/3) seems almost perfect. But Schoener (1968b), using data on predatory mammals, found a regression value of $1.4 \pm .16$, and Harestad and Bunnell (1979) give values of 1.02 for large mammalian herbivores, .92 for large mammalian omnivores, and 1.36 for large mammalian carnivores. Turner et al. (1969) find a value of .95 for lizards. These values are too big! On the other hand, might we not expect values that are "too big?" Because home ranges are not exclusive, a doubling of an individuals area does not yield twice the available food for that individual; now the area is shared with twice as many others. However, to the degree that parts of any home range are exclusive, the available food does nevertheless increase. We should therefore expect home range area to have to increase much faster than with the two-thirds power of an animal's body weight to meet its needs adequately. Harestad and Bunnell (1979) suggest, in addition, that food availability might not increase linearly with area foraged. Foods can be expected to occur in patches, larger consumers require larger patches for acceptable foraging efficiency, and the number of large patches simply does not follow size of foraging area linearly until that area becomes quite large.

One test of the opposing social stress versus energetics hypotheses can be made by comparing changes in home range size with population density. The social stress hypothesis would suggest contraction of home ranges during periods of high density due to increased, aggressive contacts; the energetics hypothesis would predict expansion due to increased local competition for food. Metzgar (1971) found no change in size with population density in the woodmouse, but these findings were in contradiction to those of Stickel (1960), who found a clear-cut decrease in home range size during times of high population density. White (1964) also found an inverse relationship between home range size and population density in this species. Similar findings were reported for the beach mouse (*Peromyscus polionotus*) (Blair, 1951), and for the harvest mouse (Dunaway, 1968). These data would appear to support the social stress hypothesis. However, this conclusion is no longer clear when we entertain the possibility—information unavailable—that high food supply might have accompanied the high population densities.

Social Groups

There are a number of reasons why grouping might raise the fitnesses of participating genotypes and thus be favored by natural selection. As with colonial living, but in exaggerated fashion, extra eyes, ears, and noses help to detect predators. Packing into a tight group may help avoid detection by predators, produce confusion in predators (as in erratically moving flocks of starlings), or provide a mass of bodies among which the selfish individual might gain protection through anonymity. In some cases, large species such as musk oxen can use their group structure to deter predators physically. A fuller discussion of the role of grouping in defense is provided in Chapter 16. "Local enhancement," the increased efficiency of foraging when operating in groups [see "Coloniality" (p. 385)], may also contribute to the value of group living. Predators feeding on large prey may also benefit from group cooperation in bringing down their quarry. Also, living in groups may convey an advantage in interspecific competition. For example, a pack of hyaenas can drive a lion from its kill, but is no match for a pride. Single hyaenas have no choice but to wait for a lion to leave. An excellent review of the advantages of living in groups may be found in Bertram (1978).

14.2 Mating Systems

We begin our discussion of social groupings with perhaps the simplest of all such assemblages, the breeding unit. We first explore the various kinds of breeding units, and under what conditions they are expected to occur. We then review the interrelationships between the nature of the breeding unit, sexual dimorphism, the sex ratio, and other life history characteristics.

The Breeding Unit

The breeding unit consists of the breeding animals and occasionally a number of other adults, subadults, or young who aid the parents in raising the latest batch of offspring. For the moment we shall ignore systems extended by "helpers" (see Chapter 15) and concentrate on the breeding individuals themselves.

Several kinds of breeding units (mating systems) occur. These are conveniently classified as shown in Table 14-4. Promiscuity may or may not involve mate choice. In oysters, for example, gametes are shed anonymously, but in at least many terrestrial vertebrates, individual preferences occur. Most mammals are promiscuous or polygynous, although some are monogamous; most passerine birds are monogamous, although many are polygynous and a few promiscuous or polyandrous. Most poikilothermic vertebrates are promiscuous, as are most invertebrates. Why? We have already commented (Chapter 11) on the improbability of parental care in species for which such activity accomplishes little. Beyond the provisioning of eggs with yolk or the protection of eggs (and sometimes early, hatched stages) in a marsupium of some sort, most invertebrates seem unlikely candidates for parental care. Exceptions are some of the predator species, such as wasps and spiders, or the social insects. Without parental care there is no advantage to social pair bonding between parental males and females (unless the male forces one or more females into a harem so as to guard his sexual investment). Indeed, with the exception of the circumstance noted above, unless both parents are somehow involved in caring for the young, there seems little reason for the female to remain with the male, and vice versa. The male's presence might well be detrimental either by his depleting needed food resources or attracting predators. Species in which paternal care of the young is of no potential use should be expected to be promiscuous (no pair bond) (or polygynous).

Zeveloff and Boyce (1980) argue that where a female's investment in her young *at birth* is small (the ratio of neonate weight to litter size is small or the gestation period is very short), more parental aid is needed during the period of parental care. Such species (rodents, lagomorphs, some carnivores) should therefore be monogamous. Species in which female investment is great by the time of birth (large offspring, long gestation period) require less aid from the father and so are more likely to be polygynous or promiscuous (primates, pinnipeds, some carnivores, ungulates.) If this reasoning seems a bit obscure, digress a step and ask why some fathers who are capable of contributing to their offsprings' welfare do so and others do not. A male that has only a small, cumulative parental investment (Chapter 11) in his offspring sacrifices minimally in the way of fitness if his offspring do without his help, and might maximize his reproductive contribution by curtailing his role as a parent in favor of inseminating other females. But a female that has invested considerably, through the risks of pregnancy and the energy drain of lactation, has much to lose if the young are abandoned. In mammals, this necessarily considerable investment by the mother assures a modicum of parental care, further lowering the cost of desertion to the male. Long gestation times and/or a high neonate weight/litter size ratio indicate particularly heavy investment of the female in her young and thus, indirectly, suggest stronger inducement for the male to leave or be distracted by other females. Hence in such cases, males are probably promiscuous or polygynous.

Of course, if the female could somehow arrange for a massive, early investment by the male, the tables would be turned. This seems a highly unlikely accomplishment for mammals, but occurs occasionally in birds. In such cases we often find, as expected, polyandry. An excellent discussion of this subject is given in S. T. Emlen and Oring (1977).

***Table 14-4* Types of Mating Systems**

Type	Subtype	Definition
Monogamy	Permanent	—
	Serial	—
Polygamy	Polygyny	One male, more than one female
	Polyandry	One female, more than one male
Promiscuity	—	Pair bond lasts only for the duration of copulations

The Verner–Willson–Orians Hypothesis

Whether both parents or only one remain to care for the young ultimately depends on cumulative parental investment and the enhancement of offspring fitness accruing from such care. But the level of investment and value of parental care are, in turn, functions of the parents' interaction with their environment. First, where polygyny is favored in the male (the male improves his reproductive contribution by gathering as many females as possible but gains either by offering some protection to them and their young, or by keeping them together, away from other males and the chance of multiple inseminations), it can, in fact, occur only if either the females somehow gain by membership in a harem or if the male has the wherewithal and the amenable environmental conditions to hold together his females and fend off other males. (For a discussion of this matter, see Kleiman, 1981.) Second, the nature of the environment, particularly with respect to forage availability, greatly affects the need for two as opposed to one parent aiding the young. The latter point was investigated by Verner (1964) and Verner and Willson (1966). Verner lists some of the advantages and disadvantages of polygyny, as opposed to monogamy, both to males and females (Table 14-5 is compressed and paraphrased from his work).

Table 14-5

Advantages of Polygyny	Disadvantages of Polygyny
To the Female	
Greater choice of mates and thus a better chance of mating with a genetically superior male	Less paternal aid in feeding the young and keeping off predators
Greater choice of territories and thus a better chance of raising young on a maximally suitable area	
To the Male	
More offspring produced (if clutch or litter size per female limits her contribution) if extra females can be inseminated	More time, energy, and perhaps risk (time vulnerable to predators) spent in courtship
	More time and energy needed for territorial defense
	Greater risk of cuckoldry by other males

In mammals the female will almost invariably have a greater, cumulative parental investment than the male at the time of her offspring's birth. Thus she has more to lose if the genetic quality of her mate, his territory, or home range (if she raises her young there) are of poor quality. She also stands to lose more than he if he deserts. She therefore gains more by breeding with a superior male in a superior area and, if possible, with one she can trust to perform at least a modicum of parental care (though see Gowaty, 1983). In short, it is very much in the interest of the female to choose her mate carefully. The male, with considerably less at stake, may on balance, contribute more genes to future generations simply by passing on as many genes to as many females as possible. Hence he is likely to be less particular in his choice of mates. Females, in a sense, become a commodity in short supply, a commodity for which males must compete. In Table 14-5, then, we should concentrate our attention on the advantages and disadvantages to the female, particularly with respect to the advantages and disadvantages of polygyny. In general, the same will be true of birds. In fish, where gestation time is nil (most species), cumulative parental investment will be more nearly equal at spawning, so parental care is less likely to fall automatically to the female. If the male has expended time in building a nest as part of a courtship procedure, he may very well have the most invested and can be expected to take over the parental responsibility (sticklebacks, for example). Generally, no cooperative care is found, so no pair bond is formed and Table 14-5 is inapplicable.

Under what circumstances might the net advantage to the female favor polygyny? Verner (1964) and Verner and Willson (1966) reasoned that high food density would make offspring care easier for one parent and thus minimize the loss due to lack of paternal help. Thus species living in highly productive areas should more often be polygynous or promiscuous. In support of this argument, Verner and Willson (1966) noted that biological productivity per unit volume in North America was generally highest in marshes, lower in fields and savannas, and lowest in forests. Presuming that this "productivity density" is an appropriate measure of food availability, polygyny and promiscuity should be relatively most prevalent among marsh-breeding birds and most uncommon in forests. Of 291 passerine species in North America, they describe 14 as being polygynous or promiscuous. Of these, 13 are marsh breeders, the fourteenth a savanna breeder. Further support comes from the observation that the wren *Troglodytes troglodytes* is monogamous where it nests on food-sparse St. Kilda Island and polygynous where it nests in food-abundant English gardens (Armstrong, 1955). Among 23 field-nesting species of African weaver finches (Ploceidae), most are polygynous or promiscuous, while of 22 forest-nesting species, most are monogamous (Crook, 1964, 1965). Furthermore, it seems likely that seed-eating species have a higher abundance of food available to them and, because seeds often form concentrated wind rows, social foraging might be expected (Chapter 16), making food discovery quicker. Feeding on concentrated food sources also would lower foraging costs. In fact, of 21 insectivorous weaver finch species, 19 nested and foraged solitarily, while of 30 seed-eating species, all 30 nested colonially or in closely adjacent territories and foraged in flocks. Sixteen of the 19 insectivorous birds were monogamous and almost all of the seed-eaters were polygynous (Crook, 1965). One glaring exception to the Verner–Willson expectation is the quelea (*Quelea quelea*), a field-nesting, seed-eating species which is monogamous. The apparently aberrant behavior may be explainable by the fact that the birds form massive breeding colonies. This results in an exaggerated depletion of food in the vicinity of the colony and necessitates foraging at considerable distances, thus greatly increasing the importance of aid, both in nest sitting and in foraging, from the male (Davis, 1952; Orians, 1961b).

In spite of its apparent success in prediction, there is some reason to be skeptical of the Verner–Willson hypothesis without some modifications. First, we have seen that the number of young produced is quite consistently and tightly related to food supply in both birds and mammals. Therefore, if productivity density is high, we should expect expansion of family size to the point where food is no longer effectively overabundant; the adjustment is likely to be in the number of young produced rather than a change in the mating system (recall Ashmole's and Ricklefs' arguments on clutch size in Chapter 11). Second in species population-limited by food abundant food promotes population growth, populations adjust to their food supply. One way out of this dilemma is to suppose (as seems quite likely true) that, at least for migratory birds, population density is mediated by winter conditions. Then summer food supply will indeed be (partly) independent of the breeding population density. This argument will not remove the first objection, though. Another possibility is that populations are limited by something other than food. The first objection is not addressed by this possibility either. Still another possibility is that factors other than productivity (such as protective cover or potential nest sites) are also important in determining mating system and that these factors may be correlated with average productivity density in an area. [For a discussion of variables used by females in choosing a territory and its relevance to this point, see Zimmerman (1966), Verner and Engelson (1970), or Pleszczynska (1978).] Finally, an idea briefly discussed by Verner and Willson (1966) and developed by Orians (1970) is that polygyny may be the result of *variance* in territory quality either in productivity or other attributes. Suppose that we plot the expected reproductive success (number of surviving offspring per female) against the quality of the territory (defined in terms of productivity or whatever factor limits the successful generation of offspring) (Fig. 14-3)

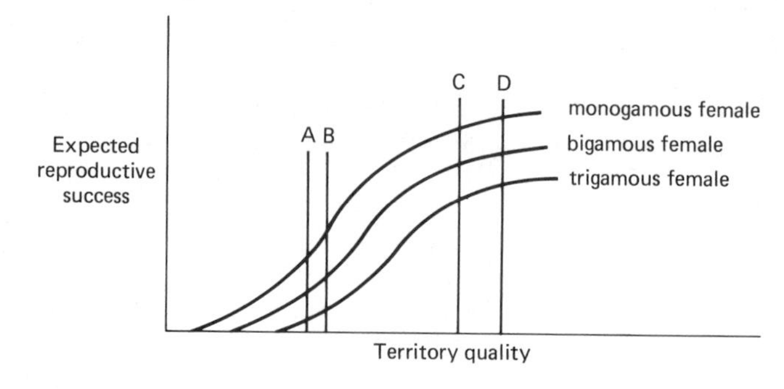

Figure 14-3 Plot of the expected reproductive success (number of successfully raised young) for monogamous, bigamous, and trigamous females as a function of the quality of the territory on which they are nesting.

[Altmann et al. (1977), suggest using reproductive success of the primary female for the abscissa, to avoid possible circularity.] Obviously, on territories (or home ranges) of higher quality, expected reproductive success is high. But also, as the number of females sharing the resident male's parental aid increases, reproductive success drops. Now consider a breeding area for which territory quality among males varies only slightly (A and B, or C and D in Fig. 14-3). Picture a female wandering about or flying over this area and assessing her likely reproductive contribution. A female choosing between males A and B should always favor B if both A and B are unmated, or if both are already mated to the same number of females. If B is mated, our female may choose between monogamy with A or bigamy with B. In the case of males A and B, or C and D, the female would always be better off choosing the unmated male. But suppose that the breeding area affords the female a much greater choice in territory quality. Examination of Fig. 14-3 will show that the expected reproductive success of a female mating bigamously with male C or even trigamously with male D exceeds that of a female mated monogamously to male A or B. Species living in areas in which territory quality is apt to be highly variable are more likely to be polygynous than are species living in more homogeneous areas.

The Orians hypothesis, as stated, is quite straightforward. But its power to provide predictions may be less than appears at first glance. It is tempting for the human observer to try to assess a territory's quality on the basis of its food productivity. But of course territory quality may mean something quite different to a bird. It may be, to some extent, a function of food. It may also be a function of the number of nest sites, the amount of protective cover, or the number of perch sites. Furthermore, even if food were known to be the overriding concern in determining quality, two territories with very different productivity might be rendered effectively almost identical if most feeding were carried out off-territory.

To the extent that food production is the important determiner of quality, and to the extent that enough feeding occurs on territory for quality differences to be of significance, the data supporting the earlier version of the Verner–Willson hypothesis are also consistent with Orians's modification. Marshes offer the possibility of extremely high variance in productivity among territories. Not only is there great heterogeneity within a marsh, but by virtue of the limited extent of marshes, some males, if all were to attempt holding territories, would be forced to poor-quality habitat *off* the marsh, making the variance in quality of territories even higher. Fields and savannas, where wind rows form and seed types and plants suitable for perching or nest building are clumped, are also quite spatially heterogeneous. Forests are relatively homogeneous.

One bit of supporting evidence for Orians's hypothesis, although a bit weak, is that older, more experienced males, who might be expected to acquire the better quality territories, do in fact have larger numbers of mates (see, e.g., Willson, 1966; Carey and Nolan, 1975). Finally, if Orians's hypothesis is correct, and if first- and last-mated females share equally in male help at nesting, reproductive success of females should be invariant with mated status (Altmann et al., 1977): A female that could expect to do better by being with another male should ideally have chosen him in the first place. If she had already mated with a male, she should choose to peacefully accept another female, and that second female should choose to join her, only if that second female cannot do better elsewhere. Male reproductive success should rise proportionately with the number of mates. Although this expectation is not always found (see below), it is realized quite exactly in the indigo bunting (*Passerina cyanea*) nesting in southern Indiana. Carey and Nolan (1975) found the number of young leaving the nest, per female, to be 1.5 for bigamous, 1.6 for monogamous females. For males the appropriate figures were 3.0 and 1.6.

But failure to find equal reproductive success for all females, both primary and secondary, does not constitute a violation of Orians's hypothesis; the ideal situation of each female equally sharing the aid of her male is seldom realized. Generally, the male lavishes the bulk of his attention on the *primary* female and her clutch. Thus we should really expect the reproductive success of *last*-mated females to be the same across all breeding units. This last observation should, in turn, cause a positive relationship between breeding unit size and number of chicks fledged per nest, a relationship which, in fact, has been documented (Goddard and Board, 1967; Haigh, 1968). An alternative explanation for this positive correlation, female cooperation, has been suggested by Altmann et al. (1977).

Mate choice does not usually lie entirely with the female. For example, the harems found among some species of pinnipeds (Bartholomew, 1970), cervids and bovids (Jarman, 1974), and the Hamadryas baboon (Kummer, 1968) probably reflect the ability of males to protect their own sexual investment by excluding other males from access to a group of females.

A more troublesome, although not common thorn in the side of Orians's hypothesis is the occasional possession of multiple territories by a single male. In the pied flycatcher (*Ficedula hypoleuca*) (Alatalo et al., 1981), males may hold more than one, nonadjacent territory. This presents a problem for a female choosing a mate because she has no way of knowing, a priori, whether a male is already mated.

We leave the discussion of Orians's model with the observation that kinship among mates of one sex in caring for the offspring of their mate may have strong potential for affecting the nature of the mating system. In the Tasmanian native hen (*Tribonyx mortierii*), the female may have one or two male mates (polyandry). Maynard Smith and Ridpath (1972) report that two males sharing their female produce roughly 1.26 or more times the number of young they might raise if singly mated. Thus the males produce only about .63 times the number of young per male as they could if mongamous. So why are they polyandrous? We take Maynard Smith and Ridpath's approach here, but argue somewhat differently. Suppose that a gene d is effectively fixed in the population and that individuals of genotype dd are passive and willing to share their female with a second male. We now introduce a gene D into the population, and suppose that this gene induces its owner (almost invariably heterozygous because the gene is very rare) to fight for the opportunity to be monogamous. A loser in such a fight must either find a new mate or fail to reproduce. Let n_1, n_2, and n_0 be the expected number of offspring produced by monogamous, bigamous, and displaced males. We now consider two scenarios. First, suppose that a D male will fight only if he knows he can win. Then, in a trait group of one female and two males, one of which carries D, the expected gain in fitness to the D gene arising from a fight is

$$\Delta W_D = \text{(contribution of the } D \text{ carrier if he alone mates)} - \text{(his contribution if he shares a female)}$$

$$= 1(n_1) - 1[\tfrac{1}{2}(1 + r)n_2] \tag{14-6}$$

where r is the coefficient of relationship between the two males. The D gene therefore fails to spread—that is, the established gene is not successfully invaded by the new form—if $\Delta W_D < 0$, or

$$n_1 < \frac{(1 + r)n_2}{2} \tag{14-7}$$

Suppose now that the D male has no information as to who will win the fight. Then

$$\Delta W_D = \frac{1}{2}\left[n_1 - \frac{(1 + r)n_2}{2}\right] + \frac{1}{2}\left[n_0 - \frac{(1 + r)n_2}{2}\right] = \frac{n_1 + n_0}{2} - \frac{(1 + r)n_2}{2} \qquad (14\text{-}8)$$

and d maintains its position if

$$n_1 < (1 + r)n_2 - n_0 \qquad (14\text{-}9)$$

If n_0 is presumed to approach zero, conditions (14-7) and (14-9) can be rewritten

$$\frac{n_1}{n_2} < \frac{1 + r}{2}$$
$$\frac{n_1}{n_2} < 1 + r \qquad (14\text{-}10)$$

But n_1/n_2 has been given as approximately 1.3. Thus bigamy can be expected to persist, in the two scenarios, if

$$1.3 < \frac{1 + r}{2} \Rightarrow r > 1.6 \qquad \text{(impossible)}$$

or if

$$1.3 < 1 + r \Rightarrow r > .3$$

Records indicated that of eight two-male breeding units where male family background was known, six involved brothers ($r = \frac{1}{2}$) and one involved a father and son ($r = \frac{1}{2}$).

Maynard Smith (1977) has expanded the Verner–Willson–Orians hypothesis. He emphasizes the role of reproductive timing. If the breeding season is very short, a male that ceases paternal care in order to acquire another female may find all other females already inseminated. His advantage to deserting the previous female is thereby lost and his behavior will only compromise the survival of the young he has already fathered. To the extent that this consideration plays a role in the evolution of breeding behavior, it is clear the sex ratio could be important in determining mating system.

Male Quality

To what degree does territory quality in the Orians model measure environmental attributes and to what degree the characteristics of the male himself? Weatherhead and Robertson (1979) argue if a male is attractive to a female, independent of his ability to gain and maintain a quality territory, the offspring of a female mated to him may inherit his attractiveness and so better attract females as well. Thus, once a male trait, for whatever reason, becomes attractive to females, it will be selectively advantageous for females to continue choosing males on this basis because they will thereby produce "sexy sons." Because it is likely that sex appeal will be related to whatever encourages male prowess at intimidating other males and acquiring a good territory, the impact of female selection of male quality will lower the "polygyny threshold." That is, the critical minimum variance in territory quality necessary to give rise to polygyny in a particular area will decrease. Heisler (1981) restates the sexy son hypothesis and provides a simple mathematical model which shows that with the proper values of sex ratio, and genetic similarity of father and son (regarding mating propensity), females ought even to tolerate a short-term loss (via current year's reproductive success) because of overcompensating gains via grandchild production. Wittenberger (1981) provides data bearing on this point. Claiming that Orians's model works, in the absence of a sexy son hypothesis, only if the ratio of reproductive success of secondary females to monogamous females equals or exceeds 1.0 (actually, at equilibrium dispersion of females on the breeding grounds, it should be equal to 1.0), he measured this ratio in six species for which data were available. The values are 1.58 for prairie warbler (*Dendroica discolor*), 1.00 for indigo buntings, .98 for lark buntings, .99 for great reed warblers, and .93 for ipswich sparrows (*Passerculus princeps*). All but one of these values are on the order of 1.0, supporting Orians's hypothesis and discrediting the sexy son hypothesis. However, the value for bobolinks (*Dolichonyx oryzivorus*) is only about .69. At least in this last species something more than territory quality might be affecting female mate choice.

A noble effort to assess the role of male quality in female choice has been made for the redwing blackbird, (*Agelaius phoeniceus*) by Yasukawa (1981). Working on a southern Indiana marsh, Yasukawa defined a number of purely arbitrary "territories"—let us call them "cells"—whose locations were unrelated to the positions of the real territories. He reasoned that if territory quality were the only factor influencing mate choice by females, parameters measuring nesting behavior and reproductive success should be independent of male turnover on these randomly chosen cells. Several years of intensive data collection, however, showed a significant rise in the among-year correlation of first-nest completion dates (relative to mean value over the years) with proportion of males returning to the cell. Inasmuch as neither the average date of first-nest completion nor the average harem size for the marsh as a whole varied significantly among years, these observations are not likely due to changes in cell attractiveness among years. Thus something other than territory quality seems to be important to the females. What is it? Yasukawa found that harem size was positively correlated with age of male, number of years the male bred previously, male wing length, and the amount of time the male spent courting. Harem size was *not* related to aggression (for definitions and measures of aggression, see Yasukawa, 1978, 1979), or time spent singing, defending, or foraging on the territory. When he tested for the importance of male age, for males with no previous reproductive experience on the marsh, that correlation disappeared. A strong positive correlation between wing length and previous experience suggested that success in territory establishment is related to physical size. But is it experience or wing size that is important in attracting females? If only males with close to the modal wing length are included in the analysis, harem size and experience are still significantly correlated. If experience is controlled in the analysis, wing size is still important. Looking at experienced males only, and controlling for wing length as well, harem size was positively correlated with courtship time. There is little question that in the redwing blackbird, females choose for male quality as well as territory quality.

The impact of mating system on sexual dimorphism may be considerable. Most species with parental care and pair bonds demonstrate predominantly female mate choice. Thus it can be expected that males experience more intense competition for mates than females, and so more rapidly and more noticeably evolve traits serving to attract the opposite sex or intimidate others of their own sex. In birds, for example, males are generally larger and more aggressive than females. This situation is reversed in polyandrous species, where we might expect the opposite, and in a few others (raptors, for example). Sexual dimorphism is likely to be particularly pronounced in polygynous species. In general, the males of species with color vision (which excludes most mammals) might be more brightly colored than their mates. And where bright colors provoke aggression (as they would be expected to do if they were used for intimidation or augmentation of agonistic displays), and serve a useful function primarily in the breeding season, they might be expected to fade or disappear in the off-season. In polygynous species, where females enjoy a wide choice of males, it can be argued that young, inexperienced males will seldom breed. In fact, quite aside from the young male's ability to garner a good territory, he may lose from the fact that females will maximize their chances of producing young with genes favoring viability if they prefer older males (Halliday, 1978). Because adult plumage might provoke aggression from other males, the young male may well do best by delaying appearance of adult plumage (Selander, 1965). Thus in many bird species, the male, although physiologically mature at 1 year of age, may still look like a juvenile. Rohwer et al. (1980) put the matter in a slightly different perspective: Noting that the yearling males really resemble females more than juveniles, and that the phenomenon is widespread among monogamous species as well (where Selander's argument is weak if applicable at all), they suggest that delayed maturation does not serve to deflect aggression per se, but rather allows the male to sneak past the adult males without recognition as a competitor, and to gain familiarity with the terrain in preparation for the next year's territorial wars. Also, the chance for a bit of cuckoldry could not help but add a bit to the adaptive value of drab feathers.

But the genetic changes favoring traits helpful in gaining a mate will not necessarily be those minimizing mortality. In fact, it is highly unlikely that characteristics favored by natural selection for one purpose will be anything but disruptive of another (but see the discussion of genetic covariance of life history traits in Chapter 12). Large antlers and huge, brightly colored tail coverts can hardly be thought of as energy efficient or of survival value against predators or the elements. A case in point is the boat-tailed grackle [*Quiscalus* (*Cassidix*) *mexicanus*]. In this species the male displays a very large tail which presumably serves to increase his likelihood of mating but, to put it a bit facetiously, serves also to get him blown away whenever a wind comes up. Selander (1965) reports that the nestling sex ratio in this species is about 1 : 1, but that the following year, just before breeding, differential male mortality has given females the edge 1.42 : 1.0. Howe (1977), looking at the common grackle (*Quiscalus quiscala*), reports a 1 : 1 sex ratio at hatching but already by fledging, a skewed ratio favoring females 1.65 : 1.0. Searcy and Yasukawa (1981) found a significant ($p < .05$) negative correlation between ratio of male survival to female survival and ratio of male size to female size for North American Icterids (blackbirds). Also, both J. M. Emlen (1973, p. 51) and Zahavi (1975) have suggested that females may be attracted to males with a "handicap," who demonstrate by successfully dealing with that disadvantage just how robust they are.

It has been argued that lower viability in men as opposed to women occurs because in human beings the male is the heterogametic sex, carrying (effectively) only one sex chromosome. Deleterious genes appearing on the X chromosome can thus more easily find expression in men. But this hypothesis is weak. In mammals and birds in general the male is less viable than the female, and in birds it is the female that is heterogametic. The sexual differences seem more likely to be the result of sexual selection acting more strongly on males.

The Sex Ratio

The discussion above leads inexorably to our next topic, the evolution of the sex ratio. The first person successfully to tackle this issue was R. A. Fisher (1958). Unfortunately, Fisher's argument is a bit obscure. Therefore, we strike out on our own, although in the same vein. If $M_i(t)$ is the number of males of age i in the population, the proportional genetic contribution of an age x male at time $t + x$ is

$$\frac{l_{mx}b'_{mx}}{\sum_i M_i(t + x)b'_{mi}}$$

where the subscript m denotes a male-specific value and l_x and b'_x, as throughout this book, denotes survivorship and fecundity at age x. Let $C_m(t + x)$ be the total number of genes at a given locus passed by *all* males in the population at $t + x$. Then the genetic contribution of one male, at age x, is

$$\frac{l_{mx}b'_{mx}}{\sum_i M_i(t + x)b'_{mi}}\, C_m(t + x)$$

and its lifetime contribution, discounted to account for its proportionately increasing or decreasing role as the population shrinks or grows, is

$$\psi_m = \sum_x \lambda^{-x} \frac{l_{mx}b'_{mx}}{\sum_i M_i(t+x)b_{mi}} C_m(t+x) \tag{14-11}$$

where λ is the population growth multiple per unit time. For a female the corresponding value is

$$\psi_f = \sum_x \lambda^{-x} \frac{l_{fx}b'_{fx}}{\sum_i F_i(t+x)b'_{fi}} C_f(t+x) \tag{14-12}$$

But in all dioeceous diploid species, each individual must receive exactly half its genome from its mother, half from its father. It follows directly that in any given breeding period the total genetic contribution by all males must equal that by all females. That is,

$$C_m(t+x) = C_f(t+x) = C(t+x)$$

The value of $C(t+x)$ is simply

$$\sum_i n_i(t+x)b'_i \tag{14-13}$$

Thus at stable age distribution, (14-11) becomes

$$\psi_m = \sum_x \left[\lambda^{-x} l_{mx} b'_{mx} \frac{\sum_i n_i(t+x)b'_i}{\sum_i M_i(t+x)b'_{mi}} \right]$$

$$= \sum_x \left[\lambda^{-x} l_{mx} b'_{mx} \frac{\sum_i n_0(t)\lambda^{x-i} l_i b'_i}{\sum_i M_0(t)\lambda^{x-i} l_{mi} b'_{mi}} \right]$$

$$= \frac{n_0}{M_0} \frac{\left(\sum_x \lambda^{-x} l_{mx} b'_{mx}\right)(1.0)}{\left(\sum_i \lambda^{-i} l_{mi} b'_{mi}\right)} = \frac{n_0}{M_0} \tag{14-14a}$$

ψ_f can be found, in similar fashion, to be

$$\frac{n_0}{F_0} \tag{14-14b}$$

(see also Goodman, 1982).

Now consider a new, mutant genotype which, when paired, produces m_0 and f_0 newborn males and females. The contribution to the ancestry of future generations by a breeding pair carrying the gene must be

$$\psi = m_0\psi_m + f_0\psi_f = n_0\left(\frac{m_0}{M_0} + \frac{f_0}{F_0}\right) \tag{14-15}$$

A simple, verbal argument can now be made. Suppose that $M_0 < F_0$, so that $\psi_m > \psi_f$. Then, in the sense of genetic contribution, a male is "worth more" than a female. Any genotype disposing a relative production of males greater than that by the average pair will be favored by natural selection. As a result, M_0 rises, and continues to do so until it equals F_0, at which point $\psi_m = \psi_f$. A 1 : 1 sex ratio has evolved.

But a breeding pair expends only so much reproductive effort. Suppose that the effort is sufficient to raise one son but that females are enough less costly that for the same price the parents could raise three. Then if the sex ratio is 1:1, so that males and females are worth the same, surely females will be favored and the sex ratio will become skewed in their favor. For situations such as this we must introduce an "energetics" constraint. Suppose that the "energy" expended on all successfully raised offspring by a pair is E, and that it costs e_m or e_f to successfully raise one son or daughter, respectively. Denote by p the age of independence. Then

$$E = e_m m_0 l_{mp} + e_f f_0 l_{fp} \tag{14-16}$$

solving for f_0 and substituting into (14-15) gives us

$$\psi = n_0 \left[\frac{m_0}{M_0} + \frac{(E - e_m m_0 l_{mp})/e_f l_{fp}}{F_0} \right]$$

$$= n_0 \left[m_0 \left(\frac{1}{M_0} - \frac{e_m l_{mp}/e_f l_{fp}}{F_0} \right) + \frac{E}{F_0 e_f l_{fp}} \right] \tag{14-17}$$

If the parenthetical term is positive (M_0 sufficiently small relative to F_0), those genotypes making the greatest genetic contribution are the ones that produce disproportionate numbers of males (large m_0). Accordingly, these genotypes become progressively more common, raising the M_0/F_0 ratio. This will continue until the parenthetical term is no longer positive. Equilibrium is reached when that parenthetical term is zero, and when $m_0/f_0 \to M_0/F_0$:

$$\frac{1}{M_0} = \frac{e_m l_{mp}/e_f l_{fp}}{F_0}$$

or

$$e_f f_0 l_{fp} = e_m m_0 l_{mp} \tag{14-18}$$

or

$$e_f f_p = e_m m_p \tag{14-19}$$

This is Fisher's general conclusion: that the sex ratio should be adjusted in such a manner that the "total parental expenditure incurred in respect of children of each sex (successfully raised to age p) shall be equal."

Simpler models concerning semelparous species have been derived (see, in particular, MacArthur, 1965). Also, for the skeptical geneticist, Leigh (1970) and Eshel (1975) provide simple, one locus–two allele models for the evolution of the sex ratio in non-age-structured populations (see also Uyenoyama and Bengtsson, 1979). In all cases, the general conclusion is that where the sexes are equally expensive to raise, the sex ratio at age p tends toward 1 : 1. Alternative models for sex ratio evolution in age-structured populations are offered by Charnov (1975) and Charlesworth (1977).

The models of MacArthur and Charnov, as well as the one presented here, predict that selection should favor the "less expensive" sex. In most terrestrial vertebrates with parental care, males are both more active and more rapidly growing than females. Therefore, they generally require more parental energy output to raise and, accordingly, should be less numerous at age p. The data from Selander and Howe mentioned above certainly support that contention for grackles. In human beings, also, the sex ratio favors females from early adulthood on; higher male mortality requires that to meet this ratio, males must be relatively common at birth (see, e.g., Hunt et al., 1965). In the oropendola of Central and South America the males are roughly twice the size of females at fledging. The observed ratio of females to males is between 5 and 10 : 1 (N. G. Smith, 1968). The cost discrepancy and thus the preponderance of females at independence should be pronounced in polygynous and promiscuous species, where the male experiences intense sexual selection.

Consideration of individual differences may provide the material for further predictions. For example, in polygynous species only the particularly robust males mate. Therefore, males born during bad times, which are likely to grow up weak, may fail to reproduce. Thus females should be favored in bad times, males in good. Offspring of healthy parents are apt to be better cared for and so grow up healthy and robust themselves. Strong, healthy, and experienced (i.e., older) parents should favor male production, especially in polygynous species. Of course, very old parents are likely to be in poor health and to produce defective or weak offspring (Chapter 12), and so should concentrate on producing females (Trivers and Willard, 1973). Trivers and Willard cite evidence of a decrease in male/female ratio with parity. They also note that deer produce fewer males under adverse conditions. Finally, larger litter size (suggesting greater energy stress and so inferior parental care per offspring) is generally correlated with a low male/female ratio. Of course, if large litter size is a consequence of abundant resources, as was argued in Chapter 11, this final observation is not applicable as evidence for Trivers and Willard's hypothesis (see also Myers, 1978).

In laboratory-raised eastern wood rats (*Neotoma floridana*) male and female offspring are equally expensive to produce. But, in the field, males disperse greater distances and can be expected to exhibit greater variance in reproductive success. Thus in times of food scarcity, poor nutrition should lower the likelihood of male success and swing the balance of selective advantage toward greater production of females. In laboratory experiments this species normally partitioned lactation energy equally among male and female offspring, but significantly biased that investment in favor of females when food was severely restricted (McClure, 1981).

Patterson and Emlen (1979) suggested that if male attributes affecting mating success are heritable, then dominant males, males with large harems, should raise disproportionate numbers of male offspring. Perhaps because females are the heterogametic sex, this expectation was not realized in the population of yellow-headed blackbirds (*Xanthocephalus xanthocephalus*) they looked at.

The possible effects of trait group selection on the sex ratio have been discussed by Wilson and Colwell (1981). These workers suggest that because one limiting factor on the number of offspring produced is the number of females present and breeding, but not the number of males, breeding groups with the sex ratio skewed in favor of females might be more productive and give rise to more dispersing individuals. In such a way (in nonmonogamously mating species) the sex ratio might evolve toward more female representation than expected on the basis of Fisher's argument. Toro (1982), using a simple kin selection model involving two alleles at a single locus, has shown that differential altruism and selfishness among sexes in the family will give rise to deviations from the Fisherian ratio (see also Bulmer and Taylor, 1980a; P. D. Taylor, 1981). For further ideas and observations on the sex ratio, see Clark, 1978; Bull and Vogt, 1979, Charnov, 1982).

Although we shall not delve into the subject here, other species with other sex-determination mechanisms, such as haplodiploidy, have also attracted a great deal of study. For the theory of sex ratio in such species, see Hamilton (1967), Oster et al. (1977), Charnov (1978), and Craig (1980).

In 1969, Ghiselin suggested that if one sex made massively greater contributions than the other either early or late in life, selection might favor a sex change. If, for example, a species displayed indeterminate growth and egg production increased markedly with size, a genotype that started out male and then at some point switched to become female might be favored. Such a change with age, or size, from male to female, does indeed occur in many species, usually those with indeterminate growth. The phenomenon is known as *protandry*, and is a form of *sequential hermaphroditism*. Species which are very strongly polygynous and which are *not* indeterminate growers might be expected to begin female and later change to male. Such *protogyny* is often found among reef fishes with fiercely maintained dominance hierarchies, in which only the most dominant males mate. These ideas were further developed by Warner (1975), who showed that it is *always* advantageous to switch sex if b'_x increases with age in one sex and, if in both sexes, then at different rates. He speculated that sequential hermaphroditism was not more widespread in nature because there often existed a prohibitive cost (in energy, time, or risk) to switching. Charnov and Bull (1977) argued that where parents lay eggs or produce young in environmental patches that differ greatly with respect to the relative advantages of being male or female, it pays to be able to change one's sex at will. Indeed, such facultative sex changes seem to be rather widespread, particularly among reef fishes. Fishelson (1970) kept a group of *Anthias squamipennis*, comprised of 10 to 20 females and one or two males, in an aquarium. When the males eventually died, he was amazed to find that one of the females, over the course of about 2 weeks, had become a male. When this fish was subsequently removed, another female changed and took its place as a male. In nature, in the Gulf of Aqaba, these fish form stationary groups of hundreds or thousands of individuals on the coral; 80 to 90% are subadult males or adult females, and a very few are active, territorial males. The ability to change sex may be an adaptation to high mortality of territorial males in a social system of pronounced dominance (see also Shapiro, 1980). *Labroides dimidiatus*, another reef fish, lives in groups of three to six females and one territorial male. Robertson (1972) watched several groups of these fish for about 2 years and found that the male was the oldest, largest, and most dominant member of the group. The older, larger female is invariably dominant to the other females and may herself defend a territory within that of the male. If the male is removed, this dominant female changes sex and takes his place. Occasionally, two large females will change, after which one will leave. If, during the change, a neighboring male takes over the missing male's territory, the changing female will stop her transformation and revert to being female. For extensive information on sex ratios and sex reversals in marine organisms, see Wenner (1972).

A particularly elegant study of sex change, in pandalid shrimp, was carried out by Charnov (1979). Return for a moment to the expression

$$\psi \propto \frac{m_0}{M_0} + \frac{f_0}{F_0}$$

As used above, the parameters represent, specifically, numbers of newborn. But the basic concept behind the expression is that a particular genotype contributes a fraction m_0/M_0 of the total male contribution, f_0/F_0 of the female contribution. Thus, in protandrous hermaphrodites, such as the pandalid shrimp, we can write

$$\psi \propto \frac{m}{M} + \frac{f}{F} \tag{14-20}$$

where the m/M stands for the fraction of total male contribution before sex reversal, and f/F the fractional female contribution after reversal. Let T be the age of reversal or any other parameter relating to reversal, and fix ψ at some constant value. Then, letting $M = M' + m$ and $F = F' + f$, where the prime denotes contribution by all *other* genotypes,

$$0 = \frac{d\psi}{dT} = \frac{\partial\psi}{\partial m}\frac{dm}{dT} + \frac{\partial\psi}{\partial f}\frac{df}{dT} = \frac{M'}{M^2}\frac{dm}{dT} + \frac{F'}{F^2}\frac{df}{dT}$$

so that $dm/df = F'm^2/M^2/M'F^2$. But at equilibrium $M/F \to M'/F' \to m/f$. Hence

$$\frac{dm}{df} \to -\frac{m}{f} \tag{14-21}$$

describes the relative values of m and f for which ψ takes a constant value. Equation (14-21) has a solution,

$$\ln m + \ln f = \ln (mf) = \ln c$$

or

$$mf = c \tag{14-22}$$

Because a given value of mf defines a given value of ψ and because a rise in mf clearly increases ψ, it follows that ψ is maximized when mf is maximized. The optimal age of sex reversal is that which maximizes the product of male and female contribution.

In pandalid shrimp individuals begin as males and switch to become females either before or immediately after first reproduction (age α). Let us define the probability of switching *before* first reproduction to be P and let $f(x)$ and $m(x)$ be female and male reproductive contribution, respectively, at age x. Then the expected male contribution of a shrimp is

$$m = (1 - P)m(\alpha)$$

and the expected female contribution is

$$f = Pf(x \geqslant \alpha) + (1 - P)f(x \geqslant \alpha + 1)$$

where $f(x \geqslant \alpha)$ is the female contribution at age α *plus* subsequent ages. To find the value of P that maximizes mf, differentiate:

$$0 = \frac{d(mf)}{dP} = (1 - 2P)m(\alpha)f(x \geqslant \alpha) - 2(1 - P)m(\alpha)f(x \geqslant \alpha + 1)$$

$$\propto (1 - P)(1 - 2Q) - P \qquad \text{so} \qquad \frac{P}{1 - P} = 1 - 2Q \tag{14-23}$$

where $Q = f(x \geqslant \alpha + 1)/f(x \geqslant \alpha)$. The value Q is

$$Q = \frac{1/l_\alpha(l_{\alpha+1}b'_{\alpha+1} + l_{\alpha+2}b'_{\alpha+2} + \cdots)}{1/l_\alpha(b'_\alpha + l_{\alpha+1}b'_{\alpha+1} + l_{\alpha+2}b'_{\alpha+2} + \cdots)} \tag{14-24}$$

where l and b' are survivorship and fecundity as usual. At this point Charnov gleaned appropriate data from the literature for mortality and fecundity values and presented justification for extrapolating age-specific values by simple formulas:

$$\text{Number of eggs } aL^C, : L = \text{ body length,}$$

L follows the Von Bertalanffy growth equation $L_t = L_{\max} - (L_{\max} - L_0)e^{-kt}$ where k, a fitted constant mortality rate is invariant with age for $x \geqslant \alpha$.

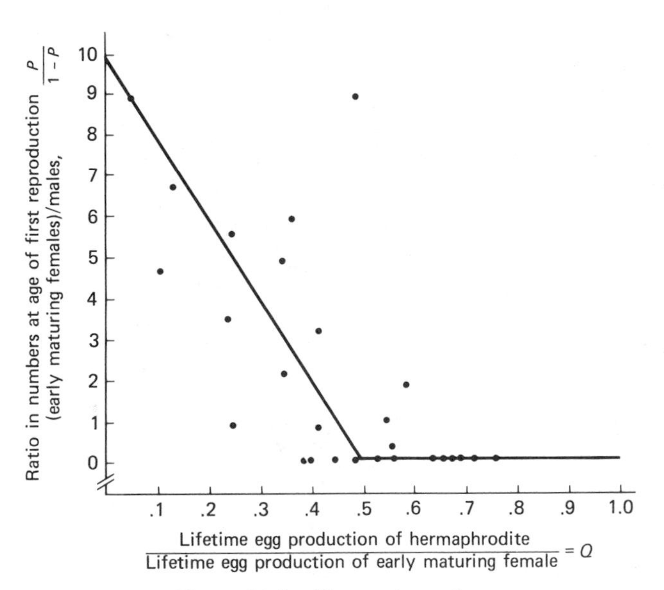

Figure 14-4 Charnov's results.

He then calculated Q for 28 populations of shrimp. The results are shown in Figure 14-4; the solid line is the expectation from theory. Except for one outlier, the fit is impressive. Charnov provides an excellent discussion of the impact of various violations of the model's simplifying assumptions, including deviation from stable age structure.

Implications for Population Dynamics

We sum up this chapter first by tying in the relationships between spacing systems, mating systems, sex ratio, and other environmental and ecological variables, then by listing some implications of spacing systems, mating systems, and sex ratio theory for population dynamics. The relationship between the various life history traits is shown in Figure 14-5.

Mating systems clearly affect the degree of inbreeding, the effective population size, and therefore the rate of evolution. First, monogamy, particularly long-term monogamy, means that the genetic composition of sperm utilized by a female has minimum variation. This, in turn, can be expected to minimize variation in the number of offspring, and this keeps effective population size slightly higher than otherwise. In addition, the strongly skewed *functional* sex ratio of polygynous species has a strongly enhancing effect on the inbreeding coefficient and thus drops the effective population size [recall from Chapter 7 that $1/n_e = 1/4n_m + 1/4n_f$, where n_e is effective population size, and n_m and n_f are the functional (breeding) numbers of males and females]. High inbreeding values tend to raise additive genetic variance (Chapter 8, see equation 8-20) and thus promote more rapid response to selection. In addition, Giesel (1972) has pointed out that the males of polygynous species display huge variance in reproductive contribution (i.e., in fitness). This, too, should lead to rapid rates of evolution. Of course, we

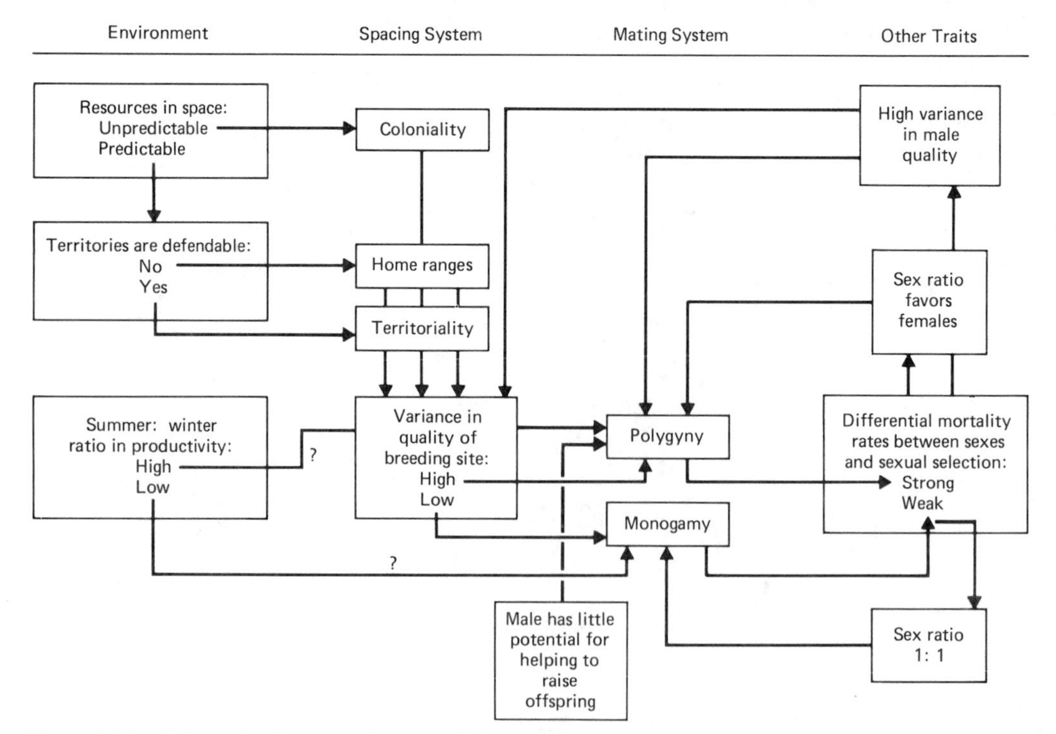

Figure 14-5 Schematic diagram of some of the relationships between environmental characteristics, spacing system, mating system, and other ecological characters.

must be careful not to make blanket statements: The differences in reproductive success among these males *may* be largely differences in age and experience—that is, nongenetic and not really fitness differences, and if genetic, reflect differences only in ability to intimidate other males and/or attract females. Successful males may be *less* fit with respect to other attributes, such as surviving.

The small effective population size of strongly polygynous species with highly skewed functional sex ratios makes such species prime candidates for interdemic selection. As such, we might expect to find group selection favoring an approach of the real sex ratio closer to the functional sex ratio (i.e., a sex ratio skewed even more toward females than expected by classic selection theory; Giesel, 1972; Clark, 1978; Bulmer and Taylor, 1980a,b; Taylor and Bulmer, 1980). In addition, optimal harem size may be less a function of female choice, or male choice, and more a function of total breeding unit success. We can write

$$R = p + b' \tag{14-25}$$

where R is the contribution to the population growth multiple of an individual with adult survival p and fecundity b'. A polygynous breeding unit with n females and one male then contributes, per individual,

$$R = \frac{p_m + np_f + nb'}{1 + n} \tag{14-26}$$

Suppose that $p_m = p_f$. Then in a system where half the males hoard two females each and half fail to breed, (14-26) becomes

$$R = \frac{1}{2}\frac{(1 + 2)p + 2b'}{1 + 2} + \frac{1}{2}\frac{(1 + 0)p + 0b'}{1 + 0} = p + \tfrac{1}{3}b' \tag{14-27}$$

In a monogamous system,

$$R = \frac{(1 + 1)p + b'}{1 + 1} = p + \tfrac{1}{2}b' \tag{14-28}$$

Clearly group selection will favor a monogamous system. But if males breed later, so that half have died by breeding age, each male will breed with two females, and

$$R = \frac{p_m + 2p_f + 2b'}{1 + 2} = \frac{\frac{1}{2}p + 2p + 2b'}{3} = \frac{5p + 4b'}{6} \tag{14-29}$$

The scheme will produce a higher group contribution than monogamy if

$$\frac{5p + 4b'}{6} > p + \tfrac{1}{2}b' \qquad \text{or if} \qquad b' > p$$

Might group selection also affect age of mating of the sexes and differential mortality rates?

There is considerable evidence that male–female differences in size, shape, and probably other characteristics preadapts the sexes to the trophic environment in different ways. Foraging niche separation might also have evolved, or be maintained by virtue of its alleviation of intraspecific competition [recall "Habitat Preference" (p. 366)]. If niche separation is exaggerated by competition [see "Balance of Intra- and Interspecific Selection Pressures" (p. 487)], high population densities will aggravate niche difference, forcing one or both sexes into more marginal habitat or habits. But the fact that the population has marginal habitat space into which it can spill means that the impact of density feedback is softened. Feedback should be less than linearly rising with population density. Of course, the degree to which spillover can take place is also a function of the density of *interspecific* competitors which may already have cornered the available spillover space. In such inscrutable ways are higher-order interactions built. Investigation of niche separation between sexes has concentrated, for obscure reasons, on woodpeckers (Kilham, 1965; Selander, 1966; Ligon, 1968; Jackson, 1970; Reller, 1972). Selander (1966) also provides data for the brown-headed nuthatch, Bewick's wren, indigo bunting, and quelea, and reports separate feeding flocks (although in apparently the same habitat) for other species. He also found greater niche divergence in island populations, where competing species are few, relative to maintain populations, suggesting strongly that the separation *is* exaggerated by intraspecific competition. Separation of niches between males and females has also been reported for lizards (Schoener, 1967) and plants (Cox, 1981).

Finally, the mating system has repercussions for the evolution of dispersal. Especially in polygynous species, mating is possible only for those males who acquire and hold high-quality territories. If younger and weaker males are to contribute genes to the future, short of catastrophic population reduction they must familiarize themselves with an area and, possibly, its denizens. Only in this way are they likely, next year, to have the wherewithal to oust a neighbor or to win the struggle for a newly vacated space. Therefore, after their initial, first-year dispersal, site fidelity is advantageous. To males already holding territories, retention of these territories also depends heavily on familiarity. They, too, should be site faithful. Site fidelity commonly referred to as *philopatry*, is difficult to document without long-term banding studies, but what information is available suggests that it is common and tenacious (Nolan, 1978; Barnard, 1979). For reviews and further references, see E. O. Wilson (1975) and Baker (1978).

Review Questions

1. Review the arguments and evidence for and against territoriality as a population-limiting factor.
2. What is the significance of territories, colonies, home ranges, and social groupings for trait group and interdemic selection? Recall that territoriality and coloniality are often seasonal, pertaining only to the breeding period, and that home ranges may slowly migrate in space.
3. Does territoriality or social grouping of individuals affect the argument for ideal free distribution? How? What about the cost (in time, energy, and risk) of movement from place to place?
4. The argument was made in this chapter that characteristics that make males attractive to females might come at the expense of viability. On the other hand, females might be attracted to characteristics that indicated high survivorship. Discuss the implications of these two alternatives for the evolution of sexual dimorphism. How might these implications affect population dynamics? How might they affect the amount of inbreeding?
5. Under what circumstances might a parent (or parents) adjust the sex ratio of its offspring late in development?

Chapter 15

Sociality: II. The Structure of Social Groups

15.1 Communal Breeding Units

In Chapter 14 we discussed mating systems which consisted of a male and one or more females—or, in a few instances, a female and several males. It has been known for some time that more complex breeding units exist. As long ago as 1940, Davis reported a rather loose mating scheme for the smooth-billed ani (*Crotophaga ani.*) This species is apparently monogamous, polygynous, or even polyandrous, often with more than one individual of *both* sexes involved. In the Galápagos mockingbird (*Nesomimus macdonaldi*), 4 to 10 individuals "defend" group territories from which nonbreeding birds are excluded (Hatch, 1966). By about 1960 it was also becoming clear that in such breeding groups not all birds were reproductively active, but that some played a role of helping the parents in various activities. Skutch (1961) defined a bird "which assists in the nesting of an individual other than its mate or otherwise a bird of whatever age which is neither its mate nor its dependent offspring" as a "helper." He went on to review the evidence to date for the existence of such helpers, and his review made it clear that helping was rather surprisingly widespread. The extended breeding unit, comprised of reproducers, offspring, and helpers, is a step above the primary family in social complexity, but simple and basic when compared to other, larger social groupings. It is also fairly well studied, and explicit attempts to explore the genetic mechanisms behind its evolution have been made. It therefore serves as a useful departure point for this chapter on the structure and evolution of social groups.

Following Skutch's paper there was a period in which the phenomenon of helping was studied extensively in a few species. Brown (1970), working with the Mexican jay (*Aphelocoma ultramarina*), discovered breeding flocks of 8 to 10 birds evenly dispersed throughout the habitat, each flock containing two or more breeding pairs plus young (yearlings) and unmated adults. Of all feeding visits to the nest, 38 to 53% were by the parents; the rest were performed by other flock members. Once the young had fledged, parents accounted for only 26% of feedings. Furthermore, at this point, parents showed no noticeable preference for their own young as opposed to the young of other breeding pairs in the flock. But while the helpers fed young birds, there was no evidence of their cooperating with nest building or incubation.

In the African bee-eater (*Merops bulocki*), birds pair monogamously for life, and often have one, two, or even three helpers which arrive 2 to 3 months before breeding. The helpers are mostly yearling males, and subordinate to the breeding male—although on occasion a "helper" may copulate with the female (Fry, 1971). Fry discovered that the average per individual output in progeny per adult was 1.1 over 36 nesting pairs and .9 for 10 breeding units with helpers present. Clearly, breeding pairs do better if they can attract helpers. But even were the helpers related by a coefficient of relationship of 1.0, their inclusive fitness would be less than if they themselves bred.

Why do helpers help? Woolfenden (1975) provides us with data for the Florida scrub jay (*Aphelocoma coerulescens*) (Tables 15-1 and 15-2), a long-lived, permanently territorial, permanently monogamous species with a short, synchronized breeding season. From Table 15-2

Table 15-1

	Number of Pairs		Number of Fledglings per Pair		Number of Young Alive at 3 Months for Pairs	
Year	*Without Helpers*	*With Helpers*	*Without Helpers*	*With Helpers*	*Without Helpers*	*With Helpers*
1969	2	5	0	2.6	0	2.0
1970	8	8	2.0	3.5	1.3	1.8
1971	6	19	1.3	2.1	1.0	1.2
1972	13	17	.3	1.6	.1	1.1
1973	18	10	1.3	1.8	.4	1.1
Total	47	59	4.9	11.6	2.8	7.2

Table 15-2

Previous breeding experience?	No	Yes	Yes	Yes	Yes
Number of helpers per pair	0	0	1	2	3–5
Number of eggs per pair	3.7	4.9	5.1	4.6	4.8
Number of fledglings per pair	0	1.2	2.1	2.2	2.2
Number alive per pair at 3 months	0	.6	1.0	1.6	1.7
Sample size	6	41	26	20	13

it seems quite clear that helpers have no influence on the number of eggs laid. Both tables indicate that the presence of helpers lowers nestling mortality, and that the advantage rises with the number of helpers. The relative advantage to having helpers (Table 15-1) can be measured by the ratio $11.6/4.9 = 2.37$ at the time of fledging, and $7.2/2.8 = 2.57$ at age 3 months. If the relative increase in this ratio from fledging to 3 months is more than a statistical artifact, it suggests that the fledglings of helped nests are more robust and probably larger. But again, although the breeding pair itself is aided by the helpers, the production of fledglings *per individual* in the breeding unit drops as group size increases. Even a coefficient of relationship of $r = 1$ is not sufficient to explain the apparently altruistic behavior of the helpers. Is there any evidence that the helpers are related to the breeding pair anyway? The answer is a clear yes. Of 78 helpers in Woolfenden's study, parentage was known for 56. Fifty-one of these helped only parents or one parent and a step-parent. Four of the remaining five helped only brothers and their mates. Then perhaps kin selection is *part* of the story. What else might account for helping behavior? Perhaps the helpers are young and inexperienced, and so might not be able to breed in any case. If so, remaining with parents (or brothers) would help them gain experience in offspring care for future benefit. The fact that it is kin that are aided allows individuals to contribute genetically a bit, even though they themselves go without breeding. (For a discussion of this line of reasoning, see West-Eberhard, 1981). This hypothesis would find support if helpers were primarily young birds (as was found by Fry, 1971). Also, because in most birds and mammals males breed somewhat later than females, this hypothesis would gain support if we were to find that most helpers, especially older helpers, were male. Woolfenden reports that of 35 yearling helpers of known sex, 18 were male; of 2-year-old helpers, 10 of 16 were male; and of 3-year-old helpers, 5 of 5 were male. Virtually all helpers are age 3 or younger; most breeding begins at age 2 to 4.

The information above, then, suggests that helpers are most likely to be found where young birds are unlikely to be able to breed successfully, and can gain experience, to their future benefit, by using their otherwise wasted efforts in helping others. The competitive loss to their genotype is minimized by directing their help toward close relatives.

Since Woolfenden's paper, the topic of helpers has burgeoned into a major field of study. Excellent reviews of the literature to 1978 are given by Brown (1978) and S. T. Emlen (1978). Emlen concentrates, among other things, on the question: Does helping really raise reproductive output? In all studies reviewed, the answer was yes; and the data continue to mount. Later work by Woolfenden (1981), building on his earlier Florida scrub jay studies, showed that the presence of helpers usually (although not always) was beneficial to parents in a variety of ways. Not only was nestling survival higher, but nests with helpers were less apt to fail, and even survival of the breeding pair was enhanced (13% annual mortality of breeders if helpers were present, 20% otherwise). Helpers were observed to feed young, mob nest predators, and help defend territorial boundaries. Koenig (1981) performed an analysis of covariance to determine the effect of (a) yearly variation in acorn crop, (b) prior group history (turnover or not since previous year), (c) territory quality (number of usable acorn storage holes or proportion of breeding seasons the site was occupied by a group), and (d) group size on the reproductive success of acorn woodpeckers (*Melanerpes formicivorus*). Although (a) and (b) were the most important factors, (d) also had an effect. Total reproductive success per group rose with number of helpers until group size reached seven or eight. Again, output *per individual* dropped with group size. Koenig reiterated the possibility that breeding groups of this sort might occur not due to any advantages accruing to the helpers, although this does not hurt, but due to shortage of suitable sites with acorn storage trees forces younger birds to forgo breeding themselves.

The question was inevitably raised as to whether the increased reproductive success of helped parents was really due to the presence of helpers, or simply because the pairs most attractive to potential helpers were those occupying the habitat most amenable to bringing up young. Alternatively, the potential helpers might be attracted to those areas because they were best able to satisfy their own foraging needs there. This awkward query was addressed directly and quite decisively for the crowned babbler (*Pomatostomus temporalis*) by Brown et al. (1982). Twenty breeding units of about the same size (six to eight) were followed. In nine, helpers were removed after completion of the first brood, so that only one yearling remained in addition to the breeding pair. The remaining 11 were left alone and simply observed. Control units raised an average of 2.4 fledglings from the second and subsequent broods. Experimental units produced only .8, a figure very close to that observed for natural groups of three.

A different perspective is offered by Ligon (1981). In the green woodhoopoe (*Phoeniculus purpureus*), nest helpers are frequently *not* offspring or full sibs. This weakens the argument for an important role by kin selection, at least for this species. In addition, Ligon found no evidence that more helpers was really any help. However, many adults never get to breed, which strengthens the experience argument for the existence of helping. Ligon also noted that all suitable habitat in a fairly large area is "owned" by a single group of 2 to 16 birds, so that new sites are rarely available unless a group somewhere dies out or becomes decimated and removable. Keeping helpers about may be a kind of extended parental care in which, in addition to the young continuing to gain experience, the parents can keep them together and send them forth in force when a potential site appears. Several sibs together seem more likely to procure a contested site successfully than does a single bird.

The nuances and *raisons d'etres* of extended breeding units are perhaps best seen by letting two of the major studies of this phenomenon speak for themselves. The white-fronted bee-eater (*Merops bullockoides*), lives in social units of two to seven individuals which cooperatively breed in burrows dug $\frac{1}{2}$ to 2 meters into earthen banks. Members of a group also roost together at night in these burrows. Seventy-two percent of all groups had more than two members and 58% had active helpers. This means that about 20 to 50% of the adult population forgoes breeding each year and, instead, helps others to breed. Helping activities include digging, building the nest chamber, incubating, feeding the young, and caring for the young after fledging (S. T. Emlen, 1981). A comparison of nest success (number of nests successfully fledging at least one young) and failure, excluding nests lost due to chance catastrophes, yields the following:

	Pairs Breeding Alone	Pairs Breeding with Helpers
Number of successful nests	22	28
Number of unsuccessful nests	38	19

Because these birds are colonial rather than territorial, differences in local habitat quality cannot be invoked to explain breeding success differences. Therefore, the significant heterogeneity ($\chi^2 = 4.67, p < .03$) of the contingency table above strongly indicates a benefit to having helpers. There is also no doubt that the helpers are related to the helped. Of 21 surviving young that Emlen followed, 13 subsequently were found helping their parents, 2 were helping a surviving parent and a step-parent, 2 were helping other, not closely related birds, 2 were roosting with parents (which were not breeding at the time), and 2 were on their own nests. Of the last two, one was being helped by its father, which had just lost his mate. There was not sufficient information to calculate whether kin selection might actually have favored helping, and there is some problem in figuring out how kin selection could initially have encouraged the origin of the habit. In territorial species we have noted a possible experiential advantage to the young remaining on home ground. As such they would be associated with close kin, providing the trait group structure for the evolution of altruistic cooperation. But these bee-eaters are colonial. Furthermore, there is little antagonism among colony members. When one nest fails, the parents often join other groups as helpers. Social dynamics are very fluid, with the closest relationships being formed on an apparent "friendship" level. "Friends" are familiar birds, past members of the same group. Thus young birds, even when remaining on the breeding grounds, could seemingly just as easily join nonrelatives as relatives. Under such circumstances kin selection should be weak. Yet relatives *do* join together preferentially. Emlen believes that altruism and kin selection are, at best, fairly unimportant aspects of the communal habit in these bee-eaters. He suggests the following rationale for the existence of helpers. This species lives in an extremely unpredictable environment with respect to the timing and intensity of rains, and also, in the dry-tropics area, food. If a poor food year is anticipated, there is a net advantage to staying home and avoiding the energy costs and risks of a likely futile reproductive effort while, at the same time, gaining experience in the fine points of offspring care. Emlen's data to show a statistically significant negative correlation between group size at the start of the breeding season and the amount of rainfall during the season. Once helping is established as an occasional pattern, there will arise secondary advantages, via inclusive fitness, to its direction toward kin.

Brown and Brown (1981) studied the gray-crowned babbler in Australia. This species lives all year round in groups of 2 to 13 or more individuals, on an all-purpose, group territory. Within each group there is typically only one breeding pair. The pattern here is for young to act as helpers to their parents for their first 2 years of life, and then attempt to breed themselves, at age 3 or 4. To determine whether helping was really helpful, Brown and Brown measured a number of territory parameters which seemed appropriate and, together with group size and age of mother, used these parameters as independent variables in a regression to determine the causes of reproductive success. Percent of territory with 20 to 30% woody vegetation, density of the woody species in question (an Acacia), and mother's age were all found to be important. So was group size. Removal experiments verified the importance of helpers (see also the article by Brown et al., 1982). Helping thus explains at least part of the observed positive correlation between number of young raised to fledging age and size of the social unit at the start of the season. This combination of approaches permitted Brown and Brown to estimate the effect of each helper; one helper effectively adds .46 fledged young to the group. The mode by which helpers affect reproductive success seems to be related primarily to the rate at which new clutches are layed. We can speculate that helping cuts the energy costs to parents of raising young and allows them to pack in more clutches per season.

A mature adult with the average (just over four) number of helpers produces 3.62 young, while a helper contributes .46. A parent is related to its real or potential offspring by a factor of $\frac{1}{2}$. A helper who is son, daughter, or sib of the offspring is also related to these offspring $\frac{1}{2}$. Thus the comparison of 3.62 and .46 is legitimate as a relative measure of genetic contribution, suggesting that helping is a pretty poor alternative to parenthood. But is it? Yearlings virtually never breed, suggesting (although by no means proving) almost total failure if they tried. If the expected number of fledglings per yearling breeder were really zero, we should be comparing not 3.62, but rather .0 with .46. Data from breeding birds of ages 2, 3, and 4 or older are given in Table 15-3. Now a helper who is also an offspring of the parents being helped, as nearly all are, is related to the offspring helped by a factor of $r = \frac{1}{2}$ if both parents are still alive, $\frac{1}{4}$ if only one parent is alive. The mean r value, then, where p_a is parental survival, is

$$r = \frac{p_a^T}{2}$$

where T is the number of years since the helper's own birth. The relationship between a breeder and its own offspring is $\frac{1}{2}$. Therefore, where F_T (Table 15-3) is the expected number of fledglings produced by a breeder of age T, the relative genetic contribution by an individual which helps rather then breeds is

$$(.46)\left(\frac{p_a^T}{2}\right) : (F_T)\frac{1}{2}$$

Brown and Brown estimate p_a to be .8. Therefore, at age 1 it is selectively advantageous to help. But by age 2, the ratio above has fallen to

$$(.46)\left(\frac{.8^2}{2}\right) : (2.0)\left(\frac{1}{2}\right) = .164 : 1.00$$

and helping is no longer a viable behavior pattern (on the basis of information given here). The fact that individuals of ages 2 and 3 often help means that there must be more to the story than is indicated. The gaining of experience is an obvious possibility.

Table 15-3

Age (yr)	Number of Individuals	Percent Breeding	Number of Fledglings per Breeder, F_T
1	88	0	0.0 (by conjecture)
2	21	19	2.0
3	34	62	2.3
4	50	100	3.6

15.2 The Evolution of Social Groups

The foregoing discussion of one class of social groupings suggests some ways in which the cohesion and cooperation of individuals might come about and be maintained through natural selection. We shall now try to be more general and talk a bit about some characteristics of populations that might be prerequisite to the evolution of sociality.

Sociality is characterized by interactions that affect each individual's contribution to its genotype's fitness. In particular, social groups constitute collections of individuals whose interactions are, on balance, cooperative. If interchanges were noncooperative, it may hardly be presumed that the individuals in question would continue to band together. Noninteractive aggregations, forming in response to availability of microhabitat, or resources, or the agencies of wind or water are not considered social. If cooperation—mutuality—is the bottom line of social grouping, sociality should not be expected unless there is some possibility of mutual benefit accruing from the acts of association. Furthermore, the potential benefit must outweigh the potential damage wrought by competitive interference. As Bertram (1978) points out, competition by grazers is passive ("a blade of grass is not worth fighting for because that is usually much more effort than finding another"), so without apparent opportunity for cooperative interaction. Thus while there is no particular disadvantage to grouping in this case, there is also no advantage—at least as respects cooperative feeding. We noted in Chapter 14 that many birds seem to benefit via increased food-finding rate by foraging in flocks. There may or may not be true cooperation here—more likely, each bird acts in an independent, selfish fashion, causing facilitation purely by chance—but the advantages of flocking bring the birds together into a situation where social interactions are likely to evolve. The same is true of groups of small mammals that huddle to keep warm. In some species, such as wolves, fitness is presumably greatly enhanced by the ability, arising from cooperative hunting, to replace mice and berries with a more substantial diet of moose, caribou, or other large game. Here there is not only the opportunity for cooperation, but almost an imperative.

Species in which individuals form passive aggregations are surely in a position to interact socially. But beetles feeding on seeds in windrows, and snakes packed into hibernacula, show little inclination to form social groups. In the latter case the reason is fairly obvious—many aggregations involve animals in inactive states. In the former it may simply be that no obvious benefit can arise from anything more complex than isolated individual effort. It is tempting to say that sociality often fails to occur because a low level of neural developement precludes it, but this would be very misleading without a much more specific definition of the form of sociality we are looking for. Although beetles do not astound us with their capacity for interpersonal discourse, the eusocial hymenoptera certainly do. Even the colonial hydrozoans are, in a sense, social. But excluding the pseudo-social organization of primitively colonial animals (and plants) such as the hydrozoans, there may be one general rule of thumb relating population parameters, as a cause, to the evolution of social behavior. In opportunistic species, evolutionary priority is on massive reproductive output and little parental care, with consequent short life. With extremely limited life spans it is unlikely that learned aspects of sociality would have a chance to develop. Also, with a deemphasis of the traits that enable individuals to get along (selection for high equilibrium densities—*K* selection), it is unlikely that cooperation is of more than minute selective value. Thus the existence of aggregations in itself is not enough to start the process of evolution toward social behavior. In equilibrium species there is competition for resources and so a situation in which strife could develop. Individuals that can form dominance hierarchies or cooperative units to deter strife are at a strong selective advantage; they can minimize "scramble" competition in which no one meets their needs, and minimize death by injury as well. It is a universal characteristic of social animals that fighting (except with weak or sick individuals, or in occasional instances of competition for mates) is largely replaced by displays within the group. In a sense, the group replaces the individual as the competitive unit. Between groups nature may retain some redness of tooth and claw. Let's look next at the mechanisms of social evolution.

Kin Selection

Perhaps the most obvious ultimate source of cooperation is kin selection. We will look briefly at two proposed examples of evolution of social traits by kin selection.

The individuals of many social species, when perceiving a potential predator, produce a so-called "alarm" call. The result of the call is that other individuals are apprised of the fact, in case they had not already noticed, that a predator is nearby, and are able to take cover, evasive action, or whatever appropriate measures are called for. But while the advantage to the social unit seems clear, it is often difficult to see any benefit to the individual caller (but see below). The case of alarm calls in birds is particularly instructive. In the presence of a ground predator the call is somewhat variable among species and often a bit raucous. The call given when an aerial predator approaches is quite uniform among species, a fairly high-pitched sliding whistle. The nature of this sound, which makes its source more difficult to locate, and the convergence on a common form by many species, suggests evolution toward a sound that will give the predator the least spatial information. It is a reasonable speculation that aerial predators, with their wider visual scan of the terrain and their greater threat to adult birds, have encouraged such evolution. The nature of the threat from ground predators makes such a sound pattern less critical. In fact, a raucous call, which is easily locatable if given by a parent some distance from the nest, may actually lure the predator away from the place where it is apt to do harm. The next logical step is to conclude, from their nature, that hawk alarm calls *do* help the predator to locate the caller (or at least did so in the past, before evolution gave them their hard-to-localize quality). Therefore, alarm calling is detrimental to the caller and constitutes altruistic behavior. Maynard Smith (1965) has proposed, therefore, that the evolution of alarm calls must have depended on the relatedness of individuals on the nesting ground. But the nature of avian dispersal, both from collections of large territories and colonies, is for offspring to leave the area, and to reappear in *other* areas, coming back to that new area fairly faithfully thereafter. It would seem, then, that relatedness would occur primarily between the parents and their offspring. We should conclude that alarm calling may be advantageous, via its effect on inclusive fitness, during the breeding season but not at other times. It is suggestive that when chaffinch (*Fringilla coelebs*) families break up in the winter, individuals cease to give alarm calls. [See also the remarks on Sherman's (1977) work, below.]

But many species continue to alarm call during nonbreeding periods. In the case of European titmice this may be reasonable because winter flocks are amalgamations of family units. But in the redwing, which also calls in the winter, this is not the case (Orians and Christman, 1968). How else might alarm calls come about? First, an alarm call might cause aggregation of the prey, and such behavior might be discouraging to the predator. Second, when all prey in a vicinity take cover, the population as a whole is less noticeable. Third, we have already seen how one joint mobbing response can be advantageous to the prey. In all these cases the disadvantages of giving up a bit of positional information is probably outweighed by the benefits of group action. Thus alarm calls may have a net beneficial effect on the caller. Charnov and Krebs (1975) suggest still another possibility. Alarm calls may convey to neighbors the information that a predator is present, but without telling them where. Might alarm calls be selfish and manipulative, allowing the caller, who *does* know where the predator is, to capitalize on the resultant confusion to position itself, without competition, in the safest place?

Within groups it is clearly advantageous to the average individual to avoid aggressive disruptions. One way in which intragroup peace can be maintained is through the setting up of dominance hierarchies. It is pretty obvious once a peck order has been established that high-ranking individuals have benefited from the scheme. But what does it gain the lowest-ranking animals to win some peace if they lose the right of access to food and mates? To some degree low-ranking individuals are young animals that will later grow into more elevated positions. Some are past reproductive age, so that in the evolutionary scheme of things, they simply do not matter. But why should a healthy mature male settle for deprivation when passing his genes is thereby prevented? In situations where deprivation is temporary, or "deprivation" an exaggeration of true events, it may benefit a subordinate, even a fully adult subordinate, to bide his time. But in many cases the term "deprivation" is quite a suitable description of life. Dittus (1977) describes a scene where a dominant toque monkey's egoism extended to actually removing food from another monkey's mouth. The "obvious" solution to this apparent dilemma is to suppose that subordinates minimize their genetic loss by being related to the dominants who get most of the food and pass most of the genes. Unfortunately, this explanation is probably unworkable. First, in instances of those hierarchical organizations where subordinates rarely breed, it is difficult to see how (say) five subordinates could possibly have inclusive fitnesses that compensated their individual losses to one dominant male. If the coefficient of relationship between individuals were $\frac{1}{2}$, for example, each male would have to increase the dominant male's output by at least twice the number of offspring that male could have raised on his own. And $\frac{1}{2}$ is probably a very generous estimate of r. If, on the other hand, interindividual strife acted to greatly reduce the reproductive output of subordinates who *did* breed, the kin-selective requirements for nonbreeding would more easily be met. If, indeed kin selection is the mechanism by which dominance hierarchies are maintained, the depression of output by subordinates due to intraspecific interference must be very considerable. Hierarchies must generally have the effect of rather severely curtailing reproductive potential of the species. Groups that maintained peace by other means would not suffer this loss and would have a marked group advantage. Unless there existed considerable mixing of groups, interdemic selection should be expected to destroy most vestiges of procrustean hierarchies rather quickly. Where hierarchical systems are more lenient, permitting low-ranking individuals to breed occasionally, this objection to the role of kin selection would be considerably weakened.

One bit of evidence of the importance of kin selection in social groups comes from studies of primates and lions where males newly establishing themselves as leader of a group may kill all the young and nursing infants, with the result that their time and effort will not promote the passage of genes other than their own. See, for example, Hrdy and Hrdy (1976) and Bertram (1978).

For further discussion and a review of theories of social evolution by kin selection, the reader is referred to West-Eberhard (1975).

Reciprocal Altruism

There are mechanisms other than kin selection that are capable of producing social cooperation. In many species, cooperation, and what looks suspiciously like altruism, occurs within groups whose members are not apparently related (among human beings, for example). One possibility is that trait group selection has occurred. Another possibility is interdemic selection acting on intergroup variance initially brought about by kin selection. There are still other possibilities. Trivers (1971) suggested that one way in which cooperation might evolve is for individuals to benefit from reciprocal altruism. Reciprocal altruism is a convenient concept because not only might it explain some observed instances of altruistic behavior among individuals, but inasmuch as it requires no genetic similarity of altruist and recipient, it might also explain the evolution of interspecific mutualism.

Unfortunately, most of the intraspecific examples Trivers provides in his paper might equally well be explainable in terms of kin selection or group selection and so are less than wholly convincing. However, two rather satisfying mechanisms for the evolution of reciprocal altruism within (or among) species have surfaced.

Schaffer (1978) asks us to consider an environment in which some element, say food supply, on which survival or eventual reproductive success depends, varies in time or space. We may, for example, picture individual consumers moving about in search of food in an area where food is patchily distributed. Suppose that "fitness" ($= \rho$) of an individual (i.e., survival or fecundity) is a function of food ingested F:

$$\rho = f(F)$$

Then the average value of ρ over time is

$$E(\rho) = E[f(F)]$$

Using the Taylor expansion [and looking, for the sake of simplicity, only at cases where var (F) is small] we obtain

$$E(\rho) = f[E(F)] + \frac{1}{2}\left[\frac{d^2 f(F)}{dF^2}\right]_{E(F)} \text{var}(F) \tag{15-1a}$$

Suppose now that individuals travel in pairs, sharing their food equally. Let F_1 and F_2 be the food eaten by individuals 1 and 2. If there is no asymmetry in foraging ability or selflessness, $E(F_1) = E(F_2) = E(F)$, and $\text{var}(F_1) = \text{var}(F_2) = \text{var}(F)$, and for each individual

$$E^*(\rho) = \text{average "fitness" if animals forage in pairs}$$

$$= f[E(F)] + \frac{1}{2}\left[\frac{d^2f(F)}{dF^2}\right]_{E(F)} \frac{1}{4}\text{var}(F_1 + F_2)$$

$$= f[E(F)] + \frac{1}{4}\left[\frac{d^2f(F)}{dF^2}\right]_{E(F)} [\text{var}(F) + \text{cov}(F_1, F_2)] \qquad (15\text{-}1b)$$

Each individual benefits from group foraging, with sharing, if

$$E^*(\rho) > E(\rho)$$

that is (from equations 15-1a and 15-1b),

$$-\frac{1}{4}(1 - \rho_{12})\left[\frac{d^2f(F)}{dF^2}\right]_{E(F)} > 0 \qquad (15\text{-}2)$$

where $\rho_{12} = \text{cov}(F_1, F_2)/\text{var}(F)$ is the correlation in the amount of food discovered by individuals 1 and 2 from place to place. Because $0 \leqslant (1 - \rho_{12}) \leqslant 2$, inequality (15-2) can be met only if

$$\left[\frac{d^2f(F)}{dF^2}\right]_{E(F)}$$

is negative. What does this mean? Picture a graph plotting f on the ordinate against F on the abscissa. It seems reasonable to suppose that survival and fecundity cannot forever increase with the amount of food eaten. Indeed, there will be a point reached where still further consumption yields diminishing returns and detracts from time beneficially spent in other activities. Thus f should rise and then fall with F. Somewhere in the vicinity of this curve's peak, the slope, df/dF, will be decreasing—$d^2f/dF^2 < 0$—and cooperative foraging is favored. If the slope is rising, $d^2f/dF^2 > 0$, then almost surely we must be at some low value of food intake at the left of the curve; $E(f)$ is small. Here we should find no advantage to cooperative foraging. Thus hungry animals will generally not be benefited by cooperation, satiated animals almost certainly will. What is happening here is that when a foraging pair encounters a dense patch of food, two individuals together can use the patch to full advantage, whereas one might not be able to do so. If this gain offsets the loss of having to share meager patches, cooperation pays. Of course, there may be some cost (squabbling, perhaps) to foraging in pairs, making the conditions for cooperation more stringent. On the other hand, in some circumstances (e.g., in wolf packs) cooperation enables more efficient foraging to take place, adding a further impetus to cooperation.

Another fascinating approach to reciprocal altruism is provided by Axelrod and Hamilton (1981). Consider the following situation. Two individuals meet under circumstances that allow them to cooperate toward a common goal. If both decide to cooperate, there is a payoff in incremental fitness of (say) ΔW_1. If individual A acts cooperatively but B does not ("defects"), A plays the sucker and experiences a fitness increment, ΔW_2, which may be negative. If B cooperates and A plays B for a sucker, A gains a lot ($\Delta W_3 > \Delta W_1$). Finally, if both players defect, the increment to A is ΔW_4, which exceeds ΔW_2 but is less than ΔW_1 or ΔW_3. The situation may be depicted as shown in Table 15-4, where we have substituted hypothetical numbers for

Table 15-4

Payoffs (fitness increments) to player A if:

		Player B	
		Cooperates	*Defects*
Player A	*Cooperates*	ΔW_1	ΔW_2
	Defects	ΔW_3	ΔW_4

$$= \begin{bmatrix} 3 & -4 \\ 5 & -1 \end{bmatrix}$$

the payoffs to A in order better to visualize their relative magnitudes. This sort of situation is known to game theoreticians as the "prisoner's dilemma." What is A's best move: to cooperate, to defect, or to mix options and choose one or the other according to a prearranged probability? Clearly, if he always cooperates, his companion can do best, competitively, by defecting, thus playing him for a fool. If he always defects, his companion ought to do likewise. In neither case will either player do very well, because we can presume that A will not continue to play his cooperative option if he is consistently punished for it. Cooperation would benefit both players, but unless cooperation can be counted on from the other, neither player should take the risk of being the sucker. This is probably a fairly typical description of the nature of two-individual interaction possibilities in real populations; it does not bode well for the evolution of cooperation. Indeed, any strategy, pure or mixed, once it became established in the population, could be successfully invaded by the pure strategy, defect. An individual who invariably defects will always increment its fitness more than others. Axelrod and Hamilton refused to be prisoners to this prognosis. Clearly, from the argument above, two players meeting only once will always do best—actually, minimize their losses—by defecting. But what if individuals meet repeatedly? If player B *does*, on one occasion, cooperate, then player A, on the subsequent encounter, knows that the possibility of a jointly cooperative effort exists, *to his benefit*. Under *this* situation might a strategy of cooperation, once set in motion, perpetuate itself? There are complications in this reasoning. If the number of future encounters is known to the players, the best move for the *last* game is to defect. But then, in the subultimate game there is no chance of cooperation next time, and the best move for this game, too, must be to defect—and so on to the very first game. That is, cooperation becomes a possibility only if the number of future encounters between the two players is unknown. Axelrod and Hamilton asked three questions:

1. What kind of a strategy can "thrive in a variegated environment composed of others using a wide variety of strategies"?
2. Under what conditions will this strategy resist invasion by other strategies?
3. How can such a strategy get an initial foothold in a population?

To find answers, they devised an ingenious scheme; they solicited ideas from a variety of biologists, game theoreticians, and computer nuts. Fourteen strategies were submitted, and each was used against a wide variety of others generated by Axelrod and Hamilton in 200-move computer simulations. The most robust (question 1) was one they dubbed "tit-for-tat"—cooperate on the first move and then do whatever your opponent did in the previous encounter. More suggestions were then solicited (62 new entries materialized) and again, tit-for-tat was always the winner. It won because (a) it was never the first to defect, (b) it was provokable into retaliation, and (c) it was immediately forgiving. A phenotypic selection simulation was then performed, each phenotype corresponding to one of a variety of game strategies. On average, the percent of wins by a particular strategy, against all others, was used as a measure of fitness for that strategy. The various fitness values were then used to increment or decrement the frequencies of the corresponding strategies each generation. Tit-for-tat invariably emerged as a noninvadable winner in the evolutionary contest—if reencounter probabilities among individuals were sufficiently high.

Let us now investigate the significance for cooperation of the result above. Suppose that both players follow the tit-for-tat strategy. Then each starts with a C (cooperate). But both next play what the other played previously. Thus the sequence over time for both players is C–C–C–––. The tit-for-tat strategy leads to consistent cooperation. But suppose that an otherwise tit-for-tat strategist cheats at the outset. Can such a variant strategy invade a population of otherwise pure tit-for-tatists? There are two ways to so cheat. Suppose that the companion, B, breaks the rules and begins with a D (defect), followed by another D. Then player A will respond with D and both players subsequently play D–D–D–––. The question now is whether this variant of the tit-for-tat strategy, which destroys cooperation, can successfully invade the true tit-for-tat approach. If the probability of a rematch between the players is p, the probability that a second interaction takes place is p; a third, p^2; and so on. The payoff to the rule breaker is thus (see Table 15-4)

$$\Delta W_3 + P\,\Delta W_4 + p^2\,\Delta W_4 + \cdots = \Delta W_3 + \frac{\Delta W_4}{1 - p} \tag{15-3}$$

The payoff to an honest tit-for-tater, on the other hand, is

$$\Delta W_1 + p\,\Delta W_1 + p^2\,\Delta W_1 + \cdots = \frac{\Delta W_1}{1 - p} \tag{15-4}$$

The variant strategy can successfully invade only if the value of (15-3) exceeds that of (15-4)—that is, if

$$p \leqslant 1 - \frac{\Delta W_1 - \Delta W_4}{\Delta W_3} \tag{15-5}$$

In the case of the numerical values given in Table 15-4, the tit-for-tat strategy, which leads to consistent cooperation, is uninvadable unless

$$p \leqslant 1 - \frac{3 + 1}{5} = .2$$

Thus sufficient probability of reencounter assures cooperation.

Suppose that player B breaks the rules in another way, playing D first and then sticking to the tit-for-tat way. Then player A plays C–D–C–D–––, and B plays D–C–D–C–––. The payoff to B in this case is

$$\begin{aligned}\Delta W_3 + p\,\Delta W_2 + p^2\,\Delta W_3 + p^3\,\Delta W_2 + \cdots &= \Delta W_3(1 + p^2 + p^2 + \cdots) \\ &\quad + \Delta W_2(p + p^3 + \cdots) \\ &= \frac{\Delta W_3 + p\,\Delta W_2}{1 - p^2}\end{aligned}$$

This value exceeds the tit-for-tat payoff (equation 15-4) only if

$$p \leqslant \frac{\Delta W_3 - \Delta W_1}{\Delta W_1 - \Delta W_2} \tag{15-6}$$

In the numerical case in Table 15-4, this is $p \leqslant \frac{2}{7}$. Again, sufficiently high probability of reencountering assures the noninvadability of the tit-for-tat strategy.

We have seen how reciprocal altruism might evolve, promoting cooperative effort within groups, and how kin selection might play a role in the evolution of sociality. Finally, it is probably a common happenstance that helping others can be personally beneficial. In breeding colonies of the western gull (*Larus occidentalis*), when one female dies, her place is generally taken by another female who cares for the orphaned young. Altruism? Possibly. But by taking over the deceased female's nesting site the new female has gained a foothold into the breeding colony for next year. In this species a female uses the same space year after year, often occupying the site to some degree for as much as 9 months of the year. The female also gains experience and a pair bond that is likely to last several years (Pierotti, 1980). As in S. T. Emlen's (1981) white-fronted bee-eaters, an opportunistic behavior pattern appears altruistically motivated and, indeed, is helpful to recipients. In such ways social organization can be prodded into ever more complex patterns.

All of these mechanisms are probably important in the evolution of social organization. One common thread is worth pointing out. Opportunism that serves altruistically is likely to be possible only where populations form groups and where individual success is dependent on membership in the group. With respect to Axelrod and Hamilton's model of reciprocal altruism, reencounters are likely—and thus evolution of cooperation by this means is possible—only where animals live in groups with small turnover. The possibility of helping oneself by helping others is possible only within groups. Altruism maintained by kin selection is possible in groups with high turnover if relatives stick together or interact preferentially with each other, but is likely to have gained an evolutionary foothold only in groups where individuals are closely related. In all cases, the mechanisms require population *viscosity*, a property describing little dispersal. Juvenile dispersal, so common in birds and mammals, nevertheless allows for parent–offspring groupings which last beyond egg laying and birth and provide a basis for social interactions to evolve by kin selection. Cohesiveness of the groups of remaining adults sets the stage for the other models of selection. Without a preadaptive tendency to group for mutual benefit and to provide parental care, it is unlikely that social organization could come about. Also, social organization is unlikely to develop to a more than elemental level in opportunistic species that depend, for their success, on highly dispersive young and who live but a short period of time.

Cooperative Subunits

In the preceding section we were not terribly kind to the position that kin selection is an important force in the evolution of social groups. But even if kin selection is of minor importance as a cause of such groups' existence (and it may be much more than this), it is unquestionably important in determining the inner working of such groups. In the Belding's group squirrel (*Spermophilus beldingi*), Sherman (1977) reports that the mating system is polygynous. As soon as a male's mates give birth, he leaves, neither defending the old territory nor exhibiting any paternal care. Often the males move considerable distances, and the evidence suggests that there is no tendency for them to move in kin groups. Thus the sedentary females form a viscous population, with neighbors generally being related, while the genetic dispersion of males is closer to random. Sherman, in over 3082 observation hours, witnessed 102 approaches of predators (dogs, foxes, coyotes, weasels, badgers, martins). The percent of instances in which various classes of individuals were observed giving alarm calls is shown in Table 15-5 next to the expected percent, based on the number of instances in which that class

Table 15-5

Class of Individual	Percent of Instances in Which Alarm Calls Were	
	Observed	*Expected*
Adult female	66	41
Adult male	7	37
Yearling female	36	17
Yearling male	4	13
Juvenile female	8	10
Juvenile male	5	9

was present. For the moment, concentrate on adults. Males clearly were much less likely to alarm-call than were females. Given the genetic distribution pattern, this is precisely what we should expect on the basis of kin selection. In the same vein, Sherman found, again consistent with kin-selection theory, that reproductive females were more likely to call if there was a living daughter, sister, mother, or granddaughter in the immediate vicinity than if no relatives were present, and females with young were more likely to call than were nonreproductives. Sherman's data also strongly suggest that alarm calling is individually disadvantageous and thus an example of true altruism. Fourteen of 107 callers were attacked by predators during his study, whereas only 8 of 168 noncallers were attacked. By a contingency table G test the difference is significant at the .025 level. Sherman further notes that in harem species of ground squirrel where the males remain with the young, males do, as expected, give their share of alarm calls.

But in more cohesive groups, how do individuals partition their altruism among the other group members? To whom are they loyal? It has often been assumed that the frequency of altruism should vary directly with the coefficient of relationship, r. A little thought, reveals a possible error of this notion (and see Altmann, 1979). Because the gain via inclusive fitness is always greatest when interaction is with the closest kin, it is invariably advantageous to lavish one's attentions on this individual until a case of diminishing returns sets in. The altruist should then turn its attention to the next most closely related individuals. (But see Hughes, 1983). Weigel (1981) has generated a simple model to describe the distribution of favors among kin and shows that the pattern expected may vary considerably as a function of the distribution of r among potential recipients, the "cost" of an altruistic act (not rigorously defined, but presumably referring to time or energy expended), the rate at which diminishing returns sets in for an altruistically beseiged recipient, and the impact of an act on the fitness of the altruist.

Data on the partitioning of altruism—in this case the aiding of fellow pigtail macaques (*Macaca nemestrina*) in agonistic encounters—has been gathered by Massey (1977). Monkeys were kept in an outdoor pen with access to an indoor area. All pairwise agonistic encounters were carefully watched, and aiding by any third individual was noted. Aiding could take the form of a chase of one of the antagonists, mounting of one of the antagonists by the newcomer (a signal of dominance that stopped further aggression), or simply blocking the interaction by bodily getting in the way. Individuals initiating an agonistic encounter did so without regard to whether the antagonist was family or not. Slightly less than half of all encounters drew in at least one additional monkey. Of 397 episodes involving aid from other animals, 309 involved help to the defender; only 88 helped the attacker. Of 30 individuals that helped defenders, 22 aided family members significantly more often than expected on the basis of no preference. Of 23 individuals observed to aid an attacker, only 13 (about half) helped primarily family members. Of all 34 individuals that at one time or another aided combatants, either attackers or defenders, 24 helped primarily their own family members. Of all 397 episodes, 88% of aid was directed toward family members. Clearly, these monkeys were often willing to risk the dangers of becoming involved in a fight to help one another. And in most cases the help was directed toward kin. Massey broke down the frequency of aiding in more detail (Table 15-6). The r-value categories represent parent–offspring or sibs; half-sibs or grandparent–grandchild; cousins; and so on. The values may exceed the expected .5, .25, .125 because of the unknown degree of

Table 15-6

	Degree of Relatedness, r		
	$\geqslant .5$ ($n = 38$)	$\geqslant .25$ ($n = 156$)	$\geqslant .125$ ($n = 481$)
Average number of aids per aggressive encounter	.149	.035	.009
Significantly different	$p < .001$ (.5 vs .25)		$p < .01$ (.25 vs .125)

inbreeding. Finally, Massey found that though both sexes received equal amounts of aid, the altruist was much more likely to be female ($p < .005$). This, it was suggested, might reflect the relative uncertainty of paternity. Another possibility is that in this species, as in many primates, the male aggressive role is directed primarily at *intergroup* conflict. Genetic predispositions may make males less likely to involve themselves in *intragroup* squabbles.

In 1970, J. M. Emlen proposed that because very young and very old individuals had low reproductive value (the arguments were actually couched in measures of selection intensity—see Chapter 12), an altruistic act directed at them would have little impact on an altruist's contribution to inclusive fitness. Also, very young and very old individuals have the least to lose by being altruistic (see also Charlesworth and Charnov, 1981, Hughes, 1983). Adding in the constraint that the very young are likely incapable of much in the way of aid giving, we derive the following expectation. Given a choice between two equally related potential recipients, an altruist should preferentially bestow its favors on the one nearest peak reproductive value, the young adult or juvenile. On the other hand, the young adults and juveniles, less likely to meet the criterion for being altruists (see Chapter 9), should be less altruistically inclined than older adults. There are few data bearing on this prediction (it works with human beings but this may be due to cultural factors). But refer to Sherman's data (Table 15-5). Note that juveniles of both sexes perform fewer calls than expected. Yearling and adult males have dispersed and so are no longer pertinent to this argument. Adult females perform disproportionate numbers of calls. It is not clear where yearling females fit. Although we do not know their reproductive value, we can conjecture that it must be near its peak. If so, the frequency of calling is unexpected high. On the other hand, juveniles have relatives ($r = \frac{1}{2}$) in the form of sibs and female parents; adults have relatives in the form of offspring sibs, and a lower probability of a surviving mother; and yearlings have all three kinds of relative about. Perhaps this swings the balance of advantage more strongly in the direction of alarm calling.

If kin selection, by directing cooperative behavior toward kin, has the capacity to mold social organization—and quite clearly it does—we must next ask how individuals recognize each other as kin. One obvious clue is familiarity. Because those populations where kin selection is apt to be an important force are viscous, kin are exactly those with whom an individual has most contact and most familiarity. A clear case in point is honeybees and many other eusocial insects, where recognition, based on hive odor, is a drastic determiner of interactions. Another possible clue is individual similarity in appearance, physical or behavioral. To the extent that kin carry the same genes and have grown up under similar environmental conditions, this is a valid alternative, or complement to familiarity. Indeed, as we saw in Chapter 9, the ultimate mechanism of trait group selection (of which kin selection is a special case) is not genetic identity by descent, but genetic identity *by type*. Thus, for genetic traits already well on their way to fixation, similarity may be nearly as appropriate a stimulus as kinship itself. There are data indicating that animals can recognize kin, even kin that have never before (since birth) been encountered. Such recognition must be based on similarity. Individuals of the primitively social sweat bee (*Lasioglossum zephyrum*) block entry to their nest by most individuals from other nests. In laboratory experiments, Greenberg (1979) found a linear relationship between the coefficient of relationship, r, and the probability with which a non-nest mate was allowed to pass. Waldman and Adler (1979) found that toad tadpoles preferentially spaced themselves near siblings rather than nonrelatives. Wu et al. (1980) separated sibling pigtail macaques from their mothers and each other at age 5 minutes, placing them in incubators until they could be placed in isolation cages. In these cages the young monkeys were isolated from each other physically, visually, and olfactorily. In the first experiment, 16 monkeys, aged 44 to 344 days, were offered a choice of playmates from two other monkeys of matched age, both strangers. One was a paternal half-sib (further controlling for early experience, even in utero), the other unrelated. A third option was an empty cage. The experimental setup was such that the subject had a chance to observe the potential playmates and the empty cage before making a choice. Two measures of preference were used: time spent oriented toward one of the three options before moving, and time spent near (but not touching) one of the options ("entry" time). On the basis of both preference criteria, subjects preferred kin. Previous experiments had shown a strong preference by monkeys for familiar individuals (playmates) over strangers. To test the relative importance of kinship versus "friendship" a second experiment offered monkeys a choice between paternal half-sibs, which, again, were complete strangers, and "friends" of matched age. Again, an empty cage was included as a third option. Although the results were not statistically significant, the subjects *still* (11 of 16) seemed to prefer kin. These kinds of tests are of extreme significance for our understanding of the nature of social evolution and deserve repeating for a variety of species.

Tightly organized social grouping has a number of implications for population genetics and population dynamics. First, increasing interdependency of group members and the emergence of groups as competitive units greatly increases the importance of both trait-group and interdemic selection as forces molding further evolutionary change. Extensive dispersal of juveniles, where it occurs, has a counter-effect, but in groups forming long-term bonds, such as wolf packs, primate troops, communal nesters with site attachment, and even house mice, such dispersal is not extensive. In fact, once group living becomes strongly established, dispersal may be deadly; admittance into new groups will be repulsed, and without the safety of the group, individuals are likely to die. Because group structure is vulnerable to interindividual conflicts whose peaceful resolution may depend on complex systems of dominance and alliance recognition and kinship ties, and because groups of ever-increasing size may have smaller, average coefficients of relationship among their members and increasing numbers of individuals and alliances to remember, there are limits to group size. Therefore, the social group life-style may have an extremely important buffering effect on population dynamics, but lead to a rather sharp-edged pattern of onset of density feedback.

15.3 Conflict Resolution

We consider two sorts of conflict. One involves decision making in adversary encounters among individuals. The second pertains to decisions of behavior priorities. We shall talk about the second only to the extent of raising an unanswered question. We shall then spend considerably more time with the first.

Consider a situation that calls for an altruistic act. For the sake of argument, suppose that alarm calling is really altruistic, carrying a not insignificant risk, and that a predator has appeared on the horizon. An individual spying the threat has two options, sound the alarm or, because that might endanger him, wait and "hope" someone else gives the alarm. Assuming that the act of alarm calling is advantageous under kin selection, too long a wait will be even more costly than an immediate response. How long should the individual wait before, if necessary, giving the call?

Close living, even with well-defined roles and dominance positions, provides opportunity for strife. The combinations of fitness increments, decrements, and kinship do not always indicate altruism; choice situations may require that to best benefit one's relatives it becomes advantageous to be selfish toward others; and individuals may make errors of commission or ommission, breaking the reciprocal altruist string of tit-for-tat cooperative acts. Sometimes, under stress, abnormal behavior might develop, or what appears aberrant and aggressive might actually be adaptive. In times of food shortage lionesses may rob cubs of food, and males may take food from both females and young. In extreme cases this can even lead to starvation of the cubs (Schaller, 1972). Does this represent a breakdown in social order? Gobbling of food and minor squabbling occur in hyaenas and wild dogs, but seldom is fighting or extreme selfishness observed. Does this mean that the latter predators are somehow more stable then lions in the face of adversity? More likely it reflects the fact that wild canid and hyaena packs include but a single mother, and so include members who are more closely related to one another than are the lions, where cubs of several females are members of a communal nursery (Bertram, 1976). It may be simply that the conditions for the advantage to altruism are more easily violated among lions for this reason. But there is also selection pressure which *favors* aggression under certain circumstances. Particularly where reciprocal altruism is critical to a group's proper functioning, individuals must be informed in no uncertain terms when they transgress the rules. Getting "angry" at a selfish individual is advantageous to the group and so, indirectly, to the angry individual.

When sources of conflict between individuals arise, how are they to be settled? This question has been rather cleverly addressed by John Maynard Smith and his co-workers (see Maynard Smith and Price, 1973; Maynard Smith, 1974; Parker, 1974; Maynard Smith and Parker, 1976). The authors analyzed conflicts as games between individuals in which the best course of action taken by one must be dependent on that followed by the other (anyone interested in a thoroughly delightful, humorous approach to game theory that is nevertheless informative and accurate is referred to Williams, 1954).

To apply the theory of games, we need first to introduce some concepts and terminology. We have already encountered the notion of "payoff" in the tit-for-tat game of Axelrod and Hamilton. It is simply a measure of the advantage or disadvantage (ideally framed in terms of fitness contribution) to a player of performing a certain act in an antagonistic encounter. It also depends, of course, on what the opponent does. We can conveniently depict the set of payoffs to a particular player with a *payoff matrix* (Table 15-7). In this case, if B makes move y, A wins 3 points if he also plays y, only 1 point if he plays x. If B plays x, A wins 2 points by playing x and 4 by playing y. In this particular example, what are the best "strategies" for A and B to follow? A "strategy" is the set of rules that governs which move a player makes. If A played x with probability 1.0, and B was sure to play y, we say that A has a "pure x strategy" and B has a "pure y strategy." If A played x with probability .3, and y with probability .7, A is said to have a "mixed strategy."

We can find A's best strategy as follows. Suppose that his opponent, B, plays a pure x strategy. Then A will always do best to play y. If B plays a pure y strategy, A will still do best always to play y. That is, A receives the greatest expected payoff by following a pure y strategy regardless of what his opponent does. Let us look at B's options. Suppose, by the nature of this contest, that B's gains are identically A's losses. Games in which one player's gains equal the other's losses are known as *zero-sum games.* Then B's interests are unequivocally to minimize the game's value to A. If A plays a pure y strategy, as we have indicated he should, B will most effectively curtail A's earnings by playing y. If A happened to follow an x strategy, B still does best to play y. Thus both players should follow pure y strategies. When both players in a game have optimal strategies that are pure, the game is said to have a *saddle point.* Saddle points may be quickly discovered, if they exist, in the following way. Examine A's moves, represented by rows 1 and 2 in Table 15-7. Choose the row with the largest minimum element (that is 3, in the second or y row). If A's best strategy is a pure one, that row is it. Now check B's choices, columns 1 and 2, and (because B wishes to minimize A's payoff) choose the column with the smallest maximum element (that is 3 in the second or y column). If the largest minimum element in A's chosen row is the same element as the smallest maximum in B's chosen column, the game has a saddle point and both players do best to follow pure strategies.

Notice that by playing y, player A maximizes his own minimum gains and simultaneously minimizes his opponent's maximum possible gains; B minimizes A's best gain and thereby also cuts his own losses. Game-theoretic strategies are conservative strategies; there is no way a player can maximize his own gains without knowledge of an opponent's strategy (which the

Table 15-7

		Player B	
		x	y
Player A	x	2	1
	y	4	3

opponent can alter if he can surmise the first player's strategy). Hence the zero-sum game solution describes the nonspeculative, safe aim of each player: A player maximizes his minimum gains or, equivalently when the payoff is negative, cuts his losses. Simultaneously, he minimizes the value of the game to his opponent.

Now look at another zero-sum game (Table 15-8). Player A, surveying this payoff matrix, finds that row y has the largest minimum element. Thus if a saddle point exists, it is at the element of value 4 in the yth row, yth column. Player B, finding the row with the smallest maximum element, however, turns to the value 5 in row x. Apparently, this game has no saddle point and the optimal strategies are mixed. We can see this more clearly perhaps by following the consequences of each player trying pure strategies. If A plays purely x, B should counter by playing x. But if B plays x, A would be foolish not to play y. However, if this happens, B ought to play y. The optimal, mixed strategies can be found as follows. Suppose that B were to use (figuratively) a random number table to decide his move, playing x with probability p', y with probability $1 - p'$, and that A played a pure x strategy. Then the expected payoff to A would be

$$p'(3) + (1 - p')(6) = 6 - 3p' \tag{15-7}$$

Table 15-8

		Player B	
		x	y
Player A	x	3	6
	y	5	4

Suppose, alternatively, that A played a pure strategy of y. In this instance, the expected payoff to A is

$$p'(5) + (1 - p')(4) = 4 + p' \tag{15-8}$$

Now if the two values in (15-7) and (15-8) are equal, it means that A's expected payoff is the same whether he plays pure x or pure y. It must therefore be the same regardless of whatever he does. But if this is so, A cannot possibly improve his performance against B. So B's strategy must be optimal. We set the two expressions equal and obtain B's optimal strategy:

$$6 - 3p' = 4 + p'$$

so that

$$p' = \tfrac{1}{2}$$

Now consider what A should do. Again, by the same reasoning, A should mix his moves in such a way that B's expected payoff is the same whether B plays pure x or pure y (and thus any mixed combination of x and y). Because this is a zero-sum game, the payoff matrix for B is the same as that for A (Table 15-8), but with all signs reversed. Let p be the probability with which A plays x. We set the

$$\text{payoffs to B} = \begin{cases} p(-3) + (1 - p)(-5) = -5 + 2p & \text{(B plays pure } x) \quad (15\text{-}9) \\ p(-6) + (1 - p)(-4) = -4 - 2p & \text{(B plays pure } y) \quad (15\text{-}10) \end{cases}$$

equal and solve to find

$$p = \tfrac{1}{4}$$

The expected payoff to A, the best he can do faced with the best defense B can muster, is (from equations 15-7 and 15-8)

$$6 - 3p' = 6 - 3(\tfrac{1}{2}) = 4\tfrac{1}{2} \qquad \text{or} \qquad 4 + p' = 4\tfrac{1}{2}$$

The expected payoff to B is (equations 15-9 and 15-10)

$$-5 + 2p = -4\tfrac{1}{2} \quad \text{or} \quad -4 - 2p = -4\tfrac{1}{2}$$

As must be the case in a zero-sum game, the expected payoff, or "value of the game," to B must be negative the value of the game to A.

A simple, quick, and dirty technique to find mixed strategies in games is as follows. To find A's strategy, look at the absolute difference in the values of the two elements in row 1, and write this value beside row 2. Then find the absolute difference in the values of the elements of row 2 and write this number beside row 1. The ratio of the two values beside rows 1 and 2 give the optimal strategy proportions of options 1 and 2. In Table 15-8, for example, we write $|3 - 6| = 3$ beside row 2, and $|5 - 4| = 1$ beside row 1. A's optimal strategy is thus $1:3$, or $\frac{1}{4}:\frac{3}{4}$. Similarly, for player B, write $|3 - 5| = 2$ beside column 2 and $|6 - 4| = 2$ beside column 1. B's optimal strategy is thus $2:2$, or $\frac{1}{2}:\frac{1}{2}$.

We are now ready to tackle some more complicated games with biological undercurrents. The following examples come from the papers cited above. Consider a situation where opponents are two individuals who are possibly about to come to blows. Two moves are possible: either (a) escalate the confrontation and continue to do so until you either win (let the value of winning be V) or get injured (value $= -D$), or (b) display but retreat if your opponent escalates. The payoff matrices are shown in Table 15-9. If both escalate, player A (or B) wins with probability $\frac{1}{2}$, and is injured with probability $\frac{1}{2}$, so the appropriate entry is $V/2 - D/2 = (V - D)/2$. We can make this costly by letting $(V - D)/2 < 0$. If player A escalates and B merely displays and retreats, A always wins. If B escalates, A loses if he retreats; if both display, A wins with probability $\frac{1}{2}$. What should A do in such encounters?

First note that because the payoff matrix for both players is the same (Table 15-9a and b are identical if the player for whom the payoff is defined is labeled at the left of the matrix); this cannot be a zero-sum game. In such cases we cannot necessarily calculate B's best strategy by setting equal A's expected gains under either pure strategy. In fact, our technique of searching for a saddle point is no longer valid; minimizing the opponent's best payoff is no longer necessarily the same as maximizing your own minimum payoff. In fact, there are now *four* possible solutions, based on whether the aim of the game is to:

1. Minimize your opponent's best payoff,
2. Maximize your own minimum payoff,
3. Hurt your opponent as much as possible, or
4. Maximize the difference between your own and your opponent's expected payoffs.

Table 15-9

		Player B	
		Escalates	*Retreats*
Player A	*Escalates*	$(V - D)/2$	V
	Retreats	0	$V/2$

(a) Payoff to A

		Player B	
		Escalates	*Retreats*
Player A	*Escalates*	$(V - D)/2$	0
	Retreats	V	$V/2$

(b) Payoff to B

		Player B	
		Escalates	*Retreats*
Player A	*Escalates*	0	V
	Retreats	$-V$	0

(c) Payoff to A – Payoff to B

Consider criterion 4 first. The appropriate payoff matrix (with respect to A's strategy) is that shown in Table 15-9c. This criterion converts the contest back into a zero-sum game with a saddle point where both players escalate. Value of the game to either player is .0.

In the case of criterion 3, player A looks at the payoff matrix for B (Table 15-9b). Whether B escalates or retreats, A always hurts B the most by escalating. The same is true for B. Thus, again, a saddle point exists. Value of the game to either player is $(V - D)/2 < .0$.

Consider criterion 2. To maximize his own minimum gains, A looks at the matrix in Table 15-9a. If A retreats, the worst he can do is .0 (if B escalates). But A stands to lose a lot if he escalates (if B also escalates). Hence A should follow a pure strategy of retreat. By the same argument, B should do likewise. Value of the game to each player is $V/2 > .0$.

Finally, look at criterion 1. To minimize B's gains, A looks at the matrix in Table 15-9b. The best B can do is 0 when A escalates. Similarly, the best A can do (Table 15-9a) is 0 when B escalates. But if B escalates, A can improve his gains by retreating. Again there is no apparent saddle point. However, note that now the aim of minimizing an opponent's gains is the same criterion as that used in (15-7) through (15-10) to calculate optimal strategy. Therefore, we can use the same procedure here. Let A's strategy be p and q. Then, looking at the payoff matrix for B (Table 15-9b), the value of the game for B playing a pure escalation strategy is

$$p\left(\frac{V - D}{2}\right) + (1 - p)V = -\frac{pV}{2} - \frac{pD}{2} + V$$

and the value of the game if B retreats is

$$p(0) + (1 - p)\frac{V}{2} = \frac{V}{2} - p\frac{V}{2}$$

Setting these expressions equal to one another, we find that

$$p = \frac{V}{D}$$

Using the quick and dirty rule (above), look at the payoff matrix for B (it is B's payoff that A wishes to minimize) and write

$$p:q = \left|V - \frac{V}{2}\right| : \left|\frac{V - D}{2} - 0\right|,$$

so that

$$p = \frac{V}{D}$$

Notice, in agreement with intuition, that as the value of victory rises, escalation becomes more likely, and that with a deepening in the agony of defeat, retreat occurs more often. The value of the game to either player is

$$-\frac{pV}{2} - \frac{pD}{2} + V \qquad \text{or} \qquad \frac{V}{2} - \frac{pV}{2} = \frac{V}{2D}(D - V) > 0$$

It is not always clear which of the four criteria is most appropriate when analyzing games. If different strategies are the phenotypic expression of different genotypes, we might expect selection to favor that genotype with the greatest *relative* fitness. If so, criterion 4 seems the one to use. The approach used by Maynard Smith and others, however, has been the standard practice of mathematicians—he used criterion 1. There is justification for this procedure, at least for some single-locus genetic models (see below). Therefore, we shall stay with this tradition in the remainder of the chapter.

Before continuing, we must ask two questions of our game-theoretic approach. First, the solutions found may represent optimal strategies, but can they evolve? Suppose that there are two game options available, as in the examples given above. If one pure strategy is determined by a given genotype, we would have to picture (for a one-locus case) a dominant allele, A, predisposing its owner to one behavioral response, the *aa* genotype performing the alternate response. Then any optimal mixed strategy can be provided by a given gene frequency. But the mixed strategy obtained applies not to individuals' strategies, but population strategies. Where subpopulations compete via the "playing" of games, group selection might act to bring about the optimal strategy in the population as a whole. Will natural selection within groups also bring about the optimal strategy? Within any group, selection will always favor that pure strategy that has, on average, the higher payoff. Therefore, unless there exists frequency-dependent selection, one allele or the other will go to fixation. Does frequency-dependent selection occur? Suppose that the frequency with which option x, as opposed to y, occurs in a population is $p:q$, and that an individual playing pure strategy x gains an average payoff of $E_x(x)$ when playing against other x strategists, $E_y(x)$ when playing against y strategists. Over many games, the average payoff to x strategists is $E_x = pE_x(x) + (1 - p)E_y(x)$. The average payoff to y-strategists is, by similar reasoning, $E_y = pE_x(y) + (1 - p)E_y(y)$. As long as $E_x > E_y$, the x-strategists are favored by natural selection and p rises in frequency. If $E_x < E_y$, then p falls. We write

$$\Delta p \text{ is of the same sign as } [pE_x(x) + (1 - p)E_y(x)] - [pE_x(y) + (1 - p)E_y(y)] \quad (15\text{-}11)$$

Clearly, selection is frequency dependent; p should come to equilibrium, $\hat{p}$, when the right side of (15-11) is zero, and the equilibrium is stable if $(\partial \Delta p/\partial p) < 0$ when $p = \hat{p}$. Solving (15-11) first, we find

$$\hat{p} = \frac{E_y(y) - E_y(x)}{[E_x(x) - E_x(y)] + [E_y(y) - E_y(x)]} \qquad 1 - \hat{p} = \frac{E_x(x) - E_x(y)}{[E_x(x) - E_x(y)] + [E_y(y) - E_y(x)]} \quad (15\text{-}12)$$

Furthermore,

$$\left(\frac{\partial \Delta p}{\partial p}\right)_{\hat{p}} = [E_x(x) - E_y(x)] - [E_x(y) - E_y(y)] = \text{denominator of (15-12)}$$

Therefore, p is stable if the denominator of (15-12) is negative, and so if $\hat{p}$ is to lie between zero and 1.0, the numerators of (15-12) must also be negative:

$$E_y(y) < E_y(x) \qquad E_x(x) < E_x(y) \quad (15\text{-}13)$$

Does the p of (15-12) agree with the game-theoretic optimum? Consider Table 15-10. We see that a saddle point exists for pure strategy x only if $E_x(x) > E_x(y)$; so that both players maximize their gains (or minimize their losses) when the other plays x. A saddle point exists at y when $E_y(y) > E_y(x)$. Thus if $E_x(x) < E_x(y)$ and $E_y(y) < E_y(x)$, a saddle point does not exist. But this (see equation 15-13) is the condition for $\hat{p}$ to be a stable mixed strategy. Suppose that a representative sample from the population, designated as hypothetical player A, plays strategy p, q. Then the payoff to another representative sample, designated B, gains $pE_x(x) + (1 - p)E_y(x)$ if it plays pure x, and $pE_x(y) + (1 - p)E_y(y)$ if it plays pure y. Setting these quantities equal, we find

$$p = \frac{E_y(y) - E_y(x)}{[E_x(x) - E_x(y)] + [E_y(y) - E_y(x)]}$$

This is identical to the $\hat{p}$ in (15-12). We conclude, then, that if there is not a saddle point, frequency-dependent natural selection acting on a single locus, with dominance, within a group,

Table 15-10

		Player B		Payoff to			**Player B**	
		x	y	A ↙ B ↘			x	y
Player A	x	$E_x(x)$	$E_y(x)$		**Player A**	x	$E_x(x)$	$E_x(y)$
	y	$E_x(y)$	$E_y(y)$			y	$E_y(x)$	$E_y(y)$

will favor the evolution of a mixed strategy which minimizes the value of the game to an average opponent. Furthermore, that strategy is an *evolutionarily stable strategy* (ESS)—that is, it is neighborhood stable; it precludes the successful invasion of an alternative strategy. Note that use of criterion 1 to solve for optimal game strategy is appropriate in single-locus models, where genotypes follow pure strategies.

Now suppose that a given genotype codes for a *mixed* strategy. Suppose that I strategists play p, q, and at least temporarily dominate the population. We introduce a new genotype whose owners are strategists playing p', q', and ask whether this new strategy can make evolutionary headway in the population. The answer is quite straightforward. If p', q' is closer than p, q to the optimal strategy, as calculated using the appropriate criterion, individuals playing J will demonstrate higher fitness than I individuals,

$$E_I(J) > E_I(I)$$

and the new genotype will increase in frequency by natural selection. Thus new strategies that come progressively closer to the optimum can successfully invade the population and increase in abundance. Suppose that I, the established strategy, is optimal. Then there is no new strategy, J, that can do better against I than I itself,

$$E_I(J) \leqslant E_I(I) \tag{15-14a}$$

Furthermore, I can do at least as well against J as J can,

$$E_J(I) \geqslant E_J(J) \tag{15-14b}$$

When conditions (15-14) are met, I is uninvadable; it is an evolutionarily stable strategy—an ESS.

Problems arise when each of the three genotypes determines a different pure strategy. In this case there are now three possible options for each player. It is hypothetically possible to achieve a ratio of genotypes such that the optimal mixed strategy is realized, but the chance that this ratio also satisfies a genetic equilibrium ($W_1 = W_2 = W_3$ when selection is frequency dependent) is vanishingly remote. Maynard Smith (1981) produced a simple genetic model, one locus, two alleles, for such a situation and came to this same conclusion (see also Slatkin, 1978). He also pointed out, though, that as the number of alleles per locus increases or the number of loci determining behavior rises, the number of phenotypic options markedly increases also. In the extreme case (recall Lande's formulations in Chapter 10), phenotypic selection models are applicable and frequency-dependent selection can result in strategies that approach the optimal game solution quite exactly.

If a single locus determines behavior and each of the three genotypes promotes a different *mixed* strategy of two options, the problem of simultaneously equilizing fitnesses and satisfying the game optimum disappears (Treisman, 1981).

If each individual can adjust its strategy appropriately (natural selection has favored adaptive plasticity), even closer adjustment to optimality might be realized. The real advantage to such behavioral plasticity is that the additional complexity engendered by many structural genes plus modifier genes acting on genotype–environment interaction make possible a tremendous array of (appropriate) responses. Thus optimal strategies do not have to evolve independently for games of different sorts played under a variety of circumstances.

Let us consider a slightly more realistic game. Suppose that both antagonists escalate, but only to a point where they perceive their probability of escaping injury to fall to (say) x. If either is injured, the other wins and the bout terminates. (A slight variant of this game, which can be analyzed identically and which more nearly fits the situation in, say, fights among fish, is to let x be the *extent* rather than the probability of damage that an organism is willing to sustain.) If a player is cautious, he may never win. But if he is too rash, he may be injured. What value of x is optimal? Suppose that player A is willing to let his x drop to x_A and that B plays $x = x_B$. Suppose that $x_B > x_A$ (player A is more of a risk taker than his opponent). Then as soon as the probability of injury rises to $1 - x_B$, B quits. The probability of no injury occurring is x_B and player A always outlasts his adversary, winning a payoff of V. But if injury occurs (probability $1 - x_B$), assuming that both participants are equally susceptible, player A will have won with probability $\frac{1}{2}$ and will have been injured with probability $\frac{1}{2}$. Player A's expected payoff is thus

$$x_B V + (1 - x_B)\frac{V - D}{2} \qquad \text{(recall that } V < D) \tag{15-15}$$

If $x_B < x_A$, player A can never win without an injury occurring. If injury does occur before A gives up (with probability $1 - x_A$ now), A gains an expected payoff of

$$(1 - x_A)\frac{V - D}{2} \tag{15-16}$$

To find the optimal, mixed strategy, suppose that B lets x_B fall to between x and $x + dx$ with probability $\phi(x)\,dx$, and that A plays a pure strategy of $x_A = k$. Then, substituting from (15-15) and (15-16), the expected payoff to A is

$$\int\left[xV + (1 - x)\frac{V - D}{2}\right]\phi(x)\,dx$$

when $x > k$, and

$$\int(1 - k)\frac{V - D}{2}\,\phi(x)\,dx$$

when $x < k$. The total payoff to A is thus

$$\psi_A = \int_k^1\left[xV + (1 - x)\frac{V - D}{2}\right]\phi(x)\,dx + \int_0^k(1 - k)\frac{V - D}{2}\,\phi(x)\,dx \tag{15-17}$$

If B's strategy is optimal, A cannot improve this payoff value; that is, ψ_A is invariant with k. Accordingly, we can solve for $\phi(x)$ by writing

$$0 = \frac{d\psi_A}{dk} = -\frac{d}{dk}\left\{\int_k^1\left[xV + (1 - x)\frac{V - D}{2}\right]\phi(x)\,dx + (1 - k)\int_0^k\frac{V - D}{2}\,\phi(x)\,dx\right\}$$

$$= -kV\phi(k) - \int_0^k\frac{V - D}{2}\,\phi(x)\,dx$$

so that

$$kV\phi(k) = \int_0^k\frac{D - V}{2}\,\phi(x)\,dx \tag{15-18}$$

Differentiating again gives us

$$V\phi(k) + kV\frac{d\phi}{dk} = \frac{D - V}{2}\,\phi(k)$$

so that

$$\frac{1}{\phi}\frac{d\phi}{dx} = \frac{3V - D}{2xV}$$

Integration yields

$$\phi(x) = cx^{\frac{D-3V}{2V}}$$

The constant c is simply the value we need such that $\int_0^1 \phi(x)\,dx = 1.0$. It is $(D - V)/2V$. Finally, then,

$$\phi(x) = \frac{D - V}{2V} x^{\frac{D-3V}{2V}} \qquad \text{where } V < D \tag{15-19}$$

Because the payoff matrix for both contestants is the same, $\phi(x)$ is optimal for both players. As expected, the optimal strategy calls for an increased willingness to risk damage [$\phi(x)$ is high for low x] if V is large, but more caution when D is large. The actual expected payoff may be calculated from (15-17). It is zero.

In passing, note that the payoff from the previous, simpler game—escalate until win or get injured with probability $p = V/D$, and display, retreat with probability $1 - V/D$—gives a higher expected payoff than that given by this game. Therefore, if animals have the capacity to play only $x = 1$ or $x = 0$ (the simple game), it pays to have confrontations. Where more fine tuning of the behavioral response is possible, it does not.

In the analyses above it was implicitly assumed that neither individual had an advantage. Conflicts began with equal chances to win or lose. But in real life this is not usually the case. One individual is larger than another, or is on home turf, and so on. In such cases the contests are "asymmetric." These more realistic games can be analyzed fairly simply. Consider the game where an individual escalates until victory or injury, or retreats. We shall now suppose that each individual entering into such a game perceives the situation differently from the other. If player A is on home ground, or is bigger, say, he views the contest in one way (label his perception of the conflict P); player B views it in an alternative way (Q). Suppose that if an individual perceives P (which occurs with frequency θ), his relative chances of winning a bout, if both players escalate, is w. If both retreat, the probability of winning is $\frac{1}{2}$. The value of winning and losing will be denoted V_P, or V_Q, D_P, or D_Q. Now let us consider two possible moves: (x) if see P, escalate until win or get injured; if see Q, retreat; and (y) if see P, retreat; if see Q, escalate. If A sees P (so that B sees Q) and both make move x, the payoff to A is V_P, for he will escalate and his opponent will retreat. If A sees Q, he will retreat and gain nothing. Thus the payoff to A of playing x, when B also plays x, is

$$\theta V_P + (1 - \theta)0 = \theta V_P$$

If A sees P and plays x, but B plays y, both escalate. The value to A is $[wV_P/(w + 1)] - [1/(w + 1)]/D_P$. If A sees Q, both retreat and A wins an amount $V_Q/2$. His expected payoff is

$$\frac{\theta(wV_P - D_P)}{w + 1} + \frac{V_Q(1 - \theta)}{2}$$

By similar reasoning, if A plays y, B plays x, the payoff to A is

$$\frac{\theta V_P}{2} + (1 - \theta)\frac{V_Q - wD_Q}{w + 1}$$

and if both play y,

$$\theta \cdot 0 + (1 - \theta)V_Q = (1 - \theta)V_Q$$

The payoff matrix to A is thus

		Player B	
		x	y
Player A	x	θV_P	$\theta \frac{wV_P - D_P}{w+1} + (1-\theta)\frac{V_Q}{2}$
	y	$\theta \frac{V_P}{2} + (1-\theta)\frac{V_Q - wD_Q}{w+1}$	$(1-\theta)V_Q$

where w = the chance that a player perceiving P wins if both escalate
V_P, V_Q = value of victory if the victor, perceives situation P, Q
D_P, D_Q = value of defeat if the loser perceives situation P, Q
θ = frequency with which a player perceives situation P

We shall analyze this game using hypothetical data for possible real-life situations.

1. Two individuals encounter each other on neutral ground. Then P, Q, which must relate to relative values of individual characteristics, affect neither V, D, nor w. Then $w = \frac{1}{2}, \theta = \frac{1}{2}$, and the payoff matrix becomes

		Player B	
		x	y
Player A	x	$\frac{1}{2}V$	$\frac{2V - D}{4}$
	yk	$\frac{2V - D}{4}$	$\frac{1}{2}V$

There is no saddle point and from considerations of symmetry it would seem that the optimal mixed strategy was $\frac{1}{2} : \frac{1}{2}$. To check this, suppose that player B plays p, $1 - p$, and A plays pure x. Then

$$\text{payoff to A} = \psi_A = p\frac{V}{2} + (1-p)\frac{2V - D}{4} = \frac{2V - D + pD}{4}$$

If A plays pure y, then

$$\psi_A = p\left(\frac{2V - D}{4}\right) + (1-p)\frac{V}{2} = \frac{2V - pD}{4}$$

Setting these two expressions equal to one another, $p = \frac{1}{2}$, as expected. The value of the game is

$$\frac{1}{2}\frac{V}{2} + \frac{1}{2}\frac{2V - D}{4} = \frac{2V - \frac{1}{2}D}{4}$$

The significant difference between this game and that described earlier (equations 15-7 to 15-10) is that now the strategies are conditional on the two antagonists viewing the environment in opposing ways. The decisions are not escalate versus retreat, but rather escalate if see P, retreat otherwise, or vice versa. Perhaps an example will make this clearer. Two individuals are foraging together when one (A) spots a particularly inviting food morsel. Both perceive that A has seen the item first. Does B allow A to proceed peacefully to eat, or does he try to reach the food first? Does A threaten B in order to try and assure prior rights, or risk losing the item if B decides to grab it? Here P designates first sighting, and the analysis above states that each individual will maximize its net gains by threatening half the time and behaving passively half the time. In the earlier game, both individuals see the food simultaneously—or at least there is no concept of prior sighting—and each decides either to fight for the item or not. The value of the conditional game is higher, meaning that a tacit recognition of asymmetry is to an individual's benefit.

2. Seeing P designates an individual's recognizing himself as larger, more aggressive, more likely to prevail in an encounter. In such a case w exceeds $\frac{1}{2}$. Furthermore, a large individual will realize a θ of greater than $\frac{1}{2}$ because he generally encounters smaller individuals. Thus the strategies of A and B will not necessarily be the same. Suppose, for illustrative purposes, that encounters occur on neutral ground ($V_P = V_Q$, $D_P = D_Q$) and that $V = 1$, $D = 2$, and $w = 1$. Then the payoff matrix is

		Player B	
		x	y
Player A	x	θ	$\frac{1 - 2\theta}{2}$
	y	$\frac{2\theta - 1}{2}$	$1 - \theta$

If $\theta < \frac{1}{4}$ (A is small or, for other reasons, not likely to prevail in an encounter), the smallest element in row one is θ, and the smallest item in row 2 is $(2\theta - 1)/2$. Because the latter is smaller than the former, A will do best to follow x. If $\frac{3}{4} > \theta > \frac{1}{4}$, the smallest element in row 1 is $(1 - 2\theta)/2$, and in the second row, $(2\theta - 1)/2$. Again the smallest element in the first row exceeds that in the second and A does best to play x. Finally, if $\theta > \frac{3}{4}$, $(1 - 2\theta)/2$ and $(1 - \theta)$ are the smallest elements in rows 1 and 2 and still x is A's best play. We conclude that a player wishing to maximize his payoff should always escalate if he perceives that he will win the fight, but retreat if he thinks that he is likely to lose.

3. P and Q designate differences in perceived values of a win or a loss. For example, an animal on its territory has much to gain by retaining its title, the same, in fact, that a challenger gains by wresting it away from him. But risk of injury in territorial squabbles is probably quite small. Because most territorial individuals spend little time challenging other territory holders, most encounters will be on home turf ($\theta > \frac{1}{2}$). The appropriate, somewhat idealized, values for such a game might be

		Player B	
		x	y
Player A	x	V	$\frac{V}{2}$
	y	$\frac{V}{2}$	0

The optimal strategy for both players is x; A territorial male will always do best to attack intruders; the intruder should back off.

In addition to its primary purpose of introducing a method for predicting strategies in conflict situations, this section points out one very important conclusion. An asymmetry in which one individual perceives that he can win a conflict, the other believes that he will lose, or simply an asymmetry of the form, "this is my territory–this is not my territory," should lead to peaceful resolution of conflicts. The critical and interesting point is that it is the *perception*, not the reality, that counts. Thus conventions are born. If, somehow, individuals in a population come to perceive, rightly or wrongly, that bright epaulets (say) conveyed an aggressive advantage, conflicts would be peacefully resolved with the more brightly endowed individual winning, the other withdrawing. The group would maintain peace while natural selection frantically favored brighter epaulets. If prior sighting of an item came to be regarded as a prior right, peaceful resolution of conflict, again, is likely to evolve.

What happens to the optimal strategies of game playing when antagonists are kin? It seems fairly straightforward to conclude that if $E_I(J)$ is the expected payoff to a J strategist playing an I strategist, the inclusive (fitness) payoff to J is

$$E_I(J) + rE_J(I)$$

where r is the coefficient of relationship. But this reasoning fails to account for the fact that related individuals are more apt to use the same strategy. If antagonists interacted nonrandomly with other individuals, the optimal approach to a game might be different. Consider the case where r, rather than indicating coefficient of relationship, measures the probability that an antagonist uses the same strategy as his opponent *by virtue of similar genotype* (identity by *type* is implicated now). Then the fitness increment (if payoffs are measured in fitness increments) to the J-strategy genotype is

$$\Delta W_J = rE_J(J) + (1 - r)\sum_I E_I(J)f(I) \tag{15-20}$$

where $f(I)$ is the frequency in the population of I strategists (see Hines and Maynard Smith, 1979). That the formulation above can alter the outcome of a game-theoretic problem is readily seen from the following example. Suppose that the *individual* payoff matrices are

Payoff to A (↙):

		Player B	
		x	y
Player A	x	-1	2
	y	1	1

Payoff to B (↘):

		Player B	
		x	y
Player A	x	-1	1
	y	2	1

A's best strategy is then, from the B payoff matrix, $|2 - 1|:|-1 - 1| = 1:2$, so that $p = \frac{1}{3}$. B's strategy is similarly $\frac{1}{3}$, and the value of the game to A (or B) is

$$\left(\frac{1}{3}\right)^2(-1) + \left(\frac{1}{3}\right)\left(\frac{2}{3}\right)(2) + \left(\frac{2}{3}\right)\left(\frac{1}{3}\right)(1) + \left(\frac{2}{3}\right)^2(1) = 1.0$$

But more accurately, because A and B may well carry similar genes ($r \neq 0$), we should write (where ΔW_J is the increment in fitness for genotype J, represented by player B)

$$\begin{aligned}\Delta W_J = {} & r[p'^2(-1) + p'q'(2) + p'q'(1) + q'^2(1)] \\ & + (1 - r)[p'p(-1) + p'q(2) + pq'(1) + qq'(1)]\end{aligned} \tag{15-21}$$

where p, q is strategy I, followed by A, p', q' is strategy J, followed by B. To find A's best strategy (best for its genotype), we look at ΔW_J for the situation in which B plays pure strategies $p' = 0$, $p' = 1$, and set them equal. Accordingly,

$$r[1(-1)] + (1 - r)[p(-1) + q(2)] = r[1(1)] + (1 - r)[p(1) + q(1)]$$

so that

$$p = \frac{1 - 3r}{3(1 - r)}$$

The value of the game to the hypothetical player A is thus

$$\frac{1 - 3r}{3(1 - r)}(-1) + \frac{1 - 3r}{3(1 - r)}\frac{2}{3(1 - r)}(2 + 1) + \left[\frac{2}{3(1 - r)}\right]^2(1) = \frac{3 - 4r - 3r^2}{3(1 - r)} < 1.0$$

So the value to an individual is less than that obtaining if he had played the game without regard to genetic similarity of his opponent. However, the value to his *genotype* is (calculated for $p' = 0$, equation 15-21), is

$$r[1(1)] + (1 - r)[p(1) + q(1)] = 1.0$$

A final word on the theory of games is necessary. All examples given above assumed that opponents encounter one another randomly. If this were not so, the conditions for stability as well as the optimal strategies themselves would be altered. This is an extremely important aspect of game-theoretic analysis in biological situations. Unfortunately, there is not adequate space to deal with it here. The reader is directed to Fagen (1980).

15.4 Group Size

In spite of the presence of peacekeeping devices such as dominance hierarchies, conventions, and the existence of reciprocal and kin-selection-based altruism, social groups are not without their insidious secrets. In gull colonies (see, e.g., Pierotti, 1980) adult males may pirate food from young ("kleptoparasitism"), attack older chicks, and injure or kill them. The only apparent advantage to the latter behavior—the dead chicks are not eaten—is to depress fitness contributions of other individuals. Such behavior might not be advantageous in a group with close kin ties, but large groups are not necessarily composed of relatives. Sherman (1981) watched conspecifics kill 8% of all pups born in his Belding's ground squirrel colony. Only nonrelatives were killed, suggesting that the practice of infanticide was directed toward increasing the killer's contribution to his genotype's relative, inclusive fitness by depressing that of others. One important function of territoriality in this species may be to keep potential killers away from young. Sherman remarks that infanticide is quite common in the ground squirrel genus *Spermophilus*.

Such strains are bound to set soft upper limits to the number of individuals that can freely or even occasionally interact. In the groove-billed ani (Vehrencamp, 1977), birds breed communally, one to four monogamous pairs laying their eggs in a single nest. As the number of laying females rises, it becomes difficult for the birds to keep track of all the eggs. Some fail to get turned, some are not properly incubated. Whether for this reason or simply as a measure to raise one's own relative output by hurting others, females often flip eggs of their nest mates out of the nest. They simply remove eggs before laying their own. This means that females that lay late will contribute a greater proportion of eggs. But late-layed eggs hatch late and so, in a single nest where there is competition for food, might be expected to be at a disadvantage in obtaining food by virtue of small size. If this is true, however, it is fairly unimportant in most breeding units. There is a clear advantage to being the last to breed, and the most dominant females enforce this order. Subordinate females, obligated to go first, counter their predicament to some degree by laying larger numbers of eggs as group size increases, and occasionally sneaking in a single, late egg. In hypothetical groups of very large size, subordinates forced to lay very early would contribute almost nothing and would find it advantageous to leave, and very late-laying females might find their own young at an extreme size disadvantage unless they raised their rate of egg destruction to genocidal levels. The social breeding group of anis simply cannot expand indefinitely.

Among foraging lions, group size must, among other things, reflect a balance between the increased efficiency of cooperative hunting and the demands of more hungry mouths. Caraco and Wolf (1975) examined hunting proficiency of lions feeding on Thomson's gazelles. Using data from Schaller (1972), they report a 15% capture rate for lone hunters, 31% for groups of two and larger. Figuring the edible weight of a gazelle per lion, the number of chases likely per time, and this hunting efficiency value, Caraco and Wolf figure that food intake per lion is maximized at a pride size of two. Single lions, and foraging groups of three, will not be able to achieve the 6-kilogram intake rate per day that is estimated to be necessary to their survival. Of course, sometimes multiple kills are made, and even in the absence of herds of larger game, prey other than gazelles are often available. The number two, then, can be taken with some room for variation. In the Serengeti during the migratory absence of larger game, lions do, in fact, usually travel in groups of one to three. Caraco and Wolf also calculated optimal pride size for the season when wildebeest and zebra were available in great numbers. Two is still the ideal number, although considerably more leeway is now possible. But groups of up to seven are often seen. This is within, but barely within, the permissible range calculated by Caraco and Wolf. There are other advantages that act to increase group size. Mutual care of loosely synchronized young in nurseries, with six to eight females, increases survival of the cubs. Also, several lions together can chase groups of hyaenas from kills.

More will be said of the influence of foraging and predation on group size in the next chapter. For the moment we simply note some further social factors encouraging or discouraging large group size. Beneficial consequences include an increase in group-competitive ability with conspecifics. Large monkey troops in general are dominant to small troops. Large wolk packs displace smaller ones. But militating against large size is increased competition for food in the small area occupied by the group at any one time. For example, birds nesting in dense colonies, like the quelea (*Quelea quelea*), a social weaver, must forage farther from the nest as a result of their concentration and, as we have already noted in Chapter 14, may have had to alter their mating system as a result. Communication systems also may break down under the strain of too many individuals, dominance rankings, and alliances to remember. Finally, consider a species in which mates of one sex are acquired outside the group and then move into the group as new members. Some human groups, such as the Yanomamo, are an example. For such groups the average relatedness among individuals, and so the genetic similarity, declines as deme size grows over time. When genetic similarity among sizable fractions of the deme drops below some critical level depending on costs and benefits of cooperation, the conditions for continued cooperative behavior under trait group selection are no longer met. We might expect strife to develop and fissioning of the group to occur along family lines. This is exactly what happens among the Yanomamo (Chagnon, 1975).

Review Questions

1. Consider a diverse array of different animal species and, for each, decide whether kin, trait group, reciprocal altruism, interdemic, or individual selection is the most important force for ongoing social evolution. Defend your answer.
2. What might be some of the advantages and disadvantages to being at the top, the bottom, or in the middle of a dominance hierarchy? Under what conditions would you expect dominance hierarchies to evolve? When we look at dominance hierarchies, are we seeing San Marco spandrels?
3. Think of concrete examples to which Axelrod and Hamilton's reciprocal altruism theory might apply.
4. Axelrod and Hamilton's tit-for-tat model is predicated on the possibility that individuals will periodically meet and interact over time. What effect might environmental fluctuations, altering the payoff matrix, have on their conclusions? What if the game varied qualitatively, from prisoners dilemma to one saddle point or the other, over time?
5. Female rhesus monkeys, as they reach maturity, assume dominance over their next oldest sister. In light of the effects of age on kin selection, does this make sense? Recall the Medawar hypothesis from Chapter 12. What are the implications of kin selection for aging?

Chapter 16

The Evolution of Prey–Predator Interactions

16.1 Predator Adaptations

It is a somewhat trivial observation that different species of animal obtain their food in different ways. The natural history of feeding and foraging is extremely rich. Methods of meeting nutrient demands range all the way from absorption of dissolved amino acids through the skin (see Stephens, 1968) to highly organized cooperative hunting behavior. The latter involves such divergent structures as wolf packs, foraging "company" ants (good reviews of ant behavior are given by Sudd, 1967, E. O. Wilson, 1971, and Oster and Wilson, 1978), and moving spread-out lines of abdims storks (*Sphenorhynchus abdimii*), walking across the African plains, driving insects before them (Jackson, 1938; cited in Friedman, 1967). Harbor porpoises (*Phocaena vomerina*) have been observed to swim in formation, driving Pacific sardines to the surface and then taking turns swimming through the concentration of food (Fink, 1959). Friedman (1967) describes wattled starlings (*Creatophora carunculata*) flocks forming a hollow cylinder within which they trap locust swarms. Poikilotherm species, including fish and sea stars (Mauzey, 1966) may feed only when food is plentiful or temperatures warm, and are energetically better off to cease feeding at other times. Some animals store food in times of abundance and can forgo extensive foraging at other times as a result. Scavenging is common. The bald eagle (*Haliaetus leucocephalus*), for example, harasses osprey on the wing, until a hapless tormentee releases the item it clutches and the eagle scoops it up. Hyaenas often approach hunting dog packs within a few yards and wait until the latter make a kill. If the hyaenas are strong enough in number, they then chase the dogs off and feed (Estes and Goddard, 1967). Predative mimicry also occurs. *Mallophora bomboides*, an asilid fly (robber fly), resembles a bumblebee, complete even to the point of mimicking the low, buzzing sound. This allows it to approach a bee, grab it from behind, and eat it (Brower et al., 1960). Among parasites, there are found an intriguing variety of techniques to facilitate continued infection. The amphipod, *Gammarus lacustris*, is normally negatively phototactic and dives when disturbed. This behavior keeps individuals away from the surface of the water body in which they live, where predators are often prevalent, and provides an escape response when attacked. When infested with the younger, immature form of the acanthocephalan parasite *Polymorphus paradoxus*, they continue to behave in this manner. But when their internal nemeses reach the adult infective stage, there is a parasite-induced change in their behavior. They skim the surface and *rise* when disturbed. This new behavior makes them almost sure targets of mallard ducks or muskrats (their definitive hosts), which ingest them incidentally while foraging in vegetation (Bethel and Holmes, 1973). Finally, although we shall resist the temptation to elaborate, one of the most fascinating and diverse groups of animals in respect to their foraging and feeding methods is the Arachnidae. The interested reader is referred to Kaston (1965).

There is no space in a book of this sort to dwell at length on natural history. So we shall get back forthwith to our attempts to discern general patterns of order in the biological smorgasbord of prey–predator interactions.

Food Preference

The seminal paper in this field is that of J. M. Emlen (1966a). His subsequent article (J. M. Emlen, 1968b), though, reiterates the same findings, is far simpler and more accurate, and will be the subject of the immediately following discussion. We consider two (or more) alternative food types for a consumer, each consisting of individual items with various food "value" to the predator. As we shall see shortly, "value" is not necessarily a parameter that can be scaled in a single dimension, but for the present, note that for carnivores (whose diet varies largely with respect to the capture and handling time of prey and not the prey items' differences in nutrient composition) and for small homeotherms (whose nutritional requirements might normally be expected to be fully met before caloric needs are satisfied), "value" can be reasonably approximated in terms of net calories obtained per unit time. Consider, then, that the items of each food class follow a frequency distribution with respect to "value" (Fig. 16-1). The ordinate describes the number of items encountered over a defined period of time, and the area under the frequency–distribution curve for (say) food species A is the total amount of A potentially available to the consumer. In Figure 16-1a, foods A and B are approximately equal in their availability to the predator, but food B, on average, consists of more valuable items than A. In Figure 16-1b, the same two foods are represented, but now type B is only, perhaps, one-half as abundant (to the consumer) as A. We now suppose that our consumer has indeed evolved to behave in an optimal manner. That is, we assume that it is reasonable to expect the consumer to feed in such a way that it eats all items of value higher than some threshold level $\hat{v}$, and none below that level. The level is set such that the area of both A and B curves *above* that point is the cumulative value actually eaten by the consumer. In Figure 16-2 we show how this "optimization" assumption translates into predictions about food preference patterns. All foods to the right of the threshold value are eaten, meaning (shaded portions of the curves) that about 80% of food B is taken, perhaps 20% of A.

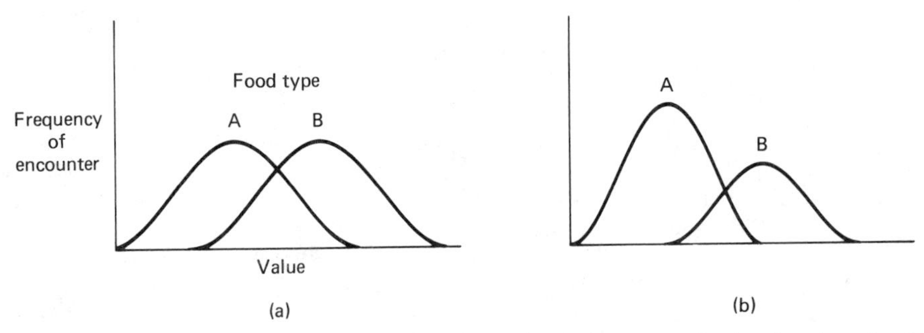

Figure 16-1 Frequency with which a foraging animal encounters foods of various food "value," plotted for two different food types. In (a) food types A and B are about equally available, in (b) there is more A available than B.

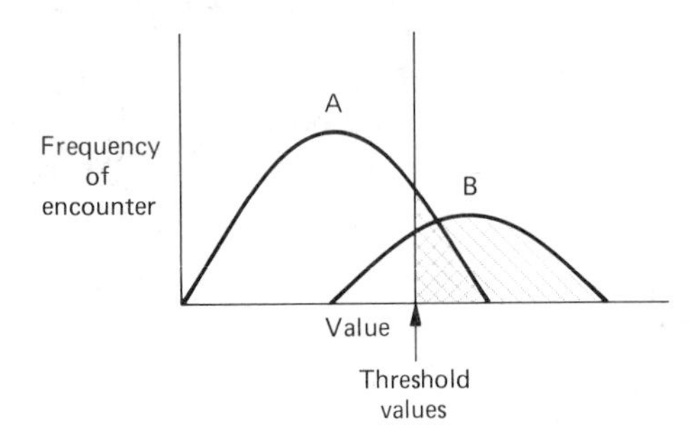

Figure 16-2 Frequency with which a foraging threshold animal encounters foods of various food "value," plotted for two different food types. If the forager feeds only on items of value greater than the threshold value $\hat{v}$, the shaded areas give the amounts of foods A, B eaten.

There are a number of basic assumptions implicit in the approach described above. First, by eating only the very best, the consumer is maximizing its food value intake per time. That presumably translates into better health, greater reserves for sustaining life and for reproduction, and hence greater fitness. We ignore the possibility of trade-offs, the chance that by feeding optimally the organism sacrifices somewhere else, with a resulting, net *decline* in fitness. We are assuming that eating an optimal diet has a magnifying effect on fitness. Second, we are assuming almost an omniscience on the part of the consumer, taking it for granted that this animal has an accurate picture in its mind of the two (or more) distribution curves, and can accurately assess the value of each food item it encounters. No one really believes a consumer will be able to do this, but we have to start somewhere and such idealistic expectations can always be relaxed as becomes necessary. Because of these exaggerated claims to consumers' abilities, the model above should be used only to make predictions of a qualitative nature.

1. If $\phi_A(v)$, $\phi_B(v)$ describe the curves in Figures 16-1 to 16-4, where v is value, then

$$V^* = \int_{\hat{v}}^{\infty} \phi_A(v)\, dv + \int_{\hat{v}}^{\infty} \phi_B(v)\, dv$$

is the total food value obtained by a foraging individual accepting items only if they lie above the critical $\hat{v}$ value. If a consumer's food value intake, V^*, is held constant, then when both foods are scarce (Fig. 16-3a), as opposed to abundant (Fig. 16-3b), $\hat{v}$ moves to the left and the relative proportions of A and B consumed are more alike. Animals that behave optimally will be more particular and so more specialized when food is common, less particular and so less specialized when food is scarce.

2. In Figure 16-4 the total area under *both* curves A and B together (representing the total amount of food encountered) is held constant, but the *relative* areas under A and B (relative amounts of foods A and B encountered) are varied. In Figure 16-4a, for example, food A makes up approximately 90% of all food (A plus B) encountered; in Figure 16-4b perhaps 70%; and so on. Suppose that V^* is constant—an animal satisfies its nutritional needs equally in all cases as depicted in Figure 16-4. Then the value of $\hat{v}$ can be found for each case, and the slashed area to the right of $\hat{v}$ can be found for both foods. These slashed areas are the amounts of A and B eaten. The amounts of (say) B eaten as a proportion of total B available can then be calculated for each case and Figure 16-5 can be constructed. There are two lessons provided by Figure 16-5. First, it is not a simple matter to describe "food preference." A single number is almost certainly inadequate in most cases without reference to both the relative and absolute abundances of the foods available. Second, the feeding response in Figure 16-5 is distinctly sigmoid, showing a "switching" in diet at some critical point as a food becomes relatively abundant. As we

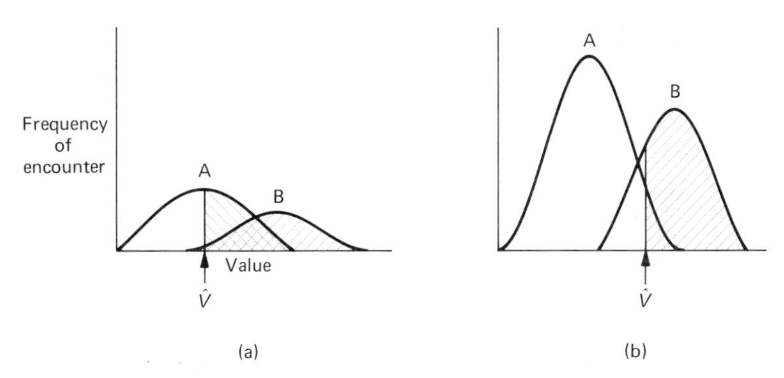

Figure 16-3 Same as Figure 16-2. In (a) both foods are scarce; in (b) both foods are of similar relative but much greater absolute availability.

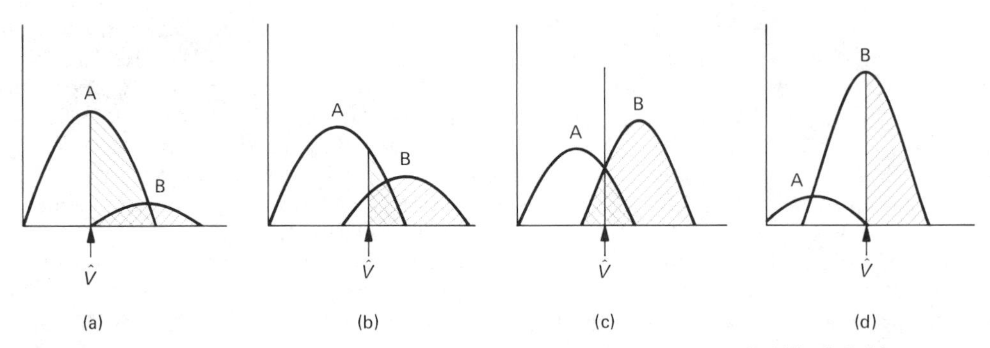

Figure 16-4 Effect on amounts of foods A and B eaten when total food availability is held constant but relative availabilities are varied.

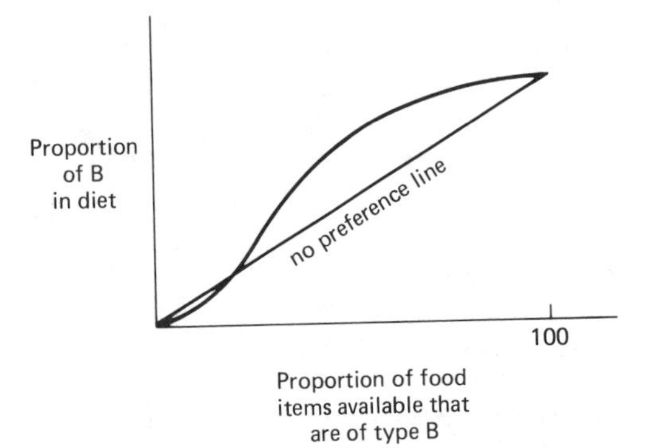

Figure 16-5 Effect on amounts of foods A and B eaten when total food availability is held constant but relative availabilities are varied.

mentioned in Chapters 4 and 5, switching, and indeed any sigmoid form of the feeding curve, is critical to the nature of the prey–predator isoclines (see Figures 4-19 and 4-20) and to the continued coexistence of competing food species under predation ["Effects of Predation on Competing Species" (p. 156)].

Before testing the most basic prediction of this model—more "valuable" foods are preferred to less valuable foods—we turn to other, more recent models. In 1974, Pulliam argued like this. Let λ_i define the availability (food density times ease of finding items) of the ith prey type, $\lambda = \sum_i \lambda_i$, and t_i be the time required to catch and eat an item of prey type i. Then the time needed to find an item of type i is proportional to $1/\lambda_i$, and the search time per item of any type is $1/\lambda$. If p_i is the proportion of type i items encountered that are eaten, and p_i and p_j are independent of what has previously been eaten, the average search time per item eaten is

$$\frac{1}{\sum_i \lambda_i p_i} = \frac{1}{\lambda^*},$$

and handling time is

$$\frac{\sum_i \lambda_i p_i t_i}{\sum_i \lambda_i p_i} = \frac{\sum_i \lambda_i p_i t_i}{\lambda^*}$$

Thus the total time spent per item, on average, is

$$\frac{1}{\lambda^*} + \frac{\sum_i \lambda_i p_i t_i}{\lambda^*} \tag{16-1}$$

But if the value (in calories) obtained from an item of type i is c_i, then the total average value intake per item is

$$\frac{\sum_i \lambda_i p_i c_i}{\lambda^*} \tag{16-2}$$

Dividing (16-2) by (16-1) we obtain the mean number of calories consumed per mean time, over all food items eaten,

$$C = \frac{\sum_i \lambda_i p_i c_i}{1 + \sum_i \lambda_i p_i t_i} \tag{16-3}$$

which Pulliam uses as a measure of diet quality.

Suppose that there are only two food types. Then

$$C_{1 \text{ and } 2} = \frac{\lambda_1 p_1 c_1 + \lambda_2 p_2 c_2}{1 + \lambda_1 p_1 t_1 + \lambda_2 p_2 t_2}$$

But because all items of a type are identical, there is no advantage to eating some but not others. Hence $p_1 = 0$ or 1, $p_2 = 0$ or 1. Thus we distinguish three kinds of diets: eat only type i, in which case

$$C_i = \frac{\lambda_i c_i}{1 + \lambda_i t_i} \tag{16-4a}$$

or eat both

$$C_{1 \text{ and } 2} = \frac{\lambda_1 c_1 + \lambda_2 c_2}{1 + \lambda_1 t_1 + \lambda_2 t_2} \tag{16-4b}$$

A pure strategy of type i is favored if $C_i > C_{i \text{ and } j}$ or, after some manipulation,

$$c_i > c_j \frac{1 + \lambda_i t_i}{t_j \lambda_i} \tag{16-5a}$$

A pure strategy of type j should occur, by similar reasoning, when

$$c_i < c_j \frac{\lambda_j t_i}{1 + \lambda_j t_j} \tag{16-5b}$$

If c_1 falls between the values in (16-5a) and (16-5b), a mixed diet is expected. The model states that:

1. If a food is "superior" and sufficiently abundant (conditions 16-5a or 16-5b), it should be eaten to the exclusion of the other food.
2. The term λ_j does not appear in (16-5a), nor does λ_i appear in (16-5b). Thus whether a food (say type i, inequality 16-5a) is eaten exclusively or not depends on *its* abundance being sufficiently high, but is independent of the abundance of the alternative food.
3. As food becomes more scarce, the diet should become more inclusive and the value of the poorest acceptable food type drops.

The predictions arising from Pulliam's model have become depressingly entangled with what appears at first glance to be a contradictory statement from Emlen's model. In Emlen's argument abundance of the inferior food certainly *does* contribute to the degree of preference for the superior food. Thus it is probably a worthwhile exercise to point out that the two approaches really address different situations. To Pulliam, a food type is a collection of identical items; to Emlen it is a set of items of variable value belonging to a given class. The class of all oat seeds might qualify as food type A in Emlen's model, while in the case of Pulliam's model, "oat seeds" contains a very large set of food types such as oat seeds of shape x, weight y, amino acid content z, and so on. When viewed in this light it can be seen that the two approaches are complementary and reinforcing and not at all contradictory. Pulliam's model says, essentially, that a food item with value v_1 should be preferred to one with value $v_2 < v_1$, and if items of value v_1 are sufficiently abundant to satisfy an animal's needs, that animal should reject items of value v_2 regardless of their abundance.

There have been other models of the same basic form as the two noted above. For example, Estabrook and Dunham (1976) constructed a model essentially similar in form to that of Pulliam. Their variant on this theme is clever, incorporating conditional probabilities of encountering an item of type j after eating one of type i. The result of this added complication is that Pulliam's prediction 2 no longer necessarily holds. Otherwise, the conclusions, as we might expect, are the same: Eat a broad diet when food is scarce; specialize when food is abundant; never eat one food but not another of higher value.

McNair (1979, 1980) has pointed out that food may not be randomly dispersed, so that elaborations such as those proposed by Estabrook and Dunham are important. He produces an elegant formulation that incorporates spatial patchiness in the food supply and also training effects (search images and increased efficiency of search and handling with practice). In the presence of either patchiness or training effects Pulliam's predictions may be seriously compromised. Further implications of McNair's work are discussed in Section 16.3.

Still another approach, utilizing Holling's disk equation as its basic premise, has been described by Marten (1973). Predictions that take into account the trade-off of time spent in feeding and in other beneficial activities are slightly different from those of the models above. Here it is possible to find total amount of food eaten rise and then fall as abundance of that food rises. Marten's paper is really more an exploration of the nature of functional response curves than of feeding preferences. Further discussion of the evolution of functional response curves can be found in Section 16.3.

Before presenting data bearing on the predictions of the models above, some comments of Schoener (1971) are in order. Schoener made some additions and refinements to the models above, but his most important contribution was to distinguish between *energy maximizers* and *time minimizers*. Assuming that the appropriate currency for food "value" is, indeed energy per time, we can identify two types of predators. Some, such as large grazers, could conceivably spend almost full time feeding, with little concern for time required in other activities. Others, which must spend considerably more time finding food, like many carnivores, or which must limit their foraging time in the interests of avoiding predation, must satisfy their energy needs in as short a time as possible. Both should eat food items that maximize their intake per unit time, but their food preferences are nevertheless slightly altered by their other needs. An energy maximizer can skip great numbers of hard-to-gather or hard-to-chew items. It can largely ignore the time lost *between* items and concentrate on the energy gained per item over time spent actually handling the item. On the other hand, a mouse, which subjects itself to higher probabilities of being predated everytime it ventures forth in search of food, must include in the evaluation of a given food class the time required to find as well as the time spent handling each item. Looked at another way, the mouse upon starting a feeding foray cannot be certain that the foray will not be interrupted without a mad dash for cover. Hence the mouse can count only on limited chances for contact with food and must view its environment as effectively less food rich. Effectively, food is scarce. Either argument leads to the conclusion that a time minimizer looks less at net energy taken in per item per handling time, and concentrates more on devouring those foods that can be found and dealt with the fastest. Thus cotton rats (*Sigmodon hispidus*) take energy-rich food such as berries when they are easily accessible, but generally concentrate on grasses, which are highly abundant and easily and quickly chewed even though energy poor (Wrazen, 1981).

Evidence that animals do, in fact, concentrate on the most energy-rich foods, after time-minimizing constraints are considered, and that they eat slightly disproportionate numbers of the poor prey when it is very common but switch readily to the better prey when it is not uncommon (Fig. 16-5), have been gathered for a wide array of species. A few examples will be given here; a few others will be presented later in the chapter in slightly different contexts. Menge (1972) studied the foraging strategy of a starfish, *Leptasterias*, and found that diet changed with season and size of the predator. Menge looked at the escape mechanisms of the prey, the calories per time yielded by a prey item and concluded that the starfish approached optimal diet choice at least in a rough sort of way. Where deviations from expectation occurred, they were in the direction expected from ease of difficulty of prey capture.

Elner and Hughes (1978) measured the time it took shore crabs (*Carcinus maenas*) to manipulate and break the shell, and then to eat individuals of the mussel *Mytilus edulis*. Shell breakage time T_b increased asymptotically with prey length to a critical point where still larger prey were ignored. If forced to eat larger prey, the crabs changed their breakage technique to the slower process of cracking the shell along the valve edges rather than crushing it. This rendered the largest mussels less valuable with respect to handling time. Energy $\hat{E}$ obtained from one prey increased with the one-third power of shell length. The calculated value of E/T_b rose with mussel size L, and then fell. The mussel size at which E/T_b peaked increased as the size of the predator increased. Elner and Hughes presented crabs of three size categories with six size classes of mussels. Assuming that the "best" prey items would be taken first, they then predicted which mussels should be taken, in order, on the basis of the E/T_b versus L curve and also from an E/T_H versus L curve, where $T_H = T_b$ plus eating time. The results are shown in Table 16-1. The observations follow, very crudely, that expected from E/T_b values, but size of prey taken is consistently low. Crabs *did* broaden their diet with respect to prey size when mussels were in short supply, became more choosy when food was plentiful. In opposition to Pulliam's prediction, the abundance of poor foods *did* influence their use in the diet. Elner and Hughes suggest that the crabs are inefficient in their scanning of the available food supply. It requires more agile intellects to behave properly.

Fish feeding behavior was studied extensively by Ivlev (1961). Plots of percent in the diet versus percent available for various prey (chironomids, amphipods, freshwater isopods, molluscs, and nonliving food) eaten by carp showed the expected patterns (Fig. 16-5). Also, using a variety of fish species as prey, and clipping parts of the tail fins to slow swimming by various amounts, Ivelv found that slower-moving prey (less handling time because they were more quickly captured) were preferred to faster-moving prey, as expected. Predators used were pike (*Esox lucius*) and perch (*Percus* spp.).

Table 16-1

Crab Size (Carapace Width in cm)	Mussel Size (cm) Expected to Be Eaten		Mussel Size (cm) Observed Eaten
	On the Basis of E/T_b	***On the Basis of E/T_H***	
5–5.5	1.5–1.75	1.75–2.0	1.25
6–6.5	2.0–2.25	2.25–2.5	1.75
7–7.5	2.25–2.75	2.75–3.0	2.25

We can argue, especially for time minimizers, that satiation permits a more relaxed approach to feeding. As feeding proceeds, need diminishes and the threshold value in Figures 16-2 to 16-4 moves to the right; the consumer should become more particular in what it deigns to eat. Psychologists have long known that hungry animals eat a broader spectrum of food items and that recently fed subjects are more persnickety. J. M. Emlen (unpublished notes) observed that both house mice and *Peromyscus maniculatus*, fed on a mixture of oats and millet seed in the laboratory showed increasing preference for the more energy rich food, millet, as the 14-hour feeding period wore on. Ivlev found abundant evidence for increased specialization on the more easily captured prey species with predator satiation.

Werner and Hall (1974) fed prey (*Daphnia magna*) of four easily distinguished size classes to bluegill sunfish. Ten fish, 70 to 80 mm in length, were allowed 24 hours to acclimate to a circular pool, and then were offered a previously determined size mixture of prey. The fish were allowed to feed for a predetermined amount of time, short enough that prey-size distribution and abundance would be minimally distorted by their behavior, and then their stomachs were examined. The visual field volume of the fish was determined and used to find the effective abundance of prey size classes (small prey are less readily observed than large). It was found that the fish saw a ratio of approximately 1 : .63 : .47 : .22 large to small prey when the true distribution was uniform. Search time and handling time per food item were measured for various concentrations of each of the four prey size classes, using isolated fish. To predict optimal diet, Werner and Hall used the expression

$$T = \frac{\text{time}}{\text{biomass eaten}} = \frac{T_S + T_H \int_{\hat{x}}^{b} f(x)\,dx}{\int_{\hat{x}}^{b} f(x)\,dx} \tag{16-6}$$

where T_S is search time, T_H handling time per item, $f(x)$ the frequency distribution of prey (in numbers, not fractions, of size x), b the largest size available (the fish can handle the largest items most efficiently), and $\hat{x}$ the lower size bound of prey eaten. Feeding will be optimal when $\hat{x}$ is adjusted so as to minimize T (this is identical in form to Pulliam's model). At effectively uniform prey distribution, and very low prey densities, all items encountered should have been eaten. Indeed, this expectation was very close to being realized—Werner and Hall reasoned that the deviations from this ideal reflected errors in their assessment of just what the fish could perceive, and used these preliminary results to "correct" their 1 : .63 : .47 : .22 estimate of effective to true prey abundance to 1 : .71 : .54 : .27. In further experiments the effective prey densities were accordingly adjusted. The fish were then tested at three density levels. Relative to the acceptance level of the second largest items, the number of the next two size classes taken are shown in Table 16-2. In the intermediate-density experiments, 80% of the fish completely ignored the smallest prey class while eating intermediate-size prey without apparent bias. At high densities the proportion of this second to smallest size class drops. Thus at low densities, essentially all items are eaten; at higher densities, the smallest size class effectively drops out; and at very high densities, the next smallest size class is also mostly rejected. The results do not fit the Emlen–Pulliam expectations very well, although qualitatively the fit is reasonable. Switching does seem to occur, but in an exaggerated form compared to theoretical expectation.

Krebs et al. (1977) tested the theories of optimal feeding in the great tit (*Parus major*) by placing mealworm pieces on a moving belt in such a way that a bird subject saw each morsel pass by a small window for .5 second, during which time it could be grabbed. Large (eight-segment), and small (four-segment) mealworm pieces were used, and presented in a regular order (example, LSLSLS . . . or LLLSLLLS . . .). If the total encounter rate was low, birds were nonselective. At higher encounter rates the smaller items were selectively ignored. In keeping with Pulliam's prediction, selectivity was independent of the encounter rate with the poorer food.

Table 16-2

	Size Class		
Prey Density	**2**	**3**	**4**
Low			
Expected	1	1	1
Observed	1	.9 ± .1	.9 ± .2
Intermediate			
Expected	1	.71	.27
Observed	1	.58 ± .02	.04 ± .01
High			
Expected	1	.71	.27
Observed	1	.23 ± .01	.05

Goss-Custard (1977a) examined size selection of worms by the redshank (*Tringa totanus*) in the field. These birds forage along shorelines, picking buried prey items from the mud. He assumed, in the standard, optimal-foraging theory spirit, that these birds should maximize the amount of worm biomass, E, collected per unit time. Measurements of handling time of worms of various sizes showed, where T_H is handling time, that E/T_H rose with size of prey. Therefore, if worms were scarce, a broader size range ought to be taken, if common, the smaller ones should be dropped from the diet. What happened was that the rate at which large worms were taken varied proportionately with their availability and that the number of smaller worms eaten per unit distance traveled in foraging increased as the number of large worms per distance dropped. Intake of small prey correlated only very poorly with small prey abundance. Predator response was thus as expected, but the lack of an all-or-none response to different prey sizes shows either a behavioral sloppiness or that factors other than worm per unit time were important in diet selection (see also Goss-Custard, 1981).

Zach (1979) watched crows (*Corvus caurinus*) picking up whelks (*Thais lamellosa*) in the British Columbia intertidal zone, carrying them into the air and dropping them in order to break them on rocks below. He divided the whelks into three size classes, and for all three, analyzed the energy content and the readiness with which their shells broke upon being dropped from different heights. He also measured the time spent by the crows carrying shells to different heights, handling the shells, and getting to and from the shore where they fed. From these data he was then able to calculate the energy obtained per unit time from whelks of different sizes, the total time and energy costs for a foraging trip, and thus which whelk sizes the crows should eat. Given the various costs involved and the fact that the whelks usually had to be dropped four or five times before breaking, Zach found that only the largest size class of snail could be used without net caloric loss to the crows. Indeed, the crows ate only the largest. He was then able to calculate the most efficient height from which the crows might drop their prey. The crows did exactly as they should.

Abbott et al. (1975) investigated food preferences in four species of finches on two of the Galápagos islands. Providing the birds with long- and short-grained rice, they found that all four species were able to crack and eat the softer short-grained variety more quickly. The smallest species was able to manipulate the grains most readily, and cracked and ate both types of rice faster than the other birds. We should expect, then, that short-grained rice should be preferred, and that the smaller finch species should show the strongest preference. In fact, the authors found no evidence whatsoever for any preferences.

Smith and Follmer (1972) offered whole and chopped acorns of white, bur, and shumard oak to gray squirrels. The first two, yielding the highest metabolizeable energy content per handling time, were clearly preferred. Shagbark hickory nuts and black walnut seeds were never eaten unless the squirrels were very hungry, suggesting that these much-harder-to-husk foods were normally avoided because their "value" was lowered by high handling time.

But Grodinski and Sawicka (1970) found no correlation between caloric richness of seeds and food preferences of mice and voles. Their subjects preferred the lower-energy deciduous tree seeds to the higher-energy conifer seeds.

Further data sets tell a similar story. The crude expectations of dietary switching, narrower diet in the midst of food plenty, and preference for foods with high-energy-per-time values are generally, but not always realized. There is clearly more to diet preferences than the simplistic hypotheses above suggest.

We shall investigate the weaknesses of our approach so far in a moment. But first the reader is reminded that the subject of dietary switching, which is commonly observed (see, e.g., the review of Murdoch and Oaten, 1975) is of great importance to the stability of prey–predator systems (Chapter 4). Second, note that the broadening of diet that accompanies drops in food availability should be expected to soften the action of density feedback—that is, we should expect smaller θ values—see Section 13.4.

Constraints on Food Preference Models

Not all the simple predictions made by the models above were borne out by the data used to test them. Other deviations from the expected can be listed. Goss-Custard (1977b) found that his redshank ought to prefer large polychaete worms not only over small worms—which was observed—but also over amphipods. Redshank, however, seem to disagree on the latter prediction. And there are many other such cases. Westoby (1978) pointed out one likely cause: Food "value" must include much more than merely energetic criteria. For carnivores, nutritional value differences among different types of foods may be fairly negligible—the simple theory is quite likely to work here. As noted earlier, the theory may well work also for animals with very high energy requirements. In these species, under most circumstances, other nutritional needs are likely to have been met by the time energy needs have been satisfied. But for a broad spectrum of consumers, especially omnivores, we must consider the importance of toxins, proteins, lipids, minerals, and so on, in our measure of "value."

Freeland and Janzen (1974) reviewed the widespread distribution of plant secondary substances that herbivores must detoxify to use. They point out that consumers must seek plants with lower concentrations of these toxins, and avoid those that produce chemicals their systems have not evolved to use. Detoxification costs may alter nutritional needs as well. The risk inherent in eating toxic foods might vary seasonally in the consumer, another factor that could influence dietary shifts independent of energetic factors. Wrazen (1981), for example, discovered that cotton rats avoided certain forbs, known to contain secondary substances, most assiduously when they were pregnant. Probably this reflects a tendency to protect the developing young, which are likely to be very sensitive to toxins.

Farentinos et al. (1981) discovered that Abert's squirrels (*Sciurus aberti*) feed heavily on cortical tissue of Ponderosa pine (*Pinus ponderosa*) twigs in winter, but concentrate their feeding heavily on certain, individual trees. Discriminant function analysis was used to determine what combinations of secondary compounds (α-pinene, β-pinene, 3-carene, and limonene—all monoterpenes) most clearly distinguished the utilized and nonutilized trees. These chemicals, primarily α-pinene, were present in significantly lower concentrations in the trees utilized as food by the squirrels.

It has been suggested that much of the observed food preferences of animals can be explained on the basis of "learned aversions" to food that produce ill side effects (Revusky and Bedarf, 1967; Gustavson et al., 1974). In the psychologists' laboratories where most tests of this hypothesis have been made, the side effects were usually produced artificially via irradiation or the addition to food or injection of lithium chloride, but natural toxins should have the same effect. Particularly among omnivores (such as the ancestors of laboratory rats), where encounters with toxins might be commonplace, learned aversions might provide an excellent guide to dietary selection.

Arnold (1981) discovered that the garter snake (*Thamnophus elegans*) has developed genetic races that specialize on eating slugs, or salamanders, or toads, and so on, and has even managed to estimate heritabilities of the chemoreceptive responses to the food-specific odors in naive newborn. Most of the prey species involved have noxious characteristics (mucous or poison glands in the skin), so specialization might be expected.

Animals eating such as to maximize energy intake in time must, nevertheless, meet certain minimum requirements for other nutrients. Recall from Chapter 2 the importance of sodium in the diet of population-stressed meadow voles. Pulliam (1975a) used a simple, linear programming approach to include nutrient constraints in his model of food preference. We will present a slightly modified version of his argument. In Figure 16-6 we plot the isoclines of fitness increment (actually energy intake for a given period of time—say 1 day) for various combinations of two foods eaten. We assume, in keeping with the models above, that only items of A, and B of energy value greater than some thresholds V_A and V_B are eaten. On the same graph we plot the set of all possible intake combinations of items above V_A and V_B. If a consumer eats a certain amount of A, it has time to eat only some other amount of B; if the amount of A eaten rises, the intake of B must drop (shaded portion of Fig. 16-6). If there were no nutrient constraints, the optimal diet would be defined where the set of possible diets reached the highest intake isocline—point P in Figure 16-6 (see also Rapport, 1971; Cody, 1974a). According to Emlen's model, that point should correspond to the consumption of all items of value exceeding $\hat{V}$. But now suppose that food A contains some critical nutrient x in concentration C_{Ax} per energy unit, and another, y, with concentration C_{Ay}. Food B contains these nutrients in concentrations C_{Bx} and C_{By}. Suppose further that the consumer requires minimal amounts, M_x and M_y of these nutrients. Then if m_A and m_B are the amounts of A and B eaten (in energy yield units),

$$m_A C_{Ax} + m_B C_{Bx} \geqslant M_x$$

so that

$$m_A \geqslant \frac{M_x}{C_{Ax}} - m_B \frac{C_{Bx}}{C_{Ax}} \tag{16-7}$$

Similarly,

$$m_A \geqslant \frac{M_y}{C_{Ay}} + m_B \frac{C_{By}}{C_{Ay}} \tag{16-8}$$

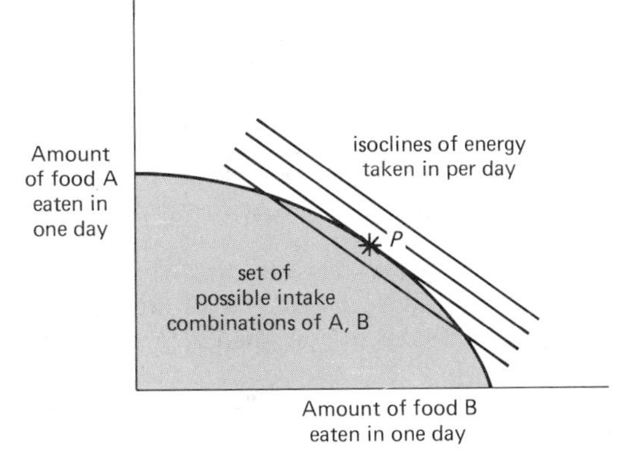

Figure 16-6

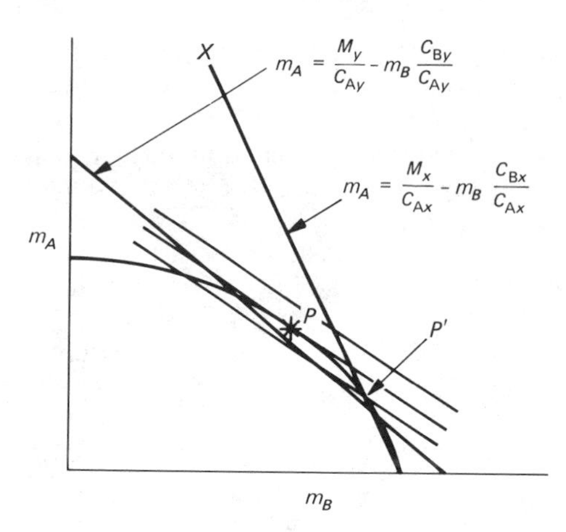

Figure 16-7 Same as Figure 16-6, but with two constraints appended. Point P is tangent to the highest energy isocline and so represents the optimal diet in the absence of the constraints. But P', although providing lower energy yield, represents the diet with the highest energy gain lying above the constraint lines.

These lines can be superimposed on Figure 16-6 as in (for example) Figure 16-7. The point P still represents the diet that maximizes energy intake per time, but it is not adequate to meet other nutritional needs. The diet that maximizes energy per time within these nutritional constraints is that given by P'. There may be additional constraints. For example, the advantage of a larger food item that can be quickly handled may disappear if the item is swallowed whole and its digestion, because of its lower surface area/volume ratio is apt to be slow. Thus fish who cannot chew may prefer smaller items than similar fish that can. Because digestion rate is proportional to the surface area of the gut, which increases less rapidly than metabolic need (which is proportional to volume) as animals get bigger, the rate at which food can be pushed through an animal may become a limiting factor in herbivores as well. Thus we might find not the energy-rich foods, but the rapidly digested foods, to be the most preferred (Westoby, 1974).

Belovsky (1978) and Belovsky and Jordan (1978) used the linear programming approach with several constraints to investigate feeding in the moose (*Alces alces*). Assuming that moose eat plant species that are most energy rich per handling time, but must also meet a minimum sodium requirement and also fit within constraints imposed by available foraging time and limited rumen capacity, these authors predicted with uncanny accuracy the preferences of this species for deciduous leaves versus forbs. The reader is referred to these papers as excellent examples of analyses successfully incorporating large amounts of information to rigorously test a simple model.

Belovsky (submitted for publication) has also applied this technique to a wide variety of other species. For the beaver (*Castor canadensis*) he calculated a gut capacity constraint based on estimated gut size and food passage rates of

$$4L + 20A \leqslant 3458 \text{ g}$$

where L and A are daily intakes of deciduous leaves and aquatic vegetation. The foraging time constraint, based on observed time required to cut and haul branches, observed cropping rate on aquatics and total time spent foraging, was

$$.49L + .19A \leqslant 269 \text{ minutes}$$

Finally, using the standard relationships between body weight and metabolic rate [see "Territoriality" (p. 388)], estimating additional activity (hauling) costs, and estimating digestive efficiencies, minimal energy requirements were deemed met when (for a 15-kg beaver)

$$2.552L + 3.28A \geqslant 1213 \text{ kcal/day}$$

A sodium constraint was also calculated but turned out to be unimportant. Then, under alternative assumptions that beavers are time minimizers or energy maximizers, the amount of

Table 16-3

	Percent of Diet Predicted If Beavers Are		Percent of Diet
	Time Minimizers	*Energy Maximizers*	Observed
Deciduous leaves	76	88	89
Aquatics	24	12	11
Energy acquired (kcal)	1213	1562	1647
Time spent foraging and feeding (min.)	186	269	286

aquatics A and leaves L that the beavers "ought" to eat were calculated (Table 16-3). Beavers seem to fit the energy maximization expectation quite well.

It may sometimes be misleading to treat needs for nutrients other than calories merely as dietary constraints. Milton (1979) analyzed leaves eaten by howler monkeys, *Alouatta palliata*, in Panama, for carbohydrate, protein, tannin, phenol, and fiber content, and determined that leaf choice was made primarily on the basis of the ratio of protein to fiber. Belovsky (1981) found that he could predict quite accurately the size and amount of twigs of various species eaten by moose by using protein plus ash content as a currency of food value. Here the ash is beneficial in that it contains useful minerals. Certain species of plant were avoided, perhaps due to the presence of toxic, secondary compounds, but some such, presumably noxious plants were eaten readily.

The question of dietary response to complementary foods with complementary nutrients has been addressed by Rapport (1980). Rapport, noting that "value" of one food in the models presented above was by the nature of the models implicitly assumed independent of what else was eaten, offered the following. As in Figure 16-6, define axes representing amounts (in bulk now rather than energy units) of two foods, A and B, eaten. Again it is hypothetically possible to define isoclines of "value" and a set of realizable feeding combinations. Again the optimal diet is defined by the point at which the latter is tangent to the highest isocline. The problem of "value" currency is further addressed by Caraco (1980), who suggests using "utilities"; utility is simply a number defined such that food item i is preferred to j when the expected utility given an item of i exceeds the expected utility given an item of j. For further discussion, the reader is referred to the primary reference.

Spatial Heterogeneity

Except with respect to the references to work by Estabrook and Dunham (1976) and McNair (1979, 1980), the discussion above ignores the possibility that food items might be other than randomly distributed and encountered. In fact, most species are clumped. Where this is so, it is pretty obvious that foraging consumers will behave most efficiently if they spend some time within patches before moving on. As any amateur naturalist can attest, this is certainly what happens. Ecologists have tried to refine this kind of observation. Tinbergen et al. (1967) asked whether feeding individuals, upon finding a camouflaged prey item, would tend to search more carefully in that area before resuming their wanderings. Where prey were clumped, such behavior would be advantageous. Spreading food items out in grid patterns in a field, they determined that this was so. Murdoch and Marks (1973) discovered that ladybird beetles, *Coccinella septempunctata*, coming upon aphids in the field, hunted extensively in the immediate area after devouring their prey, before resuming a random search pattern. Looking at the response of small mammals to seed distribution patterns, Reichman and Oberstein (1977) found that *Dipodomys merriami*, a kangaroo rat, foraged preferentially on seed clumps. On the other hand, they also discovered that *Perognathus amplus*, a pocket mouse, couldn't care less whether it foraged in seed clumps or among more widely scattered seeds. Perhaps the latter represents a divergence in foraging pattern that alleviates competition between the species (see Chapter 5)—that is, considering the species in isolation the behavior seems maladaptive, but taking into consideration all impinging criteria, it is really beneficial. Speculations such as this are not easily tested. We can, however, fine-hone our understanding of the autecological aspects of foraging behavior.

MacArthur and Pianka (1966) were the first to present a formal model of optimal patch use. Refer to Figure 16-8; let the abscissa represent the number of types of patches used by a forager, the ordinate denote time. As the number of patch types used increases, the jack of all trades–master of none problem sets in; handling (pursuit) time per item within a patch must go up. Let T_H represent the handling time experienced within patches. But also, as number of patches used rises, the time spent traveling between patches goes down. Let T_S be the between patch "search" time per item eaten. Because it is (presumably) optimal to minimize total time spent per item eaten, we can find the optimal number of patch types used, $\hat{x}$, by writing

$$0 = \Delta\,(\text{total time per item}) = \Delta T_H + \Delta T_S$$

so that $\hat{x}$ is that value for which

$$\Delta T_H = -\Delta T_S$$

Figure 16-8 can be translated into Figure 16-9, where $\hat{x}$ is marked by the point where the two curves cross. Figure 16-9 permits us to make the following predictions.

1. As food density within patches increases, animals can increasingly meet their food needs by taking only the easiest-to-catch prey. Thus T_H rises less rapidly with x, so the ΔT_H curve is lowered, shifting x to the right. Species living in areas comprised of rich patches should use a greater variety of patches.
2. If the density of patches rises, ΔT_H is unaffected, but the T_S versus x curve is lowered, so the $-\Delta T_S$ curve is lowered. Species in environments with many patches should specialize on patch types. The same is true for spatially large patches.

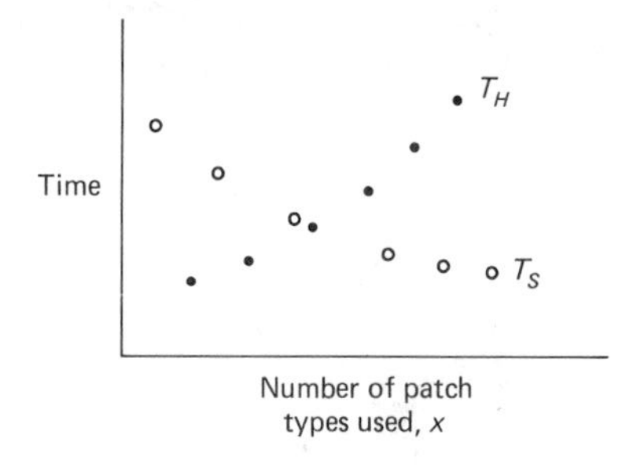

Figure 16-8 Plot of handling time per food item, T_H, and search time per food item, T_S, as functions of the number of food patch types utilized.

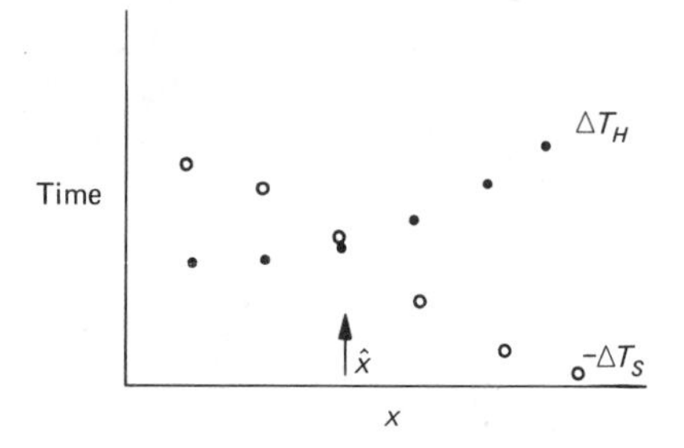

Figure 16-9 Marginal values of T_H and T_S as functions of the number of food patch types utilized (derived from Fig. 16-8).

3. If one species should invade the foraging range of another (i.e., if competition occurs for food), each should limit the number of patch types used (so as to avoid direct competition, either exploitative or interference) but maintain or, if necessary to meet their needs, even expand the variety of prey types utilized within each patch. This prediction has become known as the *compression hypothesis.*

A different sort of question has been raised by others. When a food patch is encountered, how long should the consumer remain in the patch before leaving? Gibb (1962) made extensive observations on the feeding of great tits on insect (*Ernarmonia*) larvae among the scales of Scotch pine (*Pinus sylvestris*) cones. He noticed that the birds concentrated their efforts in patches of trees that had particularly large numbers of food items in their cones. But he also noticed that within a patch of trees, the number of larvae eaten per cone increased much less rapidly, and rather erratically, with the number of larvae available per cone. He hypothesized that the birds formed an image of the number of insects a cone, on average, was likely to have in a given tree patch, and foraged until it presumed, on this basis, that it was reaching a point of diminishing returns. *Hunting by expectation* is the term used to refer to such a foraging strategy.

Krebs et al. (1974) reckoned that consumers could do better than hunt by expectation, and developed the following hypothesis. If T_S is the time spent traveling between patches, T_A and T_B are times spent in patches of type A and B, $f(T_i)$ is the cumulative food value taken at the point the forager has spent time T_i in a patch of type i, and p_A and p_B are the proportion of visits to the two patch types, then

$$\text{total time, average, per patch} = T_S + p_A T_A + p_B T_B \tag{16-9}$$

$$\text{value gained, average, per patch} = p_A f(T_A) + p_B f(T_B) \tag{16-10}$$

so that value gained per patch is

$$\frac{\bar{E}}{\bar{T}} = \frac{p_A f(T_A) + p_B f(T_B)}{T_S + p_A T_A + p_B T_B} \tag{16-11}$$

The optimal value of T_i (T_A or T_B) can be found by setting the derivative equal to zero. Then

$$\frac{df(T_i)}{dT_i} = \frac{p_A f(T_A) + p_B f(T_B)}{T_S + p_A T_A + p_B T_B} = \left(\frac{E}{T}\right)^* \tag{16-12}$$

the value of E/T when T_A and T_B are optimally chosen. What (16-12) says is that the optimal time to spend in a patch is reached when the marginal intake rate $df(T)/dT$ in that patch equals the average rate of gain per unit time over all patches, including interpatch transit time. This formulation has become known as the *marginal value theorem.*

Krebs et al. suggested that consumers would follow the marginal value theorem in choosing when to leave a patch, and could use the *giving-up time* (GUT), the time spent without finding an item, as an indicator of when that time has been reached; the time spent in fruitless search since the last item found is, on average, the reciprocal of $df(T_i)/dT_i$. Note that as habitat quality $(E/T)^*$ rises, GUT should decline. To test their hypothesis they constructed five "trees" with "pine cones." Each tree was a kind of modified hat rack, and the cones were blocks of wood into which three holes had been drilled in each of two adjacent sides. These holes served as repositories for occasional mealworm pieces. Tape was then placed over the holes so that the consumers, six hand-raised chickadees, *Parus atricapillus*, had to remove the tape to find the worms. Four cones were hung on each tree. When the experimental setup was ready, birds were introduced and allowed to forage for 5 minutes. Rich patches of cones were clearly preferred to poor patches. If the environment as a whole was poor—few mealworms per cone—GUT was longer than if the environment was mixed, as expected. Within a mixed environment, though, GUT was the same for rich and poor cone patches. It seems that the birds develop an average expectation of environmental richness and adjust their GUT accordingly. Differences among small patches of cones were probably of too small scale for the birds to respond appropriately.

The fact that the time spent in fruitless search is, only *on average*, the reciprocal of $df(T_i)/dT_i$ causes difficulties for the realization of the optimal foraging patterns. If a predator monitors time between successive captures, it is hypothetically possible for it to use that information in conjunction with fruitless search time to assess quite accurately the optimal GUT. But there is no evidence that animals do this. Poor estimation of optimal GUT may mean that alternative cues ideally less accurate than GUT should nevertheless be used to assess the point at which one patch should be left for another (e.g., see Iwasa et al., 1981).

An alternate approach to the marginal value theorem, one that avoids the difficulties of dealing with GUT is provided by McNair (1982).

The marginal value theorem was extended to more than two patch types by Charnov (1976). The intake rate expression was generalized to account for the fact that different patch types might yield food in their own unique patterns. Thus (16-11) becomes

$$\frac{\bar{E}}{\bar{T}} = \frac{\sum_i p_i g_i(T_i) - T_S E_T}{T_S + \sum_i p_i T_i} \tag{16-13}$$

where E_T is the mean energetic cost of moving between two patches, and $g_i(T_i)$ replaces the more restrictive $f(T_i)$. The result, found by differentiation, is the same as before:

$$\frac{dg_i(T_i)}{dT_i} = (E/T)^* \tag{16-14}$$

Figure 16-10 permits some additional predictions. The rate of intake is zero for time less than T_S, because the animal is still between patches and has not yet begun to feed. Then, with more time, the cumulative amount ingested rises and asymptotes as the food available is used up. The consumer should leave a patch, i, when the slope of the g_i curve, $dg_i(T_i)/dT_i$, equals $(E/T)^*$. Thus ideally (as indicated in Fig. 16-10), the time spent in a poor patch (T_i) should be less than the time spent in a rich patch. But because GUT is the reciprocal of $(E/T)^*$, it should be greater in poor patches. Also, because a patch should be left, regardless of its relative richness, when the marginal capture rate reaches a constant, all patches should be left when, at least to a crude approximation, their remaining food density drops to a constant critical level. Thus the percent of items taken from rich patches should rise with initial patch richness. If n is initial richness, n' the final richness, then percent taken should be

$$100\frac{n - n'}{n} = 100\left(1 - \frac{n'}{n}\right) \tag{16-15}$$

where n' is constant over all patches. Smith and Sweatman (1974) gathered data that are pertinent to the expectation given in (16-15). They let great tits forage in an aviary with six patches of differing food density. Food was presented in 16 × 16 arrays of food cups glued to a board and covered with aluminium foil. The birds were trained to search for food items by removing the foil cap and then introduced into the aviary and allowed to forage for 5 minutes.

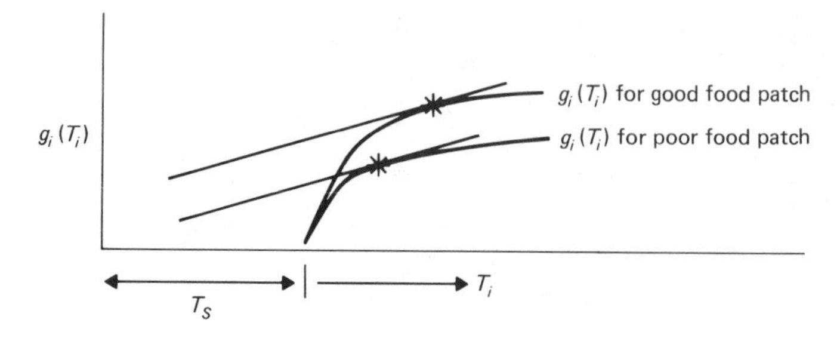

Figure 16-10 Plot of cumulative food value obtained in a food patch (i) as a function of time spent searching (T_S) for the patch and then foraging (T_i) within it.

Results are given in Table 16-4; expected percentages taken are written relative to that for the lowest food density class (four cups per patch with a piece of mealworm), using (16-15). For some reason the birds seemed to leave the rich patches too early, or the poor patches too reluctantly. For further laboratory tests of the marginal value theorem, in which *travel* time was varied, see Cowie (1977).

Table 16-4

Prey Density (Number of Cups with a Mealworm Piece per 16 × 16 Array)	Percent Taken	
	Actual	*Expected*
0	—	—
4	5	5
8	2	52.5
12	12	67.2
16	8	76.3
20	31	90.5

Field tests of the marginal value theorem have also been made. Pyke (1978) observed foraging by broad-tailed (*Selasphorus platycercus*) and rufous (*S. rufus*) hummingbirds on the scarlet gilia (*Ipomopsis aggregata*), gathering data on the number of flowers per inflorescence, the number of flowers probed per visit to an inflorescence, the time taken to move between flowers of an inflorescence, the time taken to remove nectar, and the rate of overall flower visitation. He also measured the nectar present per flower. Without belaboring the details of the methodology, results were as follows. The number of flowers probed per inflorescence (patch) before the bird left for another fit the marginal value theorem predictions extremely well. Pyke was even able to predict quite well the shape of the probability distribution of number of flowers probed per patch visit.

The marginal value theorem states that, after all energetic and time expenses are accounted for, if one patch yields a higher return than another, foragers should concentrate on that one patch. Ideally, the average intake rate from each patch used should equal that from each other patch used. Pyke (1980) found that, in keeping with this expectation, bumblebees (*Bombus flavifrons*) had identical net energy intake rates from each of two flowering patches, one of the species *Aconitum columbianum*, the other, *Delphinium barbeyi*.

Some Complicating Factors

In some insects, such as honeybees, foragers recruit other workers to join them at a rich food source. By arriving in force, a number of bees can chase off other foraging individuals. Because there are a number of foragers out at most times, finding food sources is a very efficient process in such species. But bumblebees, unlike honeybees, cannot recruit, and so are left the task of foraging and using food sources individually. The resulting foraging inefficiency generates particularly strong selective forces on bumblebees to be highly efficient at handling food sources once found. But this, in turn, is hard for any one individual to accomplish, because the types of flowers available change seasonally, and each requires a slightly different technique to remove the nectar. Bumblebees have solved their problem by "majoring." Each individual tends to develop preferences for flower types initially discovered in high frequency. These preferences then lead to practice and so increased feeding efficiency. The result is that as available flower types change over the year, new bees develop appropriate preferences and skills, and a wide variety of food sources can be efficiently tapped. "Minoring" also occurs, allowing individuals at least some leeway to switch should their major species experience a precipitous demise (Heinrich, 1978). Because different individuals are utilizing different patch types, it is not clear that the marginal value theorem holds except among patches composed of a given flower species.

Predictions of the marginal value theorem may often have to be modified when disturbances occur due to the presence of competitors. Morse (1977), for example, also working with bumblebees, discovered that workers of the small *Bombus ternarius* normally foraged on both proximal and distal parts of goldenrod flower clusters, but in the presence of the large *B. terricola*, actively avoided interference and concentrated on the distal portions. The larger species is constrained, by virtue of its body weight, to the proximal portions.

Environmental uncertainty must play a role in foraging behavior. Real (1981) enclosed a colony of bumblebees, *Bombus sandersoni*, in a mesh cage in which was placed a Lucite food board with 2304 little holes that could be filled with nectar. "Type" of flower was determined by the color of a piece of cardboard pasted to the back side of each hole. In one experiment all holes were filled with 2 μl of nectar; one half were yellow, one half blue. The bees foraged solitarily, one to a general area of the board, so there was no interference among them. Seventy-nine percent of the visits were to yellow flowers ($p < .001$ in a U-test for preference using percent of yellow versus blue over 30 trials). In a second experiment, all blue holes contained nectar (2 μl), and every third yellow hole held 6 μl of nectar. Thus the yellow and blue food supplies were again, on average, equal, but there was more uncertainty as respects the yellow. The bees shifted their preference to blue (84%, $p < .001$). Further experiments of this sort, both with the bees and with wasps (*Vespula vulgaris*), yielded the same kind of results. Variability (uncertainty) in value lowers the perceived value of a patch type to a forager. Caraco (1980) argues along similar lines, suggesting that a consumer may prefer items or patches of items that are of low mean value over others that average higher value but are also highly unpredictable. The risk of finding a patch empty, or a food item noxious or uncatchable, makes perceived expected value lower than real expected value. An example of such avoidance of risk in the field comes from the work of Morrison (1978). Morrison found that the fruit bat (*Artibeus jamaicensis*) in Panama was a "central place forager" (see Hamilton et al., 1967), flying from a home (roosting) base to food sources and then returning before setting forth again. These bats flew consistently to food sources that were *not* closest and thus most energetically valuable. The reason, apparently, is that search time can be extensive, and the bats could do best by flying farther, to known sources, than by flying short distances and having to search.

Can consumers balance two or more conflicting demands? Sih (1980) placed *Notonecta hoffmanni*, an aquatic hemipteran, into tubs, and dropped wingless *Drosophila* onto the water surface. *Notonecta* eat the flies, and the adults cannibalize their young. If the tub population of *Notonecta* is divided into inner and outer portions, and flies are provided in much greater abundance in the inner compartment, the adults, as might be expected, concentrate most of their time and all of their foraging efforts there. The first instar larvae, however, leave the inner compartment to avoid being eaten by their elders. Second and third instars are less apt to be cannibalized and divide their time about equally among the two compartments. Fourth and later instars are not cannibalized, and they stay with the adults in the more food rich area. Individuals of this species, then, do seem to be able to assess the trade-offs between the advantages of visiting rich food patches and the disadvantages of being eaten. In experiments with sticklebacks (*Gasterosteus aculeatus*), Milinski and Heller (1978) offered tubes with various numbers of *Daphnia* inside to their fish. Given a simultaneous choice between two prey populations, the fish attacked the tube with the most prey. However, if the shadow of a kingfisher—a predator to the fish—was passed over the tank immediately before the trial, the fish chose the less dense tube. The authors feel that too many prey may confuse the predator if it is simultaneously distracted by the threat of getting eaten itself. If so, it is better to attack the simpler food patch. Whatever the reason for these observations, it is clear that predation may affect patch choice by consumers.

Of course, other kinds of trade-offs occur. Can animals appropriately balance their intake of carbohydrates, protein, and so on? Richter's (1942) classic "cafeteria" experiments suggest that they can, but later workers have failed to replicate his results and the findings are now somewhat discredited.

In most tests of optimal feeding theory, subjects make consistent "errors." Are they really errors? Clearly, if a consumer is to be aware of the distribution of food types and food values, it must continually sample its environment. Thus a certain amount of "error" is necessary and adaptive (Royama, 1970; Hartwick, 1976).

We shall leave this topic with an uncomfortable query. If time between patches, or between items, is spent doing other things of benefit to the consumer in addition to searching for food, should interpatch (or interitem) time really be included in the foraging equations? Maybe foraging paths and patch-leaving criteria have evolved not to maximize foraging efficiency but best to allow for the accomplishment of other necessary tasks during the transitions between meals. At the very least, if foraging time is concurrently used as well for other purposes, and if the realization of those other purposes is influenced by the proximity of food of one type or another, linear programming may be too simplistic as a means of incorporating constraints into food preference models.

Foraging in Groups

In the preceding two chapters we alluded from time to time to the various advantages and disadvantages of living in groups. We shall briefly discuss here these considerations as they apply to predators.

Caraco (1979a,b) argues that as groups increase in membership, less time is needed per individual to scan for predators, so more time is available for feeding. On the other hand, more time and energy is lost, and more risk accrued, in aggressive encounters. He suggests that group size is a balance between these two factors. Studying the yellow-eyed junco (*Junco phaeonotus*), he recorded time spent with the head straight up or cocked (scanning for predators), the time spent in agonistic encounters, the feeding rate, and the temperature at different hours of the day for a variety of flock sizes. Mean scanning time per bird, y, does indeed vary inversely with group size, n:

$$\log_{10} y = 1.047 - .399 \log_{10} n \qquad r = -.92, \quad p < .05$$

Also, group size seems to increase the distance at which predators are sighted; larger groups flushed at greater distances than small groups when he approached them. The mean number of pecks per time (feeding rate), y, as expected, increased with group size,

$$\log_{10} y = 1.22 + .42 \log_{10} n \qquad r = .89, \quad p < .05$$

in the first of two study years, and

$$\log_{10} y = 1.25 + .56 \log_{10} n \qquad r = .87, \quad p < .05$$

in the second.

Waser (1981) has taken a different approach; more individuals find food items more readily than single individuals, but bigger groups have other, compensating disadvantages. What is optimal foraging group size? Let p_h be the probability an item (occupying a space just big enough for one item) is found by any one individual in a given period of time, and let p_r be the probability that an item exploited is replaced by reproduction, migration, or whatever other means, in one unit of time. Then if n is group size, the probability that a given item is found by none of the n individuals is

$$(1 - p_h)(1 - p_h) \cdots = (1 - p_h)^n$$

and the probability the item is found (and so eaten) is $1 - (1 - p_h)^n$. We can define a Markov process with transition probabilities, p_{ij}, where i and j denote the presence or absence of an item in any given environmental space:

$\mathbf{P} =$

	Item Present	Item Absent
Item Present	Probability that an item is not eaten, and does not die for other reasons	Probability that an item is eaten or dies for other reasons
Item Absent	Probability that an item appears in a space previously without an item	Probability that an item continues to be absent from a space

$=$

	Item Present	Item Absent
Item Present	$(1 - p_h)^n - p_d$	$1 - (1 - p_h)^n + p_d$
Item Absent	p_r	$1 - p_r$

Here p_d is the probability of a prey death by means other than predation (by the species in question). If π is the vector denoting proportion of spaces with and without items, then

$$\pi(t + 1) = \pi(t)\mathbf{P}$$

Substituting and solving for equilibrium π we obtain

$$\hat{\pi} = \hat{\pi}\mathrm{P} = \left(\frac{p_r}{p_r + 1 - (1 - p_h)^n + p_d} \quad \frac{1 - (1 - p_h)^n + p_d}{p_r + 1 - (1 - p_h)^n + p_d}\right) \tag{16-16}$$

Now, for each individual the rate of food intake is proportional to the food density (within a narrow range of the functional response curve). Thus an individual foraging in a group of size n takes food in at a rate proportional to

$$\frac{p_r}{p_r + 1 - (1 - p_h)^n + p_d} \tag{16-17}$$

The difference in food intake between a solitary feeder and a social feeder, the advantage of foraging alone, is thus the difference between expression (16-17) evaluated at $n = 1$ and $n > 1$, or

$$g = \frac{p_r}{p_r + p_h + p_d} - \frac{p_r}{p_r + 1 - (1 - p_h)^n + p_d} - C \tag{16-18}$$

where C is the cost of keeping others out of the exclusive area, plus any other costs that might accrue as a result of solitary living. The value of g rises rapidly at first with p_r, but then drops and eventually falls below zero. Thus high food renewal rates favor group foraging. Waser goes on to look at cases where renewal follows other specific patterns. For example, if renewal is not a random process, but rather a logistic one,

$$\frac{dR}{dt} = r_R R\left(1 - \frac{R}{K_R}\right) - nhR \tag{16-19}$$

where R is the amount of food. Then

$$\hat{R} = K_R\left(1 - \frac{nh}{r_R}\right)$$

so that

$$g = \frac{K_R h}{r_R}(n - 1) - C \tag{16-20}$$

In this case g drops at a decreasing rate with renewal potential (r_R). As before, high renewal rate encourages group foraging.

The primary disadvantage to group foraging seems to be interference. Beddington (1975) found that among insect predators and parasites, searching efficiency actually *dropped* when predators were in groups, apparently for this reason. The same was discovered in feral house mouse populations (Stueck and Barrett, 1978).

16.2 Prey Adaptations

Morphological Adaptations

The value of cryptic coloration, the classic anecdotal form of defense for prey, is sufficiently obvious that there seems little point in dwelling on it. Using chaffinches (*Fringilla coelebs*) and jays (*Garrulus glandarius*) as predators, DeRuiter (1952) showed that stick caterpillars mixed up with sticks on a cage floor were found only with great difficulty compared with alternative noncryptic food items. Kettlewell and his co-workers (Kettlewell, 1955b) showed that the light and dark morphs of the peppered moth were relatively hard for predators to find against light and dark backgrounds, respectively. He also showed (Kettlewell, 1955a; Kettlewell and Conn, 1977) that the moths tended to alight on backgrounds of the appropriate color. Sargent (1969) demonstrated that two patterned moth species rested in alignment with a patterned background in such a way as to make themselves maximally cryptic. He also showed that light and very dark colored underwing moths, *Catocala relicta* and *C. antinympha*, respectively, chose to land on light and very dark colored backgrounds (Sargent, 1981). Dark and heavily banded morphs of the garden snail (*Cepaea hortensis*) are found in woods, in dark habitat, whereas yellow and unbanded morphs more commonly frequent open country (although the story is much less clear for the related *Cepaea nemoralis*) (Clarke, 1960). Blair (1947) found that in the prairie deer mouse, darker coat color was correlated with darker soil color, and Dice (1947) found that owls more readily caught light-colored mice on dark backgrounds, and vice versa.

It has long been suspected that vertebrate predators form *search images*. That is, they act as if they form a mental image of a prey type and thereafter tend to find disproportionate numbers of that prey type, and overlook others. The result would be a sigmoid (type III) functional response curve for each prey type in the company of others. Switching behavior and simply optimal food preferences as predicted by Emlen [see "Food Preference" (p. 440)] would generate the same sigmoid response. But whenever a sigmoid response curve is found, uncommon morphs of a prey species will find their genotype favored by natural selection (recall the discussions of frequency-dependent predation and competitive coexistence in Chapters 5 and 6). The disproportionate predation of common prey forms has been dubbed *apostatic selection*, and it is not hard to see that it provides a kind of defense for prey species that can evolve polymorphism. To demonstrate the existence of apostatic selection, a number of workers have set out varieties of foods for a variety of consumer species. Allen and Clarke (1968) offered collections of flour and lard baits, green or brown, to wild birds and found consistent preferences by blackbirds, starlings, house sparrows, and dunnocks for the more common food type. In 1972, Allen did similar tests and obtained the same results except that when one of two foods was extremely common and the other very rare, the rare type was taken out of proportion to its relative abundance. Allen thought that in this extreme situation the rare food might stand out just by virtue of its rareness. Manly et al. (1972) and Cook and Miller (1977) gave artificial (red or blue) baits to Japanese quail and observed preferences as a function of both relative frequency and absolute abundance. Presence of a search image was inferred if

$$\alpha = \ln \frac{\text{red available/red eaten}}{\text{blue available/blue eaten}}$$

showed a significantly negative regression on frequency of red abundance. At high densities, the birds demonstrated a straightforward preference for red bait, although there was a weak but statistically significant tendency to develop a search image. At low densities, the preference for red was less marked, as we might expect from optimal feeding theory, and a slightly stronger search image was noticeable. At intermediate densities, very strong search images were observed. Soane and Clarke (1973) found evidence, using flour and lard baits scented with peppermint oil or vanilla, that house mice formed search images based on olfactory cues.

But does polymorphism in the face of apostatic selection really benefit the prey species? Harvey and Greenwood (1978) painted mussel (*Mytilus edulis*) shells baited with meat and set them out for carrion crows (*Corvus corone*). Monomorphic collections suffered higher predation than did trimorphic groups.

Under what circumstances should polymorphism evolve as an antipredator adaptation? Greenwood (1969) notes that where predators show type III functional response curves, as almost all predators should in the presence of alternative foods (see Fig. 16-5), predation is heaviest on morphs of intermediate density (Fig. 16-11). Thus in very sparse and very dense populations, polymorphism is unlikely to evolve. Polymorphism is most adaptive as a predator-defense mechanism in populations of intermediate density, when prey are important to their predators, but well below predator saturation levels.

Species whose individuals are normally brightly colored, perhaps as adaptation for attracting mates, may evolve drabness in the presence of heavy predation by visual predators. The three-spined stickleback male, in nuptual coloration, usually sports a bright red-orange belly. But in parts of western North America, black-bellied males are found. These are usually allopatric to the normal red-bellied populations. The color difference seems to have a genetic basis; both fish types raised under identical conditions breed true. Females drawn from allopatric populations of *both* types prefer the red males quite consistently, suggesting a sexual advantage to the normal genotype. On the other hand, females from black-bellied populations immediately adjacent to red-bellied populations prefer their own black-bellied males. This suggests there has been selection for isolation of the two populations, allowing for differential adaptation (see Chapter 13). Indeed, there is clear-cut hybrid inferiority; both genetic incompatability and aberrant behavior in hybrid males has been documented, and significant viability loss occurs in cross-bred populations. Field observations show that black forms are more common upstream and red forms downstream, nearer salt water. Also, the normal red form is far more tolerant of saline conditions. There is no barrier, though, to red forms swimming upstream, so why do we find only black-bellied fish there in stream systems where they occur at all? What appears to be happening is that upstream populations of pure black-bellied *Gasterosteus*, which are unique to the Chehalis River system, find a particularly voracious predator, also unique to this river system, in the western mud minnow (*Novumbra*

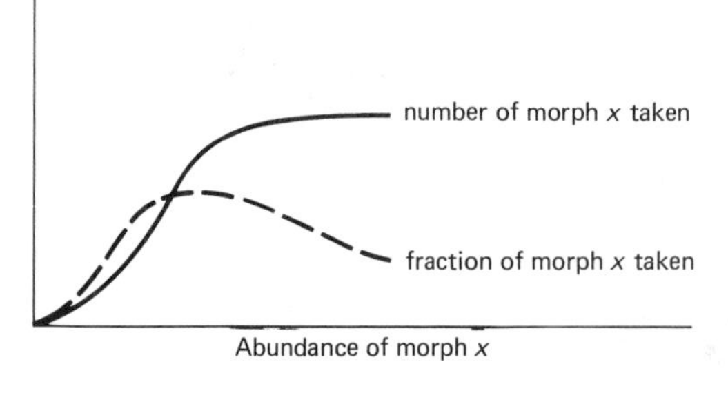

Figure 16-11 Type III functional response curve (solid line) and the corresponding fraction of a food population (morph x) taken by a predator.

hubbsi). The black-bellied males are less easily seen than are their red-bellied cousins, can fade in color when appropriate to blend with the background, and demonstrate a "striking association" with the western mud minnow. Also, laboratory tank tests on stickleback larvae, of an age when the two morphs look alike showed the black-bellied form to be better at evading predation by mud minnows (McPhail, 1969; but see Hagen et al., 1980).

Chemical Defenses

We shall return to morphological defense mechanisms shortly, in another context. But first we note another widespread technique for deterring predators, chemical defenses. Whittaker and Feeny (1975) have given a review of some of the compounds utilized by plants to discourage their getting eaten. Buttercups, *Ranunculus* spp., are distasteful to grazing animals. Several species contain protoanemonin, a substance that can lead to fatal convulsions. *Delphinium*, the larkspur, produces delphinine, a neurotoxin. Nicotine and other alkaloids are found in tobacco. Steroid cardiac glycosides cause convulsive heart attacks to any vertebrate that insists on eating plants of the genus *Digitalis*. Oleander (*Nerium oleander*) is sufficiently poisonous that one leaf is potentially lethal to human beings. Mustard oils are irritants that can cause serious tissue injury. They are also among the most potent antibiotics known from higher plants. But they are also used as *attractants* by insects that have evolved a resistance. Pyrethrins are effective insecticides occurring in *Chrysanthemum*. Some insects have evolved mixed-function oxidases in their mouths and gut that breaks down pyrethrins, but one *Chrysanthemum* species has counter-attacked by evolving lignan sesamin, an inhibitor of mixed-function oxidases. Tannins, a common constituent of many angiosperms, binds protein into indigestible complexes. Finally, some plants, mostly gymnosperms, have evolved synthetic insect hormones—α- and β-ecdysone and analogs—that may fatally accelerate insect metamorphosis. The balsam fir produces an insect juvenile hormone that prevents insect development. Janzen (1969) noted that Central American legumes appeared to use one of two possible strategies against bruchid beetles; produce either tremendous numbers of seeds (predator saturation) or a small number of large alkaloid-filled seeds.

In Colorado *Lupine* populations, alkaloid content of inflorescences was highest in species flowering at times of the season when larvae of the flower-feeding butterfly (*Glaucopsyche lygdamus*) were most common and active. Areas with high butterfly concentrations had plants with particularly high alkaloid concentrations. Also, flower populations suffering only slight butterfly damage carried three or four types of alkaloids, whereas those in butterfly dense areas often produced up to nine. A large variety of toxins makes the evolution of resistance more unlikely (Dolinger et al., 1973). A review of phenolic compounds as plant defense substances is given by Levin (1971). The alkaloid story can be found in Levin (1976).

Production by plants of secondary compounds for defense costs energy that could presumably be used for growth and reproduction. Hence the amount of toxin produced represents a balance of selection forces. It is for this reason, probably, that the lupines mentioned above, were less toxic in areas with less predation. Cates (1975) discovered that wild ginger (*Asarum caudatum*) could be classified into two genetically different forms; fast growing with high seed production, but highly sensitive to predation by slugs, and a slower-growing, less productive form more resistant to slugs. The former are more common in areas of low slug density, the latter are found in good slug habitat.

We intimated above that insects might occasionally evolve resistances to plant secondary compounds. Indeed, not only does this occur, but often the no-longer sensitive insect species will incorporate the toxins into their own bodies for use as deterrants against predators of the next trophic level. Monarch butterflies fed on milkweed (*Asclepias* spp.), their normal diet, induce vomiting in naive bluejays who at first eat them when offered, but the same insects raised on cabbage are readily eaten (Brower et al., 1968). Specific groups of toxic compounds are often found in specific plant families. This suggests that each chemical innovation led to an adaptive radiation which lasted until some enterprising insect came up with a way of neutralizing its effectiveness. One might expect that the new resistance mechanism might then lead to an adaptive radiation of insects. Ehrlich and Raven (1965) suggest that this may be one of many reasons for the immense number of observed angiosperm and insect species.

Mimicry

Although crypsis seems a reasonable sort of antipredator adaptation, it is not necessarily ideal for a potential prey item that has the chemical arsenal to protect itself adequately. When you are noxious tasting, have a painful bite or sting, or in some other way present a very unpleasant demeanor to a would be predator it may very well pay to advertise. Thus wasps are often brightly colored, as are monarch butterflies and the poison-skinned dendrobatid and atelopodid frogs of South America. Advertising colors are referred to as "aposomatic." Of course, not all noxious creatures are aposomatic. Toads (*Bufo* spp.) when disturbed, exude a bitter-tasting liquid from their parotid glands that is quite effective in dissuading initially enthusiastic but naive dogs and set them to making wry faces for some time after an encounter. Yet (most) toads are very drab in appearance. On the other side of the coin, some creatures are very brightly colored and yet quite harmless and palatable. Offhand, this would seem a rather ineffectual way to make a living. Perhaps the colors have evolved by sexual selection. That seems likely in the case of birds who, as adults, have few predators (their young are almost universally drab), and may also be true in some insects. But a more likely explanation, at least for many insects, is that the color pattern is mimetic. By closely resembling individuals of an aposomatic, noxious-tasting species, an insect may deceive a predator into responding as if it thought it was also noxious. The idea that aposomatic species might serve as models to mimicking palatable species was first suggested by Bates.

What is the evidence for *Batesian mimicry*? First there are a number of species pairs in which one is known to feed on toxic plants and to be at least mildly aposomatic, the other to be quite palatable but look very much like the former. The classic example is that of the milkweed-feeding monarch butterfly (*Danaus plexippus*) and its presumed mimic, the viceroy (*Limenitis archippus*). The latter looks almost exactly like the former, orange with black markings. It can be distinguished, superficially, by an extra black bar on its hind wings, and slightly smaller size. Platt et al. (1971) used caged, wild bluejays (*Cyanocitta cristata*) as predators, offering them viceroys (control birds), or monarchs and then viceroys. Four of 15 control birds refused to eat the viceroys; birds having previous experience with monarchs all refused the viceroys. Brower and his co-workers performed a great many experiments, offering look-alike pairs of food items, one of which was in some way noxious to various predators. Unfortunately, most of these experiments bore little relation to apparent model-mimic complexes in the wild. It is, for example, one thing to show that toads fed on stinging honeybees will thereafter show less appetite for look-alike but harmless droneflies (Brower and Brower, 1966), quite another to prove that Batesian mimicry is a viable process in nature.

Morrel and Turner (1970) spread out cardboard triangles with bits of pastry on them in a 5- by 5-yard area of suburban lawn where several bird species fed on them. Models, which had been dipped in quinine hydrochloride, were colored red. Perfect mimics were also red and poorer mimics were yellow, red with a black stripe, or yellow with a black stripe. Control pastries were dyed green. The birds learned very quickly to avoid the models, and ate significantly fewer mimics than green controls. Apparently, birds can learn, in the wild, to recognize bad-tasting but quite edible items.

Pilecki and O'Donald (1971) (see also O'Donald and Pilecki, 1970) performed similar experiments, using flour and lard "worms." Again the models were dipped in quinine. In one experiment a 1% quinine solution was used; both blue and green models were used, in roughly equal numbers. A small number of blue mimics and a large number of green mimics were set out and a fair number of yellow controls. Results are shown in Table 16-5. The effectiveness of

Table 16-5

	Blue Models	Blue Mimics	Green Models	Green Mimics	Yellow Controls
Number put out	493	80	415	438	315
Number eaten by bluejays	31	5	30	45	251

mimicry as an antipredation device is quite clear. In addition, note that when the mimic is quite rare relative to its model (80/80 + 493 = .14 for blue), only 5/80 = .06 of the mimics were eaten, but when the mimic was relatively common (438/438 + 415 = .51 for green), 45/438 = .10 of the mimics were taken. The difference is not statistically significant, but indicates the possibility of frequency-dependent (apostatic) selection favoring the rare mimic. Were frequency-dependent selection of this sort to occur in the presence of several models, we might expect frequency-dependent selection favoring several mimetic species (or several mimetic morphs within a species) with frequencies roughly that of the corresponding model forms. Indeed, such a pattern is observed in at least one model-mimic system. The mimic *Pseudacrea eurytus*, an African butterfly, has five distinct morphs, each mimicking a model species of the genus *Bematistes*. Sheppard (1959) plotted, for several locations, the frequency of the model type against the frequency of its corresponding mimic and found a good fit to a 45° line.

In 1964 Brower et al. made a valiant attempt to demonstrate Batesian mimicry in the field. Traveling to Trinidad, they used the butterflies *Parides neophilus* and *Heliconius errato*, both aposomatic and both known to be distasteful to local birds, as models. *Hyalophora promethea* was used as a control, and individuals painted to look like the models were used as mimics. Great numbers of models and (male) mimics were then released, and after some time, lured back with the aid of virgin females. The reasoning was that if mimics were protected by their resemblance to models, then more painted than control *Hyalophora* should be recaptured. Their results were statistically insignificant. Undaunted, however, they tried again (Brower et al., 1967). This time the models were *Parides anchises* females and the mimic (and control) was again *Hyalophora promethea*. This time some interesting results were obtained. The first few days of returns showed disproportionate numbers of mimics returning, suggesting that mimicry was really working. However, the last few days of recoveries favored controls. Overall, there was no difference in the return rate of mimic and nonmimic. The authors speculated that mimicry really did work, but that too many released individuals allowed the predators to learn to distinguish models and mimics. Thereafter the mimic, being more brightly colored than the control, was at a disadvantage. Smaller releases brought somewhat but not very much more satisfying results (but see Waldbauer and Sternburg, 1977).

Similar experiments were carried out several years later by Sternburg et al. (1977). These researchers painted male Promethea moths to resemble either (a) the palatable, yellow morph of the butterfly *Papilio glaucus*, or (b) the unpalatable, blue-black morph of the butterfly *Battus philenor*. Likenesses were good enough to fool the human eye at a few meters. Both models were native to the area in which the experiment was performed. A total of 436 promethea moths were released in groups of 14 or 50, with equal numbers of the two morphs. The return data, lumped over 12 days (the data are homogeneous over this period) are shown in Table 16-6. The same

Table 16-6

	Recaptures x Days After Release, x =							
Morph	*0*	*1*	*2*	*3*	*4*	*5*	*6*	***Total***
Black	54	26	12	4	2	2	2	102
Yellow	54	7	3	5	3	3	0	75

reversal of advantage found by Brower and his co-workers may well occur here, although it was not detected, but the overall results are far more satisfying. Table 16-7 summarizes the release–recapture data. The mimetic form was recaptured significantly more often than the nonmimetic form, indicating a predator-escaping advantage. Further support for the efficacy of Batesian mimicry was the observation that of all returning yellow morphs, 100% carried wing injuries apparently inflicted by birds' bills; 59% were seriously hurt (more than 2 cm^2 of wing missing). Of the black morph returnees, only 41% were injured, and only 8% seriously.

Table 16-7[a]

	Morph	
	Black	*Yellow*
Recovered	102	75
Not recovered	116	143

[a] $\chi_1^2 = 6.93, p < .01$.

The evolution of mimicry is a problem that has fascinated a number of people for some time. Some species have a mimetic and a nonmimetic morph, some have several mimetic forms. Is there a whole, coevolved complex of genes that controls the nuances of the mimetic relationship, and another "switch gene" that turns a complex on and off, or turns one complex off and another on? Has the gene complex congealed into a single "supergene" that has different allelic forms, each corresponding to another morph? It is not the business of this book to go into this question, but a recent reference that will serve as an introduction to the literature is by Charlesworth and Charlesworth (1976). A good review of both ecology and genetics of mimicry has been provided by Turner (1977).

We shall attack some ecological questions about the evolution of mimicry by perusing the work of Arnold (1978), based on a model originated by Estabrook and Jesperson (1974). Let p_M be the probability that the predator encounters a model, given its last encounter was with a mimic, and let p_I be the probability of encounter with a mimic, given the previous item was a model. Then if the likelihood of finding either potential prey type is dependent only on the identity of the previous item, we can describe the pattern of prey discovery with a Markov chain:

$$\mathbf{P} = \begin{array}{c} \\ \text{Mimic} \\ \text{Model} \end{array} \begin{array}{cc} \textbf{Mimic} & \textbf{Model} \\ \hline 1 - p_I & p_M \\ p_M & 1 - p_I \end{array}$$

Before proceeding recall that if $\boldsymbol{\pi}$ is the vector of unconditional encounter probabilities with mimics and models, then

$$\boldsymbol{\pi}(t + 1) = \boldsymbol{\pi}(t)\mathbf{P}$$

In addition, when one model has been encountered, the probability that the next item will also be a model is $1 - p_I$, the next after that $(1 - p_I)^2$, and so on. Thus, the number of consecutive models that a predator can expect, on average, to encounter is

$$1 + (1 - p_I) + (1 - p_I)^2 + \cdots = \frac{1}{1 - (1 - p_I)} = \frac{1}{p_I} \tag{16-21}$$

which we shall call $E(L_M)$, the expected "run length."

We suppose now that the predator cannot distinguish model and mimic, will readily eat a mimic, and upon mouthing a model will skip the next N food items. In their seminal paper on this topic, Estabrook and Jesperson (1974) proved that when this occurs, the proportion of mimics and models eaten will be

$$\text{proportion of mimics eaten} = \frac{\dfrac{p_M}{p_I + p_M}(1 - Z^{N+1})}{p_I(N + 1) + \dfrac{p_M}{p_I + p_M}(1 - z^{N+1})}$$

$$\text{proportion of models eaten} = \frac{p_M}{p_I(N + 1) + \dfrac{p_M}{p_I + p_M}(1 - z^{N+1})} \qquad (16\text{-}22)$$

where $z = 1 - p_I - p_M$. Thus if the "value" of a mimic to the predator is unity, and the relative "value" of a model is $-b$, the value to a predator behaving in this manner is

$$\text{"value"} = \frac{\dfrac{p_M}{p_I + p_M}(1 - z^{N+1}) - bp_M}{p_I(N + 1) + \dfrac{p_M}{p_I + p_M}(1 - z^{N+1})} \qquad (16\text{-}23)$$

where p_M = probability an encounter with a mimic is followed by an encounter with a model
p_I = probability an encounter with a model is followed by an encounter with a mimic
$z = 1 - p_I - p_M$
N = number of items the predator avoids after having sampled a model
b = value to the predator of a model − value of a mimic

Obviously, this equation is not easily put into words, but a deductive dissection of it is worthwhile.

It is in the interests of the predator to maximize the quantity in (16-23), and for the sake of argument we shall suppose that natural selection accomplishes this feat. What value of N should we expect to find? Arnold used several values of p_I, p_M, and b, and using a computer, calculated and plotted "value" as a function of N. His results indicate that N increases with $1 - p_I$. Thus, as the size of clumps of models (the run length, $L_M = 1/p_I$) increases, N also increases. This makes very good sense. If models come in large clumps it is appropriate for the predator, upon encountering a model, to skip a number of subsequent items roughly equal to the expected number of subsequent items that will be models. (In passing, consider the interesting possibility that where models form small clumps so that N_{opt} is small, predators should evolve shorter-term memories of unpleasant encounters.) If b was large (particularly noxious models), N went to infinity—it pays to eat nothing. If b was very small, $n = 0$—eat everything. When the model was bad but not awful and if p_M was sufficiently low and p_I sufficiently high, the predator could realize a net value gain, and so should feed on the model–mimic complex.

But we are interested here in what all this means to the prey species. It is to the model's interest, without doubt, to have the predator ignore the model–mimic complex. This can be accomplished by becoming extremely noxious. But this exacts a price—recall the work of Cates in the preceding section. The predator is also likely to ignore the complex if p_M is high and p_I low. But p_M is high when model and mimic are positively associated and when $(1 - p_M)$, the probability the predator encounters a mimic having just encountered a mimic, is low (i.e., when mimics are overdispersed). P_I is low when model and mimic are negatively associated and when $1 - p_I$ is high (i.e., when models are clumped). Thus models should evolve clumping behavior. Mimics will do best in opposite circumstances. Thus mimics should evolve such as to lower p_M (i.e., also to clump). The predictions with respect to species association in space are contradictory. But Arnold's calculations indicate p_I is the more critical parameter. Hence models should avoid mimics, but mimics should attempt to mix with models. This last prediction follows intuition. It seems likely that dispersion pattern may to some degree be forced on the model by virtue of the distribution of the plants on which it feeds and from which it obtains its noxious quality. The choice of association thus probably lies mostly with the mimic.

Müllerian Mimicry

Often several noxious-tasting species converge to a similar appearance. It has been suggested that similarity serves to present the potential predator with a single visual pattern, thus making it easier to remember. Also, where aposomatic coloring is found and clumping is advantageous, the advertising impact of several look-alike models is enhanced. Mimicry of this sort is termed *müllerian*. Little more will be said of müllerian mimicry here, but two observations are worth remembering. First, there is some evidence from müllerian complexes in neotropical forests that the so-called aposomatic coloration is really cryptic (Papageorgis, 1975). Second, there cannot be a clear-cut differentiation between the two types of mimicry. Müllerian mimics will inevitably vary in degree of noxiousness—is one really mimicking the other because it makes discrimination by predators more difficult or because similarity makes recognition of a common noxiousness easier? And once two species have converged, might not one, protected by the similarity, gain energetically by shifting emphasis from production or storage of toxins to greater reproductive effort?

Behavior Defense Mechanisms

To observe that prey flee from their predators is not terribly enlightening. And flight is often a tactic of last resort. Prey use a wide variety of means to avoid detection. They may use other prey as cues to predator activity, and may occupy microhabitats that make them less attractive to or less easily captured by predators. Here we will concentrate on one behavioral means of defense, namely grouping. We have already mentioned the extra eyes and ears advantage of grouping (see also Pulliam, 1973, and a review by Bertram, 1978). There are other reasons for flocking together. Brock and Riffenburg (1963) calculated the following relationship. Consider a flattened, disk-shaped school of fish with radius r_n, and containing n individuals, and suppose that a predator can detect prey at a distance, r'. Then a predator swimming randomly about will detect a school if its center is within a distance $r_n + r'$ (Fig. 16-12). This will occur per unit time with a probability proportional to $r_n + r'$. If the total number of fish is N, then the number of schools is N/n and the rate at which schools are detected is

$$\text{rate of detection} = c\frac{N}{n}(r_n + r')$$

where c is a constant. If m, the number of fish in an encountered school that are eaten, is less than n, then

$$\text{rate of feeding} = cm\frac{N}{n}(r_n + r') \tag{16-24}$$

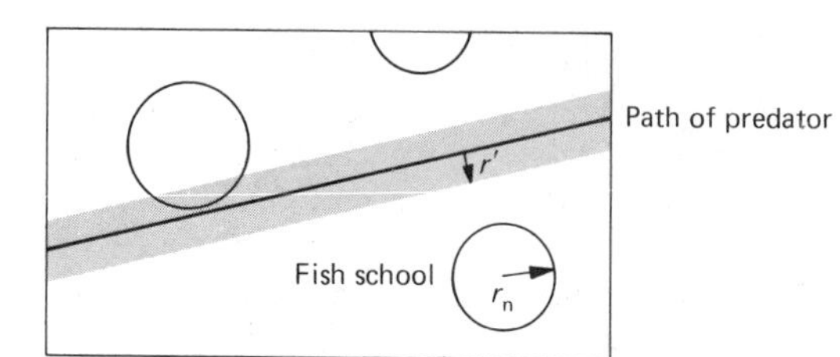

Figure 16-12 Schematic diagram of a predator sweeping a search path. If the path crosses one of the circles, representing the area of detection around the prey, the predator detects the prey.

The rate of feeding on randomly dispersed individuals is

$$\text{rate of feeding} = c(1)\frac{N}{n}(r_1 + r') \tag{16-25}$$

Schooling will be advantageous to the prey when the value of expression (16-24) is less than that of (16-25):

$$r' > \frac{mr_n - r_1}{1 - m} \tag{16-26}$$

Quite clearly, if all fish in a school are eaten on encounter with a predator ($m = n$), inequality (16-26) cannot be met and so schooling will never pay off to the prey species. If m is less than n, the advantage to schooling increases as

1. r_n declines: the school becomes tighter.
2. r' increases: the predator can detect prey at a greater distance.

The model assumes that after an encounter and feeding bout the predator must begin the search anew. If visibility is sufficiently high that the predator never loses sight of a school, there is no search involved and no advantage to schooling to avoid detection. So the grouping of antelope on the African plain is not practiced for its cryptic advantage. But where the assumptions of the model are met, there is some evidence for its legitimacy. Schooling occurs primarily in fish that serve as prey. Most large piscine predators school only when small. Also, schools tend to break up at night and in murky water where r' is small.

Treisman (1975a) takes a different approach. Suppose that p is the probability a predator detects any prey item within a certain range. Then if n prey are randomly distributed, the probability that the predator sees at least one is $p^* = 1 - (1 - p)^n$, and the probability that any one prey finds itself the intended victim is $[1 - (1 - p)^n]/n$. Now let D be the probability that the prey detects the predator while still in a position to escape. The probability the prey gets caught, then, is the probability that the prey is detected on the tth predator sighting and that the prey has not yet (in t trials) noticed the predator, summed over all t:

$$\sum_t (1 - D)^t(1 - p^*)^{t-1}p^* = \frac{p^*(1 - D)}{p^*(1 - D) + D}$$

Substituting $p^* = 1 - (1 - p)^n$ into this expression, we obtain

$$\text{probability of death} = \frac{[1 - (1 - p)^n](1 - D)}{[1 - (1 - p)^n](1 - D) + D} \tag{16-27}$$

If prey travel in groups of size n, then $p^* = p/n$, and D must be replaced by $D^* = 1 - (1 - D)^n$. The probability of death now becomes

$$\text{probability of death} = \frac{(p/n)(1 - D)^n}{(p/n)(1 - D)^n + [1 - (1 - D)^n]} \tag{16-28}$$

If a predator eats only one individual in an encountered group before it must start searching again, then grouping pays off to the prey when the value of (16-28) exceeds that of (16-27). Some rearrangement shows this condition to be

$$pD(1 - D)^{n-1} - [1 - (1 - p)^n][1 - (1 - D)^n] > 0 \tag{16-29}$$

The expression on the left falls as D rises. Thus grouping is most likely to be advantageous when D is low (individuals are unlikely, on their own, to spot a predator). The expression rises with p, indicating that grouping is increasingly likely to be advantageous when the predator is adept at finding its prey.

A more elaborate queuing theory treatment of prey grouping, considering gut capacity of the predator to control m, is offered by R. J. Taylor (1976). As in the two previous models, the assumption is made that a predator can eat only a limited number of prey and then must search for another group. The conclusion is that grouping in general is beneficial to prey.

Treisman's model can be extended to include the geometry of the group and the habitat by noting that p is proportional to $(N/n)(r_n + r')$ (see equations 16-24 and 16-25). The optimal size of the prey group may actually be less than that which minimizes expression (16-28) [with $p = (N/n)(r_n + r')$]; as group size increases there is an increasing probability of false alarms which have a disruptive effect (Treisman, 1975b).

Hamilton (1971a) points out another advantage to grouping for prey. When a predator appears, it pays an individual on the periphery of a group to dive into the center, putting others between it and the predator. The common tendency of groups to tighten up in the presence of danger suggests that this is a hypothesis to be considered seriously.

16.3 The Coevolution of Prey and Predator

Efficiency and Stability

In Chapter 4 we discussed at some length the fact that highly efficient predators destabilized prey–predator systems. It stands to reason that prey evolve morphological, chemical, and behavioral adaptations that minimize predator capture efficiency, and therefore that natural selection, acting on prey species, tends to stabilize prey–predator systems. It seems pretty obvious, on the other hand, that natural selection acting on predators will push the system toward increasing instability. But actually, the latter conclusion is not necessarily true. First, predators that learn to concentrate their efforts on the very young, old, or sick prey individuals are the ones that will be the most successful hunters. Natural selection, then, encourages efficiency at least partly by directing predators toward an "excess" prey population, the individuals that contribute the least to prey population dynamics. This kind of efficiency is more apt to be stabilizing than destabilizing. Also, predators that feed on a range of sizes of prey might be expected to show preferences for a particular size range, generally the larger prey individuals. Such a predation pattern has two effects. First, prey that are too large to be handled form an untouchable, stabilizing refugium pool. Second, as the medium-to-large items are taken, the average size of the edible class of individuals decreases, making the prey species, as a whole, a less desirable food. Predation is apt to drop in efficiency as predators are forced more and more to take those hard-to-handle, poor-energy-return items. Switches to other prey species may occur. All of these effects, too, are stabilizing. And there is good evidence that coevolution of prey and predator may alter both such as to impart stability. The European rabbit (*Oryctolagus cuniculus*) was introduced several times to Australia with a somewhat misguided zeal. The introducers got more than they bargained for. The rabbits finally took hold, from 24 pairs brought to the continent in 1859, and by 1950 had reached several hundred millions. The subsequent introduction of a rabbit virus, myxomatosis, brought the population under control—but what happened during the controlling process is of considerable interest. The virus was released in the form of a few inocculated rabbits at a number of places, and had a very rapid effect. In the population at Lake Urana, 100 infected rabbits were released and the September 1950 population of about 5000 was cut to around 50 in just 2 weeks, a 99% mortality. In the second year there was some buildup of the population, but the following summer there was a vector (mosquito) outbreak that brought another die-off, this time about 90%. Over all of Australia the first wave of mortality claimed 97 to 99%, the second year 85 to 95%, and the third only 40 to 60%. Laboratory tests showed that not unsurprisingly, the later generation rabbits had developed more resistance to the disease—natural selection works fast at a kill rate of 90% or more if the deaths are non–random. But not so expected was the finding that the myxomatosis had become less virulent. The mechanism for the latter change, though, is not hard to find. The disease produces skin lesions where the virus is picked up by mosquitos and carried to other rabbits. A rabbit infected with a mild form of myxomatosis lives longer and so provides more opportunities for mosquitos to spread the infection (Fenner et al., 1953; Fenner, 1965).

Coevolutionary processes such as that of the Australian rabbit and myxomatosis inspired Pimentel and his co-workers (Pimentel et al., 1963; Pimentel and Stone, 1968) to promote what they called a *genetic feedback* theory. Basically, these workers felt that the prey and predator would evolve such escape and exploitation patterns that only an "excess" of prey, those likely to die of other causes in any case, should fall victim. Specifically, where prey and predator cycle, the initial downturn of the prey signals a period when prey is being selected strongly for resistance or escape abilities, but the predator is enjoying a period of relative food plenty and so experiences only slight selection pressures for increased hunting and handling efficiency. Consequently, the prey, on average, evolves rapidly to become harder to catch, and its population decline is buffered. When the predator reaches its peak and begins to fall, and the prey is beginning a time of population recovery, the selection tables are turned, and the predator becomes rapidly more efficient, the prey more lax. This buffers the predator's fall. To test these ideas, Pimentel et al. (1963) arranged a series of small population cages into groups of 16 or 30, connected by $\frac{1}{4}$-inch-inner-diameter tubes. This setup mimicked a natural system of several subpopulations with limited dispersal between them. A wasp parasite, *Nasonia vitripennis*, and a host, either houseflies (*Musca domestica*) or blowflies (*Phaenicia sericata*), were then introduced. The populations were simply allowed to interact without disturbance and every few generations sample individuals were withdrawn and tested for resistance or virulence against both other "experimental" animals, or controls. In the 16-cell wasp–housefly run, wasps removed after eight generations (16 weeks) were found to lay 73.2 ± 14.3 eggs per female, on average, on experimental flies, and 99.6 ± 13.0 on control flies. Control wasps laid 77.7 ± 12.3 eggs per female on experimental flies and 131.2 ± 16.6 on control flies. Clearly, the flies had become more successful at avoiding parasitism, and also the wasps had become less virulent. The 16-cell system demonstrated two full, prey–predator cycles (about 16 generations) before going extinct. The 30-cell system, with houseflies, oscillated several times before crashing at 83 weeks. The evidence for greater fly resistance and lower wasp virulence was even more marked.

In a later experiment, Pimentel and Stone (1968) repeated the 30-cell wasp–housefly runs using control animals, or flies and wasps that had previously evolved in similar conditions. The control group showed oscillations of considerable amplitude; the experimental system fluctuated, but much more sedately.

Not all workers happily accepted Pimentel's ideas. Although genetic feedback seemed to work in an artificial wasp–fly system, Lomnicki (1971) felt it could not apply as a general rule. He argued that Pimentel had implicitly assumed that predators, or parasites, exploited prey randomly. If, in fact, the less resistant (picture a system of variably noxious plant items and an herbivore), more palatable prey individuals were preferred by the predator, the less noxious strain would disappear, and with the loss of genetic variability there would no longer be a feedback mechanism. Levin (1972) countered with two theoretical genetical models that stated the conditions for genetic feedback-induced cycles in the prey when the predator was held constant in number and genetic makeup, but the prey chose food items according to palatability. Pimentel et al. (1975) countered with experiments. Predators were houseflies and prey were 100 randomly placed vials with sugar solution and varying amounts of (noxious) NaCl. The salt concentration was determined by two additive allelic effects at each of three "loci" with no linkage. In the absence of predation, fitness of a vial genotype was set at $.9^J$, where $0 \leqslant J \leqslant 6$ was the molar concentration of salt. Thus fitness declined with salt content unless predation occurred. Flies were then introduced and allowed to feed from the vials. Definite preferences for the low-salt items were observed. After 4 days any vial with more than 50% of its "sap" gone was considered "dead." Relative viabilities of all genotypes was then assessed, the $.9^J$ factor considered, and a new generation of vials replaced the old, new gene frequencies determined by the standard selection equations (Chapter 8). Each surviving fly was replaced by two "offspring." In this way the prey was allowed to evolve, experiencing opposing selection pressures from predation and "cost" of production of salt, but prey populations remained constant. The predator population was allowed to vary in number, but did not evolve. Over 50 generations fly populations oscillated with decreasing amplitude until, by about the twentieth generation, densities were almost constant. Control populations of flies, in which the prey did not evolve, continued to fluctuate at the same amplitude over all 50 generations. By the twentieth generation the control populations were oscillating about a mean of roughly 147 flies, the experimental populations at about 53. Crowding effects were evident at high densities in the controls, but not detected in the experimental populations. Thus prey populations approached equilibrium well below carrying capacity, and at lower levels after selection had taken place. Survivorship of the prey increased from about 40% initially to about 85% at the end of the experiment, reflecting gene frequency changes for the salt-producing alleles at each locus from .12 to about .70, where they appeared to stabilize.

In a later response to Lomnicki's criticisms, Udovic et al. (1976) acknowledged that, if predators evolve the ability to discriminate and feed with strong enough preferences on the nonresistant prey, the genetic feedback process will not work. But they point out that spatial heterogeneity decreases the relative accessibility of susceptible prey and so forces the predator not to specialize strongly.

Group selection (Chapter 9) may also play a role in the evolution of prey–predator stability. Levin and Pimentel (1981) note that if a parasite is particularly virulent, it will kill its host and so kill off its own progeny. Interdemic selection, where demes consist of the collective assembly of parasites in one host, then, should favor low virulence. If virulence is associated with fecundity—as it surely must be in most cases—group selection should favor low fecundity. In combination with individual selection acting to increase fecundity, the net result should be intermediate values.

Gilpin (1975a) has developed the concept of group selection as a factor promoting prey–predator stability to a considerable degree in his book *Group Selection in Predator–Prey Communities.* Several models of varying complexity were developed; we shall deal with only one here. Suppose that the prey–predator dynamic equations are

$$\frac{dn_1}{dt} = n_1(r_1 - cn_2 + s_1 n_1 + I_1 n_1)$$

$$\frac{dn_2}{dt} = n_2(r_2 + \epsilon c n_1) \tag{16-30}$$

where n_1 and n_2 are the numbers of prey, predator, r's are intrinsic rates of increase, c is a measure of predation efficiency, ϵ measures the conversion prey biomass consumed to predator biomass, s_1 is a "cooperative" or "disoperative" term, and $I_1 < 0$ is an extra density-feedback term thrown in for the sake of generality. Consider selection at a single locus with impact on $\bar{c}$,

$$\bar{c} = p^2 c_{AA} + 2pqc_{Aa} + q^2 c_{aa} \qquad \text{where } c_{AA} > c_{Aa}, c_{aa} \tag{16-31}$$

where p and q are allelic frequencies, and suppose that selection acts only through the predator. Then

$$\frac{dp}{dt} = pq(\bar{r}_A - \bar{r}) = pq\epsilon n_1[p(c_{AA} - c_{Aa}) + q(c_{Aa} - c_{aa})] \tag{16-32}$$

Gilpin also included a term for genetic drift. If we now let predators migrate among patches at a rate proportional to the predator density at the patch of origin, allow some migration mortality, and suppose that a predator subpopulation is extinct if $n_2 < 1.0$, we can simulate the population–genetic process of several subpopulations simultaneously on a computer. Initial conditions, with respect to distribution of p values over patches, were varied, with various c_{ij} values and migration rates. The results for this, and other models, indicated that over a wide range of parameter values and initializations, group selection *can* prevail over individual selection, sending the a allele (lower predation rate) to fixation. If all subpopulations are initially very similar, lack of variance in p among groups weakens the interdemic selection process and allows A to increase. However, genetic drift generally leads, in time, to sufficiently high var(p) and turns the process around.

The Evolution of Functional Response Curves

We have discussed several reasons, both in Section 16.2 and in Chapter 4, why functional response curves might take the shape they do. Before reviewing, or proceeding, however, it is important to note that deviations from the simple type II form may come about for two entirely separate classes of reasons.

1. If the predator is faced with but a single type of food, or if we wish to consider a predator's response to all foods considered together, comparisons of amount ingested per predator over several populations of different food density seem unlikely to give us anything but a type II curve. Sigmoid curves seem reasonable only for populations responding to changing food supplies. As food density increases, the predator learns to more efficiently handle individual food items, and the response curve should rise slowly at first, and then more quickly than prey density. But during a decline in food we might expect a hysteresis effect, a disproportionately slow drop in utilization until food density becomes quite low. Evolution of the form of an average curve over all fluctuations can be viewed as follows. Use Real's (1979) formulation [same as (4-6)]

$$\text{number of items eaten per time per predator} = m_1 = \frac{aTn_1^\eta}{1 + acn_1^\eta} \qquad (16\text{-}33)$$

where T = total foraging time
a = a constant describing foraging efficiency
η = describes the sigmoid nature of m_1 versus n_1
c = a constant related to food-handling time

and ask what value of η we should expect to find. Suppose that we are dealing with a highly opportunistic prey and a much less opportunistic predator. Then the prey, relative to the predator, undergoes large density fluctuations and is found mostly at low population levels. Clearly (Fig. 16-13), it is in the interest of the predator to eat in the fashion indicated by curve A. The prey, however, will benefit by a type III curve (B). Because selection is acting, via foraging patterns on the predator, but also via defense mechanisms, on the prey, the net evolution of the functional response curve is in some doubt. One can argue that an opportunistic prey is affected primarily by density-independent factors and so predation is a relatively unimportant influence. If that is the case, selection via the predator might generally be expected to prevail, leading to a pronounced type II curve (η small). But, of course, the ultimate path of evolution depends, among other things, on the additive genetic variance of η among predators. If the predator is food-limited, η may be an important component of fitness and so have little expressed additive

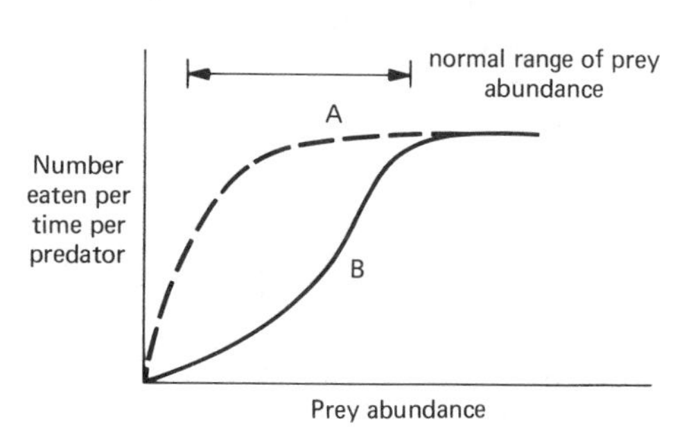

Figure 16-13 Possible functional response curves.

variance. If the predator is more opportunistic than its prey, an unlikely situation unless we consider pathogenic microbes and some parasites, the predator will be benefited if it does not eat out all its prey during predator population highs. Thus when prey abundance is low relative to predator abundance, group selection favors low feeding rates and so a type III curve.

2. If the predator has available to it alternate prey, then sigmoid functional response curves for each prey type may be expected on several bases. Continuing the foregoing line of reasoning with a prey that fluctuates more strongly than its predator, the ability to shift diet weakens the selection pressure on predators to evolve a type II response. In such cases, especially if the prey is not markedly opportunistic, a type III response is likely to occur. We have seen also that optimal food preferences can be expected to lead to type III curves (Fig. 16-5), an effect that will be exaggerated by switching or time lags in adjustment of preferences. Finally, there is the matter of search images. It is not clear why search images might have evolved. Perhaps they are the result of neural constraints. An animal may not be mentally capable of concentrating strongly on more than one type of searched-for item at a time, and so flip-flops, without internal control, back and forth between attentiveness to one or the other. Perhaps natural selection favors and fine-tunes these flip-flops with an end result encouraging use of the more readily available food sources.

There are other reasons why sigmoid functional response curves may have evolved in the context of available alternative foods. For a treatment rich in detail, but somewhat complex, the reader is referred to McNair (1979, 1980). Here we shall merely give McNair's general results. Note first that foraging animals do not necessarily encounter their foods randomly. Note also that they may not only form search images as a function of what they last encountered, but may as a result of practice (short-term learning) or shifting "mind set" express greater searching efficiency or handling ability with items as a result of just having encountered them. McNair calls these "training effects," and it is not difficult to see that they lead to sigmoid tendencies in the functional response curve. Clumping of prey also promotes type III curves. A very general expression, when all items encountered are utilized, bearing a strong similarity to the expression of Pulliam (equation 16-3), is

$$\lim_{t \to \infty} \begin{pmatrix} \text{mean energy intake per time} \\ \text{over the interval } 0, t \end{pmatrix} = \frac{(\pi_1/\gamma)e_1 + (\pi_2/\gamma)e_2}{1 + (\pi_1/\gamma)T_1 + (\pi_2/\gamma)T_2} \qquad (16\text{-}34)$$

where T_i = handling time for an item of type i

e_i = "value" of an item of type i

$\pi_i = p_{ji}/(p_{ij} + p_{ji})$

p_{ij} = probability of encountering an item of type j given an immediately previous encounter with one of type i

$\gamma = \pi_1(p_{11}s_{11} + p_{12}s_{12}) + \pi_2(p_{21}s_{21} + p_{22}s_{22})$

s_{ij} = search time for an item of type j given the previous encountered item was type i.

Following the analysis of Pulliam (equations 16-4 and 16-5) we see that a pure strategy of eating food type 1 should occur when

$$e_1 > e_2 \frac{1 + (\pi_1/\gamma)T_1}{(\pi_1/\gamma)T_2} \qquad (16\text{-}35)$$

and that a mixed diet can be expected when

$$e_2 \frac{(\pi_2/\gamma)T_1}{1 + (\pi_1/\gamma)T_2} < e_1 < e_2 \frac{1 + (\pi_1/\gamma)T_1}{(\pi_1/\gamma)T_2} \qquad (16\text{-}36)$$

Unlike the case described by Pulliam, with no training effects, the quantities π_i/γ include both p_{ij} and p_{ji} terms and are thus functions of the abundances of both foods. Therefore, in opposition to Pulliam's results, even though food types show no overlap in value whatsoever, inclusion of the poorer food will be a function of the poorer as well as of the better food's abundance.

If use of both foods varies and depends on the type of food previously encountered and also on whether that item was utilized, or if training effects reflect encounters or use of items prior to the immediately preceding one, the formulation will have to become more complicated.

Review Questions

1. Review the various criteria believed important in determining an animal's diet. For what kinds of animals might "energy per time" be an adequate measure of food value for making predictions? For what kinds of species would learned aversions be least or most important? What kinds of animals need to alter their diet because of constraints imposed by gut capacity?
2. Would you expect the relationship shown in Figure 16-5 to be a general one for most species?
3. How might a plot of percent food *A* in the diet against percent food *A* in the environment (Fig. 16-5) depend on whether food patches were large or small, rich or poor? How would the curve be affected if food species formed species-specific patches as opposed to mixed-species patches?
4. What is the importance of the shape of the curve in Figure 16-5 for understanding population dynamics?
5. Return to the section on functional response curves in Chapter 4 and "fine-tune" the models given there in light of the ideas and data presented in this chapter.
6. Refer to the section "Foraging in Groups" (p. 457) and recall the discussion of groups and social evolution in Chapter 15. To what extent might food distribution in space, and food renewal in time, be related to the nature of a species' social organization? What *general* principles can you suggest?
7. Would the various predator feeding adaptations discussed in this chapter have a stabilizing or destabilizing influence on a prey–predator system?
8. What influence might a predator's functional response curve have on the evolution of mimicry, or prey polymorphism in general [see "Morphological Adaptations" (p. 459)]? How might the evolution of mimicry or polymorphism affect the functional response curve?

Chapter 17

The Evolution of Competitive Interactions

Suppose that we were to perform the armchair experiment of introducing a variety of species, all sharing the same resources, into an isolated area and ask what might happen. Several scenarios are possible. First, it is conceivable that density-independent fluctuations depress expected population numbers sufficiently below carrying capacities that there is a shortage of resources only on rare occasion. Then some competitive exclusion might occur, but probably at a very slow rate. Perhaps periodic catastrophes accomplish the same result. In either case, some frequency dependence might be introduced, permitting indefinite coexistence of all species. Predation might also enable coexistence to occur. Finally, if the area into which the introduction took place were big enough, random codispersion plus social cohesion within species might allow continued coexistence. All of these possibilities have been discussed in some detail in Chapters 5 and 6. If we take the position that these various competition-alleviating conditions really permit coexistence of the introduced species, that few or no exclusions occur, and that those species that do die out do so largely without regard to their niche relationships to others, we must conclude that no competition-induced pattern in species composition or resource utilization occurs. Essentially, we deny a role to competition as an organizer of community structure. What differences occur between niches of the various species are serendipidous. On the other hand, if we deny that those species which persist are a random sample of those introduced, if we suppose that species that are very similar in their use of resources are less likely to coexist than very different species, we should expect to find, after some time has elapsed, that the remaining species are *not* randomly dispersed in their choice of resources or microhabitat, but are overdispersed. We should conclude that what species occur in an area and the patterns in which they exploit their environments are very much the result of competition.

The field of community ecology is currently split over which of the two approaches above is the more accurate. We shall review this controversy in more detail shortly, but for the moment let us lay some groundwork by returning to basics. What kinds of selection pressures might we expect to find in our hypothetical community?

We begin by stating a position with which many will take exception. This beginning does not represent an attempt to bias the discussion to come; rather, it is an attempt to provide some framework on which to hang the subject matter of this chapter, an endeavor to bring some organization to a very disordered topic. We assert that because various resources are differentially distributed in the environment, the resource demand patterns of their consumers will be reflected in those consumers' microhabitat distributions. Thus, knowing the spatial dispersion of a species in relation to the spatial background of resources, we should be able (if we knew how to read the information) to deduce how that species makes a living. There are other criteria that determine a species' dispersion pattern. It is not only the resources that are important but also, at least potentially, the ability to acquire those resources. If population density in a small, local area is very high, resource availability might be curtailed, either through interference or by virtue of prior consumption. Notice that both interference or resource reduction might occur either as a result of the presence of conspecifics or individuals of other species. Finally, microhabitat distribution depends on environmental characteristics unrelated (except, perhaps, statistically) to resources or population density (i.e., density-independent factors). We now reassert the existence of ideal free distribution. Populations should be expected to have evolved abilities to recognize with reasonable accuracy the suitability of a small piece of substrate, based on resources present, density feedback and density-independent factors, and the predisposition to remain preferentially in more suitable surroundings. Therefore, all individuals of a given genotype should distribute themselves in space in a manner that assures that their *expected* contributions to the genotype's fitness are equal (see Chapter 13).

The realization of ideal free distribution assures four things. First, if two species do, in fact, exploit the environment in similar enough fashion that they are potential competitors, microhabitat preference patterns should eventually evolve to separate them in space. This separation may be either patchy in nature, or occur on a macro-environmental scale. The more potential for competition, the more separation should occur. Species with no potential for competition may be randomly codispersed or, because of *indirect* interaction [see "Facilitation" (p. 161) and "Diffuse Competition" (p. 166)], either positively or negatively correlated in distribution. The areal size of the plots we would need to detect such codispersion would depend greatly on the areal extent over which the species in question interacted. Birds might be expected to exhibit separation based on a much coarser scale of microhabitat definition then, say, collembolans. The second consequence of ideal free dispersion is that the introduction of a potential competitor encourages evolution of a segregating shift in microhabitat preferences: The differences in niche between two allopatric species are likely to be exaggerated in sympatry. A third consequence of ideal free distribution is that a species' breadth of microhabitat use reflects a balance between intraspecific and interspecific competition. Recall the argument for the evolution of niche width in Chapter 13:

$$\sigma_x^2 = \frac{\hat{r}w^2 E(n)/K}{1 - E(n)/2K} \tag{13-14}$$

where σ_x^2 is phenotypic variance (niche width), $\hat{r}$ is the intrinsic rate of increase of the hypothetical optimal phenotype, $1/w^2$ measures the intensity of stabilizing selection, and $E(n)$ and K are expected population size and carrying capacity, respectively. This expression invokes *intra*specific density- (and frequency-) dependent selection only. But note that the presence of interspecific competitors lowers $E(n)$ relative K and so decreases σ_x^2. Thus the realized niche width arises from *intraspecific* competition acting to broaden the niche and *interspecific* competition (plus stabilizing selection—including, for example, selection to specialize) encouraging a narrowing of the niche. Finally, the fourth consequence of ideal free distribution is that because it reflects a balance of intra- and interspecific competition, it permits us to measure the (potential) competition intensity among species relative to that within species. In this chapter we discuss each of these four consequences from a more classical viewpoint. We end by using ideal free distribution to develop a method for assessing competition intensity in the field. The method is, in fact, more general in its import than this suggests, however, for ideal free distribution patterns carry information about far more than just relative competition pressures.

Throughout this chapter we assume that facultative shifts in distribution, resource use, or other behaviors reflect historical, evolutionary patterns. Thus, when a behavior is selectively advantageous, it is both selected for at the genetic level and, within constraints, performed "by choice."

17.1 Niche Separation and Community Organization

Niche Separation

MacArthur (1958) conducted his dissertation research on the foraging behavior of five wood warblers nesting in coniferous woods in the northeastern United States. Figure 17-1 shows the various foraging positions of these birds. The Cape May warbler (*Dendroica tigrina*)

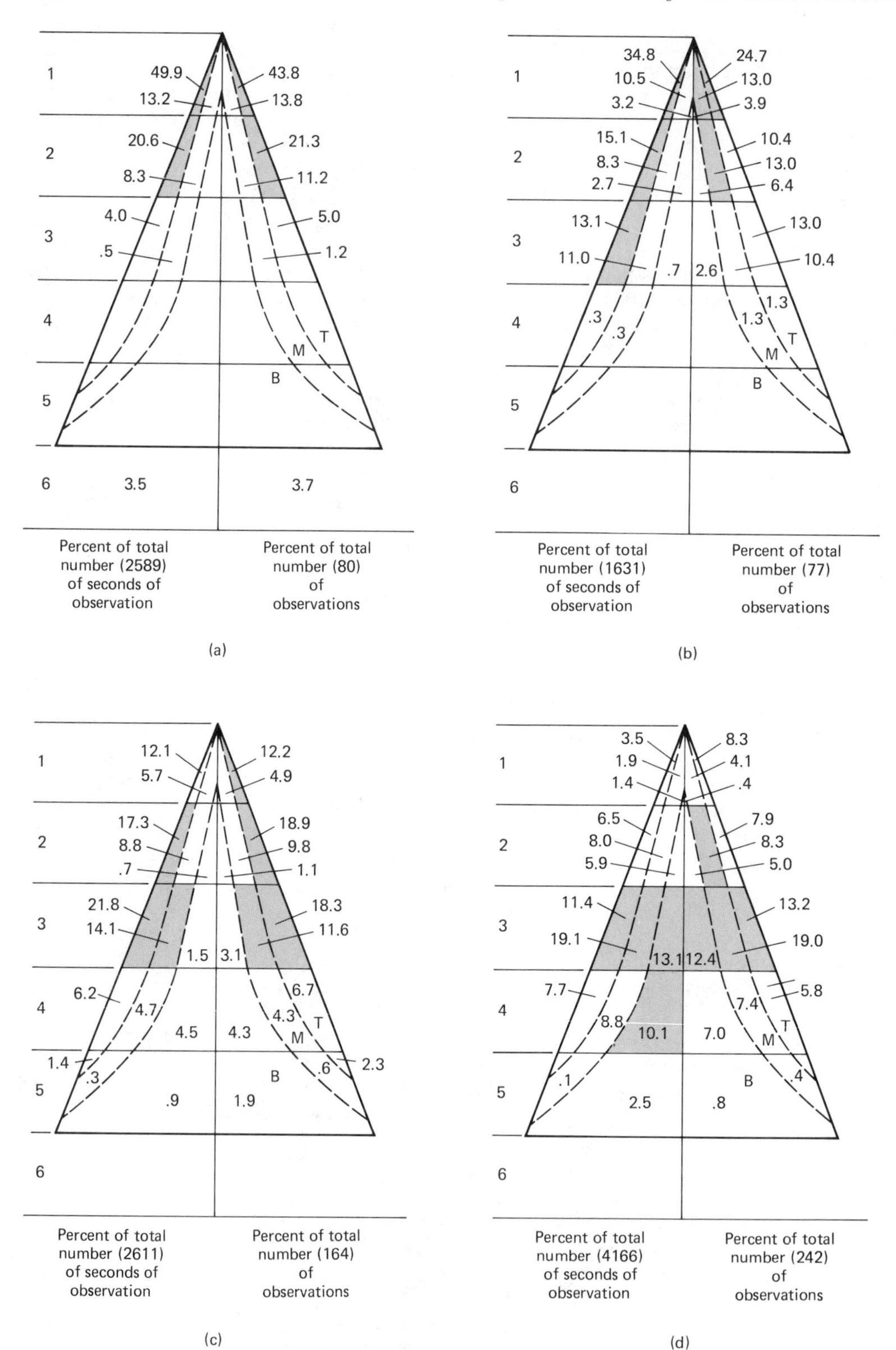
1
2
3
4
5
6
49.9
13.2
20.6
8.3
4.0
.5
43.8
13.8
21.3
11.2
5.0
1.2
T
M
B
3.5
3.7
Percent of total number (2589) of seconds of observation
Percent of total number (80) of observations
(a)
34.8
10.5
3.2
15.1
8.3
2.7
13.1
11.0
.7
2.6
.3
.3
24.7
13.0
3.9
10.4
13.0
6.4
13.0
10.4
1.3
1.3
Percent of total number (1631) of seconds of observation
Percent of total number (77) of observations
(b)
12.1
5.7
17.3
8.8
.7
21.8
14.1
1.5
3.1
6.2
4.7
4.5
4.3
1.4
.3
.9
1.9
12.2
4.9
18.9
9.8
1.1
18.3
11.6
6.7
4.3
.6
2.3
Percent of total number (2611) of seconds of observation
Percent of total number (164) of observations
(c)
3.5
1.9
1.4
6.5
8.0
5.9
11.4
19.1
13.1
12.4
7.7
8.8
10.1
7.0
.1
2.5
.8
8.3
4.1
.4
7.9
8.3
5.0
13.2
19.0
5.8
7.4
.4
Percent of total number (4166) of seconds of observation
Percent of total number (242) of observations
(d)

Figure 17-1

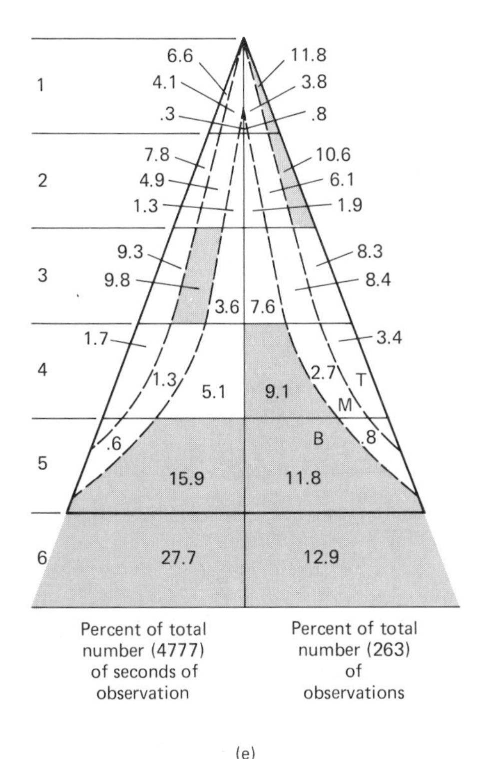

Figure 17-1 Distribution of feeding effort for five warblers nesting in northeastern coniferous forests in the United States: (a) Cape May warbler; (b) Blackburnian warbler; (c) black-throated green warbler; (d) bay-breasted warbler; (e) myrtle warbler. The zones of most concentrated activity are shaded until at least 50% of the activity is in the stippled zones. (From MacArthur, 1958.)

concentrated its efforts in the top 40% of trees, mostly on the outer branches. The Blackburnian warbler (*D. fusca*) used the top 60%, mostly the top 20%, both outer branches and closer in, although not near the trunk; the black-throated green warbler (*D. virens*) foraged as the Blackburnian except mostly in the mid 20% of the tree. The mid 20%, from the periphery to the trunk, was used by the bay-breasted warbler (*D. castanea*). Finally, the myrtle (yellow-rumped) warbler (*D. coronata*) foraged in the bottom 20% of the tree. Where the area used for feeding was similar for two species, different foraging movements were found. The Cape May moved up and down, while the Blackburnian moved radially and the black-throated green tangentially. It seems quite clear that these warblers display niche differences. Are these differences merely what we might expect were we to generate a warbler community without competitive constraints, or do MacArthur's observations show a pattern of overdispersal? The fact that lack of separation with respect to one criterion (position) was seemingly compensated by greater than average separation elsewhere (foraging mode) suggests the latter. A point of historical interest is that this investigation, and others like it, spawned a veritable storm of studies designed to look for niche differences among presumably competing species of every imaginable sort. Some of these studies were extremely elegant. Some of the high points are described below.

Schoener (1968a) studied four species of Anolis lizards in Bimini, measuring the height and size of each lizard's perch, whether it was on leaves, branches, or ground, and the color of the bark against which the lizard might be seen. In such a way a large number of possible microhabitats were defined. Schoener defined niche overlap between two species, i and j, by the quantity

$$D_{ij} = 1 - \tfrac{1}{2}\sum_i |p_{xi} - p_{yi}| \tag{17-1}$$

where p_{xi} was the proportion of species x in the ith situation. The same measure was used to measure niche separation with respect to size of prey eaten by the lizards. Schoener found that as habitat overlap increased, food size similarity decreased. This, he felt, was good evidence for competition and for the role of competition in structuring the lizard community.

Working with seven species of small mammals in the American Southwest, M'Closkey (1976) carried out an extensive live-trapping program and in 15- by 15-meter quadrats, measured 12 microhabitat variables. Discriminant function analysis showed statistically significant microhabitat differences among 20 of 21 possible species pairings. That these differences are a response to interspecific competition is supported by the fact that during the season when only permanent residents were present (some species hibernate), niche widths (the within-species variance in microhabitat parameter values) increased. Of course, this period also coincided with a time of recruitment, so the increased variety of microhabitats used *could* simply have reflected the dispersal movements of young animals. Thus competition, in concert with the ability of various small mammal species to adjust their dispersion patterns, seems likely to be important in determining both composition and niche pattern in this system.

The bumblebees *Bombus oppositus* and *B. flavifrons*, forage on larkspur (*Delphinium barbeyi*) and monkshood (*Aconitum columbianum*), two species with very similar habitat preference and flowering time (Inouye, 1978). In 15- by 20-meter plot near Gothic, Colorado, it was found that the monkshood received 32.1 bee visits every hour from *B. flavifrons*, but only 4.2 from *B. appositus*. Larkspur, on the other hand, was visited by *B. flavifrons* .7 times and by *B. appositus* 30.2 times. Thus there was a clear separation in flower use (a niche separation) by the two bee species. During an experimental period, Inouye removed *B. flavifrons* foragers from monkshood whenever he found them, reducing the visitation rate from 32.1 to .15. This procedure was rewarded by a significant decline in the preference by the other bee for its usual flower, the larkspur, and an increase in its visits to monkshood. This experiment was repeated with bees of both species being taken from their preferred flower and held 2 full days before release. The results are shown in Table 17-1. In both cases the rise is significant at the .01 level or better.

Table 17-1 **Number of Visits per Flower per Hour**

	Control	*B. flavifrons* Removed from Monkshood	Control	*B. appositus* Removed from Larkspur
On monkshood				
By *B. flavifrons*	11.40	4.36	7.35	6.85
By *B. appositus*	.20[a]	1.50[a]	.18	.23
On larkspur				
By *B. flavifrons*	12.20	12.29	4.24[a]	6.46[a]
By *B. appositus*	4.20	1.96	9.72	3.77

[a] Pairs in which the value is expected to rise in the experimental treatment if competition were keeping the bees' foraging niches separated.

A review of the data to 1973 regarding niche separations for a wide variety of species was provided by Schoener (1974a), who added the further observation that out of 81 species for which he could find data, 78% showed niche separation with respect to diet and 90% showed habitat separation. There are no data on *microhabitat* distribution for those species without clear habitat separation.

We might expect differences in habitat or resource utilization by species to be reflected in morphological differences. Thus larger food might be correlated with larger consumers, more specifically consumers possessing larger feeding apparatus. Where this correlation is realized, we should be able to assess niche separation, to some degree, on the basis of morphological divergence. There is a considerable literature supporting this contention (see, e.g., Newton, 1967, for birds, Mares and Williams, 1977, for mammals). Hutchinson (1959) suggested that just as there might be a minimum degree of niche overlap that permitted coexistence, so might there also be a minimal morphological separation between competing species, allowing them to persist. This idea has been picked up and tested by many people, but perhaps most spectacularly by Pulliam (1975b). Pulliam measured culmen length (distance from the anterior end of the nares to bill tip) in sparrows and calculated the ratios of the mean for one species to the mean for the next smallest species. Among Costa Rican sparrows the ratio is 1.11 $\pm$ .15, in southeastern Arizona, 1.09 $\pm$ .10, and for wintering populations in North Carolina 1.10 $\pm$ .05. The smallest species is Brewer's sparrow, with a culmen length of 8.8 mm. The size series thus runs 8.8, 9.7, 10.6, 11.7, 12.9, 14.2, 15.6. The largest belongs to the fox sparrow, *Passerella iliaca.* Pulliam then regressed mean seed size eaten, M, on culmen length, C, to obtain

$$M = .108 + .120C \qquad r = .814, \quad p < .01$$

The data used for the regression were gathered from birds in Maryland, but were presumed generalizable to birds elsewhere. There was no relationship between culmen length and variance in seed size. The next step was to break all seeds in three Arizona winter sites down into five size categories, x, and to measure the frequency distribution of each category, $p(x)$. Finally, the distribution of seed sizes in any hypothetical species' diet, $f_i(x)$, was presumed to be close to normal. Pulliam then used the niche overlap measure to estimate competition coefficients (Section 6.5):

$$\alpha_{ij} = \frac{\sum_x f_i(x) f_j(x)}{\sum_x f_i(x)^2} \tag{17-1}$$

and approximated carrying capacity with

$$K_i = \sum_x p(x) f_i(x) \tag{17-2}$$

The inequalities (Chapter 5) that must be met to allow competitive coexistence,

$$\frac{1}{K_i} > \frac{\alpha_{ji}}{K_j} \qquad \frac{1}{K_j} > \frac{\alpha_{ij}}{K_i} \tag{17-3}$$

were then examined for each hypothetical species pair. The results obtained for one site are shown in Table 17-2. Species rank 1 corresponded to $C = 8.8$, species rank 6 to $C = 14.2$. Table 17-2 is read as follows. The top row indicates that species 1 can coexist with species 4, 5, or 6, but not with 2 or 3; the second row indicates that species 2 can coexist with 4, 5, or 6, but not with 3. Thus the realizable species combinations are 1 and 4, 1 and 5, 1 and 6, 2 and 4, 2 and 5, and 2 and 6; all other combinations should lead to competitive exclusion. But species 3 has the largest K, and wins over any other single species or otherwise stable pair. Thus species 3 is the predicted lone inhabitant of this site. In fact, one sparrow, with culmen 10.8 (rank 3) is found,

Table 17-2

Species Rank: Smallest to Largest		Species Rank: Smallest to Largest					
		1	2	3	4	5	6
	1		−	−	+	+	+
	2			−	+	+	+
	3				−	−	−
	4					−	−
	5						−
	6						

the grasshopper sparrow, *Ammodramus savannarum.* In site 2, the results indicated that species 1 and 3, 1 and 4 could coexist, but the species 2 could successfully invade and oust either of these combinations. Here Pulliam found only the chipping sparrow, *Spizella passerina*, rank 2. In site 3 the expected species were ranks 2 and 3. Species 2 (chipping sparrow) was found, but no species 3. However, towhees (*Pipilo erythrophthalmus*, $C = 15.2$, rank 7) occurred in the area. Using the formula $n_i = K_i - \alpha_{ij}n_j$, Pulliam predicted the relative abundances of these two species to be 93:7. In actuality the ratio was 94:6.

Unfortunately, the concept that morph size ratios necessarily reflect niche difference is questionable. Not always do large species eat larger foods. In fact, Pulliam himself (Pulliam and Enders, 1971) found that within sites, five seed-eating finch species he examined, including the field sparrow (*Spizella pusilla*), the white-throated sparrow (*Zonotrichia albicollis*), and the cardinal (*Richmondena cardinalis*), showed little difference in seed-size use despite the large differences in bill size. (Perhaps this is merely a case where niche separation is realized strictly by differential habitat rather than resource use?) But for the sake of argument, suppose that body (or bill) size and food size *are* generally correlated, and that a particular minimal critical difference permits coexistence (in Pulliam's work it appeared that two species would have to differ by at least two ranks, a ratio of $1.1^2 = 1.2:1$). Then why do we find gaps in the observed series in nature? Wiens and Rotenberry (1981) looked at culmen length ratios in grassland birds and found values between adjacent species to range from 1.19 to 3.97 at one site, 1.12 to 3.22 at another, and so on. Why? Perhaps it is simply a matter that some species that *could* be present simply are not, for historical reasons. Perhaps habitat separation obviates the need for resource separation, rendering bill size differences meaningless. Or perhaps interference and environmental heterogeneity in time and space make exploitation competition a force of miniscule significance—again rendering bill size differences meaningless. But again, let us suppose that the 1.2 (or whatever) ratio *is* meaningful, that the gaps can be explained, and forge on. Why should morphological differences necessarily be expressed as ratios rather than as an arithmetic progression or some other kind of relationship? Schoener (1974b) argued that large resource items are apt to be more scarce than small items, requiring greater differences among coexisting consumers. Thus separation should increase with size and a geometric series is reasonable. Oksanen et al. (1979) argued that it would make sense for animals to view their resources in units relative to the mean size used. Thus, again, ratios would be appropriate (see also Case, 1981). But might it be that different species view their environments from different scaling perspectives? What happens if one species accomplishes niche separation from a second along one gradient, but the second views competition with the first primarily with respect to another? Horn and May (1977) made the case that the ratio (which, incidentally, seems to average somewhere around 1.28 : 1.0—see Hutchinson, 1959) may be something other than biological. They scurried about looking at various objects other than species that come in different sizes and discovered that the 1.28 : 1.0 ratio (or something very close) rather nicely described "instars" of children's bicycles, sets of kitchen skillets, ensembles of recorders, and consorts of viols. Maiorana (1978) takes a less jaundiced view, but adds sets of plates, glasses, small pewter cans, and animal figurines to the list. Roth (1981) may have made the most astute observation of all on this subject. "To replace a simple 1.3 rule with enormously complex post hoc models . . . would be no more productive than was the embellishment of commonsense notions of geocentrism with Ptolemaic epicycles."

Finally, let us return to the point that neither the existence of competition nor its role in structuring the community can be demonstrated merely by showing differences in niche placement; these differences must be nonrandom. Inference of overdispersion can be made, with caution, if niche overlap along one dimension is significantly negatively correlated with overlap along another, as with MacArthur's warblers or Schoener's lizards. If separation occurs along one gradient, separation along another may be unnecessary for coexistence to occur. Sale (1974) suggests another means to test for overdispersion of niches. He measures overlap between species i and j using the expression

$$C_{ij} = 1.0 - \tfrac{1}{2}\sum_h |p_{ih} - p_{jh}|$$

where h defines a microhabitat or segment along a gradient (although other overlap measures could also be used) for all species pairs. He then uses the observed p_{ih} values, for all h, figuratively throws them into a hat and randomly rearranges the h subscripts. After doing this for all species, he uses the values to calculate C values for a "randomized" community. Such randomized communities provide means and variances in the C values—the C values for a hypothetical, competition-free community. The competition-free ($\bar{C}_F$) values can then be compared with the true mean value, $\bar{C}_0$. If $\bar{C}_0 < \bar{C}_F$, we can infer that niches are overdispersed. Sale applied his technique to MacArthur's warbler data, calculating a $\bar{C}_0$ of .654, and a $\bar{C}_F$ of .678 $\pm$ .0143 (standard deviation). A significant value, $p < .05$, indicates that the separation that MacArthur observed is nonrandom and so, probably, preserved by competition. Application of the method to grasshopper data (14 coexisting species in Colorado) collected by Ueckert and Hansen (1972) showed a $\bar{C}_0$ of .615, a $\bar{C}_f$ of .538 $\pm$ .0057 (S.D.). In this case $\bar{C}_0 > \bar{C}_f$ ($p < .001$), indicating niche *convergence*. Perhaps these species are facilitative (?). More likely this pattern is due to similar habitat preferences, for which Sale's method does not correct.

Character Displacement and Competitive Release

Think of two species living in different areas but in similar environments, and suppose that some character related to use of a limiting resource (e.g., bill size in seed-limited birds) is similar in size and morphology for the two. Now consider what might happen in a third area where the two were sympatric. If no other competitors were present to affect the results, it is reasonable to suppose that any small difference in that character would be exaggerated by the effects of natural selection. We shall defer a defense of this statement until Section 17.2; for the moment the reader is asked to keep faith. Divergence in a trait between two species where they are sympatric has been termed *character displacement* (Brown and Wilson, 1956), and has been invoked quite commonly as evidence for the existence of exploitation competition. Brown and Wilson gave several examples, one of which we shall come back to, but there are more recent, better examples.

Dunham et al. (1979) compared the number of vertebrae and number of gill rakers among several species of suckers (catastomid fish) and regressed the values on an index of competition from other species of the same genus, correcting for a variety of environmental parameters such as latitude, longitude, elevation, stream gradient, and maximum and minimum stream discharge, by including them in the regression. In *Catastomus discobolus*, which possesses a large number of gill rakers, the number increased with competition. In *C. platyrhynchus*, which has a small number of gill rakers, the number declined as competition increased, except when competition came in the form of species with still smaller numbers. In that case the number increased. Vertebral number followed the same pattern.

Huey et al. (1974) studied two species of skinks, *Typhlosaurus lineatus* and *T. gariepensis*, on the Kalahari desert. These species are sympatric in some places, *T. lineatus* is allopatric in others. *T. lineatus* is longer, with a wider head, differences that are most pronounced in sympatry. Furthermore, the offspring of *T. lineatus* are larger in sympatry, where fewer females are reproductive. Finally, there is a dietary divergence in prey (termites) eaten for females and immatures (although not for males). It seems fairly certain that morphological divergences should arise directly from the operation of natural selection. Dietary divergence, on the other hand, might either be fairly genetically rigid or, alternatively, a facultative response. As pointed out in the introduction to this chapter, if a situation implies natural selection for niche shifts it also indicates advantages to appropriate "voluntary" shifts growing out of behavioral plasticity. The ability of organisms to make such appropriate shifts surely will have been, and will continue to be, advantageous for any species that occasionally find themselves in competitive situations. Facultative shifts are not part of Brown and Wilson's classic view of character displacement, but are effectively equivalent.

With respect to facultative character displacement, the acanthocephalan worm *Moniliformis dubius* and the cestode *Hymenolepis diminuta*, introduced into previously uninfested rats one at a time, occupy most of the host's intestine. When introduced together, though, *M. dubius* remains in the anterior portion, *H. diminuta* in the posterior portion of the gut (Holmes, 1961).

Werner and Hall (1977) have shown that in small ponds where it occurs alone, the bluegill sunfish (*Lepomis macrochirus*) feeds on "relatively large prey" associated with the vegetation. But in the presence of green sunfish, *L. cyanellus*, it shifts to feeding on smaller prey in open water.

Jared Diamond has amassed a fair number of what he feels are cases of character displacement among New Guinea birds. A review of much of his work can be found in a 1973 article in *Science*. We give two examples here. On Mount Wilhelmina, *Ptiloprora perstriata*, a honey eater, ranges from an altitude of between 1000 and 2000 meters to nearly 4000 meters. On Mount Saruwaged, a related species, *P. guisei*, occurs over a slightly attenuated range very similar to that of *P. perstriata*. They are allopatric on these mountains. But on Mount Michael, where they occur together, *P. perstriata* occurs only above 3000 meters, and *P. guisei* only below that altitude. Their ranges show almost no overlap. The New Guinea cuckoo shrikes *Coracina tenuirostris* and *C. papuensis*, where allopatric, on different islands, are of very similar size, and forage in similar habitat. Where sympatric they either diverge in weight (61 and 101 grams, respectively, on one island), or if weight differences fail to materialize (73 and 74 grams, respectively, on another island) they separate by habitat.

Mary Price (1978), working with heteromyid rodents in the American Southwest looked for facultative niche shifts arising from competition. She used physical and vegetative data to define four microhabitat types:

1. Large, open area: open space > 2 m diameter
2. Small, open area: open space .25 to 0.5 m diameter
3. Large bush: the area under a shrub > 1 m tall and > 2 m in diameter
4. Tree: the area under a *Prosopis* or *Cercidium* > 1.7 m tall

Perognathus baileyi, *P. penicillatus*, *P. amplus*, and *Dipodomys merriami* all had quite different microhabitat preferences. Two kinds of tests for niche displacement were made. First, species were kept alone or in pairs in .26- to .34-ha enclosures to see if the presence of a competitor would restrict microhabitat use, and shift it in the direction expected from the live-trap data. Specifically, when the large, large-open-area kangaroo rat was added, the pocket mice were expected to shift away from the open areas in the enclosure. If sufficient numbers of *Dipodomys* were present, the expected shift did, in fact, take place. *Dipodomys* spent more time in open areas and less in tree areas in the presence of *Perognathus*. Second, Price provided the animals with an increased amount of open area by clipping bushes 60 to 100 cm in height to the ground at about half of the trapping stations. On the basis of extensive data on microhabitat preferences, she then predicted the consequent short-term shifts due to migration in local densities of the four species. *Dipodomys* should have increased by about 1.3 individuals per meter squared of large-space habitat added. The observed figure, after 6 weeks, was 1.4 (although the change included a 9% drop in unmanipulated areas and a 21% rise in large, open areas). *Perognathus penicillatus* and *P. baileyi* were not expected to show much change, and they did not. Results for *P. amplus* were ambiguous. Price suggests that interspecific competition has a lot to do with the structuring of heteromyid communities, but notes that the observed effects seem to be mediated by microhabitat shifts (see also Rosenzweig and Winakur, 1969; Rosenzweig and Sterner, 1970; Smigel and Rosenzweig, 1974).

Although the cases of character displacement cited above are pretty convincing, not all published accounts have withstood closer scrutiny. Grant (1975), reviewing the data pertaining to one of Brown and Wilson's "classic cases" of character displacement, comes up with disturbing results. The species in question are the two Near Eastern nuthatches *Sitta tephronota* and *S. neumayer*. The former centers its range in Iran and Afghanistan, the latter in southern Turkey and Yugoslavia. The two are sympatric in parts of western Iran. Grant looked at the various characters believed involved in displacement, bill size and eye stripe. The larger bill size of *S. tephronota* is exaggerated in the zone of sympatry, as is the larger eyestripe. Regressing the size of these traits on latitude and longitude, Grant found similar trends in both species, trends sufficient to account for an apparent divergence. But could the regressions hide discontinuities? Separate analyses within and without the region of sympatry were not significantly different. Grant then collected additional data [the data cited by Brown and Wilson were from Vaurie (1951)] from museum skins, and extended the analysis to other morphological features—eye stripe intensity, weight, wing length, and tarsus length. In the new regressions were included, in addition to latitude and longitude, average July maximum temperature, average January minimum temperature, rainfall (July and January), and relative humidity (July and January). Only latitude and longitude were significant in the analyses for *any* of the characters. Again, there was no evidence for discontinuities in any of the traits except, *perhaps*, eyestripe area. But even if the apparent displacement is real for this trait, it seems more likely that it reflects a reproductive isolating mechanism (to avoid species misidentification) than a competitive shift.

The other side of the character displacement coin is *competitive release*. This term refers to the niche expansion exhibited by species upon the removal of competitors. Again, apparent cases abound. Diamond (1970) notes again with respect to the birds of New Guinea that if a species drops out from one mountain to the next, the remaining species expand their altitudinal ranges. Similar observations have been made by Terborgh and Weske (1975) working in South America. These authors gathered data on the birds of the Cordillera Vilcabamba, part of the eastern chain of the main body of the Peruvian Andes, and also on an isolated (island) massif, the Cerros del Sira, about 100 km to the east. Eighty to eighty-two percent of the bird species found in the summit zone in the (mainland) Andes were not found on the Cerros del Sira. Of those species on the Sira that lacked congeners (existing on the mainland) at higher altitudes, 71% expanded their range upward. Of those species lacking *noncongeners* at higher altitudes, 58% extended their range upward. This apparent impact (or removal of impact) by a collection of noncongeners, each contributing a small bit, has been termed "diffuse" competition [see Chapter 6 (p. 166); see also Terborgh (1971)].

Brown and Lieberman (1973) measured the variance in seed sizes (millimeter length) eaten by the kangaroo rat (*Dipodomys ordi*). Where it occurred alone, the variance was 4.56, where it occurred with the smaller, *D. merriami*, the variance dropped to .65. The difference is significant by an F test, $p < .05$.

Competition and Community Organization

Diamond (1975) feels that competitive interactions are a major driving force in community organization, and he has constructed a list of "assembly rules" that he feels determine the nature of competition communities—what combinations of species are present. These rules, somewhat paraphrased, are:

1. Only certain species combinations of all those possible really occur.
2. Permissible combinations resist invasion by species that would transform them into unstable combinations.
3. Species combinations stable on large (or species-rich) islands may not be so on small (or species-poor) islands, and vice versa.
4. On small or species-poor islands a species configuration might resist other species, whereas on a large or species-rich island those other species could be accommodated.
5. Some species pairs can never coexist.
6. Some pairs, unstable alone, may be stable when part of a larger assembly.
7. Some pairs, stable when alone, may be unstable when part of a larger assembly.

But recall from the introduction to this chapter that an alternative view exists—if competition is very slight, the configuration of species in a community and their pattern of resource use may be quite serendipidous. One strong proponent of this view is Daniel Simberloff.

Connor and Simberloff (1979) have taken direct issue with Diamond's assembly rules, spending considerable time charging that all seven are either "tautological consequences of the definitions employed, a trivial logical deduction from the stated circumstances, or a pattern which would largely be expected were species distributed randomly on the islands subject only to three constraints: (a) that each island has a given number of species, (b) that each species is found on a given number of islands, and (c) that each species is permitted to colonize islands constituting only a subset of island sizes." Their claim: None of Diamond's observations require invocation of competition. In another attack (Simberloff and Connor, 1981), the same authors take issue with the following assertion: Because, of k possible two-species combinations that can occur on islands, only a few, $x < k$, are actually found, we can conclude that species interactions dictate the existence of a limited number of stable pairs. This is the multiple domains of attraction approach that we mentioned in Chapter 6, and its validity rests on the assumption, again, that competition is a major organizing force in ecological communities. Simberloff and Connor checked a number of articles about species sets over islands or study plots, all but one of which had invoked competitive exclusion to explain the observed patterns. They were able to show that in *every* instance there was either a mathematical error or lack of "a properly framed null hypothesis," thus casting the claims into doubt. In fact, a computer simulation based on an assumption that colonization was utterly independent of the presence of other species, but which allowed different species to arrive at different rates, fit most of the data reported in the articles very well.

Strong et al. (1979) generated null hypothetical combinations of bird species (hypothetical species defined by culmen size), drawn from a mainland species pool, and compared the resulting culmen size distributions with those observed on real islands. The average ratio of culmen sizes among species was larger for the random community half the time and smaller half the time, suggesting that the real populations were merely random samples of the mainland pool (see also Simberloff and Boecklen, 1981, Pulliam, 1983).

Simberloff's objections have not gone unanswered. Referring to Strong et al.'s (1979) work, Grant and Abbott (1980) pointed out that regular patterns of culmen size should be expected only within feeding guilds. Strong et al. lumped all bird species in their analysis and so should hardly have expected to find anything other than what they did find. Grant and Abbott also claimed that Simberloff's statistics were in error. For the positive side of the ledger, Grant and Schluter (in press) noted that of six finch (*Geospiza*) species on the Galápagos, no more than five ever co-occur. The number of possible combinations of five species drawn from a pool of six is $6!/5!1! = 6$, but only one of these is ever actually found (three times). In Table 17-3 they provide similar data for all islands on which $1, 2, \ldots, 5$ species occur. These results are decidedly non-random.

Bowers and Brown (1982) tested the hypothesis that coexistence of desert rodents is independent of their size. They reasoned that were an alternative hypothesis, based on competitive interaction, correct,

Table 17-3

Number of Islands	Number of Species per Island	Number of Configurations		p
		Possible	*Observed*	
3	5	6	1	.028
3	4	15	1	.0044
8	3	20	2	$< .001$
7	2	15	4	.072
2	1	6	1	.167

1. Species in the same guild and locale should show nonrandom size relationship patterns.
2. Species in the same guild should show greater geographic range divergence if of similar than of different size.
3. Regular size distribution patterns should be diminished or obliterated when species of different guilds were lumped in analysis.

Going to the literature, Bowers and Brown collected information on the diet of desert rodents and classified them into four guilds—foliage-eating, insectivorous, granivorous, and omnivorous. Then, using p_i the frequency of occurrence of species i in each of several locations, for each of three desert habitats, they calculated expected (random) range overlaps. Tests for independence of body mass and range overlap were then carried out. Results for the granivore guild are given in Table 17-4 (using a computer-generated distribution function to test the null hypothesis). When all guilds were pooled, the corresponding significance values (p) were .870, .826, and .677. The hypothesis that coexistence is independent of body size must be rejected at the level of the guild, but cannot be rejected for all guilds lumped.

Table 17-4

		Number of Cases Where Range Overlap Was:		
Desert Type	**Body Mass Ratio**	***Greater Than Expected***	***Less Than Expected***	
Great Basin	<1.5	0	6	$p = .003$
	>1.5	15	15	
Mojave	<1.5	1	3	$p = .652$
	>1.5	5	11	
Sonoran	<1.5	0	7	$p = .047$
	>1.5	15	23	

The results of Diamond, Werner and Hall, Holmes, Terborgh, Terborgh and Weske, and Price, cited earlier, all involved niche separation by means of habitat or microhabitat differences. The earlier cited studies of MacArthur, Schoener, M'Closkey, and Inouye also implicated habitat or microhabitat separation. Schoener's (1974b) review indicated habitat separation in 90% of all studies on niche separation. Habitat separation and microhabitat separation can be the result either of evolved preference differences or facultative shifts due to interference. But the fact that the differences often seem to disappear, or at least significantly diminish when the competitor is removed, suggests that interference may be as important or even more important than exploitation competition in structuring communities. We can argue that interference is costly in time, effort, and perhaps risk, and so should evolve only to the point where the gains, as regards alleviating exploitation competition, exceed the losses. If this argument is correct, the selective advantage to interference falls off with the intensity of exploitation competition; interference should be an important phenomenon only when competition for resources is also strong. This view has some support.

Ebersole (1977) presents data showing the relationship between interspecific interference and exploitation competition in reef fish, *Eupomacentris leucostictus*. An estimate of the exploitation competition coefficient was made using the niche overlap method:

$$\alpha_{ij} = \frac{\sum_h p_{ih} p_{jh}}{\sum_h p_{ih}^2}$$

where p_{ih} is the proportion of food in the diet of species i comprised of food h. Ebersole then improved on this measure, with respect to finding an index of niche overlap impact, by weighting α_{ij} by the metabolic rate of species j:

$$\alpha_{ij}^* = M_j^{.75} \alpha_{ij}$$

where M_j is the mean body weight of species j. Finally, he calculated the correlation coefficient r of α_{ij}^* and the number of agonistic responses per trespass time by species j on species i territories and obtained

$$r = .744 \qquad p < .01$$

Clearly, interference intensity parallels his measure of exploitation competition intensity. Aggressive reaction by *E. leucostictus* to nine other species that ate its eggs was more intense.

However, to expect interference intensity necessarily to correlate with exploitation competition may be naive. The argument for correlation is not incorrect but perhaps neglects the true nature of the spandrels of San Marco. There is, for example, the possibility of mistaken identity. Lack of discrimination between conspecifics and competitors may occur so that intraspecific interference spills over to other species. They, in turn, diverge in habitat or microhabitat to avoid the squabbles. More of this will be said later in the context of interspecific territoriality. Second, interspecific interference may be a behavioral pattern that is beneficial during population highs or periods of resource shortage, when exploitation competition is intense. The evolved interference behavior then may persist in better times even if no longer advantageous, as long as it also is not very costly to the perpetrators.

Interspecific interference is clearly demonstrated in interspecific territoriality. Myrberg and Thresher (1974) described the behavior of the three-spot damselfish (*Eupomacentris planifrons*), a territorial reef fish. "Intruders" of various fish species, safely protected in a glass container, were gradually moved closer and closer to a territorial individual and the point at which an attack occurred was noted. Size of intruder made only a statistically insignificant difference, but each intruding species had a characteristic critical distance. Not surprisingly, conspecifics were attacked at the greatest distance.

Interspecific territoriality is also found among birds, where the prevailing wisdom follows the line of reasoning suggested by the fish study cited above. However, this explanation has been challenged by Murray (1971, 1976), who believes that "interspecific territoriality is generally misplaced *intraspecific* territoriality." In other words, it results from a nonadaptive overflow of intraspecific aggression. Cody (1969a,b) has argued that species showing interspecific territoriality should have evolved, and do demonstrate convergent patterns of appearance and voice. Murray retorts that the similarities came first and form the basis for the mistaken identity that allows nonadaptive interspecific territoriality to occur. He discounts much of Cody's evidence as made up of invalid examples of interspecific territoriality. (But one is tempted to suspect that if reef fishes can accurately identify other reef fish species and alter their aggressive behavior accordingly and consistently, so could birds.)

Finally, interspecific interference behavior by one individual, contributing to niche separation or, alternatively, the driving away of potential competitors or opposing interferers, might benefit immediately adjacent conspecifics. Thus, because it affects more than the actor, interference behavior must be judged by its contributions and costs for inclusive fitness. Kin selection or trait-group selection might drive the evolution of interference way beyond levels we would otherwise expect on the basis of protecting access to resources (Case and Gilpin, 1974).

On the basis of this argument we might expect strong interference competition among highly social species, those most likely to exhibit very viscous population structure. The existence of strong interference would render exploitation competition almost nonexistent.

17.2 The Evolution of Niche Overlap

The Balance of Intra- and Interspecific Selection Pressures

Inasmuch as niche overlap with respect to resources is necessary for exploitation competition to occur, and overlap in space is required for interference competition, it is reasonable to ask what patterns of niche overlap or separation of either sort might be favored by natural selection. There are a number of approaches to this question, and some very complicated arguments have been published. Roughgarden (1976) pictured individuals of two species being distributed along some niche gradient, which presumably could reflect either resource or spatial characteristics. Resource abundance, or value, varied along this gradient, falling off to either side of some point (see Fig. 17-2). Consider the simple case where the two competing species, i and j, are identical except for their mean positions $\bar{x}_i$ and $\bar{x}_j$ on the gradient. We can define the resource abundance at point x [which is the carrying capacity, $K(x)$, at point x] to follow a distribution curve of the form

$$K(x) = K_0 a \exp\left[-c\left(\frac{|x - \hat{x}|}{\sigma_K}\right)^m\right] \tag{17-4}$$

where a is a constant such that $\int_{-\infty}^{\infty} K(x)\,dx = K_0$, m is a constant, c is a function of m that controls kurtosis. Such a formulation permits us to describe an extremely broad range of curves. Roughgarden defined the species distribution curves $\phi(x)$ using the same general form. Now, if we accept the idea that α-coefficients can be calculated as suggested at the end of Chapter 6,

$$\alpha_{ij} = \frac{\int \phi_i(x)\phi_j(x)\,dx}{\int \phi_i(x)^2\,dx}$$

then, because the two species are identical except for their $\bar{x}_i$ and $\bar{x}_j$ values, we can write

$$\frac{dn_i}{dt} = r_{i_0} n_i \left[1 - \frac{n_i + \alpha(\bar{x}_i - \bar{x}_j)n_j}{K(\bar{x}_i)}\right] \tag{17-5}$$

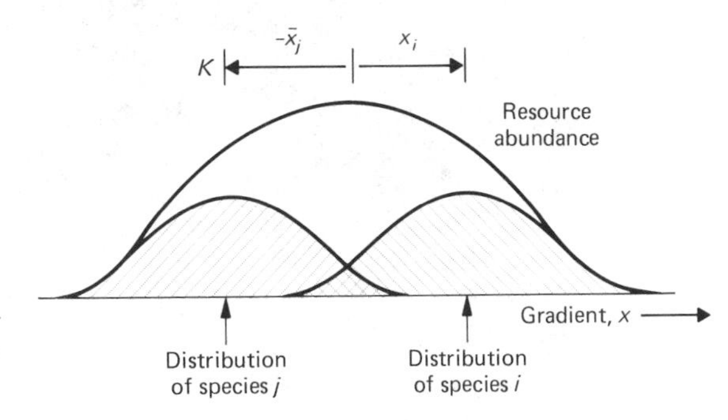

Figure 17-2 Distribution of two competing species along a resource gradient, showing the abundance of the resource.

whence, at equilibrium,

$$\hat{n}_i = \frac{K(\bar{x}_i) - \alpha(\bar{x}_i - \bar{x}_j)K(\bar{x}_j)}{1 - \alpha(\bar{x}_i - \bar{x}_j)^2} \tag{17-6}$$

(recall equation 5-6). Roughgarden then argued that, at equilibrium, maximizing mean fitness was equivalent to maximizing $\hat{n}_i$. He thus proceeded to use (17-4)–(17-6) to find the value of $\bar{x}_i - \bar{x}_j = \hat{x}$ that maximized $\hat{n}_i$. The general form he obtained was

$$\frac{\hat{x}}{\sigma} = \left\{\frac{1}{2^m c} \ln\left[\frac{1}{2}\left(\frac{2\sigma_K}{\sigma}\right)^m - 1\right]\right\}^{1/m} \tag{17-7}$$

Other models have been devised by Bulmer (1974) and by Fenchel and Christiansen (1977). Both these and that of Roughgarden have since been absorbed in a more general model by Slatkin (1980). Slatkin criticizes Roughgarden's approach for implicitly holding σ^2 fixed. He shows that when σ^2 is allowed to vary together with $\hat{x}$, multiple equilibrium $\hat{x}$ values are possible. All of the models above are mathematically difficult and all suffer from their share of simplifying assumptions. As noted by some of the authors, the degree of overlap and the value of α are very sensitive to the shape of the resource supply distribution, $K(x)$, to the shape of $\phi_i(x)$, $\phi_j(x)$, and to whether the two species view resource supply or value in the same manner—it is probably true that $K(x)$ has a different mean and variance for each competing species. Finally, we never know the shapes of these curves in real life, nor do we even know that niche overlap is a valid approach to measuring competition, especially when interference is an important component. A few of the conclusions reached by Slatkin can be simply—if not rigorously—understood by looking at Figure 17-3. Suppose that $K(x)$ is the same for both species and is distributed as shown in the outer, broader curves. The competitor distribution, or utilization curves $\phi_i(x)$ and $\phi_j(x)$ are shown in the inner curves. If natural selection acts as an efficiency expert, the total use by both species should closely fit the resource availability curve. Otherwise, resources go to waste. If we constrain $\phi(x)$ to be Gaussian, curves too close together relative to their standard deviations waste resources at the right and left extremes of the gradient (17-3a). If the curves are far apart in standard deviation units (17-3b), resources in midgradient are wasted. Figure 17-3c, d, and e demonstrate that if $\phi(x)$ shape is allowed to vary, multiple solutions occur. There seems no reason to claim that any particular curve shape or degree of separation should evolve. On the other hand, if the competitors evaluate the resource spectrum in slightly different ways, so that $K(x)$ is species specific, $\phi_i(x)$ must lie below K_i, and $\phi_i(x) + \phi_j(x)$ must lie below the higher of the two K lines (Fig. 17-4). Again, there are multiple combinations of stable species utilization curves that might evolve, shown in 17-4a, 17-4b, but now species divergence can be seen to occur as a standard rule. If species i, say, were remove, species j, to remain maximally efficient, should expand its niche.

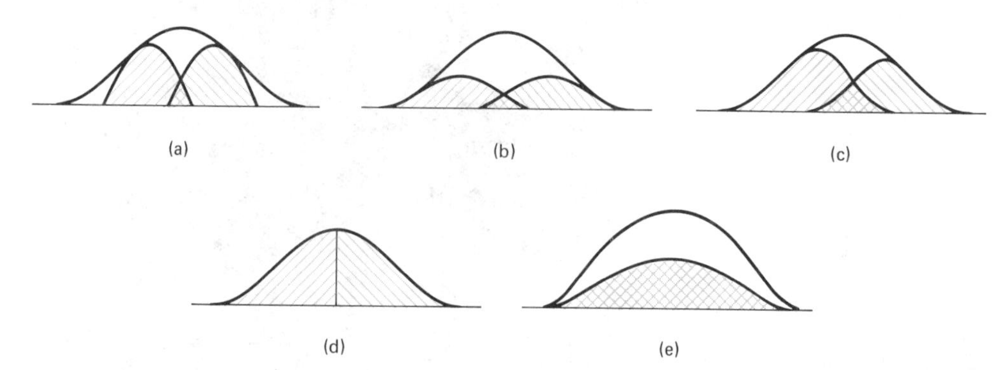

Figure 17-3 Possible, optimal solutions to resource use in two competing species. In (a) and (b) distribution curves are constrained to be Gaussian.

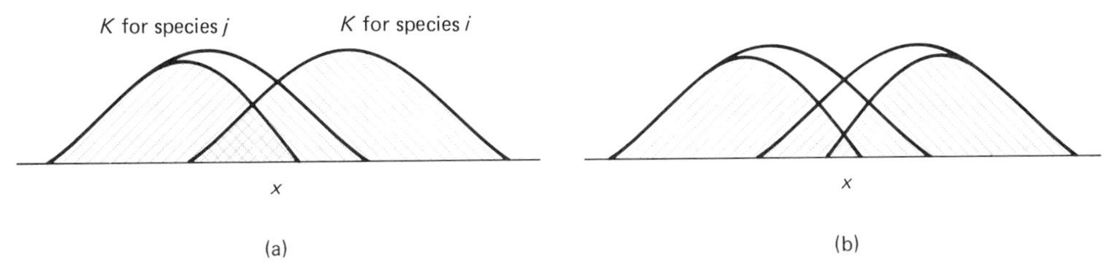

Figure 17-4 Possible optimal solutions to resource use in two competing species when each species views resource availability in a different manner.

Let us now use Figure 17-4 to explore the consequences of changing resource supply. Consider a food-limited species that experiences shortages in (say) winter. If winter shortages limit population density, then by definition, food is relatively plentiful in summer. But as we saw in Chapter 16, food plenty can be expected to result in a narrowing of diet, a pulling in of the food niche around the optimal value point on the resource gradient. Where the outer curves represent $K(x)$ in the season of food plenty, the situation pictured in Figure 17-4b might give way to that shown in Figure 17-5. Because food is plentiful, the population need not, and perhaps cannot, use it all; it pays to let much of it go and to specialize on the best items. The same argument would apply to choice of any kind of resource, including space. Thus where plastic niche responses are possible, we should expect species living in habitats with strongly variable resource supplies to be more narrow niched—more specialized—in the season of plenty than similar species occupying more constant surroundings. Also, in resource variable environments, niche shape should more faithfully conform to resource availability in the harsh than in the mild season. This last statement must be interpreted with caution, for there are a number of ways in which resource supply can vary. Suppose that all resources, X, along a hypothetical gradient, fluctuated equally so that the *shapes* of the $K(x)$ curves remained the same while only their amplitude varied. In such cases, as the niches of competitors expanded and contracted, their mean positions on the gradient would not change. The season of resource plenty would bring greater niche separation by virtue of smaller niche variances: niche overlap and consequently the ratio of inter- to intraspecific competition would drop. The voles *Clethrionomys gapperi* and *Microtus pennsylvanicus* provide a likely example. In the summer, both species occur, respectively, in woods and fields. But when winter comes in Manitoba and food supplies are low, *Clethrionomys* invades the fields, where it overlaps broadly in habitat with *Microtus*. In the spring, with renewed food productivity, it again withdraws to the woods (Iverson and Turner, 1972). On the other hand, resource flushes probably more often than not *do* alter the shape of the $K(x)$ curves. During the lean season certain resources (values of x) might be most available, and during the fat season others. Thus contraction of niches at the most productive times of year might *not* occur around the mean x values of the poor season. In fact, if the flush is constituted of resources of very similar x value, niche narrowing might even involve convergence of niches and so *greater* overlap. Smith et al. (1978) studied ground finches in the Galápagos and found that in the wet season there was no significant correlation among islands between bird biomass density and food volume, suggesting that food was not limiting then. But in the dry season, sites with low food generally also had low avian densities, indicating the importance of food at those times. When food supplies were short, niche overlap among species was least. A survey of 10 studies providing sufficient information to assess niche divergence or convergence in the lean season was compiled by Smith et al. (1978). In all 10 cases there was divergence.

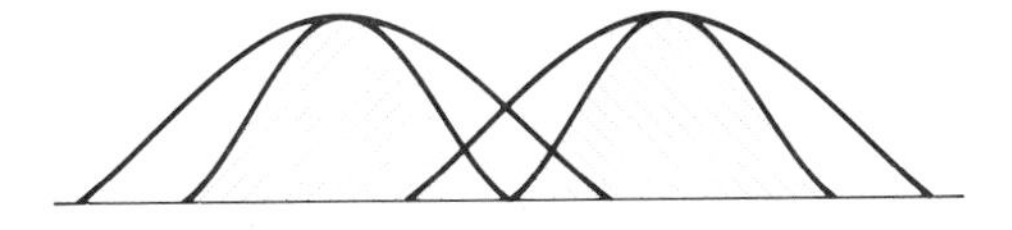

Figure 17-5 Possible response in resource utilization by two competitors when resource abundance is high.

A very different approach to the question of evolution of niche overlap has been taken by Lawlor and Maynard Smith (1976). The authors concern themselves with finite numbers (two) of resources rather than a continuum, and start with the expressions

$$r_i = \frac{1}{n_i}\frac{dn_i}{dt} = a_{i1}v_1R_1 + a_{i2}v_2R_2 - T \tag{17-8a}$$

$$\frac{1}{R_j}\frac{dR_j}{dt} = g_j(R_j) - a_{1j}n_1 - a_{2j}n_2 \tag{17-8b}$$

where r_i is the population growth rate of predator i, for $i = 1, 2$, and R_j is the amount of food resource j. The v_j's denote resource "value" (assumed equivalent for both predators), and a_{ij}'s are food preference values modifiable by natural selection. T is the metabolic cost of living. We shall discuss the fine-grained case, involving differential use of the two resources by the two predators without large-scale spatial patchiness in prey distribution. Natural selection is supposed to vary the a_{ij} values such as to maximize r_i for both predators. Thus we can write

$$0 = dr_i = \frac{\partial r_i}{\partial a_{i1}} da_{i1} + \frac{\partial r_i}{\partial a_{i2}} da_{i2}$$

so that, at equilibrium

$$\frac{da_{i1}}{da_{i2}} = -\frac{\partial r_i/\partial a_{i2}}{\partial r_i/\partial a_{i1}} = -\frac{v_2\hat{R}_2}{v_1\hat{R}_1} \qquad \text{for } i = 1, 2 \tag{17-9}$$

But also, at equilibrium $r_i = 0$, so that

$$a_{i1} = \frac{T - a_{i2}v_2\hat{R}_2}{v_1\hat{R}_1} \equiv h_i(a_{i2}) \tag{17-10}$$

For fixed $\hat{R}_1$ and $\hat{R}_2$, a_{i1} drops linearly with a_{i2}. However, as a_{i2} and so a_{i1} change, $\hat{R}_1$ and $\hat{R}_2$ also change. As use of one food increases (a_{i2} goes up), $\hat{R}_1$ and $\hat{R}_2$ are unlikely to drop very much, because of density compensation (less density feedback from other sources as $\hat{R}$ drops), so a_{i1} can be expected to drop less than a_{i2} rises. Therefore, a plot of a_{i1} against a_{i2} should be convex (Fig. 17-6). But from (17-9) we know that selection favors the a_{i1}, a_{i2} combination such that the slope of the line in Figure 17-6 is $-v_2R_2/v_1R_1$ (point *). We now draw the a_{ij} constraints for both predators on a single graph, and mark the a_{ij} values favored by selection (Fig. 17-7). If the initial a_{ij} values are interior to those shown in Figure 17-7, natural selection favors a divergence, if exterior a convergence. If we knew the functions $g_j(R_j)$ we could find $a_{i1} = h(a_{i2})$ and therefore the a_{ij} values favored by selection. These values, describing resource use by the predators, define the niches of those predators. Knowing $g_j(R_j)$ and the a_{ij} values, we could also calculate

$$\alpha_{ij} = \frac{\partial r_i/\partial n_j}{\partial r_i/\partial n_i} \tag{17-11}$$

for equilibrium conditions.

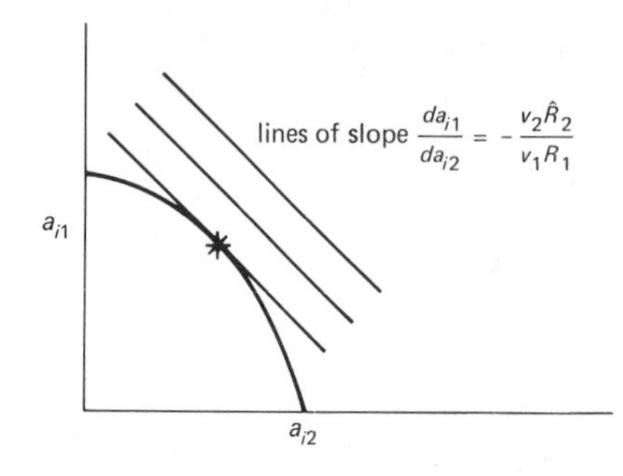

Figure 17-6 Graphical model of optimal use of resources 1 and 2 by species i.

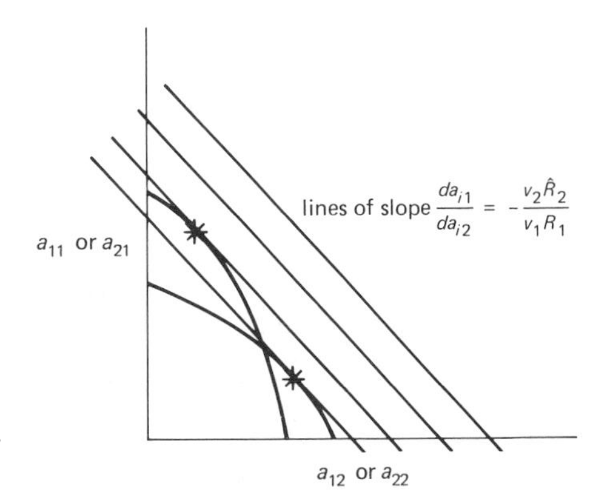

Figure 17-7 Graphical model of optimal use of resources 1 and 2 by species *i* and by species *j*.

The Genetics of Competition

We have gone blithely about trying to predict evolutionary changes in competitive relationships without examining whether the genetic machinery exists for allowing such changes. There are two kinds of evidence for a role of the genome in determining the degree of competition.

In 1962, Lerner and Dempster cited a series of interspecific competition experiments with laboratory populations of the flour beetles *Tribolium castaneum* and *T. confusum*, dating back to 1948 (Park; see also Park, 1954; Park and Lloyd, 1955), which showed a very interesting common thread. In all the experiments, one species always competitively excluded the other. But whereas the outcome of runs begun with 10 pairs of each species was almost always predictable from knowledge of temperature and relative humidity, that of experiments started with only two pairs of each species was quite unpredictable. If the species' gene pools were instrumental in deciding the winner, this makes sense, for the founder effect might be quite pronounced in very small founder populations, giving rise to high among-experiment genetic variance and so high variability in outcome. In support of this explanation, it seems that experiments using inbred populations gave more consistent results than did experiments using mixtures of inbred lines or outbred lines. Repeated tests of this sort performed by Dawson and Lerner (1966) supported these conjectures.

The second line of evidence comes from selection experiments in which the experimenter deliberately tried directly to alter the competitive viability of a species. Dawson (1967) selected for fast or slow development time (days from egg to pupa) in flour beetles and found that *both* selected lines did poorly in competition trials with controls. However, after a few generations, one of the fast-selected strains showed a significant improvement. The initial deterioration might be expected as a result of the breaking up of coadapted gene complexes by artificial selection for a trait not highly correlated with interspecific competitive viability. The improvement seemed to be due to increased cannibalism in this strain.

Park and Lloyd (1955) raised the two *Tribolium* species together for several generations, expecting to find increased competitive viability as a result of selection. But tests of selected strains against unselected stock populations indicated no improvement, and Park and Lloyd rejected the hypothesis that selection could (easily) alter competitive outcomes (see also Sokal et al., 1970; Dawson, 1972). Ayala (1966) picked the idea up again a few years later, however, and this time, working with *Drosophila*, got some encouraging results. *D. serrata* were pitted against either *D. pseudoobscura*, *D. nebulosa*, or *D. melanogaster* in bottles using the serial transfer technique (each week adults, both old and newly emerged, are transferred to a new bottle). After 6 months, at 25°C, both species, in all three experiments, were still extant. In the first, *D. pseudoobscura* was dominant at first, but then a reversal took place. In the third, *D. melanogaster* was dominant for a long time before *D. serrata* suddenly achieved the upper hand.

Pimentel et al. (1965), running experiments with houseflies and blowflies in a system of 16 small cages connected with tubes (simulating a patchy environment), found similar dynamics. Houseflies very nearly eliminated blowflies until, at week 55, the blowflies suddenly gained dominance. Eventually, the houseflies were competitively excluded. In both Pimentel's and Ayala's work, the results can be explained by the working hypothesis of Park and Lloyd. When a species is numerically dominant, most individuals it encounters are conspecifics. Thus natural selection favors traits that enhance the species' *intraspecific* competitive ability, probably at the expense of slightly different adaptations for interspecific competitive ability. The subordinate species, however, encounters mostly dominant species individuals and so experiences selection for *interspecific* competitive viability. Hypothetically, these selection differences could reverse competitive dominance. It would be interesting to measure α coefficients and examine their genetic architecture at different absolute and relative population densities.

After these early experiments, the work of Ayala and Pimentel diverged somewhat. Ayala (1972) raised *D. serrata* and *D. nebulosa* together in serial transfer populations for about 220 days, after which *D. serrata* were allowed to lay eggs used in tests. The emerging adults were placed back into an experimental situation with control unselected *D. nebulosa.* The results were compared with those of a new run of control *D. serrata* with control *D. nebulosa.* Results are shown in Table 17-5. *D. serrata* lost ground in both sets of runs, but selected populations fared much less poorly.

Table 17-5

	Mean Percent *D. serrata* over 12-Week Period	Coefficient of Regression of Percent *D. Serrata* on Time	
Experimental *D. serrata*	69.2	−1.1	Replicates
	59.7	−3.4	
	68.5	+ .2	
	54.8	−1.9	
	63.0	−1.6	
Control *D. serrata*	47.2	−4.4	Replicates
	48.3	−4.8	
	47.7	−6.6	
	48.3	−7.3	
	47.9	−5.8	
Mean difference	15.2	4.1	

Pimentel (1968) worried about mechanisms by which reversals in competitive viability might occur and came up with a variant of his genetic feedback theory (see pp. 363, 468). Consider two competing species preying on a single prey species, and suppose that the best defenses against one predator are somewhat different from those best against the other. Then if one predator becomes dominant, the prey evolves resistance to it at the expense of being able to protect itself well against the other. This gives the subordinate predator an advantage, and its numbers increase. In this way switches in competitive superiority can take place—perhaps cycling of the predators could be produced.

With the exception of Ayala's (1972) work, we have presented only indirect evidence that species selected under competitive situations improve their competitive viability. Seaton and Antonovics (1967) attempted a more rigorous demonstration, using two strains of *D. melanogaster.* Wild-type and dumpy-winged homozygotes were allowed to oviposit together for 4 days in various predetermined numbers and then removed. The larvae developed and pupated in competition with one another and when emerged, were immediately removed to avoid cross-breeding. Then the wild-type and dumpy progeny were allowed to breed separately and a sample of fertilized females was placed into new jars to repeat the process. The authors were looking for evidence of increased competitive viability in either or both of what, because of reproductive isolation, were *effectively* different species. Each generation the number of adult flies emerging was counted and plotted as shown in Figure 17-8. In only four generations there was evidence of increased competitive viability.

Clark (1979) performed similar experiments using *D. melanogaster* and *D. simulans* as competitors. He was interested primarily in the effects of competition on the frequency of a recessive, "sparkling poliert" gene, but his results are pertinent here. Three population lines were started in pint bottles, one began with 150 *D. melanogaster* and no *D. simulans*, one had 100:50, and the third 50:100. *D. melanogaster* founders were all heterozygous for the recessive allele, so $p = .50$. On day 6 the adults were removed and on day 18 the adult progeny were scored by phenotype. The *D. simulans* were discarded and a part of the *D. melanogaster* progeny (with the same genotype frequency as the total group of emergents) was used to start a new generation with new, stock, *D. simulans*. After five generations an improvement in competitive viability in the *D. melanogaster* was found, as measured by progeny output, which was independent of the sparkling gene frequency. Control populations of 50, 100, and 150 *D. melanogaster* raised alone were also tested for competitive ability and showed no improvement. That is, the improvement found in the experimentals was the result of contact with the competitor.

The most carefully carried out, most thoroughly controlled, and most exhaustive study to date is that of Sulzbach (1980) (see also Sulzbach and Emlen, 1979). Sulzbach used *D. melanogaster* strains, a wild type and a brown eye form, both obtained by multiple chain crosses between a variety of stocks from all over the world. Thus genetic variance was maximized within constraints set by available allelic types. The experimental procedure was similar to that of Seaton and Antonovics except that the experiments were run at each of two densities over many generations, and control replicates were maintained in vials, like the selected lines, as well as in

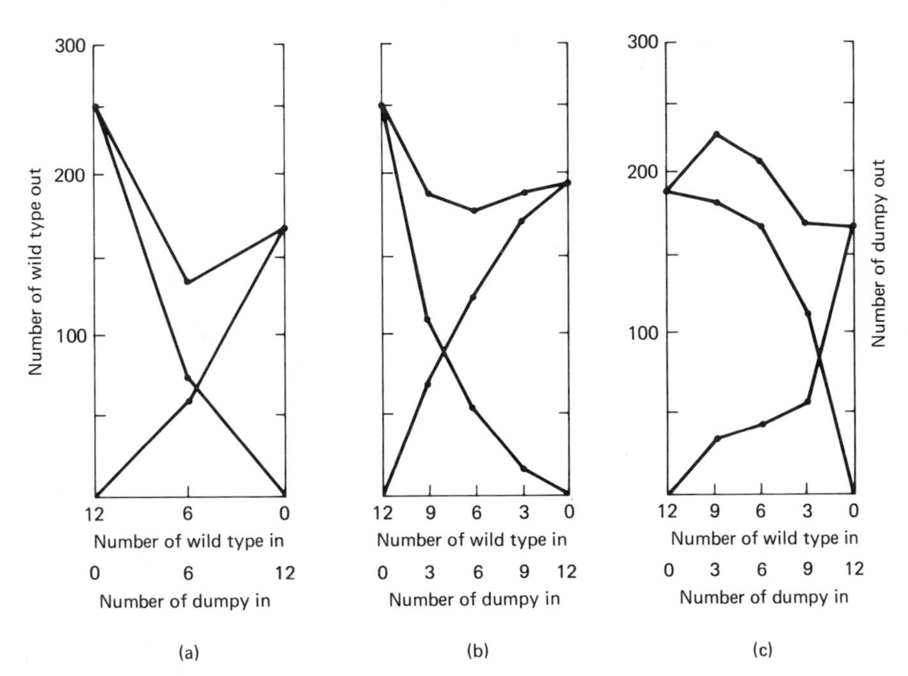

Figure 17-8 Results of selection for competitive ability: (a) stock wild type–stock dumpy; (b) stock wild type–selected dumpy; (c) selected wild type–stock dumpy.

bottles. (In Seaton and Antonovics' studies it was not clear that experimental and control flies had evolved under conditions similar except for the presence of competitors.) Reduction in the number of emerging adults (over 20 days) due to "interspecific" competition averaged about 37% for the brown-eyed strain, 48% for the wild type. So competition intensity was considerable. Selection was carried out for 21 generations in two selected lines, 23 in two others. Periodically, flies from the experimental and control lines were pitted against each other for a single generation and competitive ability assessed by looking at the proportion of emerging progeny belonging to each competitor. Selected brown-eyed flies did no better against either control or selected wild flies than did control brown-eyed flies. Selected wild types did no better than controls against either brown-eyed category. On the other hand, for all possible combinations of trials (four), over all replicates (four), 14 of the 16 series of tests showed increasing performance of the selected lines over time. Also there was a (nonsignificant) increase in the proportion of progeny with time in seven of the eight (four brown-eyed plus four wild) selected populations. Thus it appears that some changes may have occurred due to selection, but their magnitude was minute. There was a rapid improvement in the high-density selected brown-eyed strain, when competed against control wild-type flies, over the first seven generations. This is in agreement with the findings of some of the other workers. But this line had started out doing poorly, and the improvement served only to bring it up to the performance level of the brown-eyed controls. Thereafter, in spite of intense selection pressure, there was no further improvement. Sulzbach speculated that the likelihood of obtaining positive results in experiments of this sort may be very species and strain specific. He also points to the fact that competitive ability depends on a complex of subsidiary traits and, like Giesel (see references and discussion in Chapters 11 and 12), mentions the possibility of failure due to the presence of strong negative additive covariances. Perhaps competitive ability is a sufficiently important trait in *Drosophila* that past selection has brought its additive genetic variance near zero by fixation of alleles and the formation of tightly linked gene groups. [It would be interesting to repeat his experiments with irradiated flies or flies expressing frequent transpositional events (dysgenic flies, for example), and so possessing exaggerated rates of mutation and chromosomal rearrangement.] Finally, it is possibly that in the laboratory environment there was simply inadequate opportunity for niche or spatial separation and so no way that competition could have been alleviated.

17.3 Measuring Alpha Coefficients

We discussed this topic briefly in Chapter 6. Now with the benefit of considerably more knowledge, it is appropriate to return to the topic. The several problems with using niche overlap to estimate competitive intensity are (see also Lawlor, 1980):

1. Lack of knowledge of appropriate resource gradients and how to weight their contributions
2. No clear way to combine overlaps along several gradients
3. Possibility that different competing species may evaluate resources or scale resource differences in different ways
4. Lack of consideration of interference competition or facilitation

In view of these problems several authors have been inventive in suggesting alternative schemes. Schoener (1974b) notes that if K_1 is constant over different environmental patches but K_2 is allowed to vary, α_{12} could be found by regressing n_1 on n_2:

$$n_1 = K_1 - \alpha_{12}n_2 \qquad \text{where } K_1 \text{ is constant} \tag{17-12}$$

Of course, K_1 will *not* be constant, but this can be dealt with by writing K_1 as a linear combination of various environmental parameters, x_k:

$$n_1 = \sum_k b_k x_k - \alpha_{12} n_2 \tag{17-13}$$

where b_k are coefficients to be found, together with α_{12}, by applying partial regression techniques. One drawback to this method is that it does not allow for higher-order interactions. (Of course, it would be possible to include $n_1^2, n_2^2, n_1 n_2$, and so on, terms among the independent variables.) Another disadvantage is that it pertains to the Lotka–Volterra formulation rather rigidly. The advantage is that the calculated α_{12} is a *realized* α and so incorporates both exploitation and interference effects. As such, it has little theoretical value, but can be used in making predictions in real-life situations.

It is appropriate to end this book with discussion of another technique for estimating competition intensity. The technique draws on a breadth of ecological and genetical principles. It is applicable to all forms of interactions among species, including interference and facilitation as well. It can also be used to assess intermorph (intergenotypic) interactions. Finally, it does all of this by generating an empirical expression for the population growth parameter, including higher-order and interactive terms, and is not tied to the Lotka–Volterra or any other particular form. Thus it allows us to read off, directly, the nature of the density-feedback curve. The method is due to J. M. Emlen (1981); it follows directly from the concept of ideal free distribution.

We first write ρ_{gi}, the contribution to the gth genotype's growth rate in species i by an individual of that genotype and species, as a linear function of a set of environmental parameters $\{y_k\}$, including physical and physiognomic parameters $\{x_k\}$, and population density parameters n_i, $\{n_j\}$, n_i^2, $\{n_j^2\}$, and $\{n_i n_j\}$, whose values are measured in the immediate vicinity of the individual (for suggestions on the practical aspects of measuring these values, see J. M. Emlen, 1981). Where the researcher deems other combinations, such as ln x_k or n_i/n_j or $1/(n_i + n_j)$, etc. to be appropriate, these can be used as well as or in place of any of the terms above. Thus

$$\rho_{gi} = a_{gi} + \sum b_{jk} y_k \tag{17-14}$$

where the b's are so-far unknown coefficients and $a < \rho$ is an unknown constant. In vector form,

$$\rho_{gi} - a_{gi} = \mathbf{b}_{gi}^T \mathbf{y}_{gi} \tag{17-15}$$

Suppose now that we measure $\mathbf{y}$ over a large number of individuals. Then, over N samples, we will have N sets of y values and N equations of the form (17-15). Thus, we can write these collectively,

$$(\boldsymbol{\rho}_{gi}^T - a_{gi} \cdot \mathbf{1}^T) = \mathbf{b}_{gi}^T \mathbf{Y}_{gi}$$

where $\mathbf{1}^T$ is the row vector of all 1's or

$$\mathbf{z}_{gi}^T = \mathbf{b}_{gi}^T \mathbf{Y}_{gi} \qquad \text{where } \mathbf{z}_{gi}^T = \boldsymbol{\rho}_{gi}^T - a_{gi} \mathbf{1}^T \tag{17-16}$$

Because there will be differences between true z_{gi} values and those described by the simplified form $\mathbf{b}_{gi}^T\ \mathbf{Y}_{gi}$, (17-16) needs an "error" term (with mean zero). Accordingly, we rewrite (17-16) more accurately as

$$\mathbf{z}_{gi}^T = \mathbf{b}_{gi}^T \mathbf{Y}_{gi} + \boldsymbol{\epsilon}_{gi}^T \tag{17-17}$$

Now, according to the concept of ideal free distribution, the expectation of all ρ_{gi} values should be the same, say $\hat{\rho}_{gi}$. Hence we write

$$\rho_{gi} = \hat{\rho}_{gi}$$

Finally, then,

$$\mathbf{z}_{gi}^T = \mathbf{b}_{gi}^T \mathbf{Y}_{gi} + \boldsymbol{\epsilon}_{gi}^T = \hat{\mathbf{z}}_{gi} \mathbf{1}^T + \boldsymbol{\epsilon}_{gi}' : \hat{\mathbf{z}}_{gi} \mathbf{1} = \hat{\rho}_{gi} - a_{gi} \mathbf{1} \tag{17-18}$$

The goal of our calculations is now to find those coefficients b_{gk} that best fit (17-18). If we have chosen our y parameters wisely, the values of b_{gk} corresponding to the n_i, n_j, n_i^2, and so on, terms, should describe the impact of those terms on ρ_{gi}. Before proceeding, however, we must note that species i (and j) are composed of many genotypes g. Hence if δ_{gi} gives the difference between the species' value of $\hat{z}_i$ and $\hat{z}_{gi}$, we must add still another error term to (17-18):

$$\hat{z}_i \mathbf{1}^T = \mathbf{b}_{gi}^T \mathbf{Y}_{gi} + \boldsymbol{\epsilon}_{gi}^T + \boldsymbol{\delta}_{gi}^T$$

for the gth genotype. Stringing all expressions of this sort over all g into a single matrix equation, we have

$$\hat{z}_i \mathbf{1}^T = \mathbf{b}_i^T \mathbf{Y}_i + \boldsymbol{\epsilon}_i^T + \boldsymbol{\delta}_i^T \tag{17-19}$$

To solve for $\mathbf{b}_i^T$ we first rearrange,

$$\boldsymbol{\epsilon}_i^T = \hat{z}_i \mathbf{1}^T - \mathbf{b}_i^T \mathbf{Y}_i - \boldsymbol{\delta}_i^T$$

and then calculate the sums of the errors (ϵ_i) squared:

$$\begin{aligned}\boldsymbol{\epsilon}_i^T \boldsymbol{\epsilon}_i &= (\hat{z}_i \mathbf{1}^T - \mathbf{b}_i^T \mathbf{Y}_i - \boldsymbol{\delta}_i^T)(\hat{z}_i \mathbf{1}^T - \mathbf{b}_i^T \mathbf{Y}_i - \boldsymbol{\delta}_i^T)^T \\ &= \hat{z}_i^2 \mathbf{1}^T \mathbf{1} - 2\hat{z}_i \mathbf{1}^T \mathbf{Y}_i^T \mathbf{b}_i - 2\hat{z}_i \mathbf{1}^T \boldsymbol{\delta}_c + \mathbf{b}_i^T \mathbf{Y}_i \mathbf{Y}_i^T \mathbf{b}_i + 2\mathbf{b}_i^T \mathbf{Y}_i \boldsymbol{\delta}_i + \boldsymbol{\delta}_i^T \boldsymbol{\delta}_i \end{aligned} \tag{17-20}$$

Next, we use the standard least-squares technique, finding that $\mathbf{b}_i^T$ which minimizes the sum of errors squared. We differentiate with respect to all elements of the vector, $\mathbf{b}_i^T$, and set the derivatives equal to zero:

$$0 = -2\hat{z}_i \mathbf{1}^T \mathbf{Y}_i^T + 2\mathbf{b}_i^T \mathbf{Y}_i \mathbf{Y}_i^T + 2\boldsymbol{\delta}_i^T \mathbf{Y}_i^T$$

whence

$$\mathbf{b}_i^T = (\hat{z}_i \mathbf{1}^T \mathbf{Y}_i^T - \boldsymbol{\delta}_i^T \mathbf{Y}_i^T)(\mathbf{Y}_i \mathbf{Y}_i^T)^{-1} \tag{17-21}$$

The second derivatives are given by $2(\mathbf{Y}_i \mathbf{Y}_i^T)_{\text{diagonal elements}} > 0$, so we know that, as desired, (17-21) defines the $\mathbf{b}_i^T$ that minimizes sum of the errors squared. This is the vector we are looking for. The $\boldsymbol{\delta}_i^T$ term would normally be very difficult to measure. But if selection is slow the $\boldsymbol{\delta}_i^T$ term can be ignored. Therefore, from (17-15), (17-17), and (17-21),

$$\rho_i = a_i + \mathbf{b}_i^T \mathbf{y}_i + \epsilon_i = a_i + (\hat{\rho}_i - a_i)\mathbf{1}^T \mathbf{Y}_i^T (\mathbf{Y}_i \mathbf{Y}_i^T)^{-1} \mathbf{y}_i + \epsilon_i \tag{17-22}$$

This equation describes the contribution of an individual of species i to the growth rate of its species. Because the realized per individual growth rate of a population is simply the mean of the contributions of all its members, we have, finally,

$$r_i = \bar{\rho}_i = a_i + \mathbf{b}_i^T \bar{\mathbf{Y}}_i + \bar{\epsilon}_i = a_i + (\hat{\rho}_i - a_i)[\mathbf{1}^T \mathbf{Y}_i^T (\mathbf{Y}_i \mathbf{Y}_i^T)^{-1}] \bar{\mathbf{Y}}_i \tag{17-23}$$

This technique can be illustrated with a simple example. Suppose that we were examining competition between two species of land snails. After making extensive surveys on appropriate environmental parameters (say, grass cover and forb cover), and the values of n_i, n_i^2, n_j, n_j^2, and $n_i n_j$ associated in a given microhabitat plot with *each individual* of species i, we apply (17-21) and obtain

$$\mathbf{b}_i^T = (\hat{\rho}_i - a_i)[-.3 \quad .1 \quad .02 \quad -.005 \quad -.01 \quad .0 \quad .001]$$

Then, for a given individual,

$$\begin{aligned}\rho_i = a_i + (\hat{\rho}_i - a_i)[&-.3(\text{grass cover}) + .1(\text{forb cover}) \\ &+ .02n_i - .005n_i^2 - .01n_j + .001n_i n_j + \epsilon_i]\end{aligned}$$

Over the population as a whole, lumping grass cover and forb cover into a single term E_i, we have

$$r_i = a_i + (\hat{\rho}_i - a_i)[E_i + .02\bar{n}_i - .005\bar{n}_i^2 - .005 \operatorname{var}(n_i) - .01\bar{n}_j + .001\bar{n}_i\bar{n}_j + .001 \operatorname{Cov}(n_i, n_j)]$$

where the variance and covariance terms are calculated over all individuals. The population growth parameter, as well as the relative intensity of density feedback by competitors and conspecifics, clearly depends on both the dispersion of species i and the codispersion of i and j. It is, as well, both frequency and density dependent. Finally, depending on the relative population densities and codispersion pattern, the effect of species j on i may be either competitive

$$\frac{\partial r_i}{\partial \bar{n}_j} = \frac{\sum \frac{\partial \rho_i}{\partial n_i} n_j}{\sum n_j} = (\hat{\rho}_i - a_i)\left(\frac{-.01\bar{n}_j + .001\overline{n_i n_j}}{\bar{n}_j}\right)$$

$$\propto -.01 + .001 \frac{\bar{n}_i\bar{n}_j + \operatorname{cov}(n_i, n_j)}{\bar{n}_j} < 0$$

or facilitative (above derivative positive). Finally, note that species i displays a marked Allee effect, r_i rising with $.02\bar{n}_i$ at low $\bar{n}_i$ values, then falling as the magnitude of the $.02\bar{n}_i$ term is overshadowed by $-.005\bar{n}_i^2$.

By ending with this model we end in the spirit of much current research in population biology. No model will be wholly accurate or universally applicable, but (17-23) looks very promising as a tool for applying ecological–genetic theory to the real world. On the other hand, the possibly large error variance may in some circumstances render it all but useless. The experimental evidence is not yet in.

Review Questions

1. Is competition an important factor in structuring communities—that is, in determining what (kinds of) species are present and the extent to which their niches or sizes overlap? Review the arguments and the evidence.
2. If you had a lot of data on microhabitat–species associations in a small mammal community, how might you use those data to determine whether competition was occurring among the species in question? Can niche separation be used as an indicator?
3. Why might we expect intensity of exploitation competition and intensity of interference competition to be positively correlated? Under what circumstances might we expect this relationship to be reversed? Distinguish between potential and realized exploitation competition.
4. If the environment went through a particularly benign period with respect to population depression by density-independent factors, what changes in niche overlap patterns might you expect to find among competitors?
5. What is the implication of strong interference or extensive behavioral plasticity for the evolution of niche overlap?
6. Can heritability for competitive ability be measured? How?
7. Can you think of any way in which spatial structuring of populations (patchiness and codispersion of species) might affect the evolution of competition intensity?
8. How does competition intensity relate to competitive ability?
9. To what sorts of organisms and under what conditions is Emlen's method for measuring species interaction intensities (Section 17.3) most or least applicable?

Bibliography

Abele, L. G., 1976. Comparative species richness in fluctuating and constant environments: coral-associated decapod crustaceans. *Science* **192**: 461–463.

Abbott, I., L. K. Abbott, and P. R. Grant, 1975. Seed selection and handling ability of four species of Darwin's finches. *Condor* **77**:332–335.

Abrahamson, W. G., and M. Gadgil, 1973. Growth form and reproductive effort in goldenrods (*Solidago*, Compositae). *Am. Nat.* **107**:651–661.

Abrams, P. A., 1976. Niche overlap and environmental variability. *Math. Biosci.* **28**:357–372.

Abrams, P. A., 1980a. Some comments on measuring niche overlap. *Ecology* **61**:44–49.

Abrams, P. A., 1980b. Consumer functional response and competition in consumer-resource systems. *Theor. Popul. Biol.* **17**:80–102.

Abugov, R. A., and R. E. Michod, 1981. On the relationship of family structured models and inclusive fitness models for kin selection. *J. Theor. Biol.* **88**:743–754.

Acevedo, M. F., 1981. On Horn's Markovian model of forest dynamics with particular reference to tropical forests. *Theor. Popul. Biol.* **19**:230–250.

Alatalo, R. V., A. Carlson, A. Lundberg, and S. Ulfstrand, 1981. The conflict between male polygyny and female monogamy: the case of the pied flycatcher, *Ficedula hypoleuca*. *Am. Nat.* **117**:738–753.

Alexander, R. D., and G. Borgia, 1978. Group selection, altruism, and the levels of organization of life. *Annu. Rev. Ecol. Syst.* **9**:449–474.

Allard, R. W., and J. Adams, 1969. Population studies in predominantly self-pollinated species. XIII. Intergenotypic competition and population structure in barley and wheat. *Am. Nat.* **103**:621–645.

Allard, R. W., S. K. Jain, and P. L. Workman, 1968. The genetics of inbreeding populations. *Adv. Genet.* **14**:55–131.

Allen, J. A., 1972. Evidence for stabilizing and apostatic selection by wild blackbirds. *Nature* **237**:348–349.

Allen, J. A., and B. Clarke, 1968. Evidence of apostatic selection by wild passerines. *Nature* **220**:501–502.

Allison, A. C., 1955. Aspects of polymorphism in man. *Cold Spring Harbor Symp. Quant. Biol.* **20**:239–255.

Allison, A. C., 1961. Genetic factors in resistance to malaria. *Ann. N.Y. Acad. Sci.* **91**:710–729.

Altmann, S. A., 1979. Altruistic behavior: the fallacy of kin deployment. *Anim. Behav.* **27**:958–959.

Altmann, S. A., S. S. Wagner, and S. Lenington, 1977. Two models for the evolution of polygyny. *Behav. Ecol. Sociobiol.* **2**:397–410.

Anderson, P. K., 1964. Lethal alleles in *Mus musculus*, local distribution and evidence for isolation of *demes*. *Science* **145**:177–178.

Anderson, P. K., 1970. Ecological structure and gene flow in small mammals. *Symp. Zool. Soc. Lond.* **26**:299–325.

Anderson, W. W., 1971. Genetic equilibrium and population growth under density-regulated selection. *Am. Nat.* **105**:489–498.

Anderson, W. W., and C. E. King, 1970. Age-specific selection. *Proc. Natl. Acad. Sci. USA* **66**:780–786.

Anton, H., 1977. *Elementary Linear Algebra*, 2nd ed. John Wiley & Sons, Inc., New York.

Antonovics, J., 1968. Evolution in closely adjacent plant populations. V. Evolution of self-fertility. *Heredity* **23**:219–231.

Aoki, K., 1981. Algebra of inclusive fitness. *Evolution* **35**:659–663.

Aoki, K., 1982. Additive polygenic formulation of Hamilton's model of kin selection. *Heredity* **49**:163–169.

Arditi, R., J. M. Abillon, and J. V. DaSilva, 1977. The effect of a time-delay in a predator–prey model. *Math. Biosci.* **33**:107–120.

Armstrong, E. A., 1955. *The Wren.* William Collins Sons & Co. Ltd., London.

Armstrong, R. A., 1976. The effects of predator functional response and prey productivity on predator–prey stability: a graphical approach. *Ecology* **57**:609–612.

Armstrong, R. A., and M. E. Gilpin, 1977. Evolution in time-varying environments. *Science* **195**:591–592.

Armstrong, R. A., and R. McGehee, 1976. Coexistence of species competing for shared resources. *Theor. Popul. Biol.* **9**:317–328.

Armstrong, R. A., and R. McGehee, 1980. Competitive exclusion. *Am. Nat.* **115**:151–170.

Arnason, E., and P. R. Grant, 1976. Climatic selection in *Cepaea hortensis* at the northern limit of its range in Iceland. *Evolution* **30**:499–508.

Arnold, S. J., 1978. The evolution of a special class of modifiable behaviors in relation to environmental pattern. *Am. Nat.* **112**:415–427.

Arnold, S. J., 1981. The microevolution of feeding behavior. In *Foraging Behavior: Ecological, Ethological and Psychological Approaches* (A. C. Kamil and T. D. Sargent, eds.). Garland Publishing, Inc., New York, pp. 409–453.

Ashby, W. R., 1960. *Design for a Brain*, 2nd ed. Chapman & Hall Ltd., London.

Ashmole, N. P., 1963. The regulation of numbers of tropical oceanic birds. *Ibis*, **103**:458–473.

Aumann, G. D., and J. T. Emlen, 1965. Relation of population density to sodium availability and sodium selection by microtine rodents. *Nature*, **208**:198–199.

Axelrod, R., and W. D. Hamilton, 1981. The evolution of cooperation. *Science* **211**:1390–1396.

Ayala, F. J., 1966. Reversal of dominance in competing species of *Drosophila*. *Am. Nat.* **100**:81–83.

Ayala, F. J., 1969. Experimental invalidation of the principle of competitive exclusion. *Nature*, **224**:1076.

Ayala, F. J., 1971. Competition between species: frequency-dependence. *Science* **171**:820–824.

Ayala, F. J., 1972. Competition between species. *Am. Sci.* **60**:348–357.

Ayala, F. J., J. W. Valentine, D. Hedgecock, and L. G. Barr, 1975. Deep-sea asteroids: high genetic variability in a stable environment. *Evolution* **29**:203–212.

Bagenal, T. B., 1969. Relationships between egg size and fry survival in brown trout *Salmo trutta* L. *J. Fish. Biol.* **1**:349–353.

Baker, H. G., 1972. Seed weight in relation to environmental conditions in California. *Ecology* **53**:997–1010.

Baker, R. P., 1978. *The Evolutionary Ecology of Animal Migration.* Holmes and Meier Publishers, Inc., New York.

Bantock, C. R., 1975a. *Cepaea nemoralis* (L) on Steep Holm. *Proc. Malacol. Soc. Lond.* **41**:223–232.

Bantock, C. R., 1975b. Experimental evidence for non-visual selection in *Cepaea nemoralis*. *Heredity* **33**:409–412.

Barclay, H., and P. T. Gregory, 1981. An experimental test of models predicting life-history characters. *Am. Nat.* **117**:944–961.

Barnard, W. H., 1979. A population study of marsh-breeding redwing blackbirds (*Agelaius phoeniceus*) with emphasis on site fidelity, Ph.D. dissertation, Indiana University, Bloomington.

Barnes, J. R., and J. J. Gonor, 1973. The larval settling response of the lined chiton, *Tonicella lineata*. *Mar. Biol.* **20**:259–264.

Bartholomew, G. A., 1970. A model for the evolution of pinniped polygyny. *Evolution* **24**:546–559.

Beacham, T. D., 1981. Some demographic aspects of dispersers in fluctuating populations of the vole. *Microtus townsendii*. *Oikos* **36**:273–280.

Beddington, J. R., 1974. Age-distribution and the stability of simple discrete time population models. *J. Theor. Biol.* **47**:65–74.

Beddington, J. R., 1975. Mutual interference between parasites or predators and its effect on searching efficiency. *J. Anim. Ecol.* **44**:331–340.

Beddington, J. R., and C. A. Free, 1976. Age-structure effects in predator–prey interactions. *Theor. Popul. Biol.* **9**:15–24.

Beddington, J. R., C. A. Free, and J. H. Lawton, 1975. Dynamic complexity in predator–prey models framed in difference equations. *Nature* **255**:58–60.

Beddington, J. R., M. P. Hassell, and J. W. Lawton, 1976. The components of arthropod predation. II. The predator rate of increase. *J. Anim. Ecol.* **45**:165–185.

Bell, A. E., and M. J. Burris, 1973. Simultaneous selection for two correlated traits in *Tribolium*. *Genet. Res.* **21**:29–49.

Bell, G., 1978. Group selection in structured populations. *Am. Nat.* **112**:389–399.

Belovsky, G. E., 1978. Diet optimization in a generalist herbivore, the moose. *Theor. Popul. Biol.* **14**:105–134.

Belovsky, G. E., 1981. Food plant selection by a generalist herbivore, the moose. *Ecology* **62**: 1020–1030.

Belovsky, G. E. 1983. Diet optimization by beaver. *Amer. Midl. Natur.* (in press).

Belovsky, G. E., and P. A. Jordan, 1978. The time–energy budget of moose. *Theor. Popul. Biol.* **14**:76–104.

Belyea, R. M., 1952. Death and deterioration of balsam fir weakened by spruce budworm defoliation in Ontario. *J. For.* **50**:729–738.

Berthold, P., and U. Querner, 1981. Genetic basis of migratory behavior in European warblers. *Science* **212**:77–79.

Bertram, B. C. R., 1976. Kin selection in lions and in evolution. In *Growing Points in Ethology* (R. A. Hinde, ed.). Cambridge University Press, Cambridge, pp. 281–301.

Bertram, B. C. R., 1978. Living in groups: Predators and prey. In *Behavioral Ecology: An Evolutionary Approach* (J. R. Krebs and N. B. Davies, eds.). Sinauer Associates, Inc., Sunderland, Mass., pp. 64–96.

Bethel, W. M., and J. C. Holmes, 1973. Correlation of development of altered evasive behavior in *Gammarus lacustris* (Amphipod) harboring cystacanths of *Polymorphus paradoxus* (Acanthocephala) with the infectivity to the definitive host. *J. Parasitol.* **60**:272–274.

Beverton, R. J. H., and S. J. Holt, 1957. *On the Dynamics of Exploited Fish Populations*, Fish Invest. Lond. Ser. 2, 19, 533 pp.

Bidder, G. P., 1932. Senescence. *Br. Med. J.* **2**:583–585.

Birch, L. C., 1953a. Experimental background to the study of the distribution and abundance of insects. I. The influence of temperature, moisture and food on the innate capacity for increase of three grain beetles. *Ecology* **34**:698–711.

Birch, L. C., 1953b. Experimental background to the study of the distribution and abundance of insects. III. The relations between innate capacity for increase and survival of different species living together on the same food. *Evolution* **7**:136–144.

Blair, W. F., 1940a. A study of prairie deer mouse populations in southern Michigan. *Am. Midl. Nat.* **24**:273–305.

Blair, W. F., 1940b. Home ranges and populations of the meadow vole in southern Michigan. *J. Wildl. Manage.* **4**:149–161.

Blair, W. F., 1942. Size of home range and notes on the life history of the woodland deer mouse and eastern chipmunk in northern Michigan. *J. Mammal.* **23**:27–36.

Blair, W. F., 1947. Estimated frequencies of the buff and gray genes (*G*, *g*) in adjacent populations of deer mice (*Peromyscus maniculatus bairdi*) living on soils of different colors. *Contrib. Lab. Vert. Biol. Mich.* **36**:1–16.

Blair, W. F., 1951. Population structure, social behavior and environment relations in a natural population of the beach mouse (*Peromyscus polionotus leucocephalus*). *Contrib. Lab. Vert. Biol. Mich.* **48**:1–47.

Boag, P. T., and P. R. Grant, 1978. Heritability of external morphology in Darwin's finches. *Nature* **274**:793–794.

Boag, P. T., and P. R. Grant, 1981. Intense natural selection in a population of Darwin's finches (Geospizinae) in the Galápagos. *Science* **214**:82–84.

Bohren, B. B., W. G. Hill, and A. Robertson, 1966. Some observations on asymmetric correlated responses to selection. *Genet. Res.* **7**:44–57.

Bonnell, M. L., and R. K. Selander, 1974. Elephant seals: Genetic variation and near extinction. *Science* **184**:908–909.

Boonstra, R., 1978. Effect of adult townsend voles (*Microtus townsendii*) on survival of young.

Boonstra, R., and C. J. Krebs, 1979. Viability of large- and small-sized adults in fluctuating vole populations. *Ecology* **60**:567–573.

Boorman, S. A., 1978. Mathematical theory of group selection: structure of group selection in founder populations determined from convexity of the extinction operator. *Proc. Natl. Acad. Sci. USA* **75**:1909–1913.

Boorman, S. A., and P. R. Levitt, 1973. A frequency dependent natural selection model for the evolution of social cooperation networks. *Proc. Natl. Acad. Sci. USA* **70**:187–189.

Booth, W. E., 1941. Revegetation of abandoned fields in Kansas and Oklahoma. *Am. J. Bot.* **28**:415–522.

Borman, J. H., and G. E. Likens, 1979. Catastrophic disturbance and the steady state in northern hardwood forests. *Am. Sci.* **67**:660–669.

Botkin, D. B., and M. J. Sobel, 1975. The complexity of ecosystem stability. In *Some Mathematical Questions in Biology* (S. A. Levin, ed.). SIAM Institute for Mathematics and Society, Philadelphia, pp. 144–150.

Bowers, M. A., and J. H. Brown, 1982. Body size and coexistence in desert rodents: chance or community structure? *Ecology* **63**:391–400.

Boyce, M. S., 1977. Population growth with stochastic fluctuations in the life tables. *Theor. Popul. Biol.* **12**:366–373.

Boyce, M. S., 1979. Population projections with fluctuating fertility and survivorship schedules. *Proc. Summer Comp. Simul. Conf.* **10**:385–388.

Boyce, M. S., and D. J. Daley, 1980. Population tracking of fluctuating environments and natural selection for tracking ability. *Am. Nat.* **115**:480–491.

Brenchley, G. A., 1979. Community matrix models: inconsistent results using Vandermeer's data. *Am. Nat.* **113**:456–459.

Brenchley, G. A., 1981. A reply to Thomas and Pomerantz with a note on diffuse competition. *Am. Nat.* **117**:386–389.

Brenner, F. J., 1967. Seasonal correlations of reserve energy of the red-winged blackbird. *Bird-band.* **38**:195–211.

Briand, F., 1983. Environmental control of food web structure. *Ecology* **64**:253–263.

Brock, V. E., and R. H. Riffenburg, 1963. Fish schooling: a possible factor in reducing predation. *J. Cons. Intern. Explor. Mer* **25**:307–317.

Bromley, P. T., 1969. Territoriality in pronghorn bucks on the National Bison Range, Moiese, Montana. *J. Mammal.* **50**:81–89.

Brooks, J. L., and S. I. Dodson, 1965. Predation, body size, and composition of plankton. *Science* **150**:28–35.

Brower, L. P., 1961. Studies on the migration of the monarch butterfly. I. Breeding populations of *Danaus plexippus* and *D. gilippus* Berenice in south central Florida. *Ecology* **42**:76–83.

Brower, L. P., and J. V. Z. Brower, 1966. Experimental evidence of the effects of mimicry. *Am. Nat.* **100**:173–187.

Brower, L. P., J. V. Z. Brower, and P. W. Westcott, 1960. Experimental studies of mimicry. 5. The reactions of toads (*Bufo terrestris*) to bumblebees (*Bombus americanorum*) and their robberfly mimics (*Mallophora bomboides*) with a discussion of aggressive mimicry. *Am. Nat.* **94**:343–356.

Brower, L. P., J. V. Z. Brower, F. G. Stiles, H. J. Croze, and A. S. Hower, 1964. Mimicry: differential advantages of color patterns in a natural environment. *Science* **144**:183–185.

Brower, L. P., L. M. Cook, and H. J. Croze, 1967. Predator responses to artificial mimics released in a neotropical environment. *Evolution* **21**:11–23.

Brower, L. P., W. N. Ryerson. L. L. Coppinger, and S. C. Glazier, 1968. Ecological chemistry and the palatability spectrum. *Science* **161**:1349–1351.

Brown, J. H., 1971a. Mammals on mountain tops: non-equilibrium insular biogeography. *Am. Nat.* **105**:467–478.

Brown, J. H., 1971b. Mechanisms of competitive exclusion between two species of chipmunks. *Ecology* **52**:305–311.

Brown, J. H., 1973. Species diversity of seed-eating desert rodents in sand dune habitats. *Ecology* **54**:775–787.

Brown, J. H., and D. W. Davidson, 1977. Competition between seed-eating rodents and ants in desert ecosystems. *Science* **196**:880–882.

Brown, J. H., and G. A. Lieberman, 1973. Resource utilization and coexistence of seed-eating desert rodents in sand dune habitats. *Ecology* **54**:788–797.

Brown, J. H., D. W. Davidson, and O. J. Reichman, 1979. An experimental study of competitiion between seed-eating desert rodents and ants. *Am. Zool.* **19**:1129–1143.

Brown, J. L., 1964. The evolution of diversity in avian territorial systems. *Wilson Bull.* **76**:160–169.

Brown, J. L., 1969. The buffer effect and productivity in tit populations. *Am. Nat.* **103**:347–354.

Brown, J. L., 1970. Cooperative breeding and altruistic behavior in the Mexican jay, *Aphelocoma ultramarina*. *Anim. Behav.* **18**:366–378.

Brown, J. L., 1978. Avian communal breeding systems. *Annu. Rev. Ecol. Syst.* **9**:123–155.

Brown, J. L., and E. R. Brown, 1981. Kin selection and individual selection in babblers. In *Natural Selection and Social Behavior: Recent Research and New Theory* (R. D. Alexander and D. W. Tinkle, eds.). Chiron Press, New York, pp. 244–256.

Brown, J. L., E. R. Brown, S. D. Brown, and D. D. Dow, 1982. Helpers: effects of experimental removal on reproductive success. *Science* **215**:421–422.

Brown, L. E., 1966. Home range and movements of small mammals. *Symp. Zool. Soc. Lond.* **18**:111–142.

Brown, W. L., Jr., 1957. Centrifugal speciation. *Quart. Rev. Biol.* **32**:247–277.

Brown, W. L., Jr., and E. O. Wilson, 1956. Character displacement. *Syst. Zool.* **5**:49–64.

Browne, R. A., and W. D. Russell-Hunter, 1978. Reproductive effort in molluscs. *Oecologia* **37**:23–27.

Buechner, H. K., J. A. Morrison, and W. Leuthold, 1966. Reproduction in the Uganda kob with special reference to behavior. *Symp. Zool. Soc. Lond.* **15**:69–88.

Bull, J. J., and R. C. Vogt, 1979. Temperature-dependent sex determination in turtles. *Science* **206**:1186–1188.

Bulmer, M. G., 1971. The effect of selection on genetic variability. *Am. Nat.* **105**:201–211.

Bulmer, M. G., 1972. Multiple niche polymorphism. *Am. Nat.* **106**:254–256.

Bulmer, M. G., 1973. Inbreeding in the great tit. *Heredity* **30**:313–325.

Bulmer, M. G., 1974. Density-dependent selection and character displacement. *Am. Nat.* **108**:45–58.

Bulmer, M. G., 1977. Periodical insects. *Am. Nat.* **111**:1099–1117.

Bulmer, M. G., and P. D. Taylor, 1980a. Dispersal and the sex ratio. *Nature* **284**:448–451.

Bulmer, M. G., and P. D. Taylor, 1980b. Sex ratio under the haystack model. *J. Theoret. Biol.* **86**:83–89.

Burd, A. C., and W. G. Parnell, 1970. The relationship between larval abundance and stock in the North Sea herring. *Rapp. Cons. Int. Explor. Mer* **164**:30–36.

Burnett, T., 1951. Effects of temperature and host density on the rate of increase of an insect parasite. *Am. Nat.* **85**:337–352.

Burt, W. H., 1943. Territoriality and home range concepts as applied to mammals. *J. Mammal.* **24**:346–352.

Bustard, H. R., 1970. The role of behavior in the natural regulation of numbers in the gekkonid lizard, *Gehyra variegata. Ecology* **51**:724–728.

Butler, L., 1953. The nature of cycles in populations of Canadian mammals. *Can. J. Zool.* **31**:242–262.

Buzas, M. A., and T. G. Gibson, 1969. Species diversity: benthonic foraminifera in the western North Atlantic. *Science* **163**:72–75.

Cain, A. J., 1971. Colour and banding morphs in subfossil samples of the snail, *Cepaea.* In *Ecological Genetics and Evolution* (R. Creed, ed.). Blackwell Scientific Publications Ltd., London, pp. 65–92.

Cain, A. J., and P. M. Sheppard, 1954. Natural selection in *Cepaea. Genetics* **39**:89–116.

Cain, A. J., M. B. King, and P. M. Sheppard, 1960. New data on the genetics of polymorphism in the snail *Cepaea nemoralis* (L). *Genetics* **45**:393–411.

Cairns, J., Jr., M. L. Dahlberg, K. L. Dickson, N. Smith, and W. T. Waller, 1969. The relationship of fresh water protozoan communities to the MacArthur–Wilson equilibrium model. *Am. Nat.* **103**:439–454.

Calhoun, J. B., 1952. The social aspects of population dynamics. *J. Mammal.* **33**:139–159.

Cameron, A. W., 1964. Competitive exclusion between the rodent genera *Microtus* and *Clethrionomys. Evolution* **18**:630–634.

Cameron, G. N., 1973. Effec[illegible]itter size on postnatal growth in the desert wood rat. *J. Mammal.* **54**:489–493.

Cameron, G. N., 1977. Experimental species removal: demographic responses by *Sigmodon hispidus* and *Reithrodontomys fulvescens. J. Mammal.* **58**:488–506.

Campanella, P. J., 1975. The evolution of mating systems in temperate zone dragonflies (Odonata: Anisoptera). II. *Libellula luctuosa* (Burmeister). *Behaviour* **54**:278–310.

Campanella, P. J., and L. L. Wolf, 1974. Temporal leks as a mating system in a temperate zone dragonfly (Odonata: Anisoptera). I. *Plathemis lydia* (Drury). *Behaviour* **51**:49–87.

Campbell, J. H., 1982. Autonomy in Evolution. In *Perspectives on Evolution* (R. Milkman, ed.). Sinauer Associates, Inc., Sunderland, Mass., pp. 190–207.

Caraco, T., 1979a. Time budgeting and group size: a theory. *Ecology* **60**:611–617.

Caraco, T., 1979b. Time budgeting and group size: a test of theory. *Ecology* **60**:618–627.

Caraco, T., 1980. On foraging time allocation in a stochastic environment. *Ecology* **61**:119–128.

Caraco, T., and L. L. Wolf, 1975. Ecological determinants of group sizes of foraging lions. *Am. Nat.* **109**:343–352.

Carey, M., and V. Nolan, Jr., 1975. Polygyny in indigo buntings: a hypothesis tested. *Science* **190**:1296–1297.

Carleton, W. M., 1965. Food habits of two sympatric Colorado sciurids. *J. Mammal.* **47**:91–103.

Carpenter, F. L., and R. E. MacMillen, 1976. Threshold model of feeding territoriality and test with a Hawaiian honey creeper. *Science* **194**:639–642.

Carson, H. L., 1968. The population flush and its genetic consequences. In *Population Biology and Evolution* (R. C. Lewontin, ed.). Syracuse University Press, Syracuse, N.Y., pp. 123–138.

Case, T. J., 1981. Niche separation and resource scaling. *Am. Nat.* **118**:554–560.

Case, T. J., and R. G. Casten, 1979. Global stability and multiple domains of attraction in ecological systems. *Am. Nat.* **113**:705–714.

Case, T. J., and M. E. Gilpin, 1974. Interference competition and niche theory. *Proc. Natl. Acad. Sci. USA* **71**:3073–3077.

Caswell, H., 1972. A simulation study of a time lag population model. *J. Theor. Biol.* **34**:419–439.

Caswell, H., 1976. Community structure: a neutral model analysis. *Ecol. Monogr.* **46**:327–354.

Caswell, H., 1978a. Predator-mediated coexistence: a non-equilibrium model. *Am. Nat.* **112**:127–154.

Caswell, H., 1978b. A general formula for the sensitivity of population growth rate to changes in life history parameters. *Theor. Popul. Biol.* **14**:215–230.

Cates, R. G., 1975. The interface between slugs and wild ginger: some evolutionary aspects. *Ecology* **56**:391–400.

Caughley, 1966. Mortality patterns in mammals. *Ecology* **47**:906–918.

Cavalli-Sforza, L. L., and M. W. Feldman, 1978. Darwinian selection and altruism. *Theor. Popul. Biol.* **14**:268–280.

Chagnon, N. A., 1975. Genealogy, solidarity, and relatedness: limits to local group size and patterns of fissioning in an expanding population. *Yearb. Phys. Anthropol.* **19**:95–110.

Chapman, R. N., 1928. The quantitative analysis of environmental factors. *Ecology* **9**:111–122.

Charlesworth, B., 1970. Selection in populations with overlapping generations. I. The use of Malthusian parameters. *Theor. Popul. Biol.* **1**:352–370.

Charlesworth, B., 1971. Selection in density-regulated populations. *Ecology* **52**:469–474.

Charlesworth, B., 1974. Selection in populations with overlapping generations. VI. Rates of change of gene frequency and population growth rate. *Theor. Popul. Biol.* **6**:108–133.

Charlesworth, B., 1976. Recombination modification in a fluctuating environment. *Genetics* **83**:181–185.

Charlesworth, B., 1977. Population genetics, demography and the sex ratio. In *Measuring Selection in Natural Populations* (F. B. Christiansen and T. M. Fenchell, eds.). Springer-Verlag, New York, pp. 345–363.

Charlesworth, B., 1979. A note on the evolution of altruism in structured demes. *Am. Nat.* **113**:601–605.

Charlesworth, B., 1980. *Evolution in Age-structured Populations.* Cambridge University Press, Cambridge.

Charlesworth, B., and E. L. Charnov, 1981. Kin selection in age-structured populations. *J. Theor. Biol.* **88**:103–120.

Charlesworth, B., and J. T. Giesel, 1972. Selection in populations with overlapping generations. II. Relations between frequency and demographic variables. *Am. Nat.* **106**:388–401.

Charlesworth, B., and J. A. Leon, 1976. The relationship of reproductive effort to age. *Am. Nat.* **110**:449–459.

Charlesworth, D., and B. Charlesworth, 1976. Theoretical genetics of Batesian mimicry. II. Evolution of supergenes. *J. Theor. Biol.* **55**:305–324.

Charlesworth, D., and B. Charlesworth, 1979. Selection on recombination in a multilocus system. *Genetics* **91**:575–580.

Charnov, E. L., 1975. Sex ratio in an age-structured population. *Evolution* **29**:366–368.

Charnov, E. L., 1976. Optimal foraging and the marginal value theorem. *Theor. Popul. Biol.* **9**:129–136.

Charnov, E. L., 1977. An elementary treatment of the genetical theory of kin selection *J. Theor. Biol.* **66**:541–550.

Charnov, E. L., 1978. Sex ratio selection in eusocial hymenoptera. *Am. Nat.* **112**:317–326.

Charnov, E. L., 1979. Natural selection and sex change in pandalid shrimp: test of a life-history theory. *Am. Nat.* **113**:715–734.

Charnov, E. L., 1982. *The Theory of Sex Allocation*, Princeton University Press, Princeton, N.J.

Charnov, E. L., and J. Bull, 1977. When is sex environmentally determined? *Nature* **266**:828–830.

Charnov, E. L., and J. R. Krebs, 1974. On clutch size and fitness. *Ibis* **116**:217–219.

Charnov, E. L., and J. R. Krebs, 1975. The evolution of alarm calls: altruism or manipulation? *Am. Nat.* **109**:107–112.

Charnov, E. L., and W. M. Schaffer, 1973. Life history consequences of natural selection: Cole's results revisited. *Am. Nat.* **107**:791–793.

Charnov, E. L., G. H. Orians, and K. Hyatt, 1976. Ecological implications of resource depression. *Am. Nat.* **110**:247–259.

Cheung, T. K., and R. J. Parker, 1974. Effect of selection on heritability and genetic correlation of two quantitative traits in mice. *Can. J. Genet. Cytol.* **16**:599–609.

Childress, J. J., and M. H. Price, 1978. Growth rate of the bathypelagic crustacean *Ganthophausia ingens* (Mysidacea: Lophogastridae). I. Dimensional growth and population structure. *Mar. Biol.* **50**:47–62.

Chinnici, J. P., 1971a. Modification of recombination frequency in *Drosophila.* I. Selection for increased and decreased crossing over. *Genetics* **69**:71–83.

Chinnici, J. P., 1971b. Modification of recombination frequency in *Drosophila*. II. The polygenic control of crossing over. *Genetics* **69**:85–96.

Chitty, D., 1937. A ringing technique for small mammals. *J. Anim. Ecol.* **6**:36–53.

Chitty, D., 1957. Self regulation of numbers through changes in viability. *Cold Spring Harbor Symp. Quant. Biol.* **22**:277–280.

Chitty, D., and E. Phipps, 1966. Seasonal changes in survival in a mixed population of two species of vole. *J. Anim. Ecol.* **35**:313–331.

Christian, J. J., 1956. Endocrine response to population size in mice. *Ecology* **37**:258–273.

Christian, J. J., 1961. Phenomena associated with population density. *Proc. Natl. Acad. Sci. USA* **47**:428–499.

Christian, J. J., and D. E. Davis, 1956. The relationship between adrenal weight and population status of Norway rats. *J. Mammal.* **37**:475–486.

Christian, J. J., and D. E. Davis, 1964. Endocrines, behavior and populations. *Science* **146**:1550–1560.

Clark, A., 1979. The effects of interspecific competition on the dynamics of a polymorphism in an experimental population of *Drosophila melanogaster*. *Genetics* **92**:1315–1328.

Clark, A. B., 1978. Sex ratio and local resource competition in a prosimian primate. *Science* **201**:163–165.

Clark, A. B., and D. S. Wilson, 1981. Avian breeding adaptations: hatching asynchrony, brood reduction, and nest failure. *Q. Rev. Biol.* **56**:253–277.

Clarke, B., 1960. Divergent effects of natural selection on two closely-related polymorphic snails. *Heredity* **14**:423–444.

Clarke, B., 1966. The evolution of morph-ratio clines. *Am. Nat.* **100**:389–402.

Clarke, B., 1972. Density-dependent selection. *Am. Nat.* **106**:1–13.

Clarke, C. A., and P. M. Sheppard, 1966. A local survey of the distribution of the industrial melanic forms in the moth, *Biston betularia*, and estimates of the selective values of these in the industrial environment. *Proc. R. Soc B* **165**:424–439.

Cockerham, C. C., P. M. Burrow, S. S. Young, and T. Prout, 1972. Frequency-dependent selection in randomly mating populations. *Am. Nat.* **106**:493–515.

Cody, M. L., 1966. A general theory of clutch size. *Evolution* **20**:174–184.

Cody, M. L., 1968. On the methods of resource division in grassland bird communities. *Am. Nat.* **102**:107–147.

Cody, M. L., 1969a. Character convergence. *Annu. Rev. Ecol. Syst.* **4**:189–211.

Cody, M. L., 1969b. Convergent characteristics in sympatric species: a possible relation to interspecific competition and aggression. *Condor* **71**:222–239.

Cody, M. L., 1970. Chilean bird distribution. *Ecology* **51**:455–463.

Cody, M. L., 1974a. Optimization in ecology. *Science* **183**:1156–1164.

Cody, M. L., 1974b. *Competition and the Structure of Bird Communities*. Princeton University Press, Princeton, N.J.

Cohen, J. E., 1977a. Ergodicity of age structure in populations with Markovian vital rates. II. General states. *Adv. Appl. Prob.* **9**:18–37.

Cohen, J. E., 1977b. Ergodicity of age structure in populations with Markovian vital rates. III. Finite state moments and growth rate: an illustration. *Adv. Appl. Prob.* **9**:462–475.

Cohen, J. E., 1978. *Food Webs and Niche Space*. Princeton University Press, Princeton, N.J.

Cole. L. C., 1954. The population consequences of life history phenomena. *Q. Rev. Biol.* **29**:103–137.

Cole, L., 1960. Competitive exclusion. *Science* **132**:348–349.

Colwell, R. K., 1973. Competition and coexistence in a simple, tropical community. *Am. Nat.* **107**:737–760.

Colwell, R. K., and D. J. Futuyma, 1971. On the measurement of niche breadth and overlap. *Ecology* **52**:567–576.

Comfort, A., 1956. *The Biology of Senescence*. Rinehart & Co., New York.

Comins, H. N., 1982. Evolutionarily stable strategies for localized dispersal in two dimensions. *J. Theor. Biol.* **94**:579–606.

Comins, H. N., and D. W. E. Blatt, 1974. Prey–predator models in spatially heterogeneous environments. *J. Theor. Biol.* **48**:75–83.

Comins, H. N., and M. P. Hassell, 1976. Predation in multi-prey communities. *J. Theor. Biol.* **62**:93–114.

Comins, H. N., W. D. Hamilton, and R. M. May, 1980. Evolutionary stable dispersal strategies. *J. Theor. Biol.* **82**:205–230.

Connell, J. H., 1961. The influence of interspecific competition and other factors on the distribution of the barnacle, *Chthamalus stellatus*. *Ecology* **42**:710–723.

Connell, J. H., 1970. A predator–prey system in the marine intertidal region. I. *Balanus glandula* and several predatory species of *Thais*. *Ecol. Monogr.* **40**:49–78.

Connell, J. H., 1978. Diversity in tropical rain forests and coral reefs. *Science* **199**:1302–1310.

Connell, J. H., and E. Orias, 1964. The ecological regulation of species diversity. *Am. Nat.* **98**:399–414.

Connell, J. H., and R. O. Slatyer, 1977. Mechanisms of succession in natural communities and their role in community stability and organization. *Am. Nat.* **111**:1119–1144.

Connor, E. F., and D. Simberloff, 1979. The assembly of species communities: chance or competition? *Ecology* **60**:1132–1140.

Cook, L. M., and P. Miller, 1977. Density-dependent selection on polymorphic prey—some data. *Am. Nat.* **111**:594–598.

Cook, L. M., and P. O'Donald, 1971. Shell size and natural selection in *Cepaea nemoralis*. In *Ecological Genetics and Evolution* (R. Creed, ed.). Blackwell Scientific Publications Ltd., London, pp. 93–108.

Cooke, D., and J. A. Leon, 1976. Stability of population growth determined by 2 × 2 Leslie matrix with density-dependent elements. *Biometrics* **32**:435–442.

Cooke, F., G. H. Finney, and R. F. Rockwell, 1976. Assortative mating in lesser snow geese (*Anser caerulescens*). *Behav. Genet.* **6**:127–140.

Cooke, G. D., 1967. The pattern of autotrophic succession in laboratory microcosms. *Bioscience* **17**:717–721.

Cooke, S. C. A., C. Lefebvre, and T. McNeilly, 1972. Competition between metal tolerant and normal plants on normal soil. *Evolution* **26**:366–372.

Coulson, J. C., and E. White, 1958. The effect of age on the breeding biology of the kittiwake, *Rissa tridactyla*. *Ibis* **100**:40–51.

Coulson, J. C., and R. D. Wooller, 1976. Differential survival rate among breeding kittiwake gulls, *Rissa tridactyla* (L). *J. Anim. Ecol.* **45**:205–213.

Cowie, R. J., 1977. Optimal foraging in great tits (*Parus major*). *Nature* **268**:137–139.

Cox, G. W., 1968. The role of competition in the evolution of migration. *Evolution* **22**:180–192.

Cox, P. A., 1981. Niche partitioning between sexes of dioecious plants. *Am. Nat.* **117**:295–307.

Cox, W. T., 1936. Snowshoe rabbit migration, tick infestation and weather cycles. *J. Mammal.* **17**:216–221.

Craig, D. M., 1982. Group selection versus individual selection: An experimental analysis. *Evolution* **36**:271–282.

Craig, R., 1980. Sex investment ratios in social hymenoptera. *Am. Nat.* **116**:311–323.

Cramer, N. F., and R. M. May, 1972. Interspecific competition, predation, and species diversity: a comment. *J. Theor. Biol.* **34**:289–293.

Creed, E. R., 1971. Industrial melanism in the two-spot ladybird and smoke abatement. *Evolution* **25**:290–293.

Crisp, D. J., 1961. Territorial behavior in barnacle settlement. *J. Exp. Biol.* **38**:429–446.

Crook, J. H., 1961. The adaptive significance of pair formation types in weaver birds. *Symp. Zool. Soc. Lond.* **8**:57–71.

Crook, J. H., 1964. The evolution of social organization and visual communication in the weaver birds (Ploceiae). *Behav. Suppl.* No. 10.

Crook, J. H., 1965. The adaptive significance of avian social organization. *Symp. Zool. Soc. Lond.* **14**:181–218.

Crow, J. F., 1979. Gene frequency and fitness change in an age-structured population. *Ann. Hum. Genet.* **42**:355–370.

Crow, J. F., and M. Kimura, 1970. *An Introduction to Population Genetics Theory*. Harper & Row, Publishers, New York.

Crowcroft, P., and F. P. Rowe, 1958. The growth of confined colonies of the wild house mouse (*Mus musculus* L.): the effect of dispersal on female fecundity. *Proc. Zool. Soc. Lond B* **131**:357–365.

Crowley, P. H., 1981. Dispersal and the stability of predator-prey interactions. *Am. Nat.* **118**:673–701.

Culver, D. C., 1970. Analysis of simple cave communities: niche separation and species packing. *Ecology* **51**:949–958.

Culver, D. C., 1973. Competition in spatially heterogeneous systems: an analysis of simple cave communities. *Ecology* **54**:102–110.

Culver, D. C., 1974. Species packing and Caribbean and North Temperate ant communities. *Ecology* **55**:974–988.

Culver, D. C., 1981. On using Horn's Markov succession model. *Am. Nat.* **117**:572–574.

Curry, J. D., R. W. Arnold, and M. A. Carter, 1964. Further examples of variation of populations of *Cepaea nemoralis* with habitat. *Evolution* **18**:111–117.

Daley, D. J., 1979. Bias in estimating the Malthusian parameter for Leslie matrices. *Theor. Popul. Biol.* **15**:257–263.

Darevsky, I. S., 1966. Natural parthenogenesis in a polymorphic group of Caucasian rock lizards related to *Lacerta saxicola*. *J. Ohio Herpetol. Soc.* **5**:115–152.

Darling, F., 1937. *A Herd of Red Deer*. American Museum of Natural History, Garden City, N.Y.

Darling, F., 1938. *Bird Flocks and the Breeding Cycle*. Cambridge University Press, Cambridge.

Darlington, P. J., 1943. Carabidae of mountains and islands: data on the evolution of isolated faunas, and on atrophy of wings. *Ecol. Monogr.* **13**:37–61.

Davidson, D. W., 1977. Species diversity and community organization in desert seed-eating ants. *Ecology* **58**:711–724.

Davidson, D. W., 1980. Some consequences of diffuse competition in a desert ant community. *Am. Nat.* **116**:92–105.

Davidson, J., 1938. On the ecology of the growth of the sheep population in South Australia. *Trans. R. Soc. S. Aust.* **62**:141–148, 342–346.

Davidson, J., and H. G. Andrewartha, 1948. The influence of rainfall, evaporation and atmospheric temperature on fluctuations in the size of a natural population of *Thrips imaginis* (Thysanoptera). *J. Anim. Ecol.* **17**:200–228.

Davis, D. E., 1940. Social nesting habits of the smooth-billed ani. *Auk* **57**:179–218.

Davis, D. E., 1952. Social behavior and reproduction. *Auk* **69**:171–182.

Davis, D. E., and J. T. Emlen, 1948. Studies on home range in the brown rat. *J. Mammal.* **29**:207–225.

Davis, E. F., 1928. The toxic principle of *Juglans nigra* as identified with synthetic juglone, and its toxic effects on tomato and alfalfa plants. *Am. J. Bot.* **15**:620.

Davis, J. W. F., and P. O'Donald, 1976. Estimation of assortative mating preferences in the arctic skua. *Heredity* **36**:235–244.

Dawson, P. S., 1967. Developmental rate and competitive ability in *Tribolium*. II. Changes in competitive ability following further selection for development rate. *Evolution* **21**:292–298.

Dawson, P. S., 1972. Evolution in mixed populations of *Tribolium Evolution* **26**:357–365.

Dawson, P. S., and I. M. Lerner, 1966. The founder principle and competitive viability of *Tribolium*. *Proc. Natl. Acad. Sci. USA* **55**:1114–1117.

Dayton, P. K., 1971. Competition, disturbance, and community organization: the provision and subsequent utilization of space in a rocky intertidal community. *Ecol. Monogr.* **41**:351–389.

DeAngelis, D. L., 1975. Stability and connectance in food web models. *Ecology* **56**:238–243.

Deevey, E. S., 1947. Life tables for natural populations of animals. *Q. Rev. Biol.* **22**:283–314.

DeJong, G., 1976. A model of competition for food. I. Frequency-dependent viabilities. *Am. Nat.* **110**:1013–1027.

Del Moral, R., and R. G. Cates, 1971. Allelopathic potential of the dominant vegetation of western Washington. *Ecology* **52**:1030–1037.

DeLong, K. T., 1966. Population ecology of feral house mice: interference by *Microtus*. *Ecology* **47**:481–484.

Demetrius, L., 1971. Primitivity conditions for growth matrices. *Math. Biosci.* **12**:53–58.

Demetrius, L., 1975a. Natural selection and age-structured populations. *Genetics* **79**:535–544.

Demetrius, L., 1975b. Reproductive strategies and natural selection. *Am. Nat.* **109**:243–249.

Demetrius, L., 1976. Measurements of variability in age-structured populations. *J. Theor. Biol.* **63**:397–404.

Deniston, C., 1978. An incorrect definition of fitness revisited. *Ann. Hum. Genet.* **42**:77–85.

Derr, J. A., 1980. The nature of variation in life history characters of *Dysdercus bimaculatus* (Heteroptera: Pyrrhocoridae), a colonizing species. *Evolution* **34**:548–557.

DeRuiter, L., 1952. Some experiments on the camouflage of stick caterpillars. *Behaviour* **4**:222–232.

DeSteven, D., 1978. The influence of age on the breeding biology of the tree swallow, *Iridoprocne bicolor*. *Ibis* **120**:516–523.

Diamond, J. M., 1970. Ecological consequences of island colonization by south west Pacific birds. I. Types of niche shifts. *Proc. Natl. Acad. Sci. USA* **67**:529–536.

Diamond, J. M., 1973. Distribution ecology of New Guinea birds. *Science* **179**:759–769.

Diamond, J. M., 1975. Assembly of species communities. In *Ecology and Evolution of Communities* (M. L. Cody and J. M. Diamond, eds.). Harvard University Press–Belknap Press, Cambridge, Mass., pp. 342–354.

Dice, L. R., 1947. Effectiveness of selection by owls of deer mice (*Peromyscus maniculatus*) with contrast in color with their background. *Contrib. Lab. Vert. Biol. Mich.* **34**:1–20.

Dingle, H., 1972. Migration strategies of insects. *Science* **175**:1327–1333.

Dingle, H., ed., 1978. *Evolution of Insect Migration and Diapause.* Springer-Verlag, New York.

Dittus, W. P. J., 1977. The social regulation of population density and age-sex distribution in the toque monkey. *Behaviour* **63**:281–322.

Dobzhansky, T., 1950. Evolution in the tropics. *Am. Sci.* **38**:209–221.

Dobzhansky, T., and O. Pavlovsky, 1957. An experimental study of interaction between genetic drift and natural selection. *Evolution* **11**:311–319.

Dobzhansky, T., and N. P. Spassky, 1962. Genetic drift and natural selection in experimental populations of *Drosophila pseudoobscura. Proc. Natl. Acad. Sci. USA* **48**:148–156.

Dodd, A. P., 1940. The biological campaign against prickly pear. *Comm. Prickly Pear Bd.*, Brisbane, Queensland.

Dodd, A. P., 1959. The biological control of prickly pear in Australia. *Monogr. Biol.* **8**:565–577.

Dodson, M. M., 1976. Darwin's law of natural selection and Thom's theory of catastrophes. *Math. Biosci.* **28**:243–274.

Dolinger, P. M., P. R. Ehrlich, W. L. Fitch, and D. E. Breedlove, 1973. Alkaloid and predation patterns in Colorado lupine populations. *Oecologia* **13**:191–204.

Drickamer, L. C., and B. M. Vestal, 1973. Patterns of reproduction in a laboratory colony of *Peromyscus. Evolution* **20**:174–184.

Dunaway, P. B., 1968. Life history and populational aspects of the eastern harvest mouse. *Amer. Midl. Nat.* **79**:48–67.

Dunham, A. E., 1980. An experimental study of interspecific competition between the iguanid lizards, *Sceloporus merriami* and *Urosaurus ornatus. Ecol. Monogr.* **50**:309–330.

Dunham, A. E., G. R. Smith, and J. N. Taylor, 1979. Evidence for ecological character displacement in western American catastomid fishes. *Evolution* **33**:877–896.

Dybas, H. S., and M. Lloyd, 1962. Isolation by habitat in two synchronized species of periodical cicadas (Homoptera:Cicadidae:Magicicada). *Ecology* **43**:444–459.

Ebersole, J. P., 1977. The adaptive significance of interspecific territoriality in the reef fish *Eupomacentrus leucostictus. Ecology* **58**:914–920.

Edmondson, W. T., 1945. Ecological studies of sessile Rotatoria. II. Dynamics of populations and social structure. *Ecol. Monogr.* **15**:141–172.

Ehrlich, P. R., and L. C. Birch, 1967. The "balance of nature" and "population control." *Am. Nat.* **101**:97–107.

Ehrlich, P. R., and P. H. Raven, 1965. Butterflies and plants: a study in coevolution. *Evolution* **18**:586–608.

Ehrman, L., 1970. The mating advantage of rare males in *Drosophila. Proc. Natl. Acad. Sci. USA* **65**:345–348.

Einhellig, F. A., and J. A. Rasmussen, 1973. Allelopathic effects of *Rumex crispus* on *Amaranthes retroflexus*, grain sorghum and field corn. *Am. Midl. Nat.* **90**:79–86.

Eisen, E. J., and B. H. Johnson, 1981. Correlated responses in male reproductive traits in mice selected for litter size and body weight. *Genetics* **99**:513–524.

Eisen, E. J., J. P. Hanrahan, and L. E. Legates, 1973. Effects of population size and selection intensity on correlated responses to selection for postweaning gain in mice. *Genetics* **74**:157–170.

Eisenberg, R. M., 1966. The regulation of density in a natural population of the pond snail, *Lymnaea elodes. Ecology* **47**:889–906.

Elner, R. W., and R. N. Hughes, 1978. Energy maximization in the diet of the shore crab, *Carcinus maenas. J. Anim. Ecol.* **47**:103–116.

Elton, C., and M. Nicholson, 1942. The ten-year cycle in numbers of lynx in Canada. *J. Anim. Ecol.* **11**:215–244.

Emlen, J. M., 1966a. The role of time and energy in food preference. *Am. Nat.* **100**:611–617.

Emlen, J. M., 1966b. Time, energy and risk in two species of carnivorous gasropods. Ph.D. dissertation, University of Washington, Seattle.

Emlen, J. M., 1968a. Biological and cultural determinants in human behavior. *Am. Anthropol.* **69**:513–514.

Emlen, J. M., 1968b. Optimal choice in animals. *Am. Nat.* **102**:385–389.

Emlen, J. M., 1970. Age specificity in ecological theory. *Ecology* **51**:588–601.

Emlen, J. M., 1973. *Ecology: An Evolutionary Approach.* Addison-Wesley Publishing Co., Inc., Reading, Mass.

Emlen, J. M., 1975. Niches and genes: some further thoughts. *Am. Nat.* **109**:472–476.

Emlen, J. M., 1980. A phenotypic model for the evolution of ecological characters. *Theor. Popul. Biol.* **17**:190–200.

Emlen, J. M., 1981. Field estimation of competition intensity. *Theor. Popul. Biol.* **19**:275–287.
Emlen, J. T., 1952. Social behavior in nesting cliff swallows. *Condor* **54**:177–199.
Emlen, J. T., 1957. Defended area? A critique of the territory concept and of conventional thinking. *Ibis* **99**:352.
Emlen, J. T., 1978. Density anomalies and regulatory mechanisms in land bird populations on the Florida peninsula. *Am. Nat.* **112**:265–268.
Emlen, J. T., 1979. Land bird densities on Baja California islands. *Auk* **96**:152–167.
Emlen, S. T., 1968. Territoriality in the bullfrog, *Rana catesbiana. Copeia* **1968**:240–243.
Emlen, S. T., 1978. The evolution of cooperative breeding in birds. In *Behavioral Ecology: An Evolutionary Approach* (J. R. Krebs and N. B. Davies, eds.). Sinauer Associates, Inc., Sunderland, Mass., pp. 245–281.
Emlen, S. T., 1981. Altruism, kinship and reciprocity in the white-fronted bee-eater. In *Natural Selection and Social Behavior: Recent Research and New Theory* (R. D. Alexander and D. W. Tinkle, eds.). Chiron Press, New York, pp. 217–230.
Emlen, S. T., and N. J. Demong, 1975. Adaptive significance of synchronized breeding in a colonial bird: a new hypothesis. *Science* **188**: 1029–1031.
Emlen, S. T., and L. W. Oring, 1977. Ecology, sexual selection, and the evolution of mating systems. *Science* **197**:215–223.
Endler, J. A., 1973. Gene flow and population differentiation. *Science* **179**:243–250.
Endler, J. A., 1977. *Geographic Variation, Speciation and Clines.* Princeton University Press, Princeton, N.J.
Erickson, A. B., 1949. Summer populations and movements of the cotton rat and other rodents on the Savanna River Refuge. *J. Mammal.* **30**: 133–140.
Errington, P. L., 1946. Predation and vertebrate populations. *Q. Rev. Biol.* **21**:144–177, 221–245.
Errington, P. L., 1956. Factors limiting higher vertebrate populations. *Science* **124**:304–307.
Errington, P. L., 1963. *Muskrat Populations.* Iowa State University Press, Ames.
Erwin, R. W., 1977. Foraging and breeding adaptations to different food regimes in three seabirds: the common tern, *Sterna hirundo*, royal tern, *Sterna maxima* and black skimmer. *Rynchops niger. Ecology* **58**:389–397.
Eshel, I., 1972. On the neighbor effect and the evolution of altruistic traits. *Theor. Popul. Biol.* **3**:258–277.
Eshel, I., 1975. Selection on sex ratio and the evolution of sex determination. *Heredity.* **34**: 351–361.
Estabrook, G. F., and A. E. Dunham, 1976. Optimal diet as a function of absolute abundance, relative abundance, and relative value of available prey. *Am. Nat.* **110**:401–413.
Estabrook, G. F., and D. C. Jesperson, 1974. Strategy for a predator encountering a model–mimic system. *Am. Nat.* **108**:443–457.
Estes, R. D., and J. Goddard, 1967. Prey selection and hunting behavior of the African wild dog. *J. Wildl. Manage.* **31**:57–70.

Fagen, R. M., 1972. An optimal life history strategy in which reproductive effort decreases with age. *Am. Nat.* **106**:258–261.
Fagen, R. M., 1980. When doves conspire: evolution of nondamaging fighting tactics in a nonrandom-encounter animal conflict model. *Am. Nat.* **115**:858–869.
Falconer, D. S., 1954. Asymmetrical responses in selection experiments. Symp. Genet. Popul. Struct., *Ist. Genet. Univ. Pavia, Italy, Un. Int. Sci. Biol.* No. 15, pp. 16–41.
Falconer, D. S., 1960a. *Introduction to Quantitative Genetics.* The Ronald Press Company, New York.
Falconer, D. S., 1960b. Selection of mice for growth on high and low planes of nutrition. *Genet. Res.* **1**:91–113.
Farentinos, R. C., P. J. Capretta, R. E. Kepner, and V. M. Littlefield, 1981. Selective herbivory in tassel-eared squirrels: role of monoterpenes in ponderosa pine chosen as feeding trees. *Science* **213**:1273–1275.
Feder, H. M., 1966. Cleaning symbioses in the marine environment. In *Symbiosis* (S. M. Henry ed.), Vol. 1. Academic Press, Inc., New York. pp. 327–380.
Feinsinger, P., 1976. Organization of a tropical guild of nectarivorous birds. *Ecol. Monogr.* **46**:257–291.
Feldman, M. W., and J. Roughgarden, 1975. A population's stationary distribution and chance of extinction in a stochastic environment, with remarks on the theory of species packing. *Theor. Popul. Biol.* **7**:197–207.
Feldman, M. W., I. Franklin, and G. J. Thomson, 1974. Selection in complex genetic systems. I. The

symmetric equilibria of the three-locus symmetric viability model. *Genetics* **76**:135–162.

Felsenstein, J., 1971. Inbreeding and variance effective number in populations with overlapping generations. *Genetics* **68**:581–597.

Felsenstein, I., 1975. A pain in. the torus: some difficulties with models of isolation by distance. *Am. Nat.* **109**:359–368.

Felsenstein, J., 1979. Excursions along the interface between disruptive selection and stabilizing selection. *Genetics* **93**:773–795.

Fenchel, T. M., and F. B. Christiansen, 1977. Selection and interspecific competition. In *Measuring Selection in Natural Populations*, (F. B. Christiansen and T. M. Fenchel, eds.), Lect. notes in Biomath., Vol. 19. Springer-Verlag, New York, pp. 477–498.

Fenner, F., 1965. Myxoma virus and *Oryctolagus cuniculus*: two colonizing species. In *The Genetics of Colonizing Species* (H. G. Baker and G. L. Stebbins, eds.). Academic Press, Inc., New York, pp. 485–499.

Fenner, F., I. D. Marshall, and C. M. Woodroffe, 1953. Studies on epidemiology of infectious myxomatosis in rabbits. I. Recovery of Australian wild rabbits (*Oryctolagus cuniculus*) myxomatosis under field conditions. *J. Hyg.* **51**:225–244.

Findlay, C. S., and F. Cooke, 1982. Synchrony in the lesser snowgoose (*Anser caerulescens*). II. The adaptive value of reproductive synchrony. *Evolution* **36**:786–799.

Finerty, J. P., 1979. Cycles in Canadian lynx. *Am. Nat.* **114**:453–455.

Fink, B., 1959. Observations of porpoise predation on a school of Pacific sardines. *Calif. Fish Game Bull.* **45**:216–217.

Fishelson, L., 1970. Protogynous sex reversal in the fish *Anthias squamipinnis* (Teleostei, Anthiidae) regulated by the presence or absence of a male fish. *Nature* **227**:90–91.

Fisher, H. I., 1975. The relationship between deferred breeding and mortality in the Laysan albatross. *Auk* **92**:433–441.

Fisher, R. A., 1958. *The Genetical Theory of Natural Selection*, 2nd ed. Dover Publications, Inc., New York.

Fisher, R. A., A. S. Corbet, and C. B. Williams, 1943. The relation between the number of species and the number of individuals in a random sample of an animal population. *J. Anim. Ecol.* **12**:42–58.

Fisler, G. F., 1962. Homing in the California vole, *Microtus californicus*. *Amer. Midl. Natur.* **68**: 357–368.

Flanders, S. E., 1957. The complete interdependence of an ant and a coccid. *Ecology* **38**:535–536.

Fleming, T. H., and R. J. Rauscher, 1978. On the evolution of litter size in *Peromyscus leucopus*. *Evolution* **32**:45–55.

Forcier, L. K., 1975. Reproductive strategies and the co-occurrence of climax tree species. *Science* **189**:808–810.

Ford, E. B., 1964. *Ecological Genetics*. Methuen and Co. Ltd., London.

Fowler, C. W., 1981. Density dependence as related to life history strategy. *Ecology* **62**:602–610.

Fox, J. F., 1977. Alternation and coexistence of tree species. *Am. Nat.* **111**:69–89.

Freeland, W. J., and D. H. Janzen, 1974. Strategies in herbivory by mammals: the role of plant secondary compounds. *Am. Nat.* **108**:269–289.

Fretwell, S. D., 1971. *Populations in a Seasonal Environment*. Princeton University Press, Princeton, N.J.

Fretwell, S. D., and H. I. Lucas, 1969. On territorial behavior and other factors influencing distribution in birds. I. Theoretical development. *Acta Biotheor.* **19**:16–36.

Friedman, H., 1967. Avian symbiosis. In *Symbiosis* (S. M. Henry, ed.), Vol. 2. Academic Press, Inc., New York, pp. 291–316.

Fry, C. H., 1971. The social organization of bee-eaters (Meropidae) and cooperative breeding in hot-climate birds. *Ibis* **114**:1–14.

Gadgil, M., 1971. Dispersal: population consequences and evolution. *Ecology* **52**:253–261.

Gadgil, M., 1975. Evolution of social behavior through interpopulation selection. *Proc. Natl. Acad. Sci. USA* **72**:1199–1201.

Gadgil, M., and W. H. Bossert, 1970. Life historical consequences of natural selection. *Am. Nat.* **104**:1–24.

Gadgil, M., and O. T. Solbrig, 1972. The concept of *r*- and *K*-selection: evidence from wild flowers and some theoretical considerations. *Am. Nat.* **106**:14–31.

Gaines, M. S., 1971. Genetic changes in fluctuating vole populations. *Evolution* **25**:702–723.

Gaines, M. S., J. H. Myers, and C. J. Krebs, 1971. Experimental analysis of relative fitness in transferrin genotypes of *Microtus ochrogaster*. *Evolution* **25**:443–453.

Gaines, M. S., A. M. Vivas, and C. L. Baker, 1979. An experimental analysis of dispersal in fluctuating vole populations: demographic parameters. *Ecology* **60**:814–828.

Galushin, V. M., 1974. Synchronous fluctuations in populations of some raptors and their prey. *Ibis* **116**:127–134.

Gardner, M. R., and W. R. Ashby, 1970. Connectance of large dynamical (cybernetic) systems: critical values of stability. *Nature* **228**:784.

Garnett, M. C., 1981. Body size, its heritability and influence on juvenile survival among great tits, *Parus major*. *Ibis* **123**:31–41.

Garten, C. T., 1976. Relationships between aggressive behavior and genic heterozygosity in the old field mouse, *Peromyscus polionotus*. *Evolution* **30**:59–72.

Gass, C. L., 1979. Territory regulation, tenure and migration in rufous hummingbirds. *Can. J. Zool.* **57**:914–923.

Gause, G. F., 1931. The influence of ecological factors on the size of populations. *Am. Nat.* **65**:70–76.

Gause, G. F., 1932. Experimental studies on the struggle for existence, I. Mixed populations of two species of yeast. *J. Exp. Biol.* **9**:389–402.

Gause, G. F., 1934. *The Struggle for Existence*. The Williams & Wilkins Company, Baltimore.

Gause, G. F., and A. A. Witt, 1935. Behavior of mixed populations and the problem of natural selection. *Am. Nat.* **69**:596–609.

Gause, G. F., N. P. Smaragdova, and A. A. Witt, 1936. Further studies of interaction between predator and prey. *Am. Nat.* **70**:1–18.

Gerking, S. D., ed., 1978. *Ecology of Freshwater Fish Production*. John Wiley & Sons, Inc., Philadelphia.

Getty, T., 1981. Territorial behavior of eastern chipmunks (*Tamias striatus*): encounter avoidance and spatial time sharing. *Ecology* **62**:915–921.

Ghent, A. W., 1966. Studies of the behavior of the *Tribolium* flour beetles. II. Distributions in depth of *T. castaneum* and *T. confusum* in fractionable shell vials. *Ecology* **47**:355–367.

Ghiselin, M. T., 1969. The evolution of hermaphroditism among animals. *Q. Rev. Biol.* **44**:189–208.

Gibb, J. A., 1954. Feeding ecology of tits, with notes on treecreeper and goldcrest. *Ibis* **96**:513–543.

Gibb, J. A., 1960. Populations of tits and goldcrests and their food supply in pine plantations. *Ibis* **102**:163–208.

Gibb, J. A., 1962. L. Tinbergen's hypothesis of the role of specific search images. *Ibis* **104**:106–111.

Giesel, J. T., 1972. Sex ratio, rate of evolution, and environmental heterogeneity. *Am. Nat.* **106**:380–387.

Giesel J. T., 1974. Fitness and polymorphism for net fecundity distribution in iteroparous populations. *Am. Nat.* **108**:321–331.

Giesel, J. T., 1976. Reproductive strategies as adaptations to life in temporally heterogeneous environments. *Annu. Rev. Ecol. Syst.* **7**:57–79.

Giesel, J. T., 1979. Genetic co-variation of survivorship and other fitness indices in *Drosophila melanogaster*. *Exp. Gerontol.* **14**:323–328.

Giesel, J. T., and E. E. Zettler, 1980. Genetic correlations of life historical parameters and certain fitness indices in *Drosophila melanogaster*: r_m, r_s, diet breadth. *Oecologia* **47**:299–302.

Giesel, J. T., P. A. Murphy, and M. N. Manlove, 1982. The influence of temperature on genetic interrelationships of life history traits in a population of *Drosophila melanogaster*: what tangled data sets we weave. *Am. Nat.* **119**:464–479.

Gilbert, L. E., and P. H. Raven, eds., 1975. *Coevolution of Animals and Plants*. University of Texas Press, Austin.

Gill, D. E., 1974. Intrinsic rate of increase, saturation density, and competitive ability. II. The evolution of competitive ability. *Am. Nat.* **108**:103–116.

Gillespie, J. H., 1973a. Natural selection with varying selection coefficients—a haploid model. *Genet. Res.* **21**:115–120.

Gillespie, J. H., 1973b. Polymorphism in random environments. *Theor. Popul. Biol.* **4**:193–195.

Gillespie, J. H., 1974. The role of grain in the maintenance of genetic variation. *Am. Nat.* **108**:831–836.

Gillespie, J. H., 1976. A general model to account for enzyme variation in natural populations. II. Characterization of the fitness function. *Am. Nat.* **110**:809–821.

Gillespie, J. H., 1981. The role of migration in the genetic structure of populations in temporally and spatially varying environments. III. Migration modification. *Am. Nat.* **117**:223–233.

Gilpin, M. E., 1973. Do hares eat lynx? *Am. Nat.* **107**:727–730.

Gilpin, M. E., 1975a. Limit cycles in competition communities. *Am. Nat.* **109**:51–60.

Gilpin, M. E., 1975b. *Group Selection in Predator–Prey Communities*. Princeton University Press, Princeton, N.J.

Gilpin, M. E., and F. J. Ayala, 1976. Schoener's model and *Drosophila* competition. *Theor. Popul. Biol.* **9**:12–14.

Gilpin, M. E., and T. J. Case, 1976. Multiple domains of attraction in competition communities. *Nature* **261**:40–42.

Gilpin, M. E., T. J. Case, and F. J. Ayala, 1976. θ-selection. *Math. Biosci.* **32**:131–139.

Gilpin, M. E., T. J. Case, and E. A. Bender, 1982. Counterintuitive oscillations in systems of competition and mutualism. *Am. Nat.* **119**:584–588.

Glasgow, J. P., 1963. *The Distribution and Abundance of Tsetse.* Pergamon Press, Inc., Oxford.

Glass, G. E., and N. A. Slade, 1980. The effect of *Sigmodon hispidus* on spatial and temporal activity of *Microtus ochrogaster*: evidence for competition. *Ecology* **61**:358–370.

Glasser, J. W., 1978. The effect of predation on prey resource utilization. *Ecology* **59**:724–732.

Goddard, S. V., and V. V. Board, 1967. Reproductive success of red-winged blackbirds in north-central Oklahoma. *Wilson Bull.* **79**:283–289.

Goh, B. S., 1977. Global stability in many-species systems. *Am. Nat.* **111**:135–143.

Golley, F. B., and J. B. Gentry, 1964. A comparison of variety and standing crop of vegetation on a one-year and a twelve-year abandoned field. *Oikos* **15**:185–199.

Golley, F. B., K. Petrusewicz, and L. Ryszkowski, 1975. *Small Mammals: Their Productivity and Population Dynamics.* Cambridge University Press, Cambridge.

Gooch, J. L., and T. J. M. Schorf, 1972. Genetic variability in the deep sea: relation to environmental variability. *Evolution* **26**:545–552.

Goodman, D., 1974. Natural selection and a cost ceiling on reproductive effort. *Am. Nat.* **108**:247–268.

Goodman, D., 1982. Optimal life histories, optimal notation, and the value of reproductive value. *Am. Nat.* **119**:803–823.

Goss-Custard, J. D., 1977a. Optimal foraging and the size selection of worms by redshank, *Tringa totanus*, in the field. *Anim. Behav.* **25**:10–29.

Goss-Custard, J. D., 1977b. The energetics of prey selection by redshank, *Tringa totanus* (L.) in relation to prey density. *J. Anim. Ecol.* **46**:1–19.

Goss-Custard, J. D., 1981. Feeding behavior of redshank, *Tringa totanus*, and optimal foraging theory. In *Foraging Behavior: Ecological, Ethological and Psychological Approaches* (A. C. Kamil and T. D. Sargent, eds.). Garland Publishing, Inc., New York, pp. 115–133.

Gould, S. J., and R. C. Lewontin, 1979. The spandrels of San Marco and the Panglossian paradigm: a critique of the adaptionist programme. *Proc. R. Soc. Lond. B* **205**:581–598.

Gowaty, P. A., 1983. Male parental care and apparent monogamy among eastern bluebirds (*Sialia sialis*), *Am. Nat.* **121**:149–157.

Grant, P. R., 1969. Experimental studies of competition interactions in a two-species system. I. *Microtus* and *Clethrionomys* species in enclosures. *Can. J. Zool.* **47**:1059–1082.

Grant, P. R., 1971. The habitat preference of *Microtus pennsylvanicus*, and its relevance to the distribution of this species on islands. *J. Mammal.* **52**:351–361.

Grant, P. R., 1975. The classical case of character displacement. *Evol. Biol.* **8**:237–337.

Grant, P. R., 1978. Dispersal in relation to carrying capacity. *Proc. Natl. Acad. Sci. USA* **75**:2854–2858.

Grant, P. R., and I. Abbott, 1980. Interspecific competition, island biogeography and null hypotheses. *Evolution* **34**:332–341.

Grant, P. R., and T. D. Price, 1981. Population variation in continuously varying traits as an ecological genetics problem. *Am. Zool.* **21**:795–812.

Grant, P. R., and D. Schluter. *Interspecific Competition Inferred from Patterns of Guild Structure.* Princeton University Press, Princeton, N.J. (in press).

Grant, P. R., B. R. Grant, J. N. M. Smith, I. J. Abbot, and L. K. Abbott, 1976. Darwin's finches: population variation and natural selection. *Proc. Natl. Acad. Sci. USA* **73**:257–261.

Grant, V., 1966. The selective origin of incompatibility barriers in the plant genus, *Gilia*. *Am. Nat.* **100**:99–118.

Greenberg, L., 1979. Genetic component of bee odor in kin recognition. *Science* **206**:1095–1097.

Greenwood, J. J. D., 1969. Apostatic selection and population density. *Heredity* **24**:157–161.

Greenwood, P. J., P. H. Harvey, and C. M. Perrins, 1978. Inbreeding and dispersal in the great tit. *Nature* **271**:52–54.

Greenwood, P. J., P. H. Harvey, and C. M. Perrins, 1979. The role of dispersal in the great tit (*Parus major*): the causes, consequences and heritability of natal dispersal. *J. Anim. Ecol.* **48**:123–142.

Grenney, W. J., D. A. Bella and H. C. Curl, Jr., 1973. A theoretical approach to interspecific competition in phytoplankton communities. *Am. Nat.* **107**:405–425.

Griffing, B., 1956. A generalized treatment of the use of diallel crosses in quantitative inheritance. *Heredity*. **10**:31–50.

Griffing, B., 1981a. A theory of natural selection incorporating interaction among individuals. I. The modelling process. *J. Theor. Biol.* **89**:635–658.

Griffing, B., 1981b. A theory of natural selection incorporating interaction among individuals. II. Use of related groups. *J. Theor. Biol.* **89**:659–677.

Griffing, B., 1981c. A theory of natural selection incorporating interaction among individuals. III. Use of random groups of inbred individuals. *J. Theor. Biol.* **89**:679–690.

Griffing, B., 1981d. A theory of natural selection incorporating interaction among individuals. IV. Use of related groups of inbred individuals. *J. Theor. Biol.* **89**:691–710.

Grime, J. P., 1977. Evidence for the existence of three primary strategies in plants and its relevance to ecological theory. *Am. Nat.* **111**:1169–1194.

Grodinski, W., and K. Sawicka, 1970. Energy values of tree-seeds eaten by small mammals. *Oikos* **21**:52–58.

Gross, M. R., and E. L. Charnov, 1980. Alternative male life histories in bluegill sunfish. *Proc. Natl. Acad. Sci. USA* **77**:6937–6940.

Gurney, W. S. C., and R. M. Nisbet, 1975. The regulation of inhomogeneous populations. *J. Theor. Biol.* **52**:441–457.

Gustavson, C. R., J. Garcia, W. G. Hankins, and K. W. Rusiniak, 1974. Coyote predation control by aversive conditioning. *Science* **184**:581–582.

Hagen, D. W., G. E. E. Moodie, and P. F. Moodie, 1980. Polymorphism for breeding colors in *Gasterosteus aculeatus*. II. Reproductive success as a result of convergence for threat display. *Evolution* **34**:1050–1059.

Hahn, D. C., 1981, Asynchronous hatching in the laughing gull: cutting losses and reducing rivalry. *Anim. Behav.* **29**:421–427.

Haigh, C. R., 1968. Sexual dimorphism, sex ratios and polygyny in the redwinged blackbird, Ph.D. dissertation, University of Washington, Seattle.

Haigh, J., and J. Maynard Smith, 1972. Can there be more predators than prey? *Theor. Popul. Biol.* **3**:290–299.

Hairston, N. G., F. Smith, and L. B. Slobodkin, 1960. Community structure, population control and competition. *Am. Nat.* **94**:421–425.

Hairston, N. G., J. D. Allan, R. K. Colwell, D. J. Futuyma, J. Howell, M. D. Lubin, J. Mathias, and J. H. Vandermeer, 1968. The relationship between species diversity and stability: an experimental approach with protozoa and bacteria. *Ecology* **49**:1091–1101.

Haldane, J. B. S., 1932. *The Causes of Evolution.* Longmans, Green & Company Ltd., London.

Haldane, J. B. S., 1941. *New Paths in Genetics.* Harper & Row, Publishers, New York.

Haldane, J. B. S., 1955. Population genetics. *New Biol.* **18**:34–51.

Haldane, J. B. S., and S. D. Jayakar, 1962. Polymorphism due to selection of varying direction. *J. Genet.* **58**:237–242.

Hallett, J. G., and S. L. Pimm, 1979. Direct estimation of competition. *Am. Nat.* **113**:593–600.

Halliday, T. R., 1978. Sexual selection and mate choice. In *Behavioral Ecology: An Evolutionary Approach* (J. R. Krebs and N. B. Davies, eds.). Sinauer Associates, Inc., Sunderland, Mass., pp. 180–213.

Hamilton, T. H., and G. H. Orians, 1965. The evolution of brood parasitism in altricial birds. *Condor* **67**:361–382.

Hamilton, T. H., and I. Rubinoff, 1963. Isolation, endemism and multiplication of species in the Darwin finches. *Evolution* **17**:388–403.

Hamilton, T. H., R. H. Barth, Jr., and I. Rubinoff, 1961. The environmental control of insular variation in bird species abundance. *Proc. Natl. Acad. Sci. USA* **52**:132–140.

Hamilton, W. D., 1963. The evolution of altruistic behavior. *Am. Nat.* **97**:354–356.

Hamilton, W. D., 1964. The genetical evolution of social behavior, I. *J. Theor. Biol.* **7**:1–16.

Hamilton, W. D., 1966. The moulding of senescence by natural selection. *J. Theor. Biol.* **12**:12–45.

Hamilton, W. D., 1967. Extraordinary sex ratios. *Science* **155**:477–488.

Hamilton, W. D., 1970. Selfish and spiteful behavior in an evolutionary model. *Nature* **228**:1218–1220.

Hamilton, W. D., 1971. Geometry for the selfish herd. *J. Theor. Biol.* **31**:295–311.

Hamilton, W. D., 1972. Altruism and related phenomena, mainly in social insects. *Annu. Rev. Ecol. Syst.* **3**:193–232.

Hamilton, W. D., 1975. Innate social aptitudes of man: an approach from evolutionary genetics. In

Biosocial Anthropology (R. Fox, ed.). John Wiley & Sons, Inc., New York, pp. 133–155.

Hamilton, W. D., and R. M. May, 1977. Dispersal in stable habitats. *Nature* **269**:578–581.

Hamilton, W. J., III, W. M. Gilbert, F. H. Heppner, and R. Planck, 1967. Starling roost dispersal and a hypothetical mechanism regulating rhythmical and animal movement to and from dispersal centers. *Ecology* **48**:825–833.

Hanrahan, J. P., E. J. Eisen, and L. E. Legates, 1973. Effects of population size and selection intensity on short-term responses to selection for postweaning gain in mice. *Genetics* **73**:513–530.

Hansen, S. R., and S. P. Hubbell, 1980. Single-nutrient microbial competition: qualitative agreement between experiment and theoretically forecast outcomes. *Science* **207**:1491–1493.

Hardin, G., 1960. The competitive exclusion principle. *Science* **131**:1292–1297.

Harding, J., R. W. Allard, and D. G. Smeltzer, 1966. Population studies on predominantly self-pollinated species. IX. Frequency-dependent selection in *Phaseolus lunatus*. *Proc. Natl. Acad. Sci. USA* **56**:99–104.

Harestad, A. S., and F. L. Bunnell, 1979. Home range and body weight—a reevaluation. *Ecology* **60**:389–402.

Harpending, H. C., 1979. The population genetics of interactions. *Am. Nat.* **113**:622–630.

Harper, J. L., 1969. The role of predation in vegetational diversity. In *Diversity and Stability in Ecological Systems*, Brookhaven Symp. Biol. No. 22, pp. 48–61.

Harper, J. L., 1977. *The Population Biology of Plants*. Academic Press, Inc., London.

Harper, J. L., and R. A. Benton, 1966. The behavior of seeds in soil, Part 2. The germination of seeds on the surface of a water supplying substrate. *J. Ecol.* **54**:151–166.

Harper, J. L., J. T. Williams, and G. R. Sagar, 1965. The behavior of seeds in soil. Part I. The heterogeneity of soil surfaces and its role in determining the establishment of plants from seed. *J. Ecol.* **53**:273–286.

Harrison, G. W., 1979. Stability under environmental stress: resistance, resilience, persistence and variability. *Am. Nat.* **113**:659–669.

Harrison, J. L., 1958. Range of movements of some Malayan rats. *J. Mammal.* **38**:190–206.

Hartl, D. L., 1980. *Principles of Population Genetics*. Sinauer Associates, Inc., Sunderland, Mass.

Hartl, D. L., and R. D. Cook, 1973. Balanced polymorphisms of quasineutral alleles. *Theor. Popul. Biol.* **4**:163–172.

Hartwick, E. B., 1976. Foraging strategy of the black oyster catcher (*Haematopus bachmani* Audubon). *Can. J. Zool.* **54**:142–155.

Harvey, P. H., 1976. Factors influencing the shell pattern polymorphism of *Cepaea nemoralis* (L) in east Yorkshire: a test case. *Heredity* **36**:1–10.

Harvey, P. H., and P. J. Greenwood, 1978. Anti-predator defence strategies: some evolutionary problems. In *Behavioral Ecology: An Evolutionary Approach* (J. R. Krebs and N. B. Davies, eds.). Sinauer Associates, Inc., Sunderland, Mass., pp. 129–154.

Hassell, M. P., 1978. *The Dynamics of Arthropod Predator–Prey Systems*. Princeton University Press, Princeton, N.J.

Hassell, M. P., and H. N. Comins, 1976. Discrete time models for two-species competition. *Theor. Popul. Biol.* **9**:202–221.

Hassell, M. P., and R. M. May, 1973. Stability in insect host–parasite models. *J. Anim. Ecol.* **42**:693–726.

Hassell, M. P., J. H. Lawton, and J. R. Beddington, 1976. The components of arthropod predation. I. The prey death rate. *J. Anim. Ecol.* **45**:135–164.

Hastings, A., 1977. Spatial heterogeneity and the stability of predator–prey systems. *Theor. Popul. Biol.* **12**:37–48.

Hastings, A., 1978. Global stabilities of two species systems. *J. Math. Biol.* **5**:399–403.

Hatch, J. J., 1966. Collective territories in Galápagos mockingbirds, with notes on other behavior. *Wilson Bull.* **78**:198–207.

Haven, S. B., 1973. Competition for food between the intertidal gastropods *Acmaea scabra* and *A. digitalis*. *Ecology* **54**:143–151.

Hay, D. A., 1973. Genotype–environmental interaction in the activity and preening of *Drosophila melanogaster*. *Theor. Appl. Genet.* **43**:291–297.

Hayman, B. I., 1954. The theory and analysis of diallel crosses. *Genetics* **39**:789–809.

Hayne, D. W., 1949. Calculation of the size of home range. *J. Mammal.* **30**:1–18.

Hayne, D. W., and R. C. Ball, 1956. Benthic productivity as influenced by fish predation. *Limnol. Oceanogr.* **1**:162–175.

Hazlett, B. A., 1968. Effects of crowding on the agonistic behavior of the hermit crab, *Pagurus bernhardus*. *Ecology* **49**:573–575.

Heath, D. J., 1975. Colour, sunlight and internal temperatures in the land-snail, *Cepaea nemoralis* (L). *Oecologia* **19**:29–38.

Heatwole, H., 1965. Some aspects of the association of cattle egrets and cattle. *Anim. Behav.* **13**:79–83.

Heatwole, H., and R. Levins, 1972. Trophic structure, stability and faunal change during recolonization. *Ecology* **53**:531–534.

Hegman, J. P., and J. C. DeFries, 1970. Are genetic correlations and environmental correlations correlated? *Nature* **226**:284–285.

Heinrich, B., 1978. The economics of insect sociality. In *Behavioral Ecology: An Evolutionary Approach* (J. R. Krebs and N. B. Davies, eds.). Sinauer Associates, Inc., Sunderland, Mass., pp. 97–128.

Heisler, I. L., 1981. Offspring quality and the polygyny threshold: a new model for the "sexy son" hypothesis. *Am. Nat.* **117**:316–328.

Heller, H. C., 1971. Altitudinal zonation of chipmunks (*Eutamias*): interspecific aggression. *Ecology* **52**:312–319.

Henry, J. D., and J. M. A. Swan, 1974. Reconstructing forest history from live and dead plant material: an approach to the study of forest succession in south west New Hampshire. *Ecology* **55**:772–783.

Henry, S. M., ed., 1966, 1967. *Symbiosis*, 2 vols. Academic Press, Inc., New York.

Herrenkohl, L. R., 1979. Prenatal stress reduces fertility and fecundity in offspring. *Science* **206**:1097–1099.

Heydweiler, A. M., 1935. A comparison of winter and summer territories and seasonal variation of the tree sparrow (*Spizella a. arborea*). *Bird-Banding* **6**:1–11.

Hickman, J. C., 1975. Environmental unpredictability and plastic energy allocation strategies in the annual *Polygonum cascadense* (Polygonaceae). *J. Ecol.* **63**:689–701.

Hilbert, D. W., D. M. Swift, J. K. Detling, and M. I. Dyer, 1981. Relative growth rates and the grazing optimization hypothesis. *Oecologia* **51**:14–18.

Hillborn, R., 1975. The effect of spatial heterogeneity on the persistence of predator–prey interactions. *Theor. Popul. Biol.* **8**:346–355.

Hillborn, R., 1979. Some long term dynamics of predator–prey models with diffusion. *Ecol. Model.* **6**:23–30.

Hines, W. G. S., and J. Maynard Smith, 1979. Games between relatives. *J. Theor. Biol.* **79**:19–30.

Hofmann, J. E., L. L. Getz, and B. J. Klatt, 1982. Levels of male aggressiveness in fluctuating populations of *Microtus ochrogaster* and *M. pennsylvanicus*. *Can. J. Zool.* **60**:898–912.

Hogsted, G., 1980. Evolution of clutch size in birds: adaptive variation in relation to territory quality. *Science* **210**:1148–1150.

Holgate, P., 1967. Population survival and life history phenomena. *J. Theor. Biol.* **14**:1–10.

Holling, C. S., 1965. The functional response of predators to prey density and its role in mimicry and population regulation. *Mem. Entomol Soc. Can. No. 45.*

Holling, C. S., 1966. The functional response of invertebrate prey to prey density. *Mem. Entomol Soc. Can.* **48**:1–86.

Holling, C. S., 1973. Resilience and stability of ecological systems. *Annu. Rev. Ecol. Syst.* **4**:1–24.

Holmes, J. C., 1961. Effects of concurrent infections on *Hymenolepis diminuta* (Cestoda) and *Moniliformis dubius* (Acanthocephala). I. General effects and comparison with crowding. *J. Parasitol* **47**:209–216.

Holmes, T. M., R. Aksel, and J. R. Royce, 1974. Inheritance of avoidance behavior in *Mus musculus*. *Behav. Genet.* **4**:357–371.

Horn, H. S., 1966. Colonial nesting in the brewers blackbird (*Euphagus cyanocephalus*) and its adaptive significance, Ph.D. dissertation, University of Washington, Seattle.

Horn, H. S., 1968. The adaptive significance of colonial nesting in the brewers blackbird, *Euphagus cyanocephalus*. *Ecology* **49**:682–694.

Horn, H. S., 1970. Social behavior of nesting brewers blackbirds. *Condor* **72**:15–23.

Horn, H. S., 1975. Markovian properties of forest succession. In *Ecology and Evolution of Communities* (M. L. Cody and J. M. Diamond, eds.). Harvard University Press–Belknap Press, Cambridge, Mass., pp. 196–211.

Horn, H. S., 1976. Succession. In *Theoretical Ecology: Principles and Applications* (R. M. May, ed.). W. B. Saunders Company, Philadelphia, Chapter 10.

Horn, H. S., 1978. Optimal tactics of reproduction and life history. In *Behavioral Ecology: An Evolutionary Approach* (J. R. Krebs and N. B. Davies, eds.). Sinauer Associates, Inc., Sunderland, Mass., pp. 411–429.

Horn, H. S., and R. H. MacArthur, 1972. Competition among fugitive species in a harlequin environment. *Ecology* **53**:749–752.

Horn, H. S., and R. M. May, 1977. Limits to similarity among coexisting species. *Nature* **270**:660–661.

Hornocker, M. G., 1970. An analysis of mountain lion predation upon mule deer and elk in the Idaho primitive area. *Wildl. Monogr.*, Vol. 21.

Howard, R. D., 1981. Male age-size distribution and male mating success in bullfrogs. In *Natural Selection and Social Behavior: Recent Research and New Theory* (R. D. Alexander and D. W. Tinkle, eds.). Chiron Press, New York, pp. 61–77.

Howard, W. E., 1974. *The Biology of Predator Control*, Addison-Wesley Modules Biol. No. 1. Addison-Wesley Publishing Co., Inc., Reading, Mass.

Howe, H. F., 1977. Sex-ratio adjustment in the common grackle. *Science* **198**:744–745.

Howe, H. F., 1978. Initial investment, clutch size, and brood reduction in the common grackle (*Quiscalus quiscala* L.). *Ecology* **59**:1109–1122.

Hrdy, S., and D. Hrdy, 1976. Hierarchical relations among female hanuman langurs (Primates: Colobinae, *Presbytis entellus*). *Science* **193**:913–915.

Hsu, S. B., S. P. Hubbell, and P. Waltman, 1978. A contribution to the theory of competing predators. *Ecol. Monogr.* **48**:337–349.

Hubbell, S. P., 1979. Tree dispersion, abundance, and diversity in a tropical dry forest. *Science* **203**:1299–1309.

Hubbell, S. P., and P. A. Werner, 1979. On measuring the intrinsic rate of increase of populations with heterogeneous life histories. *Am. Nat.* **113**:277–293.

Hubbs, C., 1958. Geographic variation in egg complement of *Percina caprodes* and *Etheostoma spectabile*. *Copeia* **1958**:102–105.

Huey, R. B., E. R. Pianka, M. E. Egan, and L. W. Coons, 1974. Ecological shifts in sympatry: Kalahari fossorial lizards (*Typhlosaurus*). *Ecology* **55**:304–316.

Huffaker, C. B., 1958. Experimental studies on predation: dispersion factors and predator–prey oscillations. *Hilgardia* **27**:343–383.

Hughes, A. L., 1983. Kin selection of complex behavioral strategies. *Am. Nat.* (in press).

Hunt, E. E., W. A. Lessa, and A. Hicking, 1965. The sex ratios of live births in three Pacific island populations. *Hum. Biol.* **37**:148–155.

Hurlbert, S. H., 1971. The non-concept of species diversity: a critique and alternate parameters. *Ecology* **52**:577–586.

Hurlbert, S. H., 1978. The measurement of niche overlap and some relatives. *Ecology* **59**:67–77.

Hutchinson, G. E., 1959. Homage to Santa Rosalia, or why are there so many kinds of animals? *Am. Nat.* **93**:145–159.

Inglis, J., M. M. Ankus, and D. H. Sykes, 1968. Age-related difference in learning and short-term memory from childhood to the senium. *Hum. Dev.* **11**:42–52.

Inouye, D. W., 1978. Resource partitioning in bumblebees: experimental studies of foraging behavior. *Ecology* **59**:672–678.

Islam, A. B. M. M., W. G. Hill, and R. B. Land, 1976. Ovulation rate of lines of mice selected for testis weight. *Genet. Res.* **27**:23–32.

Istock, C. A., J. Zisfein, and K. J. Vavra, 1976. Ecology and evolution of the pitcher-plant mosquito. 2. The substructure of fitness. *Evolution* **30**:535–547.

Ito, Y., and Y. Iwasa, 1981. Evolution of litter size. I. Conceptual reexamination. *Res. Popul. Ecol.* **23**:344–359.

Iverson, S. L., and B. N. Turner, 1972. Winter coexistence of *Clethrionomys gapperi* and *Microtus pennsylvanicus* in a grassland habitat. *Am. Midl. Nat.* **88**:440–445.

Ivlev, V. S., 1961. *Experimental Ecology of the Feeding of Fishes*. Yale University Press, New Haven, Conn.

Iwasa, Y., M. Higashi, and N. Yamamura, 1981. Prey distribution as a factor determining the choice of optimal foraging strategy. *Am. Nat.* **117**:710–723.

Jackson, F. J., 1938. *The Birds of Kenya Colony and the Uganda Protectorate*, 3 vols. Gurney and Jackson, London.

Jackson, J. A., 1970. A quantitative study of the foraging ecology of downy woodpeckers. *Ecology* **51**:318–323.

Jackson, J. B. C., and L. Buss, 1975. Allelopathy and spatial competition among coral reef invertebrates. *Proc. Natl. Acad. Sci. USA* **72**:5160–5163.

Jansson, C., J. E. Kinman, and A. VonBromssen, 1981. Winter mortality and food supply in tits, *Parus* spp. *Oikos* **37**:313–322.

Janzen, D. H., 1966. Coevolution of mutualism between ants and acacias in Central America. *Evolution* **20**:249–275.

Janzen, D. H., 1967. Fire, vegetation structure and the ant × acacia interaction in Central America. *Ecology* **48**:26–35.

Janzen, D. H., 1968. Allelopathy by Myrmecophytes: the ant *Azteca* as an allelopathic agent of *Cecropia Ecology* **50**:147–153.

Janzen, D. H., 1969. Seed-eaters versus seed size, number, toxicity and dispersal. *Evolution* **23**:1–27.

Janzen, D. H., 1976. Why bamboos wait so long to flower. *Annu. Rev. Ecol. Syst.* **7**:347–391.

Jarman, P. J., 1974. The social organization of antelope in relation to their ecology. *Behaviour* **48**:215–267.

Jarvis, M. J. F., 1974. The ecological significance of clutch size in the South African gannet, *Sula capensis* (Lichtenstein) *J. Anim. Ecol.* **43**:1–17.

Jeffries, C., 1974. Qualitative stability and digraphs in model ecosystems. *Ecology* **55**:1415–1419.

Jenkins, O., 1948. A population study of the meadow mice (*Microtus*) in three Sierra Nevada meadows. *Proc. Calif. Acad. Sci.* (4th ser.) **26**:43–67.

Jewell, P. A., 1966. The concept of home range in mammals. *Symp. Zool. Soc. Lond.* **18**:85–109.

Jinks, J. L., J. M. Perkins, and H. S. Pooni, 1973. The incidence of epistasis in normal and extreme environments. *Heredity* **31**:263–269.

Johnson, J. S., and W. B. Heed, 1976. Dispersal of desert-adapted *Drosophila*: the Saguaro-breeding *D. nigrospiracula*. *Am. Nat.* **110**:629–651.

Johnson, L., and S. P. Hubbell, 1975. Contrasting foraging strategies and coexistence of two bee species on a single resource. *Ecology* **56**:1398–1406.

Johnson, M. P., L. G. Mason, and P. H. Raven, 1968. Ecological parameters and plant species diversity. *Am. Nat.* **102**:297–306.

Johnson, M P., and P. H. Raven, 1973. Species number and endemism: the galápagos archipelago revisited. *Science* **179**:893–895.

Johnston, D. W., and E. P. Odum, 1956. Breeding bird populations in relation to plant succession on the piedmont of Georgia. *Ecology* **37**:50–62.

Johnston, R. F., 1954. Variation in breeding seasons and clutch size in song sparrows of the Pacific Coast. *Condor* **56**:268–273.

Jones, J. S., 1973a. Evolutionary genetics and natural selection in molluscs. *Science* **182**:546–552.

Jones, J. S., 1973b. The genetic structure of a southern, peripheral population of the snail, *Cepaea nemoralis*. *Proc. Ry. Soc. Lond. B* **183**:371–384.

Jones, J. S., 1973c. Ecological genetics of a population of the snail *Cepaea nemoralis* at the northern limit of its range. *Heredity* **31**:201–211.

Jorgenson, C. D., and W. W. Tanner, 1963. The application of the density-probability function to determine the home ranges of *Uta stansburiana s.* and *Cnemidophorus tigris t. Herpetol.* **19**:105–115.

Joule, J., and D. L. Jameson, 1972. Experimental manipulation of population density in three sympatric rodents. *Ecology* **53**:652–660.

Kacser, H., and J. A. Burns, 1981. The molecular basis of dominance. *Genetics* **97**:639–666.

Karban, R., 1982. Increased reproductive success at high densities and predator satiation for periodical cicadas. *Ecology* **63**:321–328.

Karlin, S., and J. MacGregor, 1974. Towards a general theory of the evolution of modifier genes. *Theor. Popul. Biol.* **5**:59–103.

Karr, J. R., 1968. Habitat and avian diversity on strip-mined land in east-central Illinois. *Condor* **70**:348–357.

Karr, J. R., and R. R. Roth, 1971. Vegetation structure and avain diversity in several new world areas. *Am. Nat.* **105**:423–435.

Kaston, B. J., 1965. Some little-known aspects of spider behavior. *Am. Midl. Nat.* **73**:336–356.

Kawai, M., 1965. Newly-acquired pre-cultural behavior of the natural troop of Japanese monkeys on the Koshima inlet. *Primates* **6**:1–30.

Kaye, S. V., 1961. Movements of harvest mice tagged with gold-198. *J. Mammal.* **42**:323–337.

Kearsey, M. J., and K. Kojima, 1967. The genetic architecture of body weight and egg hatchability in *Drosophila melanogaster*. *Genetics* **56**:23–37.

Keith, L. B., 1963. *Wildlife's Ten-Year Cycle*. University of Wisconsin Press, Madison.

Kempthorne, O., 1973. *An Introduction to Genetic Statistics*. Iowa State University Press, Ames.

Kempthorne, O., and E. Pollak, 1970. Concepts of fitness in Mendelian populations. *Genetics* **64**:125–145.

Kendeigh, S. C., 1970. Energy requirements for existence in relation to size of bird. *Condor* **72**:60–65.

Kettlewell, H. B. D., 1955a. Recognition of appropriate backgrounds by the pale and black phases of Lepidoptera. *Nature* **175**:943–944.

Kettlewell, H. B. D., 1955b. Selection experiments on industrial melanism in the Lepidoptera. *Heredity* **9**:323–342.

Kettlewell, H. B. D., 1956. Further selection experiments on industrial melanism in the Lepidoptera. *Heredity* **10**:287–301.

Kettlewell, H. B. D., 1958. A survey of the frequencies of *Biston betularia* (L) (Lep) and its melanic forms in Great Britain. *Heredity* **12**:51–72.

Kettlewell, H. B. D., 1961. The phenomenon of industrial melanism in Lepidoptera. *Annu. Rev. Entomol.* **6**:245–262.

Kettlewell, H. B. D., and R. J. Berry, 1969. Gene flow in a cline, *Amathes glareosa* Esp. and its melanic form. *Heredity* **24**:1–14.

Kettlewell, H. B. D., and D. L. T. Conn, 1977. Further background-choice experiments on cryptic Lepidoptera. *J. Zool. Lond.* **181**:371–376.

Keyfitz, N., 1968. *Introduction to the Mathematics of Populations.* Addison-Wesley Publishing, Co., Inc., Reading, Mass.

Kilham, L., 1965. Differences in feeding behavior of male and female hairy woodpeckers. *Wilson Bull.* **77**:134–145.

Kimura, M., 1965a. Attainment of quasi-linkage equilibrium when gene frequencies are changing by natural selection. *Genetics* **52**:875–890.

Kimura, M., 1965b. A stochastic model concerning the maintenance of genetic variability in quantitative characters. *Proc. Natl. Acad. Sci. USA* **54**:731–736.

Kimura, M., and J. F. Crow, 1963. The measurement of effective population number. *Evolution* **17**:279–288.

Kimura, M., and G. H. Weiss, 1964. The stepping stone model of population structure and the decrease of genetic correlation with distance. *Genetics* **49**:561–576.

King, C. E., and W. W. Anderson, 1971. Age-specific selection. II. The interaction of *r* and *K* during population growth. *Am. Nat.* **105**:137–156.

King, C. E., and P. S. Dawson, 1972. Population biology and the *Tribolium* model. *Evol. Biol.* **5**:133–227.

Kirby, G. C., 1975. Heterozygote frequencies in small populations. *Theor. Popul. Biol.* **8**:31–48.

Kleiman, D. G., 1981. Correlations among extreme forms of monogamy. pp 332–344. In *Natural Selection and Social Behavior: Recent Research and New Theory* (R. D. Alexander and D. W. Tinkle, eds.). Chiron Press, New York.

Klomp, H., 1972. Regulation of the size of bird populations by means of territorial behaviour. *Neth. J. Zool.* **22**:456–488.

Klopfer, P. H., and R. H. MacArthur, 1961. On the causes of tropical species diversity: niche overlap. *Am. Nat.* **95**:223–226.

Kluyver, H. N., and L. Tinbergen, 1953. Territory and regulation of density in titmice. *Arch. Neerl. Zool.* **10**:265–286.

Knight-Jones, E. W., 1953. Laboratory experiments on gregariousness during settling in *Balanus balanoides* and other barnacles. *J. Exp. Biol.* **30**:584–598.

Koch, A. L., 1974a. Coexistence resulting from an alternation of density-dependent and density-independent growth. *J. Theor. Biol.* **44**:373–386.

Koch, A. L., 1974b. Competitive coexistence of two predators utilizing the same prey under constant environmental conditions. *J. Theor. Biol.* **44**:387–395.

Kodric-Brown, A., and J. H. Brown, 1978. Influence of economics, interspecific competition, and sexual dimorphism on territoriality of migrant rufous hummingbirds. *Ecology* **59**:285–296.

Koenig, W. D., 1981. Reproductive success, group size, and the evolution of cooperative breeding in the acorn woodpecker. *Am. Nat.* **117**:421–443.

Koford, C. B., 1957. The vicuña and the puna. *Ecol. Monogr.* **27**:153–219.

Kojima, K., 1971. Is there a constant fitness value for a given genotype? No! *Evolution* **25**:281–285.

Kolman, B., 1980. *Introductory Linear Algebra with Applications*, 2nd ed. Macmillan Publishing Co., Inc., New York.

Koonce, J. F., T. P. Bagenal, R. F. Carline, K. E. F. Hokausen, and M. Nagiec, 1977. Factors influencing year-class strength of percids: a summary and a model of temperature effects. *J. Fish. Res. Bd. Can.* **34**:1900–1909.

Koplin, J. R., and R. S. Hoffman, 1968. Habitat overlap and competitive exclusion in voles (*Microtus*). *Am. Midl. Nat.* **80**:494–507.

Koskimies, J., 1950. The life of the swift, *Micropus apus* (L) in relation to the weather. *Ann. Acad. Sci. Fenn., Ser. A, IV: Biol.* **12**:1–151.

Krebs, C. J., 1970. *Microtus* population biology: behavior changes associated with the population cycles in *Microtus ochrogaster* and *Microtus pennsylvanicus*. *Ecology* **51**:34–52.

Krebs, C. J., and J. H. Myers, 1974. Population cycles in small mammals. *Adv. Ecol. Res.* **8**:267–399.

Krebs, C. J., B. L. Keller, and R. H. Tamarin, 1969. *Microtus* population biology: demographic changes in fluctuating populations of *Microtus ochrogaster* and *Microtus pennsylvanicus* in southern Indiana. *Ecology* **50**:587–607.

Krebs, C. J., M. S. Gaines, B. L. Keller, J. H. Myers, and R. H. Tamarin, 1973. Population cycles in small rodents. *Science* **179**:35–41.

Krebs, C. J., I. Wingate, L. LeDuc, J. A. Redfield, M. Taitt, and R. Hilborn, 1975. *Microtus* population biology: dispersal in fluctuating populations of *M. townsendii*. *Can. J. Zool.* **54**:79–95.

Krebs, J. R., 1971. Territory and breeding density in the great tit, *Parus major* L. *Ecology* **52**:2–22.

Krebs, J. R., 1974. Colonial nesting and social feeding as strategies for exploiting food resources in the great blue heron (*Ardea herodias*). *Behaviour* **51**:99–131.

Krebs, J. R., J. C. Ryan, and E. L. Charnov, 1974. Hunting by expectation or optimal foraging? A study of patch use by chickadees. *Anim. Behav.* **22**:953–964.

Krebs, J. R., J. T. Erickson, M. I. Webber, and E. L. Charnov, 1977. Optimal prey selection in the great tit (*Parus major*). *Anim. Behav.* **25**:30–38.

Kummer, H., 1968. *Social Organization of Hamadryas Baboons: A Field Study*. Bibl. Primatol. No. 6. The University of Chicago Press, Chicago.

Lack. D., 1948a. Selection and family size in starlings. *Evolution* **2**:95–110.

Lack, D., 1948b. The significance of clutch size. Part III. Some interspecific comparisons. *Ibis* **90**:25–45.

Lack, D., 1954. *The Natural Regulation of Animal Numbers*. Oxford University Press, Oxford.

Lack, D., 1966. *Population Studies of Birds*. Clarendon Press, Oxford.

Lack, D., 1968. *Ecological Adaptations for Breeding in Birds*. Methuen & Company Ltd., London.

Lack, D., and H. Arn, 1947. Die Bedeutung der Gelegegrosse beim Alpensegler. *Ornithol. Beob.* **44**:188–210.

Lack. D., and E. Lack, 1951. The breeding biology of the swift, *Apus apus*. *Ibis* **93**:501–546.

Lacy, R. C., 1982. Niche breadth and abundance as determinants of genetic variation in populations of mycophagous Drosophilid flies (Diptera: Drosophilidae). *Evolution* **36**:1265–1275.

Land, R. B., W. R. Carr, and G. J. Lee, 1980. A consideration of physiological criteria of reproductive merit in sheep. In *Selection Experiments in Laboratory and Domestic Animals* (A. Robertson, ed.). Commonwealth Agricultural Bureaux, Slough, Buckinghamshire, England, pp. 147–160.

Lande, R., 1975. The maintenance of genetic variability by mutation in a polygenic character with linked loci. *Genet. Res.* **26**:221–235.

Lande, R., 1977. The influence of the mating system on the maintenance of genetic variability in polygenic characters. *Genetics* **86**:485–498.

Lande, R., 1979. Quantitative genetic analysis of multivariate evolution, applied to brain: body size allometry. *Evolution* **33**:402–416.

Lande, R., 1980. The genetic covariance between characters maintained by pleiotropic mutations. *Genetics* **94**:203–215.

Lande, R., 1982. A quantitative genetic theory of life history evolution. *Ecology* **63**:607–615.

Larkin, P. A., 1973. Some observations on models of stock and recruitment relationships for fishes. *Rapp. Cons. Int. Explor. Mer* **164**:316–324.

Latter, B. D. H., 1970. Selection in finite populations with multiple alleles. II. Centripetal selection, mutation, and isoallelic variation. *Genetics* **66**:165–186.

Lawler, G. H., 1965. Fluctuations in the success of year-classes of white-fish populations with special reference to Lake Erie. *J. Fish. Res. Bd. Can.*, **22**:1197–1227.

Lawlor, L. R., 1978. A comment on randomly constructed model ecosystems. *Am. Nat.* **112**: 445–447.

Lawlor, L. R., 1979. Direct and indirect effects of *n*-species competition. *Oecologia* **43**:355–364.

Lawlor, L. R., 1980. Overlap, similarity and competition coefficients. *Ecology* **61**:245–251.

Lawlor, L R., and J. Maynard Smith, 1976. The coevolution and stability of competing species. *Am. Nat.* **110**:79–99.

Lawton, J. W., M. P. Hassell, and J. R. Beddington, 1975. Prey death rates and rate of increase of arthropod predator-populations. *Nature* **255**: 60–62.

Leamy, L., 1981. The effect of litter size on fertility in *Peromyscus leucopus*. *J. Mammal.* **62**:692–697.

Lee, J. J., and D. L. Inman, 1975. The ecological role of consumers: an aggregated system view. *Ecology* **56**:1455–1458.

Lefkovitch, L. P., 1965. The study of population growth in organisms grouped by stages. *Biometrics* **21**:1–18.

Leigh, E. G., 1975. Population fluctuations, community stability and environmental variability. In *Ecology and Evolution of Communities* (M. L. Cody and J. M. Diamond, eds.). Harvard

University Press–Belknap Press, Cambridge, Mass., pp. 51–73.
Law, R., 1983. A model for the dynamics of a plant population containing individuals classified by age and size. *Ecology* **64**:224–230.
Leigh, E. G., Jr., 1970. Sex ratio and differential mortality between the sexes. *Am. Nat.* **104**:205–210.
Lerner, I. M., and E. R. Dempster, 1951. Attenuation of genetic progress under continued selection in poultry. *Heredity* **5**:75–84.
Lerner, I. M., and E. R. Dempster, 1962. Indeterminism in interspecific competition. *Proc. Natl. Acad. Sci. USA* **48**:821–826.
Lesiewski, R. C., and W. R. Dawson, 1967. A reexamination between standard metabolic rate-body weight in birds. *Condor* **69**:13–23.
Leslie, P. H., 1945. On the use of matrices in certain population mathematics. *Biometrika* **33**:183–212.
Leslie, P. H., 1959. The properties of a certain lag type of population growth and the influence of an external, random factor on a number of such species. *Physiol. Zool.* **32**:151–159.
Levene, H., 1953. Genetic equilibrium when more than one ecological niche is available. *Am. Nat.* **87**:331–333.
Levin, B. R., and W. L. Kilmer, 1974. Interdemic selection and the evolution of altruism: a computer simulation study. *Evolution* **28**:527–545.
Levin, D. A., 1971. Plant phenolics: an ecological perspective. *Am. Nat.* **105**:157–182.
Levin, D. A., 1976. Alkaloid-bearing plants: an ecogeographic perspective. *Am. Nat.* **110**:261–284.
Levin, S. A., 1972. A mathematical analysis of the genetic feedback mechanism. *Am. Nat.* **106**:145–164.
Levin, S. A., 1974. Dispersion and population interactions. *Am. Nat.* **108**:207–228.
Levin, S. A., 1977. A more functional response to predator–prey stability. *Am. Nat.* **111**:381–383.
Levin, S. A., and D. Pimentel, 1981. Selection of intermediate rates of increase in parasite–host systems. *Am. Nat.* **117**:308–315.
Levin, S. A., and L. A. Segal, 1976. Hypothesis for origin of planktonic patchiness. *Nature*, **259**:659.
Levine, S. H., 1976. Competitive interactions in ecosystems. *Am. Nat.* **110**:903–910.
Levins, R., 1968. *Evolution in Changing Environments*. Princeton University Press, Princeton, N.J.
Levins, R., 1970. Extinction. In *Some Mathematical Questions in Biology*. American Mathematical Society, Providence, R.I., pp. 75–108.
Levins, R., 1975. Evolution in communities near equilibrium. In *Ecology and Evolution of Communities* (M. L. Cody and J. M. Diamond, eds.). Harvard University Press–Belknap Press, Cambridge, Mass., pp. 16–50.
Levins, R., and D. C. Culver, 1971. Regional coexistence of species and competition between rare species. *Proc. Natl. Acad. Sci. USA* **68**:1246–1248.
Levins, R., and R. H. MacArthur, 1966. The maintenance of genetic polymorphism in a spatially heterogeneous environment: variations on a theme by Howard Levene. *Am. Nat.* **100**:585–589.
Levinton, J., 1973. Genetic variation in a gradient of environmental variability: marine bavalvia (Mollusca). *Science* **180**:75–76.
Lewis, E. R., 1976. Application of discrete and continuous network theory to linear population models. *Ecology* **57**:33–47.
Lewontin, R. C., 1965. Selection for colonizing ability. In *The Genetics of Colonizing Species* (H. G. Baker and G. L. Stebbins, eds.). pp. 77–91. Academic Press, Inc., New York.
Lewontin, R. C., 1966. Evolution and the theory of games. *J. Theor. Biol.* **1**:382–403.
Lewontin, R. C., 1977. Fitness, survival and optimality. In *Analysis of Ecological Systems* (D. J. Horn, G. R. Stairs, and R. D. Mitchell, (eds.). Ohio State University Press, Columbus, pp. 3–21.
Lewontin, R. C., and M. J. D. White, 1960. Interaction between inversion polymorphism of two chromosome pairs in the grasshopper, *Moraba scurra*. *Evolution* **14**:116–129.
Lidicker, W. Z., Jr., 1962. Emigration as a possible mechanism permitting the regulation of population density below carrying capacity. *Am. Nat.* **96**:29–33.
Lidicker, W. Z., Jr., 1965. Comparative study of density-regulation in confined populations of four species of rodents. *Res. Popul. Ecol.* **7**:57–72.
Lidicker, W. Z., Jr., 1975. The role of dispersal in the demography of small mammals. In *Small Mammals: Their Productivity and Population Dynamics*, Int. Biol. Program 5 (F. B. Golley, K. Petrusewicz, and L. Ryszkowski, eds.). Cambridge University Press, Cambridge, pp. 103–128.
Ligon, J. D., 1968. Sexual differences in foraging

behavior in two species of *Dendrocopos* woodpeckers. *Auk* **85**:203–215.
Ligon, J. D., 1981. Demographic patterns and communal breeding in the green woodhoopoe, *Phoeniculus purpureus*. In *Natural Selection and Social Behavior: Recent Research and New Theory* (R. D. Alexander and D. R. Tinkle, eds.). Chiron Press, New York, pp. 231–243.
Lloyd, J. A., and J. J. Christian, 1967. Relationship of activity and aggression to density in two confined populations of house mice (*Mus musculus*). *J. Mammal.* **48**:262–269.
Lomnicki, A., 1971. Animal population regulation by the genetic feedback mechanism: a critique of the theoretical model. *Am. Nat.* **105**:413–421.
Lord, R. D., Jr., 1960. Litter size and latitude in North American mammals. *Am. Midl. Nat.* **64**:488–499.
Loschiavo, S. R., 1968. Effect of oviposition on the egg production and longevity in *Trogoderma parabile* (Coleoptera: Dermestidae). *Can. Entomol.* **100**:86–89.
Lotka, A. J., 1925. *Elements of Physical Biology* (reprinted in 1956 by Dover, Publications, Inc., New York).
Lott, D., S. D. Scholz, and D. L. S. Lehrman, 1967. Exteroceptive stimuli of the reproductive system of the female ring dove (*Streptopelia risoria*) by the mate and by the colony milieu. *Anim. Behav.* **15**:433–437.
Lubchenco, J., 1978. Plant species diversity in a marine intertidal community: importance of herbivore food preference and algal competitive abilities. *Am. Nat.* **112**:23–39.
Luckinbill, L. S., 1973. Coexistence in laboratory populations of *Paramecium aurelia* and its predator, *Didinium nasutum*. *Ecology* **54**:1320–1327.
Luckinbill, L. S., and M. M. Fenton, 1978. Regulation and environmental variability in experimental populations of protozoa. *Ecology* **59**: 1271–1276.
Lynch, C. B., and D. S. Sulzbach. Quantitative genetic analysis of temperature regulation in *Mus musculus*. II. Diallel analysis of individual traits. *Evolution*, in press.
Lynch, M., 1978. Complex interactions between natural coexploiters, *Daphnia* and *Ceriodaphnia*. *Ecology* **59**:552–564.
Lynch, M., 1979. Predation, competition, and zooplankton community structure: an experimental study. *Limol. Oceanogr.* **24**:253–272.
Lynch, M., 1980. Predation, enrichment and the evolution of cladoceran life histories: a theoretical approach. In *Evolution and Ecology of Zooplankton Communities* (W. C. Kerfoot, ed.). University Press of New England, Hanover, N. H., pp. 367–376.

MacArthur, R. H., 1955. Fluctuations of animal populations and a measure of community stability. *Ecology* **36**:533–536.
MacArthur, R. H., 1958. Population ecology of some warblers of northeast coniferous forests. *Ecology* **39**:599–619.
MacArthur, R. H., 1959. On the breeding distribution pattern of North American migrant birds. *Auk* **76**:318–325.
MacArthur, R. H., 1965. Ecological consequences of natural selection. In *Theoretical and Mathematical Biology* (T. H. Waterman and H. J. Morowitz, eds.). Blaisdell Publishing Company, New York.
MacArthur, R. H., 1968. The theory of the niche. In *Population Biology and Evolution* (R. Lewontin, ed.). Syracuse University Press, Syracuse, N.Y., pp. 159–176.
MacArthur, R. H., 1972. *Geographical Ecology: Patterns in the Distribution of Species*. Harper & Row, Publishers, New York.
MacArthur, R. H., and J. W. MacArthur, 1961. On bird species diversity. *Ecology* **42**:594–598.
MacArthur, R. H., and E. Pianka, 1966. On optimal use of a patchy environment. *Am. Nat.* **100**:603–609.
MacArthur, R. H., and E. O. Wilson, 1963. An equilibrium model of insular zoogeography. *Evolution* **17**:373–387.
MacArthur, R. H., and E. O. Wilson, 1967. *The Theory of Island Biogeography*. Princeton University Press, Princeton, N.J.
MacArthur, R. H., H. F. Recher, and M. L. Cody, 1966. On the relation between habitat selection and species diversity. *Am. Nat.* **100**:319–125.
MacLagan, D. S., 1932. The effect of population density upon rate of reproduction with special reference to insects. *Proc. R. Soc. Lond. B* **111**: 437–454.
MacLean, G. L., 1972. Clutch size and evolution in the Charadrii. *Auk* **89**:299–324.
MacLulich, D. A., 1937. Fluctuations in the numbers of the varying hare (*Lepus americanus*). *Univ. Toronto Stud. Biol. Ser. No. 43.*
McClintock, B., 1951. Chromosomic organization and genic expression. *Cold Spring Harbor Symp. Quant. Biol.* **16**:13–47.

McClintock, B., 1956a. Controlling elements and the gene. *Cold Spring Harbor Symp. Quant. Biol.* **21**:197–216.

McClintock, B., 1956b. Intranuclear systems controlling gene action and mutation. *Brookhaven Symp. Biol.* **8**:58–74.

M'Closkey, R. T., 1976. Community structure in sympatric rodents. *Ecology* **57**:728–739.

M'Closkey, R. T., 1982. The principle of equal opportunity: A test with desert rodents. *Can. J. Zool.* **60**:1968–1972.

McClure, P. A., 1981. Sex-biased litter reduction in food-restricted wood rats (*Neotoma floridana*). *Science* **211**:1058–1060.

McClure [Randolph], P. A., J. C. Randolph, K. Mattingly, and M. M. Foster, 1977. Energy costs of reproduction in the cotton rat, *Sigmodon hispidus. Ecology* **58**:31–45.

McGovern, and C. R. Tracy, 1981. Phenotypic variation in electromorphs previously considered to be genetic markers in *Microtus ochrogaster. Oecologia* **51**:276–280.

McKechnie, S. W., P. R. Ehrlich, and R. R. White, 1975. Population genetics of *Euphydras* butterflies. I. Genetic variation and the neutrality hypothesis. *Genetics* **81**:571–594.

McLaren, I. A., 1966. Adaptative significance of large size and long life of the Chaetognath, *Sagitta elegans*, in the Arctic. *Ecology* **47**:852–855.

McNab, B., 1963. Bioenergetics and the determination of home range size. *Am. Nat.* **97**:133–140.

McNair, J. N., 1979. A generalized model of optimal diets. *Theor. Popul. Biol.* **15**:159–170.

McNair, J. N., 1980. A stochastic foraging model with predator training effects. I. Functional response, switching, and run lengths. *Theor. Popul. Biol.* **17**:141–166.

McNair, J. N., 1982. Optimal giving-up times and the marginal value theorem. *Am. Nat.* **119**:511–529.

McNaughton, S. J., 1975. Serengeti migratory wildebeest: facilitation of energy flow by grazing. *Science* **191**:92–94.

McNaughton, S. J., 1979. Grazing as an optimization process: grass–ungulate relationships in the Serengeti. *Am. Nat.* **113**:127–703.

McPhail, J. D., 1969. Predation and the evolution of a stickleback (*Gasterosteus*). *J. Fish. Res. Bd. Can.* **26**:3183–3208.

Maiorana, V. C., 1978. An explanation of ecological and developmental constants. *Nature* **273**:375–377.

Maker, W. J., 1970. The pomarine jaeger as a brown lemming predator in northern Alaska. *Wilson Bull.* **82**:130–157.

Malecot, G. M., 1969. *The Mathematics of Heredity.* W. H. Freeman and Company, Publishers, San Francisco.

Maly, E. J., 1969. A laboratory study of the interaction between the predatory rotifer, *Asplanchna*, and *Paramecium. Ecology* **50**:59–73.

Manly, B. F. J., P. Miller, and L. M. Cook, 1972. Analysis of a selective predation experiment. *Am. Nat.* **106**:719–736.

Mares, M. A., and D. F. Williams, 1977. Experimental support for food particle size resource allocation in heteromyid rodents. *Ecology* **58**:1186–1190.

Margalev, R., 1958. Temporal successions and spatial heterogeneity in phytoplankton. In *Perspectives in Marine Biology*, (*A. A. Buzzati-Traverso, ed.*), Union Int. Sci. Biol. No. 27, pp. 323–350.

Margalev, R., 1963. On certain unifying principles in ecology. *Am. Nat.* **97**:357–374.

Margalev, R., 1968. *Perspectives in Ecological Theory.* The University of Chicago Press, Chicago.

Marten, G. G., 1973. An optimization equation for predation. *Ecology* **54**:92–101.

Massey, A., 1977. Agonistic aids and kinship in a group of pigtail macaques. *Behav. Ecol. Sociobiol.* **2**:31–40.

Matessi, C., and S. D. Jayakar, 1976. Conditions for the evolution of altruism under Darwinian selection. *Theor. Popul. Biol.* **9**:360–387.

Mather, K., 1953. Genetic control of stability in development. *Heredity* **7**:297–336.

Mather, K., 1961. Competition and cooperation. In *Mechanisms of Competition*, Symp. Soc. Exp. Biol. XV, Cambridge University Press, Cambridge, pp. 264–281.

Mattson, W. J., and N. D. Addy, 1976. Phytophagous insects as regulators of forest primary production. *Science* **190**:515–522.

Mauzey, K. P., 1966. Feeding behavior and reproductive cycles in *Pisaster ochraceus. Biol. Bull.* **131**:127–144.

May, R. M., 1972. Will a large system be stable? *Nature* **238**:413–414.

May, R. M., 1973a. *Stability and Complexity in*

Model Ecosystems. Princeton University Press, Princeton, N.J.

May, R. M., 1973b. Qualitative stability in model ecosystems. *Ecology* **54**:638–641.

May, R. M., 1974. Biological populations with nonoverlapping generations: stable points, stable cycles, and chaos. *Science* **186**:645–647.

May, R. M., 1975a. Patterns of species abundance and diversity. In *Ecology and Evolution of Communities* (M. L. Cody and J. M. Diamond, eds.). Harvard University Press–Belknap Press, Cambridge, Mass., pp. 81–120.

May, R. M., 1975b. Some notes on estimating the competition matrix, α. *Ecology* **56**:737–741.

May, R. M., and G. F. Oster, 1976. Bifurcations and dynamic complexity in simple ecological models. *Am. Nat.* **110**:573–599.

May, R. M., M. P. Hassell, R. M. Anderson, and D. W. Tonkyn, 1981. Density-dependence in host–parasitoid systems. *J. Anim. Ecol.* **50**:855–865.

Maynard Smith, J., 1964. Group selection and kin selection. *Nature* **201**:1145–1147.

Maynard Smith, J., 1965. The evolution of alarm calls. *Am. Nat.* **99**:59–63.

Maynard Smith, J., 1974. The theory of games and the evolution of animal conflicts. *J. Theor. Biol.* **47**:209–221.

Maynard Smith, J., 1977. Parental investment: a prospective analysis. *Anim. Behav.* **25**:1–9.

Maynard Smith, J., 1978. Optimization theory in evolution. *Annu. Rev. Ecol. Syst.* **9**:31–56.

Maynard Smith, J., 1981. Will a sexual population evolve to an ESS? *Am. Nat.* **117**:1015–1018.

Maynard Smith, J., and G. A. Parker, 1976. The logic of asymmetric contests. *Anim. Behav.* **24**:159–175.

Maynard Smith, J., and G. R. Price, 1973. The logic of animal conflict. *Nature* **246**:15–18.

Maynard Smith, J., and M. G. Ridpath, 1972. Wife sharing in the Tasmanian native hen, *Tribonyx mortierii*: a case of kin selection? *Am. Nat.* **106**:447–452.

Maynard Smith, J., and M. Slatkin, 1973. The stability of prey–predator systems. *Ecology* **54**:384–391.

Mech, L. D., 1977. Wolf pack buffer zones as prey reservoirs. *Science* **198**:320–321.

Medawar, P. B., 1952. *An Unsolved Problem in Biology*. H. K. Lewis & Company Ltd., London.

Medawar, P. B., 1957. *The Uniqueness of the Individual*. Methuen & Company Ltd., London.

Menge, B. A., 1972. Foraging strategy of a starfish in relation to actual prey availability and environmental predictability. *Ecol. Monogr.* **42**:25–50.

Meredith, D. H., 1977. Interspecific agonism in two parapatric species of chipmunks (*Eutamias*). *Ecology* **58**:423–430.

Mertz, D. B., 1971. Life history phenomena in increasing and decreasing populations. In *Statistical Ecology* (G. P. Patil, E. C. Pielou, and W. E. Waters, eds.). Vol. 2. The Pennsylvania State University Press, University Park, pp. 361–399.

Mertz, D. B., 1972. The *Tribolium* model and the mathematics of population growth. *Annu. Rev. Ecol. Syst.* **3**:51–78.

Metzgar, L. H., 1967. An experimental comparison of screech owl predation on resident and transient white-footed mice (*Peromyscus leucopus*). *J. Mammal.* **48**:387–391.

Metzgar, L. H., 1971. Behavioral population regulation in the woodmouse, *Peromyscus leucopus*. *Am. Midl. Nat.* **86**:434–448.

Meyerriecks, A. J., 1972. *Man and Birds: Evolution and Behavior*. Pegasus, The Bobbs Merrill Company, Inc., Indianapolis.

Michod, R. E., 1972. Genetical aspects of kin selection: effects of inbreeding. *J. Theor. Biol.* **81**:223–233.

Michod, R. E., and W. D. Hamilton, 1980. Coefficients of relatedness in sociobiology. *Nature* **288**:694–697.

Milham, S., Jr., and A. M. Gittelsohn, 1965. Parental age and malformations. *Hum. Biol.* **3**:13–22.

Milinski, M., and R. Heller, 1978. Influence of a predator on the optimal foraging behavior of sticklebacks (*Gasterosteus aculeatus*). *Nature* **275**:642–644.

Miller, R. S., 1964. Larval competition in *Drosophila melanogaster* and *D. simulans*. *Ecology* **45**:132–148.

Milton, K., 1979. Factors influencing leaf choice by howler monkeys: a test of some hypotheses of food selection by generalist herbivores. *Am. Nat.* **114**:362–378.

Mimura, M., and J. D. Murray, 1978. On a diffusive prey–predator model which exhibits patchiness. *J. Theor. Biol.* **75**:249–262.

Mitter, C., and D. J. Futuyma, 1979. Population genetic consequences of feeding habits in some forest lepidoptera. *Genetics* **92**:1005–1021.

Mode, C. J., 1969. Applications of generalized multi-type age-dependent branching processes in population genetics. *Bull. Math. Biophys.* **31**:575–589.

Monro, J., 1967. The exploitation and conservation of resources by populations of insects. *J. Anim. Ecol.* **36**:531–547.

Moore, N. W., 1964. Intra and interspecific competition among dragonflies. *J. Anim. Ecol.* **33**:49–71.

Moran, P. A. P., 1953a. The statistical analysis of the Canadian lynx cycle. I Structure and prediction. *Aust. J. Zool.* **1**:163–173.

Moran, P. A. P., 1953a. The statistical analysis of the Canadian lynx cycle. I. Structure and premeteorology. *Aust. J. Zool.* **1**:291–298.

Moreau, R. E., 1944. Clutch-size: a comparative study, with special reference to African birds. *Ibis* **86**:286–347.

Morgan, P., 1975. Selection acting directly on an enzyme polymorphism. *Heredity* **34**: 124–127.

Morisita, M., 1959. Measuring of the dispersion of individuals and analysis of the distribution patterns. *Mem. Fac. Sci. Kyushu Univ., Ser. E* (*Biol.*) **2**:215–235.

Moriwaki, D., 1940. Enhanced crossing over in the second chromosome of *Drosophila ananassae*. *Jap. J. Genet.* **16**:37–48.

Morrel, G. M., and J. R. G. Turner, 1970. Experiments on mimicry. I. The response of wild birds to artificial prey. *Behaviour* **36**:116–130.

Morrison, D. W., 1978. On the optimal searching strategy for refuging predators. *Am. Nat.* **112**: 925–934.

Morrow, P. A., and V. C. LaMarch, 1978. Tree ring evidence for chronic insect suppression of productivity in subalpine *Eucalyptus*. *Science* **201**: 1244–1246.

Morse, D. H., 1977. Resource partitioning in bumble bees: the role of behavioral factors. *Science* **197**:678–679.

Moynihan, M., 1955. Some aspects of reproductive behavior in the black-headed gull, *Larus ridibundus*, and related species. *Behaviour Suppl.* No. 4.

Mukai, T., and T. Yamazaki, 1971. The genetic structure of natural populations of *Drosophila melanogaster*. X. Development time and viability. *Genetics* **69**:385–389.

Muki, T., T. K. Watanabe, and O. Yamaguchi, 1977. The genetic structure of natural populations of *Drosophila melanogaster*. XII. Linkage disequilibrium in a large, local population. *Genetics* **77**:771–793.

Mulholland, R. J., 1975. Stability analysis of the response of ecosystems to perturbations. In *Ecosystem Analysis and Prediction* (S. A. Levin, ed.). *SIAM*, Philadelphia, pp. 166–188.

Muller, C. H., 1966. The role of chemical inhibition (allelopathy) in vegetational composition. *Bull. Torrey Bot. Club* **93**:332–351.

Murdoch, W. W., 1966. Community structure, population control and competition—a critique. *Am. Nat.* **100**:219–226.

Murdoch, W. W., 1969. Switching in general predators: experiments on predator specificity and stability of prey populations. *Ecol. Monogr.* **39**:335–354.

Murdoch, W. W., 1973. The functional response of predators. *J. Appl. Ecol.* **14**:335–341.

Murdoch, W. W., and J. R. Marks, 1973. Predation of coccinellid beetles: experiments on switching. *Ecology* **54**:160–167.

Murdoch, W. W., and A. Oaten, 1975. Predation and population stability. *Adv. Ecol. Res.* **9**:1–131.

Murdoch, W. W., F. C. Evans, and C. H. Peterson, 1972. Diversity and pattern in plants and insects. *Ecology* **53**:819–830.

Murie, A., 1944. The wolves of Mount McKinley. *U.S. Dept. Interior Faunal Ser. No. 5.*

Murphy, G. I., 1968. Pattern in life history and the environment. *Am. Nat.* **102**:391–403.

Murray, B. G., Jr., 1971. The ecological consequences of interspecific territorial behavior in birds. *Ecology* **52**:414–423.

Murray, B. G., Jr., 1976. A critique of interspecific territoriality and character convergence. *Condor* **78**:518–525.

Murray, K. F., 1965. Population changes during the 1957–1958 vole (*Microtus*) outbreak in California. *Ecology* **46**:163–171.

Myers, J. H., 1978. Sex ratio adjustment under food stress: maximization of quality or numbers of offspring? *Am. Nat.* **112**:381–388.

Myers, J. H., and C. J. Krebs, 1971. Sex ratios in open and enclosed vole populations: demographic implications. *Am. Nat.* **105**:325–344.

Myers, J. P., P. G. Connors, and F. A. Pitelka, 1981. Optimal territory size and the sanderling: compromises in a variable environment. In *Foraging Behavior: Ecological, Ethological and Psychological Approaches*, (A. C. Kamil and T. D. Sargent, eds.). Garland Publishing, Inc., New York, pp. 135–158.

Myrberg, A. A., Jr., and R. E. Thresher, 1974. Interspecific aggression and its relevance to the concept of territoriality in reef fishes. *Am. Zool.* **14**:81–96.

Nakano, T., 1981. Population regulation by dispersal over patchy environment. I. Regulation by the threshold dispersal. *Res. Pop. Ecol.* **23**:1–18.

Nassar, R., 1979. Frequency-dependent selection at the Lap locus in *Drosophila melanogaster*. *Genetics* **91**:327–338.

Nassar, R., H. J. Muhs, and R. D. Cook, 1973. Frequency-dependent selection at the Payne inversion in *Drosophila melanogaster*. *Evolution* **27**:558–564.

Naumov, N. P., 1940. The ecology of the hillock mouse, *Mus musculus hortulanus*. *J. Inst. Evol. Morphol.* **3**:33–77.

Naylor, A. F., 1959. An experimental analysis of dispersal in the flour beetle, *Tribolium confusum* (Tenebrionidae). *Ecology* **40**:453–464.

Nei, M., 1967. Modification of linkage intensity by natural selection. *Genetics* **57**:625–641.

Nei, M., 1972. Genetic distance between populations. *Am. Nat.* **106**:283–292.

Nei, M., and W-H. Li, 1973. Linkage disequilibrium in subdivided populations. *Genetics* **75**: 213–219.

Neill, R. L., and E. L. Rice, 1971. Possible role of *Ambrosia psilostachya* on pattern and succession in old fields. *Am. Midl. Nat.* **88**:344–357.

Neill, W. E., 1974. The community matrix and the interdependence of the competition coefficients. *Am. Nat.* **108**:399–408.

Nevo, E., T. Shimony, and M. Libni, 1977. Thermal selection of allozyme polymorphisms in barnacles. *Nature* **267**:699–701.

Newton, I., 1966. Fluctuations in the weights of bullfinches. *Br. Birds* **59**:89–100.

Newton, I., 1967. The adaptive radiation and feeding ecology of some British finches. *Ibis* **109**:33–98.

Neyman, J., T. Park, and E. L. Scott, 1956. The struggle for existence. The *Tribolium* model: biology and statistical aspects. In *Proc. Berkeley Symp. Math. Statist. Prob.*, 3rd ed., Vol. 4. University of California Press, Berkeley, pp. 41–79.

Nice, M. M., 1937. Studies in the life history of the song sparrow. I. *Trans. Linn. Soc. N.Y.* **4**:1–247.

Nice, M. M., 1957. Nesting success in altricial birds. *Auk* **74**:305–321.

Nicholson, A. J., 1933. The balance of animal populations. *J. Anim. Ecol.* **2**:132–178.

Nicholson, A. J., and V. A. Bailey, 1935. The balance of animal populations, Part I. *Proc. Zool. Soc. Lond.* **1935**:551–598.

Nicholson, S. A., and C. D. Monk, 1974. Plant species diversity in old field succession on the Georgia piedmont. *Ecology* **55**:1075–1085.

Nikolskii, G. V., 1963. *The Ecology of Fishes.* Academic Press, Inc., New York.

Nisbet, R. M., and W. S. C. Gurney, 1976. Population dynamics in a periodically varying environment. *J. Theor. Biol.* **56**:459–475.

Nolan, V., Jr., 1978. The ecology and behavior of the prairie warbler, *Dendroica discolor*. *Ornithol. Monogr. No. 26*. pp. 1–595.

Nolan, V., Jr., and C. F. Thompson, 1978. Egg volume as a predictor of hatchling weight in the brown-headed cowbird. *Wilson Bull.* **90**:353–358.

Noy Meir, I., 1975. Stability of grazing systems: an application of predator–prey graphs. *J. Ecol.* **63**:459–481.

Nunney, L., 1980a. The stability of complex model ecosystems. *Am. Nat.* **115**:639–649.

Nunney, L., 1980b. The influence of type 3 (sigmoid) functional response upon the stability of predator–prey difference models. *Theor. Popul. Biol.* **18**:257–278.

Oakeshott, J. G., J. B. Gibson, P. R. Anderson, W. R. Knibb, D. G. Anderson, and G. K. Chambers, 1982. Alcohol dehydrogenase and glycerol-3-phosphate dehydrogenase clines in *Drosophila melanogaster* on different continents. *Evolution* **36**:86–96.

Oaten, A., and W. W. Murdoch, 1975. Functional response and stability in predator–prey systems. *Am. Nat.* **109**:289–298.

O'Connor, R. J., 1978. Brood reduction in birds: selection for fratricide, infanticide and suicide? *Anim. Behav.* **26**:79–96.

O'Donald, P., N. S. Wedd, and J. W. F. Davis, 1974. Mating preferences and sexual selection in the Arctic skua. *Heredity* **33**:1–16.

O'Donald, P. O., and C. Pilecki, 1970. Polymorphic mimicry and natural selection. *Evolution* **24**:395–401.

Odum, E P., 1969. The strategy of ecosystem development. *Science* **164**:262–269.

Odum, E. P., C. E. Connell, and L. B. Davenport, 1964. Population energy flow of three primary consumer components of old-field ecosystems. *Ecology* **43**:88–95.

Oksanen, L., S. D. Fretwell, and O. Jarvinen, 1979. Interspecific aggression and the limiting similarity of close competition: the problem of size gaps in some community arrays. *Am. Nat.* **114**:117–129.

Orians, G. H., 1961a. Social stimulation within blackbird colonies. *Condor* **63**:330–337.

Orians, G. H., 1961b. The ecology of blackbird (*Agalaius*) social systems. *Ecol. Monogr.* **31**:285–312.

Orians, G. H., 1970. On the evolution of mating systems in birds and mammals. *Am. Nat.* **103**:589–603.

Orians, G. H., and G. M. Christman, 1968. A comparative study of the behavior of redwing, tricolor, and yellow-headed blackbirds. *Univ. Calif. Publ. Zool. No. 84.*

Orlove, M. J., 1975. A model of kin selection not invoking coefficients of relationship. *J. Theor. Biol.* **49**:289–310.

Oster, G., and Y. Takahashi, 1974. Models for age-specific interactions in a periodic environment. *Ecol. Monogr.* **44**:483–501.

Oster, G., and E. O. Wilson, 1978. *Caste and Ecology in the Social Insects.* Princeton University Press, Princeton, N.J.

Oster, G. E., I. Eshel, and D. Cohen, 1977. Worker–queen conflict and the evolution of social insects. *Theor. Popul. Biol.* **12**:49–85.

Paine, R. T., 1966. Food web complexity and species diversity. *Am. Nat.* **100**:65–76.

Paine, R. T., 1969. A note on trophic complexity and community stability. *Am. Nat.* **103**:91–93.

Paine, R. T., 1971. A short-term experimental investigation of resource partitioning in a New Zealand rocky intertidal habitat. *Ecology* **52**: 1096–1106.

Paine, R. T., 1976. Size-limited predation: an observational and experimental approach with the *Mytilus–Pisaster* interaction. *Ecology* **57**:858–873.

Paine, R. T., 1980. Food webs: linkage, interaction strengths and community infrastructure. *J. Anim. Ecol.* **49**:667–685.

Paine, R. T., and R. L. Vadas, 1969. The effects of grazing by sea urchins, *Strongylocentrotus franciscanus* and *S. purpuratus*, on benthic algal populations. *Limnol. Oceanogr.* **14**:710–719.

Paltridge, G. W., and J. V. Denholm, 1974. Plant yield and the switch from vegetative to reproductive growth. *J. Theor. Biol.* **44**:23–34.

Papageorgis, C., 1975. Mimicry in neotropical butterflies. *Am. Sci.* **63**:522–532.

Park, T., 1932. Studies in population physiology: the relation of numbers to initial population growth in the flour beetle, *Tribolium confusum* Duv. *Ecology* **13**:172–181.

Park, T., 1948. Experimental studies of interspecies competition. I. Competition between populations of the flour beetles, *Tribolium confusum* Duval and *Tribolium castaneum* Herbst. *Ecol. Monogr.* **18**:265–307.

Park, T., 1954. Experimental studies of interspecific competition. II. Temperature, humidity and competition in two species of *Tribolium. Physiol. Zool.* **27**:177–238.

Park, T., and M. Lloyd, 1955. Natural selection and the outcome of competition. *Am. Nat.* **89**:235–240.

Parker, G. A., 1974. Assessment strategy and the evolution of fighting behavior. *J. Theor. Biol.* **47**:223–243.

Parrish, B. B., and A. Saville, 1965. The biology of the north-east Atlantic herring populations. *Oceanogry Mar. Biol. Annu. Rev.* **3**:323–373.

Patten, B. C., 1975. The relation between sensitivity and stability. In *Ecosystem Analysis and Prediction* (S. A. Levin, ed.). *SIAM*, Philadelphia, pp. 141–143.

Patterson, C. B., and J. M. Emlen, 1979. Variation in nestling sex ratios in the yellow-headed blackbird. *Am. Nat.* **115**:743–747.

Patterson, I. J., 1965. Timing and spacing of broods in the black-headed gull, *Larus ridibundus. Ibis* **107**:433–459.

Pearl, R., 1927. The growth of populations. *Q. Rev. Biol.* **2**:532–548.

Pearl, R., 1940. *Introduction to Medical Biometrics and Statistics*, 3rd ed. W. B. Saunders Company, Philadelphia.

Pearson, O. P., 1966. The prey of carnivores during one cycle of mouse abundance. *J. Anim. Ecol.* **35**:217–233.

Pease, J. L., R. H. Vōwles, and L. B. Keith, 1979. Interactions of snowshoe hares and woody vegetation. *J. Wild. Manage.* **43**:43–60.

Peet, R. K., 1974. The measurement of species diversity. *Annu. Rev. Ecol. Syst.* **5**:285–308.

Pennycuick, C. J., R. M. Compton, and L. Beckingham, 1968. A computer model for simulating the growth of a population of two interacting populations. *J. Theor. Biol.* **18**: 316–329.

Perrill, S. A., H. C. Gerhardt, and R. Daniel, 1978. Sexual parasitism in the green tree frog (*Hyla cinerea*). *Science* **200**:1179–1180.

Perrins, C. M., 1963. Survival in the great tit, *Parus major* L. *Proc. XIII Int. Ornithol. Congr.*, pp. 717–728

Perrins, C. M., 1965. Population fluctuations and clutch size in the great tit, *Parus major* (L). *J. Anim. Ecol.* **34**:601–647.

Perrins, C. M., and P. J. Jones, 1974. The inheritance of clutch size in the great tit (*Parus major* L.). *Condor* **76**:225–229.

Peterman, R. M., 1977. A simple mechanism that causes collapsing stability regions in exploited salmonic populations. *J. Fish. Res. Bd. Can.* **34**:1130–1142.

Petit, C., 1958. Le déterminisme génétique et psychophysiologique de la compétition sexuelle chez *Drosophila melanogaster*. *Bull. Biol. Fr. Belg.* **92**:248–329.

Pianka, E R., 1966a. Latitudinal gradients in species diversity: a review of concepts. *Am. Nat.* **100**:33–46.

Pianka, E R., 1966b. Convexity, desert lizards, and spatial heterogeneity. *Ecology* **47**:1055–1059.

Pianka, E. R., 1970. On *r*- and *K*-selection. *Am.* Nat. **104**:592–597.

Pianka, E. R., 1972. *r* and *K* selection or *b* and *d* selection? *Am. Nat.* **106**:581–588.

Pianka, E. R., 1974a. *Evolutionary Ecology*, 2nd ed. Harper & Row, Publishers, New York.

Pianka, E. R., 1974b. Niche overlap and diffuse competition. *Proc. Natl. Acad. Sci. USA* **71**: 2141–2145.

Pianka, E. R., and W. S. Parker, 1975. Age-specific reproductive tactics. *Am. Nat.* **109**:453–464.

Pielou, E. C., 1966. Species diversity and pattern diversity in the study of ecological succession. *J. Theor. Biol.* **10**:370–383.

Pielou, E. C., 1969. *An Introduction to Mathematical Ecology*. Wiley–Interscience, New York.

Pielou, E. C., 1974. *Population and Community Ecology*. Gordon and Breach, Science Publishers, Inc., New York.

Pierotti, R., 1980. Spite and altruism in gulls. *Am. Nat.* **115**:290–300.

Pikitch, E. K. Depensatory mortality as an influence on western Lake Erie yellow perch: evidence from stock-recruitment relationships (in preparation).

Pilecki, C., and P. O'Donald, 1971. The effects of predation on artificial mimetic polymorphisms with perfect and imperfect mimics at varying frequencies. *Evolution* **25**:365–370.

Pimentel, D., 1968. Population regulation and genetic feedback. *Science* **159**:1432–1436.

Pimentel, D., and F. A. Stone, 1968. Evolution and population ecology of parasite–host systems. *Can. Entomol.* **100**:655–662.

Pimentel, D., W. P. Nagel, and J. L. Madden, 1963. Space–time structure of the environment and the survival of parasite–host systems. *Am. Nat.* **97**:141–167.

Pimentel, D., E. H. Feinberg, P. W. Wood, and J. T. Hayes, 1965. Selection, spatial distribution, and the coexistence of competing species. *Am. Nat.* **99**:97–109.

Pimentel, D., S. A. Levin, and A. B. Soans, 1975. On the evolution of energy balance in some exploited victim systems. *Ecology* **56**:381–390.

Pimm, S. L., and J. H. Lawton, 1977. Number of trophic levels in ecological communities. *Nature* **268**:329–331.

Pitelka, F. A., 1957. Some aspects of population structure in the short-term cycle of the brown lemming in northern Alaska. *Cold Spring Harbor Symp. Quant. Biol.* **22**:237–251.

Platt, A. P., R. P. Coppinger, and L. P. Brower, 1971. Demonstration of the selective advantage of mimetic *Limenitis* butterflies presented to caged avian predators. *Evolution* **25**:692–701.

Pleszczynska, W. K., 1978. Microgeographical prediction of polygyny in the lark bunting. *Science* **201**:935–937.

Polednak, A. P., 1976. Paternal age in relation to selected birth defects. *Hum. Biol.* **48**:727–739.

Pollak, E., 1974. The survival of a mutant gene and the maintenance of polymorphism in subdivided populations. *Am. Nat.* **108**:20–28.

Pollard, J. H., 1973. *Mathematical Models for the Growth of Human Populations*. Cambridge Univ. Press, London.

Poole, R. W., 1972. An autoregressive model of population density change in an experimental population of *Daphnia magna*. *Oecologia* **10**: 205–221.

Poole, R. W., 1976. Stochastic difference equation predictors of population fluctuations. *Theor. Popul. Biol.* **9**:25–45.

Poole, R. W., 1978. The statistical prediction of population fluctuations. *Annu. Rev. Ecol. Syst.* **9**:427–448.

Pope, J. A., D. H. Mills, and W. M. Schearer, 1961. The fecundity of the Atlantic salmon (*Salmo salar* Linn). *Freshw. Salmon Fish. Res.* **26**:1–12.

Post, W., 1974. Functional analysis of space and related behavior in the seaside sparrow. *Ecology* **55**:564–574.

Powell, J. R., 1971. Genetic polymorphisms in varied environments. *Science* **174**:1035–1036.

Powell, J. R., and C. E. Taylor, 1979. Genetic variation in ecologically diverse environments. *Am. Sci.* **67**:590–596.

Preston, F. W., 1948. The commonness and rarity of species. *Ecology* **29**:254–283.

Preston, F. W., 1962. The canonical distribution of commonness and rarity. Part I, *Ecology* **43**:185–215; Part II, *Ecology* **43**:410–432.

Price, D. J., and C. R. Bantock, 1975. Marginal populations of *Cepaea nemoralis* (L) on the Brendon Hills, England. II. Variation in chiasma frequency. *Evolution* **29**:278–286.

Price, M. V., 1978. The role of microhabitat in structuring desert rodent communities. *Ecology* **59**:910–921.

Price, P. W., 1973. Reproductive strategies in parasitoid wasps. *Am. Nat.* **107**:684–693.

Primack, R. B., 1979. Reproductive effort in annual and perennial species of *Plantago* (Plantaginaceae). *Am. Nat.* **114**:51–62.

Prus, T., 1966. Emigrational ability and surface numbers of adult beetles in twelve strains of *Tribolium confusum* Duval and *T. castaneum* Herbst (Coleoptera, Tenebrionidae). *Ekol. Pol. A* **14**:547–588.

Pulliam, H. R., 1973. On the advantages of flocking. *J. Theor. Biol.* **38**:419–422.

Pulliam, H. R., 1974. On the theory of optimal diets. *Am. Nat.* **108**:59–74.

Pulliam, H. R., 1975a. Diet optimization with nutrient constraints. *Am. Nat.* **109**:765–768.

Pulliam, H. R., 1975b. Coexistence of sparrows: a test of community theory. *Science* **189**:474–476.

Pulliam, H. R., 1983. Ecological community theory and the coexistence of sparrows. *Ecology* **64**:45–52.

Pulliam, H. R., and F. Enders, 1971. The feeding ecology of five sympatric finch species. *Ecology* **52**:557–566.

Pyke, G. H., 1978. Optimal foraging in hummingbirds: testing the marginal value theorem. *Am. Zool.* **18**:739–752.

Pyke, G. H., 1980. Optimal foraging in bumblebees: calculation of net rate of energy intake and optimal path choice. *Theor. Popul. Biol.* **17**:232–246.

Rapport, D. L., 1971. An optimization model of food selection. *Am. Nat.* **105**:575–587.

Rapport, D. L., 1980. Optimal foraging for complementary resources. *Am. Nat.* **116**:324–346.

Ratchcke, B. J., 1976. Competition and coexistence within a guild of herbivorous insects. *Ecology* **57**:76–87.

Real, L. A., 1979. Ecological determinants of functional response. *Ecology* **60**:481–485.

Real, L., 1980. Fitness, uncertainty, and the role of diversification in evolution and behavior. *Am. Nat.* **115**:623–635.

Real, L. A., 1981. Uncertainty and pollinator–plant interactions: the foraging behavior of bees and wasps on artificial flowers. *Ecology* **62**:20–26.

Recher, H. F., 1969. Bird species diversity and habitat diversity in Australia and North America. *Am. Nat.* **103**:75–80.

Reddingius, J., and P. J. den Boer, 1970. Simulation experiments illustrating stabilization of animal numbers by spreading of risk. *Oecologia* **5**:240–284.

Reeve, E. C. R., and F. W. Robertson, 1952. Studies in quantitative inheritance. II. Analysis of a strain of *Drosophila melanogaster* selected for long wings. *J. Genet.* **57**:276–316.

Reichman, O. J., and D. Oberstein, 1977. Selection of seed distribution types by *Dipodomys merriami* and *Perognathus amplus* in the laboratory. *Ecology* **58**:636–643.

Reller, A. W., 1972. Aspects of behavioral ecology of red-headed and red-bellied woodpeckers. *Am. Midl. Nat.* **88**:270–290.

Revusky, S. H., and E. W. Bedarf, 1967. Association of illness with prior ingestion of novel foods *Science* **155**:219–220.

Richardson, A. M. M., 1974. Differential climatic selection in natural populations of the land snail, *Cepaea nemoralis. Nature* **247**:572–573.

Richmond, R. C., M. E. Gilpin, S. P. Salas, and F. J. Ayala, 1975. A search for emergent competitive phenomena: the dynamics of multi-species *Drosophila* systems. *Ecology* **56**:709–714.

Richter, C. P., 1942. Total self-regulatory functions in animals and human beings. *Harvey Lect. Ser.* **38**:63–103.

Ricker, W. E., 1954. Stock and recruitment. *J. Fish. Res. Bd. Can.* **11**:624–651.

Ricker, W. E., 1958. Handbook of computations for biological statistics of fish populations. *J. Fish. Res. Bd. Can.* **119**:1–300.

Ricklefs, R. E., 1980. Geographic variation in

clutch size among passerine birds: Ashmole's hypothesis. *Auk* **97**:38–49.

Ricklefs, R. E., 1981. The optimization of life-history patterns under density-dependence. *Am. Nat.* **117**:403–408.

Robel, R. J., 1966. Booming territory size and mating success of the greater prairie chicken (*Tympanuchus cupido pinnatus*). *Anim. Behav.* **14**:328–331.

Roberts, A., 1974. The stability of a feasible random ecosystem. *Nature* **251**:607–608.

Robertson, A., 1955. Selection in animals. *Cold Spring Harbor Symp. Quant. Biol.* **20**:225–229.

Robertson, D. R., 1972. Social control of sex reversal in a coral-reef fish. *Science* **177**:1007–1009.

Rogers, D., 1972. Random search and insect population models. *J. Anim. Ecol.* **41**:367–383.

Rogers, J. S., 1972. Measurements of genetic similarity and genetic distance. *Stud. Genet. VI, Univ. Tex. Publ. 7213*, pp. 145–153.

Rohlf, F. J., and G. D. Schnell, 1971. An investigation of the isolation by distance model. *Am. Nat.* **105**:295–324.

Rohwer, S., S. D. Fretwell, and D. M. Niles, 1980. Delayed maturation in passerine plumages and the deceptive acquisition of resources. *Am. Nat.* **115**:400–437.

Root, R. B., 1967. The niche exploitation pattern of the blue-gray gnatcatcher. *Ecol. Monogr.* **37**:317–350.

Rose, M. R., and B. Charlesworth, 1981a. Genetics of life history in *Drosophila melanogaster*. I. Sib analysis of adult females. *Genetics* **97**:173–186.

Rose, M. R., and B. Charlesworth, 1981b. Genetics of life history in *Drosophila melanogaster*. II. Exploratory selection experiments. *Genetics* **97**:187–196.

Rosenzweig, M. L., 1969. Why the prey curve has a hump. *Am. Nat.* **103**:81–87.

Rosenzweig, M. L., 1971. Paradox of enrichment: destabilization of exploitation ecosystems in ecological time. *Science* **171**:385–387.

Rosenzweig, M. L., and R. H. MacArthur, 1963. Graphical representation and stability conditions of predator–prey interactions. *Am. Nat.* **97**:209–223.

Rosenzweig, M. L., and P. Sterner, 1970. Population ecology of desert rodent communities: body size and seed husking as bases for heteromyid coexistence. *Ecology* **51**:217–224.

Rosenzweig, M. L., and J. Winakur, 1969. Population ecology of desert rodent communities: habitats and environmental complexity. *Ecology* **50**:558–572.

Roth, V. L., 1981. Constancy in the size ratios of sympatric species. *Am. Nat.* **118**:394–404.

Rothstein, S. I., 1973. The niche-variation model—is it valid? *Am. Nat.* **107**:598–620.

Roughgarden, J., 1971. Density-dependent natural selection. *Ecology* **52**:453–468.

Roughgarden, J., 1975. A simple model for population dynamics in stochastic environments. *Am. Nat.* **109**:713–736.

Roughgarden, J., 1976. Resource partitioning among competing species—a coevolutionary approach. *Theor. Popul. Biol.* **9**:388–424.

Roughgarden, J., 1978. Influence of competition on patchiness in a random environment. *Theor. Popul. Biol.* **14**:185–203.

Roughgarden, J., and M. Feldman, 1975. Species packing and predation pressure. *Ecology* **56**:489–492.

Rowen, W., 1950. Canada's premier problem of animal conservation. *New Biol.* **9**:31–57.

Royama, T., 1966. Factors governing feeding rate, food requirement, and brood size of nestling great tits, *Parus major*. *Ibis* **108**:313–347.

Royama, T., 1970. Factors governing the hunting behaviour and food selection of the great tit (*Parus major* L.). *J. Anim. Ecol.* **39**:619–668.

Sabath, M. D., 1974. Niche breadth and genetic variability in sympatric natural populations of drosophilid flies. *Am. Nat.* **108**:533–540.

Sale, P. F., 1974. Overlap in resource use, and interspecific competition. *Oecologia* **17**:245–256.

Sale, P. F., 1977. Maintenance of high diversity in coral reef fish communities. *Am. Nat.* **111**:337–359.

Salisbury, E. J., 1942. *The Reproductive Capacities of Plants.* G. Bell & Sons Ltd., London.

Salt, G. W., 1974. Predator and prey densities as controls of the rate of capture by the predator, *Didinium nasutum*. *Ecology* **55**:431–439.

Sanders, H. L., 1968. Marine benthic diversity: a comparative study. *Am. Nat.* **102**:243–282.

Sanders, H. L., and R. R. Hessler, 1969. Ecology of the deep sea benthos. *Science* **163**:1419–1424.

Sargent, T. D., 1969. Behavior adaptations of cryptic moths. III. Resting attitude of two barb-like species: *Melanopla canadaria* and *Catocola ultronia*. *Anim. Behav.* **17**:670–672.

Sargent, T. D., 1981. Anti-predator adaptations of

underwing moths. In *Foraging Behavior: Ecological, Ethological and Psychological Approaches* (A. C. Kamil and T. D. Sargent, eds.). Garland Publishing, Inc., New York, pp. 259–284.

Sawyer, S., and M. Slatkin, 1981. Density independent fluctuations of population size. *Theor. Popul. Biol.* **19**:37–57.

Schaal, B. A., and D. A. Levin, 1976. The demographic genetics of *Liatris cylindracea* Michx (Compositae). *Am. Nat.* **110**:191–206.

Schaffer, W. M., 1974a. Selection for optimal life histories: the effects of age structure. *Ecology* **55**:291–303.

Schaffer, W. M., 1974b. Optimal reproductive effort in fluctuating environments. *Am. Nat.* **108**:783–790.

Schaffer, W. M., 1978. A note on the theory of reciprocal altruism. *Am. Nat.* **112**:250–253.

Schaffer, W. M., and M. Gadgil, 1975. Selection for optimal life histories in plants. In *Ecology and Evolution of Communities* (M. L. Cody and J. M. Diamond, eds.). Harvard University Press–Belknap Press, Cambridge, Mass., pp. 142–157.

Schaffer, W. M., and M. L. Rosenzweig, 1977. Selection for optimal life histories. II. Multiple equilibria and the evolution of alternative reproductive strategies. *Ecology* **58**:60–72.

Schaller, G. B., 1972. *The Serengeti Lion.* The University of Chicago Press, Chicago.

Schoener, T. W., 1967. The ecological significance of sexual dimorphism in size in the lizard *Anolis conspersus. Science* **155**:474–477.

Schoener, T. W., 1968a. The *Anolis* lizards of Bimini: resource partitioning in a complex fauna. *Ecology* **49**:704–726.

Schoener, T. W., 1968b. Sizes of feeding territories among birds. *Ecology* **49**:123–141.

Schoener, T. W., 1971. The theory of feeding strategies. *Annu. Rev. Ecol. Syst.* **2**:369–404.

Schoener, T. W., 1974a. Competition and the form of habitat shift. *Theor. Popul. Biol.* **6**:265–307.

Schoener, T. W., 1974b. Resource partitioning in ecological communities. *Science* **185**:27–39.

Schoener, T. W., 1976. Alternatives to Lotka–Volterra competition: models of intermediate complexity. *Theor. Popul. Biol.* **10**:309–333.

Schopf, T. J. M., and J. L. Gooch, 1971. Gene frequencies in a marine ectoproct: a cline in natural populations related to sea temperature. *Evolution* **25**:286–289.

Schopka, S. A., and G. Hempel, 1973. The spawning potential of populations of herring (*Clupea harengus* L.) and cod (*Gadus morhua* L.) in relation to the rate of exploitation. *Rapp. Cons. Int. Expl. Mer* **164**:178–185.

Schroder, G. D., and K. N. Geluso, 1975. Spatial distribution of *Dipodomys spectabilis* mounds. *J. Mammal.* **56**:363–368.

Seastedt, T. R., and S. F. MacLean, 1979. Territory size and composition in relation to resource abundance in Lapland longspurs breeding in arctic Alaska. *Auk* **96**:131–142.

Searcy, W. A., and K. Yasukawa, 1981. Sexual size dimorphism and survival of male and female blackbirds (Icteridae). *Auk* **98**:457–465.

Seaton, A. P. C., and J. Antonovics, 1967. Population interrelationships. I. Evolution in mixtures of *Drosophila* mutants. *Heredity* **22**:19–33.

Selander, R. K., 1965. On mating systems and sexual dimorphism. *Am. Nat.* **105**:400–437.

Selander, R. K., 1966. Sexual dimorphism and differential niche utilization in birds. *Condor* **68**:113–151.

Selander, R. K., 1970. Behavior and genetic variation in natural populations. *Am. Zool.* **10**:53–66.

Selander, R. K., and D. W. Kaufman, 1973. Genic variability and strategies of adaptation in animals. *Proc. Natl. Acad. Sci. USA* **70**:1875–1877.

Selander, R. K., and S. Y. Yang, 1970. Biochemical genetics and behavior in wild house mouse populations. In *Contributions to Behavior–Genetic Analysis: The Mouse as a Prototype* (G. Lindzey and D. D. Thiessen, eds.). Appleton-Century-Crofts, New York, pp. 293–334.

Selander, R. K., W. Grainger Hunt, and S. Y. Yang, 1969. Protein-polymorphism and genic heterozygosity in two European subspecies of the house mouse. *Evolution* **23**:379–390.

Shapiro, D. Y., 1980. Serial female sex changes after simultaneous removal of males from social groups of a coral reef fish. *Science* **209**:1136–1137.

Shelford, V. E., 1945. The relation of snowy owl migration to the abundance of the collared lemming. *Auk* **62**:592–596.

Sheppard, D. H., 1971. Competition between two chipmunk species (*Eutamias*). *Ecology* **52**:320–329.

Sheppard, P. M., 1951. Fluctuations in the selective value of certain phenotypes in the polymorphic land snail, *Cepaea nemoralis* (L.). *Heredity* **5**:125–134.

Sheppard, P. M., 1952. Natural selection in two colonies of the polymorphic land snail, *Cepaea nemoralis. Heredity* **6**:233–238.

Sheppard, P. M., 1959. The evolution of mimicry: a problem in ecology and genetics. *Cold Spring Harbor Symp. Quant. Biol.* **24**:131–140.

Sheppard, P. M., 1960. *Natural Selection and Heredity*. Harper & Row, Publishers, New York.

Sheppe, W., 1965. Island populations and gene flow in the deer mouse (*Peromyscus leucopus*). *Evolution* **19**:480–495.

Sherman, P. W., 1977. Nepotism and the evolution of alarm calls. *Science* **197**:1246–1253.

Sherman, P. W., 1981. Reproductive competition and infanticide in Beldings ground squirrels and other animals. In *Natural Selection and Social Behavior: Recent Research and New Theory* (R. D. Alexander and D. W. Tinkle, eds.). Chiron Press, New York, pp. 311–331.

Shroder, G. D., and M. L. Rosenzweig, 1975. Perturbation analysis of competition and overlap in habitat utilization between *Dipodomys ordii* and *Dipodomys merriami*. *Oecologia* **19**:9–28.

Shuter, B. J., and J. F. Koonce, 1977. A dynamic model of the western Lake Erie walleye (*Stizostedion vitreum vitreum*) population. *J. Fish. Res. Bd. Can.* **34**:1972–1982.

Siefert, R. P., and F. H. Siefert, 1976. A community matrix analysis of *Heliconia* insect communities. *Am. Nat.* **110**:461–483.

Siefert, R. P., and F. H. Siefert, 1979. A *Heliconia* insect community in a Venezuelan cloud forest. *Ecology* **60**:462–467.

Sih, A., 1980. Optimal behavior: can foragers balance two conflicting demands? *Science* **210**:1041–1043.

Simberloff, D. S., 1969. Experimental zoogeography of islands: a model for insular colonization. *Ecology* **50**:296–314.

Simberloff, D. and W. Boecklen, 1981. Santa Rosalia reconsidered: size ratios and competition. *Evolution* **35**:1206–1228.

Simberloff, D., and E. F. Connor, 1981. Missing species combinations. *Am. Nat.* **118**:215–239.

Simberloff, D. S., and E. O. Wilson, 1969. Experimental zoogeography and islands: the colonization of empty islands. *Ecology* **50**:278–296.

Simon, C. A., 1975. The influence of food abundance on territory size in the iguanid lizard, *Sceloporus jarrovi*. *Ecology* **56**:993–998.

Simpson, C. D., 1968. Reproduction and population structure in the greater kudu in Rhodesia. *J. Wildl. Manage.* **32**:149–161.

Simpson, E. H., 1949. Measurement of diversity. *Nature* **163**:688.

Simpson, N. E., 1969. Heritabilities of liability to diabetes when age and sex at onset are considered. *Ann. Hum Genet.* **32**:283–304.

Skutch, A. F., 1949. Do tropical birds rear as many young as they can nourish? *Ibis* **91**:430–455.

Skutch, A. F., 1954. Life histories of Central American birds. *Pac. Coast Avifauna* **31**:1–448.

Skutch, A. F., 1960. Life histories of Central American birds. *Pac. Coast Avifauna* **34**:1–593.

Skutch, A. F., 1961. Helpers among birds. *Condor* **63**:198–226.

Slatkin, M., 1970. Selection and polygenic characters. *Proc. Natl. Acad. Sci. USA* **66**:87–93.

Slatkin, M., 1974. Competition and regional coexistence. *Ecology* **55**:128–134.

Slatkin, M., 1978. On the equilibration of fitnesses by natural selection. *Am. Nat.* **112**:845–859.

Slatkin, M., 1979. Frequency and density-dependent selection on a quantitative character. *Genetics* **93**:755–771.

Slatkin, M., 1980. Character displacement. *Ecology* **61**:163–177.

Slatkin, M., and R. Lande, 1976. Niche width in a fluctuating environment—density-independent model. *Am. Nat.* **110**:31–55.

Slatkin, M., and M. J. Wade, 1978. Group selection on a quantitative character. *Proc. Natl. Acad. Sci. USA* **75**:3531–3534.

Slobodkin, L. B., 1954. Population dynamics in *Daphnia obtusa* Kurz. *Ecol. Monogr.* **24**:69–88.

Slobodkin, L. B., 1961. *The Growth and Regulation of Animal Numbers*. Holt, Rinehart and Winston, New York.

Slobodkin, L. B., 1964. Experimental populations of Hydrida. *J. Anim. Ecol.* **33** (Suppl.):131–148.

Slobodkin, L. B., and H. L. Sanders, 1969. On the contribution of environmental predictability to species diversity. In *Diversity and Stability in Ecological Systems*, Brookhaven Symp. Biol. No. 22, pp. 82–95.

Slobodkin, L. B., F. E. Smith, and N. Hairston, 1967. Regulation in terrestrial ecosystems, and the implied balance of nature. *Am. Nat.* **101**:109–124.

Smigel, B. W., and M. L. Rosenzweig, 1974. Seed selection in *Dipodomys merriami* and *Perognathus penicillatus*. *Ecology* **55**:329–339.

Smith, A. T., 1980. Temporal changes in insular populations of the pika (*Ochotona princeps*). *Ecology* **61**:8–13.

Smith, C. C., 1968. The adaptive nature of social organization in the genus of tree squirrels, *Tamiasciurus*. *Ecol. Monogr.* **38**:31–64.

Smith, C. C., and D. Follmer, 1972. Food preferences of squirrels. *Ecology* **53**:82–91.

Smith, C. C., and S. D. Fretwell, 1974. The optimal balance between size and number of offspring. *Am. Nat.* **108**:499–506.

Smith, F. E., 1961. Density-dependence in the Australian thrips. *Ecology* **42**:403–407.

Smith, J. N. M., and A. A. Dhondt, 1980. Experimental confirmation of heritable morphological variation in a natural population of song sparrows. *Evolution* **34**:1155–1158.

Smith, J. N. M., and H. P. A. Sweatman, 1974. Food searching behavior of titmice in patchy environments. *Ecology* **55**:1216–1232.

Smith, J. N. M., and R. Zach, 1979. Heritability of some morphological characters in a song sparrow population. *Evolution* **33**:460–467.

Smith, J. N. M., P. R. Grant, B. R. Grant, I. J. Abbott, and L. K. Abbott, 1978. Seasonal variation in feeding habits of Darwin's ground finches. *Ecology* **59**:1137–1150.

Smith, N. G., 1968. The advantage of being parasitized. *Nature* **219**:690–694.

Smith, R. H., and R. Mead, 1974. Age structure and stability in models of prey–predator systems. *Theor. Popul. Biol.* **6**:308–322.

Smith, S. M., 1971. The relationship of grazing cattle to foraging rates in anis. *Auk* **88**:876–881.

Smith-Gill, S. J., and D. E. Gill, 1978. Curvilinearities in the competition equations: an experiment with ranid tadpoles. *Am. Nat.* **112**:557–570.

Smouse, P. E., 1976. The implications of density-dependent population growth for frequency and density dependent selection. *Am. Nat.* **110**:849–860.

Smouse, P. E., and K. M. Weiss, 1975. Discrete demographic models with density-dependent vital rates. *Oecologia* **21**:205–218.

Snell, T. W., and C. E. King, 1977. Lifespan and fecundity patterns in rotifers: the cost of reproduction. *Evolution* **31**:882–890.

Soane, I. D., and B. Clarke, 1973. Evidence for apostatic selection by predators using olfactory cues. *Nature* **241**:62–63.

Sokal, R. R., 1969. Senescence and genetic load: evidence from *Tribolium*. *Science* **167**:1732–1734.

Sokal, R. R., and C. E. Taylor, 1976. Selection at two levels in hybrid populations of *Musca domestica*. *Evolution* **30**:509–522.

Solbrig, O. T., and B. B. Simpson, 1977. A garden experiment on competition between biotypes of the common dandelion (*Taraxacum officionale*). *J. Ecol.* **65**:427–430.

Soulé, M., and B. R. Stewart, 1970. The niche-variation hypothesis: a test and alternatives. *Am. Nat.* **104**:85–94.

Sousa, W. F., 1979. Experimental investigations of disturbance and ecological succession in a rocky intertidal algal community. *Ecol. Monogr.* **49**: 227–254.

Southern, H. N., 1970. The natural control of a population of tawny owls (*Strix aluco*). *J. Zool. Lond.* **162**:197–285.

Southwick, C. H., 1955. Regulatory mechanisms of house mouse populations: social behavior affecting litter survival. *Ecology* **36**:627–634.

Southwood, T. R. E., 1961. The number of species of insect associated with various trees. *J. Anim. Ecol.* **30**:1–8.

Southwood, T. R. E., and D. J. Cross, 1969. The ecology of the partridge. III. Breeding success and the abundance of insects in natural habitats. *J. Anim. Ecol.* **38**:497–509.

Spiess, E. B., 1970. Mating propensity and its genetic basis in *Drosophila*. In *Essays in Evolution and Genetics in Honor of Th. Dobzhansky* (M. K. Hecht and W. C. Steere, eds.). Appleton-Century-Crofts, New York, pp. 315–379.

Spiess, E. B., and J. F. Kruckeberg, 1980. Minority advantage of certain eye color mutants of *Drosophila melanogaster*. II. A behavior basis. *Am. Nat.* **115**:307–327.

Spight, T. M., and J. M. Emlen, 1976. Clutch sizes of two marine snails with a changing food supply. *Ecology* **57**:1162–1178.

Stark, A. E., 1976. Hardy–Weinberg law: asymmetric approach to a generalized form. *Science* **193**:1141–1142.

Stearns, S. C., 1976. Life history tactics: a review of the ideas. *Q. Rev. Biol.* **51**:3–47.

Stearns, S. C., 1977. The evolution of life history traits. *Annu. Rev. Ecol. Syst.* **8**:145–171.

Stearns, S. C., 1980. A new view of life history evolution. *Oikos* **35**:266–282.

Steele, J. H., 1974. Spatial heterogeneity and population stability. *Nature* **248**:83.

Stefanski, R. A., 1967. Utilization of the breeding territory in the black-capped chickadee. *Condor* **69**:259–267.

Steiner, W. W. M., 1977. Niche width and genetic variation in Hawaiian *Drosophila*. *Am. Nat.* **111**:1037–1045.

Stephens, G. C., 1968. Dissolved organic matter as

a potential source of nutrition for marine organisms. *Am. Zool.* **8**:95–106.

Sternburg, J. G., G. P. Waldbauer, and M. R. Jeffords, 1977. Batesian mimicry: selective advantage of color pattern. *Science* **195**:681–683.

Stewart, F. M., and B. R. Levin, 1973. Partitioning of resources and the outcome of interspecific competition: a model and some general considerations. *Am. Nat.* **107**:171–198.

Stickel, L. F., 1960. *Peromyscus* ranges at high and low population densities. *J. Mammal.* **41**:433–441.

Stickel, W. B., 1973. The reproductive physiology of the intertidal prosobranch, *Thais lamellosa* (Gmelin). I. Seasonal changes in the rate of oxygen consumption and body component indices. *Biol. Bull.* **144**:511–524.

Streifer, W., 1974. Realistic models in population ecology. *Adv. Ecol. Res.* **8**:199–266.

Strobeck, C., 1975. Selection in a fine-grained environment. *Am. Nat.* **109**:419–425.

Strong, D. R., Jr., L. A. Szyska, and D. S. Simberloff, 1979. Tests of community-wide character displacement against null hypotheses. *Evolution* **33**:897–913.

Stubbs, M., 1977. Density dependence in the life cycles of animals and its importance in *K*- and *r*-strategies. *J. Anim. Ecol.* **46**:677–688.

Stueck, K. L., and G. W. Barrett, 1978. Effects of resource partitioning on the population dynamics and energy utilization strategies of feral house mouse (*Mus musculus* M.) populations under experimental field conditions. *Ecology* **59**:539–551.

Sudd, J. H., 1967. *An Introduction to the Behavior of Ants.* St. Martin's Press, Inc., New York.

Sulzbach, D. S., 1980. Selection for competitive ability: negative results in *Drosophila. Evolution* **34**:431–436.

Sulzbach, D. S., and J. M. Emlen, 1979. Evolution of competitive ability in a mixture of *Drosophila melanogaster* populations with an initial asymmetry. *Evolution* **33**:1138–1149.

Sutherland, J. P., 1972. Energetics of high and low populations of the limpet, *Acmaea scabra* (Gould). *Ecology* **53**:430–437.

Svardson, G., 1948. Natural selection and egg number in fish. *Rep. Inst. Freshw. Res. Drottn.* **29**:115–122.

Svardson, G., 1949. Notes on spawning habits of *Leuciscus erythrophthalmus* (L), *Abramis brama* (L), and *Esox lucius* (L). *Rep. Inst. Freshw. Res. Drottn.* **32**:102–107.

Swenson, W. A., 1977. Food consumption of walleye (*Stizostedion vitreum vitreum*) and sauger (*S. canadense*) in relation to food availability and physical conditions in Lake of the Woods, Minnesota, Shagawa Lake and western Lake Superior. *J. Fish. Res. Bd. Can.* **34**:1643–1654.

Sykes, Z. M., 1969. On discrete stable population theory. *Biometrics* **25**:285–293.

Taber, R. D., and R. F. Dasmann, 1957. The dynamics of three natural populations of the deer, *Odocoilus hemionus columbianus. Ecology* **38**:233–246.

Tait, M. J., 1981. The effect of extra food on small rodent populations. I. Deermice (*Peromyscus maniculatus*). *J. Anim. Ecol.* **50**:111–124.

Talbot, M., and C. H. Kennedy, 1940. The slave-making ant, *Formica sanguinea subintegra,* Emery, its raids, nuptual flights and nest structure. *Ann. Entomol. Soc. Am.* **33**:560–577.

Tamarin, R. H., 1977. Dispersal in island and mainland voles. *Ecology* **58**:1044–1054.

Tamarin, R. H., and C. J. Krebs, 1969. Microtine population biology. II. Genetic changes at the transferrin locus in fluctuating populations of two vole species. *Evolution* **23**:187–211.

Tanner, J. T., 1975. The stability and the intrinsic growth rate of prey and predator populations. *Ecology* **56**:855–867.

Taylor, C. E., 1975. Genetic loads in heterogeneous environments. *Genetics* **80**:621–635.

Taylor, C. E., 1976. Genetic variation in heterogeneous environments. *Genetics* **83**:887–894.

Taylor, C. E., C. Cϕndra, M. Conconi, and M. Prout, 1981. Longevity and male mating success in *Drosophila pseudoobscura. Am. Nat.* **117**: 1035–1039.

Taylor, D. L., 1973. Some ecological implications of forest fire control in Yellowstone National Park, Wyoming. *Ecology* **54**:1394–1395.

Taylor, H. M., R. S. Gourley, C. E. Lawrence, and R. S. Caplan, 1974. Natural selection of life history attributes: an analytic approach. *Theor. Popul. Biol.* **5**:104–122.

Taylor, P. D., 1981. Intra-sex and inter-sex sibling interactions as sex ratio determinants. *Nature* **291**:64–66.

Taylor, P. D., and M. G. Bulmer, 1980. Local mate competition and the sex ratio. *J. Theor. Biol.* **86**:409–419.

Taylor, R. J., 1976. Value of clumping to prey and the evolutionary response of ambush predators. *Am. Nat.* **110**:13–29.

Temin, R. G., 1966. Homozygous viability and fertility loads in *Drosophila melanogaster. Genetics* **53**:27–56.

Templeton, A. R., 1979. A frequency-dependent model of brood selection. *Am. Nat.* **114**:515–524.

Templeton, A. R., 1982. Adaptation and the integration of evolutionary forces, pp. 15–31, in R. Milkman (ed.), *Perspectives on Evolution*, Sinauer, Sunderland, Mass.

Templeton, A. R., and D. A. Levin, 1979. Evolutionary consequences of seed pools. *Am. Nat.* **114**:232–249.

Terborgh, J., 1971. Distribution on environmental gradients: theory and a preliminary interpretation of distributional patterns in the avifauna of the Cordillera Vilcabamba, Peru. *Ecology* **52**:23–40.

Terborgh, J., and J. S. Weske, 1975. The role of competition in the distribution of Andean birds. *Ecology* **56**:562–576.

Thomas, W. R. and M. J. Pomerantz, 1981. Feasibility and stability in community dynamics. *Am. Nat.* **117**:381–385.

Thomas, W. R., M. J. Pomerantz, and M. E. Gilpin, 1980. Chaos, asymmetric growth and group selection for dynamical stability. *Ecology* **61**:1312–1320.

Tilman, D., 1976. Ecological competition between algae: experimental confirmation of resource-based competition theory. *Science* **192**:463–465.

Tinbergen, L., 1960. The natural control of insects in pine wood. I. Factors influencing the intensity of predation by songbirds. *Arch. Neerl. Zool.* **13**:265–343.

Tinbergen, N., 1956. On the functions of territories in gulls. *Ibis* **98**:401–411.

Tinbergen, N., M. Impekoven, and D. Frank, 1967. An experiment on spacing-out as a defense against predation. *Behaviour* **28**:307–321.

Tinkle, D. W., 1965. Home range, density, dynamics and structure of a Texas population of the lizard *Uta stansburiana*. In *Lizard Ecology: A Symposium* (W. W. Milstead, ed.). University of Missouri Press, Columbia, pp. 5–29.

Tinkle, D. W., 1969. The concept of reproductive effort and its relation to the evolution of life histories of lizards. *Am. Nat.* **103**:501–516.

Tipton, J., 1977. *An Exaltation of Larks, or The Venereal Game*. Penguin Books, New York.

Toro, M. A., 1982. Altruism and sex ratio. *J. Theor. Biol.* **95**:305–311.

Tošić, M., and F. J. Ayala, 1981. Density- and frequency-dependent selection at the Mdh-2 locus in *Drosophila melanogaster. Genetics* **97**:679–701.

Travis, C. C., W. M. Post, D. L. DeAngelis, and J. Perkowski, 1980. Analysis of compensatory Leslie matrix models for competing species. *Theor. Popul. Biol.* **18**:16–30.

Treisman, M., 1975a. Predation and the evolution of gregariousness. I. Models for concealment and evasion. *Anim. Behav.* **23**:779–800.

Treisman, M., 1975b. Predation and the evolution of gregarious. II. An economic model for predator–prey interactions. *Anim. Behav.* **23**:801–825.

Treisman, M., 1981. Evolutional limits to the frequency of aggression between related or unrelated conspecifics in diploid species with simple Mendelian inheritance. *J. Theor. Biol.* **93**:97–124.

Trivers, R. L., 1971. The evolution of reciprocal altruism. *Q. Rev. Biol.* **46**:35–57.

Trivers, R. L., 1972. Parental investment and sexual selection. In *Sexual Selection and the Descent of Man, 1871–1971* (B. Campbell, ed.). Aldine Publishing Company, Chicago, pp. 136–179.

Trivers, R. L., and D. E. Willard, 1973. Natural selection of parental ability to vary the sex ratio of offspring. *Science* **191**:90–92.

Tuljapurkar, S. D., and S. H. Orzack, 1980. Population dynamics in variable environments. I. Long-run growth rates and extinction. *Theor. Popul. Biol.* **18**:314–342.

Turelli, M., 1977. Random environments and stochastic calculus. *Theor. Popul. Biol.* **12**:140–178.

Turelli, M., 1978. A reexamination of stability in randomly varying versus determinist environments with comments on the stochastic theory of limiting similarity. *Theor. Popul. Biol.* **13**:244–267.

Turelli, M., and D. Petry, 1980. Density-dependent selection in a random environment: an evolutionary process that can maintain stable population dynamics. *Proc. Natl. Acad. Sci. USA* **77**:7501–7505.

Turner, F. B., R. I. Jennrich, and J. D. Weintraub, 1969. Home ranges and body size of lizards. *Ecology* **50**:1076–1081.

Turner, J. R. G., 1977. Butterfly mimicry: the genetical evolution of an adaptation. *Evol. Biol.* **10**:163–206.

Udovic, J. D., D. Pimentel, and D. Nafus, 1976. The interactions between spatial heterogeneity and genetic feedback in laboratory predator–prey systems. *Oecologia* **25**:23–34.

Ueckert, D. N., and R. M. Hansen, 1972. Dietary overlap of grasshoppers on sandhill rangeland in north-eastern Colorado. *Oecologia* **8**:276–295.

United Nations, 1974. *Demographic Year Book.* United Nations, New York.

Usher, M. B., 1972. Developments in the Leslie-matrix model. In *Mathematical Models in Ecology* (J. N. R. Jeffers, ed.). Blackwell Scientific Publications Ltds., pp. 29–60.

Utida, S., 1955. Population fluctuations in the system of host–parasite interaction. *Mem. Coll. Agric. Kyoto Univ.* **71**:1–34.

Uyenoyama, M., and B. O. Bengtsson, 1979. Towards a genetic theory for the evolution of the sex ratio. *Genetics* **93**:721–736.

Uyenoyama, M., and M. W. Feldman, 1980. Theories of kin and group selection: a population genetics perspective. *Theor. Popul. Biol.* **17**:380–414.

Vance, R. R., 1980. The effect of dispersal on population size in a temporally varying environment. *Theor. Popul. Biol.* **18**:343–362.

Vandermeer, J. H., 1969. The competitive structure of communities: an experimental approach with protozoa. *Ecology* **50**:362–371.

Vandermeer, J. H., 1970. The community matrix and the number of species in a community. *Am. Nat.* **104**:73–83.

Vandermeer, J. H., 1972a. On the covariance of the community matrix. *Ecology* **53**:187–189.

Vandermeer, J. H., 1972b. Niche theory. *Annu. Rev. Ecol. Syst.* **3**:107–132.

Vandermeer, J. H., 1981. A further note on a community models. *Am. Nat.* **117**:379–380.

Van Hyning, J. M., 1973. Stock recruitment relationships for Columbia River Chinook salmon. *Rapp. Cons. Int. Explor. Mer* **164**:89–97.

Van Noordwijk, A. J., and W. Scharloo, 1981. Inbreeding in an island population of the great tit. *Evolution* **35**:674–688.

Van Valen, L., 1971. Group selection and the evolution of dispersal. *Evolution* **25**:591–598.

Vaurie, C., 1951. Adaptive differences between two sympatric species of nuthatches (*Sitta*). *Proc. 10th Int. Ornithol. Congr.*, pp. 163–166.

Vehrencamp, S. L., 1977. Relative fecundity and parental effort in communally nesting anis, *Crotophaga sulcirostris. Science* **197**:403–405.

Verner, J., 1964. Evolution of polygamy in the long-billed marsh wren. *Evolution* **18**:252–261.

Verner, J., and G. H. Engelson, 1970. Territories, multiple nest building, and polygyny in the long-billed marsh wren. *Auk* **87**:557–567.

Verner, J., and M. F. Willson, 1966. The influence of habitats on mating systems of North American passerine birds. *Ecology* **47**:143–147.

Volterra, V., 1926. Fluctuations in the abundance of a species considered mathematically. *Nature* **118**:558–560.

Wade, M. J., 1976. Group selection among laboratory populations of *Tribolium. Proc. Natl. Acad. Sci. USA* **73**:4604–4607.

Wade, M. J., 1977. An experimental study of group selection. *Evolution* **31**:134–153.

Wade, M. J., 1978a. A critical review of the models of group selection. *Q. Rev. Biol.* **53**:101–114.

Wade, M. J., 1978b. Kin selection: a classical approach and a general solution. *Proc. Natl. Acad. Sci. USA* **75**:6154–6158.

Wade, M. J., 1979. The evolution of social interactions by family selection. *Am. Nat.* **113**:399–417.

Wade, M. J., 1980. Kin selection: its components. *Science* **210**:665–667.

Wade, M. J., and F. Breden, 1981. Effect of inbreeding on the evolution of altruistic behavior by kin selection. *Evolution* **35**:844–858.

Wade, M. J., and D. E. McCauley, 1980. Group selection: the phenotypic and genotypic differentiation of small populations. *Evolution* **34**:799–812.

Waldbauer, G. P., and J. G. Sternberg, 1977. Saturnid moths as mimics: an alternative interpretation of attempts to demonstrate mimetic advantage in nature. *Evolution* **29**:650–658.

Waldman, B., and K. Adler, 1979. Toad tadpoles associate preferentially with siblings. *Nature* **282**:611–613.

Wallace, B., 1957. Influence of genetic systems on geographic distribution. *Cold Spring Harbor Symp. Quant. Biol.* **24**:193–204.

Walter, E., 1934. Grundlagen der allgemeinen fischeilichen Produktionslehre. *Handb. Binnenfisch. Mitterleur.* **4**:480–662.

Ward, P., and A. Zahavi, 1973. The importance of

certain assemblages of birds as "information centres" for food-finding. *Ibis* **115**:517–534.

Warner, R. R., 1975. The adaptive significance of sequential hermaphroditism in animals. *Am. Nat.* **109**:61–82.

Warnock, J. E., 1965. The effects of crowding on the survival of meadow voles (*Microtus pennsylvanicus*) deprived of cover and water. *Ecology* **46**:649–664.

Waser, P. M., 1981. Sociality or territorial defense? The influence of resource renewal. *Behav. Ecol. Sociobiol.* **8**:231–237.

Waser, P. M., and T. J. Case, 1981. Monkeys and matrices: on the coexistence of "omnivorous" forest primates. *Oecologia* **49**:102–108.

Watson, A., and D. Jenkins, 1968. Experiments on population control by territorial behavior in red grouse. *J. Anim. Ecol.* **37**:595–614.

Watson, M. A., 1979. Age structure and mortality within a group of closely related mosses. *Ecology* **60**:988–992.

Watt, K. E. F., 1959. A mathematical model for the effect of densities of attacked and attacking species on the number attacked. *Can. Entomol.* **91**:129–144.

Watt, K. E. F., 1968. *Ecology and Resource Management: A Quantitative Approach.* McGraw-Hill, Inc., New York.

Watt, K. E. F., 1969. A comparative study of the meaning of stability in biological systems: insect and fur bearer populations, influenza, Thai hemorrhagic fever, and plague. In *Diversity and Stability in Ecological Systems*, Brookhaven Symp. Biol. No. 22, pp. 142–150.

Weatherhead, P. J., and R. J. Robertson, 1979. Offspring quality and the polygyny threshold: "the sexy son hypothesis." *Am. Nat.* **113**:201–208.

Weatherley, A. H., 1972. *Growth and Ecology of Fish Populations.* Academic Press, Inc., New York.

Wecker, S. C., 1963. The role of early experience in habitat selection by the prairie deer mouse, *Peromyscus maniculatus bairdi. Ecol. Monogr.* **33**:307–325.

Weigel, R. M., 1981. The distribution of altruism among kin: a mathematical model. *Am. Nat.* **118**:191–201.

Weinstein, M. S., 1977. Hares, lynx and trappers. *Am. Nat.* **111**:806–808.

Wellington, N. G., 1964. Qualitative changes in populations in unstable environments. *Can. Entomol.* **96**:436–451.

Wells, L., 1970. Effects of alewife predation on zooplankton populations in Lake Michigan. *Limnol. Oceanogr.* **15**:556–565.

Wenner, A. M., 1972. Sex ratio as a function of size in marine crustacea. *Am. Nat.* **106**:321–350.

Werner, E. E., and D. J. Hall, 1974. Optimal foraging and the size selection of prey by the bluegill sunfish, *Lepomis machrochirus. Ecology* **55**: 1042–1052.

Werner, E. E., and D. J. Hall, 1977. Competition and habitat shift in two sunfishes (Centrarchidae). *Ecology* **58**:869–876.

Werner, P. A., and H. Caswell, 1977. Population growth rates and age versus stage-distribution models for teasel (*Dipsacus sylvestris* Huds.). *Ecology* **58**:1103–1111.

West-Eberhard, M. J., 1975. The evolution of social behavior by kin selection. *Q. Rev. Biol.* **50**:1–33.

West-Eberhard, M. J., 1981. Intra group selection and the evolution of insect societies. In *Natural Selection and Social Behavior: Recent Research and New Theory* (R. D. Alexander and D. W. Tinkle, eds.), Chiron Press, New York, pp. 1–17.

Westoby, M., 1974. An analysis of diet selection by large generalist herbivores. *Am. Nat.* **108**:290–304.

Westoby, M., 1978. What are the biological bases of varied diets? *Am. Nat.* **112**:627–631.

White, J. E., 1964. An index of the range of activity. *Am. Midl. Nat.* **71**:369–373.

Whiteside, M. C., and R. V. Harmsworth, 1967. Species diversity in Chydorid (*Cladocera*) communities. *Ecology* **48**:664–667.

Whittaker, R. H., and P. P. Feeny, 1975. Allelochemics: chemical interactions between species. *Science* **171**:757–770.

Wiens, J. A., 1970. Effects of early experience on substrate pattern selection in *Rana aurora* tadpoles. *Copeia* **1970**:543–548.

Wiens, J. A., 1977. On competition and variable environments. *Am. Sci.* **65**:590–597.

Wiens, J. A., and J. T. Rotenberry, 1981. Morphological size ratios and competition in ecological communities. *Am. Nat.* **117**:592–599.

Wilbur, H. M., 1972. Competition, predation and the structure of the *Ambystoma-Rana sylvatica* community. *Ecology* **53**:3–21.

Willemsen, J., 1977. Population dynamics of percids in Lake IJssel and some smaller lakes in the Netherlands. *J. Fish. Res. Bd. Can.* **34**:1710–1719.

Williams, G., 1978. *Computational Linear Algebra with Models*, 2nd ed. Allyn and Bacon, Inc., Boston.

Williams, G. C., 1957. Pleiotropy, natural selection, and the evolution of senescence. *Evolution* **11**: 398–411.

Williams, G. C., 1966. Natural selection, the costs of reproduction and a refinement of Lack's principle. *Am. Nat.* **100**:687–690.

Williams, G. C., and D. C. Williams, 1957. Natural selection of individually harmful social adaptations among sibs with special reference to social insects. *Evolution* **11**:32–39.

Williams, J. D., 1954. *The Compleat Strategist.* McGraw-Hill, Inc., New York.

Williamson G. B., 1975. Pattern and seral composition in an old-growth beech-maple forest. *Ecology* **56**:727–731.

Wills, C., and C. Miller, 1976. A computer model allowing maintenance of large amounts of genetic variability in Mendelian populations. II. The balance of forces between linkage and random assortment. *Genetics* **82**:377–399.

Willson, M. F., 1966. Breeding ecology of the yellow-headed blackbird. *Ecol. Monogr.* **36**:51–77.

Wilson, D. S., 1975. A theory of group selection. *Proc. Natl. Acad. Sci. USA* **72**:143–146.

Wilson, D. S., 1977. Structured demes and the evolution of group-advantageous traits. *Am. Nat.* **111**:157–185.

Wilson, D. S., 1980. *The Natural Selection of Populations and Communities*, Evolutionary Biology Series, Institute for Ecology, Davis, Calif.

Wilson, D. S., and R. K. Colwell, 1981. Evolution of sex ratio in structured demes. *Evolution* **35**:882–897.

Wilson, E. O., 1969. The species equilibrium. In *Diversity and Stability in Ecological Systems*, Brookhaven Symp. Biol. No. 22, pp. 38–47.

Wilson, E. O., 1971. *The Insect Societies.* Harvard University Press–Belknap Press, Cambridge, Mass.

Wilson, E. O., 1975. *Sociobiology.* Harvard University Press–Belknap Press, Cambridge, Mass.

Wilson, R. E., 1968. The role of allelopathy in old-field succession on grassland areas of central Oklahoma. *Knox College Biol. Field Stat. Spec. Publ. No. 3.*

Wittenberger, J. F., 1981. Male quality and polygyny: the "sexy son" hypothesis revisited. *Am. Nat.* **117**:329–342.

Wolda, H., 1963. Natural populations of the polymorphic landsnail, *Cepaea nemoralis* (L). *Arch. Neerl. Zool.* **15**:381–471.

Wolda, H., 1967. The effect of temperature on reproduction in some morphs of the landsnail, *Cepaea nemoralis* (L). *Evolution* **21**:117–129.

Wolda, H., 1969. Stability of a steep cline in morph frequencies of the snail, *Cepaea nemoralis* (L). *J. Anim. Ecol.* **38**:623–633.

Wolf, L. L., 1969. Female territoriality in a tropical hummingbird. *Auk* **86**:490–504.

Wolf, L. L., 1970. The impact of seasonal flowering on the biology of some tropical hummingbirds. *Condor* **72**:1–14.

Wolf, L. L., and F. G. Stiles, 1970. Evolution of pair cooperation in a tropical hummingbird. *Evolution* **24**:759–773.

Woolfenden, G. E., 1975. Florida scrub jay helpers at the nest. *Auk* **92**:1–15.

Woolfenden, G. E., 1981. Selfish behavior by Florida scrub jay helpers. In *Natural Selection and Social Behavior: Recent Research and New Theory* (R. D. Alexander and D. W. Tinkle, eds.). Chiron Press, New York, pp. 257–260.

Workman, P. L., and R. W. Allard, 1964. Population studies in predominantly self-pollinated species. V. Analysis of differential and random viabilities in mixtures of competing pure lines. *Heredity* **19**:181–189.

Wrazen, J. A., 1981. Feeding patterns of cotton rats (*Sigmodon hispidus*): seasonal, reproductive, experiential, and physicochemical factors in the selection of natural foods. Ph.D. dissertation, Indiana University, Bloomington.

Wright, J. W., and C. H. Lowe, 1968. Weeds, polyploids, parthenogenesis, and the geographic and ecological distribution of all-female species of *Cnemidophorus Copeia*, **1968**:128–138.

Wright, S., 1921. Systems of mating. III. Assortative mating based on somatic resemblance. *Genetics* **6**:144–161.

Wright, S., 1932. The roles of mutation, inbreeding, cross breeding and selection in evolution. *Proc. 6th Int. Congr. Genet.* **1**:356–366.

Wright, S., 1940. Breeding structure of populations in relation to speciation. *Am. Nat.* **74**:232–248.

Wright, S., 1945. Tempo and mode in evolution: a critical review. *Ecology* **26**:415–419.

Wright, S., 1965. Factor interaction and linkage in evolution. *Proc. R. Soc. Lond. B* **162**:80–104.

Wright, S., 1969. *Evolution and the Genetics of Populations*, Vol. 2: *The Theory of Gene*

Frequencies. The University of Chicago Press, Chicago.

Wright, S., 1980. Genic and organismic selection. *Evolution* **34**:825–843.

Wu, H. M. H., W. G. Holmes, S. R. Medina, and G. P. Aackett, 1980. Kin preference in infant *Macaca nemestrina*. *Nature* **285**:225–277.

Wu, L., and J. Antonovics, 1976. Experimental ecological genetics in *Plantago*. II. Lead tolerance in *Plantago lanceolata* and *Cynodon dactylon* from a roadside. *Ecology* **57**:205–208.

Wynne-Edwards, V. C., 1962. *Animal Dispersion in Relation to Social Behavior*. Oliver & Boyd Ltd., Edinburgh.

Yeaton, R. I., 1974. An ecological analysis of chaparral and pine forest bird communities on Santa Cruz Island and mainland California. *Ecology* **55**:959–973.

Yodzis, P., 1977. Harvesting and limiting similarity. *Am. Nat.* **111**:833–843.

Yokoyama, S., and J. Felsenstein, 1978. A model of kin selection for an altruistic trait considered as a quantitative character. *Proc. Natl. Acad. Sci. USA* **75**:420–422.

Young, H., R. L. Strecker, and J. T. Emlen, 1950. Localization of activity in two indoor populations of house mice, *Mus musculus*. *J. Mammal.* **31**:403–410.

Yasukawa, K., 1978. Aggressive tendencies and levels of a graded display: factor analysis of response to song playback in the redwinged blackbird (*Agelaius phoeniceus*). *Behav. Biol.* **23**:446–459.

Yasukawa, K., 1979. Territory establishment in red-winged blackbirds: importance of aggressive behavior and experience. *Condor* **81**:258–264.

Yasukawa, K., 1981. Male quality and female choice of mate in the red-winged blackbird (*Agelaius phoeniceus*). *Ecology* **62**:922–929.

Zach, R., 1979. Shell dropping: decision making and optimal foraging in north-western crows. *Behaviour* **68**:106–117.

Zahavi, A., 1975. Mate selection—a selection for a handicap. *J. Theor. Biol.* **53**:205–214.

Zeveloff, S. I., and M. S. Boyce, 1980. Parental investment and mating systems in mammals. *Evolution* **34**:973–982.

Zimmerman, J. L., 1966. Polygyny in the dickcissel. *Auk* **83**:534–546.

Index